Biological Magnetic Resonance
Volume 19

Distance Measurements in Biological Systems by EPR

A Continuation Order Plan is available for this series. A continuation order will bring delivery of each new volume immediately upon publication. Volumes are billed only upon actual shipment. For further information please contact the publisher.

Biological Magnetic Resonance
Volume 19

Distance Measurements in Biological Systems by EPR

Edited by

Lawrence J. Berliner
Ohio State University
Columbus, Ohio

Gareth R. Eaton

and

Sandra S. Eaton
University of Denver
Denver, Colorado

KLUWER ACADEMIC / PLENUM PUBLISHERS
NEW YORK, BOSTON, DORDRECHT, LONDON, MOSCOW

ISBN 978-1-4757-0575-1 ISBN 978-0-306-47109-4 (eBook)
DOI 10.1007/978-0-306-47109-4

Softcover reprint of the hardcover 1st edition 2000

233 Spring Street, New York, New York 10013

http://www.wkap/nl

10 9 8 7 6 5 4 3 2 1

A C.I.P. record for this book is available from the Library of Congress

To Pier Luigi Nordio

CONTRIBUTORS

Albert H. Beth • Department of Molecular Physiology and Biophysics, Vanderbilt University, Nashville, Tennessee 37232

Petr P. Borbat • Baker Laboratory of Chemistry and Chemical Biology, Cornell University, Ithaca, New York 14853

Gary W. Brudvig • Department of Chemistry, Yale University, P.O. Box 208107, New Haven, Connecticut 06520-8107

Sergei A. Dzuba • Institute of Chemical Kinetics and Combustion, Russian Academy of Sciences, Novosibirsk, Russia

Gareth R. Eaton • Department of Chemistry and Biochemistry, University of Denver, Denver, Colorado 80208

Sandra S. Eaton • Department of Chemistry and Biochemistry, University of Denver, Denver, Colorado 80208

Jack H. Freed • Baker Laboratory of Chemistry and Chemical Biology, Cornell University, Ithaca, New York 14853

Arnold J. Hoff • Department of Biophysics, Leiden University, Leiden, the Netherlands

Eric J. Hustedt • Department of Molecular Physiology and Biophysics, Vanderbilt University, Nashville, Tennessee 37232

Gunnar Jeschke • Max-Planck-Institute for Polymer Research, Postfach 3148, D-55021 Mainz, Germany

K. V. Lakshmi • Department of Chemistry, Yale University, P.O. Box 208107, New Haven, Connecticut 06520-8107

Gertz I. Likhtenshtein • Department of Chemistry, Ben-Gurion University of the Negev, P.O.B. 653, Beer-Sheva 84105, Israel

Hassane S. Mchaourab • Department of Molecular Physiology and Biophysics, Vanderbilt University School of Medicine, Nashville, Tennessee 37232

Joseph C. McNulty • Department of Chemistry and Biochemistry, University of California, Santa Cruz, California 95064

Glenn L. Millhauser • Department of Chemistry and Biochemistry, University of California, Santa Cruz, California 95064

Eduardo Perozo • Department of Molecular Physiology and Biological Physics, University of Virginia Health Science Center, Charlottesville, VA 22906

Martin Pannier • Max-Planck-Institute for Polymer Research, Postfach 3148, D-55021 Mainz, Germany

Arnold Raitsimring • Department of Chemistry, University of Arizona, Tucson, Arizona 85721

Yeon-Kyun Shin • Department of Biochemistry and Biophysics, Iowa State University, Ames, Iowa 50011

Hans W. Spiess • Max-Planck-Institute for Polymer Research, Postfach 3148, D-55021 Mainz, Germany

Wenzhong Xiao • Department of Chemistry, University of California, Berkeley, California 94270

Preface

Distance measurements in biological systems by EPR

The foundation for understanding function and dynamics of biological systems is knowledge of their structure. Many experimental methodologies are used for determination of structure, each with special utility. Volumes in this series on Biological Magnetic Resonance emphasize the methods that involve magnetic resonance. This volume seeks to provide a critical evaluation of EPR methods for determining the distances between two unpaired electrons.

The editors invited the authors to make this a very practical book, with specific numerical examples of how experimental data is worked up to produce a distance estimate, and realistic assessments of uncertainties and of the range of applicability, along with examples of the power of the technique to answer biological problems.

The first chapter is an overview, by two of the editors, of EPR methods to determine distances, with a focus on the range of applicability. The next chapter, also by the Eatons, reviews what is known about electron spin relaxation times that are needed in estimating distances between spins or in selecting appropriate temperatures for particular experiments. Albert Beth and Eric Hustedt describe the information about spin-spin interaction that one can obtain by simulating CW EPR line shapes of nitroxyl radicals. The information in fluid solution CW EPR spectra of dual-spin labeled proteins is illustrated by Hassane Mchaourab and Eduardo Perozo. Wenzhoug Xiao and Yeon-Kyun Shin show a way to determine dipolar interactions by Fourier deconvolution of nitroxyl CW EPR spectra. Glenn Millhauser describes the use of the rigid spin label abbreviated TOAC. The broadening of CW EPR of spin labels caused by paramagnetic metals reveals depth of the radical beneath the surface and the electrostatic environment, as described by Gertz Likhtenshtein. The use of T_1 and T_m effects to determine distances is described by the Eatons. Arnold Raitsimring shows how to apply the "2+1" pulse sequence for distance and spatial distribution measurements of paramagnetic centers. Double-Quantum ESR is applied to distance measurements by Petr Borbat and Jack Freed. Pulsed double electron-electron methods, sometimes called PELDOR or DEER are described by Gunnar Jeschke, Martin Pannier and Hans Spiess. K. V. Lakshmi and Gary Brudvig discuss spin-spin interactions in the photosynthetic reaction center revealed by CW EPR spectra. The distance information that is obtainable from out-of-phase electron spin echoes of photo-induced radical pairs is explained by Sergei Dzuba and Arnold Hoff.

Some related topics have been discussed in other volumes of Biological Magnetic Resonance and consequently are not included in the present volume. In particular, we call the reader's attention to the Mims and Peisach review of ESEEM in Volume 3, the review of biological applications of time domain EPR by Thomann, Dalton, and Dalton in Volume 6, the four reviews of ENDOR by Hoffman and coworkers, Hüttermann, Möbius, and Thomann and Bernardo in Volume 13, progressive saturation and saturation transfer EPR by Marsh, Páli, and Horváth in Volume 14, and of ENDOR by Makinen, Mustafi and Kasa in Volume 14. Other reviews are cited in the appropriate chapters.

Site-directed spin labeling and time-domain EPR are enabling technologies for using spin-spin interactions to estimate distances. Now that commercial pulsed EPR spectrometers are available, both the CW and pulsed EPR the techniques that are described in this volume can be applied by many researchers. We hope that this volume guides users to the technique(s) most suitable for the problem to be solved.

Sandra S. Eaton
Gareth R. Eaton
Denver, Colorado

Lawrence J. Berliner
Columbus, Ohio

Contents

Section I. Introduction

Chapter 1
Distance Measurements by CW and Pulsed EPR
Sandra S. Eaton and Gareth R. Eaton

Chapter 2
Relaxation Times of Organic Radicals and Transition Metal Ions
Sandra S. Eaton and Gareth R. Eaton

Section II. CW Measurements

Chapter 3

Structural Information from CW-EPR Spectra of Dipolar Coupled Nitroxide Spin Labels

Eric J. Hustedt and Albert H. Beth

Chapter 4

Determination of Protein Folds and Conformational Dynamics using Spin-Labeling EPR Spectroscopy

Hassane S. Mchaourab and Eduardo Perozo

Chapter 7
Depth of Immersion of Paramagnetic Centers in Biological Systems
Gertz I. Likhtenshtein

Section III. Pulsed EPR Measurements of Electron-Electron Interactions

Chapter 8
Determination of Distances Based on T_1 and T_m Effects
Sandra S. Eaton and Gareth R. Eaton

Chapter 9

Double-Quantum ESR and Distance Measurements

Petr P. Borbat and Jack H. Freed

Chapter 10
"2+1" Pulse Sequence as Applied for Distance and Spatial Distribution Measurements of Paramagnetic Centers
A. Raitsimring

Chapter 11
Double Electron-Electron Resonance
Gunnar Jeschke, Martin Pannier, and Hans W. Spiess

Section IV. Applications to Photosynthesis

Chapter 12
Electron Paramagnetic Resonance Distance Measurements in Photosynthetic Reaction Centers
K. V. Lakshmi and Gary W. Brudvig

Chapter 1

Distance Measurements by CW and Pulsed EPR

Sandra S. Eaton and Gareth R. Eaton
Department of Chemistry and Biochemistry, University of Denver, Denver, Colorado 80208

Abstract: Distances between unpaired electrons ranging from ca. 5 Å to > 70 Å can be measured by a combination of continuous wave (CW) and pulsed electron paramagnetic resonance (EPR) techniques. An overview of these techniques is provided as an introduction to the volume.

Although X-ray crystallography might seem to be the paradigm for structural studies, it is of little use for studying species in intact cells, for example. From among the large armamentarium of techniques available to the biomedical researcher, electron paramagnetic resonance (EPR) has a special power for determining distances, and deserves wider use than it has had so far. This volume, written by experts in the field, reveals the utility of EPR. A special power of EPR is its focus on the small number of unpaired electrons in the system, and hence the ability to obtain structural information in environments, such as membranes, that could confound other structural methods. Many examples are provided in this volume. The profusion of new proteins that will come as a result of the genome project create opportunities for applications of the methodology discussed in this book, well beyond what was envisioned when the relevant fundamental research began.

The focus of this book is the use of electron-electron spin-spin interaction to measure distances between two electron spins. Applications include studies of protein folding, studies of conformational changes during protein function, geometry of assembly of subunits in multi-component systems, arrangement of proteins in membranes, and structure of proteins in intact biological systems. The fact that the EPR of metals and spin labels is "blind" (with some useful exceptions delineated in chapters in this book) to most aspects of the system, and focuses on the pair of spins, gives a special power to this methodology. Spin-spin interaction can be studied in fluid or frozen solution, in crystalline, amorphous, or partially organized media, and

Biological Magnetic Resonance, Volume 19: Distance Measurements in Biological Systems by EPR, edited by Berliner et al. Kluwer Academic/Plenum Publishers, New York, 2000

over a wide range of temperatures from liquid helium temperature to room temperature, depending on the pair of spins. EPR distance measurements have been reviewed previously by us (Eaton and Eaton 1978, 1989, 1996) and by Maret (1993).

1. STRUCTURAL STUDIES OF BIOMOLECULES

The task of explaining biological function at a molecular level requires knowledge of the composition and structure of the biological system. As genome research rapidly increases the number of proteins known, the requirement for obtaining structural information will increase dramatically (Read and Wemmer, 1999). The experimental methodologies most useful for measuring distances between parts of proteins or assemblies of proteins with other species rely on scattering of electromagnetic radiation (e.g., X-ray diffraction, EXAFS, and light scattering), fluorescence energy transfer, or dipolar interactions (NMR, EPR). The most common of these methods have been X-ray diffraction, fluorescence energy transfer, NMR and EPR. Each physical method provides a different perspective of the system, and multiple methods are needed to achieve a full view (Bohr and Brunak, 1996, p. 85). There are questions that each technique can answer better than the others. A fundamental question for each system to be examined is what distance scale is most informative and which techniques are applicable. In some instances determination of even a small number of relatively long distance constraints may provide key structural information.

X-ray diffraction can yield high-resolution interatomic distances for the form of the protein that is stable in a crystal. When the sample is cryo-cooled, resolution for proteins may approach 1.2 Å. Two things that X-ray diffraction does not reveal are (a) the structure of the protein in solution or other normal environment, and (b) the electronic configuration of a metal in a metalloprotein (Bohr and Brunak, 1996, p. 34). Even in the small number of cases in which crystals have been formed of proteins that function in a membrane (ca. 20 so far), some parts have remained ill-defined by the single-crystal X-ray diffraction.

A renaissance in electron microscopy, especially using freeze-trapped conformations and image-processing is providing structures with 20 Å resolution, and in some cases near 3 Å resolution (Stowell et al., 1998). Single particles of large complexes can be studied, and when atomic-resolution structures of component parts are available from other techniques, such as X-ray diffraction or NMR, it is possible to put detailed structural information into functional context (Stowell et al., 1998).

NMR can yield an average solution structure or the structure in a solid phase based on multiple measurements of short (<5-6 Å) interproton distances. Although the distance constraints from NOE's may not be sufficient to define a unique 3-dimensional conformation, a minimum number of additional longer-range constraints make it possible to predict a conformation (Bohr and Brunak, 1996, page 85). Current activity includes efforts to achieve partial orientation of the protein, via magnetic anisotropy, to observe residual dipolar couplings that define longer-range order in the molecule (Prestegard et al., 1999), isotopic labeling and new pulse sequences, and placing proteins in micelles in low-viscosity solvents (e.g. Banci et al., 1998; Wider and Wüthrich, 1999). For ^{13}C-^{13}C distances longer than ca. 4 Å the couplings are less than ca. 120 Hz, and the standard deviation in the derived distances are about an order of magnitude larger for powder samples than for single crystals (Hodgkinson and Emsley, 1999). A new pulse sequence provides distances between methyl groups via selective NOE data (Zwahlen et al., 1998). Since many methyl groups are located in the hydrophobic core of a protein, inter-methyl constraints help to define the structure.

Fluorescence energy transfer can yield distances between naturally-fluorescent sites or added labels if all of the angles that define the relative orientations of the axes of the donor and the acceptor are known or average values are assumed.

Qualifications are needed even on these generalizations, since it is non-trivial to prepare the desired sample, whether it is a solution free of interfering light-scatterers or a crystal of a protein. There is also a fundamental tension in the field of biological structure and function measurements between "pure and well-characterized" substances and "biologically-relevant milieu." For example, to use the power of NMR to study solution structure usually involves a protein in as high concentration as it can be kept in solution with minimal other reagents and as low conductivity as possible, in a glass tube above room temperature for several days. Analogously, X-ray crystallography, which has shaped the way we think about protein secondary and tertiary structure, only reveals the structure in a crystal that usually was formed in the presence of high salt concentration, and grew slowly over several days to weeks, a time scale over which some of these same proteins are known from other studies to degrade.

1.1 Structure and Dynamics

It is also important to keep in mind that proteins are not static entities. Structure provides a vantage point from which to envision dynamics that facilitate function. While X-ray diffraction provides otherwise unobtainable detail about the stereochemistry of the protein under the conditions of the crystal, and sometimes even about the stereochemistry of transition state analogs, other distance and dynamics information is needed to create a full picture of protein function. It is well known that the detailed protein structures obtained by single-crystal X-ray diffraction are not compatible with the ingress and egress of substrates unless there is substantial motion of the protein. For example, neutron scattering measurements show that there is a lot of motion in proteins (Niimura, 1999). Increasingly, molecular mechanics and molecular dynamics calculations are being extended to biological molecules and assemblies. Molecular dynamics calculations and spectroscopy affirm large amplitude motions in proteins, even in powders, crystals, and frozen solutions (Buck and Karplus, 1999; Brooks et al., 1988; Bizzarri and Cannistraro, 1999; Schlichter, et al., 2000; Diehl et al., 1997; Arcangeli et al., 1998; Tarek and Tobias, 1999).

1.2 Rationale for Distance Measurements by EPR

Special features of EPR that provide unique power for biological structure determination are that (1) the method detects only unpaired electrons; (2) EPR can see unpaired electrons in any phase and over a wide range of temperatures; and (3) the large electron magnetic dipole (μ_e/μ_p = 658) results in very-long-range effects on EPR line shapes, on responses to pulses, and on electron spin relaxation times. EPR can be used to study proteins that cannot be crystallized for X-ray diffraction study, proteins that are too large for study by NMR in solution, and proteins in a much more natural environment than is possible for either NMR or X-ray crystallography, including in membranes and even in whole cells and tissues, environments that can confound other physical methods.

In addition to being selective and long-distance, the EPR method is also applicable to a wide range of biological species. Over sixteen different types of biological systems that exhibit electron-electron spin-spin interaction are listed in Reed and Orosz (1993), over 10,700 metal binding sites in proteins have been characterized by X-ray crystallography, and large numbers of metalloenzymes involve free radicals (Sigel and Sigel, 1994, 1999). Pairs of spins can be naturally-occurring, or one or both spins can be introduced to make the distance measurement. Site-directed mutagenesis permits introducing a nitroxyl spin label (Hubbell et al., 1998; Altenbach et al., 1989) or a metal-binding site (Voss et al., 1995a,b, 1998; Jung et al., 1995; He et al., 1995a,b; Regan, 1993; Hellinga, 1996; Lu and Valentine, 1997; Hamachi et al., 1999). These developments in site-directed introduction of paramagnetic centers, and the profusion of new proteins that will come as result of the genome project, create opportunities for applications of the EPR methodology described in this book (Read and Wemmer, 1999).

EPR spectroscopy has some limitations also. The most important limitation is that the distances are measured between paramagnetic centers. Exogenous spin labels and non-native metals have been routinely criticized as "perturbing" the "real" structure. Although researchers always have to be sensitive to the relation between their observations and the reality in a living system, EPR is not more perturbing than other physical methodologies. Any technique that involves modifying the biomolecule with the probe to be studied is subject to the concern of whether the studied features of the biomolecule are significantly perturbed by the added probe. This is a significant question for X-ray crystallography, for example, when a heavy metal is added to aid in scattering, in fluorescence energy transfer when a fluorescent label is attached, and in EPR when a diamagnetic metal is replaced by a paramagnetic metal or when a nitroxyl spin label is attached. In the cited, and other, physical techniques, the accumulated wisdom is that the cost of the perturbation is more than repaid by the information obtained that otherwise could not be obtained. One biochemist is alleged to have called this "the price of peeking." For the specific case of nitroxyl spin labels, there have been several inquiries into the extent to which the label perturbs the structure and other properties of the biomolecule. Spin-labels attached to three of the amino acids in myoglobin (His-12, Tyr-103, and Tyr-151) were shown not to affect the properties of the heme or the general conformation and stability of the protein (Postnikova, 1996). This Russian work focused on conformational properties revealed by spin label mobility, and did not mention spin-spin interactions. Similarly, studies of spin-labeled T4 lysozyme have revealed minimal perturbation of the structure (Langen et al., 2000; also see ch. 4 by Mchaourab and Perozo).

Studies of electron-nuclear distances by electron spin echo envelope modulation (ESEEM) or by electron nuclear dipolar interaction (ENDOR) are powerful tools for determining the fine details of the geometry around an unpaired electron. ESEEM has been reviewed recently by Dikanov and Tsvetkov (1992) and by McCracken (1999). Pulsed ENDOR techniques were reviewed by Grupp and Mehring (1990). ENDOR data was reviewed by Kispert and coworkers (Piekara-Sady et al., 1994; Goslar et al., 1994). Originally, it appeared to be true that ESEEM and ENDOR were applicable to different electron-nuclear coupling regimes. However, recently, Hoffman and coworkers have demonstrated the ability to measure very weak couplings to distant nuclei, extending the range of ENDOR, so that while ENDOR and ESEEM remain complementary techniques, their coupling regimes now overlap.

The interactions between electron spins and nuclear spins can also be studied by NMR, in solution and in solids (Bertini et al., 1991; see also ch. 7 by Likhtenshtein).

2. DIPOLAR INTERACTIONS

Magnetic resonance methods for estimating distances between spins (electron and/or nuclear) are based on the dipolar interactions between spins, i.e. the force of the magnetic field from one spin on the magnetic dipole of the second spin. Unless specified otherwise, all of the following section is for pairs of S = ½ paramagnetic centers. The magnetic terms considered in this discussion are tensorial quantities (see Abragam and Bleaney, 1970, sections 15.6-15.8, for refinements of these concepts). However, in practical applications it is only occasionally important to retain the full tensorial notation, and for the purposes of this introduction, we use predominantly the trigonometric relations between the vectors, and limiting conditions, to provide a feel for the magnitudes of interactions. Notation in the following discussion follows closely that of the text by Weil, Bolton and Wertz (1994), supplemented by mention of other commonly-used terms. That book provides additional background information on many of the topics introduced here.

The force on dipole 1 with magnetic moment μ_1 due to interaction with the magnetic field created by dipole 2 with magnetic moment μ_2 at distance r is (Abeles and Bos, 1967):

$$F = \frac{\mu_1 \mu_2}{r^4} \tag{1}$$

The energy of interaction is force times distance so

$$E = U = \frac{\mu_1 \mu_2}{r^3} \tag{2}$$

In the magnetic resonance experiment, the energy of interaction of two dipoles is revealed in the change in Zeeman energy of the observed spin, due to the presence of the second spin. To begin this discussion, we ignore relaxation effects, which we will consider later. The dipole-dipole interaction results in splitting of the resonance line of the observed spin. Hence, interaction with a second spin with S = ½ results in a doublet of lines for the observed spin. The vectorial nature of the interaction is conveniently simplified to a cosine dependence of the interaction energy on the relative orientation of the dipoles in an external field.

$$U_{dipolar} = \frac{\mu_1\mu_2}{r^3}\left(1 - 3\cos^2\theta\right) \tag{3}$$

where θ is the angle between the external magnetic field direction and a line joining the two dipoles. The angular dependence causes U to vary between 1 and –2 times $\frac{\mu_1\mu_2}{r^3}$. Importantly, there is an angle, sometimes called the magic angle (54.7°), at which the dipolar interaction is zero. In addition, the average over a sphere is zero. Hence, a very rapidly rotating pair of spins will not exhibit net dipolar splitting. Rapidly, in this context, means fast relative to the interaction energy expressed in frequency units. For example, the two H atoms in liquid H_2O yield only a single NMR line, whereas in a solid hydrate splitting of the NMR line is observed.

To evaluate the magnitude of the dipolar interaction in a magnetic resonance experiment we need to consider the quantization of the spins and to use the more complete form of the magnetic dipole-dipole interaction, which is

$$U_{jk} = \frac{\mu_j \cdot \mu_k}{r_{jk}^3} - 3\frac{\left(\mu_j \cdot r_{jk}\right)\left(\mu_k \cdot r_{jk}\right)}{r_{jk}^5} \tag{4}$$

In this form of the equation all terms in the numerator are vectors, and it is clear that the simplified form of the equation is valid only if the magnetic moments, μ, are isotropic. For the interaction of two protons in NMR, the isotropic assumption is fairly good, but for the interaction between two paramagnetic transition metals, or between a transition metal and an organic radical, the assumption may not be good enough for the goals of the measurement.

However, if it is assumed that the g values are isotropic, the dipolar splitting of an EPR signal can be expressed in terms of the parameter D, where D is the splitting of the signal when the interspin vector is perpendicular to the external magnetic field and –2D is the splitting when the interspin vector is parallel to the external magnetic field (Luckhurst, 1976). In a randomly oriented sample this results in a classic "Pake pattern" where, in the ideal case, the value of D can be read from the splitting between the intense perpendicular turning points. When one can identify this interaction energy in the EPR spectrum and if the point dipole approximation is valid, the distance between the two spins is calculable directly from D.

$$D\,(\mathrm{erg}) = \frac{3g^2\beta^2}{2r^3} \text{ and r is in cm.} \tag{5}$$

Conversion of units from erg to gauss gives

$$D\,(\mathrm{gauss}) = \frac{3g\beta}{2r^3} = 1.39\mathrm{x}10^4\,\frac{g}{r^3} \tag{6}$$

where r is in Å, the g value for one of the unpaired electrons is assumed to be 2.00 and the g value for the second unpaired electron is entered explicitly. D can also be given in the following units.

$$D\,(\mathrm{MHz}) = \frac{-78000}{r^3}\text{ , where r is in Å, and g is assumed to be 2,} \tag{7}$$

$$\text{and } D\,(\mathrm{mT}) = \frac{-2786}{r^3}\text{, and r is in Å, and g is assumed to be 2.} \tag{8}$$

To calibrate our thinking about the ways in which dipolar couplings are measured, it is useful to consider the actual magnitude of D for various interspin distances as shown in the following table.

Table 1. Magnitude of dipolar coupling as a function of interspin distance

r (Å)	D (gauss)
5	222
10	28
15	8.2
20	3.5
40	0.43
60	0.13

Nitroxyl radicals with normal isotopic abundance typically have frozen-solution (powder) linewidths of 6-8 gauss. Replacement of hydrogen with deuterium narrows the lines by about a factor of two. Comparison of these widths with the maximal dipolar splittings in Table 1 indicates that for distances greater than about 20 Å the dipolar interaction is small compared with typical spin label linewidths. In contrast, typical values of the spin-echo dephasing time constant, T_m, for nitroxyl radicals at temperature below about 80 K are about 2 μs, which corresponds to a spin packet linewidth of about 30 mG. Thus dipolar interactions that are small compared to CW linewidths are significant compared with spin-packet linewidth, which makes pulse techniques advantageous for measurements of longer interspin distances. In addition, the spin-lattice relaxation time constant, T_1, is much longer than T_m in typical immobilized samples of nitroxyl radicals so changes in either T_m or T_1 can be sensitive indicators of spin-spin interactions in distance regimes where CW lineshape changes are too small to detect. As the target distance becomes longer, it becomes necessary to use increasingly low sample concentrations to ensure that the intramolecular spin-spin distance of interest is significantly shorter than the average random distance between spins in the sample. A fundamental limit will be the instrumental sensitivity at these low spin concentrations.

The dipolar coupling is proportional to r^{-3} so this defines the distance dependence for techniques that measure lineshape changes and pulse techniques such as the 2+1 technique and double electron-electron resonance (DEER). Methods that measure perturbations caused by the dipolar interaction, such as changes in spin-lattice relaxation rates and the relative intensity of the half-field transition depend upon the square of the dipolar interaction and therefore vary as r^{-6}.

3. EXCHANGE INTERACTION

One unpaired electron also may influence the energy of a second unpaired electron via the exchange interaction, which is a quantum mechanical phenomenon resulting from the indistinguishability of electrons. Whenever there is orbital overlap, there is the possibility of interchanging electrons such that the linear combinations $\alpha(1)\beta(2) \pm \alpha(2)\beta(1)$ make a contribution to the wave function. These combinations are the singlet and triplet states, and the energy separation, J, between them, is written with the Hamiltonian term $H = -2JS_1 \cdot S_2$. The isotropic exchange is sometimes called Heisenberg exchange, especially when it involves electron exchange interactions between species colliding in fluid solution, and typically is much larger than anisotropic exchange at distances of interest in this book. Anisotropic exchange has the same form as the dipolar interaction and is sometimes (unfortunately) called the "pseudodipolar interaction" (Pake and Estle, 1973). The orbital overlap necessary for exchange of electrons can occur through many intervening orbitals. It is especially large when the orbitals in the usual cartoon representations, form a "W." Sometimes, as a rough approximation, it is stated that exchange interactions fall off exponentially with increasing distance between the spins. This is what would be expected for overlap of two simple atomic orbitals at large enough distances that the radial dependence of an orbital is reasonably approximated as an exponential function. However, attempts to correlate experimental exchange interactions with distance (Coffman and Buettner, 1979) yield limiting functions that are violated in enough cases that great caution should be exercised (Eaton and Eaton, 1988b). We have demonstrated resolved spin-spin splitting in CW EPR spectra due to exchange interactions over at least 12 Å, and through both saturated linkages (σ bonds) and unsaturated linkages (π bonds) (More et al., 1982). Exchange can occur through hydrogen bonds (Goodson et al., 1994; Akita et al., 1995; Eischenbroich, 1997; Yoshioka et al., 1997; Chadhuri et al., 1988). In copper dimers, exchange was found to depend strongly on the ligands bridging between the two Cu(II). J was -70 cm^{-1} through a ligand that held the two coppers 11.3 Å apart, but only -0.7 cm^{-1} through a different ligand that held the two coppers 7.8 Å apart (Chadhuri et al., 1988). Exchange interaction also can occur due to collisions between paramagnetic centers in fluid solution.

For perspective, note that the well-known high-resolution spin-spin splitting in proton NMR spectra used by organic chemists to characterize compounds in fluid solution is due entirely to intramolecular spin-spin exchange – the dipolar interaction has been averaged to zero by rapid molecular tumbling in solution and the effects of collisions are negligible. Some new NMR techniques for study of proteins and membranes recover part of the dipolar interaction via partial orientation.

For relatively long interspin distances without a short through-bond pathway it is common practice to assume that exchange interaction is negligible. For example, when spin labels are bound to positions on the peptide chain that are separated by multiple residues, this is likely to be a reasonable assumption. However, at relatively short distances it is important to use distance-determination methods that permit separation of dipolar and exchange contributions.

4. DETERMINATION OF ELECTRON-ELECTRON DISTANCES FROM DIPOLAR INTERACTION MEASURED BY EPR

The task of methodology development is to maximize sensitivity of the spectral display to the dipolar interaction and to separate the dipolar interaction from other features of the spectra and thereby extract the distance information from the spectrum. A variety of techniques have been reported for various types of samples and magnitudes of dipolar interaction as outlined in the following paragraphs and summarized in Table 2. In some cases, further details and additional examples can be found in our earlier reviews (Eaton and Eaton, 1978, 1988a,b, 1989).

Table 2. Summary of Methods for Determining Electron-Electron Distances[a]

Method	r^{-3}	r^{-6}	R_{exp} [b] (Å)	R_{max} [c] (Å)	comments	see ch. or review
Two slowly-relaxing spins, immobilized						
simulation of CW lineshape	x		6-20	20-25	longer distances with ^2H labels	Ch. 3
deconvolution of CW lineshape	x		8-25	25	longer distances with ^2H labels	Ch. 5
relative intensity of half-field transitions		x	4-12	12	longer distances at lower microwave frequencies	Ch. 6
d_1/d	x		12-22			Kokorin et al., 1972
2+1 pulse sequence	x		30-60	60		Ch. 10
pulsed ELDOR	x		15-130			Milov et al., 1998
4-pulse DEER	x		15-66	80		Ch. 11
double-quantum coherence	x		21-29	80		Ch. 9
Two slowly relaxing spins, slow tumbling						
CW fluid solution line width	x		9-21	25	requires calibration, depends on tumbling time	Ch. 4
One rapidly and one slowly-relaxing spin, immobilized						
Leigh effect		x	8-27		requires evaluation of assumptions for particular application	Eaton and Eaton, 1988a
change in T_1		x	12-40	40	longer distances for faster relaxing metals	Ch. 8, 12
change in T_m		x	12-30	40	characteristic temp. dependence	Ch. 8
selective hole burning	x		25-50			Dzuba and Kawamori, 1996
Spin-Polarized Radical Pair						
out-of-phase echo	x		25-40	40	pair must be spin-correlated	Ch. 13

[a]In this table, as in the text, slow and fast relaxation times are relative to the dipolar interaction splitting expressed in frequency units. [b] Values of r that have been measured with a particular technique. [c] Maximum values of r that has been proposed as measurable by a particular technique.

4.1 Pairwise Interaction Between Two Slowly-Relaxing Spins in an Immobilized Sample

In these cases it is assumed that there is a fixed distance between the two unpaired electrons. If the g values and/or hyperfine splittings of the two centers are anisotropic, there may be a unique orientation of the interspin vector relative to the magnetic axes of the two centers. Alternatively, there may be enough conformational flexibility of the intervening bonds that there is a distribution of relative orientations or even a random distribution of relative orientations. In selecting a technique that is appropriate for a particular system it is important to consider how well the relative orientations are defined in the sample and in the data analysis.

4.1.1 Dipolar splitting is significant compared with linewidths of spectra for corresponding monoradical

When the dipolar splitting (D) is significant relative to the linewidths of the signals in the CW spectra, analysis of the lineshape changes can be used to determine the magnitude of the dipolar interaction. A variety of approaches have been examined.

4.1.1.1 Analysis of spectral lineshape by computer simulation

Since the dipolar interaction is inherently anisotropic, computer modeling of the complete spectral lineshape can, in principle, define both the interspin distance and the orientation of the interspin vector relative to the magnetic axes of both paramagnetic centers, as well as separating dipolar and exchange interactions. Since the simulated spectral lineshapes depend upon the degree of ordering of the relative orientations of the magnetic axes, information can be obtained concerning the range of conformations present in the sample. The simulations involve multiple parameters so the uniqueness of solutions depends strongly upon the resolution of the spectra and can be substantially enhanced by fitting data at multiple microwave frequencies (Hustedt et al., 1997). This approach is discussed in Ch. 3 by Hustedt and Beth.

The separation of dipolar and exchange contributions and definition of the orientation of the interspin vector relative to the magnetic axes of the interacting spins can be facilitated by obtaining spectra as a function of orientation in oriented samples or single crystals (Damoder et al., 1983). An example of the use of one-dimensionally oriented photosynthetic membranes to separate dipolar and exchange interaction in the $S_2Y_Z^{\bullet}$ state of photosystem II is included in Ch. 12 by Lakshmi and Brudvig.

4.1.1.2 Lineshape deconvolution

If the dipolar splittings are not well resolved, it may be difficult to obtain a unique simulation of the CW lineshape. In addition, computer simulation may be a time-consuming process. Deconvolution has been proposed as a simpler alternative that does not attempt to determine the orientation of the interspin vector relative to the magnetic axes. The method assumes that there is sufficient flexibility in the linkage between the two paramagnetic centers that there is an essentially random distribution of the orientation of the interspin vector relative to the magnetic axes of the paramagnetic centers. Two methods have been suggested for application to doubly spin-labeled proteins. Both approaches rely upon comparison of the spectra for the two corresponding non-interacting mono-radicals with the spectrum for the interacting radical pair. Shin and co-workers use Fourier deconvolution of the diradical spectrum by the monoradical spectrum to determine the dipolar interaction (Rabenstein and Shin, 1995). Steinhoff et al. (1997) multiply the monoradical spectrum by a broadening function to match the diradical spectrum. The deconvolution approach is discussed in Ch. 5 by Shin. In both approaches it is necessary to have the spectra of the corresponding monoradicals in order to analyze the spectrum of the diradical.

4.1.1.3 Relative intensity of the half-field transition

Anisotropic interaction between two spins (dipolar interaction and anisotropic exchange) shifts the triplet state $m_s = \pm 1$ energy levels relative to the $m_s = 0$ level, and causes the normally forbidden transition probability between the $m_s = -1$ and $m_s = +1$ levels to become non-zero. This transition occurs at half the magnetic field required for the allowed transitions (at constant microwave frequency), and hence is called the "half-field" transition. If the anisotropic exchange is negligible, and if r > ca. 4 - 5 Å, the resonant field for the half-field transition is independent of the values of J and r. The ratio of the integrated intensity of the half-field transition to the integrated intensity of the allowed transitions is a function of r and independent of J. Therefore, isotropic and anisotropic contributions can be separated by measuring the relative intensity of the half-field transition (Eaton and Eaton, 1982, 1989; Eaton et al., 1983; Anderson et al., 1999; Steinhoff et al., 1997).

$$\text{relative intensity} = \frac{\text{integrated intensity of half-field signal}}{\text{integrated intensity of allowed transitions}} \tag{9}$$

$$\text{relative intensity} = \frac{(19.5 \pm 0.5)(9.1)^2}{r^6 \nu^2} \tag{10}$$

where r is the interspin distance in Å and ν is the microwave frequency, in GHz, at which the experiment was performed.

A matrix diagonalization study by Coffman and Pezeshk (1986) found that the numerical coefficient in eq. (10) was 19.5 + 10.9Δg, where $\Delta g = |g_{max} - g_{min}|$, for organic radicals and copper complexes and confirmed that the anisotropic exchange contribution to the anisotropic interaction was negligible for distances greater than 4 to 6 Å. Note that the relative intensity is inversely proportional to the square of the microwave frequency. Thus, longer distances can be measured at lower microwave frequencies (Eaton et al., 1983).

When the interspin distance is less than about 4 Å the position of the half-field transition depends upon the dipolar interaction and that shift can be analyzed to determine the interspin distance (Thomson, 1968). At these short distances it also becomes necessary to consider the impact of anisotropic exchange and the point dipole approximation may not be appropriate.

Since the dipolar interaction is strictly through space, it can occur between neighbors in a random mixture and does not require specific pairwise interaction. For example, a frozen solution of monomeric nitroxyl radicals will exhibit a half-field transition if the concentration is high enough that there are a significant number of near-neighbor pairs. To correct for this contribution to the intensity of the half-field transition, the relative intensity of the half-field transition should be measured as a function of concentration and extrapolated to infinite dilution to obtain the specific pairwise interaction distance (Eaton et al., 1983). Examples of the use of the intensity of half-field transitions to determine distances between spin labels are included in Ch. 6 by Millhauser.

4.1.1.4 Lineshape distortion determined by the ratio d_1/d

Broadening of the frozen solution CW EPR spectrum of a nitroxyl spin label due to dipolar interaction with a second nitroxyl changes the relative intensities of the center and outer lines of the spectrum. Kokorin et al. (1972) proposed that a ratio of peak heights, which they call d_1/d, could be used to estimate the distance between two nitroxyl radicals. The assumptions are that the dipolar spin-spin splitting is not resolved, and that the nitroxyl relaxation rate, T_1^{-1}, is slow relative to the splitting. The method has been extended to interactions between metals and nitroxyl radicals by introducing a parameter, α (Kokorin et al., 1981). In this highly parameterized method, the value of T_1 does not need to be known.

4.1.2 Dipolar interaction is small compared with linewidths of spectra for corresponding monoradical

CW techniques for measuring dipolar interaction depend upon the observation of significant broadening of the lineshape due to the interaction. Smaller dipolar interactions can be measured by pulsed techniques than by CW techniques because the distance scale for the pulse techniques is limited by the spin-packet width (proportional to the rate constant for echo dephasing), which frequently is much smaller than the inhomogeneously broadened CW linewidth. In these pulsed EPR techniques one perturbs a spectral region due to one label and observes the time-response in a spectral region due to the other label. Dipolar interaction is manifest in oscillations of spin echo intensity as a function of pulse timing. The oscillation frequency is characteristic of the interspin distance. Several pulse sequences are currently in use and it is probable that others will be developed to exploit particular features of spin-coupled systems.

The pulse techniques however face a different challenge than the CW techniques. Some of the methods require excitation of most of the spins in order to define the full dipolar broadening function. With current technology it is difficult to achieve a bandwidth that is broad enough to excite the full spectrum of an immobilized nitroxyl. This challenge becomes even greater as the spectrum is broadened by increasingly strong dipolar interaction. Thus, the pulse techniques tend to be more useful at longer distances than are accessible by CW techniques.

4.1.2.1 3-Pulse ELDOR (electron-electron double resonance)

The first of these pulse techniques was the electron-electron double resonance (ELDOR or PELDOR) technique. A spin echo is created by a two-pulse sequence at one microwave frequency. The timing of a pulse at a second microwave frequency is varied. Milov, Salikhov, and Shchirov (1981) demonstrated this technique by applying a pulse to the signal for H atoms and observing the effect on hydroquinone radicals in frozen sulfuric acid. Examples of applications of this technique have been described by Tsvetkov and co-workers (Milov and Tsvetkov, 1997; Maryasov et al., 1998; Milov et al., 1999) and are reviewed in Milov et al. (1998). Note that DEER and ELDOR are two different acronyms for an electron-electron double-resonance experiment.

4.1.2.2 2+1 sequence

Raitsiming and co-workers proposed a sequence that is similar to 3-pulse PELDOR, but with all three pulses at the same microwave frequency (Astashkin et al., 1998; Kurshev et al., 1988, 1989; Raitsimring et al., 1992;

Shigemori et al., 1998). This is denoted as the "2+1" sequence. Precautions must be taken to avoid excitation of proton nuclear modulation in this single-frequency experiment. This technique and its applications are described in Ch. 10 by Raitsimring.

4.1.2.3 4-pulse DEER (double electron-electron resonance)

In the 2+1 pulse sequence and in 3-pulse ELDOR experiments there is an inherent experimental deadtime that limits the magnitude of the dipolar interaction that can be characterized. By using a 4-pulse experiment that deadtime can be eliminated (Martin et al., 1998; Pannier et al., 2000). Comparing dipolar interactions to spin packet widths, Mims and Peisach (1981) and Spiess and coworkers (Martin et al., 1998) estimate that the distance that can be measured by ESE techniques is about 70-80 Å. This technique is described in Ch. 11 by Spiess and Jeschke.

4.1.2.4 Double-quantum coherence

Allowed double-quantum coherences (DQC) can be generated in ordered or disordered samples containing pairs of radicals and random distributions of radicals. Pake doublets obtained from DQC can be used to determine distances within pairs of radicals (Saxena and Freed, 1996, 1997; Borbat and Freed, 1999) and the decay constants yield concentrations in a distribution. The upper limit of distances between nitroxides accessible by this technique is estimated to be ca. 80 Å. In the limit of non-selective pulses the interpretation of DQC signals becomes independent of the orientation of the interspin vector relative to the hyperfine axes. The technique is described in Ch. 9 by Borbat and Freed.

4.1.3 Comparison of four distance determination methods applied to doubly spin-labeled human carbonic anhydrase II

Four doubly spin-labeled variants of human carbonic anhydrase II and corresponding singly-labeled variants were prepared by site-directed spin labelling (Persson et al, 2000). The labels are located within the hydrophobic core of the protein. The variants were characterized by catalytic activity and CD spectra. The distances between spin labels were obtained from CW spectra by analysis of the relative intensity of the half-field transition, Fourier deconvolution of lineshape broadening, and computer simulation of lineshape changes. Distances also were determined by 4-pulse DEER. For each variant, at least two methods were applicable, and reasonable agreement between methods was obtained as shown in Table 3. The useful distance ranges for the techniques examined at X-band with natural isotope abundance labels and samples that contain some singly-labeled protein were approximately: half-field transition (5 to 10 Å), lineshape simulation (up to 15 Å), Fourier deconvolution (8 to 20 Å), and DEER (>18 Å). By using multiple distance measurement techniques it is possible to more clearly define contributions from multiple conformations with different distances and contributions to the spectra from singly-labeled protein. For example, conformations of $HCAII_{121\text{-}206}$ with longer distances are clearly defined by DEER experiments but are indistinguishable from singly-labeled protein in Fourier deconvolutions or lineshape simulations. However, conformations with distances shorter than about 18 Å were not well defined by DEER using the available hardware. The DEER measurements have the advantage that the rate of damping of the dipolar modulation gives a more direct indication of the breadth of distance distributions than is currently available from CW lineshapes.

Table 3. Distances (Å) Determined from EPR Spectra

Doubly spin-labeled variant[a]	Distance between β-carbons (Å)[b]	Half-field transition	Fourier Deconvolution	Lineshape simulation	DEER
$HCAII_{67\text{-}121}$	8.8	7		7-8	-
$HCAII_{59\text{-}174}$	5.4	8	8.5-9	9-10	-
$HCAII_{121\text{-}206}$	10.9	-	16-18	14-15[c]	18 (70%)
				17-19[d]	24 (30%)
$HCAII_{67\text{-}206}$	17.9	-	17-20		20 ±1.8

[a]The subscripts denote the positions of the cysteines to which the spin labels are attached.
[b]Distance between β-carbons determined from the crystal structure (Håkansson et al., 1992).
[c]Including contribution from singly-labeled protein, [d]Assuming 100% doubly-labeled protein

4.2 Pairwise Interaction between two Slowly-Relaxing Spins on a Slowly-Tumbling Macromolecule in Fluid Solution

In fluid solution there may be substantial flexibility of the linkage between a protein and the spin label that can result in partial motional averaging of the anisotropic g and A values. However, if the overall tumbling rate of the macromolecule is slow relative to the dipolar interaction, there is little averaging of the dipolar interaction. To interpret such spectra quantitatively will require simulations that model all of the dynamics of the system. However, even without such simulations, very useful qualitative interpretation is possible. For example, the intensity in the outer wings of the CW spectra is due to the relative orientations with the largest dipolar interactions. Hence, a spectrum with greater intensity further in the wings is due to species that are closer to one another. As shown in Ch.

4 by Mchaourab and Perozo, the tertiary structure of proteins with two nitroxyl spin labels can be deduced by visual inspection of their EPR spectra and by calibrations for a particular tumbling correlation time.

4.3 Pairwise Interaction Between a Slowly-Relaxing Spin and a More Rapidly-Relaxing Spin in an Immobilized Sample

To determine whether one of the following techniques is appropriate for a particular spin-coupled pair in a particular temperature regime requires knowledge of typical relaxation rates as a function of temperature. Available information concerning spin relaxation rates for nitroxyl radicals and transition metal complexes between about 5 – 10 K and 300 K is discussed in Ch. 2 by Eaton and Eaton.

4.3.1 Impact of the rapidly-relaxing spin on the CW spectrum of the slowly-relaxing spin

The impact on the CW spectrum of a slowly-relaxing spin that is dipolar-coupled to a more rapidly-relaxing spin is temperature dependent. When the relaxation rate of the more-rapidly relaxing spin is slow compared with the dipolar splitting, broadening or splitting of the CW spectrum is observed. As temperature is increased and the relaxation rate of the rapidly relaxing center increases, the CW spectrum of the slowly-relaxing center broadens. Further increases in the relaxation rate cause collapse of the splitting due to the electron-electron coupling and then narrowing to an unbroadened spectrum. Examples of this temperature dependence are shown in Fielding et al. (1986) and Rakowsky et al. (1995) and discussed in Ch. 8.

In the temperature interval where the rapid relaxation rate is fast enough to collapse the splitting, but not fast enough to result in full narrowing, J. S. Leigh (1970) proposed that the decrease in amplitude of a nitroxide spin label spectrum could be used to determine the distance between the spin label and a paramagnetic metal. The Leigh method has been one of the most commonly applied EPR methods for estimating distances in biological system, but in many applications the assumptions of the method have not been met (Case and Leigh, 1976; Morris and Dwek, 1977; Eaton and Eaton, 1978; Eaton et al., 1979; Hyde et al., 1979; Kuo at al., 1983; Makinen and Kuo, 1983; More et al., 1984, 1985; Eaton and Eaton, 1989). There are many statements in the literature to the effect that one observes reduction in the intensity of the EPR signal without observable broadening. Actually, there is not a reduction in total integrated intensity of the CW EPR signal, but for some of the spin pairs the broadening is so great that in a normal CW spectrum with average S/N, the broad lines are not significant contributors to the signal amplitude observed in the usual first-derivative display. (Recall that in first-derivative spectra, the contribution to the height of the signal is proportional to the inverse square of the width of the signal.)

4.3.2 Impact of the rapidly-relaxing spin on spin echo dephasing of the slowly-relaxing spin

In the temperature range where the relaxation of the rapidly-relaxing spin causes broadening and partial collapse of the CW spectrum of the slowly-relaxing spin, the spin echo dephasing rate of the slowly-relaxing spin is enhanced. The enhancement of the echo dephasing can be seen clearly even in cases where the interspin distance is so long that there is little impact on the CW lineshape. The analysis of the effect of the rapidly-relaxing spin on echo dephasing is discussed in Ch. 8 by Eaton and Eaton.

4.3.3 Impact of the rapidly-relaxing spin on spin-lattice relaxation of the slowly-relaxing spin

A rapidly–relaxing spin increases the spin-lattice relaxation rate of a slowly-relaxing spin to which it is spin-coupled. The maximal effect occurs when either of two criteria is met. If the relaxation rate, T_1^{-1}, of the rapidly-relaxing spin is equal to the Larmor frequency, $2\pi\nu$, of the slowly-relaxing spin, or if T_2^{-1} of the rapidly-relaxing spin is equal to the difference between the Larmor frequencies for the two paramagnetic centers, the effect of the rapidly-relaxing spin on the slowly-relaxing spin is maximized. These effects are well known in NMR. Applications of these relaxation enhancements to distance determination are discussed in Ch. 8 by Eaton and Eaton and in Ch. 12 by Lakshmi and Brudvig.

4.3.4 Selective hole burning

In this technique a low-power selective 180° pulse is used to burn a hole in the EPR spectrum of a slowly-relaxing spin. The broadening of the hole is monitored by the FID and two-pulse echoes created by subsequent high-power pulses. The rate of broadening of the hole is caused by spectral diffusion. The fluctuations of electron spins that are dipolar coupled to the observed unpaired electron contribute to spectral diffusion. If the contributions to spectral diffusion from motion, nuclear spin flip-flops, and instantaneous diffusion are smaller than the contribution from dipolar interaction between unpaired electrons, the spectral diffusion can be used to determine the interspin distance. Application of this technique to photosynthetic systems has led to estimates of distances between 25 and 50 Å (Dzuba et al., 1993; Kodera et al., 1994; Hara and Kawamori, 1997). The technique is described and reviewed in Dzuba and Kawamori (1996).

4.4 Out-of-phase Echo for Spin-Polarized Radical Pairs

In these experiments a spin-correlated radical pair is created photochemically, followed by microwave pulses that generate a two-pulse spin echo. Crucial to the interpretation of these data is that there is phase coherence between eigenstates present in the photo-induced spin-correlated radical pair (Salikhov et al., 1992; Tang et al., 1994). The spin-spin interaction causes deep modulation in the out-of-phase echo, whereas the normal, in-phase, echo vanishes (Salikhov et al., 1992; Tang et al., 1994). Modulation of the echo can be analyzed to determine the interspin distance and the exchange interaction between the radicals in the pair. Distances in the range of 25 – 40 Å between donor and acceptors in photosystems have been measured as discussed in Ch. 13 by Dzuba and Hoff.

4.5 Collisions Between Slowly-Relaxing Label and More Rapidly-Relaxing Spin in Fluid Solution

Collisions between rapidly-relaxing paramagnets and slower-relaxing spins cause broadening of the CW EPR spectrum of the slower-relaxing spins. Likhtenshtein (1976) pioneered metal-broadening of nitroxyl spin label CW spectra to estimate the distance from the solvent-exposed surface to a spin label. These procedures are discussed in Ch. 7 by Likhtenshtein. The relative broadening effectiveness of numerous metal complexes has been quantitated in our laboratory (Yager et al., 1978; Dalal et al.; 1985a,b; Burchfield et al., 1994).

Hubbell and coworkers compared the effects of metal complexes and oxygen, taking advantage of differential partitioning between aqueous and membrane environments, to map label locations in transmembrane proteins (Hubbell et al, 1998; Sun et al., 1999), and structure of heat-shock proteins (Koteiche et al., 1998; Mchaourab et al., 1997). A series of these qualitative estimates of distance can provide a detailed understanding of the structure of the assembly including distinguishing between α-helices and β-sheets. These procedures are discussed in Ch. 4 by Mchaourab and Perozo and in Hubbell et al. (1998).

5. DISTRIBUTIONS OF DISTANCES

In many systems the distance between paramagnetic centers is not fixed at a specific value, but rather is a distribution. Even where there is a well-defined structure for a biradical or doubly-labeled protein, molecular flexibility can cause a distribution of distances that should be taken into account in data analyses. Sometimes the goal is to find the spatial distribution function. Sometimes the goal is to find a most-probable, or even closest, distance between the two centers. The difficulty is to weight a physically-reasonable distribution function with the distance dependence of the interaction. The $1/r^3$ and $1/r^6$ dependencies strongly select for the closest distances. Several papers have examined the effect of distributions of distances and orientations on the "2+1" and DEER ESE results (Astashkin et al., 1998; Milov et al., 1999; Steinhoff et al., 1997) and Ch. 9 discusses the effect on double-quantum coherence.

6. ACCURACY AND PRECISION OF DISTANCES MEASURED

The uncertainty in the precision of the distances depends upon the method used, but frequently is stated to be of the order of 1 – 2 Å. However, a major uncertainty in the interpretation is the treatment of the distributions in the distances. This distribution is often ignored, resulting in poorer accuracy than the precision would imply. Furthermore, one has to interpret the distance between paramagnetic centers in terms of the most likely orientation of the label relative to the site of attachment. Molecular structure calculations, such as Insight II (Biosym/MSI, San Diego, CA), can be useful in this part of the interpretation, although effects of solvation on conformation need to be included. For the more qualitative measures, such as d_1/d or changes in power saturation curves upon collisions, the relative distances for closely analogous systems are more accurate than the absolute distances. However, as in NMR NOE measurements, sometimes the desired geometric information can be obtained by a set of several measurements that that are only qualitatively graded as non-interacting, weakly-interacting, moderately-interacting, strongly-interacting, etc.

7. REFERENCES

Abeles, T. P. and Bos, W. G. (1967). A Dimensional Analysis of Magnetic Susceptibility Calculations. *J. Chem. Ed.* **44**, 438-441.

Abragam, A. and Bleaney, B. (1970). *Electron Paramagnetic Resonance of Transition Ions*. Oxford University Press, Oxford.

Akita, T., Mazaki, Y., and Kobayashi, K. (1995). Ferromagnetic Spin Interaction in a Crystalline Molecular Complex Formed by Inter-heteromolecule Hydrogen Bonding: a 1:1 Complex of Phenyl Nitronyl Nitroxide Radical and Phenylboronic Acid. *J. Chem. Soc. Chem. Commun.* 1861-1862.

Altenbach, C., Flitsch, S. L., Khorana, H. G., and Hubbell, W. L. (1989). Structural Studies on Transmembrane Proteins. 2. Spin Labeling of Bacteriorhodopsin Mutants at Unique Cysteines. *Biochemistry* **28**, 7806-7812.

Anderson, D. J., Hanson, P., McNulty, J., Millhauser, G., Monaco, V., Formaggio, F., Crisma, M., and Toniolo, C. (1999). Solution Structures of TOAC-Labeled Trichogin GA IV Peptides from Allowed ($g \approx 2$) and Half-Field Electron Spin Resonance. *J. Am. Chem. Soc.* **121**, 6919-6927.

Arcangeli, C., Bizzarri, A. R., and Cannistraro, S. (1998). Role of interfacial water in the molecular dynamics-simulated dynamical transition of plastocyanin. *Chem. Phys. Lett.* 291, 7-14.

Astashkin, A. V., Hara, H., and Kawamori, A. (1998) The pulsed electron-electron double resonance and "2+1" electron spin echo study of the oriented oxygen-evolving and Mn-depleted preparations of photosystem II. *J. Chem. Phys.* **108**, 3805-3812.

Banci, L., Bertini, I., Huber, J. G., Luchinat, C. and Rosato, A. (1998). Partial Orientation of Oxidized and Reduced Cytochrome b_5 at High Magnetic Fields: Magnetic Susceptibility Anisotropy Contributions and Consequences for Protein Solution Structure Determination. *J. Am. Chem. Soc.* **120**, 12903-12909.

Bertini, I., Molinari, H., and Niccolai, N., eds. (1991). *NMR and Biomolecular Structure*, VCH Publisher, Weinheim.

Bizzarri A. R. and Cannistraro, S. (1999). Molecular dynamics simulation of plastocyanin potential energy fluctuations: 1/f noise. *Physica A* **267**, 257-270.

Bohr, H. and Brunak, S., eds. (1996). *Protein Folds. A Distance-Based Approach.* CRC Press, Boca Raton, Florida.

Borbat, P. P. and Freed, J. H. (1999). Multi-quantum ESR and Distance Measurements. *Chem. Phys. Lett.* **313**, 145-154.

Brooks, C. L. III, Karplus, M., and Pettitt, B. M., eds. (1988). *Proteins: A Theoretical Perspective of Dynamics, Structure, and Thermodynamics.* Advances in Chemical Physics, Vol. 71.

Buck, M. and Karplus, M. (1999). Internal and Overall Peptide Group Motions in Proteins: Molecular Dynamics Simulations for Lysozyme Compared with Results from X-ray and NMR Spectroscopy. *J. Am. Chem. Soc.* **121**, 9645-9658.

Burchfield, J., Telehowski, P., Rosenberg, R. C., Eaton, S. S., and Eaton, G. R. (1994). A Neutral Water-Soluble Broadening Agent for Spin Label Studies, *J. Magn. Reson.* **B104**, 69-72.

Case, G. D. and Leigh, J. S., Jr., (1976). Intramitochondrial Positions of Cytochrome Haem Groups Determined by Dipolar Interactions with Paramagnetic Cations. *Biochem. J.* **160**, 769-783.

Chadhuri, P., Oder, K., Wieghardt, K., Gehring, S., Haase, W., Nuber, B., and Weiss, J. (1988). Moderately Strong Intramolecular Magnetic Exchange Interaction between the Copper(II) Ions Separated by 11.25 Å in $[L_2Cu_2(\mu\text{-terephthalato})](ClO_4)_2$ (L = 1,4,7-Trimethyl-1,4,7-triazacyclononane). *J. Am. Chem. Soc.* **110**, 3657-3658.

Coffman, R. E. and Buettner, G. R. (1979). A Limit Function for Long-Range Ferromagnetic and Antiferromagnetic Superexchange. *J. Phys. Chem.* **83**, 2387-2392.

Coffman, R. E. and Pezeshk, A. (1986a). Analytical Consideration of Eatons' Formula for the Interspin Distance between Unpaired Electrons in ESR. *J. Magn. Reson.* **70**, 21-33.

Dalal, D. P., Damoder, R., Eaton, G. R., and Eaton, S. S. (1985a). Estimates of Collision Frequencies and Solvent Effects in Collision Interactions. *J. Magn. Reson.* **63**, 327-332.

Dalal, D. P., Damoder, R., Benner, C., Eaton, G. R., and Eaton, S. S. (1985b). Metal-Nitroxyl Interactions. 44. Collision Interactions Between Transition Metal Complexes and Nitroxyl Radicals in Aqueous Solution. *J. Magn. Reson.* **63**, 125-132.

Damoder, R., More, K. M., Eaton, G. R., and Eaton, S. S. (1983). Metal-Nitroxyl Interactions. 30. Single Crystal EPR Spectra of Two Spin-Labeled Copper Porphyrins. *J. Amer. Chem. Soc.* **105**, 2147-2154.

Diehl, M., Doster, W., Petry, W., and Schober, H. (1997). Water-Coupled Low-Frequency Modes of Myoglobin and Lysozyme Observed by Inelastic Neutron Scattering. *Biophys. J.* **73**, 2726-2732.

Dikanov, S. A. and Tsvetkov, Yu. D. (1992) *Electron Spin Echo Envelope Modulation Spectroscopy*, CRC Press, Boca Raton, FL.

Dzuba, S. A., Kodera, Y., Hara, H., and Kawamori, A. (1993). The Use of Selective Hole Burning in EPR Spectra to Study Spectral Diffusion and Dipolar Broadening. *J. Magn. Reson. A* **102**, 257-260.

Dzuba, S. A. and Kawamori, A. (1996). Selective Hole Burning EPR: Spectral Diffusion and Dipolar Broadening. *Concepts in Magnetic Resonance* **8**, 49-61.

Eaton, S. S. and Eaton, G. R. (1978). Interaction of Spin Labels with Transition Metals. *Coord. Chem. Rev.* **26**, 207-262.

Eaton, S. S. and Eaton, G. R. (1982). Measurement of Spin-Spin Distances from the Intensity of the EPR Half-Field Transition. *J. Am. Chem. Soc.* **104**, 5002-5003.

Eaton, S. S. and Eaton, G. R. (1988a). Interaction of Spin Labels with Transition Metals. Part 2. *Coord. Chem. Rev.* **83**, 29-72.

Eaton, G. R. and Eaton, S. S. (1988b). EPR Studies of Long-Range Intramolecular Electron-Electron Exchange Interaction. *Acct. Chem. Res.* **21**, 107-113.

Eaton, G. R. and Eaton, S. S. (1989). Resolved Electron-Electron Spin-Spin Splitting in EPR Spectra. *Biol. Magn. Reson.* **8**, 339-397.

Eaton, G. R. and Eaton, S. S. (1996). Determination of Electron-Electron Interspin Distance by Electron Spin Echo and Saturation Recovery Electron Paramagnetic Resonance. *Current Topics in Biophysics* **20**, 5-8.

Eaton, S. S., Law, M. L., Peterson, J., Eaton, G. R., and Greenslade, D. J. (1979). Metal-Nitroxyl Interactions. 7. Quantitative Aspects of EPR Spectra Resulting from Dipolar Interactions. *J. Magn. Reson.* 33, 135-141.

Eaton, S. S., More, K. M., Sawant, B. M., and Eaton, G. R. (1983). Use of the Half-Field Transition to Determine the Interspin Distance and the Orientation of the Interspin Vector in Systems with two Unpaired Electrons. *J. Amer. Chem. Soc.* **105**, 6560-6567.

Eischenbroich, Ch., Schiemann, O., Burghaus, O., and Harms, K. (1997). Exchange Interaction Mediated by O-H···O Bonds: Synthesis, Structure, and EPR Study of the Paramagnetic Organometallic Carboxylic Acid (η^7-C_7H_7)V(η^5-C_5H_4COOH). *J. Am. Chem. Soc.* **119**, 7452-7457.

Fielding, L., More, K. M., Eaton, G. R., and Eaton, S. S. (1986). Metal-Nitroxyl Interactions. 51. Collapse of Iron-Nitroxyl Electron-Electron Spin-Spin Splitting Due to the Increase in the Electron Spin Relaxation Rate for High-Spin Iron(III) When Temperature is Increased. *J. Am. Chem. Soc.* **108**, 8194-8196.

Goodson, P. A., Glerup, J., Hodgson, D. J., Michelsen, K., and Rychlewska, U. (1994). Magnetic Exchange through Hydrogen Bonds: Structural and Magnetic Characterization of cis-Hydroxoaquachromium(III) Complexes of Tetradentate and Monodentate Ligands. *Inorg. Chem.* **33**, 359-366.

Goslar, J., Piekara-Sady, L., and Kispert, L. D. (1994). ENDOR Data Tabulations, Ch. VI in *Handbook of Electron Spin Resonance*, Vol. 1, C. P. Poole, Jr., and H. A. Farach, eds., AIP Press, American Institute of Physics, New York.

Grupp, A., and Mehring, M. (1990). Pulsed ENDOR Spectroscopy in Solids, Ch. 4 in *Modern Pulsed and Continuous-Wave Electron Spin Resonance*. L. Kevan and M. K. Bowman, eds., Wiley, New York.

Hamachi, I., Yamada, Y., Matsugi, T., Shinkai, S. (1999) Single- or dual-mode switching of semisynthetic ribonuclease S' with an iminodiacetic acid moiety in response to the copper(II) concentration. *Chem. Eur. J.* **5**, 1503-1511.

Hara, H. and Kawamori, A. (1997). A Selective Hole Burning Method Applied to Determine Distances between Paramagnetic Species in Photosystems. *Appl. Magn. Reson.* **13**, 241-257.

Håkansson, K., Carlsson, U., Svensson, M., and Liljas, A. (1992). Structure of native and apo carbonic anhydarse II and structure of some of its anion-ligand complexes. *J. Mol. Biol.* **277**, 1192-1204.

He, M. M., Voss, J., Hubbell, W. L., and Kaback, H. R. (1995a). Use of Designed Metal-Binding Sites to Study Helix Proximity in the Lactose Permease of Escherichia coli. 1. Proximity of Helix VII (Asp237 and Asp240) with Helices X (Lys319) and XI (Lys358). *Biochemistry* **34**, 15661-15666.

He, M. M., Voss, J., Hubbell, W. L., and Kaback, H. R. (1995b). Use of Designed Metal-Binding Sites to Study Helix Proximity in the Lactose Permease of Escherichia coli. 2. Proximity of Helix IX (Arg302) with Helix X (His322 and Glu325). *Biochemistry* **34**, 15667-15670.

Hellinga, H. W. (1996). Design of Metalloproteins, in *Protein Engineering, Principles and Practice*, J. L. Cleland and C. S. Craik, eds., Wiley-Liss, pp. 369-398.

Hubbell, W. L., Fross, A., Langren, R., and Lietzow, M. A. (1998). Recent advances in site-directed spin labeling of proteins. *Curr. Opinion Struct. Biol.* **8**, 649-656.

Hustedt, E. J., Smirnov, A. I., Laub, C. F., Cobb, C. E., and Beth, A. H. (1997). Molecular Distances from Dipolar Coupled Spin-Labels: The Global Analysis of Multifrequency Continuous Wave Electron Paramagnetic Resonance Data. *Biophys. J.* **72**, 1861-1877.

Hyde, J. S., Swartz, H. M., and Antholine, W. E. (1979). The Spin-Probe-Spin Label Method. Ch. 2 in *Spin Labeling II: Theory and Applications*. L. J. Berliner, ed., Academic Press, New York.

Jung, K., Voss, J., He., M., Hubbell, W. L., and Kaback, H. R. (1995). Engineering a Metal Binding Site within a Polytopic Membrane Protein, the Lactose Permease of Escherichia coli. *Biochem.* **34**, 6272-6277.

Kodera, Y., Dzuba, S. A., Hara, H., Kawamori, A. (1994). Distances from tyrosine D^+ to the manganese cluster and the acceptor iron in Photosystem II as determined by selective hole burning in EPR spectra. *Biochim. Biophys. Acta* **1186**, 91-99.

Kokorin, A. I., Zamarayev, K. I., Grigoryan, G. L., Ivanov, V. P., and Rozantsev, E. G. (1972). Measurement of the Distances between the Paramagnetic Centres in Solid Solutions of Nitroxide Radicals, Biradicals, and Spin-Labeled Proteins. *Biofizika* **17**, 34-41 (p. 31-39 in transl.)

Kokorin, A. I., and Formazyuk, V. E. (1981). New Method of Measuring Distances Between Spin Label and Paramagnetic Ions in Macromolecules. *Molek. Biol.* **15**, 930-938 (p. 722-728 in transl.)

Koteiche, H. A., Berengian, A. R., and Mchaourab, H. S. (1998). Identification of Protein Folding Patterns Using Site-Directed Spin Labeling. Structural Characterization of a β-

Sheet and Putative Substrate Binding Regions in the Conserved Domain of αA-Crystallin. *Biochem.* **37**, 12681-12688.

Kuo, L. C., Fukuyama, J. M., and Makinen, M. W. (1983). Catalytic Conformation of Carboxypeptidase A. The Structure of a True Reaction Intermediate Stabilized at Subzero Temperatures. *J. Mol. Biol.* **163**, 63-105.

Kurshev, V. V., Raitsimring, A. M., and Salikhov, K. M. (1988). Angular dependence of the dipole broadening function of the ESR line of single crystals containing paramagnetic centers with an anisotropic g tensor: Er^{3+} in $CaWO_4$. *Sov. Phys. Solid State* **30**, 239-242.

Kurshev, V. M., Raitsimring, A. M., and Tsvetkov, Yu. D. (1989). Selection of Dipolar Interaction by the "2+1" Pulse Train ESE. *J. Magn. Reson.* **81**, 441-454.

Langen, R., Oh, K. J., Cascio, D., and Hubbell, W. L. (2000) Crystal Structure of Spin Labeled T4 Lysozyme Mutants: Implications for the Interpretation of EPR Spectra in Terms of Structure. *Biochemistry* **39**, 8396-8405.

Leigh, J. S., Jr. (1970). ESR Rigid-Limit Line Shape in a System of Two Interacting Spins. *J. Chem. Phys.* **52**, 2608-2612.

Likhtenshtein, G. I. (1976). *Spin Labeling Methods in Molecular Biology*, Wiley, New York.

Lu, Y. and Valentine, J. S. (1997). Engineering metal-binding sites in proteins. *Curr. Opin. Struct. Biol.* **7**, 495-500.

Luckhurst, G. R. (1976) Biradicals as Spin Probes, *Spin Labeling: Theory and Applications*, L. J. Berliner, ed., Academic Press, N. Y., ch. 4.

Makinen, M. W. and Kuo, L. C. (1983) Spin-Label Probes of Enzyme Action, *Magn. Reson. Biol.* **2**, 53-94.

Maret, W. (1993). Detecting Metal-Metal Interactions and Measuring Distances between Metal Centers in Metalloproteins. *Methods Enzymol.* **226**, 594-618.

Martin, R. E., Pannier, M., Diederich, F., Gramlich, V., Hubrich, M., and Spiess, H. W. (1998). Determination of End-to-End Distances in a Series of TEMPO Diradicals of up to 2.8 nm Length with a New Four-Pulse Double Electron Electron Resonance Experiment. *Angew. Chem.* **37**, 2834-2837.

Maryasov, A. G., Tsvetkov, Y. D., and Raap, J. (1998). Weakly Coupled Radical Pairs in Solids: ELDOR in ESE Structure Studies. *Appl. Magn. Reson.* **14**, 101-113.

McCracken, J. (1999) Electron Spin Echo Envelope Modulation, Ch. V in *Handbook of Electron Spin Resonance*, Vol. 2, C. P. Poole, Jr., and H. A. Farach, eds., AIP Press, Springer-Verlag, New York.

Mchaourab, H. S., Berengian, A. R., and Koteiche, H. A. (1997). Site-Directed Spin-Labeling Study of the Structure and Subunit Interactions along a Conserved Sequence in the α-Crystallin Domain of Heat-Shock Protein 27. Evidence of a Conserved Subunit Interface. *Biochem.* **36**, 14627-14634.

Milov, A. D., Salikhov, K. M., and Shchirov, M. D. (1981). Application of the double resonance method to electron spin echo in a study of the spatial distribution of paramagnetic centers in solids. *Sov. Phys. Solid State* **23**, 565-569.

Milov A. D., and Tsvetkov, Yu. D. (1997). Double Electron-Electron Resonance in Electron Spin Echo: Conformations of Spin-Labeled Poly-4-Vinylpyridine in Glassy Solutions. *Appl. Magn. Reson.* **12**, 495-504.

Milov, A. D., Maryasov, A. G., and Tsvetkov, Y. D. (1998). Pulsed Electron Double Resonance (PELDOR) and Its Applications in Free Radicals Research. *Appl. Magn. Reson.* **15**, 107-143.

Milov, A. D., Maryasov, A. G., Tvestkov, Yu. D., and Raap, J. (1999). Pulsed ELDOR in spin-labeled polypeptides. *Chem. Phys. Lett.* **303**, 135-143.

Mims, W. B. and Peisach, J. (1981). Electron spin echo spectroscopy and the study of metalloproteins, *Biol. Magn. Reson.* **3**, 213-263.

More, K. M., Eaton, G. R., and Eaton, S. S. (1982). Metal-Nitroxyl Interactions. 25. Exchange Interactions via Saturated Linkages in Spin Labeled Porphyrins. *Can. J. Chem.* **25**, 1392-1401.

More, J. K., More, K. M., Eaton, G. R., and Eaton, S. S. (1984). Metal-Nitroxyl Interactions. 39. Electron-Electron Spin-Spin Splitting in the EPR Spectra of Nitroxyl Spin-Labeled Mn(II) Complexes in Fluid Solution. *J. Amer. Chem. Soc.* **106**, 5395-5402.

More, K. M., Eaton, G. R., and Eaton, S. S. (1985). Metal-Nitroxyl Interactions. 42. Spin-Spin Interaction in Frozen Solution EPR Spectra of Spin-Labeled Mn(II) Complexes. *J. Magn. Reson.* **63**, 151-167.

Morris, A. T. and Dwek, R. A. (1977). Some Recent Applications of the Use of Paramagnetic Centers to Probe Biological Systems Using Nuclear Magnetic Resonance. *Quart. Rev. Biophys.* **10**, 421-484.

Niimura, N. (1999). Neutrons expand the field of structural biology. *Curr. Opin. Struct. Biol.* **9**, 602-608.

Pake, G. E. and Estle, T. L. (1973). *The Physical Principles of Electron Paramagnetic Resonance*, 2nd Ed., W. A. Benjamin, Inc., Reading, Mass. See pages 157ff.

Pannier, M., Veit, S., Godt, A., Jeschke, G., and Spiess, H. W. (2000). Dead-time Free Measurement of Dipole-Dipole Interactions between Electron Spins, *J. Magn. Reson.* **142**, 331-340.

Persson, M., Harbridge, J. R., Hammarström, P., Mitri, R., Mårtensson, L.-G., Carlsson, U., Eaton, G. R., and Eaton, S. S. (2000). Comparison of EPR Methods to Determine Distances between Spin Labels on Human Carbonic Anhydrase II, submitted for publication.

Piekara-Sady, L., and Kispert, L. D. (1994). ENDOR Spectroscopy, Ch. V in *Handbook of Electron Spin Resonance*, Vol. 1, C. P. Poole, Jr., and H. A. Farach, eds., AIP Press, American Institute of Physics, New York.

Postnikova, G. B. (1996). Spin-Label Studies of the Conformational Properties of Heme Proteins: a Review. *Biochemistry (Moscow)* **61**, 679-693.

Prestegard, J., Tolman, J. R., Al-Hashimi, H. M., and Andrec, M. (1999). Protein Structure and Dynamics from Field-Induced Residual Dipolar Couplngs. *Biol. Magn. Reson.* **17**, in press.

Rabenstein, M. D. and Shin, Y.-K. (1995). Determination of the distance between two spin labels attached to a macromolecule. *Proc. Natl. Acad. Sci. U.S.* **92**, 8239-8243.

Raitsimring, A., Peisach, J., Lee, H. C., and Chen, X. (1992). Measurement of Distance Distributions between Spin Labels in Spin-Labeled Hemoglobin Using an Electron Spin Echo Method. *J. Phys. Chem.* **96**, 3526-3531.

Rakowsky, M. H., More, K. M., Kulikov, A. V., Eaton, G. R., and Eaton, S. S. (1995). Time-Domain Electron Paramagnetic Resonance as a Probe of Electron-Electron Spin-Spin Interaction in Spin-Labeled Iron Porphyrins. *J. Amer. Chem. Soc.* **117**, 2049-2057.

Read, R. J. and Wemmer, D. E. (1999). Biophysical Methods – Bigger, better, faster, and automatically too? Editorial overview. *Curr. Opin. Struct. Biol.* **9**, 591-593.

Reed, C. A. and Orosz, R. D. (1993). Spin Coupling Concepts in Bioinorganic Chemistry. in *Research Frontiers in Magnetochemistry*. C. J. O'Connor, ed., World Scientific, Singapore.

Regan, L. (1993). The design of metal-binding sites in proteins. *Ann. Rev. Biophys. Biomol. Struct.* **22**, 257-281.

Salikhov, K. M., Kandrashkin, Yu. E., and Salikhov, A. K. (1992). Peculiarities of free induction and primary spin echo signals for spin-correlated radical pairs, *Appl. Magn. Reson.* **3**, 199-217.

Saxena, S. and Freed, J. H. (1996). Double quantum two-dimensional Fourier transform electron spin resonance: distance measurements. *Chem. Phys. Lett.* **251**, 102-110.

Saxena, S. and Freed, J. H. (1997). Theory of double quantum two-dimensional electron spin resonance with application to distance measurements. *J. Chem. Phys.* **107**, 1317-1340.

Schlichter, J., Friedrich, J., Herenyi, L., and Fidy, J. (2000). Protein dynamics at low temperatures. *J. Chem. Phys.* **112**, 3045-3050.

Shigemori, K., Hara, H., Kawamori, A., and Akabori, K. (1998). Determination of distances from tyrosine to D and Q_A and chlorophyll$_z$ in photosystem II studied by '2+1' pulsed EPR. *Biochim. Biophys. Acta* **1363**, 187-198.

Sigel, H. and Sigel, A. (1994). Metal Ions in Biological Systems, Vol. 30, *Metalloenzymes Involving Amino Acid-Residue and Related Radicals*. Marcel Dekker, New York.

Sigel, H. and Sigel, A. (1999). Metal Ions in Biological Systems, Vol. 36, *Interelations Between Free Radicals and Metal Ions in Life Processes*. Marcel Dekker, New York.

Steinhoff, H.-J., Radzwill, N., Thevis, W., Lenz, V., Brandenburg, D., Antson, A., Dodson, G., and Wollmer, A. (1997) Determination of Interspin Distances between Spin Labels Attached to Insulin: Comparison of Electron Paramagnetic Resonance Data with the X-Ray Structure. *Biophys. J.* **73**, 3287-3298.

Stowell, M. H. B., Miyazawa, A., and Unwin, N. (1998). Macromolecular structure determination by electron microscopy: new advances and recent results. *Curr. Opin. Struct. Biol.* **8**, 595-600.

Sun, J., Voss, J., Hubbell, W. L., and Kaback, H. R. (1999). Proximity between Periplasmic Loops in the Lactose Permease of Escherichia coli as Determined by Site Directed Spin Labeling. *Biochem.* **38**, 3100-3105.

Tang, J., Thurnauer, M. C., and Norris, J. R. (1994). Electron spin echo envelope modulation due to exchange and dipolar interactions in a spin-correlated radical pair. *Chem. Phys. Lett.* **219**, 283-290.

Tarek, M. and Tobias, D. J. (1999). Environmental Dependence of the Dynamics of Protein Hydration Water. *J. Am. Chem. Soc.* **121**, 9740-9741.

Thomson, C. (1968). Electron Spin Resonance Studies of the Triplet State. *Q. Rev. Phys. Soc.* **22**, 45-74.

Voss, J., Salwinski, L., Kaback, H. R., and Hubbell, W. L. (1995a). A method for distance determination in proteins using a designed metal ion binding site and site-directed spin labeling: Evaluation with T4 lysozyme. *Proc. Natl. Acad. Sci. USA* **92**, 12295-12299.

Voss, J., Hubbell, W. L., and Kaback, H. R. (1995b). Distance determination in proteins using designed metal ion binding sites and site-directed spin labeling: Application to the lactose permease of Escherichia coli. *Proc. Natl. Acad. Sci. USA* **92**, 12300-12303.

Voss, J., Hubbell, W. L., and Kaback, H. R. (1998). Helix Packing in the Lactose Permease Determined by Metal-Nitroxide Interaction. *Biochem.* **37**, 211-216.

Weil, J. A., Bolton, J. R., and Wertz, J. E. (1994). *Electron Paramagnetic Resonance: Elementary Theory and Practical Applications*. John Wiley & Sons, Inc., New York.

Wider, G. and Wüthrich, K. (1999). NMR spectroscopy of large molecules and multimolecular assemblies in solution. *Curr. Opin. Struct. Biol.* **9**, 594-601.

Yager, T. D., Eaton, G. R., and Eaton, S. S. (1978). $(Cr(oxalate)_3)^{3-}$ as a Broadening Agent in Nitroxyl Spin Probe Studies. *J. C. S. Chem. Commun.*, 944-945.

Yoshioka, N., Irisawa, M., Mochizuki, Y., Kato, T., Inoue, H., and Ohba, S. (1997). Unusually Large Magnetic Interactions Observed in Hydrogen-Bonded Nitronyl Nitroxides. *Chem. Lett.* 251-252.

Zwahlen, C., Gardner, K. H., Sarma, S. P., Horita, D. A., Byrd, R. A., and Kay, L. E. (1998). An NMR Experiment for Measuring Methyl-Methyl NOEs in ^{13}C-Labeled Proteins with High Resolution. *J. Am. Chem. Soc.* **120**, 7617-7625.

[illegible] (1990) [illegible]

[illegible] *J. Am. Chem. Soc.* [illegible]

[illegible] (1988) [illegible] *Biochemistry* [illegible]

[illegible] (1990) [illegible]

[illegible] (1989) [illegible]

[illegible] (1990) [illegible]

[illegible] and Kalra, [illegible] (1987) [illegible] *Biochim. Biophys. Acta* **902** [illegible]

[illegible] and Kinosita, [illegible] (1990) [illegible]

[illegible] (1989) [illegible] *Chem. Phys. Lett.* [illegible]

[illegible] and Marsh, [illegible] (1989) [illegible]

[illegible] (1990) [illegible]

[illegible] (1990) [illegible] *Biochemistry* [illegible]

Chapter 2

Relaxation Times of Organic Radicals and Transition Metal Ions

Sandra S. Eaton and Gareth R. Eaton
Department of Chemistry and Biochemistry, University of Denver, Denver, Colorado 80208 USA

Abstract: Review of electron spin relaxation times for organic radicals and transition metal ions in magnetically dilute samples. Emphasis is placed on studies that have been performed as a function of temperature and that provide insight into the relaxation processes.

1. INTRODUCTION

1.1 Scope of this Chapter

Most of the methods for determining distances between spins (ch.1) are either experimentally feasible or theoretically applicable for a limited range of electron spin relaxation times. For example, analysis of the dipolar contribution to the CW lineshape to determine interspin distance requires that the relaxation rate for the interacting partner is slow enough that the dipolar splitting is not collapsed by electron spin relaxation. In addition, analysis of the lineshapes assumes that spectra are obtained under non-saturating conditions and are free from passage effects. Application of methods based on changes in relaxation times requires that the relaxation rate of the interacting partner fall within certain limits. The longest distance that can be obtained by pulsed techniques is limited, in principle, by the relaxation-determined width of an individual spin packet. Thus, knowledge of electron spin relaxation times is foundational to the methodologies presented in this book.

Biological Magnetic Resonance, Volume 19: Distance Measurements in Biological Systems by EPR, edited by Berliner et al. Kluwer Academic/Plenum Publishers, New York, 2000

Relaxation in any physical system is the recovery from a non-equilibrium state to an equilibrium state. The characteristic times are called relaxation times and the reciprocals of the relaxation times are relaxation rates. The emphasis of this chapter is on electron spin relaxation times or rates for samples in which the electron spins are magnetically dilute, as would be found for many biological systems. Although the emphasis is on the electron spins, the electrons typically are surrounded by nuclear spins and relaxation process for the nuclear spins impact the behavior of the electron spins.

Relaxation can be monitored by any physical technique that is sensitive to the relative populations of the states and has an appropriate time scale. In this chapter we emphasize the use of pulsed EPR, as opposed to CW progressive saturation, to determine relaxation times because pulsed techniques are more direct measures of relaxation. There is also an emphasis in this chapter on measurements at multiple temperatures. One might reasonably ask the relevance of low-temperature, or even variable temperature, EPR relaxation studies to biological studies. Are they not made under "unphysiological" conditions? There are two important parts to the answer to this issue. One is that relaxation studies over wide temperature ranges are needed to identify the mechanism of relaxation. Without knowledge of the mechanism, one cannot make predictions about new conditions. Second, many biological systems have strong resemblance to low-temperature glasses. Mark Chance (1993) summarized this, in a review of optical spectroscopic measurements: "The better understanding of proteins as glasses emphasizes the fact that temperature affords a continuum of evaluations which are essential to the understanding of structure-function relationships in biological systems." For example, optical studies have shown that myoglobin is glass-like below 180 K, and liquid-like from 180 K to 300 K (Ahn et al., 1995 and Shibata et al., 1996). Similarly, vibrational echo experiments display $T^{1.3}$-dependent pure dephasing between 15 and ~150 K, suggesting glass-like behavior (Rector et al., 2000).

1.1.1 Literature Covered

This chapter builds on a large literature of electron spin relaxation. To keep this chapter relatively brief, we refer to this background literature largely via some important reviews, especially those by Standley and Vaughan (1969), Abragam and Bleaney (1970), Al'tshuler and Kozyrev (1974), Bowman and Kevan (1979), and Bertini et al. (1994). Most of the papers cited in this chapter were selected to document particular features of spin relaxation, or to lead the reader to specific numerical data, with an emphasis on papers that report relaxation times measured over a significant temperature range. Relatively little emphasis is placed on the direct process (see section 4.2), since this is well covered by the reviews cited. Most references cited are after 1980. There has been no attempt to exhaustively cite the relaxation literature, but we hope we have been comprehensive in citing the literature that shapes a modern view of relaxation in biological EPR. Since a modern view owes a lot to seminal experimental and theoretical papers in many fields, we have selectively cited this literature to give credit to pioneers and to show where our ideas come from. Papers reporting studies of spin-spin interactions, and of spatial distributions of spins, are covered in other chapters and are not cited in this chapter unless they meet the preceding criteria. We will appreciate having oversights brought to our attention.

1.2 Why You Should Care about Relaxation Times

1.2.1 Experimental Design

In this section we give examples of aspects of experimental design that depend upon relaxation times. The details are provided in subsequent sections.

CW EPR. To obtain a valid CW lineshape, spectra should be recorded with a microwave power that does not cause saturation and with scan rates and modulation amplitudes that do not cause passage effects. The appropriate selection of microwave power and sweep time therefore requires knowledge of relaxation times.

ENDOR. Relaxation times determine whether a CW ENDOR experiment is feasible. Simple ENDOR models predict maximum signals when electron and nuclear spin-lattice relaxation rates are equal (Plato et al., 1981).

Pulse EPR. Relaxation times determine whether a pulsed experiment is feasible. The minimum time scale is tens of ns for current pulsed EPR. On the other extreme, long relaxation times require very long pulse repetition times. If there are multiple species with different relaxation times, pulse repetition much shorter than 5-10 times the longest T_1 will preferentially select species. In an echo-detected spectrum the choice of the interpulse spacing can be used to selectively enhance or minimize the contributions from species with different relaxation times.

1.2.2 Information That Can Be Learned from Relaxation Times

Since relaxation times are strongly dependent on the electronic structure of the paramagnetic center and on its interactions with its environment, relaxation times can provide a wide variety of information.

1. The spin-lattice relaxation rate depends upon the electronic structure of the paramagnetic center so analysis of the relaxation mechanism can tell, for example, whether there is a low-lying excited state.
2. Motions shorten spin-lattice relaxation times, T_1, by mixing spin and orbital angular momentum (spin orbit coupling, SOC), modulating zero-field splitting (ZFS), and by mixing spin and rotational angular momentum by a variety of mechanisms. Analysis can provide insight into molecular motions.
3. Dynamics of a radical can be studied by analysis of contributions to T_2 due to rotational modulation of hyperfine and Zeeman anisotropies (Freed, 1998).
4. A knowledge of electron spin relaxation permits predictions concerning potential utility of a paramagnetic species as an MRI contrast reagent or a spin probe broadening reagent.
5. The local concentration of spins is reflected in instantaneous diffusion, so it is possible to distinguish local concentrations from bulk concentrations.
6. Nuclear spins commonly determine the limiting T_m for dilute electron spins at low temperature, so spin echo decays can be used to determine local concentrations of nuclear spins.
7. The distance between a slowly relaxing spin and a more rapidly relaxing spin can be revealed in both T_1 and T_2 (T_m) measurements.
8. Methyl groups coupled to the electron spin result in shorter T_m in the intermediate averaging region. Weakly coupled methyl groups in the vicinity of the electron spin result in changes in the shape and time constant for echo decay curves. Thus, spin echo decays can be used to monitor both intramolecular and intermolecular interaction with methyl protons.

2. TERMINOLOGY

One of the most difficult aspects of the literature concerning relaxation times is the terminology that is used to describe various processes. In this section we provide brief descriptions of key terms and outline some of the issues related to their definitions.

Spin-lattice relaxation time, T_1, is the time constant for equilibration of the populations of the two electron spin Zeeman energy levels. It is defined by $\frac{dn}{dt} = -\frac{(n - n_o)}{T_1}$ where n is the difference between the populations of the m_s = +1/2 and −1/2 energy levels at time t and n_o is the equilibrium difference in populations. Analogous definitions describe the relaxation between electron spin levels when $S > ½$. T_1 is defined in the Bloch equations as the time dependence of the magnetization along the z axis. Since the magnetization along the z axis is due to the difference in the populations of the two Zeeman energy levels these two definitions are equivalent. Spin-lattice relaxation decreases the enthalpy of the electron spin system. If the EPR spectrum is a single isotropic line, with no nuclear hyperfine splitting, and is narrow enough so that all spins can be excited/detected simultaneously, then the relationship between these definitions and experimental data is straightforward. However, for most samples the interpretation of data obtained in experiments to determine T_1 is complicated by the need to account for a variety of other processes that can take spins off resonance *without* changing the net populations of the two Zeeman levels. These processes are called spectral diffusion or spin diffusion. Failure to take account of these processes when interpreting experimental recovery curves can result in interpreting a recovery time constant as T_1, when it is actually a spectral diffusion time, T_D, or some combination. The Bloch equations (unmodified) predict exponential changes in M_z, but the experimental measurements of M_z rarely yield simple exponentials. Consequently, the Bloch equations, while providing a useful conceptual basis, do not describe the full range of phenomena observed.

The word "lattice" was used in early studies because work was performed in ionic lattices. However, the usage has been generalized such that "lattice" now refers to all surrounding atoms.

Spectral Diffusion Time, T_D, is a generic term that encompasses all processes that move spin magnetization between positions in the EPR spectrum. If this process moves an excited spin outside the detection window for the experiment that is used to measure T_1 or T_2, the spectral diffusion is an apparent relaxation process. Spectral diffusion includes motion of an anisotropic paramagnetic center, electron-electron exchange, electron-nuclear cross relaxation, and nuclear spin flip-flops. It can make large contributions to CW saturation, saturation transfer, spin echo, and inversion recovery measurements. It is especially important when spectral lines overlap.

The terms spectral diffusion and spin diffusion are used interchangeably by some authors, and are used by other authors to describe distinct physical processes. One has to be careful about the context of statements in the literature concerning spectral diffusion and spin diffusion. In two important reviews, Salikhov and Tsvetkov (1979) and Bowman and Kevan (1979) distinguished spectral diffusion, defined as the change of EPR frequency of a single radical, from spin diffusion which was defined as the movement (transfer) of spin from one radical (or nucleus) to another (at a different location), which results in changes in the dipolar field at the location of a particular spin, and from cross relaxation, which describes the transfer of energy between spins with different Zeeman frequencies. In this review we will use the more general definition of the preceding paragraph.

Spectral diffusion will depend on radical concentration when the physical mechanism involves electron-electron spin coupling. However, spectral diffusion in most experimental conditions also involves electron-nuclear interactions. With the definition that spectral diffusion includes cross relaxation to transitions of neighboring molecules, then spectral diffusion will depend on radical concentration.

Nuclear Spin Diffusion Rate is the rate of $I_{1+}I_{2-}$ mutual nuclear spin flips (Anderson, 1958; Wolf, 1962; Tse and Hartmann, 1968). It is important in interpreting EPR relaxation because in most samples the magnetically-dilute electron spins are surrounded by magnetically-concentrated nuclear spins. Dipolar coupling between the electron spin and the nuclear spins makes the resonance energy for the electron spin dependent upon the spin states of the surrounding nuclei. Nuclear spin diffusion is a contribution to spectral diffusion for electron spins and to spin-echo dephasing, T_m. Because mutual nuclear spin flips can cause magnetization to move through a sample, nuclear spin diffusion is sometimes called spatial diffusion.

Spin-spin Relaxation Time, T_2, is defined by the Bloch equations as the time constant for dephasing in the x,y plane and is the rate for mutual $S_{1+}S_{2-}$ electron spin flips. As was discussed above for T_1, the measurements of $M_{x,y}$ do not often yield simple exponentials, so the Bloch equations do not provide a full description of the observed phenomena. T_2 is one of the contributors to spin-echo dephasing, but rarely dominates the echo dephasing in solid samples because of the contributions from processes involving nuclear spins. In fluid solution T_2 commonly is described in terms of the linewidth for a Lorentzian line and is attributed to rapid fluctuations in a local field (Atherton, 1993). It is also described as the contribution to the spin-spin relaxation that is caused by rotational motion of the hyperfine and Zeeman anisotropies (Goldman et al., 1972). It should be noted that these definitions of T_2 are not equivalent. A fluctuation in a local field or rotational motion can cause uncertainty in the resonance energy for the electron spin without causing an electron spin $S_{1+}S_{2-}$ flip. Spin-spin relaxation and spectral diffusion increase the entropy of the electron spin system.

In some texts T_2 is expressed as

$$\frac{1}{T_2} = \frac{1}{T_2'} + \frac{1}{2T_1} \qquad (1)$$

where the T_2' sometimes includes inhomogeneous broadening. The derivation of this expression is given in Abragam (1961, p. 443-445) where it is acknowledged that it is obtained for the special case of a single pair of states separated by the Zeeman energy. Bertini and Martini (1994a, page 52-53) discuss confusion that occurs with the use of (1).

T_1 and T_2 as defined in the Bloch equations rarely describe the experimental relaxation measurements made in EPR, because there are so many other physical mechanisms affecting the spins, as described in this chapter.

Spin Echo Dephasing Time, T_m, is a generic term encompassing all processes that cause loss of electron spin phase coherence. T_2 is one of the contributors. There are many other contributions, including, for example, librational motion of the paramagnetic species and nuclear spin diffusion. Contributions to T_m are discussed in section 5. Norris et al. (1980) point out that if the phase memory decay function is a simple exponential (i.e., $x = 1$ in Eq. (8)), then the Bloch equations can be written with T_m in place of T_2, and hence in this case one can write $T_m = T_2$. If the phase memory decay function is not a simple exponential, then T_m does not satisfy unmodified Bloch equations. Consistent with this definition, which applies, for example in the fast motion limit, as in fluid solution or when T_1 dominates T_m, we equate T_m and T_2 (and use the terms interchangeably) only when $x = 1$ (Eq. (8)).

Instantaneous Diffusion is observed in spin echo experiments when the concentration of electron spins/gauss is sufficiently high that the second pulse flips both the observed spin and a neighboring electron spin to which the observed spin is dipolar coupled. By flipping the neighboring spin the pulse changes the resonant energy of the observed spin and results in dephasing. Since this process depends upon concentration, it can be used to determine local spin concentration (Eaton and Eaton, 1993). Some authors call this process spin diffusion.

Cross-relaxation is a term that is used by some authors to describe a mutual spin flip of two unlike spins, which involves spectral features with some overlap (Bloembergen, 1949). This term includes the mutual flip of an electron and a nuclear spin that becomes allowed by dipolar interaction. When two electron spins flip simultaneously in two different transitions, it is called an allowed cross relaxation (Wenckebach and Poulis, 1973). The energy difference is transferred into the extended nuclear Zeeman system. Cross relaxation among electron spins was examined in detail by Bowman and Kevan (1977a) and by Bowman and Norris (1982). Single-step cross relaxation is assumed to be temperature-independent, but multistep cross relaxation, involving different nuclear substates of the same spin system, is expected to occur via a temperature-dependent electron nuclear dipolar relaxation (Kevan and Narayana, 1979). If the cross relaxation is to a spin whose T_1 or T_2 is temperature dependent, then the cross relaxation rate can be temperature dependent (Bowman and Norris, 1982). Cross relaxation can contribute to spectral diffusion.

<u>Saturation Factor</u>. The saturation factor is defined as $s = \frac{1}{1+\gamma^2 B_1^2 T_1 T_2}$. To record a CW spectrum with an undistorted lineshape requires that s be approximately equal to 1. S is most commonly determined by CW power saturation measurements and reflects the relief of saturation by relaxation processes. This expression is derived from the Bloch equations. If it were applied to a single-line spectrum with no nuclear hyperfine, then values of T_1 and T_2 could be obtained from s if the value of B_1 is known. Typically the saturation factor is analyzed for systems where saturation also is relieved by the range of spectral diffusion processes described in this chapter. In systems where spectral diffusion is faster than spin lattice relaxation, it would be more accurate to use T_D in place of T_1 and T_m in place of T_2 in the expression for s.

3. EXPERIMENTAL MEASUREMENTS OF RELAXATION TIMES

It is important to start with a strong caution, that one must be very careful in measuring relaxation times to avoid errors. With modern instrumentation, it is fairly straight-forward for a spectroscopist to perform, for example, a two-pulse spin echo or inversion recovery measurement. However, it may not be straightforward to determine whether the decay time is appropriately labeled as T_2 or T_1, respectively. The following sections provide some guidance in this regard. The amount of effort to be expended in the relaxation time measurement has to be budgeted with an eye on the goal for the interpretation and application of the measured time. For example, if the goal is to select reasonable operating parameters, the major concern will be for the effective time constant itself, and not its ultimate interpretation. On the other hand, if the goal is to measure a true T_1 or T_2, then a major effort will be required in some cases because of the large number of possible contributors to the experimental observables in the relaxation measurement.

Central to the discussions in this chapter are two general concepts. The first concept is that the spins that are excited by a microwave pulse are not necessarily the spins that are observed in the EPR measurement. There are a variety of spectral diffusion processes that can convert excited A spins into non-observed B spins (Fig. 1) and these processes will appear to be relaxation processes.

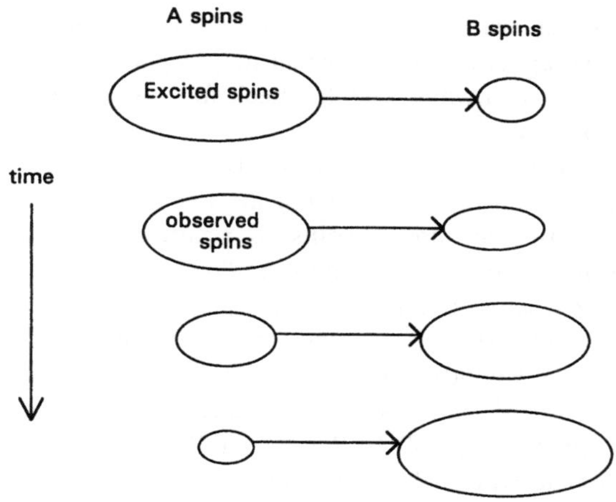

Figure 1. The spins that are excited in an experiment are called the A spins. All other spins are designated as B spins. As a function of time A spins become B spins due to spectral diffusion, which decreases the number of A spins that can be observed.

The second concept is that there are always competing relaxation pathways, and the observed relaxation rate is the sum of these rates. The fastest rate is the one that dominates the observed relaxation. For example, it is possible that the spectral diffusion rate will be fast enough relative to the electron spin relaxation rate that the observed time-domain EPR rate will actually be the spectral diffusion rate and not the spin-lattice or spin-spin relaxation rate that the methodology was intended to measure. Sometimes two or more rates are similar enough that the time-domain measurement yields a multi-parameter fit, and not a simple, single, rate constant (see, for example, Freed, 1974; Yin and Hyde, 1987; Yin et al., 1987). The challenge in designing an appropriate experiment is to pick the method that most clearly defines the time constant of interest.

For studies of electron spins in purportedly magnetically dilute samples it is very important to ensure that the spins are in fact magnetically dilute. For low-temperature measurements on compounds in solution it is necessary to use solvents that form glasses. When a solvent crystallizes solute is excluded from the lattice, which results in regions with locally high concentrations of solute, i.e. electron spins. For example, for a spin label attached to the surface of carbonic anhydrase we observed more than an order of magnitude shorter apparent T_1 in frozen water (even when the sample was thought to be "rapidly" frozen) than for a sample in 1:1 water glycerol that formed a good glass. The requirement for glassy solutions may be relaxed somewhat for paramagnetic centers *buried* in proteins because the protein environment precludes close approach of the paramagnetic centers.

If multiple species are present in solution or if there are distributions of relaxation rates, the apparent time constant may depend on the time window for the experiment. It is always wise to check for this possibility by varying the observation window.

There are many useful ways to categorize experimental methods for measuring relaxation times, and for using the impact of relaxation times to develop methodologies to induce spins to report on the nature of their environment. In this chapter we use the term "pulsed EPR" (or some variant thereof) to describe saturation recovery (SR, in its many forms), electron spin echo (ESE, ESEEM, etc.), and free induction decay – Fourier Transform (FT EPR). Reports from some labs use pulsed EPR to mean SR, and other labs mean ESE, but these are local traditions, and not in conceptual conflict. Continuous wave (CW) EPR is generally understood to encompass all methodologies in which the microwave frequency is held at a fixed frequency and power during the measurement. However, there may have been some pulsed perturbation, such as with a laser, at the start of a CW measurement. There is only a small conceptual divide between producing spin population differences by such a pulse and by a pulse of microwaves (as in SR). Magnetic field modulation is usually used during CW EPR. However, there are other types of modulation that lead to a third category of EPR, modulation spectroscopy, which encompasses methods that modulate some parameter, such as microwave frequency or amplitude, at a rate comparable to T_1^{-1}. (See the discussions below of amplitude modulation and LODESR.) Since magnetic field modulation can occur on a time scale comparable to T_1 (e.g., passage effects), the conceptual divisions are too small to achieve uniformity of nomenclature. Various combinations of the "different" types of EPR can yield unique views of spin behavior. One such example is pulsed EPR with longitudinal detection using pickup coils tuned to the pulse repetition rate (Schweiger and Ernst, 1988).

Strategies for measuring electron spin relaxation were discussed by Bowman (1993).

3.1 Which Spins are Observed?

In comparing CW and pulsed measurements of relaxation times and in comparing different pulsed methods it is important to keep in mind differences in the bandwidths of the experiments.

<u>CW bandwidth</u>. In a CW experiment or in a CW-detected saturation recovery experiment the resonator is critically coupled and the resonator Q typically is relatively high. For example, at X-band the rectangular TE_{102} resonator has Q about 3600 where $Q = \omega/\Delta\omega$. At 9.5 GHz this corresponds to a half-power bandwidth, $\Delta\omega$, of 2.6 MHz. At g = 2 there is about 2.8

MHz/G so this is about 1 Gauss (0.1 mT) at g = 2. This bandwidth means that any process that moves the resonant field by about a gauss on the time scale of the experiment can appear to be a relaxation process.

Bandwidth in pulsed experiments. In a pulsed experiment the bandwidth is determined by the smaller of two bandwidths – the bandwidth of the resonator and the excitation bandwidth that is determined by the length of the microwave pulses (Hornak and Freed, 1986; Bowman, 1990).

The Q of the resonator typically is lower for pulsed experiments than for CW experiments, which gives a larger bandwidth. For example it is fairly common in pulse experiments to overcouple a split ring or loop gap resonator to a Q of 150 which corresponds to an X-band bandwidth of about 64 MHz or 23 G at g = 2. Note, however, that saturation recovery experiments are performed with a critically-coupled resonator, so the resonator bandwidth is the same as for a CW measurement with the same resonator.

The excitation bandwidth (full width at half height), or selectivity, of a pulse is given by $\Delta B_{exc} = 1.6\ B_1 = 1.6/(2\gamma_e t_p)$ (Saalmuetter et al., 1995) where t_p is the length of the pulse. Using a 120 ns π pulse would, for example, lead to an excitation width of 3.8 G and a 20 ns π pulse would excite about 23 G. An alternate approach to estimating the bandwidth for a pulse is the following. A square pulse of length t_p yields a sinc (i.e., (sin x)/x) function in the frequency domain. The frequency separation of the first zeroes of the sinc function is $2/t_p$. Hence, a 20 ns pulse excites ca. 100 MHz or about 35 G at g = 2. This is not a uniform excitation, and the outer parts of the spectrum are turned less than the center. Bandwidth calculated based on half amplitude of an FID is $1.36/t_p$ and for an echo formed by equal power $\pi/2$ - π pulses bandwidth = $0.54/t_p$ (Bowman, 1990).

Thus, unless long selective microwave pulses are used, spin echo experiments tend to excite and observe a wider bandwidth of spins than do CW experiments.

3.2 Spin-Lattice Relaxation, T_1

It is important to distinguish between the experimentally determined response to a perturbation of the spin populations, which might be termed an "effective T_1," and the "true T_1," which is a time for transitions between particular energy levels in an energy level diagram. This distinction includes cases in which, for example, the best experimental determination of T_1 is a value that includes a contribution from spectral diffusion or a nuclear spin relaxation time. In a "T_1 measurement" the observed time will be "the sum of all possible relaxation pathways rather than the relaxation between only the observed levels" (Hyde, 1979, page 27). An attempt is made in the following discussions to indicate where measured "T_1" values may include contributions other than a true electron T_1 between two energy levels. Yin and Hyde (1989) showed that the rates of bimolecular collisions between nitroxyl radicals can be measured in saturation recovery experiments that use observing power so high that the "effective T_1" is altered. The advantage of higher observe power is higher S/N.

Small molecules, and spin labels flexibly attached to large molecules such as proteins, move over greater amplitude and at higher frequency in frozen or glassy solution than is generally assumed. For example, we have found that simple nitroxyl radicals in glassy solvent mixtures have correlation times of ca. 10^{-8} to 10^{-6} s in the liquid nitrogen temperature range (Du et al., 1995a, using the methods of Freed, 1976). This has several implications for relaxation studies. First, we need to recognize that the molecules undergo small amplitude librations many times during the spin-lattice relaxation time. Thus, spectral diffusion due to molecular reorientation is not a major contributor to the observed relaxation time if the saturating pulse is long relative to the measured recovery time. However spectral diffusion will be a major contributor if the saturating pulse is of the order of the molecular librations or tumbling time. The recovery time is measured for those molecules that happen to have an orientation such that they are in resonance at the time of observation. The set of molecules belonging to the observed population of such molecules is changing rapidly with time. Secondly, the motions of the radical relative to the solvent result in collisions between the solvent and the radical that excite vibrations in the radical. These vibrations mix spin and orbital angular momentum, providing a spin-lattice relaxation pathway. The more motion within the frozen solution, the faster the spin-lattice relaxation of the radical. Motional rates can differ because of temperature or because of the structure of the solvent and/or the radical.

Bowman and Kevan (1977a) found that T_1 for electrons trapped in methanol glass depended strongly on the microwave power used to saturate and to observe the spin system. The interpretation was that electron spins were strongly coupled to a large thermal reservoir of methyl groups undergoing tunnelling rotation at a frequency equal to the EPR resonant frequency. However, these methyl groups were only weakly coupled to the lattice, so the methyl reservoir cooled to the lattice temperature very slowly.

3.2.1 Long-pulse Saturation Recovery with CW Detection

T_1 can be measured by saturation recovery with CW detection. This was first demonstrated for nuclear spins by Bloembergen (Bloembergen, 1949). Subsequently there have been many applications to electron spins.

Bloembergen and Wang (1954) measured the change in the z magnetization following a microwave pulse by using a pickup coil outside the resonant cavity. They mentioned, but did not apply, a technique in which recovery would be monitored in a low-level microwave field after the end of the saturating microwave pulse. This method is now called saturation recovery. Weissman and coworkers (1957) manually stepped the microwave amplitude in a measurement of a 20-minute T_1 of triphenylmethyl at 1.2 K. The first saturation recovery spectrometers capable of measurements faster than those that used manual switching were described in 1958 (Davis et al.; Giordmaine et al.), 1959 (Bowers and Mims) and 1960 (Pastor and Hoskins, Pace et al.). The modern development of the field is largely due to Hyde and coworkers (Hyde, 1974; Huisjen and Hyde, 1974; Percival and Hyde, 1975, 1976; Hyde, 1979). Numerous applications of saturation recovery to organic radicals have been published by Venkataraman and coworkers (Ingam et al., 1972; Rengen et al., 1979; Fessenden et al., 1981; Venkataraman, 1982). Detailed descriptions of saturation recovery spectrometers were published by Huisjen and Hyde (1974), Percival and Hyde (1975), Mailer et al. (1985) and Quine et al. (1992). Theory of relaxation, applied to saturation recovery experiments, and illustrated with the case of nitroxyl radicals in fluid solution, was detailed by Freed (1974).

A long saturating pulse is applied to saturate the EPR transition. As soon as possible after this pump pulse the EPR signal is detected with low power continuous wave microwaves. The pulse sequences for saturation recovery and for the other methods discussed in the following sections are sketched in Fig. 2. The time between the end of the pulse and the beginning of observation is determined by switching transients and by the ring-down time of the resonator. The observation of recovery time constants that depend upon the length of the saturating pulse indicates the presence of spectral diffusion. The length of the saturating pulse is increased until a limiting value of the apparent relaxation time constant is observed. The limiting value of the recovery time constant is the best approximation to T_1. A limitation of this generalization was documented in a series of papers by Manenkov and coworkers (Manenkov et al., 1962; Manenkov and Prokhorov, 1962; Manenkov and Pol'skii, 1964). They showed that although long pulses often suppress spectral diffusion, it is possible to achieve conditions where the observed recovery is independent of the pulse width, but there is a steady state where the effect of spectral or spin diffusion is roughly balancing T_1. In this case the recovery following the pump pulse exhibits spectral or spin diffusion. Manenkov focused specifically on the importance of two paths, spin-lattice relaxation and cross relaxation to other states of neighboring multi-level paramagnetic centers, such as Cr(III), Fe(III), Nd(III) and Gd(III). The observations can be generalized to other spectral diffusion mechanisms, since the key is the relative rates of multiple mechanisms. In the Cr(III) systems studied, at various concentrations and at 1.7 and 4.2 K, cross relaxation, T_{12}, was always shorter than T_1. In the general case, not considered by Manenkov, one might have $T_{12} > T_1$, in which case the interpretation of the measured time constants might differ. Daraseliya and Manenkov (1970) showed that one could "quench" the effect of cross relaxation by sweeping rapidly through the line during the saturation time. This saturated all sublevels, so that the recovery was a true T_1. This paper also pointed out that in an inhomogeneously broadened line there could be "a spectrum of cross-relaxation times T_{12}." These ideas were applied to Fe(III) in $Al(NO_3)_3 \cdot 9H_2O$ crystals (Manenkov and Milyaev, 1970) and to Nd(III) (Daraseliya et al., 1970). The latter paper presented a theoretical basis for the prior papers, and showed that for many sets of assumptions the recovery curve will not be a simple exponential in T_1. For Nd(III) the experimental data reveal longer T_1 when the method of rapid scan through the line during saturation is used than when the recovery is deconvoluted into T_{12} and T_1. Unfortunately, the quantitative basis for this analysis was not presented.

Since the saturating microwave pulse is produced by a CW source it can, in principle, be made as long as is required to saturate the spectral diffusion processes. A disadvantage of this technique is that CW detection is less sensitive than spin-echo detection. The detection system in the CW SR experiment does not use high-frequency (e.g., 100 KHz) magnetic field modulation (which could function as a relaxation process) so it does not benefit from the noise-rejection achieved by phase-sensitive detection at the modulation frequency in the usual CW spectrometers. However, some SR spectrometers use dual-channel boxcar detection and/or low-frequency (e.g., 28 Hz) stepping of the magnetic field on and off resonance to eliminate noise at frequencies lower than the pulse repetition rate and the field-stepping rate (Percival and Hyde, 1975, Hyde, 1979). Also, it is necessary to use very low observe powers to avoid perturbing the system. An advantage of CW detection is that it can be used even when T_2 is so short, or echo envelope modulation is so deep, that it would be difficult to observe an echo. Short T_2's can arise in fluid solutions of nitroxyl radicals where T_2 is dominated by incomplete motional averaging of g and A anisotropy and T_1 typically is substantially longer than T_2.

The relative benefits of long and short saturating pulses to obtain various kinds of relaxation information have been described by Hyde and coworkers (see sections 6.1 and 7.1.4).

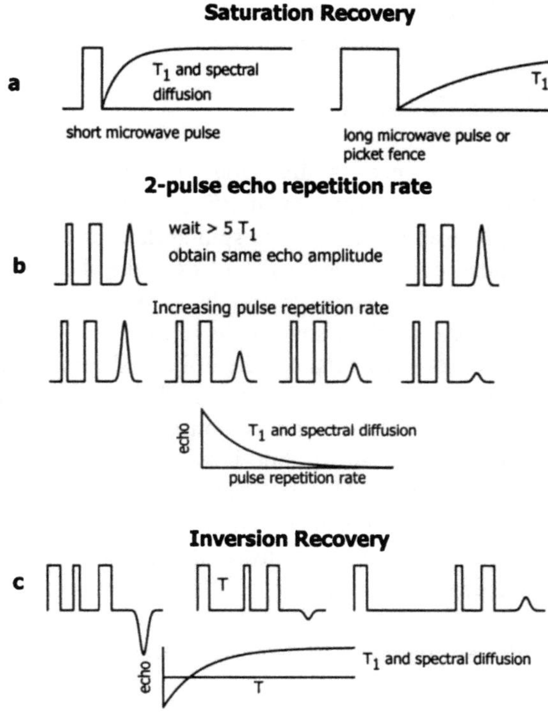

Figure 2. Pulse sequences for measurement of T_1. These sequences are discussed in section 3.2. The sketches in this figure emphasize that in most cases the "T_1" measurement is an "effective" T_1 that includes various spectral diffusion contributions. In some cases, the goal of the measurement might be to quantify the spectral diffusion terms.

3.2.2 Inversion Recovery

The inversion recovery sequence is π-T-π/2-τ-π-τ-echo (Fig. 2c). There are actually 2 FIDs and 4 echoes generated by this sequence. The first pulse inverts the spins. Relaxation occurs during the time interval T, which is stepped in the experiment. The second and third pulses generate a spin echo that monitors the magnetization that remains along the –z axis so the echo at time τ after the 3rd pulse is the one that is measured. Phase cycling can be used to eliminate unwanted echoes. If the initial π pulse inverts the spins, the echo starts out inverted and recovers through zero to an equilibrium value equal to that of the standard 2-pulse echo at time τ. This sequence is the most susceptible to effects of spectral diffusion because the inverting pulse is so short. On the other hand, a short, intense pulse excites the maximal number of spins. Competition between diffusion rates and the spectral window excited determine the impact on the measured T_1.

3.2.3 Picket Fence Excitation

The picket fence sequence is analogous to the inversion recovery sequence, but a "picket fence" (series) of $\pi/2$ pulses with spacing greater than a few times T_m is used to saturate the spin system prior to sampling its recovery with a 2-pulse echo. Successive pulses make the hole wider and wider, exceeding the inverse pulse width and the B_1 width. Pulses can be less than $\pi/2$ provided that the total turning angle is many π (Bowman, 2000). The picket fence sequence helps suppress the effects of spectral diffusion by repeatedly turning the spins toward the xy plane after intervals during which spectral diffusion occurs. If the timing is optimal for the particular spin system, the hole that is burned in the line is wide enough that the spectral diffusion time does not contribute significantly to the measured recovery time, which then can be interpreted as T_1. Imperfect saturation, resulting from various sequences of pulses, leads to results that must be fitted with a 3-parameter fit (Roscher et al., 1996).

3.2.4 Long-pulse Saturation with Spin-echo Detection

The ideal alternative to a picket fence sequence would be an initial saturating pulse that is long relative to T_D followed by two-pulse echo detection. This sequence is not feasible on spin-echo spectrometers that use a pulsed TWT due to the 10 μs limit for the on-time of the TWT. However it can be used on spectrometers with continuous duty TWT's or on spectrometers with a low-power source such as the Bruker W-band system. It has the advantage of a long saturating pulse to overcome the effects of spectral diffusion plus the sensitivity advantage of spin echo detection. Note that since the echo detection is performed with a constant 2-pulse separation, τ, this measurement should be independent of instantaneous diffusion. The measured recovery time will be dependent on spectral diffusion unless the initial saturating pulse burns a hole that is wide relative to the diffusion time, or the echo detection pulse samples a range of frequencies that is wide relative to the spectral diffusion.

3.2.5 Three-pulse Stimulated Echo

Closely related to the inversion recovery sequence is the 3-pulse "stimulated echo" sequence that is used for ESEEM. The sequence is $\pi/2$-τ-$\pi/2$-T-$\pi/2$-τ-echo, and the time T is stepped. The intensity of the τ echo would be expected to decay with the time constant T_1 if all pulses were perfect and if there were no spectral diffusion. In practice the decay is faster than T_1, so the stimulated echo sequence is not a good measure of T_1.

3.2.6 Echo Repetition Rate

Another way to determine T_1 is to measure the amplitude of a 2-pulse echo as a function of the rate of repeating the pulse sequence (Fig. 2b). If the 2-pulse echo is formed before the z magnetization that was turned into the xy plane has recovered, the echo is of lesser amplitude. If spectral diffusion can be ignored, the echo amplitude as a function of the pulse repetition rate yields T_1. A rough estimate of T_1 by this method is inherent in selecting the pulse repetition rate to use in any pulsed EPR experiment. One checks the echo amplitude as a function of pulse repetition rate to decide whether a slower or faster repetition rate should be used. This experiment typically records a smaller number of data points than in saturation recovery or inversion recovery so it is not as useful for testing for multiple relaxation processes or non-exponential decays as experiments that record a larger number of data points.

There is a possible error inherent in a rather amusing example of spin behavior. If the pulse repetition rate is very fast, and the pulse turning angle is very small, it is possible to observe an echo whose intensity decreases as the repetition rate is decreased. For example, with a ca. 0.2 mM nitroxyl solution at ca. 10 K, an echo can be observed with 60 dB attenuation of the output of a 1 KW TWT, if the pulse repetition time is less than a few hundred μs. As the repetition time is lengthened, the echo disappears, and then reappears when the repetition time becomes closer to the T_1 of the nitroxyl, which is of the order of tenths of a second at 10 K.

3.2.7 Ernst Angle

The Ernst angle can be used to optimize pulse repetition rate in pulsed EPR measurements or to determine T_1 if (a) spectral diffusion is not important, and (b) if there is not a distribution of relaxation times. $\cos\theta_{\text{Ernst}} = \exp\left(-T/T_1\right)$, where T is the reciprocal of the repetition rate. The assumptions are important. The Ernst angle for a given pulse repetition rate can be calculated if T_1 is known, or found empirically by seeking the maximum echo amplitude while varying the pulse power or repetition rate (see, for example, Fukushima and Roeder, 1981; Canet 1996; or van de Ven 1995).

3.2.8 Temperature-dependent Contribution to CW Line Width

The vector model of the electron spin polarization suggests that T_1 must always be longer than or equal to T_2, since the spin polarization along the z axis cannot return to equilibrium while there remains net polarization in the xy plane. Conversely, the net x,y polarization will be negligible if the T_1 relaxation returns the spin polarization to the z axis rapidly. Since T_1 decreases with increasing temperature, at some point the lifetime broadening due to short T_1 will dominate other contributions to the CW linewidth and then $T_1 = T_2$. For a single Lorentzian line, the derivative peak-to-peak width, ΔB_{pp}, is related to the relaxation times T_1 and T_2 and the microwave magnetic field, B_1, by (Poole and Farach, 1971, p. 267):

$$\Delta B_{pp}^2 = \frac{4}{3}\left(\frac{1}{\gamma^2 T_2^2} + B_1^2 \frac{T_1}{T_2}\right) \tag{2}$$

If $B_1 << 1$, then

$$\Delta B_{pp}^2 = \frac{4}{3}\frac{1}{\gamma^2 T_2^2}, \text{ or } \Delta B_{pp} = \frac{2}{\sqrt{3}\gamma T_2} = \frac{6.56\text{x}10^{-8}}{T_2} \text{ G}, \tag{3}$$

when T_2 is in seconds. Thus, T_2 = 66 ns causes a 1 G line width.

At low temperature EPR linewidths are determined by inhomogeneous broadening. At intermediate temperatures the linewidth is a convolution of the relaxation-broadened spin packet linewidths with the inhomogeneous broadening. The increase in linewidth due to the relaxation can be used to calculate T_2 (= T_1) using Eq. (3). An example of T_1 measured by saturation recovery merging smoothly with T_1 measured by line width is in Fig. 1 of Marshall et al. (1972). We have used the relaxation determined contributions to CW linewidths to determine relaxation rates for low-spin and high-spin Fe(III) (Budker et al., 1995; Rakowsky et al., 1995, 1998; Seiter et al., 1998). This approach is more useful when the inhomogeneously broadened line at low temperature is relatively narrow.

3.2.9 Microwave Amplitude Modulation Spectroscopy

It was recognized by Hervé and Pescia (1960) that microwave amplitude modulation at a rate close to $1/T_1$ could be used to study relaxation times. Their spectrometer was designed to modulate microwaves to a depth that saturated the spin system to varying extents and at a rate that the spin system could not exactly follow. The fundamental principle of this method is that if the microwave amplitude is changed faster than the electron spin T_1 permits the spins to follow the change, then the resulting time-dependence of the EPR signal is potentially interpretable in terms of the relaxation time (Hervé and Pescia, 1963; Pescia, 1965). Recent applications of modulation spectroscopy to relaxation of transition metals include (Atsarkin et al., 1998; Gourdon et al., 1973; Ablart and Pescia, 1980; Bourdel et al., 1981; Allam et al., 1989; Niang et al., 1995; Zinsou et al., 1996; Vernoux et al.,1996; Pescia, 1998; Pescia et al., 1999). This technique permits measurements of relaxation times as short as 10^{-9} s, which is substantially shorter than can be measured by inversion recovery or saturation recovery.

3.2.10 Pulsed Electron-electron Double Resonance (ELDOR) and 2D FT EPR

In an ELDOR experiment the first microwave pulse is applied at one position in the spectrum and a pair of pulses is then applied at a second position in the spectrum to create a spin echo for detection. Typically this experiment is performed using two microwave frequencies, but it can also be performed using a magnetic field jump between the first and second pulses. Transfer of magnetization from one position to the other is due to spectral diffusion, which can then be separated from T_1 if the equations for the kinetics of the spin response include all of the relevant pathways and enough different starting conditions are created by the pulses to provide sufficient experimental parameters to determine all of the rates.

2D ELDOR, COSY and SECSY measurements of relaxation times were reported by Lee et al. (1993), Sastry et al. (1996a,b) and by Saxena and Freed (1997). The pulse sequences are sketched in Lee et al. (1993). 2D EPR methods give enhanced resolution of structure and dynamics. Cross peaks in the 2D spectrum grow in as a function of mixing time between the 2nd and 3rd pulses, due to nuclear spin flips caused by the electron-nuclear dipolar interaction. T_1 was determined independently after fitting other parameters, but T_2 is a bit uncertain (in some cases the derived T_2 was longer than T_1) due to ambiguity in how some contributions to homogeneous line width are distributed in the model. In the absence of Heisenberg exchange, the time evolution of cross peaks is a direct measure of the nuclear spin relaxation rate (Sastry et al., 1996b).

3.2.11 Multi-Quantum EPR

T_1 can also be measured by use of multiquantum EPR (MQEPR) (Hyde et al., 1989; Sczaniecki et al., 1990, 1991; Mchaourab et al, 1991, 1993a,b,c,d; Eaton 1993; Strangeway et al, 1995). MQEPR encompasses all spectroscopies that involve excitation of an EPR transition with two or more microwave frequencies that are closely spaced relative to T_1^{-1}. This would require the microwave frequencies to be within 1 KHz if T_1 is 1 ms, and within 1 MHz if T_1 is 1 μs. Hyde and coworkers have shown both theoretically and experimentally that irradiating the spin system by ω_1 and ω_2 "leads to oscillation of the spin system at $n(\omega_1-\omega_2)$ and the production of intermodulation (IM) sidebands at $(n+1)\omega_1-n\omega_2$." The intermodulation sidebands caused by the spin system are called spin system IM sidebands to distinguish them from IM sidebands caused by nonlinearities in the microwave system of the spectrometer. Various physical mechanisms for transfer of spin population differences lead to multiquantum analogs of ELDOR and of ENDOR.

Hyde and coworkers demonstrated that in MQEPR the spectral line intensities are proportional to $T_2(T_1T_2)^n$ where n is the harmonic number. Thus, the relaxation time dependence for a 3-quantum transition (n = 1) is $T_1T_2^2$. Similarly, one predicts $T_1^2T_2^3$ for 5-quantum transitions and $T_1^3T_2^4$ for 7-quantum transitions. The scope of MQEPR experiments is presented in Table 1 of Strangeway et al. (1995). Several generations of the MQEPR spectrometer have been described in the context of applications of increasing sophistication. The theory has been presented in depth. MQEPR signals have been observed from aqueous solutions of Fremy's salt and of nitroxyl radicals at room temperature. The oxygen-sensitivity of MQEPR of nitroxyl radicals in solution, including in liposomes, and of spin-labeled peptides (cecopins) bound to liposomes, has been demonstrated. ELDOR of ^{14}N- and ^{15}N-nitroxyl radical solutions illustrate the ability to obtain saturation-transfer rates from the dependence of the MQEPR signal on the difference in the two pumping frequencies. T_{1eff} for the copper in nitrous oxide reductase was measured by observing the amplitude of the MQEPR signal as a function of the separation of the microwave frequencies. At 20 K values of 3.7-7.7 μs were estimated. The ENDOR effect has been measured by observing the MQEPR signal of tri-t-butylphenoxyl radical.

3.2.12 Longitudinally Detected ESR (LODESR)

Irradiation of a spin system with two microwave frequencies, ω_r and ω_s, near the resonant frequency, ω_0, results in oscillation of the longitudinal component of the signal at $|\omega_r - \omega_s|$ and its multiples (Chiarini et al., 1973, 1975; Colligiani et al., 1992). This EPR signal is proportional to the product T_1T_2 (Martinelli et al., 1975, 1977). Using this relation, it was shown that for aqueous Mn^{2+}, $T_1 \propto T^{-2}$ over the temperature range 104 to 182 K (Martinelli et al., 1977), and the angular dependence of T_1 was measured for a nitroxyl spin probe in a smectic A mesophase (Giordano et al., 1979). T_1 for a nitroxyl radical was found to approach about 2×10^{-8} s at 4 to 5 M in toluene (Giordano et al., 1981). This is a much shorter relaxation time than can be measured with CW or saturation recovery techniques. The LODESR spectrometer is less sensitive than a standard EPR spectrometer (Giordano et al., 1981).

3.2.13 Other Techniques

Lineshape changes in the Mossbauer spectrum were used to estimate spin-lattice relaxation of myoglobin azide at temperatures too high to measure it directly by EPR (Schultz et al., 1990). In reading the Mossbauer literature it is important to recognize that Mossbauer results typically are expressed in terms of the rates for transitions between energy levels at equilibrium whereas EPR relaxation is the net change in populations of energy levels that initially are not at equilibrium. In addition, the Mossbauer energy levels are separated by zero-field splittings rather than by the Zeeman energy. Although the relaxation processes are described by the same terminology (direct, Orbach, Raman), the expressions for the temperature dependence of the relaxation rates for these processes (Bizzari et al., 1995; Zimmerman et al., 1974) are different than for EPR measurements of T_1. Also, note that in the absence of an external magnetic field there is no distinction between T_1 and T_2.

NMR relaxation studies of H_2O in solutions of metal ions have yielded estimates of the metal T_1 from its effect on the water protons. These values are summarized in LaMar et al. (1973) and in Banci et al. (1991). For metals with $S>1/2$ there are in general $(2S+1)/2$ different relaxation times (Koenig and Brown, 1985) although the experimental data typically do not provide enough detail to obtain the individual values.

3.3 Spin Echo Dephasing, T_m

Any two pulses will yield an echo. The decay of a two-pulse echo is characterized by the time constant T_m. Frequently there is echo modulation superimposed on the decay. The modulation damps out as a function of increased interpulse spacing and this damping must be taken into account in order to get a valid value for T_m. When the emphasis is on T_m it is best to acquire data with longer low-power pulses that do not excite the modulation as effectively as high-power pulses. Alternatively, the data can be fitted by simulating the data as the product of a decay function and a modulation function. The extent to which the modulation impacts estimates of T_m depends both on the depth of the modulation and on how long the modulation lasts relative to T_m. In some systems there is a distribution of spin-lattice relaxation rates. When this is true it is important to use pulse repetition rates for the spin echo experiments that are long relative to the longest values of T_1 to ensure that all electron spin environments are appropriately sampled by the spin echoes. If shorter pulse repetition rates are used then the echo decays from the environments with the longer values of T_1 are under-sampled. Longer pulses excite a smaller bandwidth of spins and thus produce echo decays that are more sensitive to spectral diffusion processes than decays produced by shorter non-selective pulses.

3.4 Spin-Spin Relaxation, T_2

If there were no nuclear spins coupled to the electron spins, the phase memory time, T_m, would be determined by the interactions among the electron spins, provided that data are obtained under conditions where instantaneous diffusion is negligible. Following Bloembergen et al. (1948), T_2 for the electron spins then is of the order of $T_2 = \frac{hr^3}{2\pi\mu^2}$, where h = Planck's constant, r = distance between unpaired spins, and μ = electron magnetic moment. In irradiated fused SiO_2 there are so few nuclear spins that the relaxation is due to electron-electron dipole-dipole interactions (Mizuta et al., 1994). Using the spin concentration of ca. 6×10^{17} spins per cm^3 measured by ESE for a sample of irradiated fused SiO_2 (Eaton and Eaton, 1993),

$$T_2 = \frac{\left(6.626 \times 10^{-27}\,\text{erg sec}\right)\left(118 \times 10^{-8}\,\text{cm}\right)^3}{2(3.1416)\left(9.28 \times 10^{-21}\,\text{erg G}^{-1}\right)^2} \quad (4)$$

$T_2 = 2 \times 10^{-5}$ sec (= 20 μs)

This approximate calculation is in very good agreement with the measured value of ca. 25 μs, obtained by extrapolating $1/T_m$ to infinitely small turning angle to eliminate instantaneous diffusion. In a similar SiO_2 sample treated with much lower radiation dose, we have observed $T_m = T_2$ = 120 μs at room temperature, and ca. 160 μs at 77 K, where T_1 was long enough not to contribute to T_2.

The Carr-Purcell-Meiboom-Gill sequence (CPMG) (Carr and Purcell, 1954; Eliav and Freed, 1983; Kurshev and Raitsimring, 1990) is widely used in NMR to determine T_2. In this sequence a series of π pulses is used to refocus a spin echo and thereby minimize the effects of spectral diffusion. The echo amplitude is measured after each π pulse. When applied to irradiated alanine (Höfer et al., 1989), this sequence gave a longer time constant than a two-pulse echo. Interpretation of CPMG data in EPR is complicated by the fact that often one cannot excite the whole spectrum and so CPMG time constants also may be impacted by spectral diffusion. Kurshev and Raitsimring (1990) showed that under conditions of partial excitation of the spectrum, as will occur in most EPR spectra, the unwanted echoes can be suppressed relative to the desired CP echoes if the first, 90-degree, pulse is many times as long as the 180-degree pulses in the CP pulse train. It would seem that this method could be very sensitive to hole-filling spin diffusion and spectral diffusion, but definitive experiments are lacking.

When the line width increases with increasing temperature due to spin relaxation, T_2 is calculated from the temperature dependent contribution to the linewidth. Under these conditions, $T_1 = T_2$ (section 3.2.8).

Hyde and Hyde (1981) showed that the wings of symmetrical inhomogeneously-broadened lines are identical in shape to the wings that would occur if all of the Lorentzian lines were at the center, and hence define T_2.

4. PROCESSES THAT CONTRIBUTE TO T_1 IN IMMOBILIZED SAMPLES

The central idea of electron spin-lattice relaxation is that transitions between spin states occur by a process that involves giving up energy to the "lattice." In the earliest work, using crystalline solids (see reviews by Standley and Vaughan, 1969; Bowman and Kevan, 1979), the lattice literally meant the crystal lattice. More generally, by lattice we mean the "surroundings," or "environment" (not including other electron spins) of the observed spins. This environment is assumed to be in thermal equilibrium with the "bath" in which the sample is placed. The larger the molecule, the greater the role of the molecule itself and the smaller the role of the surrounding species in the "lattice". In this section we do not include the effects of a rapidly-relaxing electron spin upon a more slowly-relaxing electron spin to which it is spin-coupled, because those effects are discussed in ch. 8.

The type of transition, and hence the operator and density of states, relevant to electron spin relaxation can be of wide variety. Waller proposed one type, and Van Vleck others (direct, Raman), and later Orbach proposed another, all relevant to transition metal ions doped into crystalline lattices. The matrix elements for the transitions were outlined in Van Vleck's 1940 paper, and in later papers by Orbach. The classic work is described in Abragam (1961), Abragam and Bleaney (1970), Orton (1968), Standley and Vaughan (1969), Al'tshuler and Kozyrev (1974), Kevan and Schwartz

(1979), Poole and Farach (1971). In all of these cases, the interaction Hamiltonian was for orbit-lattice energy transfer involving electronic states and lattice phonons. The lattice vibration spectrum according to the Debye model is explained on pages 153-4 of Orton. Phonons are associated with the elastic deformations of the lattice (Kittel, 1963, 1966; Ziman, 1960). A figure comparable to Orton's 9.1 is in Abragam and Bleaney page 555 (note that the numerator here is x^3, not x^2 as in Orton). These are the now-classical results of Van Vleck and Orbach relaxation theory of transition metal ions in crystalline lattices.

4.1 Debye Temperature

The discussion of the direct, Raman, and Orbach processes are couched in the terms of the phonon spectrum of the solid matrix. Historically, the matrix was a crystal lattice, and the paramagnetic center was tightly coupled to the vibrations of the lattice. The usual treatment considers only the acoustic modes, ignoring the optical modes (Abragam, 1961, page 402). See Huang (1967) for a discussion of optical modes in electron spin relaxation. It remains convenient to continue the solid state physics nomenclature, and report an effective Debye temperature, keeping in mind the refinements discussed below.

The crystal is treated as an elastic solid with a continuum of phonons. Since there are a finite number of atoms (N) there are 3N-6 modes or frequencies in the crystal. If the crystal has s atoms (inequivalent ions in the ionic lattice model) in the primitive cell, there are 3 acoustic modes and 3(s – 1) optical modes. In the Debye model there is a cutoff frequency above which there are no more phonons. This frequency is given by

$$\nu_{max} = \frac{k\Theta_D}{h} \tag{5}$$

where k = Boltzmann's constant, h = Planck's constant, and Θ_D is the Debye temperature. Θ_D is typically of the order of a hundred to a few thousand degrees.

The concept of phonons as collective modes applies to all condensed phases, and not just to the ionic or metallic crystals invoked in textbook derivations. Because of the nature of the Debye approximation, the Debye temperature derived from experimental data depends on the type of data (e.g., heat capacity or relaxation times) and the temperature range studied.

As pointed out by Orton (p. 157) and Abragam and Bleaney (p. 551), phonons have long wavelength relative to lattice spacings. They note that the sound velocity in crystals is of the order of $3x10^3$ m s^{-1} (DiBartolo and

Krykos, 1997, p. 53) (the velocity is anisotropic, but is assumed to be isotropic in almost all calculations). At a 10 GHz phonon frequency the phonon wavelength is ($3x10^5$ cm s^{-1})/(10^{10} s^{-1}) = $3x10^{-5}$ cm, or 3000 A. This is a very long wavelength as seen by the paramagnetic ion and its neighbors. Thus, the phonon interaction picture can be viewed as a long-range interaction.

The vibrational spectra of real solids differ from the ω^2 dependence assumed in the simple Debye model. It would be better to use a more realistic spectrum than the Debye spectrum, but that would require modeling specific to a particular condensed phase. There is a large literature on phonons. Some of the experimental curves for metals look rather like the Debye spectrum, except with structure. For polymers there are several approximations to extend the Debye spectrum, and there are discussions of "optical phonons" or "hot phonons" that are due to vibrations of functional groups.

In molecular solids the Debye model is likely to be at most a crude approximation (Zahlan, 1968). However, one could argue that there probably is close to a continuum of low vibrational energies. Vibrations at higher energies or that are particularly effective in inducing relaxation are likely to appear as local vibrational modes rather than as a contributor to the Raman process.

Proteins are about 50% hydrogen atoms. Motion in proteins, even at low temperature, is an important contributor to electron spin relaxation. Collisions involving side chains are very frequent because the atom packing density in a protein is comparable to that of common organic solids (Kneller and Smith, 1994). The vibrations in proteins include low-frequency, collective modes distributed over the protein, and rigid-body diffusive motions of the side chains (Kneller and Smith, 1994). Above ~200 K there is a non-vibrational component to protein atom dynamics that has been detected by neutron scattering (Kneller and Smith, 1994). At frequencies below ca. 20 cm^{-1} the vibrational frequency distribution of myoglobin is Debye-like, with a ω^2 dependence as expected for a quasiharmonic continuum (Cusack and Doster, 1990). This results in a prediction that the Raman relaxation rate would be proportional to T^9 at low temperature. However, the vibrational density of states is less than predicted by the Debye model at >ca. 25 cm^{-1}, and above ca. 5 K the Raman relaxation rate increases at a lower power of T (Cusack and Doster, 1990).

For many crystals studied by EPR θ_D is of the order of 100 K (Abragam and Bleaney p. 556). Abragam (page 403) says that the Debye temperature is 200 to 300 K for many substances, and thus the Debye frequency is of the order of $2\pi x10^{13}$ s^{-1}. Values of effective Debye temperatures in molecular solids that we have found in the literature are in the range of 20 to 150 K

(Brown, 1979; Zhou et al., 1999). Bowman and Kevan (1977a) found 100 K in glassy aqueous NaOH. Drews et al. (1990) found Debye temperatures of 90 - 100 K for plastocyanin in aqueous solution.

4.2 Direct Process

In the Van Vleck direct process the electron spin transition occurs when there is an exact match of the spin transition energy with a phonon energy, so that there can be a direct transfer of energy from the spin system to the lattice phonon bath. So long as the temperature is much less than the Debye temperature, the phonon density of states is such that the transition probability for the direct process increases linearly with temperature. This process has been observed at temperatures below about 4 K for a variety of transition metal complexes (Scholes et al., 1971, 1984; Yim et al., 1982; Makinen and Yim, 1981; Makinen and Wells, 1987).

It is common to describe as involving a "direct process" any T_1 that has a linear dependence on temperature. One can imagine other mechanisms in which there is an exact match of energies over an extended temperature range. There could be, for example, functional group oscillatory motions in a polymer. In addition, we have observed that if samples are dissolved in solvents that crystallize (instead of forming a glass) the resulting locally-high concentration of solute can result in a weak temperature dependence of $1/T_1$ similar to that predicted by the direct process. The relaxation that is observed under these conditions may be due to intermolecular electron-electron interaction. Thus, the observation of a weak temperature dependence of $1/T_1$ at low temperatures may not always to due to the direct process for isolated spins.

4.3 Raman Process

The Raman process is a two-photon process in which the energy to be transferred to the lattice is the difference between the energies absorbed and emitted for a virtual excited state at any energy less than the Debye temperature (Orton, 1968). The Raman process would seem always to be more likely than the direct process unless there is not good coupling of the relevant vibrational spectrum with the spectrum of the lattice. The Raman process yields a "background" relaxation rate in all systems, but how important it is depends on coupling of the spins to the lattice modes available.

4.4 Orbach Process

The Orbach process also must always be available as an option. In this case the excited state involved is within the allowed range of phonon energies. The details presented in many texts will not necessarily apply, since they are specific to a crystalline lattice with a Debye phonon spectrum. However, the concept that the spin state transition occurs via a two-photon (two-phonon, in the original) process involving a specific excited state, is general. Although the Orbach process is usually written in the form

$$\frac{1}{T_1} \propto \Delta^3 e^{\left(-\frac{\Delta}{kT}\right)} \qquad (6)$$

this is actually the approximation for the limit in which $\Delta >> kT$. The general expression for the Orbach process is included in Eq. (7) (below).

Makinen has applied this relation to the estimation of ZFS terms for high-spin Co(II), since the excited state involved is above the ground state by the ZFS term $2D \sim \Delta$ (Yim et al., 1982; Makinen and Wells, 1987). Note that this treatment assumes that the excited state is still within the phonon energy range, i.e., that $\Delta < k\theta_D$ where θ_D is the Debye temperature. This should be valid for ZFS terms up to ca. kT.

4.5 Local Modes

Although most of the literature of relaxation in "rigid" media builds upon the early concepts of direct, Raman, and Orbach processes, localized lattice modes can contribute to electron spin-lattice relaxation (Klemens, 1962; Valiev and Zaripov, 1962; Castle et al., 1963; Castle and Feldman, 1965a,b; Feldman et al., 1965; Murphy, 1966) and may dominate for covalent lattices or molecular solids. Specific examples are provided in the section that summarizes experimental data.

4.6 Spectral Density Functions

Readers of this chapter who are familiar with the NMR literature may be puzzled that the relaxation processes in the preceding sections are not described in terms of spectral density functions that have the form $\frac{\tau_c}{1+\omega^2\tau_c^2}$, where τ_c is the correlation time for the perturbing process and ω is the resonant frequency for the experiment. Analysis in terms of spectral density functions assumes that the relevant motion is random and stochastic, with an autocorrelation function that decays exponentially with the characteristic time τ_c (see, for example, Carrington and McLachlan, 1967, Appendix E). In condensed phases the degrees of freedom are vibrations and oscillatory motions of the molecule as a whole. Vibrational frequencies, whether of the lattice or of the molecule under study, can be expressed in terms of normal modes, which do not behave in accordance with the autocorrelation function that leads to the above expression in terms of τ_c. Although anhármonicity effects cause mixing and finite lifetimes of phonons, τ_c should not be identified with phonon or vibrational frequencies. Oscillatory (librational) motions in solids and rotational motions in less viscous condensed phases typically make it necessary to consider the magnitude of the term $\omega^2\tau_c^2$ at X-band. If this term is significantly less than 1, the spectral density function can be approximated as τ_c.

The most commonly encountered relaxation process for which τ_c typically is slow enough that spectral density functions are used to describe the process in solid samples is the relaxation of a slowly relaxing spin by a more rapidly relaxing spin to which it is spin-coupled. This case is discussed in ch. 8. If τ_c is due to a thermally-activated process, then spectral density functions can describe the contribution to the relaxation, as proposed by Atsarkin et al. (1997). Also, in the regime where a glass is softening it may be appropriate to describe the increased importance of molecular rotations by a thermally-activated τ_c and a spectral density function.

4.7 Fitting Experimental T_1 Data

To fit the temperature dependence of $1/T_1$ over a significant temperature range one potentially needs to consider any of the processes described in the preceding paragraphs. Such a fit function would have the contributions shown in Eq. (7).

$$\frac{1}{T_1} = A_{dir}T + A_{Ram}\left(\frac{T}{\theta_D}\right)J_8\left(\frac{\theta_D}{T}\right) + A_{loc}\left[\frac{e^{\Delta_{loc}/T}}{(e^{\Delta_{loc}/T}-1)^2}\right] + A_{orb}\frac{\Delta_{Orb}{}^3}{e^{\Delta_{Orb}/T}-1} + A_{therm}\left[\frac{2\tau_c}{1+\omega^2\tau_c{}^2}\right] \tag{7}$$

where T is temperature in Kelvin

A_{dir} is the coefficient for the contribution from the direct process

A_{Ram} is the coefficient for the contribution from the Raman process

θ_D is the Debye temperature

J_8 is the transport integral, $J_8(\frac{\theta_D}{T}) = \int_0^{\theta_D/T} x^8 \frac{e^x}{(e^x - 1)^2} dx$

A_{loc} is the coefficient for the contribution from a local vibrational mode

Δ_{loc} is the energy for the local mode in units of Kelvin

A_{Orb} is the coefficient for the contribution from the Orbach process

Δ_{Orb} is the energy separation between the ground state and the excited state for the Orbach process

A_{therm} is the coefficient for the contribution from the thermally-activated process

τ_c is the correlation time for the thermally-activated process $= \tau_c^o e^{Ea/T}$

E_a is the activation energy for the thermally-activated process

The expressions for the relaxation processes are described in the following references: Raman process (Murphy, 1966), local mode (Castle and Feldman, 1965a), Orbach process (Orbach, 1961), and thermally-activated process (Atsarkin et al., 1997).

The weak temperature dependence of relaxation produced by the direct process is distinctive and typically is most important at low temperatures. The temperature dependence of $1/T_1$ produced by the Raman process also is distinctive. Over a limited temperature interval ($T < \Delta_{loc}$, Δ_{Orb}, or E_a) the temperature dependence of $1/T_1$ due to a local mode, an Orbach process, or a thermally activated process is similar and distinctions between these processes require knowledge of the electronic and vibrational structure of the paramagnetic species. For example, there are no low-lying excited states for a nitroxyl radical or for Cr(V) complexes so an Orbach process is not plausible for these systems. However, the contribution from an Orbach process needs to be considered for low-spin Fe(III). Note that more than one (indeed many) local mode and Orbach process terms may be needed to fit experimental data. For example, multiple levels have to be considered for high-spin Fe(III).

For certain cases the mathematical expressions in Eq. (7) are replaced by other terms. For example, if the two-photon Raman process involves optical phonons of energy Δ, the Raman term becomes $\text{csch}^2(\Delta/2kT)$ (Huang, 1967; Hilczer et al., 1993). If the Orbach process (sometimes called the Orbach-Aminov process) involves simultaneous transition of the spin and a local vibrational photon in doublets of excited states split by ΔE, the term becomes $\text{csch}(\Delta E/kT)$ (Hoffmann et al., 1996).

Equation (7) does not show the field dependence of the direct and Raman processes, nor does it show the different forms of these terms for non-Kramers states (see Section 4.11 and Standley and Vaughan, 1969; Abragam and Bleaney, 1970).

4.8 Modulation of Spin-orbit Coupling

The preceding paragraphs invoke vibrations in a variety of relaxation processes. Vibrations modulate the electronic energy of the paramagnetic center because of vibronic coupling. Changes in the electronic energy levels are coupled to the spins via spin-orbit coupling. Therefore, the effect of vibrations on spin-lattice relaxation depends both on spin-orbit coupling and on the vibronic coupling for a particular system. Molecular flexibility also is a factor. For a given electronic structure and magnitude of spin-orbit coupling the less rigid the molecule, the more sensitive the unpaired electron will be to motions of surrounding atoms and the more rapidly it will relax.

In the absence of spin-orbit coupling, electron g values would be isotropic and equal to the free electron g-value (Weil et al., 1994; Drago, 1992). Spin-orbit coupling arises from coupling between the spin and orbital angular momenta of the unpaired electron. Depending on the symmetry of the system, degenerate ground states can have inherent orbital angular momenta. To zeroth order approximation, non-degenerate ground states are "spin-only", but spin-orbit coupling mixes in certain symmetry-selected excited states, which creates orbital angular momentum in the ground state. The extent of the mixing is proportional to the magnitude of the spin-orbit coupling constant, λ, and inversely proportional to the energy separation between the ground and excited states (Weil et al., 1994). Thus, spin-orbit coupling has a greater impact in systems with low-lying excited states. One result of spin-orbit coupling is the deviation of g values from 2.0023. Hence, one can estimate the significance of spin-orbit coupling for a particular species based on g values. In general, spin-orbit coupling increases from left to right across the periodic table and from top to bottom. Thus, if modulation of spin-orbit coupling dominates relaxation, we expect faster relaxation for heavier elements.

The next step in the argument is to consider vibronic coupling. That is, to what extent does vibration mix with electronic states, and hence provide a means for converting spin angular momentum to vibrational energy via the orbital angular momentum? We can argue based on the symmetry of the transitions, as first pointed out by Van Vleck. See, for example, Cotton (1971, p. 280). The key is that the direct product of the wave functions, $\Psi_e H \Psi_e' \Psi_v$, contains the totally symmetric representation. This will happen whenever the direct product $\Psi_e H \Psi_e'$ contains the irreducible representation of one of the vibrational modes. In O_h for example, the vibrational modes are A_{1g}, E_g, $2T_{1u}$, T_{2g}, T_{2u}. Since the electronic transition operators have the symmetry of the unit vectors (dipole transitions), which are u, and the d-d transitions are g, this leads to the prediction that the d-d transitions are parity forbidden, but can occur due to vibronic coupling to the T_{1u} and T_{2u}

vibrational modes. The mixing occurs along all three axes in the cubic groups (e.g., O_h and T_d), but in lower symmetry environments x, y, and z may not belong to the same representation. The significant transition probabilities for d-d transitions are evidence of the importance of vibronic coupling.

Spin-lattice relaxation is only one of the manifestations of the importance of vibronic interactions. There is a large literature of vibronic interactions, in the discussion of Jahn-Teller distortions (Abragam and Bleaney Ch. 21; Ham 1965, 1968) and g-values. Leading references for the interpretation of specific systems include: Jahn-Teller distortion of d^9 Cu(II) in a cubic environment (O'Brien, 1964); interstitial Ni(I) in diamond (Isoya et al., 1990); in ferricenium cation (Prins, 1970); Cu(II) (Breen et al., 1969, Williams et al., 1969); coronene and triphenylene anions (Townsend and Weissman, 1960; McConnell, 1961). References for the temperature-dependence of g-values include discussion of Ni(I), Ni(III), Rh(II), Ru(I), Fe(I), Pd(III) (Vugman and Amaral, 1990; Vugman et al., 1992; de Abreu et al., 1992). Hoffmann et al. (1998b) argue that only for Cu^{2+} in $ZnSiF_6{\cdot}6H_2O$ is there convincing evidence that the dynamic Jahn-Teller effect influences T_1. Bacci and Cannistraro argue that the vibronic coupling mechanism effective in causing temperature dependence of g values in pseudo-tetrahedral Cu(II) complexes is not effective in square-planar complexes (Bacci, 1986; Bacci and Cannistraro, 1987a,b). This is in conflict with our spin-lattice relaxation studies of Cu(II) (Du et al., 1996). A direct comparison of Cu^{2+} and Mn^{2+} in the same single crystal host at the same orientation showed that T_1 for Cu^{2+} is about an order of magnitude faster than that for Mn^{2+} (Hoffmann et al., 1998b). The interpretation was that "g tensor modulations by the lattice phonons for Cu^{2+} are a more effective relaxation mechanism than the D tensor modulation mechanisms for Mn^{2+} ion."

4.9 Tunnelling

Bowman and Kevan (1977a,b) suggested a new relaxation mechanism involving modulation of the electron-nuclear dipolar interaction by tunnelling of atoms or molecular groups in the environment of the radical. This mechanism was important in T_1 relaxation of radicals with little g anisotropy in disordered matrices, and fit experimental data in the ca. 10 – 50 K region that could not be fit by the direct or Raman processes (Bowman and Kevan, 1977a,b, 1979; Michalik and Kevan, 1978). It is assumed that "the tunnelling rate is rapid enough to be treated as a random process" (Bowman and Kevan, 1979). For several plausible distributions of tunnelling modes, the relaxation rate was predicted to be proportional to the temperature, T. The rate is also proportional to ω^{-2}. For measurements at a given microwave frequency, the term included in the function used to fit experimental relaxation rates as a function of temperature is D^2T, where D^2 is an empirical parameter obtained by fitting the data (Bowman and Kevan, 1977a,b, 1979; Michalik and Kevan, 1978). Stinson and Stapleton (1983) also observed relaxation linearly dependent on T, and attributed it to electron spin relaxation "through localized tunnelling states created by an 'off-center' lithium impurity." This mechanism predicted a functional form that under certain limiting conditions yielded linear dependence of T_1^{-1} on T.

4.10 Fractal Relaxation

Stapleton and coworkers suggested that the spin-lattice relaxation of low-spin Fe(III) hemes could be interpreted in the framework of a fractal description of the protein (Allen et al., 1982). Subsequent work from the same lab found metalloprotein relaxation that did not fit the fractal model (Stevens and Stapleton, 1990). Subsequently, many labs have tried to fit relaxation data to a fractal model (Schultz et al. 1990; Tabti et al., 1997). We have observed that the temperature dependence of the spin-lattice relaxation rates is very similar for low-spin Fe(III) in small-molecule porphyrin complexes and in methemoglobin and metmyoglobin (Budker et al., 1995; Rakowsky et al., 1995; Zhou et al., 1999). It is unlikely that the fractal dimensions of these systems are similar, so we conclude that the fractal dimension does not determine the relaxation of heme proteins.

Early modeling of the temperature dependence of relaxation rates was expressed in terms of $1/T_1$ = coefficient(s) times T^n where integer values of n were predicted for limiting cases such as low-temperature (n = 7 or 9) and high-temperature (n = 2) limits of a Raman process. In this framework it seemed plausible to assign non-integer values of n to fractal behavior. However, when the full forms of the equations for various relaxation processes are considered, the predicted temperature dependence is not limited to integer values of the exponents. In addition, non-integer exponents can arise from overlapping contributions from processes with different temperature dependence of the relaxation rates. Thus the observation of non-integer values of n should not be taken as sufficient evidence for a fractal process. Even a demonstration that some other aspect of a system has a fractal nature does not prove that the electron spin relaxation mechanism is also fractal. Caution should be exercised in applying the fractal model to spin relaxation.

4.11 Field/frequency Dependence of Relaxation

In this section we consider the dependence of relaxation on field/frequency per se (Standley and Vaughan, 1969; Poole and Farach, 1971) as opposed to orientation dependence that arises from anisotropy of the spin system (Bowman and Kevan, 1979; Brown, 1979; Gorchester et al., 1990). Historically such a larger fraction of EPR experiments have been performed at X-band that there has been much less discussion of the frequency dependence of relaxation rates/times in EPR than in NMR. However, with the increasing availability of EPR spectrometers at frequencies other than X-band it becomes important to ask whether a particular experiment can be optimized by going to higher or lower frequency. One factor to consider is relaxation times.

Relaxation processes of isolated spins in the solid or glassy state rely on vibrations and oscillatory motions. In the limit of slow motions, the temperature dependence of relaxation is due to changes in the populations of vibrational energy levels. For example, the direct process results in a relaxation rate that is proportional to B_o^2 (non-Kramers species) or to B_o^4 (Kramers species) in the limit that $kT >> h\nu$ (Standley and Vaughan, 1969). This dependence on B_o^4 has been demonstrated (Davids and Wagner, 1964) and exploited to use higher observe power at higher frequency to improve S/N of otherwise too-slowly relaxing species (Muller et al., 1989). The first-order Raman process can vary as B_o^2 (longer relaxation time at lower field) (Marchand and Stapleton, 1974), but the more common Raman process is predicted to be independent of magnetic field. The Orbach process is independent of magnetic field, except insofar as the field may alter the value of the energy of the excited state. High magnetic fields and frequencies can bring other energy separations into resonance, such as transitions that are between states separated from the ground state by a ZFS.

For the more general case, the frequency dependence of relaxation rate depends on spectral density functions that have the form $\frac{\tau_c}{1+\omega^2\tau_c^2}$. As microwave frequency is varied, τ_c due to random motions may become comparable to ω which could cause enhanced rates of relaxation. At sufficiently high frequency the relaxation rate would then decrease. When the process that is effecting relaxation is the motion that occurs near the softening point of a glass, increasing microwave frequency could increase or decrease relaxation depending on the whether $\omega^2\tau_c^2$ is greater or less than one. When the correlation time is the relaxation time for a rapidly relaxing spin to which the slowly relaxing spin is spin-coupled, the maximum effect on the relaxation occurs when $\omega^2\tau_c^2$ is ~1 as discussed in ch. 8.

The rate of relaxation via modulation of tunneling modes in glassy matrices is predicted to depend on B_0^{-2} (Bowman and Kevan, 1977a,b). There are also special cases of apparent magnetic field dependence that involve resonant cross relaxation for two species with different g and nuclear hyperfine values (Larson and Jeffries, 1966). In some systems whose relaxation was interpreted in terms of tunnelling mechanisms, a slight decrease in relaxation rate was observed at 16.5 GHz relative to 9.5 GHz (Kurtz and Stapleton, 1980; Askew et al., 1984).

When the relaxation rate that is actually measured is dominated by spectral diffusion, and not T_1^{-1} or T_2^{-1}, as often is the case in CW power saturation measurements, the increased spectral dispersion at higher field can result in a decrease in the effective relaxation rate because the spins per gauss, and hence this contribution to spectral diffusion, decreases (Ghim et al., 1995) (see section 7.9). However, g-anisotropy has a larger effect at higher microwave frequency, since small changes in molecular orientation have a larger effect on resonant field ($dB/d\theta$ is larger). Thus, relaxation is more sensitive to faster motion at higher microwave frequencies and higher magnetic fields, and this contribution to spectral diffusion increases at higher frequency. The greater sensitivity to faster motion, and to the specific model for rotational diffusion, is fundamental to the studies of molecular motion at 250 GHz by the Freed group (Barnes et al., 1999; Budil et al., 1993; Earle et al., 1993, 1997), who have compared CW line widths at 9.5 and 250 GHz.

4.12 Effect of Pairs and Clusters

In the early literature many of the studies of relaxation focused on ions in ionic lattices. Following the statistics of Behringer (1958), at a 1% doping level of a simple cubic lattice 5% of the dopant will be in pairs and 0.4% will be in triples. In real systems charge compensation, lattice strain, etc., can cause deviations in either direction from the random distribution calculations, but the likelihood remains that significant fractions of the paramagnetic ions would exist in pairs and a non-trivial fraction would exist in triples. Note also that closed triples (where each is a neighbor of the other two) would exhibit spin frustration. Since 1% doping was then considered a relatively low level, it was important to assess the role of pairs and clusters in relaxation. At the low temperatures where most of the studies were performed, the direct process dominated. Under these conditions pairs and/or triples made major contributions to the relaxation because relaxation was faster in the pairs or triples and cross relaxation between isolated ions and the pairs/triples enhanced relaxation for all spins in the sample. The following references provide an entre to this literature: Van Vleck (1960, 1961), Abragam and Bleaney (1970, p. 552), Gill (1962), Atsarkin (1966),

Dugdale and Thorpe (1969), Harris and Yngvesson (1968), and Al'tshuler and coworkers (1975).

At the concentrations of paramagnetic species typically present in magnetically dilute samples of molecular species, the concentrations of pairs and triples present in randomly oriented samples is much lower than considered in the early studies. However these studies make the key point that when there is strong spin-spin interaction between paramagnetic ions, such as in dimers or trimers, the energy levels are different from the monomers, which can result in enhanced relaxation.

5. PROCESSES THAT CONTRIBUTE TO T_M IN IMMOBILIZED SAMPLES

As noted in section 3.1 the bandwidth for detection of spin echo decays typically is only a few gauss. Any process that takes a spin out of the observation window will eliminate it from contributing to a T_1 recovery. Any process that shifts the resonant frequency by $1/\tau$ or more will prevent the spin from refocusing to form the echo, even if it stays within the bandwidth of the detection window. Any such process constitutes a dephasing mechanism. These processes include a range of dynamic processes as enumerated in the following paragraphs.

5.1 The Shape of the Echo Decay Curve

The shape of the echo decay curve depends upon the rate of the process that dominates the dephasing (Zhidomirov and Salikhov, 1969). A fit to the full shape of the curve provides the most detailed description. Comparison with limiting forms of the expression can be performed by fitting the echo decay to the expression

$$E(2\tau) = \exp\left[-\left(\frac{2\tau}{T_m}\right)^x\right] \quad \text{where } \tau \text{ is the time between pulses.} \qquad (8)$$

The value of x is determined by the mechanism of phase memory decay and the rate, W, of the dephasing process relative to τ (Brown, 1979; Salikhov and Tsvetkov, 1979). When $W\tau \ll 1$ as is observed for nuclear spin diffusion, x is 2 to 2.5. When $W\tau$ is ~1, x can become as small as 0.5. For $W\tau \gg 1$, x = 1. When instantaneous diffusion dominates (see next paragraph), x =1. In the slow motion region, and for very short times, τ, the echo decay shape is an exponential in τ^3 (Schwartz et al., 1982).

When the echo decay curve exhibits deep modulation due to interactions with nuclei (Dikanov and Tsvetkov, 1992), the value of T_m will appear to be smaller than the true value unless proper account is taken (e.g., by simulation) of the modulation pattern.

5.2 Instantaneous Diffusion

If the local spin concentration is high enough, T_m is decreased by the effects of instantaneous diffusion (Klauder and Anderson, 1962). To separate the contribution of instantaneous diffusion, T_m can be extrapolated to infinitely small turning angles. The relevant formula is: $E(2\tau) = (\text{constant})e^{(-2b\tau)}$, where $b = \frac{1}{T_m} + \gamma\left(\frac{4\pi^2 g\beta C}{9\sqrt{3}}\right)\sin^2\left(\frac{\theta}{2}\right)$ and C is the concentration in spins/cm^3. This formula is based on the assumption that the spectrum is narrow enough that all spins are excited by the microwave pulses. An example of the application of this formula to determine the local concentration of radicals in irradiated glassy SiO_2 is given in Eaton and Eaton (1993).

Whether instantaneous diffusion is an important contributor to T_m depends on the magnitude of other contributions to T_m. In a protic environment, and for a radical with a spectrum as broad as a nitroxyl, instantaneous diffusion makes an important contribution to T_m when the radical concentration is more than a few tenths millimolar and the pulses are non-selective. Very broad spectra, such as for many metal ions, do not exhibit instantaneous diffusion effects even at millimolar concentrations. When the protons are removed, as in deuterated solvents, instantaneous diffusion is an important contribution to T_m at much lower spin concentrations than is observed in protic solvents. Examples of the effect of instantaneous diffusion on T_m are shown in section 7.1.1.

Instantaneous diffusion was a more serious problem in early experiments where instrumental sensitivity was not as good as it is today. The lower sensitivity forced early workers to use higher spin concentrations than typically are used today. For example, instantaneous diffusion was observed when the copper concentration was $3\text{x}10^{18}$ cm^{-3} but not when it was $5\text{x}10^{17}$ or lower (Breen et al., 1969; Williams et al., 1969). In the ensuing discussion of processes that impact T_m it is assumed that data were acquired under conditions such that instantaneous diffusion is negligible.

5.3 Nuclear Spin Diffusion

It has been recognized for some time that for dilute solutions of organic radicals at low temperatures ESE dephasing is dominated by nuclear spins (Gordon and Bowers, 1958; Brown, 1971, 1979; Salikhov and Tsvetkov, 1979). We have shown that this is also true for S=1/2 transition metals Cr(V) and vanadyl ion (Eaton and Eaton, 1999) and propose that it is true for most S = ½ metal complexes in the regime where T_m is much shorter than T_1. For example, S = ½ organic radicals and transition metal complexes exhibit T_m about 2 – 4 μs in normal proton-containing materials at low temperature (see below for the special effect of methyl groups). T_m is much longer in deuterated solvents, reaching ca. 100 μs. The very large difference between the effect of ^1H and ^2H on T_m makes it difficult to obtain an accurate value in deuterated solvents, because of the strong effect of small amounts of ^1H contamination. The relative effect of ^1H and ^2H (or any other nucleus) on T_m can be roughly estimated by the ratios of the magnetic moments. For example the nuclear magnetic moments of some selected spins are given in Table 1. As a rough approximation, the electron-nuclear coupling is diminished by the relative magnetic moments, and the nuclear spin diffusion is diminished by the square of the moments, so the overall effect on the electron T_m should decrease as the 3rd power of the nuclear moments. This is approximately what is observed. For the case in which the electron T_m in proton-containing solvents is ca. 2 μs at low temperature, T_m increases by about 35 times to ca. 70 μs in the analogous deuterated solvent.

Table 1. Nuclear magnetic moments and effectiveness in echo dephasing[a]

nucleus	magnetic moment in nuclear magnetons	magnetic moment relative to ^1H	relative moment to –3 power
^1H	2.793	1	1
^2H	0.8574	0.307	35
^{19}F	2.627	0.941	1.2
^{35}Cl	0.8209	0.294	39
^{37}Cl	0.6833	0.245	68

[a]data from the CRC Handbook of Chemistry and Physics

There are several practical implications of these effects of nuclear spins. Clearly, replacing ^1H by ^{19}F will not increase T_m very much (this has been confirmed in our lab). Replacing ^1H by ^2H or chlorine will increase T_m very significantly. The potential to apply some multiple-pulse techniques increases as T_m increases, so of the common solvent mixtures, ^2H-containing and chlorine-containing solvents offer the greatest spectral flexibility. Solvents with very few nuclear spins, such as CS_2 have similar advantages, if they form a glass in a mixed solvent that will dissolve the species to be studied. Note also that the longer is T_m, the more sensitive it is to other effects, from instantaneous diffusion to electron-electron spin interaction.

Exchange of spin orientation energy (mutual spin flips, $I_{1+}I_{2-} \leftrightarrow I_{1-}I_{2+}$) between two neighboring spins with essentially the same Zeeman energy is called spin diffusion (Bloembergen, 1949; Portis, 1956; Redfield, 1959; Abragam, p. 381; Suter and Ernst, 1982, 1985). These flip-flop processes modulate the electron-nuclear dipolar interaction, which provides a mechanism for echo dephasing. The large number of nuclear spins is the key to the relation between pairwise spin flips, whose rate is of the order of their dipolar interaction (~10 KHz), and T_m^{-1}, which is much faster. The average flip rate due to many nuclear spins interacting with the electron spin S can be much higher than the rate of adjacent pairs of spins (Greis et al., 1994).

The term "spin diffusion" is used because the phenomenon is described by a typical diffusion equation, which for nuclear spins is:

$$\frac{\partial p}{\partial t} = D\Delta p \tag{9}$$

where p describes the nuclear polarization and D is the diffusion constant. D is given approximately by Wa^2, where a is a distance and W the probability of a flip-flop between nearest neighbors. Abragam stated that W is ca. 10^3 sec^{-1}. We found that the dephasing of nitroxyl radicals in proton-containing solvents could be accounted for with a flip frequency of 24 to 33 Hz/M which corresponds to W about $3x10^3$ s^{-1} for typical proton concentrations (Zecevic et al., 1998). If the average distance between protons were 2.5 Å, this value of W would correspond to D = $2x10^{-12}$ cm^2s^{-1}, which is comparable to spin diffusion rates determined by NMR (Table 2). Cheung (1996) reviewed nuclear spin diffusion and found in polymers "a typical T_1 of the order of 0.1 s and ^1H spin diffusion coefficient of $6x10^{-12}$ cm^2s^{-1}."

Since the electron spin phase memory relaxation depends strongly on nuclear spin relaxation, it is important to provide some background on what is known about nuclear spin relaxation and how it fits with the observations of electron spin relaxation. The nuclear relaxation times and nuclear spin diffusion values in Table 2 are not the result of an exhaustive literature search. The references given merely provide an entre to the large literature of nuclear relaxation. Lowe and Gade (1967) estimated the spin diffusion constant to be within a factor of two or three of 0.15 $\gamma^2\hbar/a$ for a simple cubic lattice of spin-1/2 nuclei. The nuclear spin-spin relaxation time, T_{2n}, also can be used to compute the diffusion constant using D = 2 r_o^2/T_2 where r_o is the proton van der Waals radius (0.117 nm) (Asano and Takegoshi, 1998). Reynhardt and Terblanche (1996) and van Wyk et al. (1997) give the spin diffusion constant for ^{13}C relaxation in natural diamond, containing paramagnetic impurities as $D = a^2/(50\ T_{2I})$ where a is the average distance between the I spins (^{13}C).

Table 2. Nuclear Spin Diffusion

Sample	10^{12}x D (cm^2s^{-1})	Diffusion times	Reference
abundant ^{19}F or ^1H spins, typical	1		Cheung, 1996
irradiated alanine	3.7		Kevan et al., 1980; Schlick et al. 1981
alanine		1-50 μs	Brustolon and Segre, 1994
solid fluorinated steroids		<50 ms	Carss et al., 1996
alkanes	6.2		Gerstein, 1998
PEO and LDPE D_{rigid}	3.2, 2.9, 8.0		Weigand et al., 1996
D_{mobile}	0.9, 1.5, 0.5		
CaF_2 single crystal	7.1 ± 0.5 and 5.3 ± 0.3		Zhang and Cory 1998
DPPH in polystyrene		20 ± 5 μs	Boscaino et al., 1986a,b
organic polymers	7		Schmidt-Rohr and Spiess, 1994
multiphase polymers	0.8 for PS phase and 0.5 for rubbery PI phase		Wang et al., 1997
polymer blends, ^{13}C		0.5 - 0.8 s^{-1}	Henrichs and M. Linder, 1984; Linder et al., 1985
protein	4.0	0.01 to 100 ms	Kumashiro et al., 1998
natural abundance ^{13}C in rigid solids		0.01 s^{-1}	Colombo et al., 1988
crystalline solids		6-20 ms for ^{13}C, and ca. 5 for ^{14}N and ^2H	Suter and Ernst, 1985
^{31}P in crystalline solids		0.1 to 3 s^{-1}	Kubo and McDowell, 1988

It is commonly stated that spin diffusion rates are independent of temperature. However, if small amplitude motions affect the dipolar couplings among the spins, spin diffusion rates may be temperature dependent in solids (Diezmann, 1995). When T_2 is independent of temperature, the spectral diffusion process is attributed to energy-conserving spin-exchange reactions among the nuclei (Greis et al., 1994). Spin diffusion due to dipolar coupling in perdeuterated biphenyl was temperature dependent with an activation energy of 75 kJ/mol in the range 290 to 315 K, and was attributed to 180° flips of the phenyl rings (Müller and Haeberlen, 1996). The phrase "spectral spin diffusion" was used to distinguish the case in which there are resolved lines in the NMR spectrum, and energy mismatch slows down the spin diffusion process.

Here we have emphasized the impact of the nuclear spins on the electron spin. However, nuclear relaxation due to interaction with a nitroxyl radical yield estimates of intramolecular distances up to 10 – 15 Å with an accuracy of about 5% (Cinget et al., 1995).

5.3.1 The "Black Sphere" or Diffusion-barrier Radius

Since Blumberg (1960) there has been general acceptance that there is a sphere (of radius b) of nuclei that are shifted from the nuclear resonance due to the field of the paramagnetic ion that do not participate in the nuclear diffusion. Based on a comparison of echo dephasing for the nitroxyl radical tempone and a much larger trityl radical we concluded that the protons at distances closer than about 6 Å did not contribute to echo dephasing (Zecevic et al., 1998). The 6 Å value is similar to the value of 5.9 Å found by Hurrell and Davies (1971) for paramagnetic centers in CaF_2.

However, the papers by Wolfe and coworkers (King et al., 1972; Wolfe, 1973) showed that anisotropy of the electron-nuclear interaction can provide a match between energies of bulk spins and spins close to the paramagnetic center and thus provide for spin diffusion of nuclear spins within the traditional barrier radius. Nuclei as close as 3 Å to Yb^{3+} were in strong thermal contact with the bulk spins. The diffusion barrier radius is strongly affected by the anisotropy of the local dipolar field. Calculations by Horvitz (1971) show that for some cases the nuclear spin diffusion is greater inside the "critical radius" of the Bloembergen model than outside.

5.4 Dephasing by Methyl Groups in the Solvent/surroundings

Early studies of T_m focused on the distinction between protons and nuclei with smaller magnetic moment (Brown, 1979). We have observed that solvent methyl groups play a special role in echo dephasing for nitroxyl radicals (Zecevic et al., 1998) and S=1/2 transition metal complexes (Eaton and Eaton, 1999). At approximately the same total proton concentration T_m at 11 and 40 K depends both on the concentration of methyl protons and on the types of methyl protons. In methyl-containing solvents the shape of the echo decay is different from that in non-methylated solvents and fits to Eq. (8) give exponents significantly less than 2. Exponents less than 2 are characteristic of cases in which the correlation time for the dynamic process that dominates dephasing is comparable to (or shorter than) τ, the time between pulses. This indicates that in methyl-containing solvents there is an additional process that occurs more quickly than the spin-diffusion in typical proton-containing solvents. We propose that classical rotation of methyl groups in the solvent is responsible for the enhanced dephasing (Zecevic et al., 1998).

Since many organic solvents used in common mixtures that form good glasses at low temperature contain methyl groups (e.g., toluene, methyl-THF, ethanol, methanol), and many protein side chains contain methyl groups, methyl group effects on T_m likely play a greater role than had been previously recognized. In proteins both spin labels and metals may be in the vicinity of rotating $-CH_3$ and $-NH_3$ groups and other side-chain groups with correlation times in the range 10^{-7} to 10^{-12} s (Torchia, 1984). Rotation of methyl groups is observed at low temperatures, down to a few Kelvin (McBrierty, 1974). The methyl rotational correlation times of various crystalline amino acids and peptides depend strongly on packing in the crystal (Torchia, 1984). The methyl group in methionine has the lowest activation energy of any methyl group in an amino acid (Keniry et al., 1983). Decreased rotational correlation time is observed as the length of the amino acid side chain increases: e.g., 800 ps in alanine, 40 ps in leucine, and 5 ps in methionine at room temperature (Keniry et al., 1984). Above ca. 10°C large-amplitude motions of the side chain occur in proteins (Keniry et al., 1983).

5.5 Averaging of Electron-nuclear Couplings due to Rotation of Methyl Groups Within the Radical or Transition Metal Complex.

When the rate of rotation of a methyl group (τ_c^{-1}) is slow compared with the electron-nuclear spin-spin coupling to the protons of the methyl group, the coupling constants to the three protons typically are inequivalent. As the rate of rotation of the methyl group increases, the rotation causes an averaging of the three couplings, which is then a dephasing process. In the limit of fast averaging of the hyperfine couplings this process makes a negligible contribution to T_m. The maximum effect on T_m occurs in a limited temperature regime where the rate of rotation is approximately equal to the difference in the coupling constants that is averaged (ω_2-ω_1). For an activated process the rate is temperature dependent, so the methyl rotation process results in a distinctive temperature dependence of T_m. This behavior is analogous to the more familiar lineshape changes that occur in dynamic NMR, and the limiting equations that can be used to describe the electron T_m effect are the same as those used in dynamic NMR (Drago, 1977). This process causes a characteristic decrease, and then increase, in T_m as temperature is increased. This effect has been observed for nitroxyl radicals (Nakagawa et al., 1992), Cr(V) complexes (Nakagawa et al., 1992), and

Cu(II) complexes (Du et al., 1994, 1995c). Awareness of this phenomenon is particularly important to researchers who work with nitroxyl spin labels because the rotation of the ring methyl groups occurs at rates comparable to the differences in electron-nuclear couplings at temperatures between about 70 and 160 K, which results in dramatic decreases in T_m in this temperature range as discussed in section 7.1.1. Saxena and Freed (1997) pointed out that in nitroxyl radicals methyl rotation can lead to nitrogen nuclear spin flips according to Eq. (10)

$$W_n = \frac{1}{60}\left[\Delta A^H(\Omega)\right]^2 \frac{\tau_I}{1+\omega_a^2\tau_I^2} \tag{10}$$

where τ_I is the correlation time for internal motion, ω_a is the proton Larmor frequency, and $\Delta A^H(\Omega)$ "represents that portion of the methyl proton hf tensor that is averaged by the internal rotations, and it depends on the overall molecular orientation, Ω." For twelve equivalent methyl protons, Saxena and Freed found that the effective spin flip rate is about $5W_n$.

Phenyl groups undergo both small angle jumps and 180° flips (Zhao et al., 1995). Above room temperature the time for ring motion was on the order of tens to hundreds of microseconds. The effects of phenyl group motion have been observed in a doped solid containing Cu(II) (see section 7.6.7) (Du et al., 1994).

5.6 Collapse of Electron-electron Spin-spin Coupling to a More Rapidly Relaxing Partner.

The relaxation of a more rapidly relaxing partner to which a slower relaxing spin (such as a nitroxyl radical) is spin-coupled is a dephasing mechanism analogous to rotation of a methyl group. This process also results in a distinctive decrease and then increase in T_m as the rate of relaxation of the more rapidly relaxing spin passes through the regime in which the relaxation rate is comparable to the dipolar coupling. This phenomenon is described in ch. 8.

Even if the two centers have the same relaxation times, as, for example, two interacting nitroxyl radicals, modulation of the electron-electron dipolar interaction is a T_2 mechanism. Shorter T_2 for a biradical relative to a monoradical was observed by Saxena and Freed (1997), and other examples are described in other chapters in this book.

5.7 Librational Motion

For an anisotropic paramagnetic center small amplitude molecular motions that are known as librations, can result in significant changes in the resonance energy. When these changes occur on the timescale of the echo experiment the motion can contribute to echo dephasing. For a given amplitude of motion, the effect of the motion on the resonance energy is smaller near the principal axes and larger at intermediate orientations. Thus, librations result in values of T_m that are longer along the principal axes than at intermediate orientations. This phenomenon has been observed in glassy solvents for nitroxyl radicals (Dzuba et al., 1984; Paschenko et al., 2000), Cr(V) complexes (Nakagawa et al., 1992; Konda et al., 1994), copper and vanadyl complexes (Du et al., 1992) (see sections 7.1.1, 7.6.2, 7.6.7). At the same temperature this effect is greater in glassy solvents than in crystalline lattices, which is attributed to larger amplitude motion in the glasses than in crystalline solids.

5.8 Molecular Tumbling

As a glass softens there is increasing freedom of molecular tumbling. For an anisotropic paramagnetic center, tumbling moves spins off resonance and contributes to dephasing. In fluid solution this is frequently the dominant contribution to T_m. Extensive ESE studies of the effect of slow motions on T_m by Tsvetkov and coworkers reveal details of intra- and inter-molecular dynamics not visible to other physical techniques (see, for example, Dzuba et al., 1984; Tsvetkov, 1990; Dzuba et al, 1992; Toropov et al., 1998). Freed and coworkers have shown that the variations in T_2 across a spectrum are very sensitive to models of molecular orientational dynamics (e.g., Saxena and Freed, 1997, and references therein). The relation of T_2 to motion was summarized by Saxena and Freed as follows: "In the motional narrowing case $T_2 \propto \tau_c^{-1}$, where τ_c is the correlation time for the motion, so that T_2 will typically decrease with decreasing temperature. However, for the slow motional case, $T_2 \propto (\tau_c)^{\alpha}$ with $1 \leq \alpha \leq ½$ so that T_2 typically increases with decreasing temperature."

5.9 Oxygen

There is no effect on T_m, within experimental error, of degassing frozen aqueous solutions of nitroxyl radicals. Note, however, that O_2 can contribute to CW spectra at low temperature. We have observed such spectra due to O_2 trapped in the frost that forms on the outside of EPR tubes during storage in liquid-nitrogen-cooled Dewars.

6. PROCESSES THAT CONTRIBUTE TO T_1 AND T_2 IN FLUID SOLUTION

6.1 Small Molecules

For molecules in solution, the tumbling process often is a major contributor to relaxation. McConnell (1956) and McGarvey (1957) showed that relaxation times of metal ions in solution depended on the anisotropy in g, the anisotropy in the nuclear hyperfine interaction, and the correlation time that described the tumbling motion of the ion in solution. The fundamental theory of EPR line shape and relaxation for dilute solutions of free radicals was developed in papers by Freed and Fraenkel (1963) and Freed (1965), and by Kivelson and coworkers (Kivelson, 1960; Atkins and Kivelson, 1966; Wilson and Kivelson, 1966a,b,c). The full theory is complicated, and the original papers and subsequent reviews should be examined before over-interpreting phenomena in terms of simplified versions of the theory. Bertini et al. (1994) compiled many equations used to describe relaxation in solution. With the increasing emphasis on EPR frequencies other than X-band, it is important to note that most common lineshape expressions assume that the magnetic field is large relative to all anisotropies, which may not be true at low field, and that some anisotropies are small enough to be ignored relative to others, which may not be true at high field. As a practical matter, simplified equations can be very useful if the range of applicability is kept in mind. Kivelson (1960) and Wilson and Kivelson (1966a) summarized several contributions to EPR line width with the relation (11), where m is the nuclear spin quantum number:

$$T_2^{-1}(m) = \alpha + \beta m + \gamma m^2 + \delta m^3 + \kappa m^4 \quad (11)$$

Notation varies in papers that apply this equation. See Chasteen and Hanna (1972) for a convenient summary. The nuclear-spin-independent term, α, which depends on modulation of g anisotropy, hyperfine anisotropy, and molecular tumbling correlation time (variously called τ_r or τ_c) is sometimes split into two terms. One of the contributions to this term is spin-rotational interaction (see below). The coefficient of the term linear in m contains products of g and hyperfine anisotropy, and depends on cross-correlation between these. The coefficient of the term in m^2 contains hyperfine anisotropies squared. The coefficient of the m^3 term depends on electron-nuclear anisotropic dipolar and g tensor relaxation. Terms in m^3 and m^4 are usually omitted because they are small and are rather uncertain theoretically (Wilson and Kivelson, 1966a; Chasteen and Hanna, 1972).

For the specific, and common, case of ^{14}N nitroxyl radicals in dilute fluid solution, one can compare the outer lines to the center line, and rewrite (11) in terms of the line widths $\Delta B(m_n)$ as

$$\left(\frac{\Delta B(0)}{\Delta B(m_n)}\right) = 1 + Bm_n + Cm_n^2 \tag{12}$$

where the coefficients are given by

$$B = \frac{2\hbar}{\sqrt{3}g\beta}\left(\frac{4B_0}{15}\right)(b\Delta\gamma)\left(\frac{1}{\Delta B(0)}\right)\tau \tag{13}$$

$$C = \frac{2\hbar}{\sqrt{3}g\beta}\left(\frac{b^2}{8}\right)\left(\frac{1}{\Delta B(0)}\right)\tau \tag{14}$$

$$b = \frac{2}{3}\left(A_z - \frac{1}{2}\left(A_x + A_y\right)\right) \tag{15}$$

$$\Delta\gamma = \frac{\beta}{\hbar}\left(g_z - \frac{1}{2}\left(g_x - g_y\right)\right) \tag{16}$$

These expressions can also be written in terms of line heights, with the approximation that the line shapes are the same and hence the widths are inversely proportional to the square root of the heights. In this approximation expressions for B and C can be written:

$$\sqrt{\frac{I(0)}{I(+1)}} - \sqrt{\frac{I(0)}{I(-1)}} = 2B \tag{17}$$

$$\sqrt{\frac{I(0)}{I(+1)}} + \sqrt{\frac{I(0)}{I(-1)}} - 2 = 2C \tag{18}$$

Hence, from the relative line heights and values of g and hyperfine anisotropies, the tumbling correlation time, τ, can be estimated. If the tumbling rate is fast enough for these equations to be valid, then the line shape will correlate with relaxation rates, but will not be a direct measure of relaxation rates. Since there are so many contributions to the line widths of nitroxyl radicals, pulsed EPR methods are needed to measure the relaxation times. The EPR line shapes for peroxylamine disulphonate (PADS, Fremy's salt) are not encumbered by unresolved hyperfine structure, so it was possible to estimate T_1 and T_2 by CW EPR methods (Kooser et al., 1969).

For very small molecules in low-viscosity solvents, relaxation can be affected by the coupling of spin angular momentum with rotational angular momentum (Nyberg, 1967, 1969; Wilson and Kivelson, 1966; Atkins and Kivelson, 1966; Atkins, 1972; see also section 7.1.4). The spin-rotational (SR) interaction is dependent on g-anisotropy and rotational correlation time, but not on microwave frequency or magnetic field (Atkins and Kivelson, 1966; Atkins 1972).

$$\frac{1}{T_1{}^{SR}} = \frac{1}{T_2{}^{SR}} = \frac{1}{12\pi r^3} \frac{\left(\Delta g_{\parallel}^2 + \Delta g_{\perp}^2\right) kT}{\eta} \tag{19}$$

where r is the molecular radius. Robinson and Mailer and coworkers wrote this as

$$\frac{1}{T_{1e}^{SR}} = \frac{\sum_{i=1}^{3} (g_i - g_e)^2}{9\tau_c} \tag{20}$$

where $\tau_c = \dfrac{V\eta}{kT}$ and V is the hydrodynamic volume of the radical.

Spin-rotation was found to be an important relaxation mechanism for Fremy's salt (Kooser et al., 1969; Goldman et al., 1973), nitroxyl radicals (Hwang et al, 1975), and semiquinones (Leniart et al., 1975).

The role of molecular tumbling in spin-lattice relaxation for nitroxyls is discussed in section 7.1.

Lewis and Morgan (1968) summarized early studies of transition metal ions. Rubinstein et al. (1971) described the relaxation of metal ions with S >1/2 in fluid solution. Bertini and Koenig and coworkers have examined the applicable theory for various metal ions in the context of nuclear magnetic resonance dispersion (NMRD) studies (Gabor et al., 1972; Banci et al., 1986; Banci et al., 1991; Bertini and Luchinat, 1986; Bertini et al., 1985, 1986, 1989a,b, 1992a,b, 1993a,b, 1994, 1999).

6.2 Macromolecules

Koenig and Brown (1985) made the key point that macromolecules in fluid solution are tumbling so slowly that paramagnetic centers buried in the macromolecule behave as if they were in the solid state. In an NMRD study of the relaxation of Cu(II)-substituted transferrin and native (copper-zinc) superoxide dismutase it was observed that the spin-lattice relaxation rate for the Cu(II) did not depend on the rotational correlation time for the protein

(Bertini et al., 1989a). It was proposed that the electron spin relaxation occurred via vibrations of the protein that provided a "microcrystal" environment for the paramagnetic center. Results from several metalloprotein studies are included in Table 3. Our results for high-spin Fe(III) hemes (see section 9.3) suggest that relaxation times for high-spin iron in solids or glasses at room temperature are similar to values for the same protein in fluid solution. These results all suggest that relaxation of paramagnetic centers buried in proteins may be similar to results in solids and glasses, even when the protein is in solution at room temperature.

For spin-labels attached to the surface of macromolecules local motion may provide relaxation pathways that are more similar to those for freely tumbling small molecules.

6.3 Oxygen

Collision with O_2 dissolved in the solvent can be a major contributor to electron T_1 of radicals in fluid solution. With proper measurement strategies the collision rate can be measured, and the oxygen concentration can be measured (see especially the papers by Hyde and coworkers cited in section 7.1.4 and in Table 3). Kusumi et al. (1982) defined an "oxygen transport parameter,"

$$W = T_1^{-1}(\text{air}) - T_1^{-1}(N_2) \tag{21}$$

and Yin and Hyde (1987) related nitroxyl collision with O_2 to other contributors to T_1 (see section 7.1.4.).

6.4 The Ultimate Limit on Relaxation

At some high temperature, the spectral density of general "lattice" vibrations and their coupling to the spin system will be small enough that the Raman process will become less important than specific interaction with discrete energy levels. These discrete energy levels can be excited electronic states and/or local vibrational modes. Since vibrations are usually less than about 3000 cm^{-1}, and often much lower in energy, the highest frequency accessible via vibrations is ca. (3000 cm^{-1})($3x10^{10}$ cm s^{-1}) = $9x10^{13}$ s^{-1}. This is about the limit of relaxation of lanthanides in fluid solution (Burns and LaMar, 1982), where it is assumed that ZFS and spin-orbit modulation via vibrations excited by collisions are the relaxation mechanism.

7. EXPERIMENTAL DATA

Subsequent to the early studies of relaxation of transition metal and rare earth ions in crystalline lattices, summarized in Bleaney and Stevens (1953), Manenkov and Orbach (1966), Standley and Vaughan (1969), and Al'tshuler and B. M. Kozyrev (1974), most EPR relaxation studies have been of organic radicals in fluid solution and defect centers in solids. Relaxation of metals in fluid solution and in proteins has been measured via the effect of the electron on nuclear relaxation. An extensive tabulation covering papers published up to the late 1980s, with a few papers from 1990, is in the 230-page chapter by Bertini, Martini, and Luchinat (1994).

Initially, CW saturation methods were used to estimate relaxation rates for Fremy's salt and small nitroxyls at room temperature. Key papers were by Fraenkel and coworkers (Schreurs et al., 1960; Stephen and Fraenkel, 1960) and by Freed and coworkers (Eastman et al., 1969, 1970; Kooser et al. 1969; Freed 1972, 1976; Goldman et al., 1973; Schneider and Freed, 1989). Direct measurements of T_1 by saturation recovery by Rengen (1971, 1974a, 1974b) and by Hyde and coworkers (Huisjen and Hyde 1974a,b, 1976; Percival and Hyde 1975; Hyde 1974, 1998), and then of T_2 by Freed and coworkers (Gorchester et al., 1990) and by Schwartz et al. (1979; Stillman and Schwartz, 1979) solidified understanding of the rates and mechanisms of relaxation of small organic radicals in fluid solution (Freed 1972, 1976, 1998; Schneider and Freed, 1989). The direct measures of T_1 and T_2 were then used by these labs to explore motions in fluid solution and collisions with other nitroxyl radicals and with O_2 (Subczynski and Hyde, 1981; Hyde et al., 1990; Subczynski et al., 1992a,b). Dynamics of nitroxyl radicals in glassy solvents were studied by the Novosibirsk group using ESE (Tsvetkov and Dzuba, 1990; Tsvetkov, 1998) (see also section 5.7). Some of this literature was reviewed by Hyde (1974). Study of transition metal ions, except as noted above, has received less attention. Most of the understanding of the approximate magnitudes of relaxation rates of transition metal ions in solution has come from NMR studies of nuclei interacting with the paramagnetic metal (Lewis and Morgan, 1968; LaMar et al., 1973, pp. 77-81; Bertini and Luchinat, 1986, p. 77; Banci et al., 1991; Bertini et al., 1989b). Although there have been many ESEEM studies of transition metals and other paramagnetic centers, rarely have the published papers reported T_1 or T_2 values for the species studied.

In Table 3 we cite some relaxation times for organic radicals and metals as an indication of the range of information available, and the approximate numerical values for species of interest for biomedical EPR. The original papers should be consulted for methods and conditions of measurement and for interpretation of the relaxation times. Unless otherwise stated, the values cited are for fluid solution near room temperature, and were measured at X-band. Some values mentioned in the text are not included in Table 3.

Table 3. Relaxation Times for Organic Radicals and Metals[a,b]

Species	T_1 (μs)[c]	T_2 (μs)[c]	notes	Reference
solvated electron	300	0.5	2-methyl-THF at 77 K	Kevan and Narayana, 1979
organic radicals				
DPPH	0.086	0.086		Hervé, 1963
DPPH	~ 0.06	~ 0.06	crystalline	Brändle et al., 1970; Krüger, 1972
coal	0.15			Hervé, 1962
organic ion radical salt	~ 50 - 0.5 over 4 - 300K	minimum at ~60K	T_1 by inversion recovery; T_2 by FID decay	Bietsch et al., 1995
radical anion salt	1.5 at 4.5K, 0.5 at 35K, 0.8 at 60K, 0.25 at 300K	70 ns at 4.5K, 1 μs at 60K, 0.3 μs at 300K	"; $T_1 < T_2$	Krebs et al., 1990
polypyrrole, conductor and semiconductor	 1.3 to 0.3 8 to 4	 0.9 to 0.3 <0.3	inversion recovery and FID, 6 - 300K	Kanemoto et al., 1998
$(CH_3)_2CHCHOH$	1.9		function of pressure	Sueishi and Kuwata, 1989
glycine radical	0.31 s at 4 K, 0.11 s at 10 K, 100 μs at 20 K	6 at 4 K. 0.38 at 170 K	T_m exhibits minima as a function of temperature	Hoffmann et al., 1995
glycine radical	~ 9 ms at 10 K to 80 μs at 200 K	0.3-5.8; minima at 20 and 170 K	10-200 K	Gramza et al., 1997
benzene anion	0.14	0.078	-20 °C	Kooser et al., 1969
benzene anion	0.15-0.22	0.14 - 0.2	-70 to -120 °C	Rataiczak and Jones, 1972
trapped electrons and $O^{\cdot -}$ in glassy matrices	several ms to 10's of ms at ~10 K		function of solvent, deuteration and temperature	Bowman and Kevan, 1977a, b
$C_{59}N$	50	3	room temp., in 1-chloro-naphthalene	Gruss et al., 1997
$N@C_{60}$	120	50	300 K, toluene	Dietel et al., 1999
$N@C_{61}(COOC_2H_5)_2$	78	13	300 K, toluene	Dietel et al., 1999
$N@C_{66}(COOC_2H_5)_{12}$	116	24	300 K, CH_2Cl_2	Dietel et al., 1999
$N@C_{70}$	133	40	300 K, CH_2Cl_2	Dietel et al., 1999
semiquinones	1	1		Schreurs et al., 1960

Species	T_1 (μs)[c]	T_2 (μs)[c]	notes	Reference
several semiquinones	~ 2 - 140	linewidth studies	function of solvent and temperature	Rengen et al., 1971, 1974a, 1974b
semiquinones	2.1 - 6.9		function of pressure	Sueishi and Kuwata, 1989
4-carboxy-o-benzo-semiquinone	0.64±10%	0.71±7%	also study of collisions with metal ions	Basosi et al., 1984
Zn complex of above semiquinone	0.76	0.73	"	"
tropenyl	T_1/T_2 =1.05	0.44	135 - 200 °C	Kooser et al., 1969
BDPA	20	~ 0.07	in solution	Brändle et al., 1970;
	0.2	0.2	crystal	Krüger, 1972
TCNE	7			Eastman et al., 1969
TCNE	7.8	0.67	31 °C	Percival and Hyde, 1975
galvinoxyl	1.5 - 25		as function of η/T	Huisjen and Hyde, 1974a
nitroxide	100	1	77 K	Kulikov, 1974
tanol	300 - 800 at 77 K		single crystal, 10 – 300 K	Wajnberg et al., 1992
tanol	0.41			Huisjen and Hyde, 1974b
tanol, tanone	0.35 - 11.65		as function of η/T	Huisjen and Hyde, 1976
tanol, tanone	1.3	0.27	-63 °C in sec-butylbenzene	Percival et al., 1975; Percival and Hyde, 1976
nitroxides	~ 0.03 - 3		collisions with metals	Hyde and Sarna, 1978
nitroxides	0.3 - 3	0.1 - 1	solvent temperature and viscosity	Schwartz et al., 1979
tempo and cholestane	~ 0.3 - 10		~ 250 - 385 K in o-terphenyl, by LODESR	Andreozzi et al., 1999
cholestane in liquid crystal	0.068 - 0.582	0.092 - 0.442	various phases; model-dependent	Sastry et al., 1996a
perdeuterated tempone	0.446 - 0.759	0.363 - 0.532	various phases; model-dependent	Sastry et al., 1996b
spin labels	2.46		in membranes, no O_2	Kusumi et al., 1982
spin labels	0.56		same, 50% air	Kusumi et al., 1982
doxylstearic acid	2		in membranes	Subczynski and Hyde, 1981
spin-labeled stearic acid	0.94 2.54		S-band X-band	Hyde et al., 1990

Species	T_1 (μs)[c]	T_2 (μs)[c]	notes	Reference
spin-labeled stearic acids	2.2 - 5.8		in liposomes	Yin et al., 1990
spin-labeled stearic acids	1.2 - 3.9		in membranes	Subczynski et al., 1991
spin-labeled stearic acids	1.4 - 1.8		in membranes, T_{1n} = 0.7 - 0.9	Yin et al., 1987
spin-labeled stearic acids	2.8 - 7.7		in CHO cell membranes	Subczynski et al., 1992a
spin-labeled lipids	2.4 - 6.8		in membranes	Subczynski et al., 1998
spin-labeled melitin	~1.4 in solution, 3 - 4 in membrane		T_{1n} was rapid	Altenbach et al., 1989
spin-labeled retinal	6.4		rhodopsin, 0 °C	Subczynski et al., 1992b
maleimide spin-labeled hemoglobin	6.6			Huisjen and Hyde, 1974b
melanin	10-22		123-303 K	Sarna and Hyde, 1978
Inorganic radicals				
peroxylamine disulfonate	0.34	0.25	Fremy's salt	Kooser et al., 1969
"		0.41	$T_1 = T_2$	Eastman et al., 1970
"	0.33	0.25		Rataiczak and Jones, 1972
"	0.82		85% glycerol	Goldman et al., 1973
"	320±20 ns	297±2 ns	H_2O	Mailer et al., 1991
SeO_3^- radical in $(NH_4)_3H(SeO_4)_2$		0.35	single crystal, 121 K	Tritt-Goc et al., 1993
"	<μs to >ms; $T_1 \approx 3T_2$ at room temperature	0.178 to 0.596	T_1 by inversion recovery; function of temperature	Hilczer et al., 1993
$H_3N\text{-}NH^{\cdot}$ in $Li(N_2H_5)SO_4$	100 ms at 10 K to 90 μs at 170 K			Morawski et al., 1997
$[F_{Li}]^0$ center in CaO:Li	5.5 - 0.33 s		1.5 - 25 K; 9.5 and 16.5 GHz	Stinson and Stapleton, 1983
F^+ center in undoped CaO	30 s		4.2 K, 16.5 GHz	"
amorphous Si	~ 0.1 s to 140 μs		0.3 - 4.2 K; 9.5 and 16.5 GHz	Askew et al., 1984
β-alumina	~ 10 s to 25 ms		~ 1.5 - 20 K; 9.5 and 16.5 GHz	Kurtz and Stapleton, 1980
Metal ions				
aquo Ti(III)	~ 0.04 ns		NMRD	Bertini et al., 1992b

Species	T_1 (μs)[c]	T_2 (μs)[c]	notes	Reference
aquo VO^{2+} [d]	~ 0.5 ns		NMRD 5 to 35 °C	Bertini et al., 1992a Bertini et al., 1993b
VO^{2+} in KNO_3	1.4 - 12.9	1.4 - 2.4 ns	vary with m_I	Rao, 1970
VO^{2+} in $CsNO_3$	0.8 - 5.0	1.2 - 2.4 ns		
VO transferrin[d]	19 ns		NMRD, T_1=T_2=τ_c	Banci et al., 1986 Bertini et al., 1993b
1% Cr^{3+} in MgO	1 and 5.5 μs from 4.5 to 40 K	60 ns from 5 to 70 K	T_1 by inversion recovery	Sczaniecki et al., 1994
Mo^{5+} in TiO_2	100 min. 5 s		1.05 K 4.2 K	Chang and Yang, 1968
Aquo Mn(II)	~ 0.23 – 2.9 ns		NMRD, temperature and viscosity	Bertini et al., 1993a
Mn^{2+} in $(NH_4)_2Mg(SO_4)_2 \cdot 6H_2O$ single crystal	~ 0.5 ms at 20 K to 1 μs at 80 K	2 at < 30 K to 0.5 at 80 K	θ_D = 240 K	Hoffman et al., 1998b
EF-Tu-Mn(II)- GDP	2.5 ns			Wilson and Cohn,1977
Mn^{2+}, Gd^{3+}	see cited reference			Banci et al., 1986
$Fe(OH_2)_6^{3+}$	~80 ps[h]	47 ps	NMRD, EPR linewidths, depends on concentrations and pH	Sur and Bryant, 1995
$Fe(NCCH_3)_6^{3+}$	3.9 ns	44 ps		
FeF_6^{3-}		6 ns		
$FeF_5(H_2O)^{2-}$		4.4 ns		
$FeCl_4^-$	5.9 ns	1.4 ns		
$MnCl_4^{2-}$	5.6 ns	2.8 ns		
$Mn(OH_2)_6^{2+}$	2.4 ns	3 ns		
$Gd(OH_2)_{8?}^{3+}$		0.23 ns		
nitrosyl hemoproteins	1.8 at 5 K		4 - 70 K	Wajnberg et al., 1996
low-spin Fe(III) porphyrin[e]	≤ 10^{-12} s		not dependent on viscosity	Dugad et al., 1990
aquo Fe(III)[f]	$7.8x10^{-11}$ to $1.3x10^{-10}$ s		NMRD, temperature and viscosity	Bertini et al., 1993b
Fe(III) in cyto-chrome P-450	3 ns		-180 °C	Ruf and Nastainczyk, 1976
Fe^{3+} in myoglobin	~ 0.1 s to ~1 μs		1.2 - 4 K	Scholes et al., 1971
Fe^{3+} in myoglobin azide	~ 0.5 s to 50 μs		1.5 - 10 K	Colvin and Stapleton, 1985
Heme a in cytochrome c oxidase	~ 250 ms 21 μs		1.5 K 9.2 K	Scholes et al., 1984
Co(II) HCA II	τ_c=3.3 $x10^{-11}$ s[g]			Banci et al., 1986
aquo Ni(II)	T_1=T_2=2.9 ps		NMR	Friedman et al., 1979

Species	T_1 (μs)[c]	T_2 (μs)[c]	notes	Reference
Ni^{2+}-centered bacterio-chlorophyll	> 100 ps		transient absorption at 780 nm	Musewald et al., 1999
Cu(II)-doped triglycine selenate	275 at 15 K 1.7 at 83 K		T_1 by inversion recovery and SR; θ_D=168 K	Hoffmann et al., 1996a
Cu(II)-doped triglycine selenate		0.2 - 1.2; 0.402 at 55K	T_m min. at 55 K due to glycine motion	Hoffmann et al., 1996b
$Cu(H_2O)_6^{2+}$ in $(NH_4)_2Mg(SO_4)_2\cdot 6H_2O$ single crystal	~50 at 12 K to 0.25 at 55 K	~0.4 at 20 K to 0.08 at 53 K	Jahn-Teller effect	Hoffmann et al., 1998a
Cu^{2+} in many crystalline hosts			review	Hoffmann and Goslar, 1998
Cu_A in cytochrome c oxidase	~250 ms 1.1 - 2.1 ms		1.5 K 9.2 K	Scholes et al., 1984
Cu^{2+} in galactose oxidase	8.4 ns		NMRD	Knowles et al., 1995
Cu(II) SOD	4.6 to 1.8 ns		NMRD, 0 to 25 °C	Bertini et al., 1985
Cu amine complexes	1.2 ns for NH_3 complex		NMRD	Szczepaniak and Bryant, 1991
Cu,ZnSOD	1.8 - 2.9 ns		NMRD, T_1=T_2=τ_c	Banci et al., 1986; Bertini et al., 1989a
Cu BCA II	1.9 ns		"	Banci et al., 1986
Cu BCA II N_3^-	2.6 ns		"	Banci et al., 1986
Cu BCA II HCO_3^-	2.1 ns		"	Banci et al., 1986
Cu transferrin	5.7 - 6.4 ns		"	Banci et al., 1986; Bertini et al., 1989a
Cu plastocyanin	0.2 - 0.8		NMR	Bertini et al., 1999
Cu plastocyanin	~ 0.1 s - 20 μs		1.41 - 15.6 K	Drews et al., 1990
Ag atoms	~ 2 ms at 10 K to 100 μs at 50 K		several solvents	Michalik and Kevan, 1978
Gd^{3+} complexes		~10^{-10} s	Based on line width; longer at 34 GHz than at 9 GHz	Reuben, 1971
$Gd(EGTA)(H_2O)^-$		~ 10^{-10} 2×10^{-9}	9 GHz 2 mm	Aime et al., 1997
Gd^{3+} complexes	estimated, 10^{-6} - 10^{-11} s	~ 10^{-9} - 10^{-11}s	strongly field-dependent	Powell et al., 1993; Clarkson et al., 1998; Tóth et al., 1998, 1999
Ln, except Gd	7.8×10^{-13} s		NMR	Burns and LaMar, 1982

Species	T_1 (μs)[c]	T_2 (μs)[c]	notes	Reference
$LnDOTP^{5-}$	2.8 - 15.4 $\times 10^{-13}$ s		NMR	Ren and Sherry, 1996
Nd^{3+}in YAG, Yb^{3+} in CaF_2			~ 1.5 - 12 K	Aminov et al., 1998
Dy^{3+}	Eq. given; ~11 ns at 10K		3.7 - 20 K; two Orbach processes	Kurkin et al., 1991
Eu^{2+}(aq)		2×10^{-9} to 3×10^{-10} s; longer times are at higher field	As a function of temperature and magnetic field; based on line width and fitting NMR	Caravan and Merbach, 1997; Caravan et al., 1999
La@C_{82}	~ 1 - 6	~ 0.5 - 1.5	173 - 283 K, CS_2 m_I dependence	Okabe et al., 1995

Numbers preceeded by "~" were in most cases read from a graph and are only approximate.
[a]The original papers should be consulted for conditions of measurement and interpretation of the relaxation times. Unless otherwise stated, the values cited are for fluid solution near room temperature. NMRD indicates that the metal electron spin relaxation time was estimated by fitting to the nuclear magnetic resonance dispersion data. Some of the papers reported T_1 as a function of O_2 concentration – the value in the table is for no O_2.
[b]Values are in units of microseconds unless other units are specified
[c]Abbreviations used in the table are: BCA, bovine carbonic anhydrase; BDPA, α,γ-bisdiphenylene-β-phenylallyl; DOTP, 1,4,7,10-tetraazacyclododecane-1,4,7,10-tetrakis(methylene phosphonate); DPPH, 2,2-di(4-*tert*-octylphenyl)-1-picrylhydrazyl; EF-TuMnGDP, Mn(II) guanosine diphosphate complex with elongation factor Tu; HCA, human carbonic anhydrase; SOD, superoxide dismutase; TCNE, tetracyanoethylene, tanol (also called tempol), 1-oxyl-2,2,6,6-tetramethyl-4-hydroxypiperidine; tanone (also called tempone) 1-oxyl-2,2,6,6-tetramethyl-4-hydroxypiperidone. Sometimes the abbreviations for the various nitroxyl radicals are written all in capital letters and sometimes they are written lower case.
[d]Bertini et al. (1992a,1993b) (Banci et al., 1986) have found magnetic field dependence of electron spin relaxation for VO^{2+} in water but not when it is in a protein.
[e]The limiting value was derived from NMR ^1H T_1 values for low-spin metMbCN (Mb denotes myoglobin), high-spin metMbH_2O, and dicyano[tetrakis(4-sulfophenyl)porphine]iron(III) (Dugad et al., 1990).
[f]It was concluded by Bertini et al. (1993b) that the electron spin relaxation of high-spin Fe(III) in proteins is different from that observed in the aquo complex.
[g]See the cited reference for assignment that $\frac{1}{\tau_c} \approx \frac{1}{\tau_s} = \frac{1}{3}\left(\frac{1}{T_{1e}} + \frac{2}{T_{2e}}\right)$ and that this should be the case for systems in which an anisotropic term in the spin Hamiltonian is dominant over the Zeeman term. Four-coordinate Co(II) relaxation time is ca. 10^{-11} s and 5 or 6 coordinate Co(II) relaxation times are in the range 10^{-12} – 10^{-13} s (Banci et al., 1986).
[h]τ_{SO} is the electron spin relaxation time at zero field.

The emphasis in this section is on experimental results for molecular species that have been studied over a wide enough temperature range or other conditions to provide insight into possible mechanisms for relaxation.

Even in the limit of long saturating pulses that appear to overcome spectral diffusion it is our experience that few experiments result in recovery curves that can be fit to a single exponential. When we have fit data for immobilized samples to the sum of two exponentials the temperature dependence of the two components appear to be similar to the temperature dependence of the single component fit. We propose that in many samples there are distributions of relaxation times. Multiple contributions to a relaxation curve sometimes can be identified with physical mechanisms – such as rotation and collisions with oxygen. In some samples it is reasonable to model the distribution. For example, the dipolar interaction between a rapidly relaxing spin and a more slowly relaxing spin results in spin-lattice relaxation that is orientation dependent. In this case it is best to analyze the full shape of the saturation recovery curve in terms of the specific model of the distribution as described in ch. 8. A distribution of relaxation times can also be modeled as a stretched exponential, Kohlrausch or Weibull function (Dishon et al., 1990; Weibull, 1951), although that approach is not taken here. In the following discussion relaxation times were obtained by fitting experimental data to a single exponential, unless noted otherwise.

7.1 Nitroxyls

7.1.1 T_m

For Fremy's salt (peroxylaminedisulfonate) T_m remains nearly constant from low temperatures up to about 120 K as shown in Fig. 3. If D_2O:glycerol-d_8 is used as the solvent instead of H_2O:glycerol, T_m becomes substantially longer (Fig. 4). Thus we conclude that the limiting value of T_m in this low temperature regime is dominated by nuclear spin diffusion as discussed in section 5.3. In the deuterated solvent mixture, where protons are not present to dominate the dephasing, instantaneous diffusion plays a much greater role than in proton-containing solvents, even at the same concentration of paramagnetic centers. For example, compare the decay curves shown in Fig. 4. When a large pulse turning angle is used for the Fremy's salt sample in deuterated solvent, the echo has decayed almost to baseline within about 20 μs (Fig. 4b), but when a small turning angle pulse is used the decay lasts for about 60 μs (Fig. 4c).

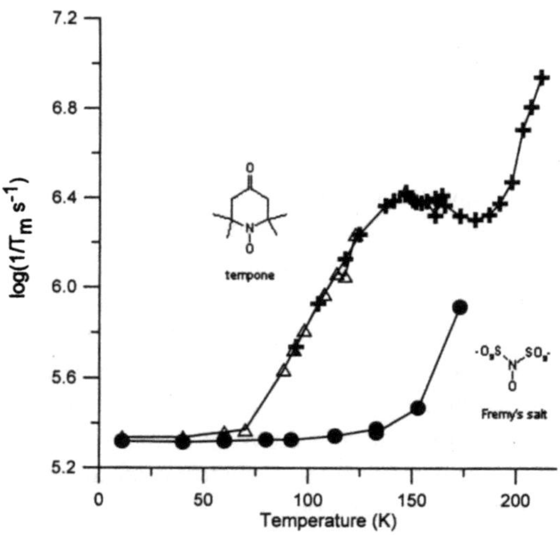

Figure 3. Temperature dependence of spin echo phase memory relaxation rate, $1/T_m$, for 0.1 mM tempone (+), 0.3 mM tempone (Δ), or 0.3 mM Fremy's salt in 1:1 water:glycerol (•) at X-band. Echo decays were recorded in the center of the spectrum.

For tempone and for nitroxyls that are typically used for spin labeling, T_m is similar to that for Fremy's salt at low temperatures. In this low temperature regime the echo dephasing for nitroxyl radicals depends not only the concentration of protons, but also on the type of protons (Zecevic et al., 1998). Dephasing is much faster in solvents that contain methyl groups than in solvents without methyl groups. The faster spin echo dephasing for spin labels buried inside carbonic anhydrase than for spin labels on the surface of the protein is attributed to the higher concentration of methyl groups in hydrophobic regions than on the surface (Lindgren et al., 1997).

For nitroxyls that contain methyl groups, spin echo dephasing is dominated by another process at temperatures between about 80 and 200 K (Fig. 3). This enhanced dephasing is not seen for Fremy's salt, which does not contain methyl groups. The increase and then decrease in $1/T_m$ as temperature is increased is attributed to rotation of the ring methyl groups at rates comparable to the electron-nuclear coupling (see section 5.5). The assignment of this enhancement to rotation of the nitroxyl ring methyl groups is supported by ENDOR studies that demonstrate that methyl group rotation is slow on the EPR timescale below about 80 K and fast on the EPR timescale above about 200 K (Barbon et al., 1999). Analysis of the temperature dependence of $1/T_m$ for tempone yielded an activation energy for methyl group rotation of 2.0 kcal/mol (Nakagawa, 1992). Dzuba and coworkers obtained 2.0 to 2.3 kcal/mol for methyl group rotation in several nitroxyls (Dzuba et al. 1981, 1984; Shushakov 1986, 1989; Tsvetkov and Dzuba, 1990).

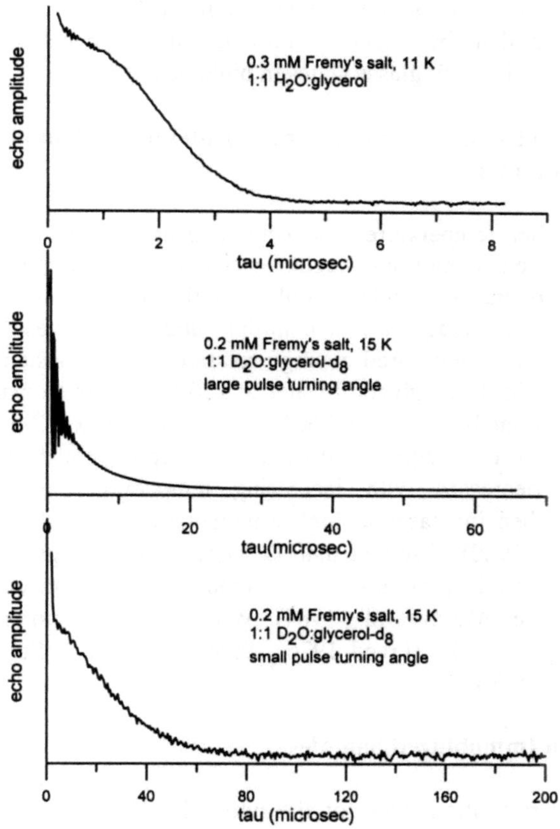

Figure 4. Spin echo decays for Fremy's salt at X-band, recorded in the center of the spectrum. a) 0.3 mM solution in 1:1 H_2O:glycerol, 11 K. b) 0.2 mM solution in 1:1 D_2O:glycerol-d_8, 15 K, with 90° turning angle for the first pulse. c) same solution as in part b but with an 11° turning angle for the first pulse. Note that deuterium echo modulation is excited by the hard pulse used in part b, but not with the softer pulses used in part c.

Dzuba, Tsvetkov, and coworkers (Dzuba et al., 1984; Shushakov et al., 1986; Dzuba and Tsvetkov, 1988; Dzyuba et al., 1989; Tsvetkov, 1990; Tsvetkov and Dzuba, 1990; Samoilova et al., 1991; Dzuba, 1992; Kudryashov et al., 1993; Dzuba et al., 1992, 1995; Dzuba, 1996; Toropov et al.,1998) showed that nitroxyl T_m is sensitive to slow motion of the nitroxyl molecule, which they termed "superslow rotations" (Dzuba et al., 1984) and "orientational vibrational motions (librations)" (Dzuba, 1992). In solvents such as ethanol at 77 K nitroxyl T_m is orientation dependent. At this temperature the contribution to dephasing is not dominated by methyl group rotation and the orientation dependence of T_m has been attributed to molecular motion (Maresch et al., 1992; Dubinskii et al., 1994). Dzuba et al. (1993) proposed that the effect of nitroxyl motion on echo-induced EPR spectra be used to identify glassy states in biological samples.

7.1.2 T_m and T_2 in Regimes where Librational or Tumbling Motion Dominates

At still higher temperatures the echo dephasing rate ($1/T_m$) increases further as molecular motions begin to average anisotropic interactions. Librations and overall molecular tumbling motion in glassy solution are fast enough to begin to average various couplings and anisotropies. For example, Freed and coworkers measured average rotational time constants of $\sim 10^{-7}$ s for Fremy's salt in 85:15 glycerol:water at 220 K. In frozen water at 200 K the value was about 10^{-7} s and at 220 K it was about 2.5×10^{-9} s. These times are of the order of the reciprocal of hyperfine coupling constants. As motion increases T_m decreases with increasing temperature, goes through a minimum and then increases at higher temperatures. Freed and coworkers (Stillman et al., 1980) observed that T_m became longer again above about 300 K. For the small Fremy's salt T_2 is about 0.3 μs in aqueous solution at room temperature (More et al., 1984). When tumbling motion dominates echo dephasing $T_m = T_2$. (Freed 1976, 1998; Stillman and Schwartz, 1979; Schwartz et al., 1979).

7.1.3 T_1 in Immobilized Samples

Spin-lattice relaxation rates for the nitroxyl radical tempone in sucrose octaacetate and in 1:1 water:glycerol between about 10 and 120 K are shown Fig. 5. In each solvent the relaxation rate over most of the temperature range can be fit with a Raman process. The Debye temperature that was required to fit the data in 1:1 water:glycerol ($\theta_D = 112$ K) is substantially higher than for sucrose octaacetate ($\theta_D = 67$ K) (Zhou et al., 1999), which is consistent with the qualitative expectation that a hydrogen-bonded matrix such as water:glycerol would be more rigid than a matrix such as sucrose octaacetate. Toward the higher temperature end of the range shown in Fig. 5 the experimental data are above the curves calculated for the Raman process, which is attributed to a contribution from a thermal process that arises from slight softening of the glass.

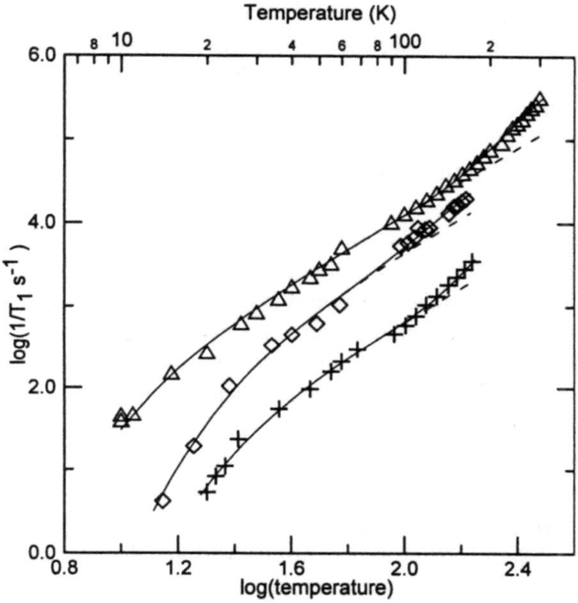

Figure 5. Temperature dependence of X-band spin-lattice relaxation rates for 0.2 mM Nycomed symmetrical-trityl radical in 1:1 water:glycerol (+), 1 mM tempone in 1:1 water:glycerol (◊), and 1 mM tempone in glassy sucrose octaacetate (Δ). The data for tempone were obtained in the center of the spectrum. The solid lines through the data are the fits obtained using Eq. (7) and the parameters in Table 5. The dashed lines are the curves calculated using only the Raman contribution to the relaxation rates. (Figure reproduced from Zhou et al., 1999).

T_1 of nitroxyl radicals depends on the solvent and position in the EPR line as well as on temperature (Dzuba et al., 1984, 1985; Tsvetkov and Dzuba, 1990). Extensive studies of T_1 for nitroxyl radicals as a function of solvent and position in the line at 100 K found the following trends (Du et al., 1995a). T_1 is longer when the external magnetic field is along the molecular z axis and shorter when the magnetic field is in the perpendicular plane for nitroxyl in glassy solutions or doped into single crystals. Deuteration of the nitroxyl, deuteration of the solvent, or replacement of natural-abundance nitrogen by ^{15}N in the nitroxyl caused less than 10% change in T_1, which indicates that inter- or intra-molecular electron-nuclear couplings are not the dominant factor in nitroxyl spin-lattice relaxation in the temperature regimes examined. The longest T_1 values for nitroxyl radical at 100 K occur in water:glycerol solution. Other hydrogen-bonding systems also result in long relaxation times. Much shorter values of T_1 occur in solvents such as decahydro-naphthalene, di-o-xylylethane, polystyrene, and methyltetrahydrofuran-methanol mixtures. The relaxation times correlate well with values of A_{zz} for the nitroxyl radical. Freed and coworkers showed that A_{zz} is a measure of molecular motion (Freed, 1976). We conclude that the shorter relaxation times occur in solutions in which there is faster molecular motion. Indeed, A_{zz} is more temperature-dependent (even down to 100 K) in these solvents than in water:glycerol. The solvents that permit greater motion also are the solvents for which lower Debye temperatures are expected.

Unlike the orientation dependence observed for T_m, the orientation dependence observed for T_1 is not that expected for librations that change the resonance field for the radical. Instead, it is interpreted as the effect of the symmetry of the vibrations that effect relaxation. Treating the C_2NO unit of the nitroxyl as having local C_{3v} symmetry, it can be shown that there are twice as many modes for perpendicular relaxation as for axial (parallel) relaxation. Similarly, approximating the molecular orbital containing the unpaired electron as a p orbital on N, spin-orbit coupling mixes in excited states when the magnetic field is in the perpendicular plane, but not when it is along the z axis. Thus, vibrations in the perpendicular (x,y) plane are more effective in modulating spin-orbit coupling and are therefore more effective in causing relaxation than vibrations along the z axis (Du et al., 1995a).

For nitroxyl radicals in polycarbonate at 110 K, $T_{1N} = 82 \pm 13$ μs and $T_{1e} = 23 \pm 4$ μs (Saalmueller et al., 1995). G-tensor fluctuations were stated to be the dominant factor in determining the anisotropy of electron spin relaxation. Below T_g, and for mixing times in the pulse sequence of less than 200 μs, the predominant cause of magnetization transfer was assigned as nuclear spin relaxation.

T_1 and T_1^{eff} of tempo in supercooled ortho-terphenyl was measured by LODESR in the temperature range where $T_1 >> \tau >> \omega_0^{-1}$(Cianflone et al., 1998). T_1 was about 3 times longer than T_1^{eff}, and neither was strongly dependent on m_I. Andreozzi et al. (1999), also using LODESR, determined that at the lower range of temperatures studied the rotational motion affecting relaxation in supercooled o-terphenyl was activated, while at the higher temperatures the relaxation is exponential.

7.1.4 T_1 in Fluid Solution

The early saturation recovery literature concerning radicals in solution has been reviewed fairly comprehensively (Rengen et al., 1972; Hyde, 1974, 1979; Venkataraman, 1982; Mailer et al., 1991).

For nitroxyl radicals in solution, the interpretation of the measured "T_1" has to take account of the ratio of the nitrogen nuclear spin-flip rate to the electron spin-flip rate, denoted b in Goldman et al. (1973): $b = W_N/W_e$. Formulae relating the electron T_1 to b for various cases are given in Goldman et al. (1973). If $b \ll 1$, then $T_1^{-1}(\pm) = T_1^{-1}(0) = 2W_e(0)$. In the other extreme, with $b \gg 1$, $T_1^{-1}(\pm) = T_1^{-1}(0) = 6W_e(0)$. If nonsecular terms are not negligible, $T_1(1) \neq T_1(-1)$. In the water:glycerol mixtures, b ranged from 0.24 (i.e., $W_N < W_e$) for $\tau_R = 3x10^{-11}$ s to 24 for $\tau_R = 1.18x10^{-9}$ s. Over the same range of τ_R, $T_1^{-1}(0)$ decreased from $1.22x10^6$ s^{-1} to $0.836x10^6$ s^{-1} and W_e decreased from $0.365x10^6$ s^{-1} to $0.135x10^6$ s^{-1}. T_1 for perdeuterated tempone was studied by CW saturation in several deuterated solvents from fast motional narrowing ($\tau_R \sim 10^{-12}$ s) to the rigid limit ($\tau_R \sim 10^{-6}$ s) (Hwang et al., 1975). Changing the temperature from 80.4 °C to –92.6 °C increased τ_R in perdeuterated toluene from $2.44x10^{-12}$ s to $2.78x10^{-10}$ s, increased b from 0.00538 to 16.7, decreased $T_1^{-1}(0)$ from $3.89x10^6$ s^{-1} to $0.623x10^6$ s^{-1}, and decreased W_e from $1.90x10^6$ s^{-1} to $0.0695x10^6$ s^{-1}. This selection of data from Hwang et al. (1975) illustrates the strong temperature dependence of both the ^{14}N nuclear spin relaxation and the electron spin relaxation, and the impact of the ^{14}N relaxation on the apparent electron T_1 measured with CW saturation methods. In this and other papers from the Freed lab there is extensive discussion of the relaxation data in terms of the details of liquid state dynamics. 2D ELDOR and COSY measurements of cholestane and tempone in various phases of a liquid crystal yielded T_1, T_2, W_N, and model-dependent rotational parameters (Sastry et al., 1996a,b). For example, at 57°C the homogeneous T_2 for perdeutero-tempone was significantly shorter for $m_I = +1$ than for $m_I = -1$ and 0.

The measured saturation recovery time constant can, depending on experimental conditions, include T_1, the nuclear T_{1n}, motion of the molecule, Heisenberg exchange between the studied radicals, and Heisenberg exchange with O_2 present in solution. Consequently, the recovery is usually multiexponential, only one term of which is the electron T_1 (Freed, 1979; Hyde 1979; Sugano et al., 1987; Beth and Robinson, 1989). The energy level diagram, including ^{14}N nuclear spin terms, for a nitroxyl radical, and the associated detailed rate equations are presented in Yin and Hyde (1987). The general solution yields three exponentials, with time constants:

$$\begin{aligned}
\tau_1^{-1} &= 2W_e + 2\omega_x \\
\tau_2^{-1} &= 2W_e + 2\omega_x + W_n + K_e[SL] \qquad (22) \\
\tau_3^{-1} &= 2W_e + 2\omega_x + 3W_n + K_e[SL]
\end{aligned}$$

W_e and W_n are the electron and nuclear relaxation rates (e.g., $2W_e = T_{1e}^{-1}$), K_e is the rate constant for Heisenberg spin exchange between nitroxyl radicals in units of $M^{-1}s^{-1}$, and ω_x is the rate of collisions between O_2 and nitroxyl radicals. This paper also compares the relative benefits of short-pulse and long-pulse saturation recovery for sorting out contributions of the several rates to the observed recovery.

The first direct measurement of the nitrogen nuclear T_{1n} for a spin label (Hyde et al., 1984) revealed that T_{1e} and T_{1n} were both of the order of a microsecond. For CTPO T_{1e} decreased from 4.5 to 0.55 μs and T_{1n} increased from 0.7 to 1.2 μs as the temperature of the heavy mineral oil solvent increased from -39 to +52 °C. T_{1n} is generally shorter than T_1 for spin labels in membranes (Yin et al., 1987, 1988; Yin and Hyde, 1987). Yin and Hyde (1987) used short-pulse SR, in which the pulse (0.1 to 0.5 μs and 2.0 to 3.5 G) is short relative to the nitrogen relaxation time. T_{1n} was about 0.7 to 0.9 μs, and the other three measured relaxation times were from ca. 0.3 to 1.7 μs at 54 °C and somewhat longer at 27 °C. For nitroxyls in membranes at higher temperatures the relaxation times are all of comparable magnitudes, while at lower temperatures the nitrogen relaxations are the fastest. In the long-pulse (ca. 5 μs) SR method the nitrogen relaxation is equilibrated and the recovery is electron relaxation and Heisenberg exchange (Kusumi et al., 1982).

Table 4. Relaxation of degassed Fremy's salt in aqueous solution.

T (°C)	9	24	34
$T/\eta \times 10^{-2}$	2.08	3.25	3.78
$T_1 \times 10^7$ s	4.5±0.5	4.1±0.35	3.4±0.3
$T_2 \times 10^7$ s	4.7±0.2	4.1±0.2	3.5±0.1

T_1 depends strongly on the tumbling correlation time (Tables 3, 4). For the small Fremy's radical in fluid solution T_1 is approximately equal to T_2 and is about 0.3 μs (Table 4). Goldman et al. (1973) extended the study of Fremy's salt to glycerol-water mixtures. When a spin label is attached to a small porphyrin T_1 in degassed organic solvents increases to about 0.6 μs (More et al., 1984). Spin labels in various membranes or liposomes at various temperatures exhibited T_1 in the range of 0.1 - 6 μs. For slowly-tumbling spin labels near room temperature, T_1 is in the range 0.5 - 6 μs, depending on temperature and oxygen level (Ashikawa et al., 1994; Kusumi et al., 1982; Subczynski et al., 1989, 1992b; Yin et al., 1990). Yin et al. (1990) measured T_{1e} = 2 - 6 μs for n-doxyl stearic acid spin labels. Single-exponential signals were obtained after 5 μs pump pulses. Subczynski et al. (1992b) observed T_1 = 4.0 μs in air and 6.4 μs in N_2 at 0 °C and closer to 5 μs at 20 °C for spin labeled rhodopsin in membranes. These values of T_1 were described as typical of those for slowly-tumbling spin labels, based on their prior work. Similar relaxation times were observed by Subczynski et al.

(1998). Thus, the general trend is toward longer T_1 as tumbling slows, which is the opposite of what is observed for T_2. In the presence of oxygen, Heisenberg exchange dominates the O_2-broadening and T_1-shortening effect (Hyde et al., 1990). Nitroxyl radicals exhibit shorter T_1 than common organic π-radicals (Percival and Hyde, 1976).

ELDOR using ^{14}N, ^{15}N-nitroxides revealed vertical and lateral motion in membranes (Feix et al., 1984). Haas et al. (1993) showed how T_{1e} can be obtained from the CW power saturation curve of nitroxyl radicals at all rotational rates. In the very slow motional regime, saturation-recovery ELDOR was used to determine T_{1e} and T_{1n} as well as the rotational correlation time (Haas et al., 1992).

Robinson and Mailer proposed that the relaxation of nitroxyl radicals can be described in terms of rotational dynamics (spin-rotation interactions), electron-nuclear dipolar (END) interactions, spin diffusion interactions with solvent nuclei, and collisions with oxygen (Mailer et al., 1992; Haas et al., 1993; Robinson et al., 1994):

$$\frac{1}{T_{1e}} = \frac{1}{T_{1e}^{SR}} + \frac{1}{T_{1e}^{END}} + \frac{1}{T_{1e}^{SD}} + \frac{1}{T_{1e}^{Ox}} \tag{23}$$

The details of these terms are given in (19) and (20) and as follows.

Robinson and Mailer and coworkers extended Abragam's treatment of the electron-nuclear dipolar interaction to yield a relaxation rate for the electron:

$$\frac{1}{T_{1e}^{END}} = \frac{1}{6}\sum_{i=1}^{3}(a_i - \bar{a})^2 \frac{\tau_c}{1+(\omega_e\tau_c)^2} \tag{24}$$

Note that if the END mechanism is important, the relaxation is dependent on the electron Larmor frequency. It has been stated that the END mechanism dominates when the spin-orbit coupling is small. Spin diffusion, by which the electron magnetization is relaxed by the bulk solvent, was expressed as:

$$\frac{1}{T_{1e}^{SD}} = 8.5\rho\left(CD_{SD}^3\right)^{1/4} \qquad C = \alpha\frac{\tau_d}{1+(\omega_e\tau_d)^{3/2}} \qquad D_{SD} = 2rW \tag{25}$$

where ρ is the density of protons, W is the proton spin-flip rate, r is the distance between protons, τ_d is the relative solvent-probe translational diffusion time, which is assumed to be approximately equal to τ_c, the rotational correlation time. Combining these expressions,

$$\frac{1}{T_{1e}^{SD}} = R_{1e,max}^{SD}\left(\frac{2\omega_e\tau_c}{1+(\omega_e\tau_c)^{3/2}}\right)^{1/4} \tag{26}$$

Robinson et al. state that D_{SD} is not known, so they use $R^{SD}{}_{1e,max}$, the maximum possible relaxation rate, as an adjustable parameter. A similar expression can be written for the nuclear T_1, replacing ω_e by ω_n. The maximum electron and nuclear rates, R, are the same.

In Mailer et al. (1992) the relative importance of the various contributions is shown for an experimental data set. Spin diffusion dominates for $\tau_c > 10^{-9}$ s, and spin rotation dominates for $\tau_c < 10^{-11}$ s. In the intermediate range, with τ_c = ca. 10^{-10} s, the END mechanism is also important. If this analysis is correct the spin diffusion mechanism would be strongly affected by isotopic substitution in the intermediate tumbling regime. Note that Lloyd and Pake (1954) showed that substituting D_2O for H_2O did not affect line width or CW saturation of Fremy's salt at 60 MHz (30 G), and Wilson and Kivelson (1966c) showed that deuteration and changing from toluene to CS_2 did not affect the line widths of vanadylacetylacetonate. Wilson and Kivelson ruled out "the interaction between the electron spin and the proton magnetic moment on toluene as an effective perturbation in causing spin relaxation." The impact of solvent deuteration on nitroxyl T_1 at higher frequencies would be an important test of the proposed mechanisms. Some unpublished results obtained by J.-J. Yin in Hyde's lab showed that T_1 was systematically longer for perdeutero-tempone in D_2O:glycerol-d8 (3:7) than for the protiated analogs, at X-band. The difference was very small at 10-35 °C, and the difference increased to about 20% at -30°C (J.-J. Yin, private communication, 12/1/99).

The END and SD terms in this mechanism predict a frequency dependence of T_1 when these contributions dominate. When the spin rotation term dominates, T_1 would be field independent. Hyde found that the oxygen relaxivity was the same at X-band and S-band. There is very little information on relaxation times of nitroxyl spin probes as a function of frequency. The Lloyd and Pake (1954) value of the relaxation time for Fremy's salt was 0.5 μs, which is close to the X-band value (More et al., 1984) and suggests at most a weak dependence on magnetic field. Hyde and coworkers (1990) observed a decrease in T_1 from X-band to S-band for a spin-labeled stearic acid in a membrane. At that time the results were described as "unexpected." The mechanism of Robinson et al. can explain the frequency dependence of relaxation of nitroxyls observed by Hyde and coworkers. The possibility of vibrational modulation of spin-orbit coupling analogous to solid state mechanisms does not appear to have been included in these analyses. At this stage it appears that extensive interplay of theory and experiment will be needed to sort out the mechanisms of the most important contributors to the relaxation of nitroxyl radicals in fluid solution at various microwave frequencies.

Recently, Robinson and coworkers (1999) showed that apparent inconsistencies between approximate treatments of the m_I dependence of relaxation for nitroxyl radicals in the fast-motion limit are removed by extending the operator method of Abragam. In agreement with experiment, they predict that T_1 for nitroxyl radicals is independent of m_I. T_1 in fluid solution does not depend on m_I (Robinson et al., 1999).

7.2 Other Organic Radicals

Early studies of relaxation of organic radicals in solution were summarized by Rengen et al. (1972). Spin lattice relaxation of semiquinones in solution, measured by saturation recovery, was found to be independent of nuclear spin state, and to be proportional to τ_R over the temperature and viscosity range studied (Rengen et al., 1972, 1974a). Relaxation was attributed to modulation of the g-tensor and to spin-rotation interaction, with spin-rotation being the major contributor, even in the $\omega_e\tau_R < 1$ region. The slope of T_1 as a function of temperature did not change at the freezing point of the solvent, suggesting that the mechanism of relaxation involved the same type of motion across the temperature range (Rengen et al., 1972, 1974a). It was pointed out that the activation energy (~1.7 kcal mol^{-1}) was similar to that found for nuclear spin lattice relaxation in acetone (1.5 ± 0.2 kcal mol^{-1}). The linewidth was due largely to viscosity-dependent modulation of g anisotropy (Rengen et al., 1974a). These observations may be relevant to the interpretation of nitroxyl relaxation in solution.

In the context of a quantitative test of ENDOR theory, T_1 was measured as a function of temperature for semiquinone radicals formed from p-benzoquinone, duraquinone, and 2,5-dimethyl-p-benzoquinone (Leniart et al., 1975). Figures in the paper present T_1 vs T/η for the semiquinones in ethanol and dimethoxyethane. T_1 values of the order of 2 to 12 μs were observed for various conditions in the temperature range -80 to 0 °C, with longer T_1 at lower temperatures and at lower concentrations. This paper also emphasizes the importance of keeping the concentration low enough that Heisenberg exchange does not contribute to the apparent T_1.

Viscosity, temperature, concentration, and solvent dependence of T_1 were measured by Prabhananda and Hyde (1986) for several semiquinones. T_1 depended on the semiquinone substituents, but not on concentration of the semiquinone from ca. 0.18 to 1.8 mM. T_2^* and T_1 were independent of microwave frequency (9 and 1.15 GHz) for tetrachlorosemiquinone in ethanol at room temperature. Spin-rotational interaction was identified as the dominant relaxation mechanism.

The temperature dependence of T_1 for the Nycomed sym-trityl radical in 1:1 water:glycerol between 20 and 160 K (Fig. 5) can be fit with a Raman process. The unpaired electron is located predominantly near the center of this large molecule, which largely constitutes its own lattice. The relaxation rate and Debye temperature exhibit much less solvent dependence than for smaller molecules (Zhou et al., 1999). There is no experimental data on frequency dependence of T_1 for the Nycomed sym-trityl probe, but there is the observation from the NCI lab (Murugesan et al., 1997) that the relaxation time is long enough (>1 μs) at room temperature and 300 MHz to obtain good FIDs.

Relaxation times for benzene anion radical and tropenyl radical are summarized in Table 3. The short relaxation times for benzene anion radical and tropenyl radical were attributed to the orbitally-degenerate ground state of these species. Kooser et al. (1969) found larger line widths at 35 GHz than at 9.1 GHz, and attributed the difference to a g-tensor mechanism that yielded a contribution to the line width of the form bB_0^2.

T_1 for C_{60} anions in glassy solvents and solutions is strongly temperature dependent and does not change significantly on going from glassy to liquid solution (Eaton et al., 1996). The data can be fit with a Raman process at low temperature and a thermally-activated process at higher temperatures (Fig. 6) (Zhou et al., 1999).

Radicals formed by encapsulation of nitrogen atoms in C_{60} (denoted N@C60) have provided some of the narrowest-line EPR spectra. The N ground state is $^4S_{3/2}$, the hyperfine splitting is isotropic, and in the solid matrix (largely C_{60}) there is broadening of the EPR lines from X-band to Q-band, attributed to dispersion of g-values (Murphy et al., 1996). T_1 of N@C_{60} was about 200 μs in polycrystalline C_{60} at room temperature (Weiden, et al., 1999). In CS_2 solution, CW EPR linewidths of N@C_{60} were determined by the magnetic field inhomogeneity. The central line exhibited ESE decay constant of 120 μs, corresponding to homogeneous linewidth of 2.5 kHz. (ca. 1 mG), and T_1 was found to be equal to T_2 (Knapp et al., 1997). When the symmetry is lowered by substitution, forming N@$C_{61}(COOEt)_2$, the 3-fold degeneracy of the lines is removed in the solid, and the resulting splitting can be simulated using D_{zz} = 2 G and E = 0.13 G. In solution, the EPR of this species looks similar to that of unreacted N@C_{60} (Pietzak et al., 1997). Demonstration that the relaxation of N@C_{60} in solution was a thermally-activated process, attributed to collisional modulation of ZFS, was provided by comparison of relaxation at 9.5 and 95 GHz, using Eq. (27) for S = 3/2 (l = 1,2) (Knapp et al., 1998a):

$$\frac{1}{T_1(\ell)} = \frac{12}{15} D^2 \left(1 + \frac{\eta^2}{3}\right) \frac{\tau}{1 + (\ell\omega\tau)^2} \tag{27}$$

where D is the ZFS parameter, η is the asymmetry parameter, and τ is the time of the fluctuating process. T_2 was independent of frequency, but T_1 was about 4 times longer at 95 GHz than at 9.5 GHz (512 vs. 128 μs). This indicates that the correlation time satisfies the condition $\omega\tau \approx 1$ at 9.5 GHz. Additional relaxation time data are in Table 3.

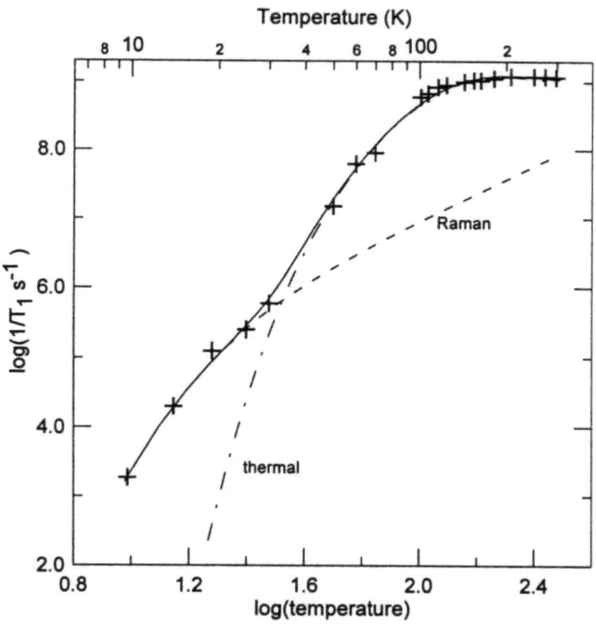

Figure 6. Temperature dependence of X-band spin-lattice relaxation rates in the perpendicular plane for 0.3 mM C_{60}^- anion in methyl-THF (+). The solid line through the data is the fit obtained using Eq. (7) and the parameters in Table 5. The contributions to the relaxation from the Raman process (- - -) and the thermally activated process (_ . _ . _) are shown separately. (Figure reproduced from Zhou et al., 1999).

P@C_{60} also is in the $^4S_{3/2}$ ground state. Each of the ^{31}P (I=1/2) hyperfine lines is further split into three lines due to transitions within the S = 3/2 multiplet (Knapp et al., 1998b). This is the first observation of such lines in solution. In toluene solution, τ was ca. 5 to 32 ps from 190 to 300 K, values consistent with the model of collision-induced deformations. At room temperature, T_1 by inversion recovery was 1 μs and T_2 = 0.5 and 0.8 μs from the linewidths of the FT EPR spectrum (Knapp et al., 1998b).

The relaxation properties of Lu@C_{60} at temperatures close to the freezing point of the solvents, up to ca. 250 K, are determined by the viscosity of the solvent. At higher temperatures additional thermally-activated relaxation processes lead to line broadening (Knapp et al., 1998).

T_1 of La@C_{82} in toluene solution, from 180 to 300 K, was due largely to spin-rotational interaction, and T_1 was close to T_2 (Rübsam et al., 1995). T_1 was 0.7 μs at room temperature and 2.1 μs near the freezing point of toluene.

The "T_1" of photosynthetic reaction centers, measured with a 180-90-180 pulse sequence, was found by Gast and coworkers (1982) to have a fast (4 ms) component, which was attributed to spin diffusion, with a time short compared to T_1. The longer time (100 ms to 5 s) depended on sample history, including freezing history. "We have the impression that T_1 is related to strain within the frozen sample as indicated by the number of cracks."

7.3 Radicals Produced by Irradiation

7.3.1 Irradiated Fused SiO_2

Castle and Feldman (1965a, 1965b) analyzed spin-lattice relaxation rates for the E' defect in crystalline and vitreous quartz between 4.2 and about 250 K in terms of the direct process and two local modes. At room temperature T_1 was found to be approximately independent of microwave frequency between 1 and 9 GHz and to increase by about 50% at 19.4 GHz (Ghim et al., 1995). Preliminary results indicate a further increase in T_1 at 95 GHz. At each microwave frequency T_1 is longer near the parallel axis and shorter in the perpendicular plane of the g-axis system, which is interpreted in terms of coupling to local vibrational modes of the defect center. The low-power T_m (avoiding instantaneous diffusion) is consistent with the magnitude expected for electron-electron spin-spin interaction, since there are very few nuclear spins in the lattice (Eaton and Eaton, 1993) (see section 3.4).

The relaxation process for atomic hydrogen in fused silica between 2 and 100 K could be modeled with either an Orbach process or a local mode (Feldman et al., 1966). The data for atomic hydrogen in fused silica demonstrate that the similarity in temperature dependence predicted by some relaxation processes within finite temperature intervals requires that assignments be based not only on the temperature dependence of the relaxation rates, but also on the plausibility of the parameters obtained by fitting to various models.

7.3.2 Malonic Acid

Nechtschein and Hyde (1970) used pulsed ELDOR to measure $T_1 = 40 \pm 5$ μs and T_x (cross relaxation) = 300 ± 30 μs in a sample of irradiated malonic acid. Freed and coworkers (Lee et al., 1993) measured 2D FT EPR of irradiated (1 Mrad) malonic acid. In their simulations they used W_e (= $1/2T_{1e}$) = 1.67×10^4 s^{-1}, $W_n = (1/2T_{1n}) = 7.14\times10^3$ s^{-1}, $W_x = 6.17\times10^3$ s^{-1}, T_{2e} = 900 ns, and T_{2n} = 22 μs. They found that T_m was orientation dependent.

7.3.3 Alanine

Preliminary measurements of T_1 of irradiated crystalline alanine showed a small dependence on frequency (S vs. X) and on isomer (DL vs. L) (Nakagawa et al., 1993). Substantial difference was observed between apparent values of T_1 for irradiated L-alanine obtained by inversion recovery and saturation recovery at microwave frequencies between 1.8 and 19.4 GHz (Ghim et al., 1996). For each experiment the data exhibited strong deviation from single exponential fits, which indicates major contributions from spectral diffusion, consistent with the observations of Brustolon and Segre (1994). The rapid rotation of the methyl groups may be an efficient spectral diffusion process. T_1 for irradiated L-alanine was measured as a function of temperature by LODESR. The room temperature value was ca. 48 μs (Angelone et al., 1993). Further work is needed to define the true T_1 as a function of frequency.

7.3.4 Irradiated Organic and Inorganic Systems

Tsvetkov (1983) reviewed spectrometer design, data analysis, and interpretation of the distributions of radicals produced by irradiation of organic and inorganic systems.

7.4 Radicals in Polymers

Radicals in polymers are formed during synthesis and by thermal and mechanical chain scission, and have CW spectra about 6 G wide (Kinell et al., 1973; Ranby and Rabek, 1977; Boyer and Keinath, 1978; Woodward and Bovey, 1980). The relaxation times, at room temperature, of a stable radical produced during production of a commercial polymer were $T_{1e} = 580 \pm 12$ μ s, $T_{2e} = 1.74 \pm 0.04$ μs, and $T_{1n} \approx 350$ μs (Hubrich et al., 1995). T_m is constant over the entire spectrum.

7.5 Triplets

There is a large literature on the sub-field of photo-excited triplet states (Schmidt, 1989). These systems have the additional complexity that lifetimes of states can be limited by intersystem crossing. The smaller the singlet-triplet separation the shorter the relaxation time. This is analogous to the observation in transition metal systems that relaxation times are shorter when there are low-lying excited states. For example, low-spin Fe(III) (S=1/2) has much faster relaxation than does Cr(V) (S=1/2) for this reason.

Spin-lattice relaxation was faster in glassy 3-methyl pentane than in crystalline n-hexane "in agreement with the general observation that spin-lattice relaxation rates in disordered solids are higher than in crystalline systems" (Trinbgali and Brenner, 1998). The direct process dominated from 1.75 to 4.2 K, and there was anisotropy of relaxation.

Phase coherence of phosphorescent triplet spins depends strongly on the magnetic moments of the nuclei on the guest as well as on the host molecules (van't Hof and Schmidt, 1979). In photo-excited triplet states, the transfer of the electron spin polarization to the nuclear spin system takes place in the order of a few hundred ns (Schmidt et al., 1992), and the spin-spin relaxation is dominated by a fast spin-lattice relaxation induced by a thermally-activated exchange process (van Kooten and Schmidt, 1985). "Peculiar spin dynamics" were observed in the photoexcited triplet state of pentacene in benzoic acid crystal (Ong et al., 1994).

7.6 Transition metals with S = ½

7.6.1 Vanadyl

At low temperature spin echo dephasing for vanadyl ion in 1:1 water:glycerol (Fig. 7) is similar to that for nitroxyl radicals with T_m about 4 μs and an exponent in Eq. (8) of 2.4 (Eaton and Eaton, 1999). Solvents that contain methyl groups enhance the rate of dephasing analogous to what was observed for nitroxyl radicals, and replacement of H_2O:glycerol by D_2O:glycerol-d_8 results in a dramatic lengthening of the echo dephasing time (Fig. 7). In proton-containing solvents, T_m measured along a principal axis exhibits very little temperature dependence up to about 100 K. In deuterated solvents where T_m is longer, echo dephasing becomes strongly temperature dependent above about 40 K because T_1 becomes comparable to T_m and dominates the echo dephasing (Fig. 7). At 50 K in glassy solvents echo dephasing is faster for intermediate orientations than along the principal axes, which indicates a significant contribution from librational motion (Du et al., 1992). For vanadyl porphyrin T_m also is approximately independent of temperature at low temperature and is attributed to nuclear spin diffusion (Du et al., 1996).

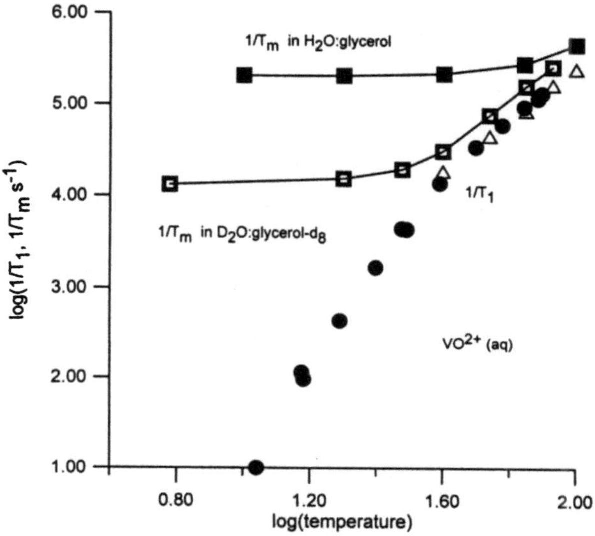

Figure 7. Temperature dependence of relaxation rates for VO^{2+} (aq) at X-band (9.4 GHz): $1/T_m$ for 1.2 mM solution in 1:1 H_2O:glycerol, (■); $1/T_m$ for 0.3 mM solution in D_2O:glycerol-d_8, (□); $1/T_1$ obtained by long-pulse saturation recovery for 0.9 mM solution in 1:1 H_2O:glycerol, (•); $1/T_1$ obtained by inversion recovery for 0.3 mM solution in D_2O-glycerol-d_8, (Δ). The lines through the data for $1/T_m$ connect the data points. Data were recorded for the $m_I = -1/2$ line. (Figure reproduced from Eaton and Eaton, 1999).

For vanadyl ion in 1:1 water:glycerol values of T_1 measured by long-pulse saturation recovery with CW detection or inversion recovery were in good agreement at temperatures above about 40 K (Fig. 7). However, below about 40 K values of T_1 measured by inversion recovery were systematically shorter than values measured by long-pulse saturation recovery, which demonstrated the presence of spectral diffusion processes (Eaton and Eaton, 1999).

Above about 20 K the relaxation rates for VO^{2+} in 1:1 water:glycerol were faster than for a vanadyl porphyrin (Fig. 8). The relaxation rates for VO^{2+} and for the vanadyl porphyrin (Du et al., 1996) were modeled with contributions from the Raman process and a local mode. For the vanadyl porphyrin it was necessary to include a contribution from the direct process to match the data at lower temperatures (Zhou et al., 1999). Faster relaxation for VO^{2+} than for the vanadyl porphyrin may occur because the vibrational modes in the aqua complex are "softer" than for the metalloporphyrin.

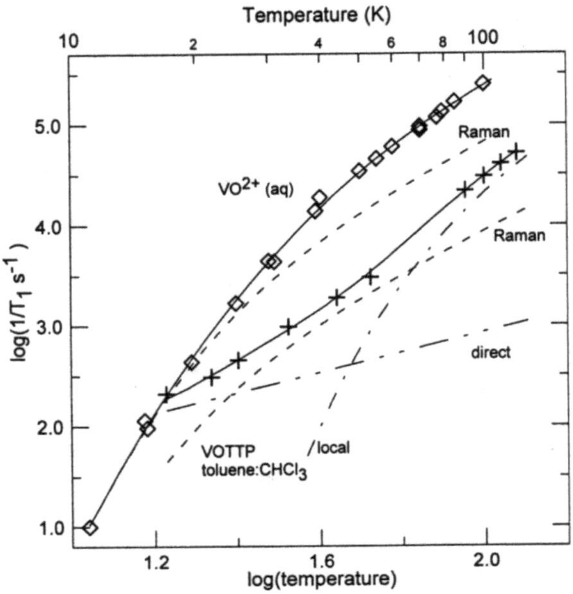

Figure 8. Temperature dependence of X-band spin-lattice relaxation rates for 1 mM VOTPP-COOH in 2:1 toluene:$CHCl_3$ (+) and for 1.2 mM VO^{2+} in 1:1 water:glycerol (◊) for the m_I = -1/2 perpendicular transitions. The experimental data for VO^{2+} are the same as in Figure 7. The solid lines through the data are the fits obtained using Eq. (7) and the parameters in Table 5. The contributions from individual processes to the relaxation for mM VOTPP-COOH are: direct (_ . . _ . . _), Raman (- - -), and local mode (_ . _ . _). (Figure reproduced from Zhou et al., 1999).

7.6.2 Cr(V)

For Cr(V) complexes T_m at low temperature is dominated by nuclear spin diffusion and is much longer in deuterated solvent than in proton containing solvents (Eaton and Eaton, 1999). In proton-containing solvent, T_m for complexes that do not contain methyl groups exhibits little temperature dependence until the temperature is high enough that the lattice softens and there is onset of motional averaging of g and A anisotropy (Konda et al., 1994; Eaton and Eaton, 1999). For complexes that contain methyl groups, rotation of the methyl groups dominates echo dephasing between about 80 and 160 K. Analysis of the effect of methyl rotation on T_m found barriers to rotation in the range of 1.0-1.5 kcal/mole (Nakagawa et al., 1992). For chromium(V) nitrido porphyrins at 100 K, spin echo dephasing was faster in glassy solvent than in crystalline solvent and faster at intermediate orientations than along the principal axes. Both of these observations indicate that librational motion had a significant impact on dephasing at this temperature.

To fit the temperature dependence of $1/T_1$ for the chromium nitrido porphyrins (Zhou et al., 1999) or chromium nitrosyls (Carruthers et al., 1999) required contributions from the Raman process as well as a local mode (Fig. 9). Relaxation rates are faster in the perpendicular plane than along the principal axis for the axially symmetric Cr(V) nitrido porphyrins (Konda et al., 1994) and for $Cr(NO)(CN)_5^-$, but not for chromium nitrosyls with lower symmetry (Du et al., 1995b; Carruthers et al., 1999). In a Cr(V) complex prepared from isotopically enriched ^{53}Cr (I=3/2) T_1 at 100 K was independent of m_I, which indicates that spin-lattice relaxation was not dominated by modulation of the nuclear hyperfine interaction (Du et al., 1995b).

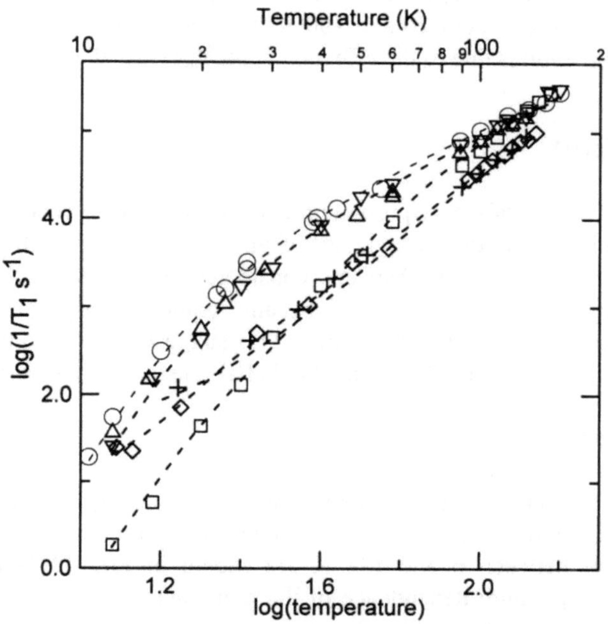

Figure 9. Temperature dependence of X-band electron spin lattice relaxation rates, $1/T_1$, at the $g_\perp$ turning point in the CW spectra in chromium nitrosyl complexes, compared with values for chromium(V) complexes: 0.9 mM $Cr(NO)(CN)_5^{3-}$, (□); 0.26 mM $Cr(NO)(EHBA)^+$, (∇); 0.47 mM $Cr(NO)(EHBA)_2$, (Δ); 0.39 mM $Cr(NO)(H_2O)_5^{2+}$, (O); 1 mM CrNTTP (nitridochromium(V) tetratolylporphyrin), (+); 1 mM $CrO(EHBA)_2^-$, (◊). CrNTTP was dissolved in toluene and the other species were dissolved in 1:1 water:glycerol. The dashed lines are the fits of the data to Eq. (7) obtained with the parameters shown in Table 5. EHBA is 2-ethyl-2-hydroxybutyrate. (Figure reproduced from Carruthers et al., 1999).

7.6.3 Mo(V)

Spin lattice relaxation times of oxo(5,10,15,20-tetratolyl-porphinato)molybenum(V) complexes, O=Mo(TTP)X, in glassy solution, also had faster $1/T_1$ when the magnetic field was in the plane perpendicular to the molecular plane (Husted et al., 1995). As in other cases studied that exhibited similar dependence on position in the spectrum, this was attributed to relaxation via spin-orbit coupling by local vibrational modes. The orientation-dependent values of $1/T_1$ are approximately the same for isotopes with I = 0 or I = 5/2, which indicates that the Mo nuclear hyperfine interaction is not a major factor in the electron spin-lattice relaxation rate (Husted et al., 1995). For the isotopes with nuclear spins $1/T_m^{\prime}$ was shorter at orientations at which the resonant field is more strongly dependent on small changes in molecular orientation. $1/T_m$ was approximately independent of temperature between about 20 and 100 K. It became temperature dependent at higher temperatures due to the effect of T_1 on T_m (Husted et al., 1995).

7.6.4 Fe(II)

^{1}H NMR of crystals of 1-propyltetrazole complexes of iron(II) exhibited a sharp minimum in the propyl side chain T_1 at 8 K, which cannot be explained by any of the mechanisms considered (Bokor et al., 1996). It is possible that at 8 K the metal T_1^{-1} became equal to the nuclear Larmor frequency, giving a resonant interaction. This suggestion is in contrast to the assumption of Bokor et al. (1996) that the Fe(II) relaxation time was 10^{-13} s.

7.6.5 Low-spin Fe(III)

Spin-lattice relaxation rates for low-spin Fe(III) in a spin-labeled mutant of sperm whale myoglobin bound to either imidazole or cyanide in 1:1 water:glycerol between 6 and 138 K are shown in Fig. 10 (Zhou et al., 1999). The temperature dependence of the dominant contribution between 6 and 18 K is consistent with the Raman process. However, to fit the lowest temperature data it is necessary to invoke a contribution from the direct process. An additional contribution is required to fit the relaxation rates above 30 K. The fit-lines shown in Fig. 10 were obtained by assigning the high-temperature process as thermally-activated. Comparable agreement between calculated and experimental relaxation rates could be obtained by assigning the high-temperature process as Orbach or as a local mode with approximately the same characteristic energy. The characteristic energies required to fit the data for the imidazole and cyanide complexes are similar to the value required for the small molecule complex $FeTTP(Im)_2^+$.

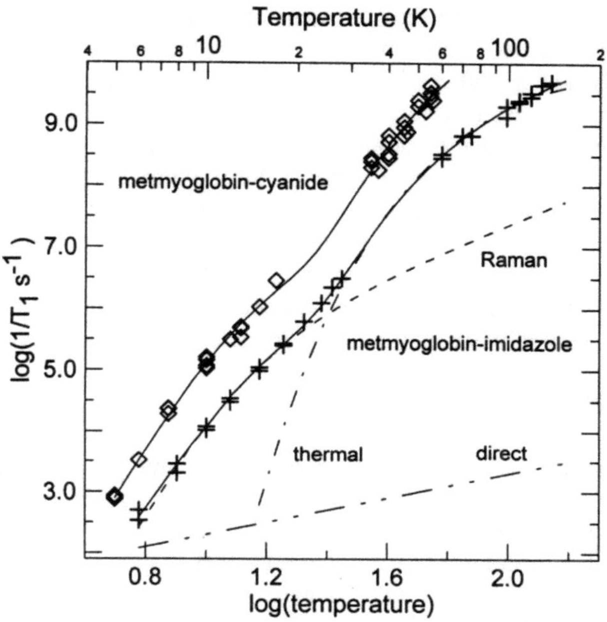

Figure 10. Temperature dependence of X-band spin-lattice relaxation rates for 1 mM imidazole adducts of metmyoglobin variants: R-Mb-V66C-Im and R-Mb-K98C-Im (+) and 1 mM cyanide adducts: R-Mb-V66C-CN, R-Mb-K98C-CN, and horse heart myoglobin cyanide (◊) in 1:1 water:glycerol. The solid lines through the data are the fits obtained using Eq. (7) and the parameters in Table 5. The contributions from individual processes to the relaxation for the imidazole adducts are: direct (_ . . _ . . _), Raman (- - -), and thermal mode (_ . _ . _). (Figure reproduced from Zhou et al., 1999).

Based solely on the fits to the experimental data, it is difficult to judge whether the high-temperature process observed for the low-spin Fe(III) should be assigned to a thermal process, an Orbach process, or a local vibrational mode. If the process is Orbach then the value obtained for Δ_{Orb} should reflect the energy separation between the ground state and a low-lying electronic excited state. The EPR g-values for low-spin Fe(III) can be analyzed to determine the energy of the low-lying excited state (Palmer, 1979, 1983) in units of λ, the spin-orbit coupling constant. Application of this approach to $FeTTP(Im)_2^+$, R-Mb-Im, and R-Mb-CN gives energies of 2.23λ, 1.68 λ and 0.90 λ, respectively (Zhou et al., 1999). Since it is likely that λ is similar for the three low-spin hemes, these calculations indicate that the energies for the low-lying electronic states are quite different for the three systems. Since the characteristic energies for the high-temperature process were similar for the three systems it is not reasonable to assign this energy as the energy for the excited state as required if we assign the process as Orbach. Assignment to a local mode would require a vibrational frequency of 165 to 185 cm^{-1}. Resonance Raman spectra of heme proteins indicate that the Fe-N stretching frequency for the bond to the distal histidine is typically about 200 cm^{-1} (Spiro, 1983). Thus, assignment of the high-temperature process to a local vibrational mode appears to be plausible. At room temperature, rotation around the Fe-N bond to coordinated imidazole is a relatively facile process in model Fe(III) porphyrinates (Momot and Walker, 1997), which suggests that low-amplitude oscillations might occur with a relatively low barrier at lower temperatures. This process is a possible candidate for a thermally-activated relaxation process. Comparison with results for other low-spin heme systems will be needed to distinguish between a thermally-activated process and a local vibrational mode.

7.6.6 Fe-S proteins

By measuring T_1 as a function of temperature for a 2-iron, 3-sulphur protein, it was determined that the best fit was an Orbach process with Δ = 350 K (250 cm^{-1}). The most likely excited state was due to antiferromagnetic coupling of two irons in the Fe-S protein. High-spin Fe(III) and high-spin Fe(II) couple to give a first excited state at 3J, so this study revealed that J = 83 cm^{-1}, a value not available from other measurements (Gayda et al., 1976).

Gayda et al. (1979) studied spin-lattice relaxation for a 2-iron-2-sulfur protein between 1.25 and 130 K. To fit the data several processes were required: below 3 K a phonon-bottleneck, between about 3 and 30 K the Raman process, and near 100 K an Orbach process.

7.6.7 Cu(II)

Since Cu(II) is one of the spin systems most extensively studied by EPR, one might expect that its relaxation properties would be "well known." The information concerning electron spin relaxation times of Cu(II), as is the situation for most metals, both historically and conceptually divides into three categories: ionic solids at liquid He temperatures, fluid solution at room temperature, and an incompletely explored intermediate region. A detailed survey of Cu(II) is presented to illustrate the incompleteness of knowledge of relaxation times.

There is a substantial solid state physics literature, almost exclusively T_1, not T_2 , mostly below 20K, and mostly in ionic lattices. The observations are summarized fairly comprehensively by Standley and Vaughan (1969) and by Al'tshuler and Kozyrev (1974). Al'tshuler and coworkers (1975) observed in a copper(II) system that below 4 K, T_1 was faster by about a factor of 3.5 at 9 GHz than at 28.8 GHz. This is the reverse of predictions for the direct process. In the temperature range of the direct process (4.3 K) there was a dependence of about a factor of 4 on nuclear spin quantum number for $g_{\parallel}$ but not for $g_{\perp}$. An angular dependence of $1/T_1$ was observed in the x,z plane. In other planes there were too many overlaps, leading to cross relaxation, to get a complete picture. The angular dependence was attributed to the mechanism of Larson and Jeffries (1966). In the range $4 < T < 12$ K, the relaxation rate was fitted by an Orbach-Aminov process with $D = 13\ cm^{-1}$.

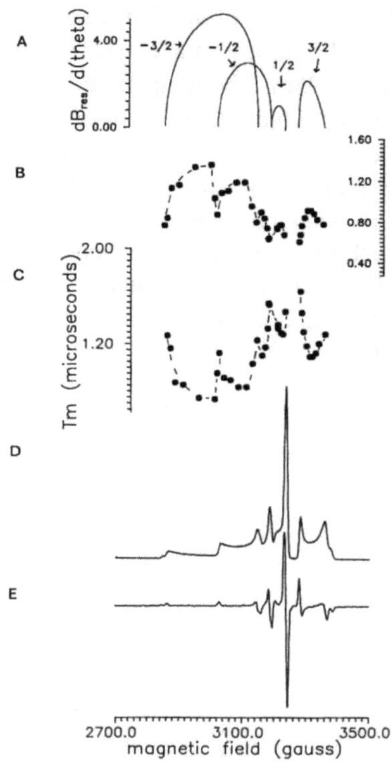

Figure 11. (A) Plots of $dB_{res}/d\theta$ in gauss/degree for $Cu(dtc)_2$ for each of the four nuclear spin states of the ^{63}Cu ion as a function of magnetic field. Resonant fields were calculated to first order using g and A values obtained by simulation of the CW EPR spectrum and a microwave frequency of 9.20 GHz. Parts B – E contain data for $Cu(dtc)_2$ in glassy 2:1 toluene: chloroform; (B) $1/T_m$ at 50 K as a function of magnetic field; (C) T_m at 50 K as a function of magnetic field. (D) Field-swept echo-detected spectrum at 100 K obtained with 30 and 60 ns pulses and a constant τ = 140 ns; (E) CW EPR spectrum obtained at 100 K with 1.0 G modulation amplitude and 0.10 mW microwave power.

Between liquid He temperatures and room temperature, there is not much information available. It is difficult to obtain accurate values for T_m for Cu(II) in complexes with nitrogenous ligands or distant nitrogens in the ligands because of deep nitrogen modulation. However, typical 2-pulse decay curves last for a few microseconds and the decay rates are not strongly dependent on temperature at low temperature (Du et al., 1996), so it seems likely that the decays are dominated by nuclear spins, analogous to what has been demonstrated for nitroxyl radicals, Cr(V), and vanadyl complexes. At 50 to 100 K, values of T_m for $Cu(dtc)_2$ (dtc denotes diethyldithiocarbamate anion) in glassy solution are longer along the principal axes and shorter at intermediate orientations. The pattern in $1/T_m$ correlates well with $dB/d\theta$, the change in resonance field per degree of angular displacement (Fig. 11). This correlation indicates that librations make significant contributions to dephasing (Du et al., 1992; 1995c). The temperature dependence of $1/T_m$ for several copper complexes in which there is weak coupling between the copper and a methyl group is shown in Fig. 12. Averaging of inequivalent Cu-proton couplings by rotation of the methyl group causes significant enhancement of echo dephasing between about 85 and 130 K (Du et al., 1994). Enhancement of echo dephasing for the phenyl-substituted complex bis(diphenyldithiophosphinate)Cu(II) is attributed to dynamics of the phenyl groups that modulate coupling between the copper unpaired electron and protons on the phenyl ring (Fig. 12).

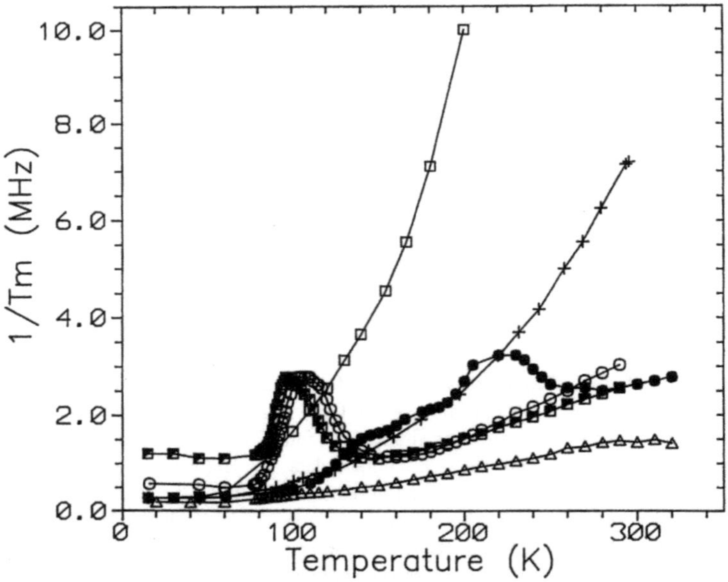

Figure 12. **Temperature dependence of $1/T_m$ for the perpendicular lines in the spectra of VOTTP doped in ZnTTP (Δ), CuTTP doped into ZnTTP (+), AgTTP doped into H_2TTP (□), $Cu(dtc)_2$ doped into $Ni(dtc)_2$ (half-filled square), $Cu(Et_2dtp)_2$ doped into $Ni(Et_2dtp)_2$ (Et_2dtp is diethyldithiophosphinate) (O), and $Cu(Ph_2dtp)_2$ doped into $Ni(Ph_2dtp)_2$ (Ph_2dtp is diphenyldithiophosphinate) (•). (Figure reproduced from Du et al., 1994).**

Spin-lattice relaxation for Cu(II) is strongly dependent upon the coordination environment. Examples of the temperature dependence of $1/T_1$ for a square planar 4-coordinate Cu(II) complex and a six-coordinate complex are shown in Fig. 13. The data were modeled with a Raman process and a local mode (Zhou et al., 1999). The difference in relaxation rates between the four- and six-coordinate complexes is not due exclusively to the differences in coordination number because a more rigid six-coordinate complex exhibited substantially slower relaxation than for the complex for which data are shown in Fig. 13 (Burchfield et al., 1997). This comparison is an example of the important role of molecular rigidity in relaxation rates in solids and glasses.

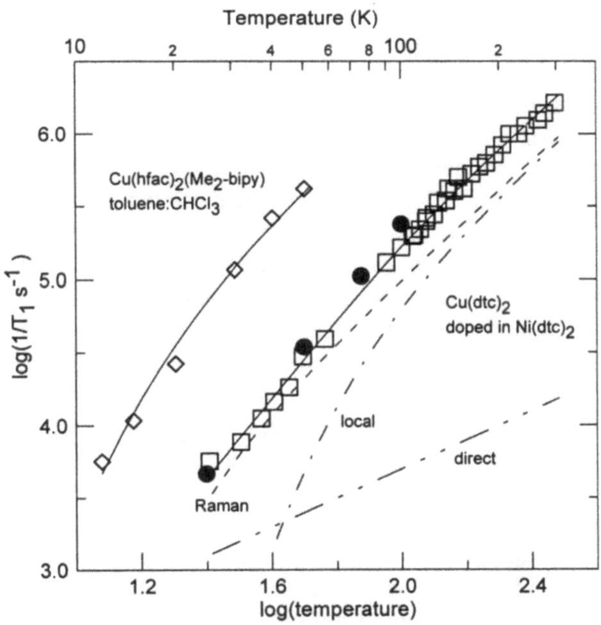

Figure 13. Temperature dependence of X-band spin-lattice relaxation rates for ~ 0.2% $Cu(dtc)_2$ doped into solid $Ni(dtc)_2$ (+) and for 1 mM $Cu(hfac)_2(Me_2\text{-bipy})$ in 2:1 toluene:$CHCl_3$ (◊) for the $m_I = ½$ perpendicular transitions. The solid lines through the data are the fits obtained using Eq. (7) and the parameters in Table 5. The contributions from individual processes to the relaxation for $Cu(dtc)_2$ are: direct (_ . . _ . . _), Raman (- - -), and local mode (_ . _ . _). (Figure reproduced from Zhou et al., 1999).

In glassy solution and doped into the diamagnetic Ni(II) crystalline hosts, $Cu(dtc)_2$ exhibits T_1 dependence on position in the line analogous to that described above for nitroxyl radical. In powdered crystalline hosts and in a single crystal host $1/T_1$ increased monotonically from the parallel to the perpendicular direction and is independent of m_I. $Cu(dtc)_2$ takes the geometry of the host lattice, approximately square planar with D_{2h} symmetry in $Ni(dtc)_2$ and distorted five-coordinate in $Zn(dtc)_2$. Relaxation in the Zn host is about two to eight times faster than in the Ni host and is less orientation dependent, which is attributed to a lower symmetry in the Zn host (Du et al., 1995c).

Cu(II) in inorganic glasses (Stösser et al., 1999) exhibits concentration-dependence of T_m that was attributed to aggregation of Cu in the glass. A strong decrease in T_m as the temperature increased from ca. 5 to 35 K was attributed to decreases in T_1 of Cu-Cu aggregates. Maximum T_m values found were about twice as long at 3.8 GHz as at 9.2 GHz for the same sample and temperature. Orientation dependence of T_m was less at 3.8 GHz than at 9 GHz.

The NMR literature reports metal ion T_1 values inferred from nuclear relaxation times. The first studies of Cu(II) in aqueous solution were by Bloembergen and Morgan (1961). The early work was summarized in the LaMar, Horrocks, and Holm book (1973). The major concern in this work was the impact of the metal T_1 on the observability of the NMR signal. More recently, the interest in MRI contrast reagents has led to much more detailed analysis of electron spin relaxation times in solution at room temperature, and especially of the magnetic field/RF frequency dependence. The extensive NMR dispersion (NMRD) work of Koenig and of the Italian group has resulted in a much more detailed understanding of electron T_1 in solution. These NMRD results are summarized in the recent book by Banci, Bertini, and Luchinat (1991). Table 4-3 in that book provides some estimates for several metals. Small Cu(II) molecules and Cu(II) proteins are discussed on pages 125-130 of that book. The dependence of Cu(II) T_1 on tumbling correlation time is presented in their Figure 4-4. Use of any of these values should be guided by detailed consideration of the bonding environment of the Cu(II). For example, the aqueous Cu(II) ion is pseudo-octahedral d^9 and thus exhibits rapid Jahn-Teller rearrangements, resulting in very fast T_1, but a low-symmetry complex would not exhibit the Jahn-Teller distortions, and might have a longer relaxation time. Also, the effect of the Cu(II) relaxation on other species will depend on the accessibility of that species to the metal, and hence on the coordination environment.

Koenig and coworkers (Gaber et al., 1972) stated that their NMRD results for a copper protein could be interpreted only in terms of a dependence of the Cu(II) relaxation time on magnetic field, being 5×10^{-10} s at low field, and longer at >750 G.

Based on the available data we conclude that for simple Cu(II) complexes in low-viscosity solution near room temperature, the Cu(II) T_1 is probably ca. 10^{-9} s and dependent on coordination geometry and rigidity.

7.6.8 Ag(II)

In glassy solutions or doped solids T_m for AgTTP is approximately independent of temperature up to about 40 K (Fig. 12, Du et al., 1996) and is attributed to nuclear spin diffusion. The spin-lattice relaxation rate for AgTTP is substantially faster than for CuTTP, consistent with the expectation of greater spin-orbit coupling and therefore more efficient modulation of spin-orbit coupling by molecular vibrations (Du et al., 1996). Above about 40 K T_1 makes a substantial contribution to T_m in doped solid samples (Fig. 12). In glassy solvents increased librations and rotations also contribute to T_m as temperature is increased.

7.7 Transition metals with S >1/2

7.7.1 Cr(III)

In ruby, Kyhl and Nageswara-Rao (1967) attributed the fastest process observed (in the microsecond range) by the pulse saturation recovery method to "spectral spin diffusion." The intermediate rate process was cross relaxation (3 ms), and the slowest rate was T_1 = 125 ms.

For $Cr(ox)_3^{3-}$ in 1:1 water:glycerol the temperature dependence of $1/T_1$ between 15 and 70 K (Fig. 14) can be modeled with a dominant contribution from a Raman process and smaller contributions from an Orbach process at low temperatures and a local mode at higher temperatures (Zhou et al., 2000).

7.7.2 Fe(III)

We have examined the spin-lattice relaxation rates for four non-heme Fe(III) proteins (Klug et al, 1998; Gaffney et al., 1998) at temperatures between 5 and 20 K. The zero-field splittings parameter D for these systems is less than 1 K (1K = 0.7 cm^{-1}). The relaxation rates for Fe(III) enterobactin bound to iron protein A (FepA) are typical of what we observed for these systems (Fig. 14). The temperature dependence of $1/T_1$ could not be modeled with a single process. We propose that it is due to a combination of the Orbach process (involving the ZFS) and the Raman process (Zhou et al., 2000).

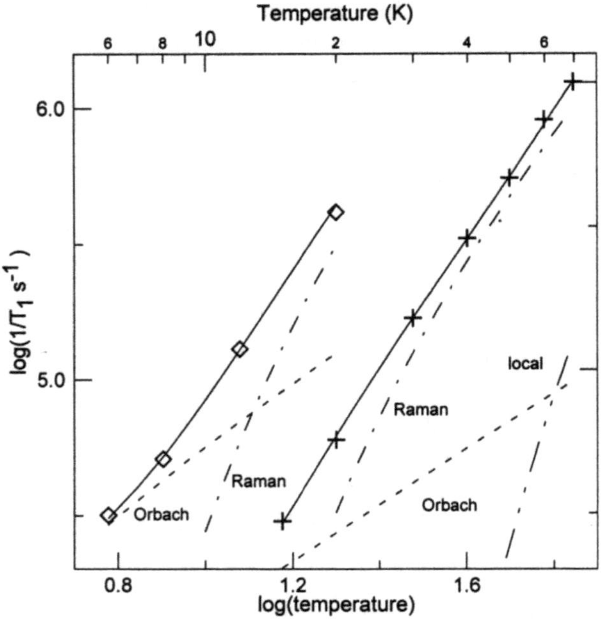

Figure 14. Temperature dependence of X-band spin-lattice relaxation rates for 1.0 mM $Cr(ox)_3^{3-}$ in 1:1 water:glycerol (+) measured at the $g_\perp$ turning point and for 0.2 mM FeEnt in 1:1 water:glycerol (◊) (Klug et al., 1998). The solid lines through the data are the fits obtained using Eq. (7) and the parameters in Table 6. The individual contributions are: Orbach process involving excited states at 2D and 6D (— — —), Raman process (— - — -) and local mode (— - - - — - -), (Figure reproduced from Zhou et al., 2000).

We also have examined T_1 for three iron porphyrins in organic solvents between 5 and 118 K and four heme proteins between 5 and 150 to 298 K (Zhou et al., 2000). The temperature dependence of the relaxation rates is characteristic of Orbach processes. The zero-field splittings for these complexes are in the range of 7 to 15 K. To analyze the relaxation rates over temperatures ranging from 5 K to room temperature it is necessary to include the Orbach process involving the excited states at both 2D and 6D (Fig. 15). For the Fe(III) in methemoglobin the relaxation rates extrapolated to room temperature are similar to results obtained by NMR in fluid solution. This observation supports the proposal that relaxation for metal ions in large macromolecules is largely determined by the protein vibrations and is not strongly impacted by the phase change of surrounding solvent from glass to liquid (section 6.2).

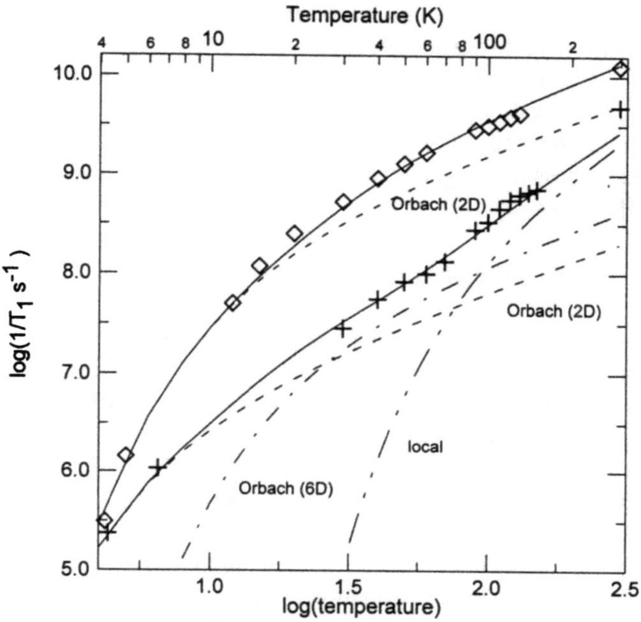

Figure 15. Temperature dependence of X-band spin-lattice relaxation rates for the $m_s = \pm 1/2$ transitions for Hb-F (+) and for Hb-H_2O (◊) in 1:1 water:glycerol (Budker et al., 1995). The solid lines through the data are the fits obtained using Eq. (7) and the parameters in Table 6. The individual contributions to the relaxation are: Orbach process involving excited state at 2D (— — —), Orbach process involving excited state at 6D (— - — -), and local mode (— - - — - -). The data points at 298 K are values of T_{1e} obtained by analysis of NMR spin-lattice relaxation rates for water protons in solutions of Hb-F or Hb-aquo (Budker et al., 1995). (Figure reproduced from Zhou et al., 2000).

Fast spectral diffusion was observed in saturation recovery of high-spin heme proteins in the temperature range 1.5 to 2.9 K by comparing species in ^{17}O-enriched and normal water (Wells et al., 1992).

7.7.3 Co(II)

Makinen and coworkers applied continuous wave microwave power saturation EPR to ascertain the zero-field splittings (ZFS) in Co(II) proteins (Yim et al., 1982; Makinen et al., 1985, Kuo and Makinen, 1985, Makinen and Wells, 1987).

7.7.4 Ions in Fluid Solution

The following results have been obtained by NMR in fluid solution. T_1 for Mn^{2+} increased with frequency (from 14 to 60 MHz proton NMR frequencies), and for V^{2+} T_1 increased from 2.7 to 60 MHz NMR frequency (Bloembergen and Morgan, 1961). For Ni^{2+}, Co^{2+}, and Fe^{2+} there was no frequency dependence up to 60 MHz. For Fe^{3+}, Mn^{2+} (S=5/2), and Cr^{3+}(S=3/2), relaxation is dominated by modulation of the quadratic ZFS interaction (Rubinstein et al., 1971). For these species there is a decrease in $1/T_{1e}$ with increasing magnetic field, by about a factor of 5 from 30 to 100 MHz NMR frequency.

7.8 Lanthanides and Actinides

The relaxation behavior of lanthanides and actinides, mostly doped into crystalline salts, was intensively studied early in the development of EPR, especially at Oxford. Summaries are in the reviews by Bleaney and Stevens (1953), Stevens (1969), and by Standley and Vaughan (1969). The book by Al'tshuler and Kozyrev (1974) summarized the early literature, and provided the functions that fit T_1^{-1}. The reported relaxation times of lanthanides are strongly dependent on the host lattice.

Many of the concepts of direct, Raman, and Orbach relaxation processes, and their experimental confirmation were outgrowths of these studies, building on the earlier work of Gorter, Waller and Van Vleck. In recent years Gd(III) has received the most attention (Stephens 1989), in large part because of its importance as a contrast reagent in MRI, but also because as an S-state f^7 ion it can be studied by a wide range of EPR techniques over a wide temperature range. The relaxation behavior in fluid solution is especially complicated, as has been shown by the studies of Clarkson et al. (1998), Powell et al. (1993, 1996), Rast et al., (1999), and Tóth et al. (1998, 1999). However, the primary relaxation mechanism is modulation of the ZFS.

The T_1 of Dy^{3+} at 77 K is nearly equal to ω^{-1} at 9 GHz, which makes it an effective relaxer of other spins (Sarna at al., 1976; Hyde and Sarna, 1978; Antholine et al., 1978). Gd^{3+} is more effective than Dy^{3+} in solution at 20°C (Hyde and Sarna, 1978). There have been several applications of broadening of radical EPR spectra by proximity to Dy^{3+} in attempts to determine distances in biological systems (Blum et al., 1981; Innes and Brudvig, 1989; Isogai et al., 1990). Unfortunately, very little information concerning the Dy^{3+} relaxation times exists. Blum et al. (1981) used the temperature-dependent width of the Dy^{3+} CW EPR line to estimate that $T_1 \approx 10^{-10}$ s at 10 K, but this value does not seem consistent with the reported effect on line broadening of a nearby slowly-relaxing species, so caution should be exercised in using this number.

Discrete lanthanide and actinide molecular complexes, especially organometallic bonding interactions, often can stabilize special oxidation states and electronic energy configurations. EPR as a diagnostic tool is revealing that some of these complexes have long enough relaxation times to be studied at room temperature. The developing consensus is that longer relaxation times occur when the unpaired electron is in an orbital of largely d character. In general, d orbitals have less orbital angular momentum than do f orbitals (Parry at al., 1999).

The electron spin T_1 of Er^{3+} doped into a rare earth metaphosphate glass was dominated by the Orbach process, with crystal field splitting $\Delta = 102$ K (Goudemond et al., 1997), based on monitoring the ^{31}P nuclear spin relaxation. At ca. 20 K with a 1% Er-doped sample the nuclear T_1 was ca. 10^{-2} sec. A similar study of 1% Nd^{3+} in a rare earth metaphosphate glass interpreted the Nd^{3+}electron T_1 at 10 KG as being dominated by the direct process below 5 K and by the Orbach process with $\Delta = 97$ K above 5 K. T_1^{-1} increased from 10^2 s^{-1} at 3 K to 10^{11} s^{-1} at 100 K (Goudemond et al., 1995).

7.9 Frequency Dependence of Relaxation

There are very few actual comparisons of the frequency dependence of relaxation times. Some data are cited in table 7 of Prisner's review (1997). He summarized relaxation times from the literature and his own measurements, comparing X-band and W-band (95 GHz) results for T_1 and T_2. In some cases relaxation times are longer at higher frequency, in some cases they are about the same, and in some cases relaxation times get shorter at higher frequency. The few data points span such widely different radicals that to find a pattern will require identifying the relaxation mechanisms for each type of sample. Caution must be exercised in interpreting the frequency dependence of apparent relaxation times based on CW experiments if differences in B_1 have not been taken into account. Field sweeps at higher magnetic fields also may be faster, leading to larger passage effects. Less spectral diffusion at high frequency may give rise to effects that are interpreted as showing longer relaxation times.

7.9.1 T_1 for Immobilized Samples

The T_1 of F-centers in irradiated LiD was estimated to be ca. 1 s at 136 and 182 GHz, but ca. 2.8 s at 71 GHz (Bouffard et al., 1980). The exact temperature was not stated, but the context indicates it was between 0.43 K and 1.9 K, so the direct process was likely to dominate the electron spin lattice relaxation, and the relative values are consistent with expectations for the direct process. At 50 to 100 K T_1 for tempone doped into 4-hydroxy-piperidinol and for Cu(II) doped into $Ni(dtc)_2$ are essentially the same at X-band and W-band (Weber et al., unpublished data), consistent with dominant contributions from the Raman process. Relaxation rates for vanadyl porphyrin doped into zinc porphyrin below about 50 K are somewhat faster at W-band than at X-band, which suggests a larger contribution from the direct process at the higher microwave frequency. Hoffman reported (Davoust et al., 1996) that T_1 is shorter at Q-band than at X-band at 2-4 K for ferredoxin and nitrile hydratase.

Mizoguchi and coworkers (1984) found that relaxation in polyacetylene as a function of frequency from 5 to 450 MHz fit the square root of frequency relation,

$$\frac{1}{T_1} \propto \frac{1}{\sqrt{\omega}} \tag{28}$$

The data at X-band obtained by others also lies on their curve. A review of the relaxation of F-centers includes discussion of field-dependence of T_1 (Baldacchini, 1993).

Witowski and coworkers (Witowski et al., 1995, 1997; Strutz et al., 1992) showed that in the temperature range where the direct process dominates, electron spin relaxation rate increases at higher magnetic fields are more accurately simulated by replacing the Debye approximation by a more realistic description of the phonon distribution. The rate in CdMnTe increases as $B^{1.5}$ to ca. 15 T and as B^5 at higher fields.

7.9.2 T_1 in Fluid Solution

Hyde et al. (1990) observed that T_1 for a stearic acid spin label at 27°C was 0.94 μs at S-band and 2.54 μs at X-band. Robinson et al. provided a model consistent with this result (section 7.1.4). Furthermore, the molecular tumbling motions could cause deformations of the nitroxyl radical, modulating the mixing of spin and orbital angular momentum.

The studies of relaxation of semiquinones by Prabhananda and Hyde (1986) yield an important insight. The g-anisotropy of tetrachlorosemiquinone and tetrabromosemiquinone are as large as the g-anisotropy of many nitroxyl radicals. Thus, the spin-rotation contributions to T_1 should be similar for the same molecular size for nitroxyl radicals and these halogenated semiquinone radicals. Prabhananda and Hyde (1986) wrote the dependence of T_1 on rotation due to g modulation as:

$$\frac{1}{T_1(g\,\mathrm{mod})} = \sum_i (g_i - g_s)^2 \frac{1}{5g_s} \frac{\omega_0^2 \tau_\theta}{1 + \omega_0^2 \tau_\theta^2} \tag{29}$$

where g_s is the isotropic g value and τ_θ is the rotational correlation time. If $\omega_0\tau_\theta >> 1$, this contribution to T_1 is not dependent on microwave frequency. For small molecules, such as the semiquinones, in fluid solution where $\omega_0\tau_\theta << 1$, T_1 is predicted to be shorter at higher microwave frequency. However, the observation that T_1 was nearly equal at 1.15 GHz and 9 GHz indicated that this was not an important contribution to T_1 for these radicals.

Magnetic field/frequency dependence of electron spin relaxation of metal ions in solution involves several relaxation mechanisms. For the case in which the electron T_1 is determined by fluctuations in ZFS caused by collisions with solvent molecules, the functional form is (Bertini et al., 1986):

$$\frac{1}{T_1} = \frac{1}{5\tau_{SO}}\left\{\frac{1}{1+\omega_s^2\tau_v^2} + \frac{4}{1+4\omega_s^2\tau_v^2}\right\} \tag{30}$$

where the electronic relaxation at zero field, τ_{SO}, is proportional to the quadratic ZFS, ω_s is the electron Larmor frequency, and τ_v is the correlation time for the collision process. At some value of ω_s the electron T_1^{-1} will no longer decrease as predicted by this equation, but will instead become dominated by the rotational correlation time or the lifetime of labile adducts of the metal (Bertini et al., 1986).

The case of Ti(III) in aqueous solution (Bertini et al., 1992b) highlights the need to focus on the relevant relaxation mechanism. Bertini and coworkers found that the relaxation time for aqueous Ti(III) decreased from 5.7×10^{-11} s to 3.2×10^{-11}s with increasing temperature (5 – 35 °C), whereas prior studies of VO^{2+}, Mn^{2+}, Cr^{3+}, and Fe^{3+} showed the opposite temperature dependence. They noted that a decrease of relaxation time with increasing temperature had also been observed for a Mn(III)porphyrin. The decrease of the Ti(III) T_1 with increase in temperature, and independence of magnetic field, was attributed to the dominance of an Orbach process for this species (Bertini et al., 1992b).

7.9.3 T_m in Immobilized Samples

Spin echo dephasing for tempone at 11 and 40 K is the same at X-band and W-band, which is consistent with dephasing by nuclear spins in the surrounding solvent (Weber et al., unpublished results).

Dzuba et al. (1992) presented a formula based on Abragam and on Ernst et al., that included the adiabatic rate constant

$$\frac{1}{T_2^a} = \tau_c \left(\begin{array}{l} \frac{\gamma^2}{4} H_0^2 \left\langle \left(g_{zz} - \langle g_{zz} \rangle \right)^2 \right\rangle + m_I^2 \frac{\left\langle \left[(A \cdot A_0) - A_0^2 \right]^2 \right\rangle}{A_0^2} \\ + \gamma H_0 m_I \frac{\left\langle \left(g_{zz} - \langle g_{zz} \rangle \right) \left(A \cdot A_0 - A_0^2 \right) \right\rangle}{A_0} \end{array} \right) \tag{31}$$

This formula is the basis for the statement that the phase memory rate constant depends on magnetic field squared. Dzuba et al. (1992) stated that this formula was valid only if τ_c is between 10^{-8} and 10^{-6} s. If it is faster, "the model does not describe the experiment." If it is slower, "the Redfield approximation is no longer valid." This work was subsequently quoted (Onischuk et al., 1998) as stating that $1/T_2$ depends on magnetic field squared if the radicals have rapid librations with low amplitude, and suggested application of this to α-Si:H particles. They did not actually have data at more than one frequency, and were relying on line widths, not direct measure of T_2.

Cutter, Moll, and coworkers (Kutter et al., 1995; Moll et al., 1999) found longer T_m at 604 GHz than at 9.5 GHz for a sample of tempone diluted 1:30 in polystyrene. T_1 was measured to be 10 μs with a three-pulse saturation, spin-echo-detection sequence. Measurements were made in the range 4 – 13 K. The authors pointed out that at 604 GHz the upper Zeeman level is almost empty, making possible very few flip-flop processes, and hence resulting in slow decay rates (Kutter et al., 1995). It is likely that these numerical values reflect some instantaneous diffusion and spectral diffusion, due to the high spin concentrations in the samples.

8. IMPLICATIONS FOR RECORDING CW SPECTRA OF NITROXYLS AT ~100 K

Techniques for determination of distances between nitroxyl spin labels that rely upon analysis of CW lineshapes require that spectra are recorded under conditions that are not saturating, which means that the saturation factor, $s = \frac{1}{1+\gamma^2 B_1^2 T_1 T_2}$ should be close to 1. For a nitroxyl radical at 100 K T_1 is about 200 μs and T_m is about 1 μs. If one uses a standard rectangular TE_{102} cavity, B_1 is given approximately by $B_1 = 0.02\sqrt{QP}$ with the power, P, in watts and Q about 3600 (More et al., 1984). Thus, a power of 1 μW (ca. 53 dB attenuation of 200 mW) would give a B_1 = 0.0012 G. For the nitroxyl relaxation times this would correspond to s = 1/(1+0.07), which is reasonably close to unsaturated.

Based on Bloch (1946), Hyde (1960), Weger (1960), Poole (1967, p. 700), Mailer (1973), the conditions for "normal" slow passage CW EPR spectra are

$$\frac{d\delta}{dt} << \frac{1}{\sqrt{T_1 T_2}} \text{ , where } \delta = \frac{B_o - B^*}{B_1} \tag{32}$$

$B_0 - B^*$ is the difference between the resonant field and the actual value. If T_1 = 200 μs and T_m = 1 μs, the right hand side of (32) is $7x10^4$ s^{-1}. The left hand side of (32) can be estimated for an incident power of 20 or 1 μW, which corresponds to B_1 of ca. 0.0054 or 0.0012 G, respectively, in a TE_{102} cavity. If we assume that the main B_o field is swept at 1 G/s,

$$\begin{aligned} \frac{d\delta}{dt} &= \frac{1}{0.0054\,G}(1\,G/s) = 185 \text{ s}^{-1} \text{ and} \\ &\quad \frac{1}{0.0012\,G}(1\,G/s) = 833 \text{ s}^{-1} \text{, respectively} \end{aligned} \tag{33}$$

These would satisfy the slow-passage criterion. However, in a field-modulated spectrometer, the sweep of the B_0 magnetic field usually is slow relative to the modulation field $\omega_m B_m$. For example, with 1 G modulation at 100 KHz, $\omega_m B_m = 2\pi x10^5$G/s, and the left hand side of (33) becomes

$$\frac{\omega_m B_m}{B_1} = \frac{2\pi(10^5\,Gs^{-1})}{0.0054\,G} = 1x10^8 \text{ and} \frac{2\pi(10^5\,Gs^{-1})}{0.0012\,G} = 5x10^8 \text{ s}^{-1} \tag{34}$$

Clearly, this is much larger than the right hand side of (32), and slow passage conditions are not achieved. This may be a surprising result, since many CW EPR spectra of nitroxyl spin labels are recorded under similar conditions. Note that the nitroxyl T_2 defines a spin packet width of ca. 0.065 G and the T_1 defines a rate limit of ca. 800 s^{-1}. If the modulation amplitude is made equal to the spin packet width and the rate is slow relative to 800 s^{-1} (e.g., 100 s^{-1}), then

$$\begin{aligned} \frac{\omega_m B_m}{B_1} &= \frac{2\pi(100x0.0065\,Gs^{-1})}{0.0054\,G} = 756 \text{ and} \\ &\quad \frac{2\pi(100x0.0065\,Gs^{-1})}{0.0012\,G} = 3.4x10^3 \text{ s}^{-1} \end{aligned} \tag{35}$$

which is now smaller than the right-hand side of (32). It is difficult to achieve slow-passage CW EPR spectra of nitroxyl spin labels at or below liquid nitrogen temperature. This conclusion is relaxed somewhat by the occurrence of spectral diffusion, which relieves saturation. When spectral diffusion is significant, T_1 in Eq. (32) should be replaced by T_D, the time constant for spectral diffusion.

As described by Hyde (1960) and Weger (1960), and references cited therein, Portis explained that under conditions of rapid passage the signal 90° out of phase from the absorption can look like an undifferentiated absorption signal instead of the derivative of the dispersion. Since the dispersion signal does not saturate as readily as the absorption signal, it is advantageous to record the dispersion signal when relaxation times are long. For example, at very low temperature, T_1 becomes long enough that even for transition metal complexes it is not possible to obtain an unsaturated slow-passage CW EPR spectrum. When the saturated EPR signal is detected with the spectrometer tuned to dispersion mode, the response approximates the absorption spectrum. When EPR lines with different relaxation times overlap, the lines are brought into passage conditions at different microwave powers, so this method provides a means of selectively observing different transitions (Hales et al., 1989). The ENDOR effect on such a line appears as an absorption rather than a derivative display (DeRose and Hoffman, 1995). Gaffney and Silverstone (1993) published additional examples, including an in-phase second harmonic spectrum.

9. SUMMARY OF TRENDS IN T_M AND T_1 FROM EXPERIMENTAL DATA.

9.1 Spin Echo Dephasing and T_2 for S = 1/2

At low temperatures T_m for S = ½ species is dominated by nuclear spin diffusion and is much shorter than T_1. It is dependent upon the concentration and types of nuclear spins. In the absence of proton or fluorine spins T_m is longer than in proton-containing solvents and may be determined by electron-electron dipolar interaction at which point we call it T_2. T_m increases very slowly with increasing temperature unless there are dynamic effects such as averaging of coupling to inequivalent nuclei or collapse of spin-spin coupling to a more rapidly relaxing electron spin. Librational motion can decrease T_m and has its greatest impact at positions in the spectrum where the resonance condition is most sensitive to orientation. Typically this causes T_m to be shorter at intermediate orientations and longer along the principal axes. At some temperature, as temperature is increased, the electron T_1 becomes short enough to begin to decrease T_m. Gradually T_m and T_1 approach each other. The temperature at which $T_1 \sim T_m$ depends on how short T_1 is. For example, it occurs at ca. 20 K for low-spin Fe(III) but near room temperature for isolated organic radicals. If T_1 becomes short enough, this may happen before the softening point of the matrix. For species with anisotropic g and/or A values molecular rotation causes averaging of the anisotropies, which decreases T_m. These effects are observed as a glass softens.

In fluid solution $T_m = T_2$. If a spin-1/2 center is not interacting with another paramagnetic center, T_2 will increase with increasing temperature as increasing rates of molecular tumbling more effectively averages anisotropy (the fast averaging limit), provided that collisions do not shorten T_m (Stillman et al., 1980). If another paramagnetic center is present and collision interactions dominate T_2, T_2 becomes shorter with increasing temperature due to increasing collision frequencies. This latter case is common, and provides information on Heisenberg exchange, leading to applications such as oxymetry and study of dynamics of membranes. T_1 decreases with increasing motion and when there is complete motional averaging of anisotropies T_2 becomes equal to T_1.

9.2 Spin-lattice Relaxation for S = ½

The available data all point to modulation of spin-orbit coupling as the dominant contribution to spin-lattice relaxation in immobilized samples. It had been proposed that modulation of nuclear hyperfine interaction could contribute to orientation-dependent spin-lattice relaxation (Aleksandrov and Zhidomirov, 1961). We have not observed any cases in solids or glassy solvents in which relaxation is dependent on m_I, which leads to the conclusion that modulation of electron-nuclear hyperfine is not a major contributor to T_1 (Eaton and Eaton, 1996). The general picture of spin lattice relaxation in immobilized samples is that it is dominated by molecular vibrations. Vibrations of the molecule exchange energy with the lattice and modulate the spin-orbit coupling that mixes spin and orbital angular momentum. In solution rotation provides additional modulation processes. In solution in some cases it is necessary to include the effects of collisions with oxygen and/or other paramagnetic species.

Table 5. Contributions to Spin-lattice Relaxation Determined by Fitting to Eq. (7)

sample[a]	lattice	Temp range (K)	g values[b]	direct A_{dir}	Raman A_{Ram}, θ_D	local A_{loc}, Δ_{loc} [c]	Thermal A_{therm}, E_a[c], τ_c^0
organic radicals							
trityl	H_2O:glycerol	22 - 173	2.005		$5.3x10^3$, 105		$5.4x10^{15}$, 1050, $5.5x10^{-12}$ [d]
trityl	9:1 n-PrOH: MeOH	12 - 132	2.005		$3.4x10^3$, 80		
tempone	H_2O:glycerol	14 - 200	2.009, 2.005, 2.002		$4.2x10^4$, 112		$2.9x10^{15}$, 600, $5.5x10^{-12}$ [d]
tempone	decalin	10 - 130	"		$3.9x10^4$, 67		
tempone	sucrose octaacetate	11 - 300	"		$2.3x10^4$, 51		$6.1x10^{16}$, 1150, $4.5x10^{-12}$
tempone	9:1 n-PrOH: MeOH	12 - 124	"		$2.4x10^4$, 54		
C_{60}^-	MeTHF	10 - 298	1.995, 2.000		$6.0x10^7$, 95		$6.8x10^{19}$, 330, $3.3x10^{-12}$
C_{60}^-	toluene: acetonitrile	9 - 217	1.994, 1.999		$1.6x10^7$, 75		$3.3x10^{19}$, 370, $3.9x10^{-12}$
C_{60}^{3-}	DMSO	6 - 134	2.008, 2.000, 1.995		$1.3x10^7$, 50		$2.6x10^{19}$, 334, $3.1x10^{-12}$
3d transition metals							
VOTTP-COOH	ZnTTP solid	41 - 260	1.984, 1.964	11.5	$6.0x10^4$, 100	$4.0x10^5$, 350	$5.5x10^{16}$, 950, $6.5x10^{-12}$ [d,e]
VOTTP-bipy	toluene: $CHCl_3$	17 - 120	1.984, 1.966	8.5	$6.5x10^4$, 100	$6.5x10^5$, 350	
VO^{2+} (aq)	H_2O:glycerol	11 - 100	1.980, 1.937		$7.4x10^5$, 120	$7.9x10^5$, 185	
$Cr(NO)(CN)_5^{3-}$	H_2O:glycerol	10-150	2.002, 1.970		$1.3x10^5$, 135	$1.3x10^6$, 310	
$Cr(NO)(H_2O)_5^{2+}$	H_2O:glycerol	10-160	1.998, 1.916		$9.1x10^5$, 110	$2.0x10^5$, 360[f]	
$Cr(NO)(EHBA)^+$	H_2O:glycerol	12-160	1.995, 1.914		$8.6x10^5$, 120	$2.8x10^5$, 360[f]	
$Cr(NO)(EHBA)_2$	H_2O:glycerol	12-130	1.996, 1.917		$6.6x10^5$, 110	$5.6x10^5$, 360[f]	
$CrO(EHBA)_2^-$	H_2O:glycerol	10 - 150	1.971, 1.972, 1.983	1.3	$1.3x10^5$, 115	$8.7x10^5$, 370	
CrNTPP	tol:THF	18 - 130	1.994, 1.956	4.8	$1.2x10^5$, 135	$4.3x10^5$, 320	$8.0x10^{16}$, 750, $6.5x10^{-12}$ [d]
$Cu(dtc)_2$	$Ni(dtc)_2$ solid	26 - 298	2.018, 2.083	30	$5.8x10^5$, 90	$5.7x10^5$, 240	
$Cu(dtc)_2$	toluene: $CHCl_3$	25 - 140	2.019, 2.087	20	$3.8x10^5$, 65		$2.5x10^{17}$, 650, $5.5x10^{-12}$

sample[a]	lattice	Temp range (K)	g values[b]	direct A_{dir}	Raman A_{Ram}, θ_D	local A_{loc}, Δ_{loc} [c]	Thermal A_{therm}, E_a [c], τ_C^0
CuTTP	ZnTTP solid	30 - 200	2.045, 2.190	19.5	$4.8x10^5$, 120	$1.5x10^6$, 250	$9.7x10^{17}$, 950, $5.5x10^{-12}$
CuTTP	toluene: $CHCl_3$	20 - 135	2.049, 2.203	14.5	$5.5x10^5$, 110	$3.4x10^5$, 250	$8.6x10^{17}$, 650, $5.5x10^{-12}$
ZnTTPbipy-Cu(hfac)$_2$	toluene: $CHCl_3$	15 - 100	2.063, 2.288		$1.2x10^7$, 125[g]		
Cu(hfac)$_2$-(Me$_2$-bipy)	toluene: $CHCl_3$	10 - 50	2.075, 2.277		$6.0x10^6$, 70[g]		
FeTPP(Im)$_2^+$	toluene: $CHCl_3$	11 - 130	1.50, 2.13, 2.61		$1.2x10^7$, 80		$0.8x10^{20}$, 240, $5.5x10^{-12}$ [h]
R-Mb-Im	H_2O:glycerol	6 - 138	1.44, 2.22, 3.02	20	$1.2x10^8$, 82		$3.0x10^{20}$, 265, $6.5x10^{-12}$ [h]
R-Mb-CN	H_2O:glycerol	5 - 60	~1., 1.72, 3.48	28	$9.0x10^8$, 78		$4.5x10^{21}$, 275, $6.5x10^{-12}$ [h]
4d transition metals							
MoOTTP-OEt	toluene: $CHCl_3$	20 - 140	1.967, 1.971		$1.8x10^5$, 120[i]	$3.6x10^6$, 350	$1.9x10^{18}$, 750, $6.5x10^{-12}$
AgTTP	H_2TTP, solid	28 - 115	2.040, 2.110		$1.3x10^5$, 120[i]	$1.1x10^7$, 270	
AgTTP	toluene: $CHCl_3$	28 - 114	"			$1.3x10^7$, 275	

[a] Abbreviations used in this table: CrNTTP, nitrido chromium(V) complex of 5,10,15-20-tetratolylporphyrin; dtc, diethyldithiocarbamate; EHBA, ethylhydroxybutyrate; hfac, hexafluoroacetylacetonate; R-Mb-CN, spin-labeled metmyoglobin cyanide; R-Mb-Im, spin-labeled metmyoglobin imidazolate; tempone, 4-oxo-2,2,6,6-tetramethylpiperidin-1-yl; trityl, Nycomed sym-trityl; TTP, 5,10,15,20-tetratolylporphyrin; VOTTP-bipy, vanadyl complex of 5,10,15-tri-tolyl-20-(4'-(4-methyl)bipyridyl)porphyrin; VOTTP-COOH, vanadyl complex of 5-(4-carboxy)-10,15,20-tri(tolyl)porphyrin

[b] For complexes with axial symmetry g values are given in the order $g_\perp$, $g_\parallel$. [c] Energies in units of Kelvin. [d] Contribution from thermal process is small and has observable impact only at temperatures above about 130 K. [e] Alternate fitting is with 2nd local mode with $A=5.5x10^6$ and Δ_{loc} = 950 K. [f] Contribution from local mode makes a small contribution at the highest temperatures examined. [g] Data can also be fit with a local mode instead of a Raman process

[h] Comparable fit to the data also could be obtained with either an Orbach process or a local mode. [i] Dominant contributions are from local mode and/or thermal process. There is substantial uncertainty in the parameters for the Raman process.

9.2.1 Temperature Dependence of T_1

The general pattern is that T_1 for most S=1/2 samples is strongly temperature dependent. The functional form for the temperature dependence depends upon the processes that dominates relaxation in a particular temperature region. As a glass softens, T_1 for a small molecules rapidly becomes shorter, until it approaches the fluid solution value. Results that we have obtained by fitting the experimental temperature dependence of T_1 using Eq. (7) are summarized in Table 5 (Zhou et al., 1999).

Direct process. For the samples for which data are summarized in Table 5 the direct process makes a relatively small contribution to the relaxation. As discussed in section 4.2 the direct process is expected to be a more important contributor at lower temperatures. Its effects are only detectable at the lowest temperatures for which data were examined.

Raman process. For the organic radicals trityl and tempone the Raman process is the dominant contribution to the relaxation over most of the temperature range for which data were obtained. For transition metal complexes it is necessary to include processes in addition to the Raman process. For all of the samples studied, the Debye temperature is between 50 and 135 K. These values are in a reasonable range based on comparison with literature data, such as the Debye temperature for crystalline ice which is about 200 K (Whalley and Labbé, 1969). Lower values are expected for organic solvents because the vibrational modes are softer. In 1:1 water:glycerol the Debye temperatures for the relatively small molecules tempone (112 K), VO^{2+} (120 K), and $CrO(EHBA)_2^-$ (115 K) are similar. These Debye temperatures are higher than were observed for small molecules in organic solvents: tempone in decalin (67 K), tempone in 9:1 n-propanol:methanol (54 K), $Cu(dtc)_2$ in 2:1 toluene:$CHCl_3$ (65 K) and $Cu(hfac)_2(Me_2\text{-bipy})$ in 2:1 toluene:$CHCl_3$ (70 K). The difference in θ_D values between 9:1 n-propanol:methanol and 1:1 water:glycerol is smaller for the bulky trityl radical (80 K vs. 105 K) than for tempone (54 K vs. 112 K). The unpaired electron of the trityl radical is largely localized on the central carbon, which is insulated from the solvent by the rest of the molecule. The trityl molecules is sufficiently large (m.w. = 1151 g/mole) that θ_D is strongly influenced by the molecule itself, with some effect of the solvent. The tempone radical is much smaller and the unpaired electron is in close proximity to the solvent, which makes θ_D more solvent dependent.

For the d^1 and d^9 metalloporphyrins θ_D is in the range of 110 to 135 K, independent of whether the host is a polycrystalline diamagnetic metalloporphyrin or 2:1 toluene:$CHCl_3$. The small range in θ_D for the metalloporphyrins suggests that, analogous to trityl, the porphyrin largely defines the motions that contribute to the Raman process at the paramagnetic center. In contrast to these metalloporphyrins, the Debye temperatures for the low-spin Fe(III) porphyrins, whether small molecule or protein, are in the range of 78-82 K. These lower values suggest that the additional axial ligands in these complexes have a substantial impact on the motions that contribute to the Raman process. These Debye temperatures are somewhat higher than the θ_D = 60 K obtained for two iron-sulfur ferredoxins (Gayda et al., 1979; Bertrand et al., 1982). Debye temperatures between 56 and 65 K were observed for alkyl tin derivatives of DNA (Barbieri et al., 1995).

If modulation of the spin orbit coupling is a dominant contribution to the spin lattice relaxation, we would expect the coefficients in Eq. (7) to increase with increasing spin orbit coupling. The coefficients of the Raman process listed in Table 5 are plotted as a function of the g anisotropy (defined as $g_{max} - g_{min}$) in Fig. 16. The g anisotropy is used as a reporter of spin orbit coupling. There is a clear trend to larger coefficients as g anisotropy increases. Some of the spread in coefficients is attributed, at least in part, to differences in molecular flexibility.

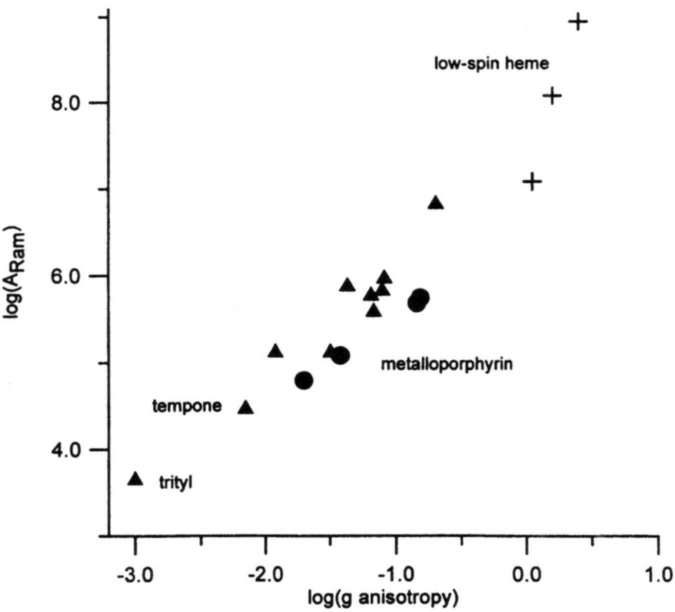

Figure 16. Dependence of the coefficients for the Raman process from Table 5 on the g anisotropy for organic radicals and metal complexes (▲), metalloporphyrins (•), and low-spin hemes (+).

Local modes. For most of the transition metal complexes an additional process with a characteristic energy of 185 to 350 K (130 to 240 cm^{-1}) was required to fit the data. These processes are tentatively assigned to local modes. The coefficients, A_{loc}, were of the order of 10^5 for the 3d transition metals and increased to $4x10^6$ to $1x10^7$ for the 4d Mo(V) and Ag(II), respectively. The increase in A_{loc} with increasing principal quantum number is consistent with larger spin-orbit coupling for heavier transition metals.

Thermally activated process. For several of the samples an additional contribution to the relaxation was observed at the highest temperatures for which data are available. This contribution was treated as a thermally-activated process that may arise from the onset of molecular rotation as the glass softens. Comparable fit to the data could be obtained with an additional local mode or an Orbach process. The Orbach process is considered less likely because sufficiently low-lying excited states are not known to exist for most of the systems. To distinguish between a local mode, a thermal process, and an Orbach process, based on the temperature dependence of relaxation, requires experimental data over an extended temperature range up to and beyond the characteristic energy.

9.2.2 Orientation Dependence of T_1

At low temperature, where T_1 is dominated by the direct process T_1 is strongly orientation dependent. At higher temperatures where the Raman process or local modes dominate, T_1 in samples with axial symmetry typically is longer along the molecular z axis than in the perpendicular plane. We attribute this orientation dependence to the symmetry of vibrational modes.

9.2.3 Dependence of T_1 on Electronic Configuration

The domination of spin-lattice relaxation by spin-orbit coupling leads to the following generalizations. 1) T_1 will decrease in the order 3d > 4d > 5d, other factors being equal. 2) Relaxation will be faster for non-centrosymmetric complexes (including tetrahedral) than for centrosymmetric complexes (including octahedral). 3) Ignoring other effects of excited states, relaxation will become faster as the symmetry is decreased. 4) Relaxation will increase as g values become further from 2.0023. 5) Relaxation will decrease as the character of the orbital containing the unpaired electron changes from predominantly metal to predominantly ligand.

9.3 Spin-lattice Relaxation for S > ½

In principle, relaxation for S>1/2 species may include (a) modulation of spin-orbit coupling, as for S = 1/2 species; (b) modulation of ZFS, which can occur via molecular vibrations (including distortions caused by collisions in solution); (c) mixing of excited states into the ground state; and (d) thermal population of excited states. Coefficients of relaxation processes obtained for Fe(III) are summarized in Table 6. At temperatures from 15 to 70 K for $Cr(ox)_3^{3-}$ and from 5 to 30 K for non-heme iron proteins with zero-field splitting, D, less than 1 K, the Raman process made a dominant contribution to spin-lattice relaxation. For heme proteins with larger zero-field splittings the Orbach process dominates (Zhou et al., 2000). In Fig. 17 the coefficients for the Orbach process are plotted as a function of the zero-field splitting. Within the set of iron porphyrins the sum of the coefficients for the two Orbach processes increases as 2D increases as expected if modulation of the zero-field splitting dominates the relaxation. Similarly, within the series of heme proteins there is a general trend toward larger coefficients as 2D increases. However, for comparable values of 2D the coefficients for the heme proteins are much smaller than for the iron porphyrins (Table 6 and Fig. 17). Modulation of the zero-field splitting by motion of the axial ligands is likely to be a significant contribution to the relaxation. In the heme proteins, the axial ligands are hydrogen bonded (Deatherage et al., 1976; Leci et al., 1995) to water molecules or protein side chains, which substantially restricts the motional freedom of the ligands compared with that in the small molecules. We propose that this is another example of the importance of molecular flexibility in spin relaxation rates.

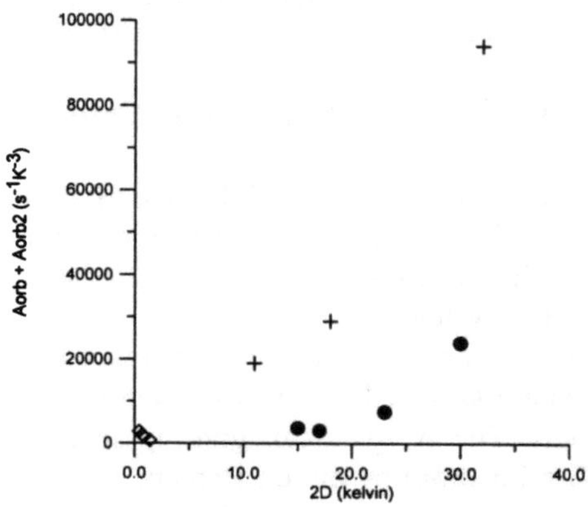

Figure 17. Dependence of the combined coefficients of the Orbach process (A_{orb} + A_{orb2}) (Table 6) obtained by fitting the temperature dependence of $1/T_1$, on the zero-field splitting, 2D, for the series of Fe(III) compounds: non-heme iron with small zero-field splittings (◊), iron porphyrins (+), and heme proteins (•).

Table 6. Relaxation Processes for High-spin Complexes[a]

sample	lattice	Temp range (K)	Orbach, A_{Orb}, A_{Orb2}, 2D	Raman, A_R, θ_D	local, A_{loc}, Δ_{loc}
$Cr(ox)_3^{3-}$	H_2O:glyc	15 - 70	$3.6x10^2$ [b], 2.0	$1.2x10^7$, 85	$1.0x10^7$ [b], 300
$Fe(ox)_3^{3-}$	H_2O:glyc	7 - 30	$1.4x10^3$, $1.4x10^3$ [c], 0.43	$4.7x10^7$, 65	
FeTfoxalate	H_2O:glyc	5 - 20	$1.0x10^3$, $1.0x10^3$ [c], 0.70	$1.6x10^7$, 52	
$FeTfCO_3$	H_2O:glyc	5 - 20	$7.6x10^2$, $7.6x10^2$ [c], 0.86	$2.7x10^7$, 52	
FeEnt	H_2O:glyc	6 - 20	$3.4x10^2$, $3.4x10^2$ [c], 1.4	$2.3x10^7$, 52	
FeTTPF	toluene	4.5 - 118	$2.0x10^3$, $1.7x10^4$, 11		
FeTTPCl	toluene	4.6 - 117	$1.5x10^4$, $1.4x10^4$, 18		
FeTTPBr	toluene	4.5 - 118	$9.4x10^4$, $1.0x10^2$, 32		
Mb-F	H_2O:glyc	4.8 - 298	$1.9x10^3$, $1.9x10^3$, 14		$4.5x10^9$ [b], 1000
Mb-F[d]	H_2O:glyc	2 - 4	$6.1x10^3$, na, 17.5		
Hb-F	H_2O:glyc	4.2 - 150	$2.4x10^3$, $5.8x10^2$, 17		$2.7x10^9$ [b], 300
Mb-formate	H_2O:glyc	4.6 - 298	$6.4x10^3$, $1.1x10^3$, 23		$2.9x10^{10}$ [b], 700
$Mb-H_2O$[d]	H_2O:glyc	2 – 4	$9.9x10^3$, na, 26.3		
$Hb-H_2O$	H_2O:glyc	4.3 - 150	$2.0x10^4$, $3.9x10^3$, 30		

[a]Energies (2D, θ_D, Δ_{loc}) are in kelvin. [b]Process makes small contribution in temperature range examined so parameters are very uncertain. [c] A_{Orb} was set equal to A_{Orb2} in the fitting procedure. [d] To be consistent with the form of equation (7), values of the coefficient A_{Orb} reported in units of s^{-1} by C. P. Scholes, R. A. Isaacson, and G. Feher, *Biochim. Biophys. Acta*, **244**, 206-210 (1971) were converted to s^{-1} K^{-3} by dividing by $(2D)^3$.

9.4 Summary of Relaxation Mechanisms

The temperature dependence of T_1 and T_m due to a range of processes is summarized in Fig. 18. The temperature range for a particular process will shift depending upon the system and only a subset of these phenomena may be present in a particular system.

Processes, pathways, and mechanisms of relaxation outlined in the preceding paragraphs are sketched in Fig. 19. Microwave energy is absorbed resonantly by A spins. The A spins can give up spin energy to the lattice thermal motions (T_1 relaxation) by direct, Raman, and /or Orbach processes. The spin angular momentum of the A spins is mixed with other angular momenta by spin-orbit coupling or spin-rotation coupling. The time-dependence of the mixing of angular momenta, which is the mechanism of relaxation, occurs by thermal motions modulating zero-field splitting (ZFS), mixing of excited states (e.g., by conformational changes), and rotational angular momentum. The A spins communicate with B spins by dipolar and exchange interactions. There is always a dipolar interaction with other spins in the sample. There may also be exchange interactions if there is overlap of electron spin density. Exchange interactions can result in both T_1 and T_2 processes. For example, time-dependent exchange between a pair of nitroxyl radicals, whether in a transitory collision or due to thermally-agitated close proximity in a rigid lattice, can result in relaxation. If the electron exchange occurs between molecules with the same nuclear spin state (e.g., both molecules have ^{14}N in the $m_I = +1$ spin state) there will be no net change, but if the exchange occurs between molecules with different m_I there is a T_2 effect. Exchange or dipolar interaction with a different paramagnetic species can have both T_1 and T_2 effects. A very common example is the relaxation effect of collisions with O_2 in solution. Interaction with a paramagnetic metal ion also provides a relaxation pathway. Overall motion of a molecule can take spins off resonance, which is a spectral diffusion process. If the spin concentration is high enough instantaneous diffusion can contribute to the observed T_m. In summary, any process that takes spins off resonance is a T_m mechanism. Any process that changes the net populations of the electron spin states is a T_1 process.

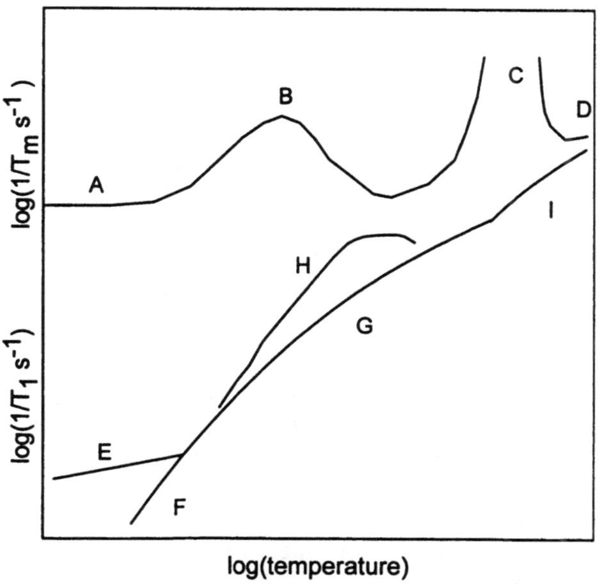

Figure 18. Schematic representation of temperature dependence of relaxation rates observed when T_m is dominated by: A, nuclear spins in the surroundings; B, collapse of couplings to methyl groups or rapidly relaxing spin; C, averaging of g- or A-anisotropy; D, T_1; and when T_1 is dominated by: E, direct process and/or high spin concentration; F, G, Raman or Orbach process; H, interaction with more rapidly-relaxing electron spin; I, local mode or thermally-activated process.

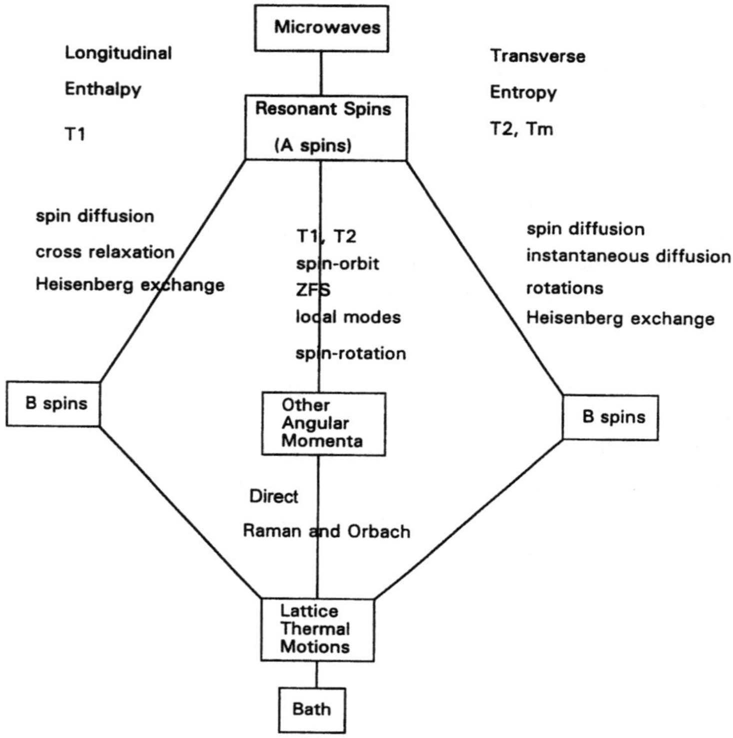

Figure 19. Summary of processes that contribute to spin-lattice relaxation, spin-spin relaxation, and spectral diffusion.

9.5 Estimating Relaxation Times of Species or Environments for which Experimental Values are not Available

This chapter has attempted to comprehensively survey the scope of available data and current understanding of electron spin relaxation times. The reader will surely be impressed by how little is known! Not all papers that have reported relaxation times have been cited, and the early literature has been invoked via references to earlier reviews, but we hope that we have not overlooked any relaxation time measurements that would change the general conclusions of this chapter. This state of the art leaves the biomedical spectroscopist in a difficult position. The other chapters in this book illustrate the enormous power of EPR for measuring distances in biological systems, but many of the methodologies require at least a good estimate of the electron T_1 and T_2 or T_m. For organic radicals well-isolated from other paramagnetic species, and for single metals in reasonably high symmetry environments, analogy with the data presented in this chapter would give a good first guess of the relaxation times. However, for interacting spins, multiple metal sites, and unusual coordination environments (i.e., most of the interesting cases!), direct measurement of relaxation times is needed. Fortunately, pulsed EPR spectrometers are now available commercially.

10. ACKNOWLEDGMENT

The support of this work by National Institutes of Health Grants GM21156 and RR13338 is gratefully acknowledged. Helpful critique of this chapter was provided by Drs. James S. Hyde, Michael K. Bowman, and Jack H. Freed. We are especially grateful to Dr. Bowman for extensive discussions that helped clarify several parts of this Chapter.

11. REFERENCES

Ablart, G. and Pescia, J., (1980) Magnetic field dependence of spin-lattice relaxation in three iron group salts. *Phys. Rev. B* **22**, 1150-1162.

Abragam, A. (1961) *The Principles of Nuclear Magnetism*. Oxford University Press, London.

Abragam, A., and Bleaney, B. (1970) *Electron Paramagnetic Resonance of Transition Ions*. Oxford University Press, London.

Ahn, J. S., Kitagawa, T., Kanematsu, Y., Nishikawa, Y., and Kushida, T., (1995). Glass transition of Zn-substituted myoglobin probed by absorption and site-selective fluorescence spectroscopies. *J. Luminesc.* **64**, 81-86.

Aime, S., Barge, A., Borel, A., Botta, M., Chemerisov, S., Merbach, A. E., Müller, U., and Pubanz, D. (1997) A Multinuclear NMR Study on the Structure and Dynamics of Lanthanide(III) Complexes of the Poly(amino carboxylate) $EGTA^{4-}$ in Aqueous Solution. *Inorg. Chem.* **36**, 5104-5112.

Aleksandrov, I. V., and Zhidomirov, G. M. (1961) Calculation of the Spin-Lattice Relaxation Time for Radicals in Molecular Crystals. *J. Exptl. Theoret. Phys. USSR* **41**, 127-132. *Sov. Phys. JETP Engl. Transl.* **14**, 94 (1962).

Allam, L., Ablart, G., Pescia, J., Clement, S., and Renard, J. P. (1989) Effect of diamagnetic doping on the magnetic resonance in a quasi-one-dimensional system tetramethylammonium manganese trichloride. *Phys. Rev. B* **40**, 8654-8665.

Allen, J. P., Colvin, J. T., Stinson, D. G., Flynn, C. P., and Stapleton, H. J. (1982) Protein conformation from electron spin relaxation data. *Biophys. J.* **38**, 299-310.

Altenbach, C., Froncisz, W., Hyde, J. S., and Hubbell, W. L. (1989) Conformation of spin-labeled melittin at membrane surfaces investigated by pulse saturation recovery and continuous wave power saturation electron paramagnetic resonance. *Biophys. J.* **56**, 1183-1191.

Al'tshuler, S. A. and Kozyrev, B. M. (1974) *Electron Paramagnetic Resonance in Compounds of Transition Elements*, 2nd ed., Halsted-Wiley, Jerusalem.

Al'tshuler, S. A., Kirmse, R., and Solov'ev, B. V. (1975) Spin-lattice relaxation of exchange-coupled Cu^{2+}-Cu^{2+} pairs and single Cu^{2+} ions in crystals of zinc(II) bis(diethyldithiocarbamate). *J. Phys. C: Solid State Phys.* **8**, 1907-1920.
Aminov, L. K., Kurkin, I. N., and Lukoyanov, D. A. (1998) Some Peculiarities of Spin-Lattice Relaxation of Impurity Rare-Earth Ions in Crystals, Caused by the Structure Defects. *Appl. Magn. Reson.* **14**, 447-456.
Anderson, P. W. (1958) Absence of Diffusion in Certain Random Lattices. *Phys. Rev.* **109**, 1492-1505.
Andreozzi, L., Faetti, M., Hvala, M., and Giordana, M. (1999) Short-time rotational relaxation in supercooled and glassy o-terphenyl: a nonlinear electron spin resonance study by using molecular probes with different symmetries. *Philosph. Mag. B* **79**, 1993-2003.
Angelone, R., Forte, C., and Pinzino, C. (1993) Relaxation Time Measurements by Longitudinally Modulated ENDOR Spectroscopy on Irradiated L-Alanine Single Crystals. *J. Magn. Reson. A* **101**, 16-22.
Antholine, W. E., Hyde, J. S., and Swartz, H. M. (1978) Use of Dy^{3+} as a Free Radical Relaxing Agent in Biological Tissues. *J. Magn. Reson.* **29**, 517-522.
Asano, A., and Takegoshi, K. (1998). Polymer Blends and Miscibility, in *Solid State NMR of Polymers*, I. Ando and T. Asakura, eds., Elsevier, Amsterdam.
Ashikawa, I, Yin, J.-J., Subczynski, W. K., Kouyama, T., Hyde, J. S., Kusumi, A. (1994) Molecular Organization and Dynamics in Bacteriorhodopsin-Rich Reconstituted Membranes: Discrimination of Lipid Environments by the Oxygen Transport Parameter Using a Pulse ESR Spin-Labeling Technique. *Biochem.* **33**, 4947-4952.
Askew, T. R., Muench, P. J., Stapleton, H. J., and Brower, K. L. (1984) Electron Paramagnetic Resonance and Relaxation of Amorphous Silicon Below 1 K. *Solid State Commun.* **49**, 667-670.
Atherton, N. W. (1993) *Principles of Electron Spin Resonance*, Prentice Hall, New York, ch. 1.
Atkins, P. W. (1972) Spin-Rotation Interaction, in *Electron Spin Relaxation in Liquids*, L. T. Muus and P. W. Atkins, eds., Plenum, New York.
Atkins, P. W. and Kivelson, D. (1966) ESR Linewidths in Solution. II. Analysis of Spin-Rotational Relaxation Data. *J. Chem. Phys.* **44**, 169-174.
Atsarkin, V. A. (1966) On the Mechanism of Paramagnetic Spin-Lattice Relaxation at Low Temperatures. *Soviet Phys. JETP* **22**, 106-113.
Atsarkin, V. A., Demidov, V. V., and Vasneva, G. A. (1997) Metal-insulator transition in RbC_{60} polymer fulleride studied by ESR and electron-spin relaxation, *Phys. Rev. B* **56**, 9448-9453.
Atsarkin, V. A., Demidov, V. V., and Vasneva, G. A. (1998) Electron Spin-Lattice Relaxation in Noncommon Metals $Yba_2Cu_3O_x$ and Rb_1C_{60}. *Appl. Magn. Reson.* **15**, 323-331.
Bacci, M. (1986) Jahn-Teller Effect in Five-Coordinated Copper(II) Complexes. *Chem. Phys.* **104**, 191-199.
Bacci, M., and Cannistraro, S. (1987a) Temperature Dependence of the g Values in Blue Copper Protein EPR Spectra. *Chem. Phys. Lett.* **133**, 109-112.
Bacci, M., and Cannistraro, S. (1987b) A Vibronic Coupling Approach for the Interpretation of the g-Value Temperature Dependence in Type-I Copper Proteins. *J. Chem. Soc. Faraday Trans. I* **83**, 3693-3700.
Baldacchini, G., Botti, S., Grassano, U. M., and Luty, F. (1993) Spin-Orbit Coupling, Spin-Lattice Relaxation, and Spin-Memory Studies of F-Center-Molecular-Ion Pairs in Alkali Halides. *Il Nuovo Cimento* **15D**, 207-227.
Banci, L., Bertini, I., and Luchinat, C. (1986) Electron Relaxation. *Magn. Reson. Rev.* **11**, 1-40.

Banci, L., Bertini, I., and Luchinat, C. (1991) *Nuclear and Electron Relaxation*, VCH, Weinheim.

Barbieri, R., Ruisi, G., Silvestri, A., Giullani, A. M., Barbieri, A., Spina, G., Pieralli, F., and Giallo, F. D. (1995). Dynamics of tin nuclei in alkyltin(IV)-deoxyribonucleic acid condensates by variable-temperature tin-119 Mossbauer spectroscopy. *J. C. S. Dalton Trans*, 467-475.

Barbon, A., Brustolon, M., Maniero, A. L., Romanelli, M., and Brunel, L.-C. (1999) Dynamics and spin relaxation of tempone in a host crystal. An ENDOR, high field EPR and electron spin echo study. *Phys. Chem. Chem. Phys.* **1**, 4015-4023.

Barnes, J. P., Liang, Z., Mchaourab, H. S., Freed, J. H., and Hubbell, W. L. (1999) A Multifrequency Electron Spin Resonance Study of T4 Lysozyme Dynamics. *Biophys. J.* **76**, 3298-3306.

Basosi, R., Sealy, R. C., Kalyanaraman, B., and Hyde, J. S. (1984) Interaction of Semiquinones with Paramagnetic Metal Ions in Aqueous Solution. Relaxation and Complex Formation. *J. Magn. Reson.* **59**, 41-49.

Behringer, R. E. (1958) Number of Single, Double, and Triple Clusters in a System Containing Two Types of Atoms. *J. Chem. Phys.* **29**, 537-539.

Bertini, I., and Luchinat, C. (1986) *NMR of Paramagnetic Molecules in Biological Systems.* Benjamin/Cummings Publ. Menlo Park, CA.

Bertini, I., Briganti, F., Luchinat, C., Mancini, M., and Spina, G. (1985) The Electron-Nucleus Dipolar Coupling in Slow Rotating Systems. 2. The Effect of g Anisotropy and Hyperfine Coupling when $S = \frac{1}{2}$ and $I = 3/2$. *J. Magn. Reson.* **63**, 41-55.

Bertini, I., Briganti, F., and Luchinat, C. (1986) Solvent Proton Nuclear Magnetic Relaxation Dispersion (NMRD) in Solutions of Paramagnetic Macromolecules. In *Advanced Magnetic Resonance Techniques in Systems of High Molecular Complexity*, Birkhäuser, Boston, 165-195.

Bertini, I., Luchinat, C., Brown, R. D., III, and Koenig, S. H. (1989a) Relaxation of the electronic spin moment of copper(II)-macromolecular complexes in solution. *J. Am. Chem. Soc.* **111**, 3532-3536.

Bertini, I., Banci, L., and Luchinat, C. (1989b) Proton Magnetic Resonance of Paramagnetic Metalloproteins. *Methods Enzymol.* **117**, 246-263.

Bertini, I., Xia, Z., and Luchinat, C. (1992a) Solvent Water ^{1}H NMRD Study of Oxovanadium(IV) Aquò Ion. *J. Magn. Reson.* **99**, 235-246.

Bertini, I., Luchinat, C. and Xia, Z. (1992b) Electronic Relaxation of the Titanium(III) Hexaqua Complex Detected by Solvent Water ^{1}H NMRD Spectroscopy. *Inorg. Chem.* **31**, 3152-3154.

Bertini, I., Briganti, F., Xia, Z., and Luchinat, C. (1993a) Nuclear Magnetic Relaxation Dispersion Studies of Hexaaquo Mn(II) Ions in Water-Glycerol Mixtures. *J. Magn. Reson. A* **101**, 198-201.

Bertini, I., Capozzi, F., and Luchinat, C., and Xia, Z. (1993b) Nuclear and Electron Relaxation of $Fe(OH_2)_6^{3+}$. *J. Phys. Chem.* **97**, 1134-1137.

Bertini, I., Martini, G., and Luchinat, C. (1994a) Relaxation, Background and Theory, in *Handbook of Electron Spin Resonance*. Poole, C. P., Jr., and Farach, H. A. eds, American Institute of Physics, New York, 51-77.

Bertini, I., Martini, G., and Luchinat, C. (1994b) Relaxation Data Tabulation, in *Handbook of Electron Spin Resonance*. Poole, C. P., Jr., and Farach, H. A. eds, American Institute of Physics, New York, 79-310.

Bertini, I., Ciurli, S., Dikiy, A., Gasanov, R., Luchinat, C., Martini, G., and Safarov, N. (1999) High-Field NMR Studies of Oxidized Blue Copper Proteins: The Case of Spinach Plastocyanin. *J. Am. Chem. Soc.* **121**, 2037-2046.

Bertrand, P., Gayda, J.-P., and Rao, K. K. (1982). Electron spin-lattice relaxation of the (4Fe-4S) ferredoxin from Bacillus stearothermophilus. Comparison with other iron proteins. *J. Chem. Phys.* **76**, 4715-4719.

Beth, A. H. and Robinson, B. H. (1989) Nitrogen-15 and Deuterium Substituted Spin Labels for Studies of Very Slow Rotational Motion. *Biol. Magn. Reson.* **8**, 179-253 (see pages 234-248).

Bietsch, W., Takagi, S., and von Schütz, J. U. (1995) Pulsed ESR on the Organic Radical Salt: 3,3'-Diethyl-4,4'-Dimethyl-2,2'-Thiazolocyanine-$[TCNQ]_2$. *Appl. Magn. Reson.* **9**, 517-525.

Bizzarri, A. R., Iakovleva, O. A., Parak, F. (1995) Spin-lattice relaxation in Mossbauer spectra of metmyoglobin: investigation of crystals, water, and water-glycerol solutions, *Chem. Phys.* **191**, 185-194.

Bleaney, B. and Stevens, K. W. H. (1953) Paramagnetic Resonance. *Reports Prog. Phys.* **16**, 108-159.

Bloch, F. (1946) Nuclear Induction. *Phys. Rev.* **70**, 460-485.

Bloembergen, N. (1949) On The Interaction of Nuclear Spin in a Crystalline Lattice. *Physica* **15**, 386-426.

Bloembergen, N., and Morgan, L. O. (1961) Proton Relaxation Times in Paramagnetic Solutions. Effects of Electron Spin Relaxation. *J. Chem. Phys.* **34**, 842-850.

Bloembergen, N. and Wang, S. (1954) Relaxation Effects in Para- and Ferromagnetic Resonance. *Phys. Rev.* **93**, 72-83.

Bloembergen, N., Purcell, E. M., and Pound, R. V. (1948) Relaxation Effects in Nuclear Resonance Absorption. *Phys. Rev.* **73**, 679-712.

Blum, H., Cusanovich, M. A., Sweeney, W. V., and Ohnishi, T. (1981) Magnetic Interactions between Dysprosium Complexes and Two Soluble Iron-Sulfur Proteins. *J. Biol. Chem.* **256**, 2199-2206.

Blumberg, W. E. (1960) Nuclear Spin-Lattice Relaxation Caused by Paramagnetic Impurities. *Phys. Rev.* **119**, 79-84.

Bokor, M., Marek, T., and Tompa, K. (1996) Solid-State NMR of 1-Propyltetrazole Complexes of Iron(II) and Zinc(II). 1. ^{1}H Spin-Lattice Relaxation Time. *J. Magn. Reson. A* **122**, 157-164.

Boscaino, R., Gelardi, F. M., and Mantegna, R. N. (1986a) Spectral Diffusion and Saturation Kinetics in Inhomogeneous Systems. *J. Magn. Reson.* **70**, 251-261.

Boscaino, R., Gelardi, F. M., and Mantegna, R. N. (1986b) Experimental Detection of Spectral Diffusion by the Saturation Transient Method. *J. Magn. Reson.* **70**, 262-269.

Bouffard, V., Roinel, Y., Roubeau, P., and Abragam, A. (1980) Dynamic nuclear polarization in ^{6}LiD. *J. Physique* **41**, 1447-1451.

Bourdel, D., Ablart, G., Pescia, J., Clement, S., and Renard, J. P. (1981) Electron-spin-relaxation in a one-dimensional system. *Phys. Rev. B* **23**, 1339-1341.

Bowers, K. D., and Mims, W. B. (1959) Paramagnetic Relaxation in Nickel Fluosilicate. *Phys. Rev.* **115**, 285-295.

Bowman, M. K. (1990) Fourier transform electron spin resonance, in *Modern Pulsed and Continuous Wave Electron Spin Resonance*, L. Kevan and M. K. Bowman, eds., Wiley, N. Y., 1-42, especially p. 28-29.

Bowman, M. K. (1993) Strategies for Measurement of Electron Spin Relaxation, in *Magnetic Resonance of Carbonaceous Solids*, R. E. Botto and Y. Sanada eds., *Adv. Chem. Ser.* **229**, Chapter 5.

Bowman, M. K. (2000) unpublished comments to the Eatons, February 2, 2000.

Bowman, M. K., and Kevan, L. (1977a) Electron Spin-Lattice Relaxation Mechanisms of Radiation Produced Trapped Electrons and Hydrogen Atoms in Aqueous and Organic

Glassy Matrices: Modulation of Electron Nuclear Dipolar Interaction by Tunnelling Modes in a Glassy Matrix. *Disc. Faraday Soc.* **63**, 7-17.

Bowman, M. K., and Kevan, L. (1977b) An electron spin-lattice relaxation mechanism involving tunneling modes for trapped radicals in glassy matrixes. Theoretical development and application to trapped electrons in γ-irradiated ethanol glasses. *J. Phys. Chem.* **81**, 456-461.

Bowman, M. K., and Kevan, L. (1979) Electron Spin-Lattice Relaxation in Nonionic Solids, in *Time Domain Electron Spin Resonance*, L. Kevan and R. N. Schwartz, eds., Wiley, New York, pages 68-105.

Bowman, M. K. and Norris, J. R. (1982) Cross Relaxation of Free Radicals in Partially Ordered Solids. *J. Phys. Chem.* **86**, 3385-3390.

Boyer, R. F. and Keinath, S. E., eds. (1978) *Molecular Motion in Polymers by ESR : papers presented at the Eighth Midland Macromolecular Meeting held between August 21-25, 1978*, Published for MMI Press by Harwood Academic Publishers.

Brändle, R., Krüger, G. J., Müller-Warmuth, W. (1970) Pulsed Spectroscopic Investigation of Electron Spin Relaxation in Free Radicals. *Z. Naturforsch.* **25A**, 1-11.

Breen, D. P., Krupka, D. C., and Williams, F. I. B. (1969) Relaxation in a Jahn-Teller System. I. Copper in Octahedral Water Coordination. *Phys. Rev.* **179**, 241-254.

Brown, I. M. (1971) Electron Spin Echoes and Spectral Diffusion in Organic Radical Solids. *J. Chem. Phys.* **55**, 2377-2384.

Brown, I. M. (1979) Electron Spin-Echo Studies of Relaxation Processes in Molecular Solids, in *Time Domain Electron Spin Resonance*, L. Kevan and R. N. Schwartz, eds., Wiley, New York, chapter 6

Brustolon, M., and Segre, U. (1994) Electron Spin-Lattice Relaxation Time and Spectral Diffusion in γ-irradiated L-Alanine. *Appl. Magn. Reson.* **7**, 405-413.

Budil, D. E., Earle, K. A., and Freed, J. H. (1993) Full Characterization of the Rotational Diffusion Tensor by Electron Paramagnetic Resonance at 250 GHz. *J. Phys. Chem.* **97**, 1294-1303.

Budker, V., Du, J.-L., Seiter, M., Eaton, G. R., and Eaton, S. S. (1995) Electron-Electron Spin-Spin Interaction in Spin-Labeled Low-Spin Methemoglobin. *Biophys. J.* **68**, 2531-42.

Burchfield, J. M., Du, J.-L., More, K. M., Eaton, S. S., and Eaton, G. R. (1997) Enhancement of electron spin relaxation rates of metalloporphyrins due to interaction with a faster relaxing metal bound to an appended bipyridyl. *Inorg. Chim. Acta* **263**, 23-33.

Burns, P. D. and La Mar, G. N. (1982) Proton Spin Relaxation for the Nonlabile Coordinated Chelate in Lanthanide Shift Reagents. *J. Magn. Reson.* **46**, 61-68.

Canet, D. (1996) *Nuclear Magnetic Resonance: Concepts and Methods*. Wiley, Chichester, 1996, page 106.

Caravan, P., and Merbach, A. E. (1997) An extreme water exchange rate: the europium(II) aqua ion. *Chem. Commun.* 2147-2148.

Caravan, P., Tóth, É., Rockenbauer, A., and Merbach, A. E. (1999) Nuclear and Electronic Relaxation of $Eu^{2+}_{(aq)}$: An Extremely Labile Aqua Ion. *J. Am. Chem. Soc.* **121**, 10403-10409.

Carr, H. Y. and Purcell, E. M. (1954) Effects of Diffusion on Free Precession in Nuclear Magnetic Resonance Experiments. *Phys. Rev.* **94**, 630-638.

Carrington, A. and McLachlan, A. D. (1967) *Introduction to Magnetic Resonance with Applications to Chemistry and Chemical Physics.* Harper & Row, Publishers, New York.

Carruthers, L. M., Closken, C. L., Link, K. L., Mahapatro, S. N., Bikram, M., Du, J.-L., Eaton, S. S., and Eaton, G. R. (1999) Electron spin relaxation in chromium-nitrosyl complexes. *Inorg. Chem.* **38**, 3529-3534.

Carss, S. A., Scheler, U., Harris, R. K., Holstein, P., and Fletton, R. A. (1996) ^{19}F NMR of Proton-Containing Solids. *Magn. Reson. Chem.* **34**, 63-70.

Castle, J. G. Jr., Feldman, D. W., Klemens, P. G., and Weeks, R. A. (1963) Electron Spin-Lattice Relaxation at Defect Sites; E' Centers in Synthetic Quartz at 3 Kilo-Oersteds. *Phys. Rev.* **130**, 577-588.

Castle, J. G., Jr., and Feldman, D. W. (1965a) Resonance modes at defects in crystalline quartz, *Phys. Rev. A* **137**, 671-673.

Castle, J. G., Jr. and Feldman, D. W. (1965b) Temperature dependence of paramagnetic relaxation at point defects in vitreous silica, *J. Appl. Phys.* **36**, 124-128.

Chance, M. R. (1993) Low-Temperature Optical Spectroscopy: Metalloprotein Structure and Dynamics. *Meth. Enzymol.* **226**, 97-118.

Chang, T., and Yang, G. C. (1968) Temperature Dependence of Paramagnetic Relaxation of Mo^{5+} in TiO_2. *J. Chem. Phys.* **48**, 2546-2549.

Chasteen, N. D. and Hanna, M. W. (1972) Electron Paramagnetic Resonance Line Widths of Vanadyl(IV) α-Hydroxycarboxylates. *J. Phys. Chem.* **76**, 3951-3958.

Cheung, T. T. P. (1996) Spin Diffusion in Solids, *Encyclopedia of NMR* **7**, 4518-24.

Chiarini, F., Martinelli, M., Rolla, F. A., and Santucci, S. (1973) Periodic Longitudinal Magnetization of a Spin System Irradiated with Two Transverse Radio-Frequency Waves. *Phys. Lett.* **44A**, 91-92.

Chiarini, F., Martinelli, M., Pardi, L., and Santucci, S. (1975) Electron-spin double resonance by longitudinal detection: Line shape and many-quantum transitions. *Phys. Rev. B* **12**, 847-852.

Cianflone, F., Francia, F., and Leporini, D. (1998) Measurement of the Longitudinal Relaxation Time by Continuous-Wave, Nonlinear Electron Spin Resonance Spectroscopies. *J. Magn. Reson.* **131**, 86-91.

Cinget, F., Fries, P. H., Greilich, U., and Vottero, P. J. A. (1995) Electron-Nuclear Relaxation Time T_{1e} for Intramolecular Distance Evaluation. *Magn. Reson. Chem.* **33**, 260-272.

Clarkson, R. B., Smirnov, A. I., Smirnova, T. I., Kang, H., Belford, R. L., Earle, K., and Freed, J. H. (1998) Multi-frequency EPR determination of zero-field splitting of high spin species in liquids: Gd(III) chelates in water. *Mol. Phys.* **95**, 1325-1332.

Colligiani, A., Giordano, M., Leporini, D., Lucchesi, M., Martinelli, M., Pardi., L., and Santucci, S. (1992) Longitudinally Detected Electron Spin Resonance: Recent Developments. *Appl. Magn. Reson.* **3**, 107-129.

Colombo, M. G., Meier, B. H., and Ernst, R. R. (1988) Rotor-Driven Spin Diffusion in Natural-Abundance ^{13}C Spin Systems. *Chem. Phys. Lett.* **146**, 189-196.

Colvin, J. T. and Stapleton, H. J. (1985) Fractal and spectral dimensions of biopolymer chains: Solvent studies of electron spin relaxation rates in myoglobin azide. *J. Chem. Phys.* **82**, 4699-4706.

Cotton, F. A. (1971) *Chemical Applications of Group Theory, 2nd Ed.*, Wiley, New York.

Cusack, S., and Doster, W. (1990) Temperature dependence of the low frequency dynamics of myoglobin. Measurement of the vibrational frequency by ineleastic neutron scattering. *Biophys. J.* **58**, 243-251.

Daraseliya, D. M. and Manenkov, A. A. (1970) Quenching of Cross Correlation in Inhomogeneously Broadened EPR Lines. *JETP Letters* **11**, 224-226 (337-339 in Russian)

Daraseliya, D. M., Epifanov, A. S., and Manenkov, A. A. (1970) Relaxation in Inhomogeneously Broadened EPR Lines. *Soviet Physics JETP* **32**, 244-249 (**59**, 445-456 in Russian).

Davids, D. A., and Wagner, P. E. (1964) Magnetic Field Dependence of Paramagnetic Relaxation in a Kramers Salt. *Phys. Rev. Lett.* **12**, 141-142.

Davis, C. F. Jr., Strandberg, M. W. P., and Kyhl, R. L. (1958) Direct Measurement of Electron Spin-Lattice Relaxation Times. *Phys. Rev.* **111**, 1268-1272.

Davoust, C. E., Doan, P. E., and Hoffman, B. M. (1996) Q-Band Pulsed Electron Spin-Echo Spectrometer and Its Application to ENDOR and ESEEM. *J. Magn. Reson. A* **119**, 38-44.

De Abreu, C. D., Pinhal, N. M., and Vugman, N. V. (1992) Temperature Dependence of g Values of Fe(I) and Ru(I) Cyanide Complexes in a KCl Host Lattice. *J. Magn. Reson.* **100**, 588-592.

Deatherage, J. F., Loe, R. S., and Moffat, K. (1976) Structure of fluoride methemoglobin. *J. Mol. Biol.* **104**, 723-728.

DeRose, V. J. and Hoffman, B. M. (1995) Protein Structure and Mechanism Studied by Electron Nuclear Double Resonance Spectroscopy. *Meth. Enzymol.* **246**, 554-589; see page 567.

DiBartolo, B. and Kyrkos, S., eds. (1997) *Spectroscopy and Dynamics of Collective Excitations in Solids*, NATO ASI series B, **356**, 53.

Dietel, E., Hirsch, A., Pietzak, B., Waiblinger, M., Lips, K., Weidinger, A., Gruss, A., and Dinse, K.-P. (1999) Atomic Nitrogen Encapsulated in Fullerenes: Effects of Cage Variations. *J. Am. Chem. Soc.* **121**, 2432-2437.

Diezemann, G. (1992) High-field spin-lattice relaxation of methyl groups: relation to neutron scattering. *J. Phys. Condensed Matter* **4**, 9153-9180.

Dikanov, S. A., and Tsvetkov, Y. D. (1992) *Electron Spin Echo Modulation (ESEEM) Spectroscopy.* CRC Press, Boca Raton, Florida.

Dishon, M., Bendler, J. T., and Weiss, G. H. (1990) Tables of the Inverse Laplace Transform of the Function $e^{-s^{\beta}}$, *J. Res. Natl. Inst. Stand. Technol.* **95**, 433-467.

Drago, R. S. (1977) *Physical Methods in Chemistry*, Saunders, Philadelphia, p. 252.

Drago, R. S. (1992) *Physical Methods for Chemists*, Saunders, Philadelphia, p. 578-594.

Du, J.-L., More, K. M., Eaton, S. S., and G. R. Eaton, G. R. (1992) Orientation Dependence of Electron Spin Phase Memory Relaxation Times in Copper(II) and Vanadyl Complexes in Frozen Solution. *Israel J. Chem.* **32**, 351-355.

Du, J.-L., Eaton, G. R., and Eaton, S. S. (1994) Effect of Molecular Motion on Electron Spin Phase Memory Times for Copper(II) Complexes in Doped Solids. *Appl. Magn. Reson.* **6**, 373-8.

Du, J.-L., Eaton, G. R., and Eaton, S. S. (1995a) Temperature, Orientation, and Solvent Dependence of Electron Spin-Lattice Relaxation Rates for Nitroxyl Radicals in Glassy Solvents and Doped Solids. *J. Magn. Reson. A* **115**, 213-221.

Du, J.-L., Eaton, G. R., and Eaton, S. S. (1995b) Electron-Spin-Lattice Relaxation in Natural Abundance and Isotopically Enriched Oxo-chromium(V)bis(2-hydroxy-2-ethylbutryate). *J. Magn. Reson. A* **115**, 236-40.

Du, J.-L., Eaton, G. R., and Eaton, S. S. (1995c) Temperature and Orientation Dependence of Electron Spin Relaxation Rates for Bis(diethyldithiocarbamato) Copper(II). *J. Magn. Reson. A* **117**, 67-72.

Du, J.-L., Eaton, G. R., and Eaton, S. S. (1996) Electron Spin Relaxation in Vanadyl, Copper(II), and Silver(II) Porphyrins in Glassy Solvents and Doped Solids. *J. Magn. Reson. A* **119**, 240-246.

Dubinskii, A. A., Maresch, G. G., and Spiess, H.-W. (1994) Two-dimensional electron paramagnetic resonance spectroscopy of nitroxides: Elucidation of restricted molecular motions. *J. Chem. Phys.* **100**, 2437-48.

Dugad, L. B., La Mar, G. N., and Unger, S. W. (1990) Influence of Molecular Correlation Time on the Homonuclear Overhauser Effect in Paramagnetic Proteins. *J. Am. Chem. Soc.* **112**, 1386-1392.

Dugdale, D. E., and Thorpe, J. S. (1969) Spin-lattice relaxation of Cr:Al acetylacetonate. *J. Phys. C (Solid St. Phys.) Ser. 2,* **2,** 1376-1383.

Dzuba, S. A. (1992) Echo-induced EPR spectra of nitroxides: Study of molecular librations. *Pure & Appl. Chem.*. **64,** 825-831.

Dzuba, S. A. (1996) Librational motion of guest spin probe molecules in glassy media. *Phys. Lett. A* **213,** 77-84.

Dzuba, S. A., and Tsvetkov, Yu. D. (1988) Magnetization Transfer in Pulsed EPR of ^{15}N Nitroxides: Reorientational Motion Model of Molecules in Glassy Liquids. *Chem. Phys.* **120,** 291-298.

Dzuba, S. A., Salikhov, K. M., and Tsvetkov, Yu. D. (1981) Slow rotations ($\tau \geq 10^{-5}$s) of methyl groups in radicals studied by pulse ESR spectroscopy. *Chem. Phys. Lett.* **79,** 568-572.

Dzuba, S. A., Maryasov, A. G., Salikhov, K. M., and Tsvetkov, Yu. D. (1984a) Superslow rotations of nitroxide radicals studied by pulse EPR spectroscopy. *J. Magn. Reson.* **58,** 95-117.

Dzuba, S. A., Maryasov, A. G., Salikhov, K. M., and Tsvetkov, Yu. D. (1984b) Superslow Rotations of Nitroxide Radicals Studied by Pulse EPR Spectroscopy. *J. Magn. Reson.* **58,** 95-117.

Dzuba, S. A., Shushakov, O. A., and Tsvetkov, Yu. D. (1985) Spin-Lattice Relaxation in Nitroxyl Radicals. *J. Structural Chem.* **26,** 480-482. (186-188 in Russian)

Dzyuba, S. A., Suvernev, A. A., and Temkin, S. I. (1989) Nitroxyl Radical Rotational-Mobility Anisotropy Effects in Magnetization-Transfer Electron Spin Echo. *Theor. Exptl. Chem.* **25,** 596-601. (Note that in some papers the name Dzuba is also transliterated as Dzyuba.)

Dzuba, S. A., Tsvetkov, Yu. D., and Maryasov, A. G. (1992) Echo-induced EPR spectra of nitroxides in organic glasses: model of orientational molecular motions near equilibrium position. *Chem. Phys. Lett.* **188,** 217-222.

Dzuba, S. A., Golovina, Ye. A., and Tsvetkov, Yu. D. (1993) Echo-Induced EPR Spectra of Spin Probes as a Method for Identification of Glassy States in Biological Objects. *J. Magn. Reson. B* **101,** 134-138.

Dzuba, S. A., Watari, H., Shimoyama, Y., Maryasov, A. G., Kodera, Y., and Kawamori, A. (1995) Molecular Motion of the Cholestane Spin Label in a Multibilayer in the Gel Phase Studied Using Echo-Detected EPR. *J. Magn. Reson. A* **115,** 80-86.

Earle, K. A., Budil, D. E., and Freed, J. H. (1993) 250-GHz EPR of Nitroxides in the Slow-Motional Regime: Models of Rotational Diffusion. *J. Phys. Chem.* **97,** 13289-13297.

Earle, K. A., Moscicki, J. K., Polimeno, A., and Freed, J. H. (1997) A 250 GHz ESR study of o-terphenyl: Dynamic cage effects above T_c. *J. Chem. Phys.* **106,** 9996-10015.

Eastman, M. P., Kooser, R. G., Das, M. R., and Freed, J. H. (1969) Studies of Heisenberg Spin Exchange in ESR Spectra. I. Linewidth and Saturation Effects. *J. Chem. Phys.* **51,** 2690-2709.

Eastman, M. P., Bruno, G. V., and Freed, J. H. (1970) ESR Studies of Heisenberg Spin Exchange. III. An ELDOR Study. *J. Chem. Phys.* **52,** 321-327.

Eaton, G. R. (1993) "New and Notable": A new EPR methodology for the study of biological systems. *Biophys. J.* **64,** 1373-4.

Eaton, S. S. and Eaton, G. R. (1993) Irradiated Fused-Quartz Standard Sample for Time-Domain EPR. *J. Magn. Reson. A* **102,** 354-6. Note that in this paper the magnetic field range in Figure 1 should be 3263 to 3273 G, not 2869 to 2879 G as published.

Eaton, S. S. and Eaton, G. R. (1996) Electron spin relaxation in discrete molecular species, *Current Topics in Biophysics* **20,** 9-14.

Eaton, G. R. and Eaton, S. S. (1999) Solvent and temperature dependence of spin echo dephasing for chromium(V) and vanadyl complexes in glass solution. *J. Magn. Reson.* **136**, 63-68.

Eaton, S. S., Kee, A., Konda, R., Eaton, G. R., Trulove, P. C., and Carlin, R. T. (1996) Comparison of electron paramagnetic resonance line shapes and electron spin relaxation rates for C_{60}^{-} and C_{60}^{3-} in 4:1 toluene:acetonitrile and dimethylsulfoxide. *J. Phys. Chem.* **100**, 6910-6919.

Eliav, U. and Freed, J. H. (1983) Multipulse sequences in electronic-spin echoes, *Rev. Sci. Instrum.* **54**, 1416-1417.

Feix, J. B., Popp, C. A., Venkataramu, S. D., Beth, A. H., Park, J. H., and Hyde, J. S. (1984) An Electron-Electron Double-Resonance Study of Interactions between [^{14}N]- and [^{15}N]Stearic Acid Spin Label Pairs: Lateral Diffusion and Vertical Fluctuations in Dimyristoylphosphatidylcholine. *Biochem.* **23**, 2293-2299.

Feldman, D. W., Castle, J. G. Jr., and Murphy, J. (1965) Spin Relaxation of Atomic Hydrogen in CaF_2: Evidence for Local Modes. *Phys. Rev.* **138 A**, 1208-1216.

Feldman, D. W., Castle, J. G., Jr., and Wagner, G. R. (1966). Spin relaxation of atomic hydrogen in fused silica: temperature dependence. *Phys. Rev.* **145**, 237-240.

Fessenden, R. W., Hornak, J. P., and Venkataraman, B. (1981) Electron spin-lattice relaxation of transient free radicals. *J. Phys. Chem.* **74**, 3694-3704.

Freed, J. H. (1965) Theory of Saturation and Double Resonance Effects in ESR Spectra. *J. Chem. Phys.* **43**, 2312-2332.

Freed, J. H. (1972) Electron Spin Resonance. *Ann. Rev. Phys. Chem.* **23**, 265-310.

Freed, J. H. (1974) Theory of Saturation and Double Resonance in Electron Spin Resonance Spectra. VI. Saturation Recovery. *J. Phys. Chem.* **78**, 1155-1167.

Freed, J. H. (1976) Theory of slow tumbling ESR spectra for nitroxides in *Spin Labeling Theory and Applications*, L. J. Berliner, ed., Academic Press, New York, ch. 3.

Freed, J. H. (1979) Theory of ESR Saturation Recovery in Liquids and Related Media, in Kevan, L., and Schwartz, R. N., eds., *Time Domain Electron Spin Resonance*, Wiley-Interscience, New York, pages 31-66.

Freed, J. H. (1998) Linewidths, lineshapes, and spin relaxation in the one and two-dimensional ESR of organic radicals and spin labels, in *Foundations of Modern EPR*, World Scientific, Singapore, 658-683.

Freed, J. H., and Fraenkel, G. K. (1963) Theory of Linewidths in Electron Spin Resonance Spectra. *J. Chem. Phys.* **39**, 326-348.

Friedman, H. L., Holz, M., and Hertz, H. G. (1979) EPR relaxations of aqueous Ni^{2+} ion. *J. Chem. Phys.* **70**, 3369-3383.

Fukushima, E. and Roeder, S. B. W., 1981, *Experimental Pulse NMR A Nuts and Bolts Approach.* Addison-Wesley Publishing Co., Reading, Massachusetts, pp. 446-448.

Gaber, B. P., Brown, R. D., Koenig, S. H., and Fee, J. A. (1972) Nuclear Magnetic Relaxation Dispersion in Protein Solutions. *Biochim. Biophys. Acta* **271**, 1-5.

Gaffney, B. J. and Silverstone, H. J. (1993) Simulation of the EMR Spectra of High-Spin Iron in Proteins. *Biol. Magn. Reson.* **13**, 1-57.

Gaffney, B. J., Eaton, G. R., and Eaton, S. S. (1998) Electron spin relaxation rates for high-spin Fe(III) in iron transferrin carbonate and iron transferrin oxalate. *J. Phys. Chem. B* **102**, 5536-5541.

Gast, P., Mushlin, R. A., and Hoff, A. J. (1982) Nonuniform Transfer of Electron Spin Polarization in Reaction Centers of the Photosynthetic Bacterium Rhodopseuomonas sphaeroides. *J. Phys. Chem.* **86**, 2886-2891.

Gayda, J.-P., Gibson, J. F., Cammack, R., Hall, D. O., and Mullinger, R. (1976) Spin Lattice Relaxation and Exchange Interaction in a 2-Iron, 3-Sulphur Protein. *Biochim. Biophys. Acta* **434**, 154-163.

Gayda, J.-P., Bertrand, P., Deville, A., More, C., Roger, G., Gibson, J. F., and Cammack, R. (1979) Temperature dependence of the electronic spin-lattice relaxation time in a 2-iron-2-sulfur protein. *Biochim. Biophys. Acta* **581**, 15-26.

Gerstein, B. C. (1998) ^{1}H NMR, in *Solid State NMR of Polymers*, I. Ando and T. Asakura, eds., Elsevier, Amsterdam.

Geschwind, S., ed., (1972) Electron Paramagnetic Resonance, Plenum Press, New York.

Ghim, B. T., Eaton, S. S., Eaton, G. R., Quine, R. W., Rinard, G. A., and Pfenninger, S. (1995) Magnetic Field and Frequency Dependence of Electron Spin Relaxation Times of the E' Center in Irradiated Vitreous Silica. *J. Magn. Reson. A* **115**, 230-235.

Ghim, B. T., Du, J.-L., Pfenninger, S., Rinard, G. A., Quine, R. W., Eaton, S. S., and Eaton, G. R. (1996) Multifrequency Electron Paramagnetic Resonance of Irradiated L-Alanine. *Appl. Radiat. Isot.* **47**, 1235-1239.

Gill, J. C. (1962) Spin-Lattice Relaxation of Chromium Ions in Ruby. *Proc. Phys. Soc.* **79**, 58-68.

Giordano, M., Lazzareschi, A., Pardi, L., Santucci, S., and Veracini, C. A. (1979) Spin-Probe Longitudinal Relaxation in a Smectic A Mesophase. *J. Magn. Reson.* **33**, 1-10.

Giordano, M., Martinelli, M., Pardi, L., and Santucci, S. (1981) Dependence of the longitudinal spin relaxation time on concentration for a nitroxyl radical. *Mol. Phys.* **42**, 523-531.

Giordmaine, J. A., Alsop, L. E., Nash, F. R., and Townes, C. H. (1958) Paramagnetic Relaxation at Very Low Temperature. *Phys. Rev.* **109**, 302-311.

Goldman, S. A., Bruno, G. V., Polnaszek, C. F., and Freed, J. H. (1972) ESR study of anisotropic rotational reorientation and slow tumbling in liquid and frozen media. *J. Chem. Phys.* **56**, 716-735.

Goldman, S. A., Bruno, G. V., and Freed, J. H. (1973) ESR studies of anisotropic rotational reorientation and slow tumbling in liquid and frozen media. II. Saturation and nonsecular effects. *J. Chem. Phys.* **59**, 3071-3091.

Gorchester, J., Millhausser, G. L., and Freed, J. H. (1990) Two-Dimensional Electron Spin Resonance, in *Modern Pulsed and Continuous-Wave Electron Spin Resonance*, L. Kevan and M. K. Bowman, eds., Wiley, New York, 119-194.

Gordon, J. P. and Bowers, K. D. (1958) Microwave Spin Echoes from Donor Electrons in Silicon. *Phys. Rev. Lett.* **1**, 368-369.

Goudemond, I. P., Keartland, J. M., Hoch, M. J. R., Senin, H. B., and Saunders, G. A. (1995) Paramagnetic Ion Relaxation in Rare-earth Doped Phosphate Glasses via ^{31}P NMR. *Bull. Magn. Reson.* **17**, 234-235.

Goudemond, I. P., Keartland, J. M., Hoch, M. J. R., and Saunders, G. A. (1997) Model for nonexponential nuclear relaxation in highly doped magnetic glasses. *Phys. Rev. B* **56**, R8463-R8466.

Gourdon, J. C., Vigouroux, B., and Pescia, J. (1973) Measurement of Spin-Lattice Relaxation Times Shorter than 10^{-8} s. Application to $CuSO_4 \cdot 5H_2O$ and Gd Metal. *Phys. Lett.* **45A**, 69-70.

Gramza, M., Hilczer, W., Goslar, J., and Hoffmann, S. K. (1997) Electron Spin Relaxation and ESEEM Spectroscopy of the Glycine Radical in Diglycine Nitrate Single Crystal. *Acta Chem. Scand.* **51**, 556-561.

Greis, J. W., Angerhofer, A., Norris, J. R., Scheer, H., Struck, A., and von Schütz, J. U. (1994) Spectral diffusion and ^{14}N quadrupole splittings in absorption detected magnetic resonance hole burning spectra of photosynthetic reaction centers. *J. Chem. Phys.* **100,** 4820-4827.

Gruss, A., Dinse, K.-P., Hirsch, A., Nuber, B., and Reuther, U. (1997) Photolysis of $(C_{59}N)_2$ Studied by Time-Resolved EPR. *J. Am. Chem. Soc.* **119,** 8728-8729.

Haas, D. A., Sugano, T., Mailer, C., Sugano, T., and Robinson, B. H. (1992) New Developments in Pulsed Electron Paramagnetic Resonance: Direct Measurement of Rotational Correlation Times from Decay Curves. *Bull. Magn. Reson.* **14,** 35-40.

Haas, D. A., Sugano, T., Mailer, C., and Robinson, B. H. (1993) Motion in Nitroxide Spin Labels: Direct Measurement of Rotational Correlation Times by Pulsed Electron Double Resonance. *J. Phys. Chem.* **97,** 2914-2921.

Haas, D. A., Mailer, C., and Robinson, B. H. (1993) Using nitroxide spin labels. How to obtain T_{1e} from continuous wave electron paramagnetic resonance spectra at all rotational rates. *Biophys. J.* **64,** 594-604.

Hales, B. J., True, A. E., and Hoffman, B. M. (1989) Detection of a New Signal in the EPR Spectrum of Vanadium Nitrogenase from Azotobacter vinelandii. *J. Amer. Chem. Soc.* **111,** 8519-8520.

Ham, F. S. (1965) Dynamical Jahn-Teller Effect in Paramagnetic Resonance Spectra: Orbital Reduction Factors and Partial Quenching of Spin-Orbit Interaction. *Phys. Rev.* **138,** A1727-1737.

Ham, F. S. (1968) Effect of Linear Jahn-Teller Coupling on Paramagnetic Resonance in a ^2E State. *Phys. Rev.* **166,** 307-321.

Harris, E. A., and Yngvesson, K. S. (1968a) Spin-lattice relaxation in some iridium salts. I. Relaxation of the isolated $(IrCl_6)^{2-}$ complex. *J. Phys. C (Proc. Phys. Soc.) Ser. 2,* **1,** 990 – 1010.

Harris, E. A., and Yngvesson, K. S. (1968b) Spin-lattice relaxation in some iridium salts. II. Relaxation of nearest neighbor exchange-coupled pairs. *J. Phys. C (Proc. Phys. Soc.) Ser. 2,* **1,** 1011-1023.

Henrichs, P. M., and Linder, M. (1984) Carbon-13 Spin Diffusion in the Determination of Intermolecular Structure in Solids. *J. Magn. Reson.* **58,** 458-461.

Hervé, J. (1963) Measurement of Electronic Spin-Lattice Relaxation Times by Rapid Modulation of the Saturation Factor, in *Paramagnetic Resonance: Proceedings of the First International Conference held in Jerusalem, July 16-20, 1962*, W. Low, ed., vol. II, pp. 689-697, Academic Press, NY.

Hervé, J. and Pescia, J. (1960) Résonance Paramagnétique – Mesure du temps de relaxation T_1 par modulation du champ radiofréquence H_1 et détection de la variation d'aimantation selon le champ director. *Compt. Rend.* **251,** 665-667.

Hervé, J. and Pescia, J. (1963) Résonance Magnétique Électronique – Théorie phénoménologique de la mesure du temps de relaxation T_1, utilisant un champ radiofréquence modulé en amplitude. Vérification expérimentale sur le diphényl-picryl-hydrazyl. *Compt. Rend.* **255,** 2926-2928.

Hilczer, W., Hoffmann, S. K., Goslar, J., Tritt-Goc, J., and Augustyniak, M. (1993) Electron Spin Echo Studies of Spin-Lattice and Spin-Spin Relaxation of SeO_3^- Radicals in $(NH_4)_3H(SeO_4)_2$ Crystal. *Solid State Commun.* **85,** 585-587.

Höfer, P., Holczer, K., and Schmalbein, D. (1989) Characterization of Gamma-ray Irradiated Powder Alanine Samples by Pulsed EPR. *Appl. Radiat. Isot.* **40,** 1233-1235.

Hoffmann, S. K. and Goslar, J. (1998) Electron Spin Relaxation of Copper(II) Ions in Diamagnetic Crystals. *Appl. Magn. Reson.* **14,** 293-303.

Hoffmann, S. K., Gramza, M., and Hilczer, W. (1995) Molecular Dynamics of Diglycine Nitrate Studied by Phase Memory Relaxation Time of Glycine Radical. *Ferroelectrics* **172,** 431-435.

Hoffmann, S. K., Hilczer, W., and Goslar, J. (1996a) EPR, Electron Spin-Lattice Relaxation, and Debye Temperature of Cu(II)-Doped Triglycine Selenate Crystal. *J. Magn. Reson. A* **122**, 37-41.

Hoffman, S. K., Hilczer, W., and Goslar, J. (1996b) Electron Spin Echo Studies of Flipping Type Minimum in Phase Memory Time of Cu(II) Ions in Triglycine Selenate Crystal at Low Temperatures. *Solid State Commun.* **100**, 449-452.

Hoffmann, S. K., Goslar, J., Hilczer, W., Augustyniak, M. A., and Marciniak, M. (1998a) Vibronic Behavior and Electron Spin Relaxation of Jahn-Teller Complex $Cu(H_2O)_6^{2+}$ in $(NH_4)_2Mg(SO_4)_2 \cdot 6H_2O$ Single Crystal. *J. Phys. Chem. A* **102**, 1697-1707.

Hoffmann, S. K., Augustyniak, M. A., Goslar, J., and Hilczer, W. (1998b) Does the Jahn-Teller effect influence electron spin relaxation? Electron paramagnetic resonance and electron spin echo studies of the Mn^{2+} doped $(NH_4)_2Mg(SO_4)_2 \cdot 6H_2O$ single crystal and comparison with Cu^{2+} data. *Mol. Phys.* **95**, 1265-1273.

Hornak, J. and Freed, J. H. (1986) Spectral Rotation in Pulsed ESR Spectroscopy. *J. Magn. Reson.* **67**, 501-518.

Horvitz, E. P. (1971) Nuclear Spin Diffusion Induced by Paramagnetic Impurities in Nonconducting Solids. *Phys. Rev. B* **3**, 2868-2872.

Huang, C.-Y. (1967) Optical Phonons in Electron Spin Relaxation. *Phys. Rev.* **154**, 215-219.

Hubrich, M., Maresch, G. G., and Spiess, H. W. (1995) Application of Pulsed ENDOR to the Study of Radicals in a Liquid-Crystalline Copolyester. *J. Magn. Reson. A* **113**, 177-184.

Huisjen, M. and Hyde, J. S. (1974a) Saturation recovery measurements of electron spin-lattice relaxation times of free radicals in solution. *J. Chem. Phys.* **60**, 1682-1683.

Huisjen, M. and Hyde, J. S. (1974b) A Pulsed EPR Spectrometer. *Rev. Sci. Instrum.* **45**, 669-675.

Hurrell, J. P., and Davies, E. R. (1971) Nuclear Induced Electron-Spin-Echo Decay in Solids. *Solid State Communications* **9**, 461-463.

Husted, R., Du, J.-L., Eaton, G. R., and Eaton, S. S. (1995) Temperature and Orientation Dependence of Electron Spin Relaxation in Molybdenum(V) Porphyrins. *Magn. Reson. Chem.* **33**, S66-S69.

Hwang, J. S., Mason, R. P., Hwang, L.-P., and Freed, J. H. (1975) Electron Spin Resonance Studies of Anisotropic Rotational Reorientation and Slow Tumbling in Liquid and Frozen Media. III. Perdeuterated 2,2,6,6-Tetramethyl-4-piperidone-N-Oxide and An Analysis of Fluctuating Torques. *J. Phys. Chem.* **78**, 489-511.

Hyde, J. S. (1960) Magnetic Resonance and Rapid Passage in Irradiated LiF. *Phys. Rev.* **119**, 1483-1495.

Hyde, J. S. (1974) Paramagnetic Relaxation. *Ann. Rev. Phys. Chem.* **25**, 407-435.

Hyde, J. S. (1979) Saturation Recovery Methodology, in Kevan, L., and Schwartz, R. N., eds., *Time Domain Electron Spin Resonance*, Wiley-Interscience, New York.

Hyde, J. S. (1998) Saturation Recovery, Chapter I.1 in *Foundations of Modern EPR*, Eaton, G. R., Eaton, S. S., and Salikhov, K. M. eds., World Scientific, Singapore.

Hyde, J. S. and Hyde, D. A. (1981) Determination of T_2 from Analysis of Wings of Symmetrical Inhomogeneous Lines. *J. Magn. Reson.* **43**, 137-140.

Hyde, J. S. and Sarna, T. (1978) Magnetic interactions between nitroxide free radicals and lanthanides or Cu^{2+} in liquids. *J. Chem. Phys.* **68**, 4439-4447.

Hyde, J. S., Froncisz, W. and Mottley, C. (1984) Pulsed ELDOR Measurement of Nitrogen T_1 in Spin Labels. *Chem. Phys. Lett.* **110**, 621-625.

Hyde, J. S., Sczaniecki, P. B. and Froncisz, W. (1989) The Bruker Lecture. Alternatives to Field Modulation in Electron Spin Resonance Spectroscopy. *J. Chem. Soc. Faraday Trans. I*, **85**, 3901-12.

Hyde, J. S., Yin, J.-J., Feix, J. B., and Hubbell, W. L. (1990) Advances in spin label oximetry. *Pure & Appl. Chem.* **62**, 255-260.

Innes, J. B. and Brudvig, G. W. (1989) Location and Magnetic Relaxation Properties of the Stable Tyrosine Radical in Photosystem II. *Biochem.* **28**, 1116-1125.

Isogai, Y., Itoh, S., and Nishimura, M. (1990) Location of D^+ and distribution of surface charges in Photosystem II. *Biochim. Biophys. Acta* **1017**, 204-208.

Isoya, J., Kanda, H., and Uchida, Y. (1990) EPR studies of interstitial Ni centers in synthetic diamond crystals. *Phys. Rev. B* **42**, 9843-9852.

Kanemoto, K., Yamauchi, J., and Adachi, A. (1998) Electron Spin Relaxation Studies of Conducting Polypyrroles: The Difference of the Relaxation Processes Between Highly Conducting and Semiconducting Polypyrroles. *Solid State Commun.* **107**, 203-207.

Keniry, M. A., Rothgeb, T. M., Smith, R. L., Gutowsky, H. S., and Oldfield, E. (1983) Nuclear Magnetic Resonance Studies of Amino Acids and Proteins. Side-Chain Mobility of Methionine in the Crystalline Amino Acid and in Crystalline Sperm Whale (Physeter catodon) Myoglobin. *Biochem.* **22**, 1917-1926.

Keniry, M. A., Kintanar, A., Smith, R. L., Gutowsky, H. S., and Oldfield, E. (1984) Nuclear Magnetic Resonance Studies of Amino Acids and Proteins. Deuterium Nuclear Magnetic Resonance Relaxation of Deuteriomethyl-Labeled Amino Acids in Crystals and in Halobacterium halobium and Escherichia coli Cell Membranes. *Biochem.* **23**, 288-298.

Kevan, L., and Narayana, P. A. (1979) Disordered Matrices, in *Multiple Electron Resonance Spectroscopy*. Dorio, M. M. and Freed, J. H., eds., Plenum Press, New York, pp. 229-259.

Kevan, L., and Schwartz, R. N., eds., (1979) *Time Domain Electron Spin Resonance*, Wiley-Interscience, New York.

Kevan, L., Schlick, S., Toriyama, K., and Iwasaki, M. (1980) Matrix ENDOR of the Protonated Carboxylic Anion Radical in γ-Irradiated L-Alanine. Simulation Using a General Matrix ENDOR Line-Shape Model and Single Crystal Data. *J. Phys. Chem.* **84**, 1950-4.

Kinell, P.-O., Rånby, B. G., and Runnström-Reio,V., eds. (1973) ESR Applications to Polymer Research; Proceedings Nobel Symposium (22nd : 1972 : Södergarn). Wiley, New York.

King, A. R., Wolfe, J. P., and Ballard, R. L. (1972) NMR of Nuclei near a Paramagnetic Impurity in Crystals. *Phys. Rev. Lett.* **28**, 1099-1102.

Kittel, C. (1963) *Quantum Theory of Solids*. Wiley, New York. See chapters 2 and 3.

Kittel, C. (1966) *Introduction to Solids State Physics*, 3rd ed., Wiley, New York, Chapter 5.

Klauder, J. R., and Anderson, P. W. (1962) Spectral Diffusion Decay in Spin Resonance Experiments. *Phys. Rev.* **125**, 912-932.

Klemens, P. G. (1962) Localized Modes and Spin-Lattice Interactions. *Phys. Rev.* **125**, 1795-1798. See correction in *Phys. Rev.* **138 A**, 1217 (1965).

Klug, C. S., Eaton, S. S., Eaton, G. R., and Feix, J. B. (1998) Ligand-induced conformational change in the ferric enterobactin receptor FepA as studied by site-directed spin labeling and time-domain EPR. *Biochemistry* **37**, 9016-9023.

Knapp, C., Dinse, K.-P., Pietzak, B., Wailbinger, M., and Weidinger, A. (1997). Fourier transform EPR study of $N@C_{60}$ in solution. *Chem. Phys. Lett.* **272**, 433-437.

Knapp, C., Weiden, N., and Dinse, K.-P. (1998a) EPR investigation of endofullerenes in solution. *Appl. Phys. A. Mater. Sci. Process.* **66**, 249-255.

Knapp, C., Weiden, N., Käss, H., Dinse, K.-P., Pietzak, B., Waiblinger, M.,, and Weidinger, A. (1998b) Electron paramagnetic rsonance study of atomic phosphorus encapsulated in [60]fullerene. *Mol. Phys.* **95**, 999-1004.

Kneller, G. R., and Smith, J. C. (1994) Liquid-like Side-chain Dynamics in Myoglobin. *J. Mol. Biol.* **242**, 181-185.

Knowles, P. F., Brown, R. D., III, Koenig, S. H., Wang, S., Scott, R. A., McGuirl, M. A., Brown, D. E., and Dooley, D. M. (1995) Spectroscopic Studies of the Active Site of Galactose Oxidase. *Inorg. Chem.* **34**, 3895-3902.

Koenig, S. H., and Brown, R. D. III (1985) Relaxation of Solvent Protons and Deuterons by Protein-Bound Mn^{2+} Ions, Theory and Experiment for Mn^{2+} - Concanavalin A. *J. Magn. Reson.* **61**, 426-439.

Konda, R., Du, J.-L., Eaton, S. S., and Eaton, G. R. (1994) Electron Spin Relaxation Rates for Nitridochromium(V) Tetratolylporphyrin and Nitridochromium(V) Octaethylporphyrin in Frozen Solution. *Appl. Magn. Reson.* **7**, 185-193.

Kooser, R. G., Volland, W. V., and Freed, J. H. (1969) ESR Relaxation Studies on Orbitally Degenerate Free Radicals. I. Benzene Anion and Tropenyl. *J. Chem. Phys.* **50**, 5243-5257.

Krebs, M., von Schütz, J. U., and Wolf, H. C. (1990) Pulsed ESR Studies on Low Dimensional Organic Conductors. *Congress Ampere on Magnetic Resonance and Related Phenomena, Stuttgart, 1990*, page 503-504.

Krüger, G. J. (1972) Electron Spin Relaxation in Solutions of Organic Free Radicals. *Adv. Molec. Relax. Proc.* **3**, 235-251.

Kubo, A., and McDowell, C. A. (1988) ^{31}P Spectral Diffusion. *J. Chem. Phys.* **89**, 63-70.

Kudryashov, A. N., Dzuba, S. A., Samoilova, R. I., Markaryan, G. A., Lunina, Ye. V., and Tsvetkov, Yu. D. (1993) Investigation of the Reorganizational Dynamics of Nitroxides Adsorbed on Surfaces Using Echo-Induced EPR Lineshapes. *J. Magn. Reson. A* **105**, 204-208.

Kulikov, A. V. (1974) Determination of the distance between the spins of the spin label and paramagnetic center in spin-labeled proteins from the saturation curve parameters of the EPR spectra of the labels at 77 K. *Molek. Biol. (Moscow)* **10**, 132-141 (p. 109 in transl.)

Kumashiro, K. K., Schmidt-Rohr, K., Murphys, O. J., III, Ouellette, K. L., Cramer, W. A., and Thompson, L. K. (1998) A novel tool for probing membrane protein structure: solid state NMR with proton spin diffusion and X-nucleus detection, *J. Am. Chem. Soc.* **120**, 5043-5051.

Kuo, L. C., and Makinen, M. W. (1985) Ground State Splitting of High-Spin Co^{2+} as a Probe of Coordination Structure. 2. The Ligand Environment of the Active Site Metal Ion of Carboxypeptidase A in Ester Hydrolysis. *J. Am. Chem. Soc.* **107**, 5255-5261.

Kurshev, V. V. and Raitsimring, A. M. (1990) Carr-Purcell Train in the Conditions of Partial Excitation of Magnetic Resonance Spectrum. *J. Magn. Reson.* **88**, 126-129.

Kurtz, S. R., and Stapleton, H. J. (1980) Effects of disorder on electron-spin relaxation in β-alumina: A prototype glass. *Phys. Rev. B* **22**, 2195-2205.

Kusumi, A., Subczynski, W. K., and Hyde, J. S. (1982) Oxygen transport parameter in membranes deduced by saturation recovery measurements of spin-lattice relaxation times of spin labels. *Proc. Natl. Acad. Sci. USA* **79**, 1854-1858.

Kutter, C., Moll, H. P., van Tol, J., Zuckermann, H., Maan, J. C., and Wyder, P. (1995) Electron-Spin Echoes at 604 GHz Using Far Infrared Lasers. *Phys. Rev. Lett.* **74**, 2925-2928.

Kyhl, R. L., and Nageswara-Rao, B. D. (1967) Effect of Al^{27} on Electron Cross Relaxation in Ruby. *Phys. Rev.* **158**, 284-287.

LaMar, G. N., Horrocks, W. DeW. Jr., and Holm, R. H., eds., (1973) *NMR of Paramagnetic Molecules*, Academic Press, New York, pages 288, 568, 611, 638.

Larson, G. H., and Jeffries, C. D. (1966) Spin Lattice Relaxation in Some Rare Earth Salts. II. Angular Dependence, Hyperfine Effects, and Cross Relaxation. *Phys. Rev.* **145**, 311-324.

Leci, E., Brancaccio, A., Cutruzzola, F., Allocatelli, C. T., Tarricone, C., Bolognesi, M., Desideri, A., and Ascenzi, P. (1995) Formate binding to ferric wild type and mutant myoglobins. Thermodynamic and X-ray crystallographic study. *FEBS Lett.* **357**, 227-229.

Lee, S., Patyal, B. R., and Freed, J. H. (1993) A two-dimensional Fourier transform electron-spin resonance (ESR) study of nuclear modulation and spin relaxation in irradiated malonic acid. *J. Chem. Phys.* **98**, 3665-89.

Leniart, D. S., Conner, H. D., and Freed, J. H. (1975) An ESR and ENDOR study of spin relaxation of semiquinones in liquid solution. *J. Chem. Phys.* **63**, 165-199.

Lewis, W. B., and Morgan, L. O. (1968) Paramagnetic Relaxation in Solutions, in *Transition Metal Chemistry*, R. L. Carlin, ed., Dekker, New York, 33-112.

Linder, M., Hendrichs, P. M., Hewitt, J. M., and Massa, D. J. (1985) Use of carbon-carbon nuclear spin diffusion for the study of the miscibility of polymer blends. *J. Chem. Phys.* **82**, 1585-1598.

Lindgren, M., Eaton, G. R., Eaton, S. S., Jonsson, B.-H., Hammarström, P., Svensson, M., and Carlsson, U. (1997). Electron spin echo decay as a probe of aminoxyl environment in spin-labeled mutants of human carbonic anhydrase II, *J. C. S. Perkin 2*, 2549-2554.

Lingam, K. V., Nair, P. G., and Venkataraman, B. (1972) Spin-lattice relaxation studies on semiquinone ions. *Proc. Ind. Acad. Sci. A* **76**, 207-220.

Lloyd, J. P., and Pake, G. E. (1954) Spin Relaxation in Free Radical Solutions Exhibiting Hyperfine Structure. *Phys. Rev.* **94**, 579-591.

Lowe, L. J., and Gade, S. (1967) Density-matrix Derivation of the Spin-Diffusion Equation. *Phys. Rev.* **156**, 817-825.

Mailer, C. and Taylor, C. P. S. (1973) Rapid Adiabatic Passage EPR of Ferricytochrome c: Signal Enhancement and Determination of the Spin-Lattice Relaxation Time. *Biochim. Biophys. Acta* **322**, 195-203.

Mailer, C., Danielson, J. D. S., and Robinson, B. H. (1985) Computer-controlled pulsed electron-paramagnetic-resonance spectrometer. *Rev. Sci. Instrum.* **56**, 1917-1925.

Mailer, C., Haas, D. A., Hustedt, E. J., Gladden, J. G. and Robinson, B. H. (1991) Low Power Electron Paramagnetic Resonance Spin-Echo Spectroscopy. *J. Magn. Reson.* **91**, 475-496.

Mailer, C., Robinson, B. H., and Haas, D. A. (1992) New Developments in Pulsed Electron Paramagnetic Resonance: Relaxation Mechanisms of Nitroxide Spin Labels. *Bull. Magn. Reson.* **14**, 30-35.

Makinen, M. W., and Yim, M. B. (1981) Coordination environment of the active-site metal ion of liver alcohol dehydrogenase. *Proc. Natl. Acad. Sci. USA* **78**, 6221-6225.

Makinen, M. W., and Wells, G. B. (1987) Application of ESR Saturation Methods to Paramagnetic Metal Ions in Proteins. In *Metal Ions in Biological Systems* **22**, 129-206.

Makinen, M. W., Kuo, L. C., Yim, M. B., Wells, G. B., Fukuyama, J. M., and Kim, J. E. (1985) Ground State Splitting of High-Spin Co^{2+} as a Probe of Coordination Structure. 1. Dependence of the Splitting on Coordination Geometry. *J. Am. Chem. Soc.* **107**, 5245-5255.

Manenkov, A. A. and Milyaev, V. A. (1970) Paramagnetic Relaxation Processes in $Al(NO_3)_2 \cdot 9H_2O:Fe^{3+}$ Single Crystals at Helium Temperatures. *Soviet Physics JETP* **31**, 427-428 (**58**, 796-799 in Russian)

Manenkov, A. A. and Orbach, R., eds., (1966) *Spin-Lattice Relaxation in Ionic Solids*, Harper & Row, Publ. New York. Contains reprints of many of the early papers.

Manenkov, A. A. and Pol'skii, Yu. E. (1964) Relaxation Processes in the Paramagnetic Resonance of Gd^{3+} in CaF_2. *Soviet Physics JETP* **45**, 1425-1429 (1963) (**18**, 985-987 (1964) in transl.)

Manenkov, A. A. and Prokhorov, A. M. (1962) Spin-Lattice Relaxation and Cross-Relaxation Interactions in Chromium Corundum. *Soviet Physics JETP* **42**, 75-83 (54-59 in translation).

Manenkov, A. A., Milyaev, V. A. and Prokhorov, A. M. (1962) The Relaxation Time of Cr^{3+} and Fe^{3+} Ions in Single Crystals of Rutile. *Soviet Physics Solid State* **4**, 388-391 (280-283 in translation).

Maresch, G. G., Weber, M., Dubinskii, A. A., and Spiess, H. W. (1992) 2D-ELDOR detection of magnetization transfer of nitroxides in disordered solid polymers. *Chem. Phys. Lett.* **193**, 134-140.

Marchand, R. L., and Stapleton, H. J. (1974) Observation of an $H^2T^7 \sin^2\theta$ Raman spin-lattice relaxation rate in a neodymium salt. *Phys. Rev. B* **9**, 14-21.

Marshall, S. A., Nistor, S. V., and Serway, R. A. (1972) Spin-Lattice Relaxation-Time Measurements of Trivalent Iron in Single-Crystal Calcite. *Phys. Rev. B* **6**, 1688-1689.

Martinelli, M., Pardi, L., Pinzino, C., and Santucci, S. (1975) Dependence of relaxation times of longitudinally detected paramagnetic resonance. *Solid State Commun.* **17**, 211-212.

Martinelli, M., Pardi, L., Pinzino, C., and Santucci, S. (1977) Electron-spin double resonance by longitudinal detection. II. Signal dependence on relaxation times. *Phys. Rev. B* **16**, 164-169.

McBrierty, V. J. (1974) N.m.r. of solid polymers: a review. *Polymer* **15**, 503-520.

McConnell, H. M. (1956) Effect of Anisotropic Hyperfine Interactions on Paramagnetic Relaxation in Liquids. *J. Chem. Phys.* **25**, 709-711.

McConnell, H. M. (1961) Spin-Orbit Coupling in Orbitally Degenerate States of Aromatic Ions. *J. Chem. Phys.* **34**, 13-16.

McGarvey, B. R. (1957) Line Widths in the Paramagnetic Resonance of Transition Ions in Solution. *J. Phys. Chem.* **61**, 1232-1237.

Mchaourab, H. S. and Hyde, J. S. (1993a) Continuous wave multiquantum electron paramagnetic resonance spectroscopy. III. Theory of intermodulation sidebands. *J. Chem. Phys.* **98**, 1786-96.

Mchaourab, H. S. and Hyde, J. S. (1993b) Dependence of the Multiple-Quantum EPR Signal on the Spin-Lattice Relaxation Time. Effect of Oxygen in Spin-Labeled Membranes. *J. Magn. Reson.* **B101**, 178-184.

Mchaourab, H. S., Christidis, T. C., Froncisz, W., Sczaniecki, P. B. and Hyde, J. S. (1991) Multiple-Quantum Electron-Electron Double Resonance. *J. Magn. Reson.* **92**, 429-33.

Mchaourab, H. S., Christidis, T. C. and Hyde, J. S. (1993) Continuous wave multiquantum electron paramagnetic resonance spectroscopy. IV. Multiquantum Electron-Nuclear Double Resonance. *J. Chem. Phys.* **99**, 4975-85.

Mchaourab, H. S., Pfenninger, S., Antholine, W. E., Felix, C. C., Hyde, J. S. and Kronek, P. M. H. (1993) Multiquantum EPR of the mixed valence copper site in nitrous oxide reductase. *Biophys. J.* **64**, 1576-9.

Michalik, J., and Kevan, L. (1978) Temperature Dependence of Electron Spin-Lattice Relaxation of Radiation-Produced Silver Atoms in Polycrystalline Aqueous and Glassy Organic Matrices. Importance of Relaxation by Tunnelling Modes in Disordered Matrices. *J. Magn. Reson.* **31**, 259-270.

Mims, W. B. (1972). Electron Spin Echoes, in *Electron Paramagnetic Resonance*, S. Geschwind, ed., Plenum Press, N. Y., ch. 4.

Mims, W. B. (1998) Electron Spin Echo Experiments Then and Now, Chapter I.3 in *Foundations of Modern EPR*, Eaton, G. R., Eaton, S. S., and Salikhov, K. M. eds., World Scientific, Singapore.

Mizoguchi, K., Kume, K., and Shirakawa, H. (1984) Frequency Dependence of Electron Spin-Lattice Relaxation Rate at 5-450 MHz --- New Evidence of One Dimensional Diffusive Motion of Electron Spin (Neutral Soliton). *Solid State Commun.* **50**, 213-218.

Mizuta, Y., Kohno, M., and Fujii, K. (1994) Relaxation Mechanism of Free Radicals in Silica Glass Studied by Pulsed-ESR Spectroscopy. *Jpn. J. Appl. Phys.* **33**, 1885-1889.

Moll, H. P., Kutter, C., van Tol, J., Zuckerman, H., Wyder, P. (1999) Principles and performance of an electron spin echo spectrometer using far infrared lasers as excitation sources. *J. Magn. Reson.* **137**, 46-58.

Momot, K. I. and Walker, F. A., (1997) Investigation of rotation of axial ligands in six-coordinate low-spin iron(III) tetraphenylporphyrinates: measurement of rate constants from saturation transfer experiments and comparison to molecular mechanics calculations. *J. Phys. Chem. A* **101**, 2787-2795.

Morawski, P., Hoffmann, S. K., Hilczer, W., and Goslar, J. (1997) ESR and Electron Spin Echo Studies of Spin-Lattice Relaxation of Hydrazinium Radical in $Li(N_2H_5)SO_4$ Single Crystal. *Acta Physica Polonica A* **91**, 1121-1129.

More, K. M., Eaton, G. R., and Eaton, S. S. (1984) Determination of T_1 and T_2 by simulation of EPR power saturation curves and saturated spectra, application to spin-labeled iron porphyrins. *J. Magn. Reson.* **60**, 54-65.

Muller, F., Hopkins, M. A., Coron, N., Grynberg, M., Brunel, L. C., and Martinez, G. (1989) A high magnetic field EPR spectrometer. *Rev. Sci. Instrum.* **60**, 3681-3684.

Müller, A., and Haeberlen, U. (1996) Temperature-dependent spectral spin diffusion in solid state deuteron magnetic resonance. *Chem. Phys. Lett.* **248**, 249-254.

Muromtsev, V. I., Shteinshneider, N. Ya., Safronov, S. N., Golikov, V. P., Kuznetsov, A. I., and Zhidomirov, G. M. (1975) Spin-lattice relaxation of iminoxyl radicals in frozen solutions. *Fiz. Tverd. Tela* **17**, 813-816. [p.517 in transl.]

Murphy, J. (1966) Spin-lattice relaxation due to local vibrations with temperature-independent amplitudes. *Phys. Rev.* **145**, 241-247.

Murphy, T. A., Pawlik, Th., Weidinger, A., Höhne, M., Alcala, R., and Spaeth, J.-M. (1996) Observation of Atomlike Nitrogen in Nitrogen-Implanted Solid C_{60}. *Phys. Rev. Lett.* **77**, 1075-1078 (1996).

Murugusan, R., Cook, J. A., Devasahayam, N., Afeworki, M., Subramanian, S., Tschudin, R., Larsen, J. A., Mitchell, J. B., Russo, A., and Krishna, M. C. (1997) In Vivo Imaging of a Stable Paramagnetic Probe by Pulsed-Radiofrequency Electron Paramagnetic Resonance Spectroscopy. *Magn. Reson. Med.* **38**, 409-414.

Musewald, C., Gilch, P., Hartwich, G., Pöllinger-Dammer, F., Scheer, H., and Michel-Beyerle, M. E. (1999) Magnetic Field Dependence of Ultrafast Intersystem-Crossing: A Triplet Mechanism on the Picosecond Time Scale? *J. Am. Chem. Soc.* **121**, 8876-8881.

Nakagawa, K., Candelaria, M. B., Chik, W. W. C., Eaton, S. S., and Eaton, G. R. (1992) Electron-spin relaxation times of chromium(V). *J. Magn. Reson.* **98**, 81-91.

Nakagawa, K., Eaton, S. S., and Eaton, G. R. (1993) Electron Spin Relaxation Times of Irradiated Alanine, *Appl. Radiat. Isotop.* **44**, 73-76.

Nechtschein, M. and Hyde, J. S. (1970) Pulsed Electron-Electron Double Resonance in an $S = ½, I = ½$ System. *Phys. Rev. Lett.* **24**, 672-674.

Niang, A., Ablart, G., Pescia, J., Servant, Y., Duplessix, R., Chanh, N. B., Parys, T., and Misra, S. K. (1995) Electron Paramagnetic Resonance and Spin-Lattice Relaxation in Two-Dimensional Systems. *Appl. Magn. Reson.* **8**, 319-333.

Norris, J. R., Thurnauer, M. C., and Bowman, M. K. (1980) Electron Spin echo Spectroscopy and the Study of Biological Structure and Function. *Adv. Biol. Med. Phys.* **17**, 365-416.

Nyberg, G. (1967) Spin-rotational relaxation in solution E.S.R. *Mol. Phys.* **12**, 69-81.

Nyberg, G. (1969) E.S.R. spectrum of the amino-cuprate ion. *Mol. Phys.* **17**, 87-90.

O'Brien, M. C. M. (1964) The dynamic Jahn-Teller effect in octahedrally coordinated d^9 ions. *Proc. Roy. Soc. A* **281**, 323-339.

Okabe, N., Ohba, Y., Suzuki, S., Kawata, S., Kikuchi, K., Achiba, Y., Iwaizumi, M. (1995) Fourier transform EPR studies of metallofullerene ($La@C_{82}$) in CS_2 solution. *Chem. Phys. Lett.* **235**, 564-569.

Ong, J.-L., Sloop, D. J., and Lin, T.-S. (1994) Peculiar Spin Dynamics of the Photoexcited Triplet State of Pentacene in a Benzoic Acid Crystal: an ESE Study. *Appl. Magn. Reson.* **6**, 359-371.

Onischuk, A. A., Samoilova, R. I., Strunin, V. P., Chesnokov, E. N., Musin, R. N., Bashurova, V. S., Maryasov, A. G., and Panfilov, V. N. (1998) EPR Investigation of α-Si:H Aerosol Particles Formed under Silane Thermal Decomposition. *Appl. Magn. Reson.* **15**, 59-94.

Orbach, R. (1961). On the theory of spin-lattice relaxation in paramagnetic salts, *Proc. Phys. Soc. (Lond.)* **77**, 821-826.

Orton, J. W. (1968) *Electron Paramagnetic Resonance*, Gordon and Breach, N. Y.

Pace, J. H., Sampson, D. F., and Thorp, J. S. (1960) Spin-Lattice Relaxation Times in Ruby at 34.6 kMc/sec. *Phys. Rev. Lett.* **4**, 18-19.

Pake, G. E. and Estle, T. L. (1973) *The Physical Principles of Electron Paramagnetic Resonance*, 2nd ed., W. A. Benjamin Inc. Reading, MA.

Palmer, G. (1979) Electron paramagnetic resonance of hemoproteins, in *The Porphyrins*, vol IV, D. Dolphin ed, Academic Press, N.Y., ch. 6, pp. 313-353.

Palmer, G. (1983) Electron paramagnetic resonance of hemoproteins, in *Iron Porphyrins*, Part II, A. B. P. Lever and H. B. Gray, ed., Addison Wesley, Reading, MA, ch. 2, pp. 43-88.

Parry, J. S., Cloke, F. G. N., Coles, S. J., and Hursthouse, M. B. (1999) Synthesis and Characterization of the First Sandwich Complex of Trivalent Thorium: A Structural Comparison with the Uranium Analogue. *J. Am. Chem. Soc.* **121**, 6867-6871.

Paschenko, S. V., Toropov, Yu. V., Dzuba, S. A., Tsvetkov, Yu. D., and Vorobiev, A. Kh. (1999) Temperature dependence of amplitudes of librational motion of guest spin-probe molecules in organic glasses. *J. Chem. Phys.* **110**, 8150-8154.

Pastor, R. C. and Hoskins, R. H. (1960) Paramagnetic Resonance in Charred Dextrose. *J. Chem. Phys.* **32**, 264-269.

Percival, P. W. and Hyde, J. S. (1975) Pulsed EPR Spectrometer II. *Rev. Sci. Instrum.* **46**, 1522-1529.

Percival, P. W. and Hyde, J. S. (1976) Saturation-Recovery Measurements of the Spin-Lattice Relaxation Times of Some Nitroxides in Solution. *J. Magn. Reson.* **23**, 249-257.

Percival, P. W., Hyde, J. S., Dalton, L. A., and Dalton, L. R. (1975) Molecular and applied modulation effects in electron electron double resonance. V. Passage effects in high resolution frequency and field swept ELDOR. *J. Chem. Phys.* **62**, 4332-4342.

Pescia, J. (1965) La mesure des temps de relaxation spin-réseau très courts. *Ann. Phys.* **10**, 389-406.

Pescia, J. (1998) Microwave Amplitude Modulation in EPR and Short T_1 Measurement, in *Foundations of Modern EPR*, S. S. Eaton, K. M. Salikhov, and G. R. Eaton, eds., World Scientific Publ., Singapore.

Pescia, J., Misra, S. K., Zaripov, M., and Servant, Y. (1999). Spin-lattice relaxation in the polymer resin P4VP doped with transition ions Cu^{2+}, Cr^{3+}, Mn^{2+}, and Gd^{3+} possessing weak spin-orbit coupling. *Phys. Rev. B* **83**, 1866-1869.

Pietzak, B., Wailbinger, M., Murphy, T. A., Weidinger, A., Höhne, M., Dietel, E., and Hirsch, A. (1997) Buckminsterfullerene C_{60}: a chemical Faraday cage for atomic nitrogen. *Chem. Phys. Lett.* **279**, 259-263.

Plato, M., Lubitz, W., and Möbius, K. (1981) A Solution ENDOR Sensitivity Study of Various Nuclei in Organic Radicals. *J. Phys. Chem.* **85**, 1202-1219.

Poole, C. P. Jr., (1967) *Electron Spin Resonance*, Wiley-Interscience, New York.

Poole, C. P. Jr., and Farach, H. A. (1971) *Relaxation in Magnetic Resonance*, Academic Press, New York.

Portis, A. M. (1956) Spectral Diffusion in Magnetic Resonance. *Phys. Rev.* **104**, 584-588.

Powell, D. H., Merbach, A. E., González, G., Brücher, E., Micskei, K., Ottaviani, M. F., Köhler, K., von Zelewsky, A., Grinberg, O. Ya., and Lebedev, Y. S. (1993) Magnetic-Field-Dependent Electronic Relaxation of Gd^{3+} in Aqueous Solutions of the Complexes $[Gd(H_2O)_8]^{3+}$, [Gd(propane-1,3-diamine-N,N,N',N'-tetraacetate)$(H_2O)_2]^-$, and [Gd(N,N'-bis[N-methylcarbamoyl)methyl]-3-azapentane-1,5-diamine-3,N,N'-triacetate)(H_2O)] of Interest to Magnetic-Resonance Imaging. *Helv. Chim. Acta* **76**, 2129-2146.

Powell, D. H., Dhubhghaill, O. M. N., Pubanz, D., Helm, L., Lebedev, Y. S., Schlaepfer, W., and Merbach, A. E. (1996) Structure and Dynamic Parameters Obtained from ^{17}O NMR, EPR, and NMRD Studies of Monomeric and Dimeric Gd^{3+} Complexes of Interest in Magnetic Resonance Imaging: An Integrated and Theoretically Self-Consistent Approach. *J. Am. Chem. Soc.* **118**, 9333-9346.

Prabhananda, B. S. and Hyde, J. S. (1986) Study of molecular motions in liquids by electron spin relaxation: Halogenated p-semiquinone anions in alcohols. *J. Chem. Phys.* **85**, 6705-6712.

Prins, R. (1970) Electronic structure of the ferricenium cation. Electron spin resonance measurements of the cations of ferrocene derivatives. *Mol. Phys.* **19**, 603-620.

Prisner, T. F. (1997) Pulsed High-Frequency/High-Field EPR. *Adv. Magn. Optic. Reson.* **20**, 245-299.

Quine, R. W., Eaton, S. S. and Eaton, G. R. (1992) Saturation recovery electron paramagnetic resonance spectrometer. *Rev. Sci. Instrum.* **63**, 4252-4262.

Rakowsky, M. H., More, K. M., Kulikov, A. V., Eaton, G. R., and Eaton, S. S. (1995) Time-Domain Electron Paramagnetic Resonance as a Probe of Electron-Electron Spin-Spin Interaction in Spin-Labeled Low-Spin Iron Porphyrins. *J. Am. Chem. Soc.* **117**, 2049-57.

Rakowsky, M. H., Zecevic, A., Eaton, G. R., and Eaton, S. S. (1998) Determination of high-spin iron(III)-nitroxyl distances in spin-labeled porphyrins by time-domain EPR. *J. Magn. Reson.* **131**, 97-110.

Ranby, B. G. and Rabek, J. F. (1977) *ESR Spectroscopy in Polymer Research*. Springer-Verlag, New York.

Rao, K. V. S. (1970) Relaxation Times of VO^{2+} ion Doped in KNO_3 and $CsNO_3$ Single Crystals, *Chem. Phys. Lett.* **5**, 438-440.

Rast, S., Fries, P. H., and Belorizky, E. (1999) Theoretical study of electronic relaxation processes in hydrated Gd^{3+} complexes in solution. *J. Chim. Phys.* **96**, 1543-1550.

Rataiczak, R. D. and Jones, M. T. (1972) Investigation of the CW Saturation Technique for Measurement of Electron Spin-Lattice Relaxation: Application to the Benzene Anion Radical. *J. Chem. Phys.* **56**, 3898-3911.

Rector, K. D., Thompson, D. E., Merchant, K., and Fayer, M. D. (2000) Dynamics in globular proteins: vibrational echo experiments. *Chem. Phys. Lett.* **316**, 122-128.

Redfield, A. G. (1959) Spatial Diffusion of Spin Energy. *Phys. Rev.* **116**, 315-316.

Ren, J. and Sherry, A. D. (1996) NMR Determination of Crystal Field Parameters and Electron-Spin Correlation Times for the $LnDOTP^{5-}$ Complexes. *J. Magn. Reson. B* **111**, 178-182.

Rengen, S. K., Khakhar, M. P., Prabhananda, B. S., and Venkataraman, B. (1972) Electron Spin-Lattice Relaxation in Organic Free Radicals in Solution. *Pure & Appl. Chem.* **32**, 287-305.

Rengen, S. K., Khakhar, M. P., Prabhananda, B. S., and Venkataraman, B. (1974a) Study of molecular motions in liquids by electron spin lattice relaxation measurements. I. Semiquinone ions in hydrogen bonding solvents. *Pramana* **3**, 95-121.

Rengen, S. K., Khakhar, M. P., Prabhananda, B. S., and Venkataraman, B. (1974b) Study of Molecular Motions in Liquids by Electron Spin Lattice Relaxation Measurements. II. 2,5-

Di-tert-butylsemiquinone Ion in Acetonitrile and Tetrahydrofuran. *J. Magn. Reson.* **16**, 35-43.
Rengen, S. K., Bhagat, V. R., Sastry, V. S. S., and Venkataraman, B. (1979) Magnetic Field-Pulsed ELDOR Spectroscopy. *J. Magn. Reson.* **33**, 227-240.
Reuben, J. (1971) Electron Spin Relaxation in Aqueous Solutions of Gadolinium(III)-Aquo, Cacodylate, and Bovine Serum Albumin Complexes. *J. Phys. Chem.* **75**, 3164-3167.
Reynhardt, E. C. and Terblanche, C. J. (1996) ^{13}C Relaxation in Natural Diamond. *Bull. Magn Reson.* **18** (1-2), 145-146.
Robinson, B. H., Haas, D. A., and Mailer, C. (1994) Molecular Dynamics in Liquids: Spin-Lattice Relaxation of Nitroxide Spin Labels. *Science* **263**, 490-493.
Robinson, B. H., Reese, A. W., Gibbons, E., and Mailer, C. (1999) A Unified Description of the Spin-Spin and Spin-Lattice Relaxation Rates Applied to Nitroxide Spin Labels in Viscous Liquids. *J. Phys. Chem. B* **103**, 5881-5894.
Roscher, A., Emsley, L., and Roby, C. (1996) The Effect of Imperfect Saturation in Saturation-Recovery T_1 Measurements. *J. Magn. Reson. A* **118**, 108-112.
Rubinstein, M., Baram, A., and Luz, Z. (1971) Electronic and nuclear relaxation in solutions of transition metal ions with S = 3/2 and 5/2. *Mol. Phys.* **20**, 67-80.
Rübsam, M., Plüschau, M., Schweitzer, P., Dinse, K.-P., Fuchs, D., Rietschel, H., Michel, R. H., Benz, M., Kappes, M. M. (1995) 2D EPR investigation of endohedral La@C_{82} in solution. *Chem. Phys. Lett.* **240**, 615-621.
Ruf, H. H., and Nastainczyk, W. (1976) Binding of a Metyrapone Spin Label to Microsomal Cytochrome P-450. *Eur. J. Biochem.* **66**, 139-146.
Saalmueller, J. W., Long, H. W., Maresch, G. G., and Spiess, H. W. (1995) Two-Dimensional Field-Step ELDOR. A Method for Characterizing the Motion of Spin Probes and Spin Labels in Glassy Solids. *J. Magn. Reson. A* **117**, 193-208.
Salikhov, K. M. and Tsvetkov, Yu. D. (1979). Electron spin-echo studies of spin-spin interactions in solids, in *Time Domain Electron Spin Resonance*, ed. by L. Kevan and R. N. Schwartz, Wiley, N. Y., ch. 7.
Samoilova, R. I., Milov, A. D., and Tsvetkov, Yu. D. (1991) Use of the Electron Spin Echo Effect for the Study of Radical Diffusion in Porous Media. *Sov. J. Chem. Phys.* **8**, 898-909.
Sarna, T., Hyde, J. S., and Swartz, H. M. (1976) Ion-Exchange in Melanin: An Electron Spin Resonance Study with Lanthanide Probes. *Science* **192**, 1132-1134.
Sarna, T. and Hyde, J. S. (1978) Electron spin-lattice relaxation times of melanin. *J. Chem. Phys.* **69**, 1945-1948.
Sastry, V. S. S., Polimeno, A., Crepeau, R. H., and Freed, J. H. (1996a) Studies of spin relaxation and molecular dynamics in liquid crystals by two-dimensional Fourier transform electron spin resonance. I. Cholestane in butoxy benzylidene-octylaniline and dynamic cage effects. *J. Chem. Phys.* **105**, 5753-5772.
Sastry, V. S. S., Polimeno, A., Crepeau, R. H., and Freed, J. H. (1996b) Studies of spin relaxation and molecular dynamics in liquid crystals by two-dimensional Fourier transform electron spin resonance. II. Perdeuterated tempone in butoxy benzylidene-octylaniline and dynamic cage effects. *J. Chem. Phys.* **105**, 5773-5791.
Saxena, S., and Freed, J. H. (1997) Two-Dimensional Electron Spin Resonance and Slow Motions. *J. Phys. Chem. A* **101**, 7998-8008.
Schlick, S., Kevan, L., Toriyama, K., and Iwasaki, M. (1981) The importance of nuclear spin diffusion as an angularly independent relaxation mechanism controlling the matrix ENDOR response of radicals in molecular crystals. *J. Chem. Phys.* **74**, 282-7.

Schmidt, J. (1989) Spin-Lattice and Spin-Spin Relaxation in Photo-Excited Triplet states in Molecular Crystals, in *Relaxation Processes in Molecular Excited States*, J. Funfschilling, ed., Kluwer Academic Publishers.

Schmidt, J., van den Heuvel, D. J., Henstra, A., Lin, T.-S., and Wenckebach, W. Th. (1992) Polarizing Nuclear Spins via Photo-Excited Triplet States. *Israel J. Chem.* **32**, 165-172.

Schmidt-Rohr, K., and H. W. Spiess (1994) *Multidimensional Solid-State NMR and Polymers*. Academic Press, N. Y.

Schneider, D. J., and Freed, J. H. (1989) Calculating Slow Motional Magnetic Resonance Spectra: A User's Guide, in *Biological Magnetic Resonance*, Vol. 8, Berliner, L. J., and Reuben, J., eds., Plenum Press, N. Y. ch. 1.

Scholes, C. P., Isaacson, R. A., and Feher, G. (1971) Determination of the zero-field splitting of Fe^{+3} in heme proteins from the temperature dependence of the spin-lattice relaxation rate. *Biochim. Biophys. Acta* **244**, 206-210.

Scholes, C. P., Janakiraman, R., Taylor, H., and King, T. E. (1984) Temperature Dependence of the Electron Spin-Lattice Relaxation Rate from Pulsed EPR of Cu_A and Heme a in Cytochrome c Oxidase. *Biophys. J.* **45**, 1027-1030.

Schreurs, J. W. H., Blomgren, G. E., and Fraenkel, G. K. (1960) Anomalous Relaxation of Hyperfine Components in Electron Spin Resonance. *J. Chem. Phys.* **32**, 1861-1869.

Schulz, C. E., Brandon, S., and Debrunner, P. G. (1990) Spin Fluctuation Rates in Myoglobin Azide: Comparison of Mossbauer and EPR Results. *Hyperfine Interactions* **58**, 2399-2404.

Schwartz, R. N., Jones, L. L., and Bowman, M. K. (1979) Electron Spin-Echo Studies of Nitroxide Free Radicals in Liquids. *J. Phys. Chem.* **83**, 3429-3434.

Schwartz, L. J., Stillman, A. E., and Freed, J. H. (1982) Analysis of electron spin echoes by spectral representation of the stochastic Liouville equation, *J. Chem. Phys.* **77**, 5410-5425.

Schweiger, A. and Ernst, R. R. (1988) Pulsed ESR with Longitudinal Detection. A Novel Recording Technique. *J. Magn. Reson.* **77**, 512-523.

Sczaniecki, P. B., Hyde, J. S., and Froncisz, W. (1990) Continuous Wave Multi-Quantum EPR Spectroscopy. *J. Chem. Phys.* **93**, 3891-8.

Sczaniecki, P. B., Hyde, J. S. and Froncisz, W. (1991) Continuous Wave Multi-Quantum EPR Spectroscopy II: Spin-System Generated Intermodulation Sidebands. *J. Chem. Phys.* **94**, 5907-16.

Seiter, M., Budker, V., Du, J.-L., Eaton, G. R., and Eaton, S. S. (1998) Interspin distances determined by time domain EPR of spin-labeled high-spin methemoglobin, *Inorg. Chim. Acta* **273**, 354-366.

Shibata, Y., Kurita, A., and Kushida, T. (1996) Transient hole-burning spectroscopy of Zn-protoporphyrin-substituted myoglobin. *J. Luminesc.* **66&67**, 13-18.

Shushakov, O. A., Dzyuba, S. A., and Tsvetkov, Yu. D. (1986) Dielectric β-relaxation in supercooled liquids and mobility of nitroxyl radicals from pulsed ESR data. *Khim. Fiz.* **5**, 1340-1345 (p. 2713 in transl.).

Shushakov, O. A., Dzyuba, S. A., and Tsvetkov, Yu. D. (1986) Superslow Rotation of Nitroxyl Radicals in Aqueous Glycerol Solution. *Sov. J. Chem. Phys.* **3**, 2713-2723.

Shushakov, O. A., Dzyuba, S. A., and Tsvetkov, Yu. D. (1989) Barriers for internal rotation of methyl groups screening nitroxide paramagnetic fragments as determined by an electron spin-echo technique. *J. Struct. Chem.* **30**, 75-80.

Spiro, T. G. (1983) The Resonance Raman Spectroscopy of Metalloporphyrins and Heme Proteins, in *Iron Porphyrins* Part Two, A. B. P. Lever and H. B. Gray, eds., Addison-Wesley, Reading. MA. Pp. 89-159, especially p. 109.

Standley, K. J. and Vaughan, R. A. (1969) *Electron Spin Relaxation Phenomena in Solids*. Plenum Press, New York.

Stephen, M. J. and Fraenkel, G. K. (1960) Theory of Saturation of Electron Spin Resonance Spectra, *J. Chem. Phys.* **32**, 1435-1443.
Stephens, E. M. (1989) Gadolinium as an EPR Probe, in *Lanthanide Probes in Life, Chemical and Earth Sciences: Theory and Practice*, Bünzli, J.-C. G., and Choppin, G. R., eds., Elsevier, Amsterdam, pp. 181-217.
Stevens, K. W. H. (1969) Paramagnetic Resonance, *Reports Prog. Phys.* **30**, 189-226.
Stevens, S. B., and Stapleton, H. J. (1990) Electron-spin-lattice relaxation in Yb^{3+} -doped silicate glass. *Phys. Rev.* **42**, 9794-9801.
Stillman, A. E., and Schwartz, R. N. (1979) Theory of Spin Echoes in Nonviscous and Viscous Liquids, in *Time Domain Electron Spin Resonance*, Kevan, L., and Schwartz, R. N., eds., Wiley, New York, ch. 5.
Stillman, A. E., Schwartz, L. J., and Freed, J. H. (1980) Direct determination of rotational correlation time by electron-spin echoes. *J. Chem. Phys.* **73**, 3502-3503; erratum (1982) *J. Chem. Phys.* **76**, 5658.
Stinson, D. G., and Stapleton, H. J. (1983) Li tunnelling effects on spin-lattice relaxation rates of a color center in CaO:Li. *Phys. Rev. B* **27**, 5386-5393.
Stösser, R., Sebastian, S., Scholz, G., Willer, M., Jeschke, G., Schweiger, A., and Nofz, M. (1999) Pulse EPR Spectroscopy of Cu^{2+}-Doped Inorganic Glasses. *Appl. Magn. Reson.* **16**, 507-528.
Strangeway, R. A., Mchaourab, H. S., Luglio, J. R., Froncisz, W., and Hyde, J. S. (1995) A general purpose multiquantum electron paramagnetic resonance spectrometer. *Rev. Sci. Instrum.* **66**, 4516-28.
Strutz, T., Witowski, A. M., and Wyder, P. (1992) Spin-Lattice Relaxation at High Magnetic Fields. *Phys. Rev. Lett.* **68**, 3912-3915.
Subczynski, W. K., and Hyde, J. S. (1981) The diffusion-concentration product of oxygen in lipid bilayers using the spin-label T_1 method. *Biochim. Biophys. Acta* **643**, 283-291.
Subczynski, W. K., Hyde, J. S., and Kusumi, A. (1989) Oxygen permeability of phosphatidylcholine-cholesterol membranes. *Proc. Natl. Acad. Sci. USA* **86**, 4474-4478.
Subczynski, W. K., Hyde, J. S., and Kusumi, A. (1991) Effect of Alkyl Chain Unsaturation and Cholesterol Intercalation on Oxygen Transport in Membranes: A Pulse ESR Spin Labeling Study. *Biochem.* **30**, 8578-8590.
Subczynski, W. K., Hopwood, L. E., and Hyde, J. S. (1992a) Is the Mammalian Cell Plasma Membrane a Barrier to Oxygen Transport? *J. Gen. Physiol.* **100**, 69-87.
Subczynski, W. K., Renk, G. E., Crouch, R. K., Hyde, J. S., and Kusumi, A. (1992b) Oxygen diffusion-concentration product in rhodopsin as observed by a pulse ESR spin labeling method. *Biophys. J.* **63**, 573-577.
Subczynski, W. K., Lewis, R. N. A. H., McElhaney, R. N., Hodges, R. S., Hyde, J. S., and Kusumi, A. (1998) Molecular Organization and Dynamics of 1-Palmitoyl-2-oleoylphosphatidylcholine Bilayers Containing a Transmembrane α-Helical peptide. *Biochem.* **37**, 3156-3164.
Sueishi, Y., and Kuwata, K. (1989) Pressure Effects on the Electron Spin Relaxation of Several Radicals in Solution. *Chem. Phys. Lett.* **160**, 640-643.
Sugano, T., Mailer, C., and Robinson, B. H. (1987) Direct detection of very slow two-jump processes by saturation recovery electron paramagnetic resonance spectroscopy. *J. Chem. Phys.* **87**, 2478-2488.
Sur, S. K., and Bryant, R. G. (1995) Nuclear- and Electron-Spin Relaxation Rates in Symmetrical Iron, Manganese, and Gadolinium Ions. *J. Phys. Chem.* **99**, 6301-6308.
Suter, D., and Ernst, R. R. (1982) Spectral spin diffusion in the presence of an extraneous dipolar reservoir. *Phys. Rev. B* **25**, 6038-6041.

Suter, D., and Ernst, R. R. (1985) Spin diffusion in resolved solid-state NMR spectra. *Phys. Rev. B* **32**, 5608-5627.

Szczepaniak, L. S., and Bryant, R. G. (1991) Proton nuclear magnetic relaxation in aqueous copper(II) amine chelate complexes. *Inorg. Chim. Acta* **184**, 7-11.

Tabti, T., Goldman, M., Jacquinot, J.-F., Fermon, C., and Le Goff, G. (1997) Relaxation without spin diffusion in fractal systems: Polymers in glassy solutions. *J. Chem. Phys.* **107**, 9239-9251.

Torchia, D. A., (1984) Solid State NMR Studies of Protein Internal Dynamics. *Ann. Rev. Biophys. Bioeng.* **13**, 125-144.

Toropov, Yu. V., Dzuba, S. A., Tsvetkov, Yu. D., Monaco, V., Formaggio, F., Crisma, M., Toniolo, C., and Raap, J. (1998) Molecular Dynamics and Spatial Distribution of TOAC Spin-Labelled Peptaibols Studied in Glassy Liquid by Echo-Detected EPR Spectroscopy. *Appl. Magn. Reson.* **15**, 237-246.

Tóth, É., Connac, F., Helm, L., Adzamli, K., and Merbach, A. E. (1998) ^{17}O-NMR, EPR, and NMRD Characterization of [Gd(DTPA-BMEA)(H_2O)]: A Neutral Contrast Reagent. *Eur. J. Inorg. Chem.* 2017-2021.

Tóth, É., Helm, L., Kellar, K. E., and Merbach, A. E. (1999) Gd(DTPA-bisamide)alkyl Coplymers: A Hint for the Formation of MRI Contrast Reagents with Very High Relaxivity. *Chem. Eur. J.* **5**, 1202-1211.

Townsend, M. G., and Weissman, S. I. (1960) Possible Symptom of the Jahn-Teller Effect in the Negative Ions of Coronene and Triphenylene. *J. Chem. Phys.* **32**, 309-313.

Trinbgali, A. E., and Brenner, H. C. (1998) Spin-lattice relaxation and ODMR line narrowing of the photoexcited state of pyrene and polycrystalline Shpol'skii hosts and glassy matrices. *Chem. Phys.* **226**, 187-200.

Tritt-Goc, J., Goslar, J., Hilczer, W., Hoffmann, S. K., and Augustyniak, M. A. (1993) Electron Spin Echo Envelope Modulation Analysis of SeO_3^- Radical in $(NH_4)_3H(SeO_4)_2$ Single Crystal. *Acta Physica Polonica A* **84**, 1131-1141.

Tse, D., and Hartmann, S. R. (1968) Nuclear Spin-Lattice Relaxation via Paramagnetic Centers without Spin Diffusion. *Phys. Rev. Lett.* **21**, 511-514.

Tsvetkov, Yu. D. (1983) Electron Spin Echo Study of the Steric Characteristics of the Formation and Reactions of Radicals in a Solid Phase. *Russ. Chem. Rev.* **52**, 866-880. (1514-1537 in Russian)

Tsvetkov, Yu. D. (1990) Physicochemical Applications of the Electron Spin Echo Method. *Bulletin Acad. Sci. USSR, Div. Chem. Sci.* **39**, 2158-2172.

Tsvetkov, Yu. D. (1998) Electron Spin Echo in Siberia. Chapter I.4 in *Foundations of Modern EPR*, Eaton, G. R., Eaton, S. S., and Salikhov, K. M. eds., World Scientific, Singapore.

Tsvetkov, Yu. D. and Dzyuba, S. A. (1990) Pulsed ESR and molecular motions. *Appl. Magn. Reson.* **1**, 179-194.

Valiev, K. A., and Zaripov, M. M. (1962) On the theory of spin-lattice relaxation in liquid solutions of electrolytes. *Soviet Phys. JETP* **14**, 545-549.

Van de Ven, F. J. M. (1995) Multidimensional NMR in Liquids – Basic Principles and Experimental Methods. Wiley-VCH, New York, page 109.

Van Kooten, J. F. C., and Schmidt, J. (1985) Spin-spin and spin-lattice relaxation in the triplet state of AB-pairs in isotopically mixed crystals of naphthalene. *Mol. Phys.* **55**, 351-361.

Van't Hof, C. A., and Schmidt, J. (1979) The effect of spectral diffusion on the phase coherence of phosphorescent triplet spins. *Mol. Phys.* **38**, 309-320.

Van Vleck, J. H. (1960) The Puzzle of Spin-Lattice Relaxation at Low Temperatures, in *Quantum Electronics*, C. H. Townes, ed., Columbia Univ. Press, N.Y.

Van Vleck, J. H. (1961) Recent Developments in Spin-Lattice Relaxation, in *Advances in Quantum Electronics*, J. R. Singer, ed., Columbia Univ. Press, N.Y.

Van Wyk, J. A., Reynhardt, E. C., High, G. L., and Kiflawi, I. (1997) The dependences of ESR line widths and spin-spin relaxation times of single nitrogen defects on the concentration of nitrogen defects in diamond. *J. Applied Phys. D. Applied Physics* **30**, 1790-1793.

Venkataraman, B. (1982) Time Resolved Electron Spin Resonance Spectroscopy. *Current Science* **51**, 397-400.

Vernoux, D., Zinsou, P. K., Zaripov, M., Ablart, G., Pescia, J., Misra, S. K., Rakhmatullin, R., and Orlinskii, S. (1996) Electron spin lattice relaxation of Yb^{3+} and Gd^{3+} ions in glasses. *Appl. Magn. Reson.* **11**, 493-498.

Vugman, N. V., and Amaral, M. R., Jr. (1990) Influence of localized anharmonic vibrations on electron paramagnetic resonance spectroscopy. *Phys. Rev. B* **42**, 9837-9842.

Vugman, N. V., Grillo, M. L. Netto, and Jain, V. K. (1992) Dynamic effects in Ni(III) and Pd(III) cyanides in KCl and NaCl studied by EPR spectroscopy. *Chem. Phys. Lett.* **188**, 419-422.

Wajnberg, E., and Alves, O. C. (1996) Spin-Lattice Relaxation of Denatured Nitrosyl Hemoproteins. *J. Magn. Reson. B* **113**, 119-124.

Wajnberg, E., Alonso, A., Nascimento, O. R., and Tabak, M. (1992) Spin-lattice relaxation of a nitroxide radical in a single crystal. *Chem. Phys. Lett.* **199**, 111-118.

Wang, J., Jack, K. S., and Natansohn, A. L. (1997) Spin diffusion and spin-lattice relaxation in multiphase polymers. *J. Chem. Phys.* **107**, 1016-1020.

Weber, R., Harbridge, J., Eaton, G. R., and Eaton, S. S., unpublished results.

Weger, M. (1960) Passage Effects in Paramagnetic Resonance Experiments. *The Bell System Technical Journal* **39**, 1013-1112.

Weibull, W. (1951) A Statistical Distribution Function of Wide Applicability. *J. Appl. Mechanics: Trans. ASME Ser. E* **18**, 293-297.

Weiden, N., Käss, H., Dinse, K.-P. (1999) Pulse Electron Paramagnetic Resonance (EPR) and Electron-Nuclear Resonance (ENDOR) Investigation of $N@C_{60}$ in Polycrystalline C_{60}. *J. Phys. Chem. B.* **103**, 9826-9830.

Weigand, F., Demco, D. E., Blümich, B., and Spiess, H. W. (1996) Spatially Resolved NMR Spin Diffusion in Solid Polymers. *J. Magn. Reson. A* **120**, 190-200.

Weil, J. A., Bolton, J. R., Wertz, J. E. (1994). *Electron Paramagnetic Resonance*, Wiley, N. Y., 100-105, 213-234.

Weissman, S. I., Feher, G., and Gere, E. A. (1957) The Spin Relaxation Time of Triphenylmethyl at Low Temperatures. *J. Am. Chem. Soc.* **79**, 5584-5585.

Wells, G. B., Yim, M. B., and Makinen, M. W. (1992) The influence of oxygen-17 enriched oxygen-donor ligands on the electronic spin relaxation behavior of paramagnetic metal ions. *Mol. Phys.* **75**, 1285-1300.

Wenckebach, W. Th., and Poulis, N. J. (1973) Spin Dynamics in Paramagnetic Crystals. , 17th Congress Ampere, V. Hovi, ed., North-Holland Publ., pp. 120-142.

Whalley, E. and Labbé, H. J. (1969) Optical Spectra of Orientationally Disordered crystals. III. Infrared Spectra of the Sound Waves. *J. Chem. Phys.* **51**, 3120-3127.

Williams, F. I. B., Krupka, D. C., and Breen, D. P. (1969) Relaxation in a Jahn-Teller System. II. *Phys. Rev.* **179**, 241-254.

Wilson, G. E., Jr., and Cohn, M. (1977) Magnetic Resonance Studies of the Manganese Guanosine Di- and Triphosphate Complexes with Elongation Factor Tu. *J. Biol. Chem.* **252**, 2004-2009.

Wilson, R., and Kivelson, D. (1966a) ESR Linewidths in Solution. I. Experiments on Anisotropic and Spin-Rotational Effects. *J. Chem. Phys.* **44**, 154-168.

Wilson, R., and Kivelson, D. (1966b) ESR Linewidths in Solution. III. Experimental Study of the Solvent Dependence of Anisotropic and Spin-Rotational Effects. *J. Chem. Phys.* **44**, 4440-4444.

Wilson, R., and Kivelson, D. (1966c) ESR Linewidths in Solution. IV. Experimental Study of Anisotropic and Spin-Rotational Effects in Copper Complexes. *J. Chem. Phys.* **44**, 4445-4452.

Witowski, A. M. and Bardyszewski, W. (1995) Spin-Lattice relaxation in high magnetic fields – beyond the Debye approximation. *Solid State Commun.* **94**, 9-12.

Witowski, A. M., Kutter, C., and Wyder, P. (1997) Spin-Lattice Relaxation in High Magnetic Fields: A Tool for Electron-Phonon Coupling Studies. *Phys. Rev. Lett.* **78**, 3951-3954.

Wolf, E. L. (1966) Diffusion Effects in the Inhomogeneously Broadened Case: High-Temperature Saturation in the F-Center Electron Spin Resonance. *Phys. Rev.* **142**, 555-569.

Wolfe, J. P. (1973) Direct Observation of a Nuclear Spin Diffusion Barrier. *Phys. Rev. Lett.* **31**, 907-910.

Woodward, A. E., and Bovey, F. A., eds. (1980) *Polymer Characterization by ESR and NMR*, American Chemical Society, Washington, D.C.

Yim, M. B., Kuo, L. C., Makinen, M. W. (1982) Determination of the Zero-Field Spltting Constants of High-Spin Metalloproteins by a Continuous-Wave Microwave Saturation Technique. *J. Magn. Reson.* **46**, 247-256.

Yin, J.-J., and Hyde, J. S. (1987) Spin-Label Saturation-Recovery Electron Spin Resonance Measurements of Oxygen Transport in Membranes. *Z. Physikal. Chem Neue Folge* **153**, 57-65.

Yin, J.-J., and Hyde, J. S. (1989) Use of high observing power in electron spin resonance saturation-recovery experiments in spin-labeled membranes. *J. Chem. Phys.* **91**, 6029-6035.

Yin, J.-J., Pasenkiewicz-Gierula, M., and Hyde, J. S. (1987) Lateral Diffusion of lipids in membranes by pulse saturation recovery electron spin resonance. *Proc. Natl. Acad. Sci. USA* **84**, 964-968.

Yin, J.-J., Feix, J. B., and Hyde, J. S. (1988) Solution of the nitroxide spin-label spectral overlap problem using pulse electron spin resonance. *Biophys. J.* **53**, 525-531.

Yin, J.-J., Feix, J. B., and Hyde, J. S. (1990) Mapping of collision frequencies for stearic acid spin-labels by saturation-recovery electron paramagnetic resonance. *Biophys. J.* **58**, 713-720.

Zahlan, A. B., editor (1968) Excitons, Magnons and Phonons in Molecular Crystals. *Proceedings of an International Symposium Held at the Physics Department of the American University of Beirut, Lebanon*. Cambridge University Press. See page 65 for phonon density of states curves.

Zecevic, A., Eaton, G. R., Eaton, S. S. and Lindgren, M. (1998) Dephasing of electron spin echoes for nitroxyl radicals in glassy solvent by non-methyl and methyl protons. *Mol. Phys.* **95**, 1255-1263.

Zhang, W., and Cory, D. G. (1998) First Direct Measurement of the Spin Diffusion Rate in a Homogeneous Solid. *Phys. Rev. Lett.* **80**, 1324-1327

Zhao, J., Chin, Y. H., Liu, Y., Jones, A. A., Inglefield, P. T., Kambour, R. P., and White, D. M. (1995) Deuterium NMR Study of Phenyl Group Motion in Glassy Polystyrene and a Blend of Polystyrene with Polyphenylene Oxide. *Macromolecules* **28**, 3881-3889.

Zhidomirov, G. M. and Salikhov, K. M. (1969) Contribution to the theory of spectral diffusion in magnetically diluted solids. *Sov. Phys. JETP* **29**, 1037-1040.

Zhou, Y., Bowler, B., Eaton, S. S., and Eaton, G. R. (1999) Electron spin relaxation rates for $S = \frac{1}{2}$ molecular species in glassy matrices and magnetically dilute solids at temperatures between 10 and 300 K. *J. Magn. Reson.* **139**, 165-174.

Zhou, Y., Bowler, B., Eaton, G. R., and Eaton, S. S. (2000) Electron Spin-Lattice Relaxation Rates for High-Spin Fe(III) Complexes in Glassy Solvents at Temperatures between 6 and 298 K. *J. Magn. Reson.* in press.

Ziman, J. M. (1960) *Electrons and Phonons. The Theory of Transport Phenomena in Solids.* Oxford University Press, London. See especially Chapter 1.

Zimmerman, R., Spiering, H., and Ritter, G. (1974) Mossbauer spectra of the tetrakis-(1,8-naphthyridine)iron(II)perchlorate in external magnetic fields. Evidence of slow relaxation in paramagnetic iron(II). *Chem. Phys.* **4**, 133-141.

Zinsou, P. K., Vergnoux, D., Ablart, G., Pescia, J., Misra, S. K., and Berger, R. (1996) Temperature and concentration dependences of the spin-lattice relaxation rate in four borate glasses doped with Fe_2O_3. *Appl. Magn. Reson.* **11**, 487-492.

Chapter 3

Structural Information from CW-EPR Spectra of Dipolar Coupled Nitroxide Spin Labels

Eric J. Hustedt and Albert H. Beth
Department of Molecular Physiology and Biophysics
Vanderbilt University
Nashville, TN 37232

Abstract: Site directed spin labeling has emerged as a powerful technique for determining structural features of proteins. By incorporating two spin labeled side chains, distances between elements of secondary structure can be determined by quantitation of spin-spin interactions between the probes. Recent advances in methods for extracting both the distance between spin labeled probes and their relative orientations are providing the capability for testing and refining accurate structural models for a wide variety of proteins. These methods take advantage of multifrequency EPR and global non-linear data analysis tools. The capabilities of current analytical methods are described for three relevant models of spin-spin interactions and directions for further refinement of these methods for future applications are described. Reliable structural information can be obtained for interelectron distances ranging from 5 to 20 Å and up to 25 Å using perdeuterated probes.

1. OVERVIEW OF SITE DIRECTED SPIN LABELING

Since its introduction by Hubbell and coworkers (Altenbach et al., 1990) site directed spin labeling (SDSL) has emerged as a valuable approach for determining the structures, at the level of the backbone fold, of a wide range of soluble and membrane bound proteins. The general approach of SDSL, its capabilities and limitations, and the proteins that have been investigated have been discussed in a number of recent reviews (Millhauser, 1992;

Biological Magnetic Resonance, Volume 19: Distance Measurements in Biological Systems by EPR, edited by Berliner et al. Kluwer Academic/Plenum Publishers, New York, 2000

Hubbell and Altenbach, 1994b; Hubbell et al., 1998; Feix and Klug, 1999) and in several chapters in this volume. More recently, there has been an increasing emphasis on determining tertiary and quaternary structural features of proteins through the observation of dipolar interactions between two nitroxide spin labels either introduced on the same monomer, on different monomers of an oligomeric protein, or on different subunits in a complex assembly of proteins. Dipolar interactions between nitroxides are observable for interspin distances up to 25 Å using CW-EPR (Rabenstein and Shin, 1995). As discussed in subsequent chapters in this volume, considerably longer distances can be measured using pulsed EPR methods.

The range of distances measurable by CW-EPR provides a useful ruler for determining the spatial arrangement of known elements of secondary structure in a protein. This fact, together with the high sensitivity of electron paramagnetic resonance (EPR) spectroscopy, the ability to determine structural properties of proteins in their native environments (i.e. in solution, in membranes, etc), and the ability to observe the time course of structural transitions on msec and longer timescales (Shin et al., 1993; Qu et al., 1997; Sienkiewicz et al., 1999) have all contributed to the emergence of double site directed spin labeling (DSDSL) as the method of choice for many different structural problems which are not amenable to characterization using classical techniques such as nuclear magnetic resonance or X-ray crystallography. The applications of DSDSL reported to date have demonstrated the utility and flexibility of the approach and have opened the door for a vast array of future applications as discussed at length in recent reviews (Millhauser, 1992; Hubbell and Altenbach, 1994b; Hubbell et al., 1998; Feix and Klug, 1999; Hustedt and Beth, 1999) and by other authors in this volume.

As applications of DSDSL have emerged, it has become apparent that there is a need to expand capabilities for extracting accurate distance and orientation information from CW-EPR data obtained from spin-spin coupled probes. The accuracy and precision of the measurement of these parameters will depend on factors that are intrinsic to the biological system under investigation, factors that are subject to experimental manipulation, and the methods employed to analyze the data. The average distance between the probes and the distance distribution, the relative orientation of the probes and the uniqueness of this orientation, the global rotational dynamics of the labeled protein, and the local dynamics of the spin probe all have a profound effect upon the CW-EPR lineshapes observed for dipolar coupled spin labels as discussed in a previous review (Hustedt and Beth, 1999). This last statement contains both good news and bad news, depending upon one's point of view. Specifically, in theory, it should be possible to extract a remarkable amount of structural information (both distance and orientation) from experimental data obtained from dipolar coupled nitroxide spin labels. However, extraction of a distance from a spectrum can be difficult due to the complex way in which orientation and dynamics affect the lineshape.

The factors cited in the previous paragraph are, for the most part, intrinsic to the problem under investigation. There are additional factors, which are under direct experimental control that also determine the accuracy and precision of distance and orientation parameters. These include the use of multiple and high (greater than X-band) microwave frequencies, the use of perdeuterated and ^{15}N versus ^{14}N nitroxides, the nature of the nitroxide side chain employed, and the use of lowered temperature or high viscosity solutions to slow rotational dynamics.

If one is interested in measuring the dipolar coupling between residues *j* and *k* in a particular protein, then ideally the CW-EPR spectra of three different mutant proteins should be obtained: 1) a spin labeled mutant at residue *j*; 2) a spin labeled mutant at residue *k*; 3) the doubly spin labeled mutant at residues *j* and *k*. The presence or lack of dipolar coupling can be inferred by comparing the CW-EPR spectrum of the doubly labeled mutant to the sum of the spectra of the two singly labeled mutants (the spatially isolated nitroxides; e.g. Mchaourab et al., 1997b). In the case of a homooligomeric protein complex, dipolar coupling between the same residues on different monomers can be inferred by comparing the CW-EPR of the labeled protein with the spectrum of a mixture of labeled protein with excess wild type protein (e.g. Berengian et al., 1997). In many applications, the simple inspection of experimental data is sufficient to answer fundamental structural questions. The observation of spectral broadening or splitting due to dipolar coupling will provide a qualitative answer as to whether two elements of secondary structure are near one another in a single protein or whether two protein surfaces are adjacent in a protein-protein complex. Many examples of this level of analysis are present in the literature. Hubbell and coworkers have proposed a scheme whereby spin-spin interactions are classified as close (<10 Å), intermediate (10 – 15 Å), or long (>15 Å) with respect to the interprobe separation (Hubbell et al., 1998).

Perozo and coworkers have used a phenomenological parameter which is defined in terms of ratios of spectral amplitudes, to characterize the relative proximity of spin probes (Perozo et al., 1998). Approximate methods for determining internitroxide distances from the EPR spectra of dipolar coupled nitroxides using deconvolution (Rabenstein and Shin, 1995) or convolution (Steinhoff et al., 1997) methods have also been developed. These approaches can provide valuable structural insights in many experimental systems. However, the EPR spectrum of dipolar coupled probes contains a wealth of structural and dynamics information which, in principal, can be extracted and thereby provide extremely powerful constraints for constructing and refining structural models. When constructing structural models from a small number of distances, it is possible that the structure will be underdetermined and a number of different structures may satisfy the distance restraints. Therefore, it is advantageous to extract the greatest amount of information from each DSDSL pair in order to obtain the most accurate structural models using the fewest number of pairs. The advantages of obtaining combined distance and orientation information for building models is now well established in NMR (Tjandra and Bax, 1997) and EPR (Hanson et al., 1998). Therefore, the development of spectroscopic and analytical methods in order to obtain the maximum distance and orientation constraints from a single DSDSL pair is an important goal.

In this chapter, the capabilities of current analytical methods for extracting interprobe distances and relative orientations are discussed for three cases that are relevant to a wide range of DSDSL applications. In addition, directions for further improvement of analytical methods are discussed within the framework of experimental applications. The interested reader is referred to previous reviews for additional discussions of the general area of DSDSL and on the use of spin-spin interactions to obtain structural information (Eaton and Eaton, 1989; Hubbell and Altenbach, 1994b; Hubbell and Altenbach, 1994a; Millhauser et al., 1995; Hubbell et al., 1998; Feix and Klug, 1999; Hustedt and Beth, 1999).

2. THE THREE CASES

The majority of DSDSL studies reported to date can be grouped into one of three limiting cases (Hustedt and Beth, 1999).

Case 1: The nitroxides are rigidly bound to the protein or protein complex and adopt a unique internitroxide distance and orientation with respect to each other. The protein has a global rotational correlation time of approximately 1 μsec or longer. This assures that the EPR spectrum at conventional microwave frequencies is insensitive to global motions.

Case 2: The nitroxides adopt a static distribution of orientations and distances with respect to each other. Neither the global rotation of the protein, nor significant local rotational motions of the nitroxides occur on a timescale less than 1 μsec.

Case 3: The nitroxide side chains are undergoing significant local motions or the protein itself is undergoing global rotational diffusion on a timescale to which linear EPR is sensitive.

Ideally, all DSDSL studies would fall into case 1 since analytical methods for extracting accurate and precise distance and orientation information have already been developed for this case (Hustedt et al., 1997). However, global rotational diffusion, internal protein dynamics, side chain dynamics, and inherent orientational disorder will mean that many, if not most, experimental situations will fall under case 2 or case 3. Given the potential wealth of structural information that can be extracted for cases 2 and 3, there is strong motivation to develop methods for analyzing data for these cases as well. There is also strong motivation for adjusting experimental conditions when appropriate to reduce the extent to which rotational dynamics can reduce the resolution of resolved dipolar splittings. Factors that determine which of the three cases applies include the correlation time for the global rotational diffusion of the protein or protein complex, the timescale and amplitude of any internal protein dynamics, and the timescale and amplitude of local probe motion. The global rotational diffusion time can be increased by freezing (Rabenstein and Shin, 1995), cooling to near liquid nitrogen temperature in solvent systems that form a glass (Hanson et al., 1998), or by adding agents such as glycerol or sucrose to significantly increase the viscosity (Beth et al., 1984). All of the methods will also slow internal protein and local probe motions without necessarily altering any distribution of interprobe distance and orientation. In this event, small soluble proteins which in aqueous solution at temperatures above 0° C would fall into case 3 can be converted to case 2 or possibly case 1. Binding a molecule to a solid matrix has been shown to be extremely effective at slowing global rotational diffusion, however it may (Hustedt and Beth, 1999) or may not (Hustedt et al., 1995) slow local dynamics depending on circumstances that are difficult to predict *a priori*. Any method used to slow protein or probe dynamics does have the potential to perturb the structure of the system being studied. Examining the same system under different experimental conditions can be used to check for such perturbations.

There are two factors that determine the extent of local probe motion and the extent to which the nitroxide side chain will adopt a distribution of orientations relative to the protein backbone. These are the chemical structure of the spin label (and thus the nature of the linkage between the spin label and the protein) and the location of the probe in the secondary and tertiary structure of the protein (Mchaourab et al., 1996; Mchaourab et al., 1999). Of all of the various spin labels that can be used to modify cysteine residues, the most commonly used is 3-(methanesulfonylthiomethyl)-2,2,5,5-tetramethylpyrrolidin-1-yloxy, or methane thiosulfonate spin label (MTSSL; Berliner et al., 1982). Extensive studies by Hubbell and coworkers (Mchaourab et al., 1996; Mchaourab et al., 1999) have shown that the MTSSL disulfide linkage to the protein backbone is more rigid than that of other types of spin labels and presumably, MTSSL adopts a smaller range of relative orientations with respect to the backbone. The 4-methyl derivative of MTSSL has been shown to have an even lower degree of mobility due to increased steric interactions of the additional methyl group (Mchaourab et al., 1996). Finally, an intrinsically spin labeled unnatural amino acid, TOAC, which when incorporated into peptides is rigidly linked to the C_α carbon of the backbone has proven to have a wide range of applications (Hanson et al., 1996a; Hanson et al., 1996b; Hanson et al., 1998; Anderson et al., 1999; Monaco et al., 1999a; Monaco et al., 1999b). In addition to those mentioned above, there are a wide variety of probes available that exhibit differing steric and flexibility properties (Hideg and Hankovszky, 1989) and which can be used to provide an additional test of structural models.

2.1 Distinguishing Between the Three Cases

Given the EPR spectra of the singly spin labeled mutants at site *j*, S_j, and site *k*, S_k, and the EPR spectrum of the doubly spin labeled mutant, S_{j+k}, how does one determine which of these cases applies? There is extensive knowledge of the effects of rotational dynamics on nitroxide lineshapes (McCalley et al., 1972; Schneider and Freed, 1989; Mchaourab et al., 1996; Mchaourab et al., 1999). Inspection of the two spectra of the singly labeled mutants, should give a clear indication of whether the labels are undergoing significant rotational dynamics. If it appears they are, then case 3 applies. If not, the two singly labeled spectra should be analyzed using procedures designed for fitting rigid limit nitroxide spectra (Fajer et al., 1990; Hustedt et al., 1993). Good fits to the spectra S_j and S_k using a rigid limit model will confirm that either case 1 or case 2 applies, and the fits will give values for the **A**- and **g**-tensors which will be necessary for further analysis of the spectrum of the doubly labeled mutant.

Distinguishing between case 1 and case 2 will largely depend on the degree to which resolved splittings can be observed. The observation of sharp, well-resolved splittings in the spectrum of the double mutant, S_{j+k}, is an indication that case 1 may apply. As discussed below, a broad distribution of orientations will eliminate these well-resolved splittings resulting in a significant broadening of the dipolar coupled spectrum. The detection of dipolar coupling induced broadening will require a direct comparison of the spectrum of the double mutant, S_{j+k}, with the sum of the spectra of the single mutants, $S_j + S_k$. It should be emphasized that the extent to which well-resolved splittings versus broadening can be observed is strongly dependent on the isotopic composition of the nitroxides. As discussed below, the resolution of splittings can be greatly enhanced by using perdeuterated nitroxides. Many perdeuterated spin labels, including MTSSL, are now commercially available making their routine use possible.

When interprobe distances exceed ~20 Å, the major effect on the lineshape is overall broadening of the lineshape. This situation offers considerable challenge in distinguishing between case 1 and case 2. At these larger distances, it may be advantageous to adopt time-domain techniques for determining distances. The optimal approach to combining time domain and CW techniques remains to be determined in the intermediate (~20 to 30 Å) distance range.

3. THEORETICAL BASIS

The spin Hamiltonian for a pair of dipolar coupled nitroxides is given by,

$$\hat{\mathscr{H}} = \beta_e \vec{H}_0 \cdot \mathbf{g}_1 \cdot \hat{S}_1 + \beta_e \vec{H}_0 \cdot \mathbf{g}_2 \cdot \hat{S}_2 - \omega_n (I_{1z} + I_{2z}) + \gamma_e \hat{I}_1 \cdot \mathbf{A}_1 \cdot \hat{S}_1 + \gamma_e \hat{I}_2 \cdot \mathbf{A}_2 \cdot \hat{S}_2 + \hat{S}_1 \cdot \mathbf{D} \cdot \hat{S}_2 + J \hat{S}_1 \cdot \hat{S}_2 \quad (1)$$

where β_e is the Bohr magneton, γ_e is the electron gyromagnetic ratio, $\mathbf{g}_1$ and $\mathbf{g}_2$ are the tensors defining the interaction of the electron spin of nitroxide 1 ($\hat{S}_1$) and nitroxide 2 ($\hat{S}_2$) with the applied magnetic field ($\vec{H}_0$), ω_n is the Larmor frequency of the nitrogen nucleus, $\mathbf{A}_1$ and $\mathbf{A}_2$ are the hyperfine tensors defining the interaction of the nitrogen nuclear spins ($\hat{I}_1$ or $\hat{I}_2$) with $\hat{S}_1$ or $\hat{S}_2$, $\mathbf{D}$ is the dipolar coupling tensor, and J is the scalar exchange interaction. Two fundamental properties serve to make nitroxides ideal spin labels for SDSL studies in general and DSDSL studies in particular.

1) The unpaired electron density is highly localized to the N-O bond of the nitroxide and the interelectron distances to be observed are considerably larger than the size of the molecular orbital. As a result, the point dipole approximation is valid for the unpaired electron and the position of the unpaired electron can be taken to be a point dipole along the N-O bond approximately equidistant from the nitrogen and oxygen (Mustafi et al., 1991).

2) The principal axes of the **g**- and **A**-tensors of a spin label are essentially coincident and are well-defined within the molecular frame of the nitroxide, with g_{xx} along the N-O bond and g_{zz} parallel to the nitrogen p_z orbital (Lajzerowicz-Bonneteau, 1976). As a result, the determination of the relative orientation of two nitroxides with respect to the principal axis frames of their respective magnetic tensors is equivalent to a determination of the relative orientation of these two nitroxides with respect to their molecular frames.

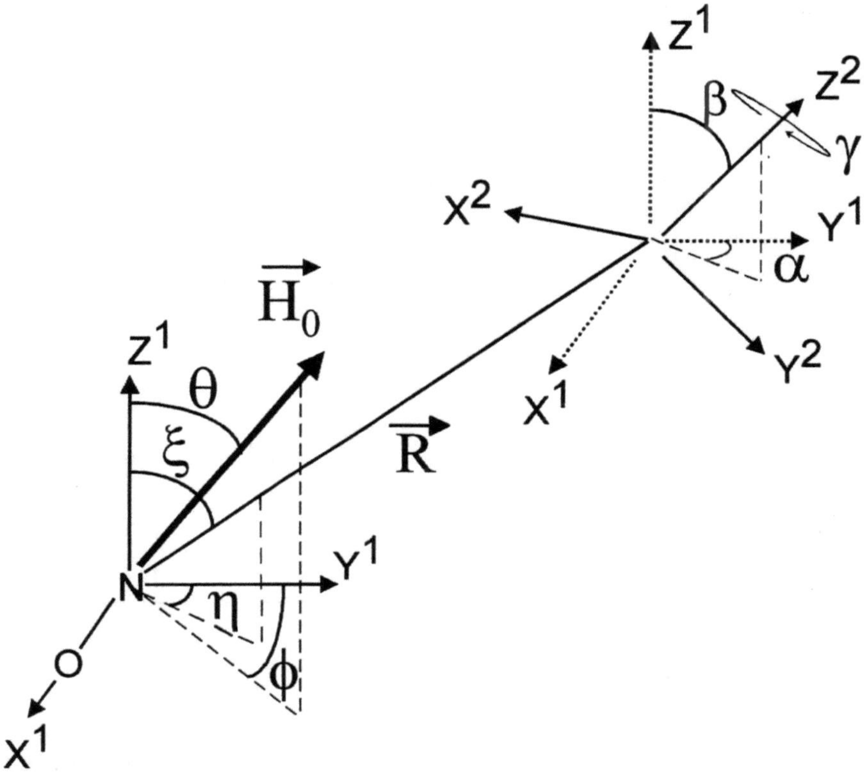

Figure 1. Diagram illustrating the two angles (ξ,η) relating the interelectron vector, R, to the axis frame defined by nitroxide 1, and the three angles (α,β,γ) relating the orientation of nitroxide 2 to nitroxide 1. The angles θ and φ define the orientation of the external magnetic field, H_0, with respect to the axis frame of nitroxide 1. For simplicity, the interelectron vector is drawn from nitrogen to nitrogen. In actuality, the position of the electron point dipole is approximately half way along the nitrogen - oxygen bond of each nitroxide.

The elements of the various tensors in the Hamiltonian in the laboratory frame are determined by three different Euler angle rotations as shown in Figure 1. Using the high field approximation so that all non-secular terms can be ignored and also under the assumption that the small nuclear Zeeman terms can be neglected the effective Hamiltonian becomes,

$$\begin{aligned}\hat{\mathcal{H}} = {} & g_{1zz}\ \beta_e H_0\ S_{1z} + g_{2zz}\ \beta_e H_0\ S_{2z} \\ & + \gamma_e\left(A_{1xz}\ I_{1z} S_{1z} + A_{1yz}\ I_y S_z + A_{1zz} I_{1z} S_{1z}\right) \\ & + \gamma_e\left(A_{2xz} I_{2x} S_{2z} + A_{2yz} I_{2y} S_{2z} + A_{2zz} I_{2z} S_{2z}\right) \\ & - \frac{(D+J)}{4}\ S_{1+} S_{2-} - \frac{(D+J)}{4}\ S_{1-} S_{2+} + (D+J)\ S_{1z} S_{2z}\end{aligned} \qquad (2)$$

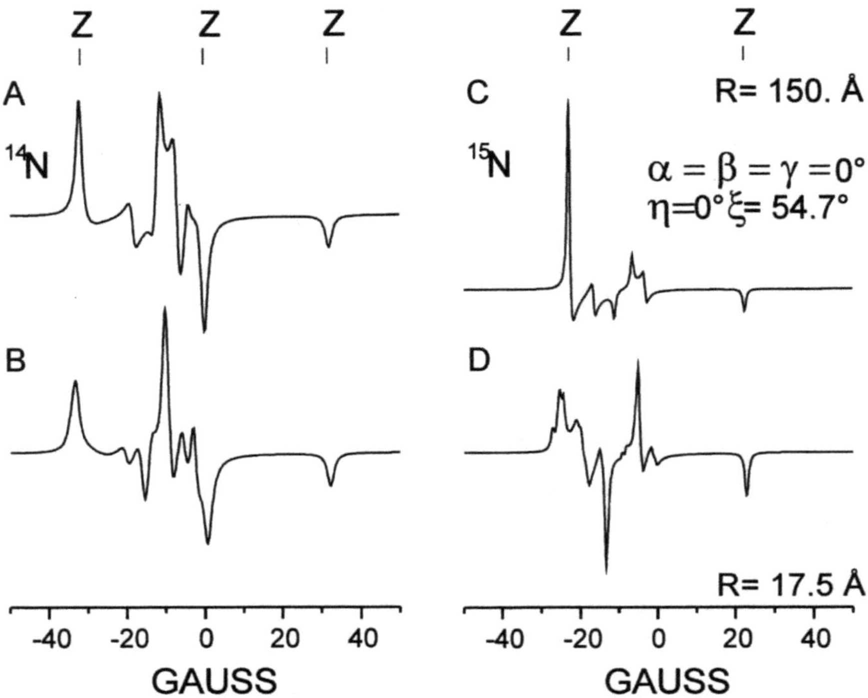

Figure 2. Calculated X-band CW-EPR spectra for dipolar coupled [^{14}N,D] (A and B) and [^{15}N,D] nitroxides (C and D) assuming $\alpha = \beta = \gamma = \eta = 0°$ and $\xi = 54.7°$. The upper spectra (A and C) were calculated for R = 150 Å ≈ ∞, the lower spectra (B and D) were calculated for R = 17.5 Å. The *z*-turning points of the spectra are labeled at top.

Finally, by neglecting the low probability "forbidden" transitions for which the change in nuclear quantum number for either of the nitroxide nitrogen nuclei is non-zero, the CW-EPR spectrum of dipolar coupled nitroxides in the rigid limit is calculated as a sum of Lorentzian lines centered at the calculated resonant field values as a function of the angles θ and ϕ (for additional details see (Hustedt et al., 1997). The resulting spectrum, $S(R,\Omega,\Gamma,\sigma)$, is a function of the interelectron distance, R, the interelectron geometry, $\Omega(\xi,\eta,\alpha,\beta,\gamma)$, and the Lorentzian and Gaussian contributions to the linewidth, Γ and σ respectively. Examples of spectra calculated from this Hamiltonian are shown in Figure 2. These spectra provide an interesting illustration of how the complex interplay between the anisotropic **A-**, **g-**, and **D**-tensors can be used to extract detailed information on the interprobe geometry. In these examples α, β, γ, and η are all equal to 0°, while the angle ξ is set to the magic angle. At the *z*-turning points ($\theta = 0°$), the effective dipolar coupling is proportional to $(1-3\cos^2\xi)=0$. However, at the *x*- and *y*-turning points the dipolar coupling is determined by $(1-3\cos^2\mu)\neq 0$, where μ is a function of ξ,θ, and ϕ (Hustedt et al., 1997). Even at this non-optimal geometry, $\xi = 54.7°$, resolved dipolar splittings can be observed at the *x*- and *y*-turning points of an isotropic powder pattern spectrum. In fact, there is no interprobe orientation at which there is a "blind spot" to the dipolar coupling for a randomly oriented sample. Instead, those regions of the spectrum that show maximal and minimal dipolar splittings are precisely that aspect of the data that contains information on the interprobe orientation.

The procedure described above assumes that the two nitroxides adopt a static, unique orientation with respect to each other. To treat cases where either the nitroxides are rotationally immobile and adopt a static distribution of orientations and distances with respect to each other, or the nitroxides are rotationally mobile, alternate methods for simulating spectra must be employed. Approximate methods have been developed to treat the case where the two nitroxides are immobile but do not adopt a unique interelectron orientation (Rabenstein and Shin, 1995; Steinhoff et al., 1997). These methods assume that the distribution of interelectron orientations is isotropic and that the dipolar coupled spectrum can be calculated as the sum of the two experimentally determined spectra of the isolated nitroxides convoluted with a Pake pattern. An alternative approach is the direct simulation of spectra in which there is a distribution of interelectron distances and orientations $P(R,\Omega)$. In this case the spectrum can be calculated by summing over a finite number of orientations and distances.

$$S(\langle R\rangle,\langle\Omega\rangle,P(R,\Omega),T_{2e},\sigma) = \sum_{i=1}^{N} S(R_i,\Omega_i,T_{2e},\sigma)P(R_i,\Omega_i) \tag{3}$$

The sum in Equation 3 can correspond to a small number (two or more) of discrete orientations and distances assuming that the two spin labels adopt discrete orientations corresponding to well-defined local minima. Alternatively, the Ω_i and R_i can correspond to a finite set of orientations and distances on a grid fine enough to represent a continuous distribution of orientations and distances. It should be emphasized that for two spin labels tethered to a peptide backbone, the distribution in side chain orientations will potentially result in a correlation between distance and interprobe orientation.

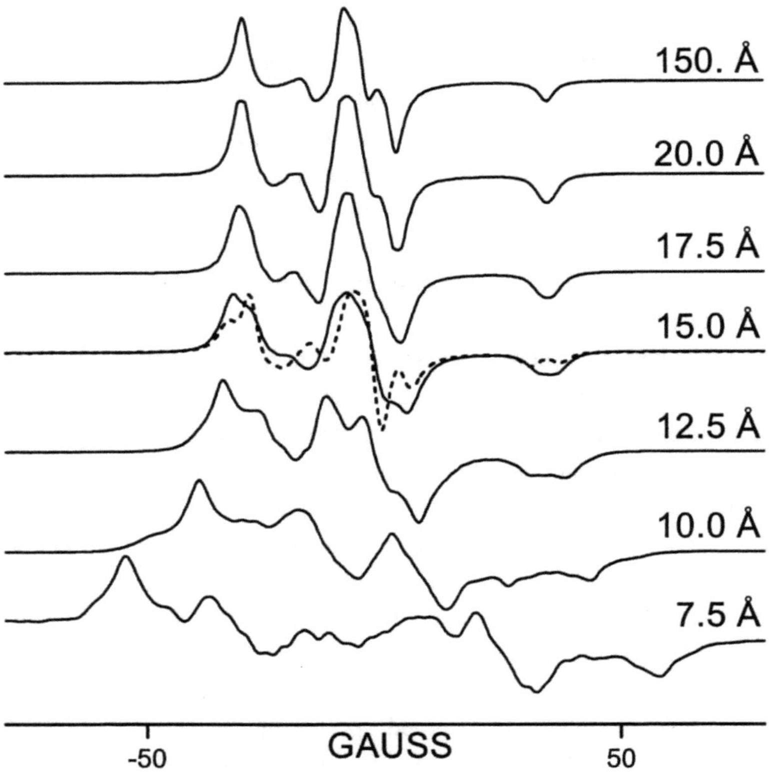

Figure 3. Calculated X-band CW-EPR spectra for dipolar coupled [^{14}N,H] nitroxides assuming a fixed distance, *R*, and an isotropic distribution of relative orientations between the two nitroxides. The dashed line is a calculated spectrum assuming a single relative orientation between the two nitroxides ($\alpha = \beta = \gamma = \eta = 0°$, $\xi = 90°$, and $R = 15$ Å). The Lorentzian linewidth for all calculations is 2 Gauss.

A simple model for treating orientational disorder between two nitroxides is to assume that they are held at a fixed distance *R* and that there is an isotropic distribution of relative orientations between them. This model is somewhat artificial in that it ignores any correlation between orientation and distance distributions mentioned above, but it gives an indication of the effect of orientational distribution on the resulting EPR spectrum. Figure 3 shows EPR spectra calculated according to Equation 3 assuming an isotropic distribution of all 5 angles (ξ, η, α, β, γ). The effect of a static distribution of distances and orientations is to cause shifts in the splittings at different turning points in the spectrum. As the width of the distribution of distances or orientations increases, the splittings, which are clearly evident at a single orientation, will be washed out and a broadened spectrum will be observed.

An equally important consideration is the effect of rotational dynamics on the EPR spectra of dipolar coupled nitroxides. For distances in the range of 10 to 20 Å, maximal dipolar splittings of 55 to 7 Gauss will be observed in a CW-EPR spectrum. This is the same range as the rigid limit widths of the different manifolds for ^{14}N and ^{15}N nitroxides at X-band. Thus, the timescale of global rotational dynamics at which changes in the X-band lineshape of an isolated nitroxide occur is the same as that which will average dipolar splittings. This timescale can be defined as follows. For isolated nitroxides at X-band, the rigid limit corresponds to correlation times greater than 500 to 1000 nsec. As the rotational correlation time decreases, the overall widths of the manifolds will decrease while features within each manifold will broaden. The three manifolds of a ^{14}N nitroxide (or the two manifolds of a ^{15}N nitroxide) will collapse to three single lines (or two for a ^{15}N nitroxide) at a rotational correlation time of 1 to 2 nsec. For shorter correlation times, the fast motion limit, the linewidths will decrease with correlation time. In the same way, rotational motions with correlation times less than 1 μsec will cause the broadening and ultimately the collapse of splittings due to dipolar coupling. For dipolar coupled nitroxides undergoing unrestricted rotational diffusion, in the fast motion limit, the traceless **D** tensor will average to zero.

Hustedt and coworkers adapted the method of Robinson and coworkers to simulate the EPR spectra of dipolar coupled nitroxides undergoing global rotational diffusion (Hustedt et al., 1997). Robinson and coworkers (Robinson et al., 1992) calculated CW-EPR spectra of spatially isolated nitroxides as the Fourier transform of a free induction decay (FID) generated by solving the time dependent Liouville equation for the spin Hamiltonian whose molecular orientation variables are stochastically modulated according to classical Brownian dynamics trajectories. With this method, the calculation of the FID for a given Brownian trajectory and the calculation of the trajectory itself are not mathematically linked. Simulations of CW-EPR spectra of dipolar coupled nitroxides undergoing rigid body global rotational diffusion with a fixed internitroxide distance and orientation assuming axial **A**- and **g**-tensors have been used to validate this method by demonstrating that: 1) in the limit of large interspin distances, this algorithm gives results which are in excellent agreement with simulations for single nitroxides calculated using programs developed by Freed and coworkers (Schneider and Freed, 1989); and 2) in the limit of long rotational correlation times, this algorithm gives results which are in excellent agreement with simulations calculated for dipolar coupled nitroxides in the rigid limit. Example calculations are shown in Figure 4. To date, this approach has only been used for the simple model described above. However, the algorithm for calculating spin dynamics that has already been developed can be used regardless of the complexity of the model used to describe the rotational dynamics of the spin labels.

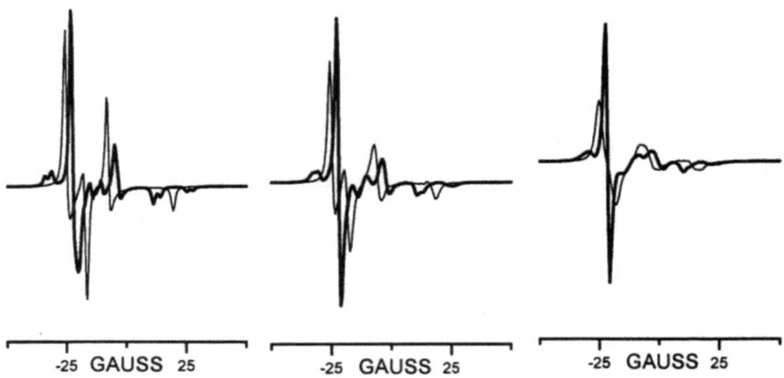

Figure 4. Simulated X-band EPR spectra of dipolar coupled [^{15}N,D] nitroxides undergoing rigid body isotropic rotational diffusion for interspin distances of 100 Å (thin lines) and 15 Å (thick lines). Rotational correlation times are 1μsec (left), 125 nsec (middle), and 20 nsec (right). Axial A- and g-tensors are used. $\alpha = \beta = \gamma = \xi = \eta = 0°$.

These early studies provide the basis for the further development of methods to simulate the effects of local side chain dynamics on DSDSL spectra. Using appropriate models of side chain dynamics, it should be possible to directly fit data to determine both the average interspin distance and the effective relative orientations of the interacting spin labels provided that the amplitudes of side chain dynamics is not so large as to substantially average the dipolar tensor. The apparent order parameters for the MTSSL side chain located in a globular protein at a variety of different positions in the tertiary structure strongly suggest that at most labeling sites the motion of the spin label will be sufficiently constrained so that dipolar coupling can be observed. For the case of isolated nitroxides, Steinhoff and Hubbell (Steinhoff and Hubbell, 1996) adapted the method of Robinson and coworkers (Robinson et al., 1992) to the calculation of EPR spectra for nitroxide side chain dynamics. Stochastic dynamics simulations of the MTSSL label in a polyleucine α–helix trimer were used to generate effective single particle potential energy functions for the nitroxide spin label. Using these effective potentials, Brownian trajectory calculations were used to generate CW-EPR spectra. These simulations were in qualitative agreement with representative spectra obtained from SDSL studies of bacteriorhodopsin (Steinhoff and Hubbell, 1996). This same approach, using effective single particle potential energy functions for both nitroxides, can be adapted to treat the case of dipolar coupled nitroxides.

4. ANALYSIS OF EXPERIMENTAL DATA FOR CASE 1

Analysis of experimental data for case 1 involves determining the interelectron separation and the five unique angles relating the interacting spin labels as shown in Figure 1. Unfortunately, there are not any straightforward approaches for determining any one of these six independent parameters from simple inspection of experimental data due to the complex way in which each contributes to the observed EPR lineshape of a disordered sample. Therefore, methods have been developed to automate the fitting of experimental data that are based upon global analysis and simulated annealing (Hustedt et al., 1997). Global analysis was adopted to simultaneously analyze data obtained at multiple microwave frequencies and to rigorously test the accuracy of and increase the precision of the structural parameters determined. The method of simulated annealing (Press et al., 1992) was adopted to find the global minimum in the complex χ^2 surface.

Global analysis was originally developed to simultaneously analyze multiple data sets from fluorescence experiments (Beechem et al., 1991; Beechem, 1992). As adapted to fit EPR data, it has proven extremely useful because of its extreme flexibility for allowing a subset of the parameters to vary from one data set to another (such as H_0 in a global analysis of data collected from a sample containing dipolar coupled nitroxides at multiple frequencies) while others are required to be the same for each data set (such as the **A**-, **g**-, and **D**-tensors). In addition, the subset of the parameters whose values are known can be fixed at specific values (H_0 is an example), while others are varied to minimize χ^2 ($\alpha,\beta,\gamma,\xi,\eta$, and R). The original global analysis, which relied on the Marquardt-Levenberg algorithm to perform the non-linear least squares analysis, has proven highly useful in analyzing a wide variety of EPR data (Hustedt et al., 1993; Hustedt and Beth, 1995; Hustedt et al., 1995; Hustedt and Beth, 1996). When fitting EPR spectra of dipolar coupled nitroxides, the χ^2 surfaces exhibit numerous local minima in which the Marquardt-Levenburg method becomes trapped. As a result, the analysis of the complex χ^2 surfaces as a function of $\alpha,\beta,\gamma,\xi,\eta$, and R required the incorporation of a simulated annealing algorithm into the program to make the analysis of data corresponding to case 1 routinely possible (Hustedt et al., 1997).

The capabilities of the simulated annealing approach for analysis of data for case 1 have been demonstrated in previous studies on glyceraldehyde-3-phosphate dehydrogenase (GAPDH; Hustedt et al., 1997), on model biradicals (Hustedt and Beth, 1999), and on dilabeled model peptides (Hanson et al., 1998). Calvo and coworkers have adopted this same approach to determine the separation and relative orientations of quinone radicals in the photoreaction center complex from Rb. sphaeroides (Calvo et al., 1999).

The study by Millhauser and coworkers on model peptides labeled with TOAC has shown that it is possible to recover reasonable values for R and the five angles even in the presence of strong exchange coupling (Hanson et al., 1998). Results from this same study on 16 residue peptides doubly labeled at positions 4 and 7 or 4 and 8 has clearly shown that the orientation information in combination with the measured distance can be used to build detailed structural models (Hanson et al., 1998). More work will be required to define the limitations of the analytical methods when the spin labels are very close (e.g. < 5 Å). However, based upon the studies published to date, it appears that reliable structural information can be obtained at least from the 5 Å range out to a distance of approximately 20 Å. At interspin distances greater than approximately 20 Å, there are no resolved spectral features in the CW-EPR spectrum that arise from dipolar coupling even for case 1 when using normal isotope spin labels. Though no previous work has focused upon the longer distance range, it is likely that one can obtain a reasonable estimate of the interprobe separation using the computational tools described above but that the orientational variables will not be determined with good precision. Resolved splittings can be observed out to distances of 25 Å when perdeuterated probes are utilized (Beth et al., 1984; Park and Trommer, 1989; Hustedt et al., 1997).

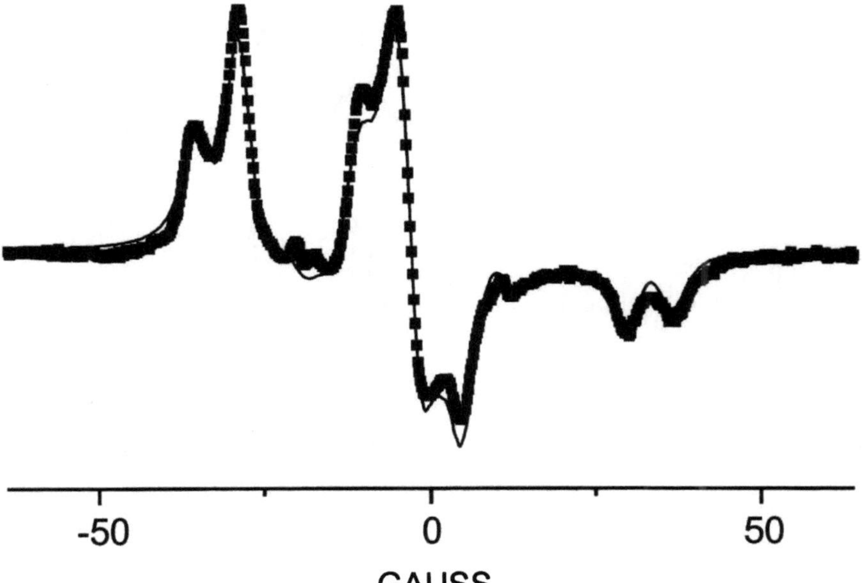

Figure 5. The EPR spectrum (squares) of the MTSSL labeled F138C mutant of HSP 27 analyzed according to case 1. The best fit (solid lines) parameters obtained are ξ= 81°, η= 92°, α= 249°, β= 174°, γ= 76°, and R= 13.9 Å. Spectra were obtained at 20° C in 20 mM MOPS, 50 mM NaCl, 0.1 mM EDTA, pH 7.2

An example of a situation that naturally falls into case 1 comes from the work of Mchaourab and coworkers (Mchaourab et al., 1997a) on the small heat shock protein HSP 27. This protein forms large multimeric complexes with molecular weights greater than 500 kD which is sufficient to reduce the global rotational diffusion rate of the complexes to the slow motion limit for X-band CW-EPR. SDSL studies of HSP 27 at residues 133 through 144 has demonstrated that this region forms a β–sheet and the observation of strong dipolar coupling between residues in different subunits labeled at the same site demonstrates that equivalent strands from different subunits are in close proximity across the subunit interface. Given the large molecular weight of these proteins and the possibility of strong steric interactions between side chains on adjacent subunits, it is not surprising that, as demonstrated by the results in Figure 5, the EPR spectrum of the MTSSL labeled F138C mutant of HSP 27 can be successfully analyzed according to case 1.

4.1 Factors Influencing the Precision of the Determined Parameters

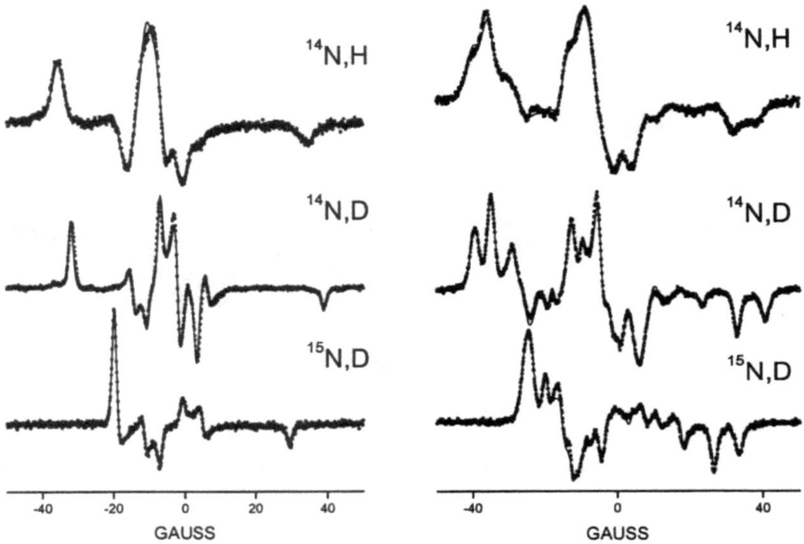

Figure 6. X-band EPR spectra (dots) of the SL-NAD$^+$/GAPDH complex at substoichiometic (left) and nearly full stoichiometric labelling (right) as a function of the isotopic composition of the nitroxides. The best fit (solid lines) Lorentzian, Γ, and Gaussian, σ, components to the linewidth of the spectra of the substoichiometrically labeled samples are as follows: [^{14}N,H] Γ= 2.1 Gauss and σ= 0.5 Gauss; [^{14}N,D] Γ= 0.5 Gauss and σ= 0.5 Gauss; and [^{15}N,D] Γ= 0.5 Gauss and σ= 0.5 Gauss. All spectra were collected at 4° C from polycrystalline ammonium sulfate precipitated samples, essentially identical spectra were observed from soluble samples in 50% (w/w) glycerol , 5 mM sodium phosphate, pH 7.5 (see Beth et al., 1984, for additional details).

The benefit of the use of isotopically substituted probes is demonstrated in Figures 6 and 7. The dramatic decrease in the Lorentzian linewidth due to perdeuteration is clearly evident from the X-band EPR spectrum of N^6-[$^{14}N,H_{17}$]-SL-NAD$^+$, N^6-[$^{14}N,D_{17}$]-SL-NAD$^+$, and N^6-[$^{15}N,D_{17}$]-SL-NAD$^+$ bound substoichiometrically to GAPDH. The effect of perdeuteration on the ability to resolve splittings due to dipolar coupling is demonstrated by the spectra of N^6-[$^{14}N,H_{17}$]-SL-NAD$^+$, N^6-[$^{14}N,D_{17}$]-SL-NAD$^+$, and N^6-

[$^{15}N,D_{17}$]-SL-NAD$^+$ stoichiometrically bound to GAPDH. The χ^2 curves in Figure 7 quantitatively demonstrate the effect of perdeuteration on the precision with which parameters can be determined. These curves were obtained by fixing the values of ξ or η to specific values while allowing all other parameters to vary in order to minimize χ^2. The horizontal lines define the 95% confidence intervals for the parameter in question as estimated from the F-statistic (Hustedt et al., 1993). Note that for all three nitroxides, the best fit values for the parameters ξ and η are essentially identical while the precision with which these parameters are determined is dramatically increased by perdeuteration.

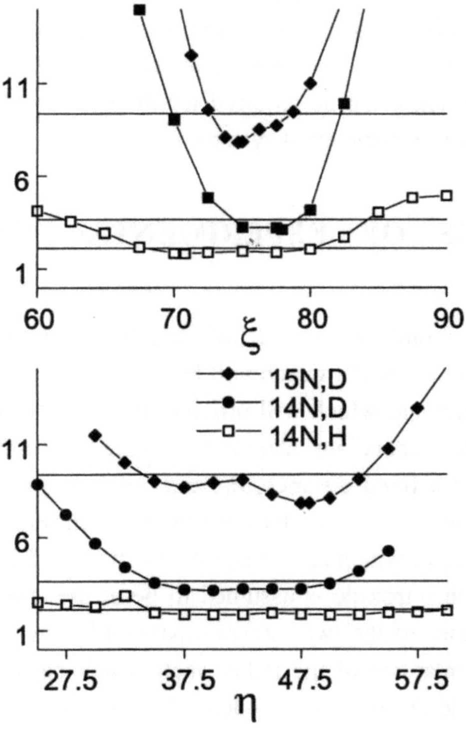

Figure 7. χ^2 curves showing the values of χ^2 for fixed values of ξ (top) and η (bottom) allowing all other parameters to vary. The horizontal lines define the 95% confidence intervals for the parameter in question as estimated from the F-statistic. The best fit values of ξ and η are as follows: [$^{14}N,H$] ξ= 70° and η= 50° (open circles); [$^{14}N,D$] ξ= 78° and η= 39° (closed circles); and [$^{15}N,D$] ξ= 75° and η= 48° (closed diamonds).

The importance of analyzing data obtained at multiple microwave frequencies has been previously demonstrated (see Figure 9 of Hustedt et al., 1997). As H_0 increases, the **g**-tensor begins to dominate the spectral dispersion while the **A**- and **D**-tensors are field independent. The resulting increased separation of the x- and y-turning points at high fields leads to increased precision in the determination of certain parameters. Additional calculations of χ^2 surfaces for data obtained at X-band alone and Q-band alone has further demonstrated the advantages of performing global analysis of data obtained at multiple microwave frequencies. For the case of the SL-NAD^+/GAPDH complex, data obtained at X-band is better than data obtained at Q-band at precisely defining the parameters R, ξ, and β; data obtained at Q-band is better at precisely defining the parameters η, α, and γ. Based on these results, it is clear that the combined analysis of multifrequency data will often be required for the precise determination of all parameters for other experimental systems as well.

5. ANALYSIS OF EXPERIMENTAL DATA FOR CASE 2

Currently work is underway to identify the forms of the distance and orientation distribution, $P(R,\Omega)$, which will best model the true distance and orientation distribution and which will provide the best fits to experimental spectra. Three such models are shown in Figure 8. In Model A, the nitroxides are held at a fixed distance, R_0, and an isotropic distribution of orientations is assumed. In this model the fitting parameters will be R_0 and the exchange coupling, J, which can be significant at short distances ($R_0 < 10$ Å). In Model B, each nitroxide is assumed to be at the end of a tether of length R_T and the bases of the two tethers separated by the distance R_0. All possible relative orientations of the tether vectors will be included. In this case, the internitroxide distance will be dependent on the orientations of the two tethers. In this model the fitting parameters will be R_0 and R_T. Finally in Model C, the tether is assumed to have a limited range of orientational freedom. In the simplest case, the tethers will be restricted to adopt a random orientation within a cylindrically symmetric, hard walled cone.

Molecular dynamics simulations have been performed for MTSSL placed at three different sites, with different degrees of tertiary contact, within a trimer of polyleucine helices (Steinhoff and Hubbell, 1996).

These simulations showed that in the absence of tertiary contacts the nitroxide can adopt a wide range of orientations. In molecular dynamic simulations of MTSSL with significant tertiary contact, the nitroxide was restricted to narrow regions of orientation space (Steinhoff and Hubbell, 1996). At sites with tertiary contacts, the nitroxide orientation may be restricted to two (or more) narrow local minima. This situation can be addressed by adapting model C to include two narrow cones or, to simplify the model, to include only two well-defined orientations. Another approach that can be considered is direct modeling of distributions of dihedral angles about the C_α-C_β-S-S-C-SL tether of MTSSL. The development of these additional models will be guided by the ongoing molecular dynamic studies on the orientation of spin labels at selected sites within myosin by Fajer and coworkers (Sale et al., 1999), the ongoing x-ray crystallographic studies on MTSSL labeled proteins by Hubbell and others (unpublished results), and our own molecular dynamics studies of MTSSL orientation in peptides and T4 lysozyme mutants.

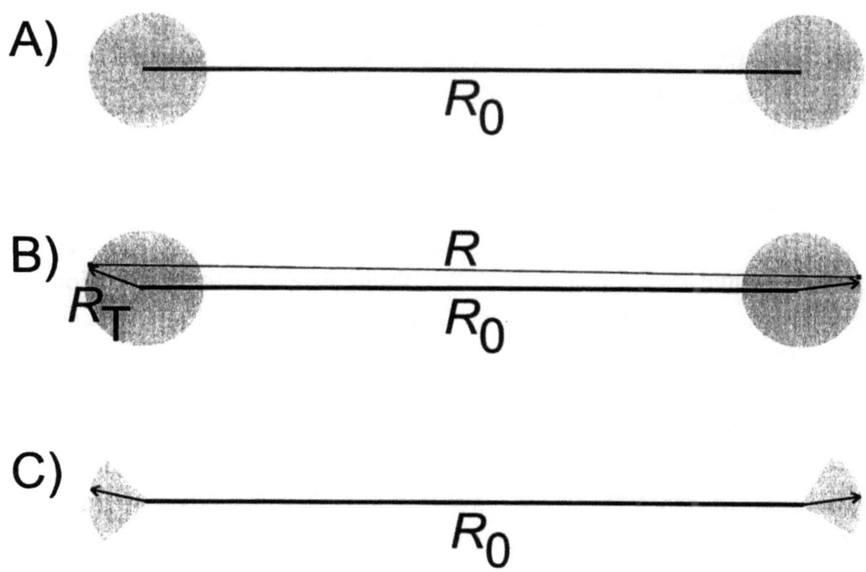

Figure 8. Three models for treating case 2.

To test methods for treating case 2, alanine-rich "4K" peptides which have been shown to adopt helical conformations under a wide range of conditions (Fiori et al., 1993a; Fiori et al., 1993b; Martinez and Millhauser, 1993; Fiori et al., 1994; Fiori and Millhauser, 1995; Rabenstein and Shin, 1995) have been investigated. Figure 9 shows the EPR spectra at X- and Q-band of two such peptides,

4K-C11	AAAAKAAAAKCAAAKAAAAKA
4K-C11,12	AAAAKAAAAKCCAAKAAAAKA

labeled at the cysteine residues with MTSSL. In 70% glycerol at -30° C, the EPR spectra of the singly labeled 4K-C11 peptide, are rigid limit spectra, demonstrating that under these conditions both the global rotational dynamics of the peptide and local motions of the label at the slow motion limit of CW-EPR. Global analysis of the X- and Q-band spectra (top panel) of the MTSSL labeled 4K-C11 peptide were used to determine the **A**- and **g**-tensors that were used in the subsequent analysis of the spectra of 4K-C11,12 and other doubly labeled 4K peptides. It is assumed that the **A**- and **g**-tensors for 4K-C11 are representative of all possible singly labeled peptides and that the **A**- and **g**-tensors are not strongly dependent on position within the peptide.

It is expected that in the absence of tertiary interactions the spin labels will adopt a range of orientations with respect to the peptide backbone and each other and that the EPR spectrum of the doubly labeled 4K-C11,12 peptide should correspond to case 2 with the extent of the distribution of orientations and distances unknown. Overlaid on the data in Figure 9 are fits assuming that the labels adopt a single interelectron distance and orientation with respect to each other as in case 1 (middle) and fits assuming that the labels adopt a single distance but an isotropic orientation distribution (bottom). While neither fit matches the data, both models give approximately the same interelectron distance. Notably, while the unique orientation model overestimates splittings that are evident in the spectra, the isotropic orientation model underestimates the degree of splitting evident in the spectrum. These observations suggest that what is required is a model which allows for a restricted distribution of orientations (and distances) as would be expected for the orientational disorder of the spin label side chains. The fact that both the fits assuming a single orientation and those assuming

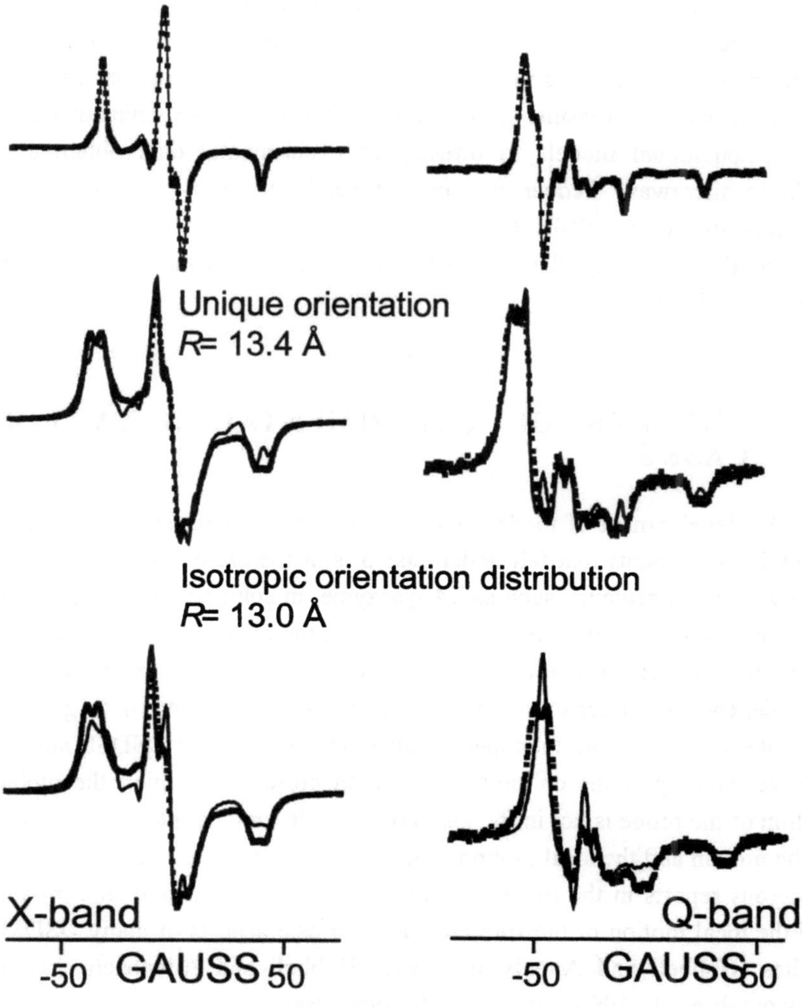

Figure 9. X-band (left) and Q-band (right) EPR spectra (dots) of the singly spin labeled peptide 4K-C11 (top) and the doubly spin labeled peptide 4K-C11,12 (middle and bottom spectra are the same) in 70% glycerol at -30° C. The overlaid simultaneous fits (solid lines) to the spectra for the 4K-C11 peptide are to a rigid limit powder pattern model. The overlaid simultaneous fits to the spectra from the 4K-C11,12 peptide assume a unique orientation and distance (middle) or a single distance and an isotropic distribution of orientations (bottom).

an isotropic orientation distribution give similar interprobe distances and that neither fits the data well (cf. Figures 5 and 6) strongly suggests that the spectra contain information on the orientation and distance distribution and that $<R>$, $<\Omega>$, and $P(R,\Omega)$ can all be determined. The observation of this anisotropic orientation distribution in this peptide without any tertiary contacts is encouraging given that such contacts should further constrict $P(R,\Omega)$. These initial results provide a strong motivation for development of the computational models to directly fit experimental data obtained at multiple microwave frequencies and, ultimately, to define the capabilities and limitations of CW-EPR to provide structural information (both the average distance and relative orientation) under experimental conditions where case 2 applies.

6. ANALYSIS OF EXPERIMENTAL DATA FOR CASE 3

The development of models and methods for the analysis of data from DSDSL experiments that fall under case 3 is in the preliminary stages. For relatively small proteins, such as T4 lysozyme, in solution where the global rotational motion is near the fast motion limit, the effect of dipolar coupling between nitroxides on linewidths can be treated using an approach based on time dependent perturbation theory (Mchaourab et al., 1997b) to give an estimate of the average interspin distance. However, most DSDSL studies involve larger proteins or membrane bound proteins in which the global motion of the probe is not in the fast motion limit, and the internal dynamics of the protein and the local motion of the probe are restricted. It is clear from numerous reports in the literature that the internal dynamics of the protein and the local motion of the spin label are intrinsic aspects of many DSDSL studies (Hubbell and Altenbach, 1994b; Hubbell and Altenbach, 1994a; Mchaourab et al., 1996; Perozo et al., 1998; Barnes et al., 1999; Feix and Klug, 1999; Mchaourab et al., 1999; Perozo et al., 1999). The methods described previously for the calculation of EPR spectra of dipolar coupled nitroxides undergoing rigid body isotropic rotational dynamics (see Figure 4) can be adapted to treat virtually any conceivable dynamics model. The question then is what models should be used to treat data from DSDSL studies of proteins. The dynamics (not attributable to global rotational dynamics of the protein) of the MTSSL can be attributed to flexibility of the peptide backbone and bond angle rotation about the C_α–C_β-S-S-C-SL bonds. The effect of the latter is largely determined by restrictions imposed by steric interactions determined by the protein structure (for review see Hubbell et al., 1996).

Molecular dynamics simulations have been used to model the dynamics of MTSSL depending on the degree of tertiary contact (Steinhoff and Hubbell, 1996). X-band EPR studies have been used to give qualitative descriptions of the dynamics of MTSSL and other cysteine specific spin labels (Mchaourab et al., 1996; Mchaourab et al., 1999). Combined X-band and high frequency (250 GHz) have been used to propose a detailed model of the internal and local dynamics of MTSSL at two different sites in T4 lysozyme (Barnes et al., 1999). ENDOR (Mustafi et al., 2000), and X-ray crystallography have been used to determine the static distribution of orientations of MTSSL. These early studies provide an important starting point for the development of computational algorithms to explicitly treat the effects of local probe motions on the CW-EPR lineshapes of dipolar coupled nitroxides. Once developed, these algorithms can be employed to analyze data from a wide range of DSDSL studies on proteins. These advances are of obvious importance given the potential of EPR to address structural questions on a wide range of proteins under physiologically relevant conditions.

7. CONCLUSIONS

The intent of this review has been to define the experimental conditions and the analytical methods of data analysis that can be used to obtain the maximum structural information from DSDSL studies. Qualitative interpretation of DSDSL can be useful to determine how structural changes relate to protein function. This is particularly true if a large number of labeled double cysteine mutants can be obtained. On the other hand, highly detailed structural models can be built from a single pair of labels when the distance and orientation information contained in DSDSL spectra are combined as has been demonstrated for case 1. Continued effort is needed to define the capabilities and limitations of CW-EPR spectroscopy as a method for the determination of structural parameters from DSDSL experiments corresponding to case 2 and case 3.

8. AVAILABILITY OF COMPUTER PROGRAMS

All of the computer programs described in this review for the analysis of data from DSDSL experiments can be obtained from Eric J. Hustedt (eric.hustedt@mcmail.vanderbilt.edu).

9. ACKNOWLEDGEMENTS

We wish to thank Dr. Hassane S. Mchaourab for insightful discussions and for reviewing this manuscript prior to submission. This work has been supported in part by NIH R37 HL34737 (A.H.B.).

10. REFERENCES

Altenbach, C., T. Marti, H. G. Khorana, and W. L. Hubbell, 1990, Transmembrane protein structure: spin labeling of bacteriorhodopsin mutants. *Science* **248**: 1088 .

Anderson, D. J., P. Hanson, J. McNulty, G. Millhauser, V. Monaco, F. Formaggio, M. Crisma, and C. Toniolo, 1999, Solution structures of TOAC-labeled trichogin GA IV peptides from allowed (g approximate to 2) and half-field electron spin resonance. *J. Am. Chem. Soc.* **121**: 6919.

Barnes, J. P., Z. Liang, H. S. Mchaourab, J. H. Freed, and W. L. Hubbell, 1999, A multifrequency electron spin resonance study of T4 lysozyme dynamics. *Biophys. J.* **76**: 3298.

Beechem, J. M., 1992, Global analysis of biochemical and biophysical data. *Methods Enzymol.* **210**: 37.

Beechem, J. M., E. Gratton, M. Ameloot, J. R. Knutson, and L. Brand, 1991, The global analysis of fluorescence intensity and anisotropy decay data: second generation theory and programs in *Topics in Fluorescence Spectroscopy* (Lakowicz, J. R., Ed.), Plenum Press, New York.

Berengian, A. R., M. P. Bova, and H. S. Mchaourab, 1997, Structure and function of the conserved domain in αA-crystallin. Site-directed spin labeling identifies a β-strand located near a subunit interface. *Biochemistry* **36**: 9951.

Berliner, L. J., J. Grunwald, H. O. Hankovsky, and K. Hideg, 1982, A novel reversible thiol-spcific spin label: papain active site labeling and inhibition. *Anal. Biochem.* **119**: 450.

Beth, A. H., B. H. Robinson, C. E. Cobb, L. R. Dalton, W. E. Trommer, J. J. Birktoft, and J. H. Park, 1984, Interactions and spatial arrangement of spin-labeled NAD^+ bound to glyceraldehyde-3-phosphate dehydrogenase. *J. Biol. Chem.* **259**: 9717.

Calvo, R., W. Hofbauer, F. Lendzian, W. Lubitz, M. L. Paddock, E. C. Abresch, R. A. Isaacson, M. Y. Okamura, and G. Feher, 1999, EPR study of the molecular and electronic structure of the $Q_A^{\cdot-}$ $Q_B^{\cdot-}$ state of the photosynthetic reactions center in Rb. Spaeroides R26 at 35 and 94 GHz. *Talk at the 22nd International EPR Symposium.*

Eaton, G. R., and S. S. Eaton, 1989, Resolved electron-electron spin-spin splittings in EPR spectra in *Biological Magnetic Resonance* (Berliner, L. J., & Reuben, J., Eds.), Plenum Press.

Fajer, P., R. Bennet, C. Polnazek, E. Fajer, and D. Thomas, 1990, General method for multiparameter fitting of high-resolution EPR spectra using a simplex algorithm. *J. Magn. Reson.* **88**: 111 .

Feix, J. B., and C. S. Klug, 1999, Site-Directed spin labeling of membrane proteins and peptide-membrane interactions in *Spin Labeling: The New Millenium* (Berliner, L. J., Ed.) pp 251.

Fiori, W. R., K. M. Lundberg, and G. L. Millhauser, 1994, A Single Carboxy-Terminal Arginine Determines the Amino- Terminal Helix Conformation of an Alanine-Based Peptide. *Nat. Struct. Biol.* **1**: 374.

Fiori, W. R., G. V. Martinez, and G. L. Millhauser, 1993a, Length Dependence in Alanine-Based Peptides - the 3(10) Helix to Alpha-Helix Transition. *Biophys. J.* **64**: A378.

Fiori, W. R., S. M. Miick, and G. L. Millhauser, 1993b, Increasing Sequence Length Favors Alpha-Helix Over 3(10)-Helix in Alanine-Based Peptides - Evidence For a Length-Dependent Structural Transition. *Biochemistry* **32**: 11957.

Fiori, W. R., and G. L. Millhauser, 1995, Exploring the Peptide 3(10)-Helix-Reversible-Arrow-Alpha-Helix Equilibrium With Double-Label Electron-Spin-Resonance. *Biopolymers* **37**: 243.

Hanson, P., D. J. Anderson, G. Martinez, G. Millhauser, F. Formaggio, M. Crisma, C. Toniolo, and C. Vita, 1998, Electron spin resonance and structural analysis of water soluble, alanine-rich peptides incorporating TOAC. *Mol. Phys.* **95**: 957.

Hanson, P., G. Martinez, G. Millhauser, F. Formaggio, M. Crisma, C. Toniolo, and C. Vita, 1996a, Distinguishing helix conformations in alanine-rich peptides using the unnatural amino acid TOAC and electron spin resonance. *J. Am. Chem. Soc.* **118**: 271.

Hanson, P., G. Millhauser, F. Formaggio, M. Crisma, and C. Toniolo, 1996b, ESR characterization of hexameric, helical peptides using double TOAC spin labeling. *J. Am. Chem. Soc.* **118**: 7618.

Hideg, K., and O. H. Hankovszky, 1989, Chemistry of spin-labeled amino acids and peptides. Some new mono- and bifunctionalized nitroxide free radicals in *Biological Magnetic Resonance* (Berliner, L. J., & Reuben, J., Eds.) pp 427, Plenum Press, New York.

Hubbell, W. L., and C. Altenbach, 1994a, Investigation of structure and dynamics in membrane proteins using site-directed spin labeling. *Curr. Opin. Struct. Biol.* **4**: 566.

Hubbell, W. L., and C. Altenbach, 1994b, Site-directed spin-labeling of membrane proteins in *Membrane Protein Structure* (White, S. H., Ed.) pp 224, Oxford, New York.

Hubbell, W. L., A. Gross, R. Langen, and M. A. Lietzow, 1998, Recent advances in site-directed spin labeling of proteins. *Curr. Opin. Struct. Biol.* **8**: 649.

Hubbell, W. L., H. S. Mchaourab, C. Altenbach, and M. A. Lietzow, 1996, Watching proteins move using site-directed spin labeling. *Structure* **4**: 779.

Hustedt, E. J., and A. H. Beth, 1995, Analysis of saturation transfer electron paramagnetic resonance spectra of a spin-labeled integral membrane protein, band 3, in terms of the uniaxial rotational diffusion model. *Biophys. J.* **69**: 1409.

Hustedt, E. J., and A. H. Beth, 1996, Determination of the orientation of a band 3 affinity spin-label relative to the membrane normal axis of the human erythrocyte. *Biochemistry* **35**: 6944 .

Hustedt, E. J., and A. H. Beth, 1999, Nitroxide spin-spin interactions: Applications to protein structure and dynamics. *Annu. Rev. Biophys. Biomol. Struct.* **28**: 129.

Hustedt, E. J., C. E. Cobb, A. H. Beth, and J. M. Beechem, 1993, Measurement of rotational dynamics by the simultaneous nonlinear analysis of optical and EPR data. *Biophys. J.* **64**: 614.

Hustedt, E. J., J. J. Kirchner, A. Spaltenstein, P. B. Hopkins, and B. H. Robinson, 1995, Monitoring DNA dynamics using spin-labels with different independent mobilities. *Biochemistry* **34**: 4369.

Hustedt, E. J., A. I. Smirnov, C. F. Laub, and A. H. Beth, 1997, Molecular distances from dipolar coupled spin-labels: the global analysis of multifrequency continuous wave electron paramagnetic resonance data. *Biophys. J.* **74**: 1861.

Lajzerowicz-Bonneteau, J., 1976, Molecular structure of nitroxides in *Spin Labeling: Theory and Applications* (Berliner, L. J., Ed.) pp 239, Academic Press, New York.

Martinez, G. V., and G. L. Millhauser, 1993, Alpha-Helices and 3(10)-Helices in Aqueous-Solution May Readily Be Distinguished Using FTIR Spectroscopy. *Biophys. J.* **64**: A378.

McCalley, R. C., E. J. Shimshick, and H. M. McConnell, 1972, The effect of slow rotational motion on paramagnetic resonance spectra. *Chem. Phys. Lett.* **13**: 115.

Mchaourab, H. S., A. R. Berengian, and H. A. Koteiche, 1997a, Site-directed spin-labeling study of the structure and subunit interaction along a conserved sequence in the α-crystallin domain of heat shock protein 27. Evidence of a conserved subunit interface. *Biochemistry* **36**: 14627.

Mchaourab, H. S., T. Kálai, K. Hideg, and W. L. Hubbell, 1999, Motion of spin-labeled side chains in T4 lysozyme: Effects of side chain structure. *Biochemistry* **38**: 2947.

Mchaourab, H. S., M. A. Lietzow, K. Hideg, and W. L. Hubbell, 1996, Motion of spin-labeled side chains in T4 lysozyme. Correlation with protein structure and dynamics. *Biochemistry* **35**: 7692.

Mchaourab, H. S., K. J. Oh, C. J. Fang, and W. L. Hubbell, 1997b, Conformation of T4 lysozyme in solution. Hinge-bending motion and the substrate -induced conformational transition studied by site-directed spin labeling. *Biochemistry* **36**: 307.

Millhauser, G. L., 1992, Selective placement of electron resonance spin labels: new structural probes for peptrides and proteins. *Trends Biochem. Sci.* **17**: 448.

Millhauser, G. L., W. R. Fiori, and S. M. Miick, 1995, Electron spin labels. *Methods Enzymol.* **246**: 589.

Monaco, V., F. Formaggio, M. Crisma, C. Toniolo, P. Hanson, G. Millhauser, C. George, J. R. Deschamps, and J. L. Flippen-Anderson, 1999a, Determining the occurrence of a 3(10)-helix and an alpha-helix in two different segments of a lipopeptaibol antibiotic using TOAC, a nitroxide spin-labeled C-alpha-tetrasubstituted alpha- amino acid. *Bioorg. Med. Chem.* **7**: 119.

Monaco, V., F. Formaggio, M. Crisma, C. Toniolo, P. Hanson, and G. L. Millhauser, 1999b, Orientation and immersion depth of a helical lipopeptaibol in membranes using TOAC as an ESR probe. *Biopolymers* **50**: 239.

Mustafi, D., H. Joela, and M. W. Makinen, 1991, The effective position of the electronic point dipole of the nitroxyl group of spin labels determined by ENDOR spectroscopy. *J. Magn. Reson.* **91**: 497 .

Mustafi, D., A. Sosa-Peinado, V. Gupta, D. Gordon, and M. W. Makinen, 2000, Endor structure of spin-labeled methylmethanethiol-sulfonate in solution and bound to TEM-1-β-lactamase. *Biophys. J.* **78**: 382A.

Park, J. H., and W. Trommer, 1989, Advantages of ^{15}N and deuterium spin probes for biomedical electron paramagnetic resonance investigations in *Biological Magnetic Resonance* (Berliner, L. J., & Reuben, J., Eds.) pp 547, Plenum Press, New York.

Perozo, E., D. M. Cortes, and L. G. Cuello, 1998, Three-dimensional architecture of a K+ channel: implications for the mechanism of ion channel gating. *Nat. Struct. Biol.* **5**: 459.

Perozo, E., D. M. Cortes, and L. G. Cuello, 1999, Structural rearrangements underlying K^+-channel activation gating. *Science* **285**: 73.

Press, W. H., B. P. Flannery, S. A. Teukolsky, and W. T. Vetterling (1992) *Numerical Recipes in FORTRAN: the art of scientific computing.*, 2 ed., Cambridge University Press, New York

Qu, K., J. L. Vaughn, A. Sienkiewicz, C. P. Scholes, and J. S. Fetrow, 1997, Kinetics and motional dynamics of spin-labeled yeast iso-1-cytochrome c: 1. Stopped-flow electron paramagnetic resonance as a probe for folding/unfolding of the C-terminal helix spin-labeled at cysteine 102. *Biochemistry* **36**: 2884.

Rabenstein, M. D., and Y.-K. Shin, 1995, Determination of the distance between two spin labels attached to a macromolecule. *Proc. Natl. Acad. Sci. USA* **92**: 8239 .

Robinson, B. H., L. J. Slutsky, and F. P. Auteri, 1992, Direct simulation of continuous wave electron paramagnetic resonance spectra from Brownian dynamics trajectories. *J. Chem. Phys.* **97**: 2609.

Sale, K. L., K. A. Sharp, and P. G. Fajer, 1999, Predictions of spin label orientation in the crystal structure of proteins. *Biophys. J.* **76**: A34.

Schneider, D. J., and J. H. Freed, 1989, Calculating slow motional magnetic resonance spectra: a user's guide in *Biological Magnetic Resonance* (Berliner, L. J., & Reuben, J., Eds.) pp 1, Plenum Press, New York.

Shin, Y.-K., C. Levinthal, F. Levinthal, and W. L. Hubbell, 1993, Colicin E1 binding to membranes: time-resolved studies of spin-labeled mutants. *Science* **259**: 960.

Sienkiewicz, A., A. M. d. C. Ferreira, B. Danner, and C. P. Scholes, 1999, Dielectric resonator-based flow and stopped-flow EPR with rapid field scanning: a methodology for increasing kinetic information. *J. Magn. Reson.* **136**: 137.

Steinhoff, H.-J., and W. L. Hubbell, 1996, Calculation of electron paramagnetic resonance spectra from Brownian dynamics trajectories: application to nitroxide side chains in proteins. *Biophys. J.* **71**: 2201.

Steinhoff, H.-J., N. Radzwill, W. Thevis, V. Lenz, D. Brandenburg, A. Antson, G. Dodson, and A. Wollmer, 1997, Determination of interspin distances between spin labels attached to insulin: comparison of the electron paramagnetic resonance data with the X-ray structure. *Biophys. J.* **73**: 3287.

Tjandra, N., and A. Bax, 1997, Direct measurement of distances and angles in biomolecules by NMR in a dilute liquid crystalline medium. *Science* **278**: 1111.

Chapter 4

Determination of Protein Folds and Conformational Dynamics Using Spin-Labeling EPR Spectroscopy

Hassane S. Mchaourab[1] and Eduardo Perozo[2]

[1]*Department of Molecular Physiology and Biophysics*
Vanderbilt University School of Medicine
Nashville, TN 37232

[2]*Department of Molecular Physiology and Biological Physics*
University of Virginia Health Sciences Center
Charlottesville, VA 22906

1. INTRODUCTION

The near completion of a large number of genome sequencing projects is ushering in a new perspective on how research in biological and biomedical sciences will be conducted in the future, both at the conceptual and experimental levels. The rush is on to transform the molecular blueprints of life into global views of the biochemical and physiological circuitry that interconnect to form entire organisms. The evolutionary histories, now available as multiple sequence alignments, will be complemented with pictures of evolving structures. In this post genomic era, the process of scientific discovery will be the result of large-scale parallel measurements to answer questions formulated from a genomic perspective.

Within this new conceptual framework, efforts for high throughput generation of structures have emerged in the context of the current avalanche of genome sequence data. This approach, termed structural genomics (Burley *et al.*, 1999; Kim, 1998; Montelione and Anderson, 1999), aims to establish a structural catalog of the entire complement of unique protein folds in a cell using efficient protein expression and purification, advanced structural techniques and computational analysis of sequence and structure.

Biological Magnetic Resonance, Volume 19: Distance Measurements in Biological Systems by EPR, edited by Berliner et al. Kluwer Academic/Plenum Publishers, New York, 2000

The emphasis is on the collective analysis of a list of target proteins rather than individual ones, where the choice of the target set is based on a biological or medical problem (Terwilliger *et al.*, 1998).

In the last few years, the feasibility of structural genomics has been discussed in workshops and in the published literature. Issues and concerns include the choice of the target list, an estimate of a realistic throughput, the cost per generated structure and the time needed for completion (Gaasterland, 1998a; Gaasterland, 1998b). At the fundamental level, the pilot projects already underway appear to establish that the technologies needed for such an endeavor are in place. Perhaps the most valid criticism centers around the ability of such initiatives with their focus on X-ray crystallography and NMR spectroscopy to provide the structural information required for understanding the molecular basis of biological function. Because of reasons of speed, competitiveness and efficiency, the various projects are likely to adopt an approach that gravitates towards easily crystallizable proteins and/or proteins of molecular weights smaller than 30 kDa. This essentially excludes most membrane proteins, key control points in cellular communication and control, which constitute about 25% of proteins encoded by a genome and 50% of the pharmaceutically important receptors. Furthermore, signaling and transduction involve conformational transitions and molecular recognition and interaction events. Thus, the development of a dynamic view of the cell requires complementary techniques that are able to explore the structure and dynamics of membrane proteins and molecular assemblies with appropriate time resolution.

Reporter group techniques provide alternative approaches to address these aspects of structural genomics (Burr and Koshland, 1964). In these techniques, where information is obtained from analysis of the properties of a conformationally and environmentally sensitive probe, proteins are typically studied in their natural milieu and few constraints are imposed on the ordering or size of the sample. More importantly, reporter groups can provide a highly localized view of the conformational dynamics of the polypeptide chain with exquisite temporal and spatial resolution (Ogawa and McConnell, 1967; Ogawa *et al.*, 1968; Wahl and Weber, 1967). While early applications of spin labeling by McConnell and coworkers focused on the dynamic aspect, it was implicit that the conformation-dependent spectroscopy can also be exploited to obtain structural information at biologically relevant resolution.

Much of the promise of spin labeling, however, remained unfulfilled until site-directed mutagenesis was developed into a routine laboratory technique. The ability to routinely introduce spin label probes into protein sequences allowed the systematic study of the structural determinants of the spin label spectral parameters in protein model systems of known X-ray structure. In parallel, the many contributions to the development of new spectroscopic approaches through novel excitation and detection schemes and advances in theory and instrumentation were critical for the renaissance of spin labeling (Eaton and Eaton, 1989; Hanson *et al.*, 1996; Hustedt *et al.*, 1997; Robinson *et al.*, 1994, Hyde, 1979; Saxena and Freed, 1996). In fact, the modern version of spin labeling, also known as site-directed spin labeling and developed by Wayne L. Hubbell and his colleagues (Hubbell and Altenbach, 1994; Hubbell *et al.*, 1998; Hubbell *et al.*, 1996), is the synthesis of multiple contributions key among them are the development of the loop-gap resonator (Froncisz and Hyde, 1982), spin-labeling relaximetry (Hyde and Subczynski, 1989) and motional analysis of CW-EPR spectra (Freed and Schneider, 1989).

At present, spin labeling can be used to determine both folding patterns of proteins and biologically relevant conformational dynamics. Although the purpose of this chapter is not to summarize the history of SDSL, it is important to briefly reference the contributions that led to the maturation of the technique. Early work by Altenbach et al. on Bacteriorhodopsin (bR) established the use of SDSL to determine sequence-specific secondary structure and solvent accessibility (Altenbach *et al.*, 1990). SDSL studies of Rhodopsin and Colicin E1 demonstrated the ability of SDSL to resolve triggered conformational transitions and along with the work on Bacteriorhodopsin suggested that the introduction of the nitroxide does not lead to debilitating effects on the protein structure (Farrens *et al.*, 1996; Shin *et al.*, 1993). Systematic investigation of the issue of perturbation by the nitroxide was quantitatively addressed by Mchaourab et al using T4 Lysozyme (Mchaourab *et al.*, 1996), where it was shown that nitroxides can be introduced almost at will on the protein surface with little if any changes in stability and structure. Furthermore, the changes in the free energy of folding were of the same order of magnitude as those observed with a mutation to a natural amino acid at the same site (Matthews, 1995). These results are consistent with what is known concerning the response of protein structures in general to amino acid substitution from the work of Brian Matthews and his colleagues (Matthews, 1995). Studies on Colicin E1, Diphtheria Toxin, Annexins and other proteins for which X-ray structures are known further validate these conclusions (Langen *et al.*, 1999; Oh *et al.*, 1996; Salwinski and Hubbell, 1999).

The early work on BR and other protein model systems set the stage for the widespread application of SDSL to determine local secondary structure and fingerprints of the tertiary fold from systematic nitroxide scanning. However, realizing that these structural constraints are not enough to restrict conformational space to a manageable set of structures, interest has shifted to use internitroxide proximities with the goal of obtaining the global features of the fold. This is the second phase in the renaissance of spin labeling. In its essence, the goal is to duplicate the success of high resolution

NMR in using spatial restraints to obtain protein structure. While this task has not been achieved, the advances have been phenomenal. This chapter describes predominantly the work carried out in our laboratories.

2. SPATIAL RESTRAINTS AND PROTEIN CONFORMATIONAL SPACE

In NMR, the principal source of geometric information used in structure determination is obtained from short-range interproton distances. Conformational space is then searched to simultaneously satisfy the experimentally derived spatial restraints (Nuclear Overhauser Enhancements (NOEs) and J-couplings) and holonomic constraints of atom connectivity, bond lengths, and bond angles. Plausible models can be generated in either Cartesian or dihedral angular space using a variety of computational methods, known as distance geometry algorithms, usually refined by restrained molecular dynamics methods (Havel, 1991). The success of the NMR approach is based on the fact that the number of experimentally determined spatial restraints and their accuracy is sufficient to overcome the degrees of freedom of the polypeptide chain. This relation was established empirically, i.e. it was confirmed by the solution of a structure and its close resemblance to that obtained from x-ray diffraction. The question of the number of restraints needed and their accuracy is fundamental to the effort of the SDSL community in attempting to use SDSL to obtain protein structures.

Unlike NMR however, in SDSL the spectroscopist is not presented with a map of inter-residue proximities. Rather, the task is to select a minimal set of sites where inter-nitroxide distance measurements are to be performed. This obviously limits the number of distance-based restraints available and strongly encourages a rational approach in the design of these doubly labeled mutants. In principle, because the strength and distance dependence of the dipolar coupling in interacting spin-label pairs allows for distance determinations up to 20 Å using continuous wave methods, fewer constraints are needed to generate plausible structures using distance geometry methods. Generally, the proximities between the nitroxide moieties can be interpreted to within 3 Å (see Chapter 5) and in cases with well resolved dipolar splittings to within 0.1 Å (see chapter 3). Unfortunately, it is not clear how soft these restraints should be if translated to reflect the α-carbon separation. In other words, the uncertainty due to the flexible link between the spin label and the protein increases the number of degrees of freedom in the problem. In an extreme case, one can assume a constant link between the ring and the backbone and an isotropic distribution of the nitroxide ring relative to the α-carbon. In general, analysis of the mobility of the commonly used spin label side-chain R1 (see Figure 1) reveals that such an assumption is not always justifiable and introduces further uncertainty to the problem (see below).

The paragraph above outlines two separate yet related problems, namely the need of a general methodology for the selection of the double mutants and the structural interpretation of inter-nitroxide distances. In our laboratories, we took advantage of the tremendous restriction on the conformational space available to a polypeptide chain as a consequence of the topological rules that govern the assembly of proteins in general (Chothia, 1984; Chothia and Finkelstein, 1990; Chothia *et al.*, 1997), to simplify these two problems. Protein structures are assembled from recurrent building blocks consisting of simple combinations of secondary structures. These combinations, known as supersecondary structures or motifs, are characterized by unique geometric arrangements of two or more secondary structures. The frequent occurrence of these motifs is the result of the fact that only two basic types of secondary structure exist and their pairwise packing arrangement has to satisfy steric and hydrogen bonding constraints. Thus, a defined task in SDSL is to design the minimal number of double mutants able to identify the presence of these motifs. Furthermore, the interpretation is based on the presence of a pattern of proximities thus alleviating the problems arising from the context-dependent conformational preferences of the nitroxide.

Supersecondary structures are also subject to packing rules. As a result of the restricted basis set, structure space is quite limited when expressed in terms of folds. This degeneracy is best appreciated by the fact that the 4000 protein structures in the protein data bank as of 1996 corresponded to only 327 distinct folds (Chothia *et al.*, 1997). It is estimated that the 100,000 or so human genes belong to only about 3000 distinct fold families. For instance, all β-sheet proteins tend to have one of few distinct overall folds. Each of these folds can be reduced to a unique pattern of proximities, the presence of which can be tested using suitable nitroxide pairs.

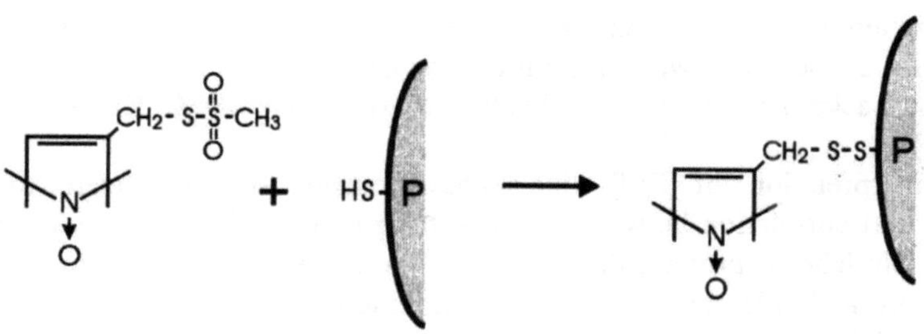

Figure 1. Chemical structure of the methanethiosulfonate spin label (MTSSL) and its reaction with a protein sulfhydryl group to generate side chain R1

Although the topological rules for supersecondary structures have been derived from analyses of water-soluble proteins, a recent survey of helical membrane proteins suggests further restrictions on the available conformational space of these proteins (Bowie, 1997). Because of the energetic penalty of having unsatisfied hydrogen bonds within the low dielectric of the membrane, there are only two types of structural motifs found in the intra-membrane regions of integral membrane proteins: α-helical bundles or β-barrels. Thus, an outline of a general methodology emerges where the spin labeler follows the hierarchy of the protein structure in an iterative process that is combined with molecular modeling. In its essence, this approach can be considered an experimental version of the purely computational fold recognition protocols.

3. EXPERIMENTAL STRATEGIES IN PROTEIN FOLD DETERMINATION

3.1 Structural Information Derived from EPR Analysis of Spin Labeled Proteins

The structure of the most commonly used nitroxide side chain, hereafter referred to as R1, is shown in Figure 1. SDSL involves the introduction of R1 into protein sequences via cysteine mutagenesis followed by derivatization of the reactive cysteine with the sulfhydryl selective spin label I shown also in Figure1. EPR analysis of spin labeled mutants yields three types of structural information: 1) nitroxide dynamics and motional freedom from spectral line-shapes, 2) solvent accessibility of the attached nitroxide through collisional relaxation methods and 3) distances between pairs of nitroxides through dipole-dipole interactions. The first two are local parameters in the sense that they are interpreted in terms of structure in the vicinity of the probe while inter-nitroxide proximities reflect global features of the backbone fold (see (Hubbell and Altenbach, 1994; Hubbell *et al.*, 1998)).

Interpretation of EPR line-shapes depends on knowledge of the structural correlation between the reporter group and its local environment. In a spin-labeled protein, the relation between the spectral line-shape, local structure and side chain dynamics is extremely complex, given the number of possible rotations around bonds that connect the label to the protein backbone. Line-shape analysis therefore depends on the structure of the spin-label itself, and on the motional model assumed for the internal rotation of the side chain. Mchaourab et al. have recently established the basis for spectral interpretation in a study of spin-labeled T4 lysozyme, a protein well characterized structurally. Spin label dynamics can be quantified through the lineshape parameter, ΔHo, or the spectral second moment (Mchaourab *et al.*, 1996), both semi-empirical measures derived from the spectral line-shape, determined by the degree of averaging of the anisotropic **g** and hyperfine tensors, respectively. Their values increase as the frequency and amplitude of nitroxide motions is reduced. ΔHo has been thoroughly calibrated using T4 lysozyme as a model system, and it is currently used as a standard measure of spin-probe dynamics. This type of measurement is particularly useful, since it has been found that mobility of the spin label allows the classification of a site into one of three structural classes: exposed, buried or in tertiary contacts. Mobility at solvent-exposed sites in α-helices or loops reflect, in addition to rotational isomerization around the tethering bonds, large amplitude segmental flexibility modes, while being weakly dependent on the interaction between side chains (Mchaourab *et al.*, 1996; Steinhoff and Hubbell, 1996).

Collision of nitroxides with fast relaxing radicals such as oxygen and metal ion complexes, causes spin exchange that effectively shortens the spin-lattice relaxation time T_1 of the nitroxide (Hyde and Subczynski, 1989). This effect can be measured either by continuous wave (CW) power saturation techniques or by saturation recovery methods. Collision frequency is directly proportional to the accessibility of the paramagnetic reagent to the nitroxide radical, and is defined as

$$\Pi = \frac{\Delta P_{1/2}(X)}{P_{1/2}(DPPH)} \frac{\Delta H_o(DPPH)}{\Delta H_o}$$

Where $P_{1/2}(DPPH)$ is the microwave power that saturates the signal relative to that without saturation for a 2,2-diphenyl-1-picrylhydrazyl crystal, $\Delta P_{1/2}(X) = [P_{1/2}(X) - P_{1/2}(N_2)]$, the difference in $P_{1/2}$ of a given sample exposed to a fast-relaxing paramagnetic species (like O_2 or NiEdda) and N_2. ΔHo is an estimate of probe mobility, as described above. This Π parameter can be used to deduce secondary and tertiary structure information, which is derived from the periodic behavior of a series of sequential mutants in the presence of relaxing agents (see below). Combination of accessibility data with spin label dynamics can be used effectively to determine the overall location of a specific residue in a soluble protein fold, or within the membrane-protein-water interface for membrane proteins.

Quantitative information on inter-spin distances can be obtained from through-space dipole-dipole interactions between nitroxide groups in doubly spin-labeled mutants. Direct estimation of distances is achieved under rigid limit conditions (low temperature), as has been shown by a number of groups (Hustedt *et al.*, 1997; Rabenstein and Shin, 1995). These dipolar interactions are reflected as a general broadening of the spectrum, and the extent of this broadening can be used to estimate actual distances with some precision (Rabenstein and Shin, 1995). In general, any structural conclusion originating from the determination of inter-spin distances should be based on the presence of patterns of proximities rather than single distances. Consequently, the impact of the uncertainty in the position of the nitroxide ring relative to the α-carbon is minimized.

3.2 Methods for Secondary Structure Assignment

The first step in a successful EPR-based fold assignment is to properly assign secondary structure information along the linear sequence of amino acids. General secondary structure assignments can be obtained in favorable cases simply by observation of the mobility or accessibility profile in a given protein segment. A quantitative approach used to estimate secondary and super-secondary structure from a one-dimensional data set might be needed to obtain more subtle information, including deviations from ideal behavior and to define the ends of a strand or a helix.

One of such quantitative approaches is based on a Fourier transform method first described by Cornette et al (Cornette *et al.*, 1987). The data set used for these calculations can be a given residue specific property, like hydrophobicity, variability in a sequence alignment, or any of the experimentally determined structural parameters described above. To estimate the periodic behavior of a given sequence property, a discrete Fourier transform (DFT) power spectrum, P(ω), of a given amino acid segment is calculated as a function of the angle between two adjacent side chains (ω). In the DFT, for a given window of residue properties along an amino acid sequence (any finite sequence of numbers $\{h_k\}_{k=0}^{h=-1}$ and angular frequency ω measured in radians per residue), the power spectra P(ω) is defined as:

$$P(\omega) = [X(\omega)]^2 + [Y(\omega)]^2$$

$$X(\omega) = \sum_{k=0}^{l-1} U_k - \overline{U} \cos(k\omega)$$

$$Y(\omega) = \sum_{k=0}^{l-1} U_k - \overline{U} \sin(k\omega)$$

Where U_k is the value of a given residue-specific property (Mobility, ΔHo^{-1}, Collision Frequency, Π_x), while k is the position of the residue in the linear sequence. $\bar{U}$ is the average value of U_k over the length of the segment or residue window. P(ω) gets evaluated for the value $\omega = \hat{\omega}$ that maximizes P(ω) and is interpreted as the characteristic frequency of $\{U_k\}_{k=0}^{U=-1}$.

In order to minimize the influence of spurious frequency components, unrelated to the natural frequency components in protein secondary structural elements Donnelly et al (Donnelly *et al.*, 1994) introduced two types of data correction. To filter out low frequency components derived from the uneven distribution of data intensities, a linear regression of the U_k profile (U_k vs. residue #) is subtracted from each amino acid window segment. Additionally, outliers are corrected by using a discontinuous function:

$$U_k = \begin{Bmatrix} U_{max} & if\ U_k \geq \overline{U} \\ U_{min} & if\ U_k \leq \overline{U} \end{Bmatrix}$$

where

$$U_{max} = \max[U_k] \quad for \quad |-U_k| \quad \geq \quad |+2\sigma|$$

$$U_{min} = \min[U_k] \quad for \quad |-U_k| \quad \leq \quad |-2\sigma|$$

U_{max} and U_{min} are defined as the value of $\bar{U}$ ± twice its standard deviation over the length of the sequence segment; any value of U_k beyond those limits is automatically set to U_{max} or U_{min}. These modifications produce cleaner spectra without affecting other individual frequency components in the power spectrum.

Figure 2 illustrates the correspondence between the periodicity of a given residue environmental property and the secondary structure of a defined protein segment within a three-dimensional fold. As an example, we have chosen two membrane proteins with both intra- and extra-membrane regions, but with different elements of secondary structure: α-hemolysin, which forms a complex β-barrel (Fig 2A, left panels) and the mechanosensitive channel MscL, which forms an α-helical bundle (Fig 2B, left panels). In each case, residue solvent accessibility was calculated for a segment with defined secondary structure, exposed to either water or lipid (using the program DSSP, Kabsch and Sander, 1983). Each profile was plotted against residue number (central panels) and power spectra were calculated for each of these segments, shown as the right-most panels in Figure 2. In each case, the main frequency component of the spectrum was observed squarely within the angular space expected for α-helices or β-strands.

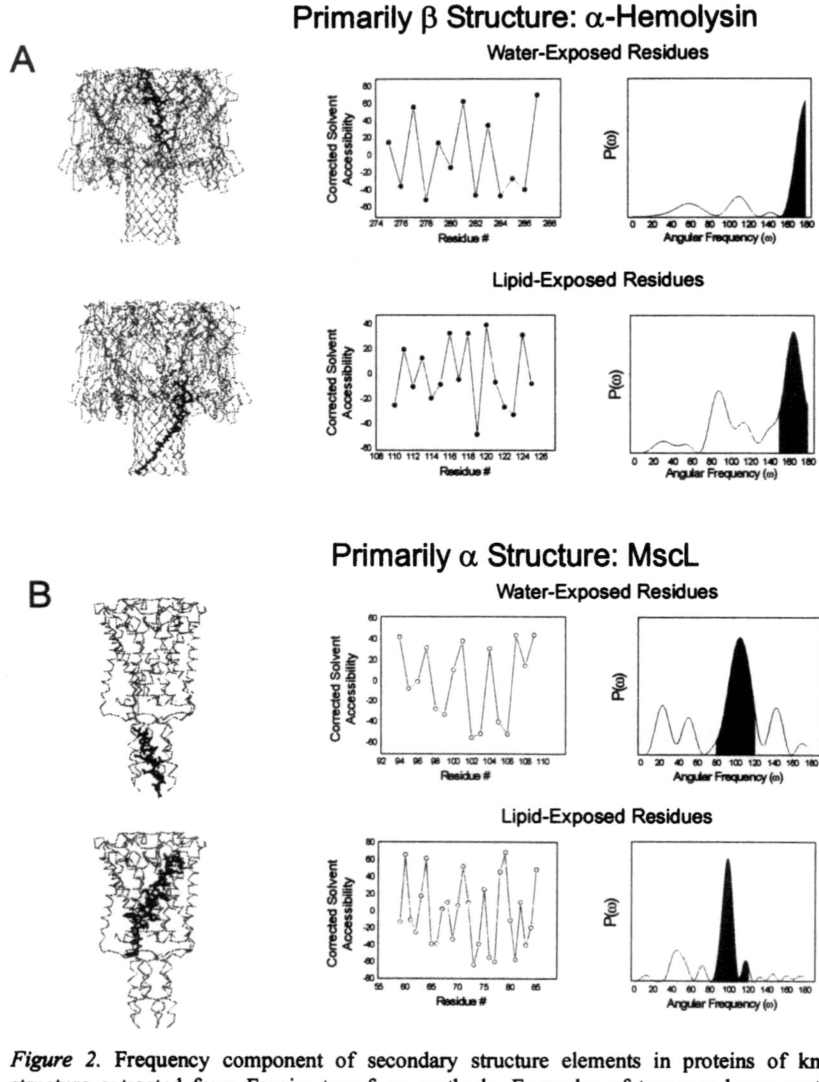

Figure 2. Frequency component of secondary structure elements in proteins of known structure extracted from Fourier transform methods. Examples of two membrane proteins with primarily β-barrel (α-hemolysin, A) or α-helical bundle (the MscL channel, B) structures are shown. In each example, elements of secondary structure in water-exposed or lipid-exposed regions of each molecule have been analysed independently. In each case the left-most panel shows the region of the molecule under analysis, the center panel a solvent accessibility profile and the right-most panel its corresponding power spectrum.

The significance of the α-helical or β-sheet periodicities derived from a defined peak in a given power spectrum can be estimated from the so-called periodicity index (PI) (Cornette *et al.*, 1987). PI represents the value of P (ω) in either the α-helical region of angular space (θ= 80-120°) or the β-strand region of angular space (θ= 150-180°), relative to the total area of the entire angular space of the spectrum (0-180°). αPI can be computed according to:

$$\alpha PI = \frac{\frac{1}{40}\int_{80^\circ}^{120^\circ} P(\omega)d\omega}{\frac{1}{180}\int_{0^\circ}^{180^\circ} P(\omega)d\omega}$$

While βPI is calculated from:

$$\beta PI = \frac{\frac{1}{30}\int_{150^\circ}^{180^\circ} P(\omega)d\omega}{\frac{1}{180}\int_{0^\circ}^{180^\circ} P(\omega)d\omega}$$

sequence segments with PI values greater than two are considered to be α-helical (Cornette *et al.*, 1987).

Using a sliding window approach (Rees *et al.*, 1989), PI estimation can be extended to help detect probable limits of a given secondary structure element, which will then serve as the basis for secondary structure assignment. This is illustrated in Figure 3. Starting from a given residue environmental profile (corrected for low frequency components), a window of analysis is defined (typically 7-11 aa long) and a power spectrum is calculated with the data string contained in that window. From that spectrum, a PI is calculated for a given frequency range (for example, β-sheet) and assigned to the mid-point of the window of analysis. This procedure is repeated until the entire protein segment of interest is fully analyzed. Based on the known threshold of significance (PI=2), this periodicity profile can help establish the beginning and end of helices, strands and even loops, and thus refine secondary structural assignments.

There are some important drawbacks in the use of Fourier analysis to estimate and assign secondary structures in proteins of unknown structure. The method is limited in resolution to the smallest usable size of the scanning window, which is about two turns of a helix (7 residues) and two full cycles in a β-strand (5 residues). Because of this, it will be very difficult to detect either short stretches of secondary structure, or short loops connecting two secondary structure elements. Clearly, this is a more serious problem when studying soluble proteins than with integral membrane proteins. This approach also requires the presence of two distinct solvent-accessible surfaces within the same structure. Either isolated secondary structure elements, or those tightly packed in the interior of a protein or an oligomeric complex will show little or no contrast for a given structural parameter and thus no significant frequency component in a power spectrum.

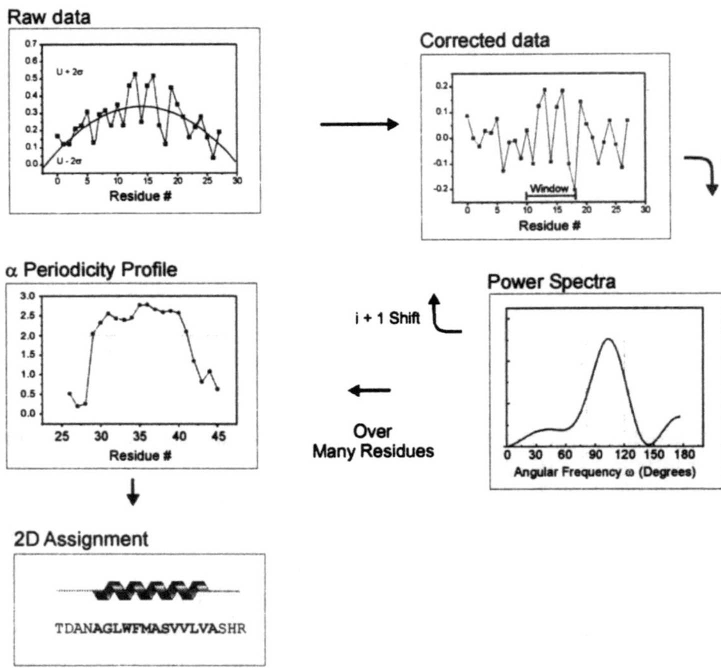

Figure 3. Use of windowed periodicity analysis for secondary structure assignment. Initially, the raw environmental parameter profile (Raw data) is corrected by subtraction of low frequency components (Corrected data). A window of analysis is then defined and a power spectrum calculated (Power spectra). The α periodicity index (α-PI) is then calculated for each spectrum and the result assigned to the residue at the center of the window. This window is then moved one residue at a time and the overall result (α-Periodicity profile) can be used to assign a stretch of secondary structure onto a linear sequence of amino acids (within 2 residues).

3.3 Methods for the Analysis of Inter-Nitroxide Proximities

Spin-spin interactions between nitroxides occur through two mechanisms: spin exchange and dipole-dipole interactions. Through-bond spin exchange is negligible for most systems of interest since the two sites in a pair are usually separated by more than three amino acids in the primary structure. Through-space exchange has been modeled as an exponentially decaying function with an effective range less than 10 Å. Beyond this distance the dipolar interaction dominates. In general, the dipole-dipole interaction is a function of the spatial separation of the two nitroxides and the frequency and amplitude of the motion averaging the orientation of the spin vector relative to the magnetic field. Doubly labeled mutants can be divided into three different categories depending on the degree of order and the presence or absence of motion. The first category corresponds to the limiting case obtained when the two nitroxides adopt a unique orientation relative to each other in the rigid limit. Such conditions are achievable for spin labels introduced at buried sites. As demonstrated in the chapter 3, a rigorous solution for the spin Hamiltonian of the system can be obtained yielding the distance between the two nitroxides and five Euler angles describing their relative orientation and that of the interspin vector relative to the magnetic field (Hustedt *et al.*, 1997).

The second category consists of pairs of nitroxides that adopt a static distribution of distances and orientations relative to each other. This is perhaps the most frequently encountered case in SDSL as any doubly labeled system will reduce to this case in the solid state. If the distribution of relative orientations is isotropic, the interaction can be analyzed by the deconvolution/convolution method introduced by Rabenstein and Shin ((Rabenstein and Shin, 1995), for details see chapter 5). The same approach provides an adequate approximation if the two nitroxides are partially oriented relative to each other.

The third category consists of pairs where either the interspin vector or the nitroxide side chains are dynamic on the EPR timescale. Global reorientation of proteins with molecular mass of less than 20 KDa in aqueous solution at room temperature partially averages the static dipolar interaction resulting in a relaxation mechanism. Theoretical treatment of this case was carried out by Abragam (Abragam, 1961) and later by Rassat in biradicals (Michon and Rassat, 1974). Mchaourab et al. demonstrated this mechanism in the context of T4L (Mchaourab *et al.*, 1997b). Using pairs of labels introduced at sites with known separation it was demonstrated that the modulation of the dipolar interaction by the global tumbling of the protein results in a homogenenous broadening of the EPR spectrum. Using this mechanism, reasonable fits were obtained and the distances experimentally determined were in the range expected based on the X-ray structure as shown in Figure 4.

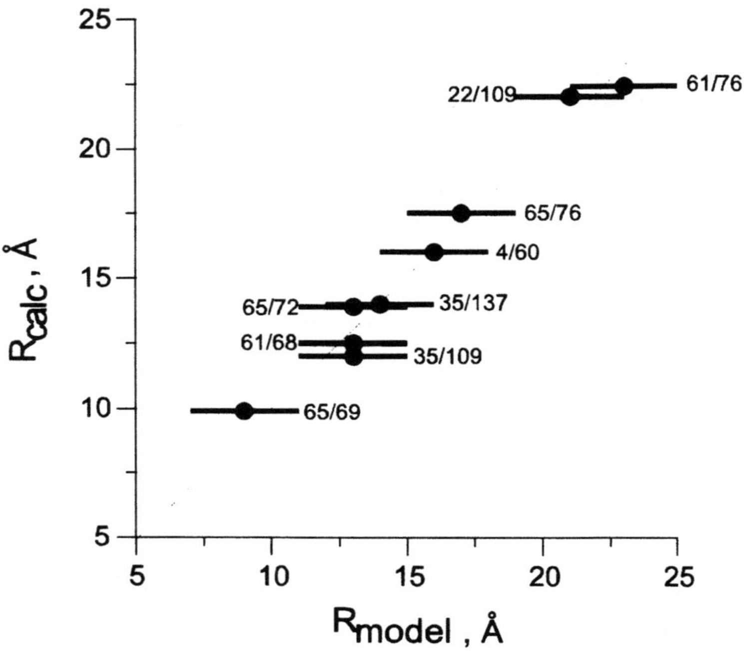

Figure 4. Plot of interspin distance calculated from the dynamic dipolar mechanism versus Rmodel, estimated using a model of R1 where the disulfide bond lays along the backbone. The horizontal bars indicate the probable range of distances determined from modeling. The solid line has a slope of one. (Reproduced from Mchaourab et al., 1997).

For proteins with molecular mass larger than 30KDa and membrane proteins, the motion of the inter-spin vector is too slow on the EPR time scale and the only possible source of averaging is the local motion of the nitroxide side chains. Although the time scale of this motion is in the right range, it is not clear that the amplitudes are large enough to average the 55-7 G dipolar spitting expected in the 10-20 Å distance range. In fact, studies of R1 (Fig. 1) in T4L and other proteins reveal that the rotational isomerization of side chain R1 occurs exclusively at the last two bonds tethering the ring to the disulfide bonds and that the amplitude of motion is restricted due to the internal structure of the side chain. Consequently the broadening of the EPR spectrum when compared to the reference state of the spectral sum of the single mutants can be interpreted semi quantitatively to rank distances into three broad categories: close (<10 Å), intermediate (10-15 Å) and far (15-20 Å).

3.4 Quaternary Structure Packing and Symmetry from Inter Subunit Spin Coupling

The assembly of subunits in oligomeric proteins can result in the proximity of nitroxide side chains introduced at a single site of a subunit near the symmetry axis. Spectral broadening is deduced from comparing the spectrum of the fully labeled oligomer to that of a spin-diluted oligomer. In addition to identifying the subunit interfaces, analysis of the spin-spin interaction determines the symmetry relating the subunits and characterizes the detailed packing of the interacting secondary structural elements. In water-soluble proteins with reversible folding reactions, spin dilution can be achieved by refolding the spin labeled oligomer with excess cysteine-less WT subunits. For cases where the sites are accessible, underlabeling, i.e. using sub-stoichiometric amount of the spin label is adequate. For buried sites where excess spin label has to be used, a mixture of spin label and a diamagnetic sulfhydryl specific reagent is a superior way to achieve spin dilution (Gross *et al.*, 1999).

The number of interacting spins reflects the symmetry of the interface. This number can be experimentally determined through dilution of an oligomer showing spin-spin interactions with increasing amount of unlabeled WT subunits. This results in subpopulations of oligomers where the interface has one or more unlabeled subunits. If the nitroxide side chains are separated by less than 13 Å, the spectral contribution of oligomers with more than one spin label at the interface is so broad as to be negligible compared to the oligomers containing a single spin. As a result, the normalized EPR spectrum reflects the fractional population of interfaces with a singly labeled subunit. For a given stoichiometry of unlabeled to labeled subunits N, the fractional population of interfaces with one spin-labeled subunit can be calculated from the binomial distribution and is a function of the total number of subunits that make up the interface. Thus, by fitting the dependence of the experimentally determined fractional population of singly labeled oligomers on N, the type of symmetry can be determined (Langen *et al.*, 1998). Once an interface has been identified, the packing of the secondary structure can be deduced using nitroxide pairs following the same strategy as in monomeric proteins.

4. DETERMINATION OF PROTEIN FOLDING PATTERNS AND QUATERNARY STRUCTURE PACKING

4.1 α-Crystallin as an Example of Water soluble proteins

4.1.1 The folding pattern of the conserved domain in the αA-crystallin subunit.

In the last couple of years, a systematic SDSL study was carried out of the α-crystallin domain of α-crystallin and serves to illustrate the use of this approach to determine protein folding patterns (Koteiche *et al.*, 1998; Koteiche and Mchaourab, 1999).

The main protein components of the vertebrate lens are a class of water-soluble proteins called the α-crystallins (for review see (Horwitz, 1993)). αA and αB-crystallin, two distant gene-product subunits, associate to form oligomeric structures of more than 40 subunits. Both subunits are members of the small heat-shock protein (sHSP) superfamily of which three other members are expressed in eukaryotic species (de Jong *et al.*, 1998). Small heat-shock proteins are part of a universal protein machinery that protects cells and organisms from the toxic and pathogenic consequences of protein aggregation. While their in-vivo role remains enigmatic, sHSPs recognize and bind unfolding proteins with unique efficiency preventing protein aggregation. The binding results in stable complexes and sHSPs do not appear to be ATP-powered engines like other heat shock proteins, nor are they able to refold their bound substrate.

The chaperone-like function of α-crystallin is critical for the lens, a tissue under constant stress and where little protein turnover occurs. Lens transparency is the result of short-range spatial order between lens protein, an order that can be disturbed if protein aggregation occurs. Thus, it is believed that α-crystallin, in addition to participating in the ordered protein matrix, delays the onset of scattering by constantly binding and sequestering proteins of compromised structures or stability.

While all sHSPs assemble into oligomeric structures of 9 to 40 subunits, the symmetries and sizes are divergent across the superfamily (Chang *et al.*, 1996;Berenegian *et al.*, 1999). Many sHSPs from archeal and bacterial species have well ordered quaternary structures while those of mammalian sHSPs are variable with some constantly exchanging subunits (Bova *et al.*, 1997). Compelling evidence suggests that the flexible and dynamic nature of these oligomers is essential to chaperone function. Thus, the divergence in oligomeric assembly might be a reflection of a divergent functional mechanism.

The conformational heterogeneity of the mammalian oligomeric structure of mammalian sHSPs makes them intrinsically unsuitable for high resolution X-ray crystallography. Cryo-electron microscopy studies confirmed that human αB-crystallin has an irregular and variable quaternary structure (Haley *et al.*, 1998). At 500 KDa, these proteins are too large for structure determination by high resolution NMR. Therefore, they are ideal candidates for investigation using reporter group techniques.

The common sequence characteristics in the superfamily are focused in an 80-100 amino acid stretch in the C-terminal part of the sequence, dubbed for historical reasons as the α-crystallin domain. Sequence-similarity in this domain is highest between sHSPs of the same species, while in distant members of the superfamily the similarity can be marginal. This divergence in sequence, in addition to the diverse molecular mass, abundance and expression, only increases the enigmatic nature of these proteins. Because of its ubiquitous presence, the α-crystallin domain was the initial focused of SDSL studies.

4.1.2 Secondary structure assignment and solvent accessibility

Following the hierarchal approach outlined above, the first step is to use nitroxide scanning to determine the sequence-specific secondary structure and solvent accessibility. As an illustration, the results of the nitroxide scan of Koteiche et al. (Koteiche *et al.*, 1998) are shown in Figure 5. The data shows the variation in accessibility to NiEdda (Π_{NiEdda}) as a function of residue number along the α-crystallin domain of α-crystallin Π, the accessibility parameter is proportional to the collision frequency between the nitroxide side chain and the paramagnetic reagent. The main feature of the plot in Figure 5 is a recurrent variation of Π with a periodicity of two. This is the hallmark of anisotropically solvated β-strands with the face of higher accessibility in contact with the aqueous phase. The strands are connected via turns and loops as deduced from the absence of regular periodic patterns. In the 60-65 and 70-80 sequences, the contrast in Π was shallow and the variations do not reveal unequivocally any regular periodicity. This is an example of the effect of the quaternary structure. If one face of the segment of secondary structure is in contact at a subunit interface with the other face buried in the fold of a subunit, then the lack of asymmetry in the environment can result in a non-periodic variation of accessibility or mobility. In this case, patterns of internitroxide proximities can be used to determine the backbone conformation. In a β-strand, the Cα carbon of residues i and i+2 and those of i and i+4 are separated by 6.6 Å and 13.2 Å respectively. In contrast, the signature of an α-helix is the close proximity between side chains of residues i and i+3 and i and i+4 (Koteiche and Mchaourab, 1999).

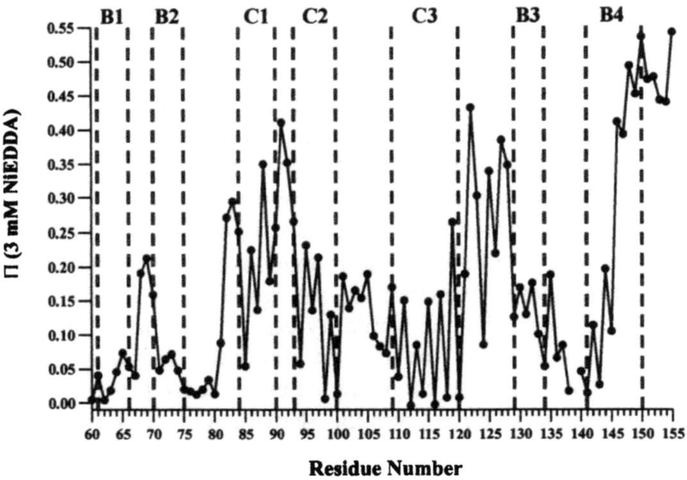

Figure 5. Π (NiEDDA) plotted versus residue number along the α-crystallin domain. (Reproduced from Koteiche et al., 1998)

4.1.3 Packing of secondary structure

Once sequence-specific secondary structure is obtained, the next step in the hierarchy is to determine the pairwise packing of segments adjacent in the sequence. For β-strands, the most frequently observed motifs connecting two strands are: (1) Side-to side antiparallel packing with hydrogen-bonded backbone, i.e. a β-hairpin, (2) face-to-face antiparallel packing in a Greek key arrangement where the strands are not coplanar, i.e. in different sheets (3) the αβα motif where the two strands are parallel and connected by an α-helix. In Figure 6, the occurrence of motifs 1 and 2 in protein folds is illustrated. Each motif is characterized by a distinct geometric arrangement; therefore, they can be distinguished from patterns of proximities in sets of nitroxide pairs. For instance, in a β-hairpin motif, residues at the N-terminus of one strand are in close proximity to residues at the C-terminus of the second strand. In motif 2, where normally each strand belongs to a different sheet, this is only true for residues on the packed faces of the strands. Residues on the more exposed surfaces point their side chains in opposite direction. In contrast, in motif 3 where the strands are parallel, residues at the N-terminus of each strand are in close proximity. Furthermore, the presence of a β-α-β motif can be recognized from the accessibility pattern since the first two motifs have intervening turns and loops instead of helices. It is noted that the differences in the expected patterns of proximities should not be influenced by the net orientation of the nitroxide group relative to the α-carbon. Therefore, the nitroxide pairs can be designed to distinguish between these motifs.

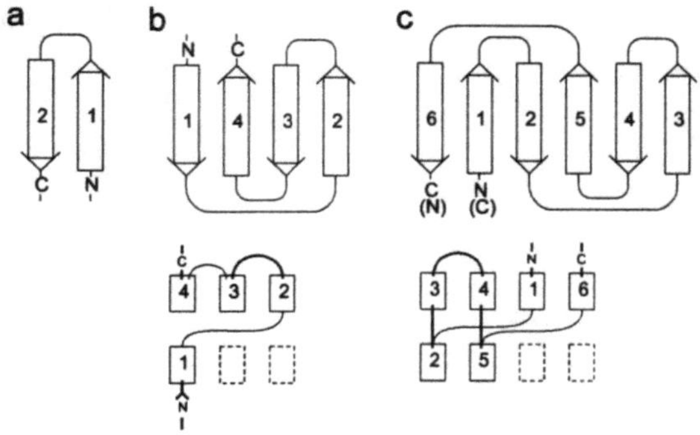

Figure 6. A. topology diagram of a b-hairpin motif. B. topology diagram and side view of a Greek key motif. C. topology diagram and side view of a jelly roll motif. Arrows and squares are different view of the β-strands.

Figure 7 is an illustration of this approach. Strands C1 and C2 of α-crystallin span residues 84-91 and 93-101 in α-crystallin according to the profile of Figure 5. To test the local packing of the two strands a set of nitroxide pairs with one member on the exposed surface of each strand was constructed. Because the accessibility variations do not indicate the presence of a helix, the set was selected to differentiate between motifs one and two. As shown in Figure 7, strong spin-spin interactions are observed where expected based on a hairpin configuration. This is deduced from the broadening of the double mutant spectrum relative to that of the sum of the single mutants normalized to the same number of spins and the appearance of intensities separated by more than the hyperfine splitting. The pairs 88/98 and 90/99 were designed such that weak spin-spin interactions were expected based on an antiparallel arrangement of the strands. The pair 88/98 is particularly instructive since the 88/97 pair shows extensive broadening. In a β-hairpin, the nitroxide groups at residues 88 and 98 are separated by at least 15-20 Å due to the projection of the side chain in opposite directions.

In contrast, if the two strands adjacent in the primary sequence do not form a β-hairpin, this pattern of proximity is not observed as illustrated in Figure 8. Strands C3 and B3 (residues 129-133) form such a pair. The EPR spectra indicate that K119 at the C-terminus of C3 is not in proximity to residues at the N-terminus of B1. Either the strands are on different sides of the same sheet or each strand is in a different sheet. In general, every pair of strands that are adjacent in the primary structure are tested for proximity in the tertiary structure.

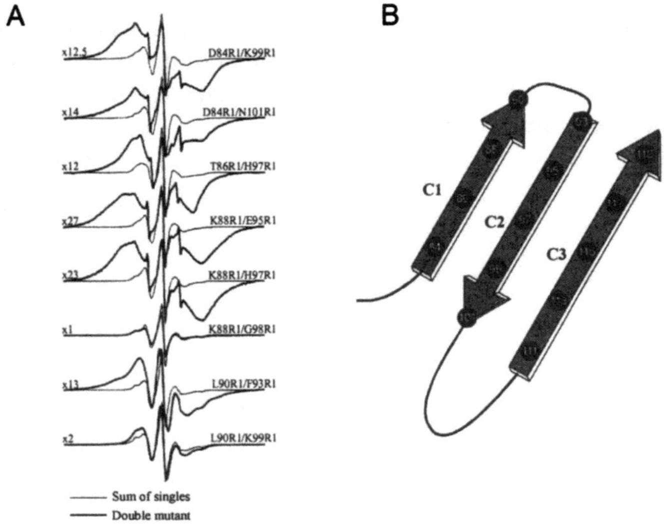

Figure 7. A. Room temperature EPR spectra of the indicated double mutants (heavy trace) and the sum of the corresponding single mutants (light trace). The mutants were selected to test a hairpin packing of C1 and C2 in αA-crystallin. The spectra were normalized to represent the same number of spins. The spectrum of each double mutant was scaled with respect to the corresponding sum of singles to show details of the lineshape. The scaling factor is shown at the left of each spectrum. B. Topology diagram of the αA-crystallin between residues 83 and 120. β-strands, represented by arrows, are sequentially named C1, C2, and C3. Sites used to derive geometric constraints are indicated by the solid circles. The lines representing the turns and loops are not drawn to proportions.

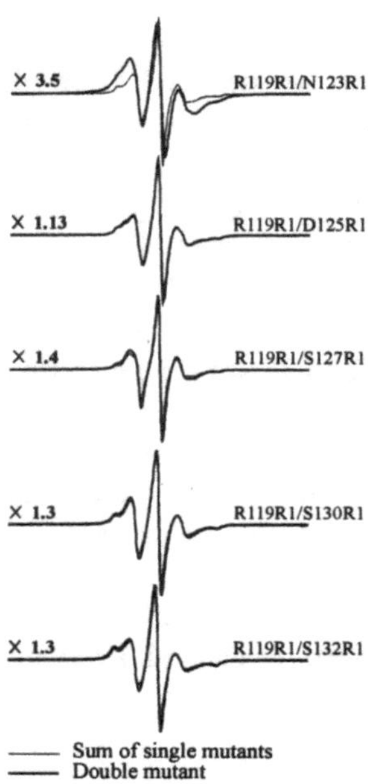

Figure 8. EPR spectra of nitroxide pairs selected to probe the proximity of strand C3 to B3, two strands adjacent in the sequence of αA-crystallin. The light trace is the sum of the single mutant spectra and the heavy trace is the double mutant spectrum. Both spectra were recorded with a 200 G scan width, normalized to the same number of spins and the doubly labeled spectrum is scaled to show the details of the lineshape. (reproduced from Koteiche and Mchaourab, 1999)

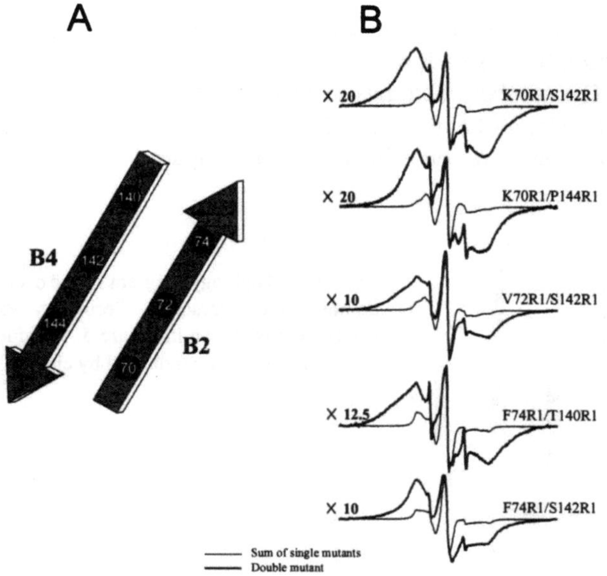

Figure 9. A. Topology diagram of the packing between B2 and B4. Residues indicated by the solid circles are along the more exposed surfaces of the strands and are pointing out of the plane of the figure. B. EPR spectra of nitroxide pairs with one member in B2 and the other in B4. The light trace is the sum of the single mutant spectra and the heavy trace is the double mutant spectrum. Both spectra were recorded with a 200 G scan width, normalized to the same number of spins and the doubly labeled spectrum is scaled to show the details of the lineshape. (reproduced from Koteiche and Mchaourab, 1999)

This approach can also be used to identify strands that are adjacent in the 3D structure but not in the sequence. The possibility of such arrangement is usually deduced from the exhaustive proximity analysis of pairs of strands. For instance, strands B2 spanning residues 70 through 76 and strand B4, spanning residues 140-146, are not sequential yet were suspected of being neighbors in the three-dimensional structure. Evidence suggested that strands B1, B2, B3, and B4 form a separate β-sheet consisting of two β-hairpins: the hairpin formed by B1 and B2 and the hairpin formed by B3 and B4. One of the possible arrangements of the sheet has strands B2 and B4 antiparallel as depicted in Figure 9. This topological model predicts a pattern of proximity similar to that expected if the two strands formed a β-hairpin, with side-by-side packing and a hydrogen-bonded backbone. The EPR spectra shown in Figure 9 demonstrate strong interactions between residues at the N-terminus (C-terminus) of one strand with residues at the C-terminus (N-terminus) of the other strand consistent with such a structure.

Systematic application of this approach not only identifies motifs of secondary structure, but also allows the complete structural characterization of the β-sheets. The number of strands in a sheet, their relative direction and their connectivity along the polypeptide chain can be determined. Two sheets were mapped out in α-crystallin as shown in Figure 10.

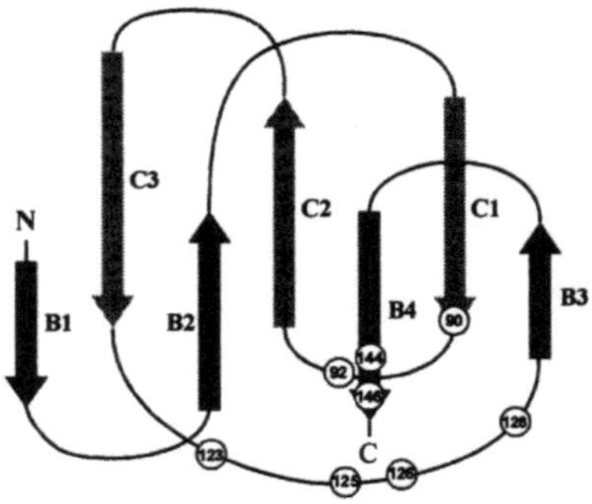

Figure 10. Topology diagram of the α-crystallin domain in αA-crystallin. Secondary structure assignment is shown in Figure 5. Residues used in nitroxide pairs are indicated by circles.

4.1.4 Overall folding pattern of the α-crystallin domain

To determine the packing pattern of the two sheets, Koteiche et al. took advantage of the general rules that govern the assembly of all β-sheet proteins (Chothia, 1984; Chothia and Finkelstein, 1990). In general, these proteins form layer structures with the strands arranged in two sheets that pack face-to-face. The relative orientation of the two sheets defines two general classes of structures (Janin and Chothia, 1980). Aligned packing results in a β-sandwich where the two sheets are independent and one sheet is at an angle of ~30° relative to the plane of the other. In orthogonal packing on the other hand, the two sheets are at 90° and in one corner or in two diagonally opposed corners, one strand crosses from one sheet to the second without interruption. Considering that the α-crystallin domain in the distantly related small heat-shock protein Hsp16.5 from *Methanoccocus Jannaschii* Hsp16.5 has the former folding pattern, it was logical to test whether the strands of αA form a β-sandwich as shown in Figure 10. Furthermore, the accessibility and internitroxide proximities do not suggest that one of the edge strands crosses between the sheets.

Because the most characteristic difference between orthogonal and aligned packing is the relative orientation of the two sheets, there are two possible SDSL approaches to distinguish between the two folds. The first is based on the relative proximities of residues along the packing interface between the two sheets. The second takes advantage of the predicted proximities between inter-strand turns and loops of the two sheets in the aligned β-sandwich. In the orthogonal packing, the loops and turns of the two sheets are distributed over the four edges of the barrel.

The use of nitroxide pairs along the packing interface proved difficult. Most double substitutions resulted in more than 100% increase in the apparent molecular mass. Furthermore, in some of the pairs the sum of the single spectra had a different lineshape than that of the doubly labeled mutant indicating structural rearrangement presumably because of the non-additive effects of the double mutations (data not shown). In total, 10 pairs of buried double mutants were constructed.

The second approach was considerably simpler and the selection of the pairs was facilitated by the structure of Hsp16.5. If the two sheets are aligned then it is expected that the long two-turn segment connecting strands C3 and B3, the turn connecting strands C1 and C2 and the 146-150 dynamic region will be clustered together at one end of the structure. A direct approach to test this model is to construct pairs with one member in each of the loops. The pairs were selected based on the Hsp16.5 structure although the strategy can be applied in the absence of a homologous structure. In such case, it is expected that more pairs will have to be constructed.

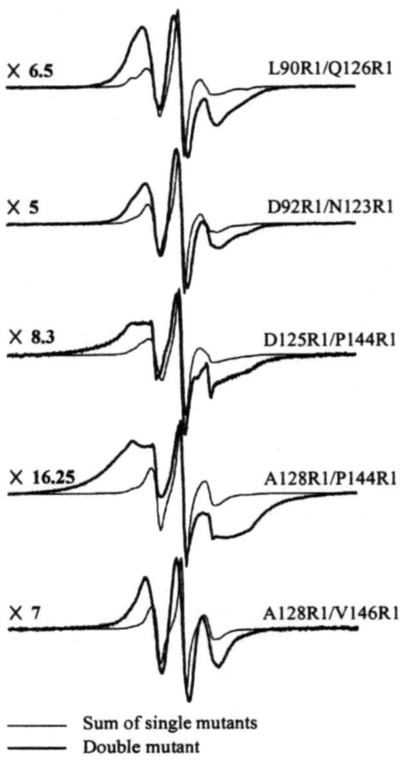

Figure 11. EPR spectra of nitroxide pairs selected to probe the overall fold of the α-crystallin domain. The light trace is the sum of the single mutant spectra and the heavy trace is the double mutant spectrum. Both spectra were recorded with a 200 G scan width, normalized to the same number of spins and the doubly labeled spectrum is scaled to show the details of the lineshape. (reproduced from Koteiche and Mchaourab, 1999).

The EPR spectra of the double mutants, superimposed on the sum of the singles, are shown in Figure 11. Residue L90R1 is on the exposed surface of the edge strand C1 projecting away from the packing interface of the two sheets. Therefore, the proximity between L90R1 and Q126R1 and D92R1 and N123R1 constrains the 121-128 segment to extend from one end to another across the sheet composed of C1, C2 and C3. From the proximities in these two pairs, the N-terminus of strand B3 is localized near the C-terminus of strand C1.

An antiparallel, face-to-face packing of C1 and B3 is deduced from the pairs D125R1/P144R1 and A128R1/P144R1. The former suggests an overlap between the 144-150 segment and the 121-128 segment. This is inconsistent with a perpendicular packing of C1 and B3 that results in the C-terminus of strand B4 being projected over the middle of strand C1. Because B2 is side-by-side antiparallel to B4, then the segment 76-83 has to cross from near the middle of C3 back to C1. Perhaps, this is the most convincing steric evidence against an orthogonal packing.

4.1.5 Determination of Quaternary Packing Interactions

A comparative study of the oligomer symmetry in sHSPs from different species illustrates the application of spin-spin interaction to explore proteins quaternary structure. These studies provided the first experimental evidence indicating that the structural plasticity of the α-crystallin domain has significant impact on the symmetry and size of the oligomeric structure in the sHSP. In the initial SDSL study of α-crystallin, Berengian *et al.* targeted residues 109-120 constituting strand C3 (Berengian *et al.*, 1997). The rationale was that this is the region of highest homology between the two α-crystallin isoforms. Furthermore, in mammalian Hsp27, the equivalent position to E113 in the sequence alignment was reported to readily form disulfide bonds with a cysteine of a neighboring subunit. Spin-spin interactions were observed at residues 110-113 at the N-terminus of strand C3 strongly suggesting a subunit interface. Mchaourab *et al.* demonstrated that this subunit interface is conserved in mammalian Hsp27. Spin-spin interactions were observed at the N-terminus of the 134-142 sequence in Hsp27, the equivalent to C3 in the sequence alignment (Mchaourab *et al.*, 1997a) (Figure 12).

To determine the symmetry of the interface, Berengian *et al.* titrated Hsp27 C137R1 and αA R136R1 with increasing amounts of the cysteine-less wild type protein (WT*) (Berengian *et al.*, 1999). This is illustrated in Figure 13 where the addition of 0.25 molar equivalent WT* results in a composite spectrum where the amplitude of the central line is dominated by singly labeled oligomers. The dependence of the fractional population of singly labeled oligomers, determined from the EPR spectra, on the ratio of unlabeled to labeled subunits revealed a dimeric interface both in Hsp27 and α-crystallin.

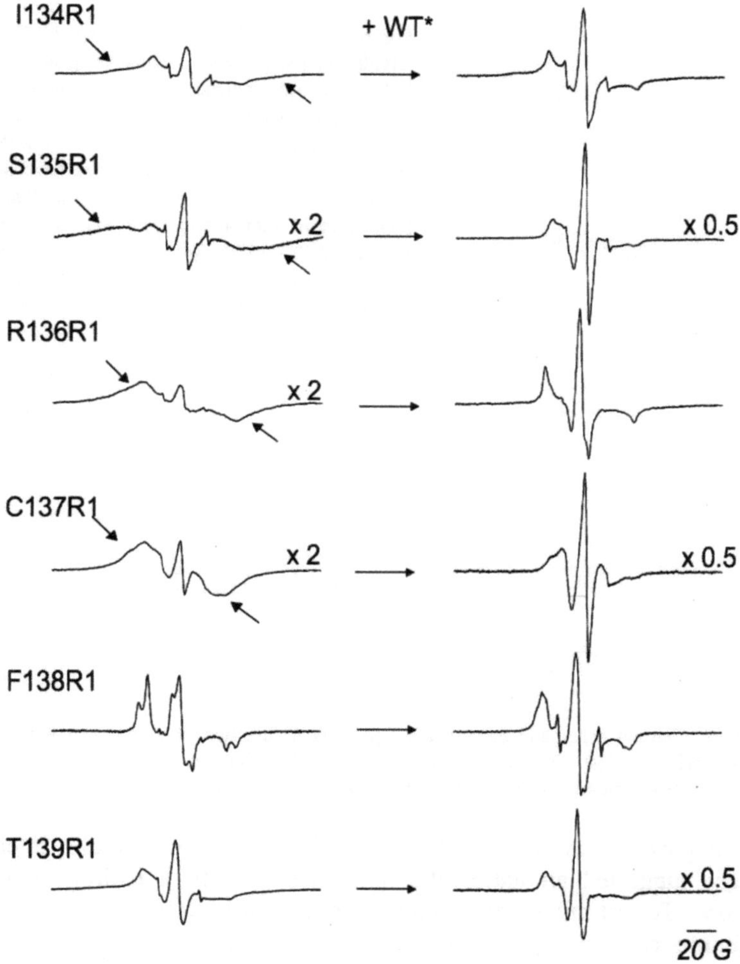

Figure 12. Room temperature EPR spectra of singly labeled Hsp 27 mutants in the presence and absence of excess unlabeled WT. All spectra were recorded with a 200 G scan width. They were normalized by double integration to represent the same number of spins. The broadened spectra were scaled to show the detailed of the lineshape. The scaling factor is shown to the right of each spectrum. (Reproduced from Mchaourab et al., 1997).

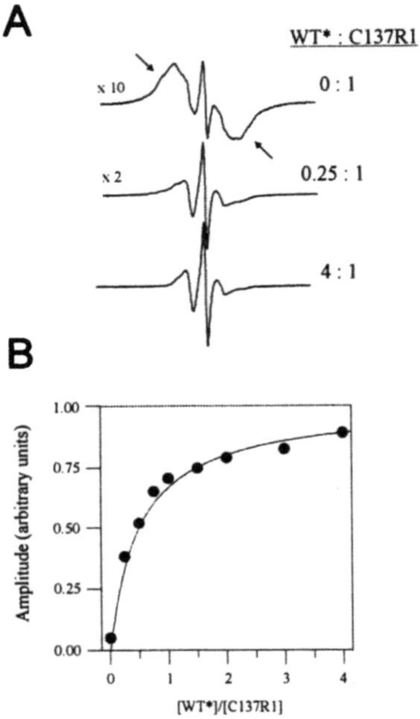

Figure 13. A. EPR spectra of Hsp 27 C137R1 refolded in the presence of increasing amounts of WT*. Arrows indicate spectral features arising from dipole-dipole broadening of the spectrum. Where appropriate, a scaling factor is indicated to the left of the spectrum. All spectra have a scan width of 200 G. B. Increase in the fractional population of monomeric spins versus the molar ratio of WT*:C137R1 (•). The solid line is the theoretical increase calculated from the binomial distribution for a dimer. (Reproduced from Berengian et al., 1999)

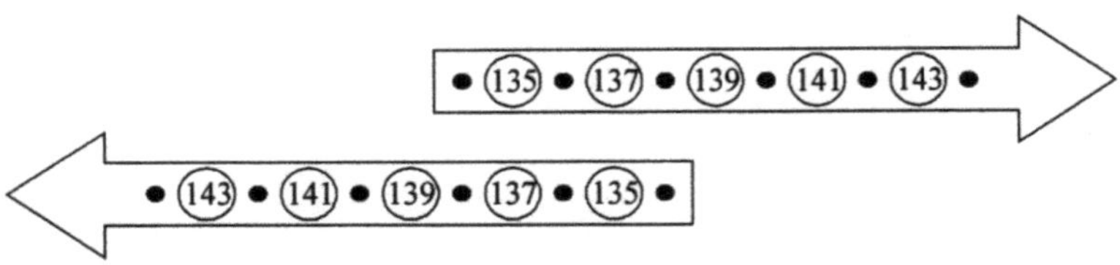

Figure 14. Cartoon schematic depicting the packing of the two □-strands at the isologous subunit interface of Hsp 27. Open circles indicate residues that are more solvent exposed and filled circles represent buried sites. (reproduced from Berengian et al., 1999)

The presence of β-strands at a two-fold symmetric subunit interface strongly suggested an antiparallel arrangement with a hydrogen-bonded backbone. To test this model, mixed oligomers were formed from two types of subunits each having a spin label at a different residue. On a statistical basis, 50% of the refolded oligomers consist of mixed subunits of the two R1 labeled mutants. Spectral subtraction was used to separate the 25% contribution from homo-oligomers of each mutant. The pattern of interaction was found to be consistent with the model of Figure 14. Alternatively, double mutants could have been constructed to avoid the spectral subtraction.

Finally, the equivalent region in *M. Tuberculosis* Hsp16.3, a bacterial sHSP with an ordered oligomeric structure, was probed with the purpose of exploring the extent to which the interface and its associated two-fold symmetry were conserved in distant sHSP (Berengian *et al.*, 1999). In contrast to the mammalian sHSP, spin-spin interactions were observed at a single residue near the C-terminus of the strand. Furthermore, gel filtration chromatography indicates that Hsp16.3 has an equilibrium intermediate consisting of a trimer populated in the presence of 1M Gnd-HCl. Under these conditions, panel a of Figure 15 shows that the spectral features arising from spin-spin interactions are eliminated indicating that the observed spin-spin interaction at residue S103 is due to the assembly of trimers. The complete unfolding curve is shown in panel b. Two cooperative transitions are reported by R1: the first arising from an oligomer to a trimer and the second from a trimer to an unfolded monomer. When S103R1 was titrated with increasing amounts of WT, the increase in the fractional population of singly labeled interfaces follows that expected from a three-fold symmetric interface (panel d). Unlike mammalian sHSP, Hsp16.3 is a trimer of trimers.

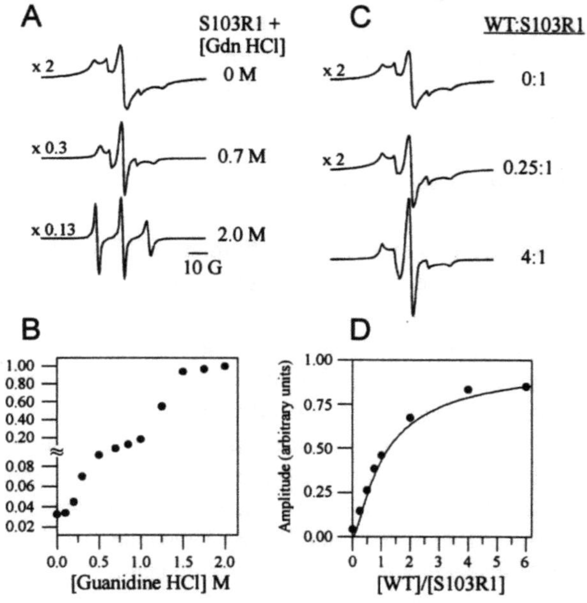

Figure 15. A. EPR spectra of Hsp 16.3 S103R1 in the trimeric and unfolded states B. Gdn HCl denaturation curve obtained from measurement of the change in the high-field line amplitude of S103R1 plotted versus the molar concentration of Gdn HCl. C. refolded in the presence of increasing amounts of WT. Where appropriate, a scaling factor is displayed to the left of the spectrum. All spectra have a scan width of 160 G. D. Increase in the fractional population of monomeric spins versus the molar ratio of WT:S103R1 (•). The solid line is the theoretical increase calculated from the binomial distribution for a trimer. (reproduced from Berengian et al., 1999)

4.2 Membrane proteins

The power of SDSL in the determination of structural parameters in proteins is perhaps ideally suited to the study of membrane proteins. This class of proteins represents a particular challenge for other structural approaches. Their size, requirement for a detergent micelle or lipid vesicle, and propensity to form oligomeric complexes, tends to preclude the application of solution NMR methods, while solid state NMR methods are yet to deliver for systems larger than 30 kDa. At the same time, despite the impressive advances in x-ray crystallography of membrane proteins, obtaining high-resolution crystals remains a daunting task in the majority of cases. The use of spin-spin coupling between pairs or systems of spins has been very productive in defining general patterns of inter-residue proximity (see above), and there are a number of examples of this approach in the literature. However, determination of de-novo folds in membrane proteins seems to be limited by the difficulty in generating and expressing a relatively complete set of spin-labeled mutants. The following examples highlight recent advances in the use of spin-spin dipolar couplings in the determination of tertiary or quaternary fold information from site-directed spin labeled membrane proteins.

4.2.1 HIV envelope glycoprotein gp41

Membrane fusion by simian (SIV) and human (HIV) immunodeficiency viruses is mediated by the envelope glycoproteins gp120 and gp41. Upon membrane binding, the ectodomain of gp41 is thought to undergo a major conformational rearrangement that ultimately leads to the fusion of the viral and host membranes, thus initiating viral infection. Full-length gp41 contains a fusion peptide at its N-terminus, followed by a "leucine zipper" or heptad repeat region and a heavily glycosilated portion immediately before a putative transmembrane segment at its C-terminus. In the first application of the Fourier deconvolution method to determine inter-spin distances, Rabenstein and Shin (Rabenstein and Shin, 1996) analyzed the tertiary structure of the gp41 ectodomain in its prefusogenic state.

The study was based on their ability to reconstitute the protease-resistant core of gp41, based on the interaction between peptides from the heptad repeat (hr), and from the region immediately adjacent to the transmembrane segment (bp). These two peptides were synthesized with cysteines at either three (hr) or two (bp) different positions, and each cysteine containing peptide was labeled using an iodoacetamido spin label. Equimolar mixtures of each spin-labeled peptide were analyzed for dipolar interactions in the frozen state using peptide mixtures with only one spin-labeled peptide at a time as a reference. Since the gp41 ectodomain assembles as a trimer, there was some concern that individual spin-labeled peptides could be close enough to interact across monomers. However, this potential error was calculated to produce at most a 9 % underestimation in the value of the calculated distance.

Based on a matrix of five unique inter-peptide distances, a tertiary and quaternary packing model was generated, in which hr and bp were proposed to be anti-parallel, forming a trimeric coiled-coil. This model was found to be in agreement with specific tertiary contacts deduced from mobility measurements of the spin-labeled peptides. The model was also consistent with recent higher resolution models based on X-ray crystallography (Chan *et al.*, 1997; Weissenhorn *et al.*, 1997) or multidimensional NMR measurements (Caffrey *et al.*, 1998).

4.2.2 *KcsA*, the Streptomyces lividans K^+ channel

Potassium channels are integral membrane proteins that catalyze the translocation of K^+ ions across the membrane with exquisite selectivity and at near diffusion-limited rates. The *Streptomyces lividans* K^+ channel (*KcsA*) is a 160-residue protein that forms homotetrameric channels closely related to the pore domain of larger voltage-dependent channels (Schrempf *et al.*, 1995). When purified and reconstituted in lipid bilayers, *KcsA* forms K^+-selective channels with a single channel conductance close to 140 pS and selectivity properties identical to those of other eukaryotic K^+ channels (Cuello *et al.*, 1998; Heginbotham *et al.*, 1998; Schrempf *et al.*, 1995).

These properties, and the fact that *KcsA* is readily expressed in *E. coli* at milligram levels made this channel an ideal target for high-resolution structural studies. Crystallographic studies led to the first high-resolution structure of *KcsA* by Doyle et al. (Doyle *et al.*, 1998), an achievement that has lent a firm structural foundation to more than three decades of functional work on K^+ channels. The channel is formed by the association of four subunits that contribute equally to form a water-filled pore. Each subunit has two transmembrane segments, TM1 in the periphery of the complex, and TM2, lining the permeation path. Towards the extracellular face of the channel is the selectivity filter, were ions are dehydrated by an extended chain that includes the so-called signature sequence (Heginbotham *et al.*, 1994), present in all known K^+ channels. The structure does not contain information on either the N- or the C- terminus, since a truncated version of the channel is required to obtain high quality crystals.

An extensive SDSL study of *KcsA* was carried out by Perozo and co-workers (Perozo *et al.*, 1998) with the idea of obtaining structural dynamic information of the two transmembrane segments, and to generate a three-dimensional architecture of the full-length channel in its native environment.

A total of sixty-six single cysteine mutants were prepared and analyzed for residues 22 - 52 and residues 86 - 120 of *KcsA*, where the two putative transmembrane segments are thought to be located. Using the type of discrete Fourier transform method described above (Cornette *et al*., 1987; Donnelly *et al.*, 1994), angular frequency information was extracted from residue environmental profiles from each of the scanned protein segments and used in the assignment of secondary structure elements (Perozo *et al.*, 1998). Solvent accessibility information derived from either O_2 or NiEdda collision frequencies was in excellent agreement with the crystal structure, suggesting that TM1 has a large lipid-exposed surface, while TM2 is more protected from membrane lipid.

Because *KcsA* is a tetrameric complex and SDSL was carried out placing one cysteine per monomer, it is expected that only those positions close to the four-fold axis of symmetry will show considerable dipolar coupling. Strong broadening was in fact observed every three or four residues on one face of the C-terminal half of TM2 (in residues A108, T112, V115 and G116), with significantly weaker spin-spin coupling detected on the N-terminal half of the helix (Figure 16). This broadening disappears when channels are under-labeled so that on average, there is only one spin label per tetramer, clearly indicating that this effect is a consequence of strong spin-spin interactions. Due the tetrameric stoichiometry of *KcsA*, spin-spin coupling can originate either from interactions between next-neighbors or from diagonally related subunits. Thus, in an ideally labeled *KcsA*, four spin labels would be contributing to the spectral broadening. Under these conditions, a qualitative estimate of the magnitude of spin-spin interactions can be obtained from the ratio of amplitudes of the central resonance line (M=0) between the under-labeled and fully labeled mutant, both normalized to the total number of spins in the sample. This interaction parameter Ω, was used to survey patterns of spin coupling along the length of TM2 and deduce possible inter-helical orientations.

These results have been recently confirmed by Gross et al (Gross *et al.*, 1999), in a similar SDSL analysis of TM2 in *KcsA*. The extent of the spin-spin interaction in residues along the lower portion of TM2 of *KcsA* was quantified using a new method based on mixtures of structurally related paramagnetic and diamagnetic labels. Actual distances were calculated assuming that coupling between spin labels in neighboring subunits dominated spectral broadening and most were found to be in general agreement with Cβ-Cβ values derived from the crystal structure. Nonetheless, some differences in the completeness of labeling were reported for some of TM2 residues compared to those of Perozo et al (Perozo *et al.*, 1998).

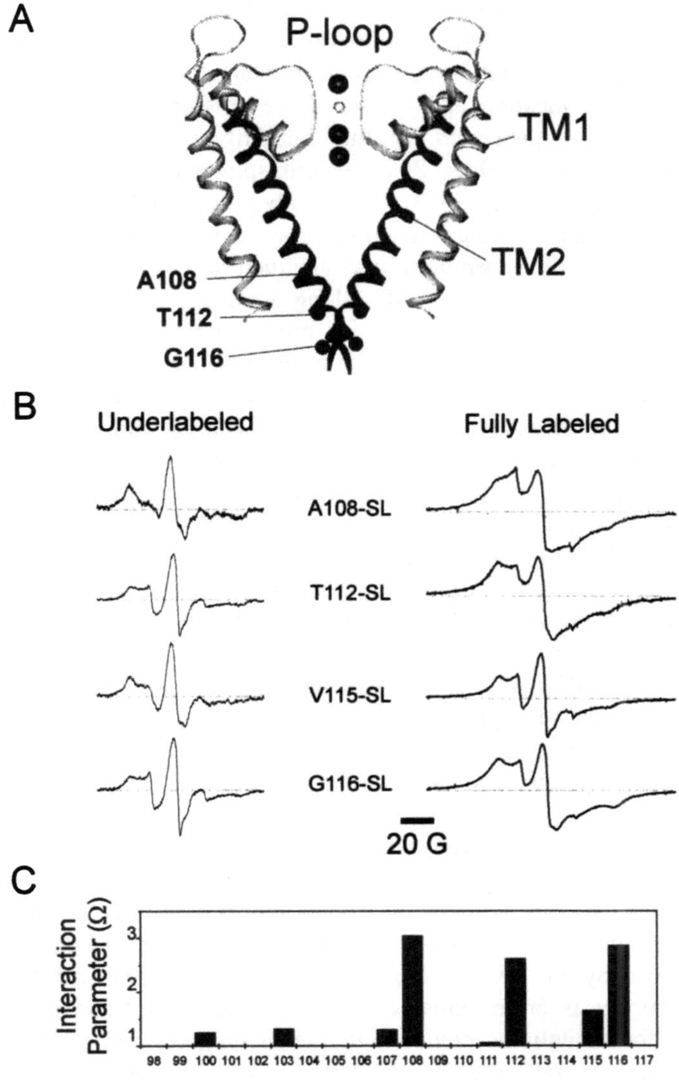

Figure 16. A. Location of some of the interacting residues in the crystal structure. For clarity, two of the subunits have been eliminated. B. EPR spectra of residues showing strong spin-spin interactions in the second transmembrane segment of the K^+ channel KcsA. The spectra at left were from channels underlabeled with the MTS-spin label at a 1:10 label:channel molar ratio. The spectra at right were obtained from channels labeled using the standard protocol (5:1 molar ratio). C. Estimation of the spin-spin coupling for the C-terminal end of TM2 using interaction parameter Ω.

A three-dimensional packing model of the transmembrane segments of *KcsA* was constructed by Perozo and coworkers (Perozo *et al.*, 1998) based on the values for individual residue environmental parameters, their periodic behavior, and the location and extent of the spin-spin interactions. The packing model was generated by first building α-helices according to canonical Phi-Psi angles values. Residues with spin-spin interactions in TM2 were then placed within 12-14 Å of each other, according to conservative estimates of inter-spin distances from other systems and in agreement with the overall configuration seen in the crystal structure. TM1 was subsequently placed around the perimeter of TM2, along a narrow path of immobilized residues and following known rules for transmembrane helix packing. A tetramer was finally generated by imposing fourfold symmetry on each TM1/TM2 monomer. General inconsistencies were evaluated by calculating individual residue accessibilities (using DSSP, (Kabsch and Sander, 1983) with a 1.5 Å probe) from the tetrameric packing model, and comparison with the mobility and accessibility profiles derived from EPR data.

Overall, secondary structure estimates from EPR data in TM1 and TM2 (Perozo *et al.*, 1998) as well as packing models based on dipolar coupling data from TM2 (Gross *et al.*, 1999; Perozo *et al.*, 1998) tend to agree remarkably well with the crystallographic data. Nonetheless, apparent differences in the precise location of the narrowest point within the TM2 bundle (position 107 in the crystal structure or 108 from EPR data) could originate from small changes in the conformation of the truncated channels (used to obtain high-resolution crystals) or as a consequence of the uncertainty in the precise orientation of the spin label.

4.2.3 Pre-synaptic SNARE complex

The determination of a three-dimensional model of the SNARE complex by Shin and collaborators constitutes a fine example of the applicability of EPR spectroscopy in the determination of tertiary/quaternary folds. The SNARE complex is an essential component of the fusogenic machinery responsible for excitation-secretion coupling in synapses and of membrane trafficking in general. SNARE is formed by the association of the presynaptic plasma proteins SNAP25 and syntaxin-1 with the synaptic vesicle protein VAMP1 (or VAMP2), also known as synaptobrevin. SNAP-25 can be further divided into N- and C-terminal domains, which can be expressed or synthesized separately. All these proteins form a stable complex that is required, and may be sufficient for membrane fusion (Weber *et al.*, 1998).

Poirier et al (Poirier *et al.*, 1998) used SDSL and dipolar coupling-based distance determinations to investigate the three-dimensional arrangement of the N- and C-terminal domains of SNAP-25 within the SNARE complex. A set of cysteine mutants was generated for each of the individual components of the SNARE complex, expressed in bacteria as fusion proteins and labeled with a MTS spin label (Fig. 1). The SNARE complex was obtained by combining either wild-type or spin-labeled SNAP-25[N], SNAP-25[C] and syntaxin-1A with wild-type VAMP-1 or spin-labeled VAMP2.

Out of a total of 22 complexes containing two spin labels, 7 did not show spin-spin interactions, suggesting that these residues were separated by more than 25 A. The remaining 15 double-labeled samples displayed a range of dipolar coupling strengths that allowed distances from 7 to 22 A to be determined. Figure 17A shows a schematic representation of all of the individual components of the SNARE complex, highlighting the residues individually spin-labeled and their respective inter-spin distances calculated by the Fourier deconvolution method (Rabenstein and Shin, 1995). To confirm and further constrain three-dimensional packing models, Poirier and co-workers also correlated the extent of side-chain immobilization in the singly labeled components with the predicted location of positions a or *d* in a coiled coil structure (although not performed in a systematic way).

This information was then used to generate a three-dimensional model of the SNARE complex. The model was grounded on established evidence suggesting that the entire complex was assembled as a four-helix coiled coil (Weimbs *et al.*, 1997). Distance data placed each of the individual helices in a parallel configuration relative to the others. Based on the multiple close distances among residues in SNAP-25[N] and SNAP-25[C], these were assigned neighboring positions in the bundle, while syntaxin-1A and SNAP-25[N] were also next- neighbors. The final position in the bundle was given to VAMP2. An initial rough model was built from four straight, parallel helices, using the experimentally determined distances to manually define inter-helix registers. The final structure (Figure 17B) was obtained by applying energy minimization and restrained simulated annealing to this initial structure using a parabolic energy function to enforce the inter-helical distance constraints. The integrity of each helix was preserved during the annealing steps by imposing a standard α-helical pattern of hydrogen bonding.

Note however that the distances obtained from dipolar couplings were not imposed directly in the annealing steps. Rather, they were used as a general blueprint for crude model building while the actual distance constraints were derived from Cα-Cα patterns of known coiled-coils. It should be pointed out that this proposed model of the SNARE complex turned out to be very close to the crystal structure determined almost simultaneously by Sutton et al. (Sutton *et al.*, 1998). Although there was a relatively large root-mean-squared deviation between the EPR-derived distances and those calculated from the model (2.8 A RMS), the general agreement between the X-ray and EPR-based structures is remarkable indeed.

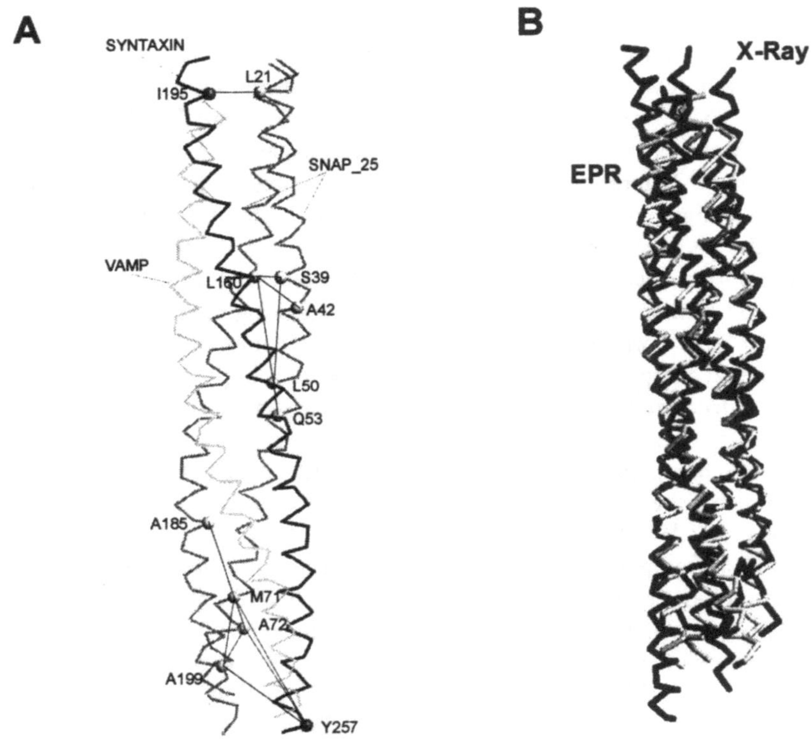

Figure 17. The three-dimensional structure of the pre-synaptic SNARE complex as derived from site-directed spin labeling. **A.** Proposed model showing individual spin-labeled residues (spheres) and their distance relationships. Each helix in the bundle is depicted with a different gray level. **B.** Backbone superposition between the EPR structure and the corresponding crystal structure.

4.2.4 Lactose Permease

One of the most exhaustively studied transport systems, the lactose permease is an integral membrane protein responsible for the stoichiometric translocation of β-galactosides across the membrane using the energy accumulated in the transmembrane H^+ gradient (Kaback and Wu, 1999). A great deal of evidence indicates that the Lac permease is composed of 12 transmembrane helices and that both N- and C termini face the cytoplasm. Despite the fact that the lac permease can be easily obtained in milligram quantities, detergent solubilized, and routinely reconstituted into lipid vesicles with full activity, the road towards a working three-dimensional model of lac permease has been slow and painful.

Most of the structural studies of Lac permease have come from the collaborative efforts of Kaback, Hubbell and co-workers, using a number of reporter group biophysical approaches to bypass the current lack of quality crystals from native or even engineered Lac permease constructs. SDSL studies have been carried out in an attempt to obtain experimental evidence for specific helix packing models for the 12 transmembrane segments of Lac permease (reviewed by (Kaback and Wu, 1999)). These have include nitroxide scanning studies (He *et al.*, 1996; Ujwal *et al.*, 1995; Voss *et al.*, 1996; Voss *et al.*, 1997), distance estimations from doubly-labeled protein (He *et al.*, 1997; Sun *et al.*, 1999; Wang *et al.*, 1998; Wu *et al.*, 1996; Zhao *et al.*, 1999) and newly developed methods based on engineering metal biding sites (Voss *et al.*, 1995b) to measure paramagnetic metal binding or to compute distances through metal-nitroxide dipolar coupling (Voss *et al.*, 1995a).

The first set of distances from the Lac permease were obtained by Voss et al (Voss *et al.*, 1995a), who used engineered metal binding sites to obtain distances to site-directed spin-labeled sites. Distances from a metal binding site located in the loop between helices III and IV and positions 103, 111 and 128 of helix IV were found to increase progressively, and this increase was consistent with an α-helical conformation for this transmembrane segment.

Distance measurements have also been obtained from spin-labeled sites on different transmembrane segments. Wu et al (Wu *et al.*, 1996) determined inter-spin distances in doubly spin-labeled lac permease for residues along helices V and VII or V and VIII. Three independent distances were estimated between positions 148 in helix V, and 275 in helix VIII, or 226 and 228 in helix VII. These distances, together with additional experiments on chemical crosslinking were then used to deduce the helical packing of this region of the molecule. He et al (He *et al.*, 1997) confirmed the spatial proximity between Arg 302 in helix IX and Glu 269 in helix VIII by two independent approaches. First, it was demonstrated that histidine mutations at positions 269 and 302 generate a high-affinity Mn^{++} binding site.

Measurement of spin-spin interactions in doubly spin-labeled permease at positions 269 and 302 confirmed their proximity. Residues 273 (Helix VIII) and 299 (Helix XI) were also found to be spatially close from measurement of spin-spin interactions in doubly spin-labeled lac permease (Wang *et al.*, 1998).

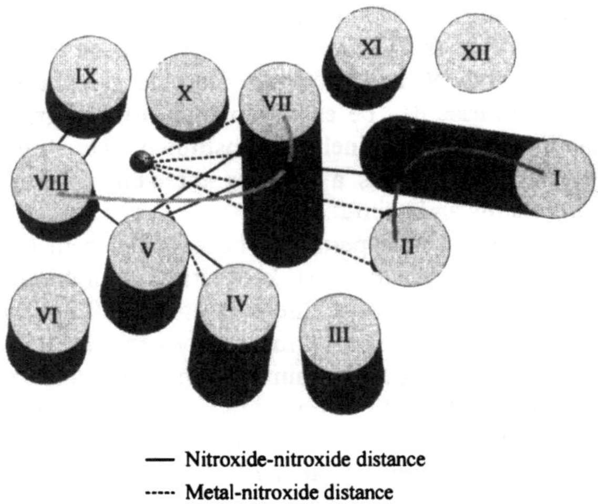

Figure 18. Proposed helix packing model for the lactose permease. Although a variety of techniques have been used to reach this particular model, only data derived from inter-nitroxide and metal-nitroxide spin-spin interactions are highlighted.

More recently, Sun et al (Sun *et al.*, 1999), used doubly spin-labeled lac permease mutants to verify results from thiol crosslinking experiments that place periplasmic loops I/II in close proximity to loops VII/VIII and XI/XII. Zhao et al (Zhao *et al.*, 1999) using doubly spin-labeled lac permease showed that residues Glu126 (Helix IV) and Arg144 (Helix V) are spatially close. This confirms earlier mutagenesis studies suggesting these residues form an anion pair.

So far, the best example of the use of a distance matrix to support a specific helix-packing model in Lac permease comes from the work of Voss et al (Voss *et al.*, 1998). A high-affinity metal binding site was created by His mutagenesis at residues 302 (Helix IX) and 325 (Helix X) and used as a reference point to estimate distances to positions in helices II, IV, and VII. Distances were obtained by contrasting dipolar broadening of individually spin-labeled sites in the presence of Cu(II) as paramagnetic metal, or Zn(II) as control. From this study, a general packing scheme was proposed (Fig 18), with helix VII in the core, surrounded in clockwise order by helices II, V, VIII, IX, X and XI. Based on these and a number of other complementary biophysical measurements an overall view of the packing of transmembrane segments in Lac permease is starting to emerge (Kaback and Wu, 1999).

5. DETECTION OF CONFORMATIONAL DYNAMICS IN SPIN-LABELED PROTEINS

5.1 Conformational dynamics in proteins

The functional behavior of most proteins is intimately related to their conformational dynamics. Thus, a thorough understanding of structure-function relations in a given protein requires a molecular description of the types and extent of the protein movements underlying function, including catalysis, transport, cellular locomotion, regulation of activity, and formation of protein assemblies. X-ray and NMR structures have made it clear that proteins undergo a wide variety of motions, whether as a consequence of a triggered action (ligand binding, protein-protein contact, physical stimuli) or as the result of the natural fluctuations that usually occur around a flexible linker between domains. Closer examination reveals that the types and dynamic behavior of these movements can be hierarchically divided according to four major groups (Figure 19): Side-chain rotamers (ps-ns), reorientation of secondary structure elements (ns), large domain movements (ns-μs), and whole subunit rearrangements in cooperative oligomeric complexes (μs-ms).

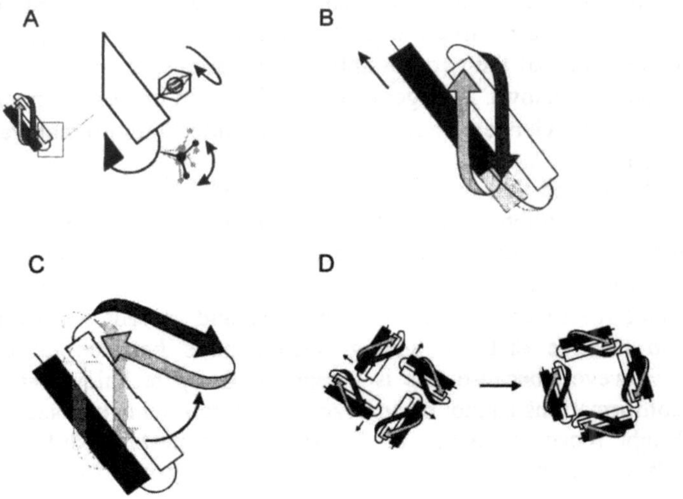

Figure 19. The types of molecular motions in proteins. **A.** Side-chain rotamers, **B.** Reorientation of secondary structure elements, **C.** Large domain motions, **D.** Whole subunit rearrangements.

Among the types of mechanisms of motion mentioned above, by far, the best understood is that occurring among protein domains. Gerstein and co-workers have reviewed and classified domain motions from a growing database of proteins with known three-dimensional structures (Gerstein *et al.*, 1994). They concluded that most domain motions ultimately originate from two basic mechanisms, hinge bending or shear motions. In hinge bending, motion occurs due to changes in a few main-chain torsion angles along a localized region of the protein connecting two interacting modules (strands, β-sheets, α-helices, or combinations of them). These two modules are normally not constrained by tertiary packing interactions, and the transition from "open" to "closed" positions is accompanied by the exclusion of water molecules from the inter-domain space. In contrast, interactions among tertiary structure elements are severely constrained due to packing interactions, precluding large changes in main-chain torsion angles. Hence, shear motions have been observed mostly among closely packed regions of proteins. Shear involves small changes in side-chain torsion angles among interdigitating side-chains, without large changes in main chain conformation. These can be observed either parallel or perpendicular to the interface of closely packed segments of a polypeptide chain. An argument is also made that because these types of protein rearrangements are usually fast, transition between the two conformations cannot involve large energy barriers. This is particularly relevant in the analysis of equilibrium fluctuations, since the relatively weak forces involved in crystal packing are known to favor one of the two possible conformations, as in the case of the T4 lysozyme (see below). A large database of protein motions is maintained on the web by Gerstein and co-workers (http://bioinfo.mbb.yale.edu/MolMovDB).

It can be argued that one of the most powerful aspects of site-directed spin labeling as a structural approach, is its potential to discern the types of protein motions described above. Information on the relative movement of secondary structure elements, domains or whole subunits can be obtained by studying patterns of change in probe mobility and solvent accessibility or changes in relative distance among residue pairs between domains or subunits. However, because this is a reporter group technique, side-chain rotamer conformations cannot be analyzed. In the following two sections, we will highlight recent advances in the analysis of protein conformational dynamics based on the study of inter-spin dipolar couplings for spontaneous (T4 Lysozyme) or triggered rearrangements (several membrane proteins). Triggered conformational changes in Myosin and other motor proteins have been extensively covered elsewhere (Fajer, 2000) and will not be considered here.

5.1.1 Equilibrium fluctuations

Protein conformational equilibria are the results of the thermally activated interconversion of the structure between multiple states of similar energies. These motions differ in their amplitude, time scale and biological consequence. Thermally activated equilibrium fluctuations in the structure can facilitate access of a substrate to an active site cleft, binding of an effector to an allosteric site and transport of ions through channels. Because thermally activated equilibria involve states of similar energies, lattice forces may be sufficiently large to select a particular state in a crystal, obscuring the existence of others. In some cases, crystallization of the same protein in two different crystal forms allows the observation of multiple conformations. Dynamics can also be inferred from the occurrence of disorder in the crystal and analysis of thermal factors.

Conformational equilibria in solution can be explored by SDSL as long as the interconversion rate is within the time scale of the employed EPR technique. Since the dynamic range of EPR covers over 7 orders of magnitude, it provides a suitable window for many protein motions of interest. Both the mobility of the nitroxide side chain and the distance between the nitroxide and a second paramagnetic center can be used to detect conformational changes. The structural determinants of these observables as well as their resolution makes them ideal to detect large scale concerted motion of large segments of the protein molecule. The population of multiple structures results in a heterogeneous environment in the vicinity of the nitroxide that is reflected in the EPR spectral lineshape as multiple motional states. Similarly, if the dynamics involve rigid body motion of a secondary structural element it will modulate the proximity of an attached nitroxide to a nearby reference paramagnetic center. Although the contribution of the backbone flexibility to the mobility of the nitroxide is not well understood, in principle backbone fluctuations are also detectable. Because conformational equilibria involve motions that are constantly occurring in solutions, the spectral properties are ensemble averages with each state with distinct spectral characteristics contributing to a different degree.

In practice, detailed quantitative interpretation of EPR data arising from such ensembles are yet to be developed. At the fundamental level, multicomponent EPR spectra can arise from different conformations of the spin label itself. Furthermore, the addition of parameters such as the number of intermediate states and the mobility and/or internitroxide distance in each state makes the linear CW-EPR spectrum underdetermined.

The problem is considerably simplified if the structure of one of the conformers is known. This provides a reference state against which spectral observables can be compared. An ideal situation arises if the states of known structure can be selectively populated via changes in the physicochemical conditions or by binding of a ligand or a substrate. In this case, reference EPR spectra can be obtained and the sign of the changes in mobility or interresidue distances when multiple states are populated can be interpreted in terms of the nature, extent and magnitude of the movement involved in the equilibria.

5.1.2 Hinge-bending motion of T4 Lysozyme (T4L)

In the context of exploring the relation between the lineshape of side chain R1 and the structure of T4 Lysozyme, it became apparent that at some sites the deduced mobility of R1 was not consistent with what is expected based on the X-ray structure. The sites in question are clustered together in one region of the protein at the interface of helices A and C. Two lines of evidence suggested that this region is inherently flexible. First, analysis of crystallographic thermal factor in the WT molecule led to the hypothesis that one domain undergoes a rotation and displacement relative to the other one. In the context of this motion helix A moves relative to helix C. Second, mutants of T4L I3P and M6I in helix A, where residue isoleucine 3 is substituted by a proline and residue methionine 6 is replaced by an isoleucine crystallize in different conformations that are related to the WT and to each other by a bend about a hinge in the molecule suggesting that these conformers are of similar energy. In one of the conformers the C-terminal domain undergoes a rotation and a displacement relative to the N-terminal domain resulting in a repacking of the interface between helices A and C. In the X-ray structure of the WT, the active site cleft, located at the interface between the two domains, is occluded and inaccessible to the substrate. Thus, from a catalysis perspective, hinge bending motion that opens up the active site allowing substrate binding must be an integral aspect of the function.

Except for the crystallographic analysis of T4L mutants, direct experimental verification of hinge bending remained elusive. A direct approach to test whether the conformation of T4L in solution more closely resembles the open state of M6I rather than it does the WT structure is to measure inter-residue distances for selected pairs of sites that distinguish between the two conformations. Mchaourab *et al.* (1997b) designed pairs such that one member of the pair was in the N-terminal region (residues 4, 22, 35) and the other in a region that undergoes relative displacement upon hinge bending (residues 60, 61, 64, 71, 109 and 137). Furthermore, rather than relying on absolute distances, the sites were selected to provide a distinctive pattern of distance changes unique to the close-open transformation. This approach has the advantage that the hinge-bending hypothesis can be tested by observing the sign of the relative displacement between pairs. As in fold determination, the use of the patterns approach minimizes the impact of the lack of accuracy in the measured distance due to the uncertainty in the position of the nitroxide relative to the α-carbon.

Given that conformational equilibria may be strongly temperature and solvent dependent, the use of low temperature is problematic. Therefore, internitroxide proximities were determined at room temperature where the correlation time of the interspin vector is estimated to be about 6 ns. In this regime, the rotational motion of the protein modulates the dipolar interaction and leads to a mechanism of relaxation as described in section 2. Examination of the EPR spectra at room temperature revealed that the observed pattern of proximities was inconsistent with the X-ray structure of the WT. However, because of the uncertainty in the position of the nitroxide relative to the α-carbon and the possibility of multiple conformation of the nitroxide itself the detailed interpretation of the data is problematic. In order to characterize the most populated state(s) in solution, locking the protein in a reference state, the open or closed states, is required. For T4L, it was found that the mutation T26E in the active site results in a covalent enzyme-substrate intermediate. X-ray analysis reveals that the structure of this intermediate in the crystal has a closed conformation similar to that of the WT within the resolution of SDSL (Kuroki *et al.*, 1993). Therefore, proximities were examined in this reference state.

5.1.3 T4L Conformation in the presence of substrate

The pairs shown in Figure 20 provide a fingerprint of the state populated in solution in the presence of substrate. The heavy trace shows EPR spectra for the indicated double mutants, while the light trace shows the sum of the spectra for the corresponding single mutants, both in the T26E background. In all but D22R1/T109R1, a significant broadening and concomitant loss of intensity is observed in the EPR spectra of the double mutants compared to the sums of the spectra of the corresponding single mutants, demonstrating spin-spin interactions between the nitroxides. Qualitative interpretation of the results reveals that they are consistent with the known crystal structure of the T26E variant, i.e. the closed conformation. For example, the inter-spin distance between the nitroxides in D22R1/T109R1 is estimated from molecular modeling to be about 21 Å, near the limit of detection for the dipolar interactions (see below), consistent with the small spectral broadening. For D22R1/R137R1 and F4R1/V71R1, the extreme broadening of the double mutants spectra relative to the sum of the singles is consistent with molecular modeling suggesting that the nitroxide groups are near van der Waal's contact. The distances between the nitroxide moieties in

K35R1/T109R1, K35R1/R137R1 and F4R1/K60R1 are estimated from molecular modeling to be 14, 13, and 17 Å, respectively, consistent with the observed intermediate degree of spectral broadening.

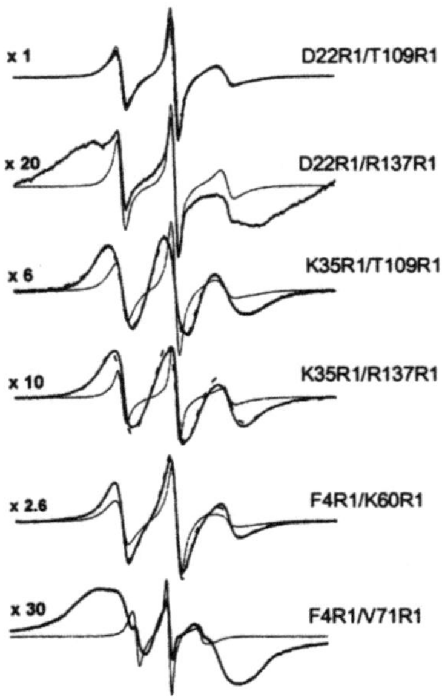

Figure 20. Room temperature EPR spectra of spin labeled mutants of T4L with bound substrate (i.e. introduced in the T26E background). The light trace is the sum of the single mutant spectra. The heavy trace is the double mutant spectrum normalized to the same number of spins as the corresponding sum of singles, then scaled for convenience of presentation. The scaling factor is shown to the left of the spectra. The superimposed dashed lines is the simulated spectra based on the method described in section 3.3.

5.1.4 T4L conformation in the absence of substrate

When the same double mutants are introduced in the WT background, i.e., with no trapped substrate, a large conformational change is detected relative to the substrate-bound state. In Figure 21A the spectra of the double mutants with (heavy trace) and without (light trace) bound substrate are compared. Figure 21B makes the same comparison for the sum of the spectra of the corresponding single mutants and serves to demonstrate that overall no significant changes are associated with substrate binding. In all cases except K35R1/T109R1, the differences between spectra of a given double mutant due to substrate binding clearly exceed those in the sum of the spectra of the single mutants. Furthermore, the changes in the double mutant spectra involve new features that cannot arise from changes in mobility. Thus, the spectral differences due to the presence of substrate in the double mutants are not a result the differences in side chain mobilities. Rather they are a consequence of spin-spin interactions and hence changes in inter-spin distances.

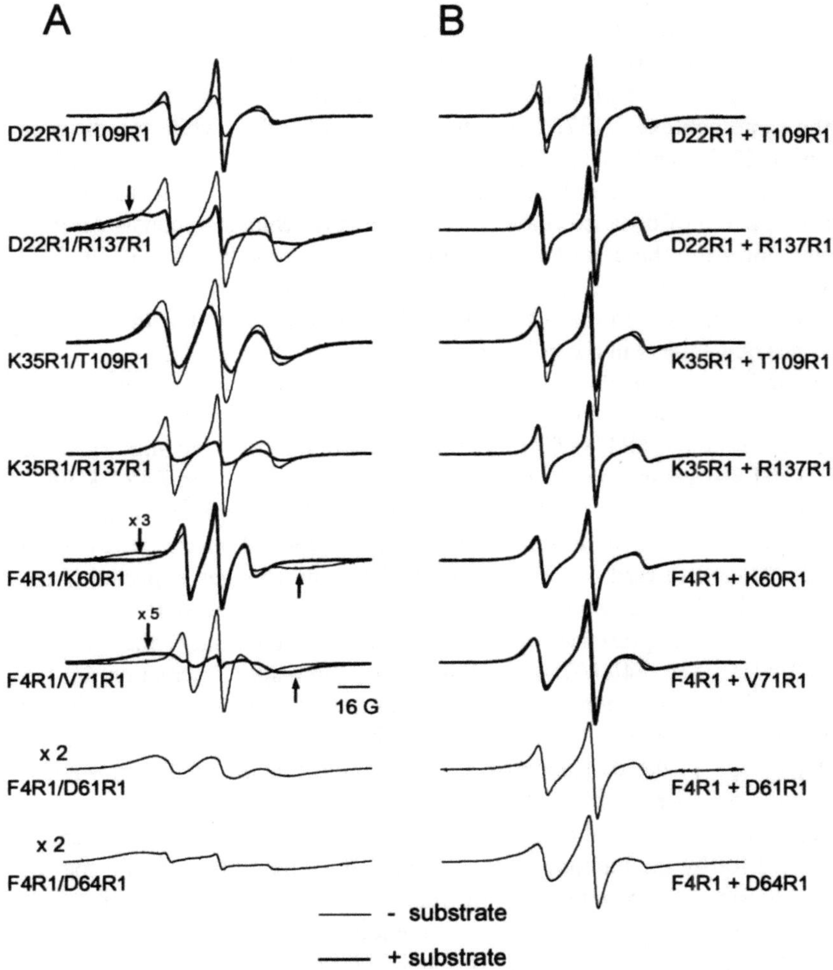

Figure 21. Room temperature EPR spectra of spin labeled mutants of T4L with and without bound substrate. A. EPR spectra of the indicated double mutants of T4L with (heavy trace) and without (light trace) bound substrate; B. The sum of the EPR spectra of the single mutants corresponding to A, with (heavy trace) and without (light trace) bound substrate. For each pair, the spectra are normalized to represent the same number of spins, except where noted by a scaling factor. The protein with bound substrate is the T26E mutant. (reproduced from Mchaourab et al., 1997)

Thus, the sign of the distance changes can be interpreted in terms of structural differences between the substrate- bound and substrate-free forms. The increase and decrease in spectral widths of D22R1/T109R1 and D22R1/R137R1, respectively, relative to the substrate-bound closed state, indicate a movement of T109R1 toward, and R137R1 away from, D22R1. In the same context, R137R1 moves away from K35R1, while the distance between K35R1 and T109R1 changes little, but could increase slightly. The triangle consisting of residues D22R1, K35R1, and R137R1 serves to show that the pattern of distance changes upon substrate removal is consistent with a transition from the closed to a more open state resembling that of the M6I mutant. Thus, these data are in accord with a global rigid body rotation of one domain relative to the other upon substrate binding, rather than limited local rearrangements in the polypeptide.

In addition to the above changes near the active site cleft, removal of the substrate alters the packing of helices A and C, monitored by the pairs F4R1/V71R and F4R1/K60R1. The interaction between the nitroxides in F4R1/V71R decreases while the interaction between those in F4R1/K60R1 increases as evidenced by the appearance of a broad spectral component indicated by arrows in Figure 21A. Thus, residue F4 moves away from residue 71 and toward 60. These are precisely the changes expected if the open state is the state populated in solution while the closed state is the state populated in the presence of the substrate. Additionally, the mutants F4R1/D61R1 and F4R1/D64R1 show strong spin-spin interaction in the absence of substrate, confirming the proximity of residue 4R1 to residues D61 and D64 near the N-terminus of helix C. Collectively, these results strongly support the existence of an open conformation of the enzyme in solution in the absence of substrate.

Unlike the spectra obtained in the presence of substrate, which can be accounted for by the interaction of two spins at a single distance, the spectra in the absence of substrate are more complex and suggest a conformational heterogeneity that may arise from an equilibrium mixture of states. In addition to the broad component in the F4R1/K60R1 spectrum (arrows), a narrow component suggests at least two populations, one with the spins in close proximity and the other with the spins at a much larger distance. The broad component, corresponding to an open state, is dominant in terms of spin population. In the spectra of D22R1/R137R1 and F4R1/D61R1, the existence of multiple components is deduced from the presence of intensities in the wings separated by nearly 100 gauss although the central features of the spectrum are relatively narrow.

5.1.5 Analysis of side chain dynamics

Hinge bending motions also result in repacking of tertiary contacts particularly at the interface of helices A and C. Such changes can be detected in the dynamics of single nitroxide side chains, and hence are reflected in the EPR spectral lineshape (Mchaourab *et al.*, 1996). Sites 4 and 71 were selected to represent regions of the molecule that undergo significant changes in tertiary interaction upon transition from the closed to open state. F4 moves from a solvent-exposed site in the closed state to a site buried at the contact face of the A and C helices in the open state. Concomitantly, V71 moves in the opposite sense, from a buried site at the contact face of these helices to a solvent-exposed site.

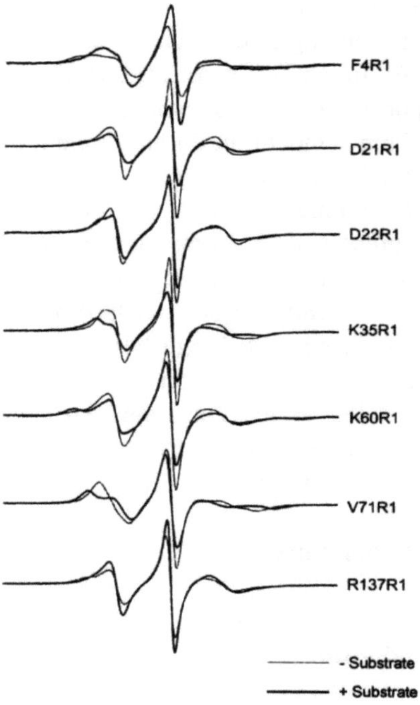

Figure 22. EPR spectra of spin labeled single cysteine mutants of T4L with (heavy trace) and without (light trace) bound substrate. The protein with bound substrate is the T26E mutant. All spectra were recorded in the presence of 30 % sucrose with 100 G field scans. (reproduced from Mchaourab et al., 1997)

Figure 22 shows the EPR spectra in 30% sucrose for a set of mutants containing single nitroxide side chains at the indicated positions, both in the T4L-substrate complex and in T4L alone. The addition of sucrose is necessary to eliminate the contribution of the global tumbling of T4L to the motional averaging. Comparison of the EPR spectra of F4R1 and V71R1 in the T4L-substrate complex and T4L in solution demonstrate that spectral changes in fact take place upon removal of substrate: the spectrum of F4R1 in the T4L-substrate complex is characteristic of a rapid anisotropic motion of a single solvent-exposed spin population, while that in T4L has multiple components, one of which is highly immobilized resulting from tertiary contact interaction. On the other hand, the spectrum of V71R1 in the T4L-substrate complex indicates a highly immobilized state, consistent with the expected buried location. In T4L, the spectrum is that of a mobile, solvent exposed site, reflecting the open state.

Substrate binding is also expected to affect the dynamics of loops near the active site cleft. Sites 21, 22 and 35 were selected to monitor these loops. Each of these sites show EPR spectra that reflect significantly less mobility in the T4L-substrate complex than in T4L. This is likely to be the result of damped backbone fluctuations, because such fluctuations appear to be the dominant contribution to mobility at these sites (Mchaourab *et al.*, 1996).

Residue 137 is at the N-terminus of the short helix I. In general, the spectrum of R1 has been found to have a characteristic two-component at such sites, apparently due to hydrogen bonding interactions with backbone amide residues (Mchaourab *et al.*, 1996). In the T4L-substrate complex, only a single component is observed as opposed to the two-component spectrum observed in solution. This may result from competition for backbone H-bond formation between the R1 side chain and the nearby substrate.

Residue 60 is located near the N-terminus of helix C. In the T4L-substrate complex, a typical two-component N terminal spectrum is observed for K60R1. In T4L, the spectrum reflects an increase in the population of the more mobile state. Faber and Matthews (1990) observed changes in backbone conformation associated with hinge bending around residues 13, 59 and 80, and this effect may be the origin of the observed spectral difference. Single R1 side chains at locations not expected to undergo changes in tertiary contact interactions upon substrate removal show no corresponding spectral changes (data not shown).

5.2 Triggered structural transitions

5.2.1 Rhodopsin and Bacteriorhodopsin

Bacteriorhodopsin (bR) is a light-driven proton pump found in specialized membrane patches from the extremophile *Halobacterium salinarium*. Its structure is now known at a resolution of 1.55 Å (Luecke *et al.*, 1999) but its general features have been known for a number of years both from electron crystallography (Henderson *et al.*, 1990) and X-ray studies (Pebay-Peyroula *et al.*, 1997). This wealth of structural information made bR the ideal candidate for the development of site-directed spin labeling (Altenbach *et al.*, 1990). BR is a seven transmembrane segment protein with a covalently attached chromophore (all-trans retinal, in the dark) cradled within an extracellularly-facing cavity.

Rhodopsin (Rho), is a visual protein-chromophore complex (with 11-cis retinal) found in the retina of a wide variety of animals. It serves as a light-activated switch that initiates a G-protein based signal transduction cascade and ultimately leads to changes in the membrane potential of the photoreceptor. As in bR, Rho also contains seven transmembrane segments, and although it was previously thought that bR could serve as an ideal structural model for Rhodopsin, electron crystallography studies have shown that the overall packing and helix orientation of Rho is significantly different (Unger *et al.*, 1997). In both cases, light absorption produces chromophore isomerization (to 13-cis retinal in bR and to all-trans in Rho) leading to conformational changes that promote either transmembrane proton translocation (in bR) or binding to a G-protein α-subunit (in Rho).

It is to the understanding of these conformational changes that SDSL and EPR spectroscopy have made significant contributions. While it has been clear for a number of years that these light-induced conformational changes can be detected by monitoring changes in spin label mobility or solvent accessibility at specific positions (Altenbach *et al.*, 1999a; Altenbach *et al.*, 1999b; Altenbach *et al.*, 1996; Farahbakhsh *et al.*, 1993; Rink *et al.*, 1997; Steinhoff *et al.*, 1994), distinguishing between specific models of motions requires the monitoring of changes in inter-spin distances. In a seminal study, Farrens et al made use of six double-cysteine mutants to probe the nature of light-triggered inter-helical movements in Rho (Farrens *et al.*, 1996). Dipole-dipole interactions were observed at low temperature, between a reference position in helix C (residue 139) and residues 248-252 along helix F. Upon illumination, a *decrease* in the extent of the dipolar coupling broadening was found between residue 139 and residues 248, 251 and 252, an *increase* in broadening was observed for the pair 139-250, while little or no change was detected in the pair 139-249. These results were interpreted in terms of a rigid-body movement of helix F in reference to helix C, with a simultaneous rotation and tilt of helix F away from helix C during Rho activation. Farrens et al (Farrens *et al.*, 1996) speculated that because of this type of movement the conformation of the E-F loop could change to recognize transducin (the G-protein) and Rhodopsin kinase and thus initiate the phototransduction cascade.

Light-induced conformational changes have also been measured recently in bR by Thorgeirsson et al (Thorgeirsson *et al.*, 1997), who used double cysteine mutants to follow changes in inter-spin distances for residues in loop EF, relative to loops AB and CD. It was found, by a combination of time-resolved and low temperature EPR measurements that upon illumination the EF loop moves away from both the AB and DC loops in a transient fashion (about 5 Å change in inter spin distance). This movement temporarily opens a cavity in the extracellular face of the molecule that may change access to the Schiff base for proton uptake. This loop movement might be a consequence of an outward tilt of helix F, as observed in electron diffraction studies (Subramaniam and Henderson, 1999), and coincides well with the types of motion proposed for Rhodopsin helix F by Farrens et al (Farrens *et al.*, 1996).

5.2.2 *KcsA* Potassium Channel

The structure of *KcsA* shows a remarkably efficient blueprint for ion permeation: a short selectivity filter where carbonyl oxygens are likely to interact with dehydrated ions, followed by an energetically favorable water-filled cavity, capped at the intracellular end by a closely packed bundle of helices restricting ion flow in the closed state (Doyle *et al.*, 1998). Gating mechanisms, however, are harder to deduce from direct examination of the *KcsA* structure. A description of the molecular basis of ion channel gating requires an understanding of the energy transduction machinery that converts multiple forms of physical stimuli (i.e. voltage, ligand binding, force, etc) into the kind of protein movements that ultimately leads to ion permeation.

The discovery that *KcsA* can be activated by changes in proton concentration (Cuello *et al.*, 1998) set the stage for the analysis of the types of protein motions that underlie activation gating in K^+ channels. Thus, Cuello and co-workers showed using both radiotracer flux experiments and single channel current measurements in planar lipid bilayers that the equilibrium between open and closed states in *KcsA* could be shifted towards the open state by lowering pH. Although initially the proton site was thought to be extracellular, recent studies clearly show that channel opening is induced by intracellular changes in pH, and that a pH gradient is not necessary for channel activation (Heginbotham *et al.*, 1999). Earlier results from voltage dependent channels suggested that channel opening was associated with global conformational changes in the intracellular face of the channel (Armstrong, 1971; Liu *et al.*, 1997). The first direct demonstration of the types of movements that may be associated with channel opening came from Perozo et al. (Perozo *et al.*, 1998), who observed large pH-dependent changes in inter-subunit spin-spin coupling at selected residues along the C-terminal end of the second transmembrane segment TM2 (Figure 23). From the crystal structure, these residues are known to be located at the narrowest section of the internal helix bundle, close to the fourfold axis of symmetry (See Figure 16). In each case, spectra obtained at low pH showed a reduction in the magnitude of the inter-subunit dipolar coupling, leading to the suggestion that channel opening was a consequence of a widening in the inner vestibule of the channel (Perozo *et al.*, 1998).

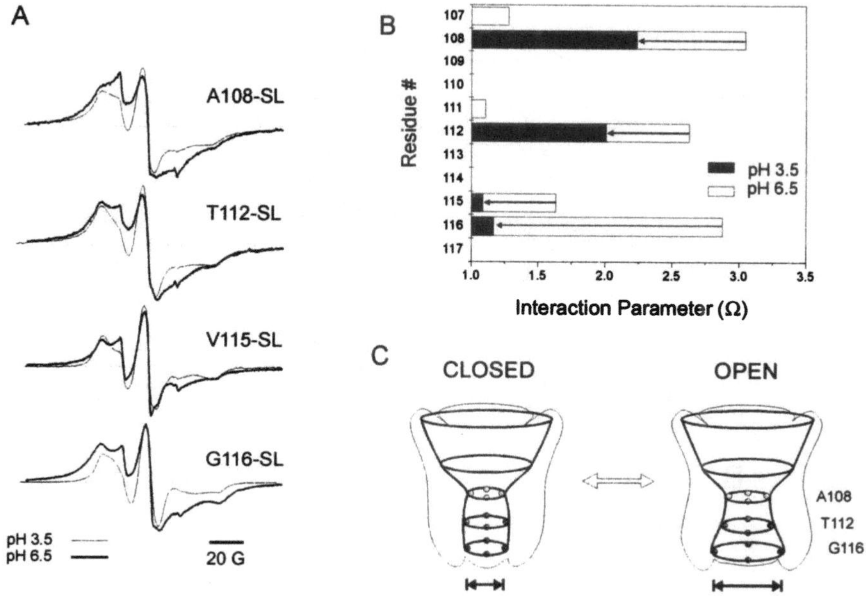

Figure 23. Spin-Spin interactions in TM2 and detection of the opening of the intracellular vestibule of the channel associated with channel gating. **A.** EPR spectra of interacting residues from the C-terminal end of TM2 under conditions stabilizing the closed state (pH 6.5, thick line) or the open state (pH 3.5, thin line). **B.** The interaction parameter Ω calculated relative to the underlabeled spectra for the closed conformation (open bars) and the open conformation (filled bars). The length of the arrow represents the difference in spin-spin interaction parameter between the two conformations. **C.** Cartoon representation of the opening of the internal vestibule during channel gating, and the relative positions of the residues with the strongest spin-spin interactions (reproduced from Perozo et al., 1998)

A more thorough analysis of these movements was subsequently carried out by studying changes in mobility and dipolar coupling along the main structural elements in the channel: the two transmembrane segments TM1 and TM2, as well as the regions flanking the selectivity (Perozo *et al.*, 1999). The changes in spectral line shape were analyzed using discrete Fourier transform methods (see section 2b, above) to detect possible patterns that could indicate specific types of movements in each of the TM segments originating from acidification. An example of this analysis for the "internal" transmembrane segment TM2 is shown in Figure 24, where changes in probe mobility (ΔΔHo) point to dramatic changes in helix-helix contacts (leftmost panes), while the remarkable periodicity in the changes of the per-residue proximity among subunits (Ω) are an indication of helix movements relative to the fourfold axis of symmetry. The majority of the residues along TM2 showed a reduction in the extent of spin-spin coupling as measured by the Ω parameter, indicating a translational movement or helical tilt away from the

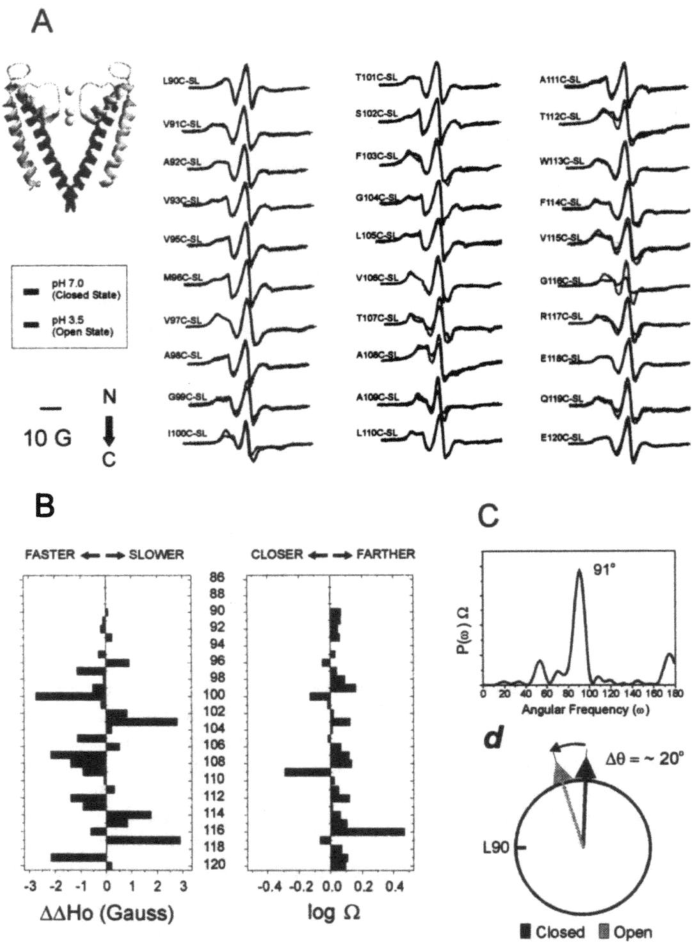

Figure 24. Conformational rearrangements in the internal transmembrane segment of KcsA (TM2). **A**. Black spectra were taken at pH 7.2 where the channel is mostly closed, grey spectra were obtained at pH 3.5, which favors the open conformation. The scale represents 15 Gauss for the spectra marked by an asterisk. The top left shows a model of KcsA (minus two subunits), highlighting the region of the channel where the spectra were obtained (dark). **B**. Profile of the changes in structural parameters associated with probe mobility (ΔΔHo, left panel) and spin-spin dipolar coupling (Ω parameter, right panel). Numbers in the center correspond to residue positions along TM1. **C**. Fourier transform power spectra of the proximity parameter Ω. The calculated peak angular frequency is 91° with a periodicity index = 3.71. **D**. Changes in dipolar coupling moment upon channel gating. Each vector is the sum of individual spectral amplitudes, normalized to the total number of spins in the sample, calculated relative to residue L90 and plotted in polar coordinates viewed extracellularly.

symmetry axis that would indicate an increase in the diameter of the internal vestibule of the channel. The power spectrum obtained from the TM2 Ω profile revealed a sharp peak centered at 91°; this was taken to suggest that the movement of TM2 occurred as a rigid-body movement with no major changes in secondary structure. Vector analysis of these changes in the context of a helical wheel projection indicated that the pattern of change in inter-subunit proximity could be explained, in addition to the helical tilt, by a rotation of the individual TM2 helices in a counterclockwise direction (Perozo *et al.*, 1999).

Similar analysis of the spectral changes in the "external" transmembrane segment TM1 also revealed helix rotations and tilts compatible with a rigid-body movement of the entire helix. However, because TM1 is located further away for the axis of symmetry, changes in inter-subunit residue proximity are more subtle and harder to detect than in TM2. For this reason, and given that less than half of TM1 residues are involved in any extensive tertiary contacts, changes in probe mobility (ΔΔHo) are a very sensitive indicator of helix tilting and rotation. As in TM2, these data suggested a counterclockwise rotation of TM1 upon channel gating. Interestingly, data on the types of conformational changes occurring near the selectivity filter showed no significant changes in either probe mobility or spin-spin interaction in the extracellular vestibule of the channel (Perozo *et al.*, 1999). However, spin labels located at the C-terminal end of the pore helix reported small changes in spectral line shape suggestive of possible conformational rearrangements that could affect the energetics of ion coordination at the selectivity filter.

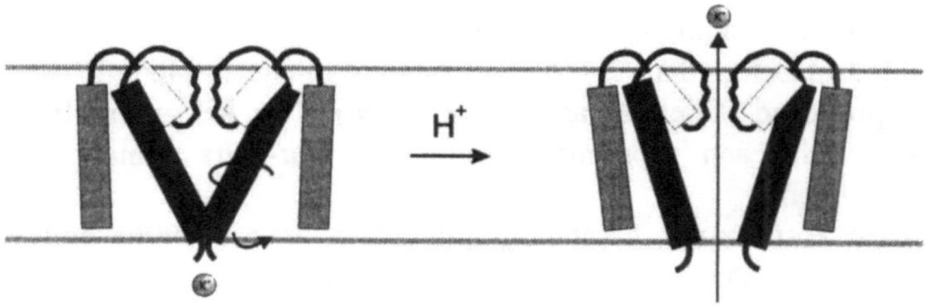

Figure 25. A model of activation gating for the KcsA K^+ channel. TM1s are light, TM2s are dark, the pore helices are while. Initiated by a proton-dependent step, TM2 rotates in a counterclockwise direction and away from the permeation pathway, opening the channel. In the channel periphery, TM1 mirrors the rotation of TM2 but tilts inwards, towards the permeation pathway. These movements are depicted here much larger than expected, for illustration purposes.

The overall picture of K^+ channel gating that emerges from these studies is summarized in Figure 25. Viewed from the extracellular side of the channel this model proposes that upon lowering the pH, TM1 and TM2 undergo a counterclockwise rotation and a simultaneous movement away from the symmetry axis of the tetramer, opening the intracellular side of the permeation path. Although the data are currently limited, it was also proposed that movements in or around the selectivity filter are small, but cannot be ruled out as having a role in activation gating. Taken together, the results on the conformational rearrangements responsible for *KcsA* opening serve as an example of the type of conclusions that can be obtained by studying patterns and periodicities derived from large sets of qualitative distance data.

5.2.3 Aspartate Receptor

Another example of the power of SDSL in discriminating between mechanistic models underlying specific protein actions comes from the studies of Koshland, Shin and coworkers on the bacterial aspartate receptor (Ottemann *et al.*, 1998; Ottemann *et al.*, 1999). Encoded by the *tar* gene, the aspartate receptor (AR) regulates a kinase cascade composed of the CheA kinase, the CheW coupling protein and CheY, the flagellar motor regulator. Upon binding of aspartate, a conformational change in the receptor leads to a decrease in the rate of CheA autophosphorylation and a subsequent dramatic reduction in the levels of phosphorylated CheY.

The AR consists of a periplasmic domain, responsible for ligand binding, a transmembrane domain and an effector domain, were interactions with members of the signaling cascade occur (Falke *et al.*, 1997). The crystal structure of both the periplasmic domain of the AR (Milburn *et al.*, 1991) and the cytoplasmic domain of a related receptor (Kim *et al.*, 1999) have been obtained at better than 2.7 Å, and models of the full-length receptor have been proposed (Chervitz and Falke, 1996). The AR assembles as a homodimer, with each monomer forming a four-helix bundle in the periplasmic domain.

Evidence form x-ray crystallography (Yeh *et al.*, 1996) and EPR measurements (Ottemann *et al.*, 1998) suggested that binding of aspartate to the receptor is accompanied by small movements in the helix 1 relative to helix 4 in a piston-like motion. A recent report by Ottemann and co-workers (Ottemann *et al.*, 1999) addressed the precise mechanism by which aspartate binding at the extracellular domain of the receptor propagates to its intracellular end. A strategy was developed to distinguish among five possible models of signaling by determining the pattern of change in intrasubunit distances in the transmembrane segments of the AR. The five models considered were Association-Dissociation, Scissor, Seesaw, Piston and Rotation.Spectra were obtained for receptors doubly labeled at positions 10/207, 10/210, 10/215, located at the intracellular portion of the membrane-embedded regions of the receptor and 39/179 located towards its extracellular side. Because the receptor is a dimer, the fully labeled complex contains four spin labels and thus dipolar coupling could derive either from within a subunit or between subunits. To avoid these problems, spin-labeled receptor was mixed with an equimolar amount of unlabeled receptor and allowed to exchange, bringing down the total amount of receptor with four spin labels to about 1/8 of the total receptor population. Upon aspartate binding, inter-spin distances decreased for the 10/207 pair, while increasing for all others, suggesting that a downward piston movement of TM2 relative to TM1 was responsible for aspartate signaling. These movements are quite small, with changes in the order of 1 Å for all pairs studied, suggesting that little or no amplification of the signal occurs due to the aspartate-induced conformational change.

6. CRITICAL PERSPECTIVE

As a technique, site-directed spin labeling is now emerging as a viable alternative for the determination of global folds and conformations at the backbone level in a variety of systems. The work on α-crystallin, KscA channel, Snare proteins and the Lac permease are perhaps the most extensive attempts to obtain static structural information and together support the contention that one can derive enough environmental and distance restraints to determine the packing structure of the protein.

Nevertheless, it is apparent that, despite the recent advances in molecular biology and protein purification, this brute-force approach requires a large investment of time and resources even when applied to a single protein. Currently, determination of secondary structure and solvent accessibility requires a cysteine scan along the entire sequence. An alternative would be a more limited and targeted use of nitroxide scanning that takes advantage of secondary structure prediction using neural networks particularly for simple geometries, i.e. membrane proteins, where the complexity is limited by the nature of the environment. Future development in this area should include increasing use of automatization and robotics in performing mutagenesis and low-level protein expression and purification. A most tantalizing notion involves the generation of an engineered organism capable of incorporating spin-labels and other non-natural mutagenesis in vivo, but that requires the directed evolution of both a tRNA synthetase and an uptake system.

Simplification in the number of double mutants is likely to result from multiple improvements in methodology and instrumentation. A more quantitative analysis of the CW-EPR spectrum to obtain both distance and orientation information can reduce the number of double mutants required to characterize the pairwise packing of helices (chapter 3). Equally important will be the development of computational approaches to include side-chain accessibility/collision frequency as a structural constraint, in conjunction with the more traditional distance constraints. A better understanding of the general conformational preferences of the side chain R1 and the origin of the restricted motion observed on the surfaces of helices will also play an important role in structure calculations. Perhaps most promising in this context is the use of time domain EPR methods, in particular DEER (chapter Eaton 8 and 11) and multiple quantum coherences (chapter 9), both of which extend the range of distances that can be measured to 40-50 A. However, their widespread use will require considerable investment in instrumentation development to bring the sensitivity to a more practical range.

ACNOWLEDGEMENT. This work was supported by grants from the National Eye Institute, EY12018 and EY12683 to H.M and from the National Institute of General Medical Sciences, GM54690 and GM57846 to E.P. Support of the McKnight endowment for neurosciences to E.P. is greatly acknowledged.

7. REFERENCES

Abragam, A. (1961). *Principles of Nuclear Magnetism*. Oxford: Oxford University Press.

Altenbach, C., Cai, K., Khorana, H. G. and Hubbell, W. L. (1999a) Structural features and light-dependent changes in the sequence 306-322 extending from helix VII to the palmitoylation sites in rhodopsin: a site-directed spin-labeling study. *Biochemistry* **38**, 7931-7.

Altenbach, C., Klein-Seetharaman, J., Hwa, J., Khorana, H. G. and Hubbell, W. L. (1999b). Structural features and light-dependent changes in the sequence 59-75 connecting helices I and II in rhodopsin: a site-directed spin-labeling study. *Biochemistry* **38**, 7945-9.

Altenbach, C., Marti, T., Khorana, H. G. and Hubbell, W. L. (1990). Transmembrane protein structure: spin labeling of bacteriorhodopsin mutants. *Science* **248**, 1088-92.

Altenbach, C., Yang, K., Farrens, D. L., Farahbakhsh, Z. T., Khorana, H. G. and Hubbell, W. L. (1996). Structural features and light-dependent changes in the cytoplasmic interhelical E-F loop region of rhodopsin: a site-directed spin-labeling study. *Biochemistry* **35,** 12470-8.

Armstrong, C. M. (1971). Interaction of tetraethylammonium ion derivatives with the potassium channels of giant axons. *Journal of General Physiology* **58,** 413-37.

Berengian, A. R., Bova, M. P. and Mchaourab, H. S. (1997). Structure and function of the conserved domain in αA-crystallin. Site-directed spin labeling identifies a beta-strand located near a subunit interface. *Biochemistry* **36,** 9951-7.

Berengian, A. R., Parfenova, M. and Mchaourab, H. S. (1999). Site-directed spin labeling study of subunit interactions in the α-crystallin domain of small heat-shock proteins. Comparison of the oligomer symmetry in αA-crystallin, HSP 27, and HSP 16.3. *Journal of Biological Chemistry* **274,** 6305-14.

Bova, M. P., Ding, L. L., Horwitz, J. and Fung, B. K. (1997). Subunit exchange of αA-crystallin. *Journal of Biological Chemistry* **272,** 29511-7.

Bowie, J. U. (1997). Helix packing in membrane proteins. *Journal of Molecular Biology* **272,** 780-9.

Burley, S. K., Almo, S. C., Bonanno, J. B., Capel, M., Chance, M. R., Gaasterland, T., Lin, D., Sali, A., Studier, F. W. and Swaminathan, S. (1999). Structural genomics: beyond the human genome project. *Nature Genetics* **23,** 151-7.

Burr, M. and Koshland, D. E. (1964). Use of "reporter groups" in structure-function studies of proteins. *Biochemistry* **52,** 1017-1024.

Caffrey, M., Cai, M., Kaufman, J., Stahl, S. J., Wingfield, P. T., Covell, D. G., Gronenborn, A. M. and Clore, G. M. (1998). Three-dimensional solution structure of the 44 kDa ectodomain of SIV gp41. *EMBO Journal* **17,** 4572-84.

Chan, D. C., Fass, D., Berger, J. M. and Kim, P. S. (1997). Core structure of gp41 from the HIV envelope glycoprotein. *Cell* **89,** 263-73.

Chang, Z., Primm, T. P., Jakana, J., Lee, I. H., Serysheva, I., Chiu, W., Gilbert, H. F. and Quiocho, F. A. (1996). Mycobacterium tuberculosis 16-kDa antigen (Hsp16.3) functions as an oligomeric structure in vitro to suppress thermal aggregation. *Journal of Biological Chemistry* **271,** 7218-23.

Chervitz, S. A. and Falke, J. J. (1996). Molecular mechanism of transmembrane signaling by the aspartate receptor: a model. *Proceedings of the National Academy of Sciences of the United States of America* **93,** 2545-50.

Chothia, C. (1984). Principles that determine the structure of proteins. *Annual Review of Biochemistry* **53,** 537-72.

Chothia, C. and Finkelstein, A. V. (1990). The classification and origins of protein folding patterns. *Annual Review of Biochemistry* **59,** 1007-39.

Chothia, C., Hubbard, T., Brenner, S., Barns, H. and Murzin, A. (1997). Protein folds in the all-beta and all-a classes. *Annual Review of Biophysics & Biomolecular Structure* **26,** 597-627.

Cornette, J. L., Cease, K. B., Margalit, H., Spouge, J. L., Berzofsky, J. A. and DeLisi, C. (1987). Hydrophobicity scales and computational techniques for detecting amphipathic structures in proteins. *Journal of Molecular Biology* **195,** 659-85.

Cuello, L. G., Romero, J. G., Cortes, D. M. and Perozo, E. (1998). pH-dependent gating in the Streptomyces lividans K^+ channel. *Biochemistry* **37,** 3229-36.

de Jong, W. W., Caspers, G. J. and Leunissen, J. A. (1998). Genealogy of the α-crystallin--small heat-shock protein superfamily. *International Journal of Biological Macromolecules* **22,** 151-62.

Donnelly, D., Overington, J. P. and Blundell, T. L. (1994). The prediction and orientation of α-helices from sequence alignments: the combined use of environment-dependent substitution tables, Fourier transform methods and helix capping rules. *Protein Engineering* **7,** 645-53.

Doyle, D. A., Cabral, J. M., Pfuetzner, R. A., Kuo, A., Gulbis, J. M., Cohen, S. L., Chait, B. T. and MacKinnon, R. (1998). The structure of the potassium channel: molecular basis of K^+ conduction and selectivity. *Science* **280,** 69-77.

Eaton, G. R. and Eaton, S. S. (1989). Resolved electron-electron spin-spin splittings in EPR spectra. In *Biological Magnetic Resonance*, vol. 8 (ed. L. J. Berliner and J Reuben), pp. 340-397. New York: Plenum.

Faber, H. R. and Matthews, B. W. (1990). A mutant T4 lysozyme displays five different crystal conformations. *Nature* **348,** 263-6.

Fajer, P. G. (2000) EPR in Peptide and Protein Analysis. In *Encyclopedia of Analytical Chemistry* (ed. R.A. Meyers) John Wiley and Sons, Ltd. *In press.*

Falke, J. J., Bass, R. B., Butler, S. L., Chervitz, S. A. and Danielson, M. A. (1997). The two-component signaling pathway of bacterial chemotaxis: a molecular view of signal transduction by receptors, kinases, and adaptation enzymes. *Annual Review of Cell & Developmental Biology* **13,** 457-512.

Farahbakhsh, Z. T., Hideg, K. and Hubbell, W. L. (1993). Photoactivated conformational changes in rhodopsin: a time-resolved spin label study. *Science* **262,** 1416-9.

Farrens, D. L., Altenbach, C., Yang, K., Hubbell, W. L. and Khorana, H. G. (1996). Requirement of rigid-body motion of transmembrane helices for light activation of rhodopsin. *Science* **274**, 768-70.

Freed, J. H. and Schneider, D. J. (1989). Calculating slow motional magnetic resonance spectra: A user's guide. In *Biological Magnetic Resonance*, vol. 8 (ed. L. J. a. R. Berliner, J.), pp. 1-76. New York: Plenum.

Froncisz, W. and J.S., H. (1982). The loop-gap resonator: A new microwave lumped circuit ESR sample structure. *J. Magn. Reson.* **47,** 515-522.

Gaasterland, T. (1998a). Structural genomics taking shape. *Trends in Genetics* **14,** 135.

Gaasterland, T. (1998b). Structural genomics: bioinformatics in the driver's seat. *Nature Biotechnology* **16,** 625-7.

Gerstein, M., Lesk, A. M. and Chothia, C. (1994). Structural mechanisms for domain movements in proteins. *Biochemistry* **33,** 6739-49.

Gross, A., Columbus, L., Hideg, K., Altenbach, C. and Hubbell, W. L. (1999). Structure of the *KcsA* potassium channel from Streptomyces lividans: a site-directed spin labeling study of the second transmembrane segment. *Biochemistry* **38,** 10324-35.

Haley, D. A., Horwitz, J. and Stewart, P. L. (1998). The small heat-shock protein, αB-crystallin, has a variable quaternary structure. *Journal of Molecular Biology* **277,** 27-35.

Hanson, P., Millhauser, G., Formaggio, F., Crisma, M. and Toniolo, C. (1996). Esr Characterization of Hexameric, Helical Peptides Using Double Toac Spin Labeling. *Journal of the American Chemical Society* **118,** 7618-7625.

Havel, T. F. (1991). An evaluation of computational strategies for use in the determination of protein structure from distance constraints obtained by nuclear magnetic resonance. *Progress in Biophysics & Molecular Biology* **56,** 43-78.

He, M. M., Sun, J. and Kaback, H. R. (1996). Cysteine-scanning mutagenesis of transmembrane domain XII and the flanking periplasmic loop in the lactose permease of EScherichia coli. *Biochemistry* **35,** 12909-14.

He, M. M., Voss, J., Hubbell, W. L. and Kaback, H. R. (1997). Arginine 302 (helix IX) in the lactose permease of Escherichia coli is in close proximity to glutamate 269 (helix VIII) as well as glutamate 325. *Biochemistry* **36,** 13682-7.

Heginbotham, L., Kolmakova-Partensky, L. and Miller, C. (1998). Functional reconstitution of a prokaryotic K^+ channel. *Journal of General Physiology* **111,** 741-9.

Heginbotham, L., LeMasurier, M., Kolmakova-Partensky, L. and Miller, C. (1999). Single streptomyces lividans K^+ channels: functional asymmetries and sidedness of proton activation. *Journal of General Physiology* **114,** 551-60.

Heginbotham, L., Lu, Z., Abramson, T. and MacKinnon, R. (1994). Mutations in the K^+ channel signature sequence. *Biophysical Journal* **66,** 1061-7.

Henderson, R., Baldwin, J. M., Ceska, T. A., Zemlin, F., Beckmann, E. and Downing, K. H. (1990). Model for the structure of bacteriorhodopsin based on high-resolution electron cryo-microscopy. *Journal of Molecular Biology* **213,** 899-929.

Horwitz, J. (1993). The function of α-crystallin. *Investigative Ophthalmology & Visual Science* **34,** 10-22.

Hubbell, W. L. and Altenbach, C. (1994). Site-directed spin labeling of membrane proteins. In *Membrane protein structure. Experimental approaches* (ed. S. H. White). New York, NY: Oxford Univ. Press.

Hubbell, W. L., Gross, A., Langen, R. and Lietzow, M. A. (1998). Recent Advances in Site-Directed Spin Labeling of Proteins. *Current Opinion in Structural Biology* **8,** 649-656.

Hubbell, W. L., Mchaourab, H. S., Altenbach, C. and Lietzow, M. A. (1996). Watching proteins move using site-directed spin labeling. *Structure* **4,** 779-83.

Hustedt, E. J., Smirnov, A. I., Laub, C. F., Cobb, C. E. and Beth, A. H. (1997). Molecular Distances From Dipolar Coupled Spin-Labels - the Global Analysis of Multifrequency Continuous Wave Electron Paramagnetic Resonance Data. *Biophysical Journal* **72,** 1861-1877.

Hyde, J. S. and Subczynski, W. K. (1989). Spin label oxymetry. In *Biological Magnetic Resonance*, vol. 8 (ed. L. J. a. R. Berliner, J.), pp. 399-425. New York: Plenum.

Janin, J., . and Chothia, C. (1980). Packing of α-helices onto beta-pleated sheets and the anatomy of α/β proteins. *J. Mol. Biol* **143,** 95-128.

Kaback, H. R. and Wu, J. H. (1999). What to do while awaiting crystals of a membrane transport protein and thereafter. *Accounts of Chemical Research* **32,** 805-813.

Kabsch, W. and Sander, C. (1983). Dictionary of protein secondary structure: pattern recognition of hydrogen-bonded and geometrical features. *Biopolymers* **22,** 2577-637.

Kim, K. K., Yokota, H. and Kim, S. H. (1999). Four-helical-bundle structure of the cytoplasmic domain of a serine chemotaxis receptor. *Nature* **400,** 787-92.

Kim, S. H. (1998). Shining a light on structural genomics. *Nature Structural Biology* **5,** 643-5.

Koteiche, H. A., Berengian, A. R. and Mchaourab, H. S. (1998). Identification of protein folding patterns using site-directed spin labeling. Structural characterization of a beta-sheet and putative substrate binding regions in the conserved domain of a A-crystallin. *Biochemistry* **37,** 12681-8.

Koteiche, H. A. and Mchaourab, H. S. (1999). Folding pattern of the a-crystallin domain in aA-crystallin determined by site-directed spin labeling. *Journal of Molecular Biology* **294,** 561-77.

Kuroki, R., Weaver, L. H. and Matthews, B. W. (1993). A covalent enzyme-substrate intermediate with saccharide distortion in a mutant T4 lysozyme. *Science* **262,** 2030-3.

Langen, R., Cai, K., Altenbach, C., Khorana, H. G. and Hubbell, W. L. (1999). Structural features of the C-terminal domain of bovine rhodopsin: a site-directed spin-labeling study. *Biochemistry* **38,** 7918-24.

Langen, R., Isas, J. M., Luecke, H., Haigler, H. T. and Hubbell, W. L. (1998). Membrane-mediated assembly of annexins studied by site-directed spin labeling. *Journal of Biological Chemistry* **273,** 22453-7.

Liu, Y., Holmgren, M., Jurman, M. E. and Yellen, G. (1997). Gated access to the pore of a voltage-dependent K^+ channel. *Neuron* **19,** 175-84.

Luecke, H., Schobert, B., Richter, H. T., Cartailler, J. P. and Lanyi, J. K. (1999). Structure of bacteriorhodopsin at 1.55 A resolution. *Journal of Molecular Biology* **291,** 899-911.

Matthews, B. W. (1995). Studies on protein stability with T4 lysozyme. *Advances in Protein Chemistry* **46,** 249-78.

Mchaourab, H. S., Berengian, A. R. and Koteiche, H. A. (1997a). Site-directed spin-labeling study of the structure and subunit interactions along a conserved sequence in the α-crystallin domain of heat-shock protein 27. Evidence of a conserved subunit interface. *Biochemistry* **36,** 14627-34.

Mchaourab, H. S., Lietzow, M. A., Hideg, K. and Hubbell, W. L. (1996). Motion of spin-labeled side chains in T4 lysozyme. Correlation with protein structure and dynamics. *Biochemistry* **35,** 7692-704.

Mchaourab, H. S., Oh, K. J., Fang, C. J. and Hubbell, W. L. (1997b). Conformation of T4 lysozyme in solution. Hinge-bending motion and the substrate-induced conformational transition studied by site-directed spin labeling. *Biochemistry* **36,** 307-16.

Michon, J. and Rassat, A. (1974). Rotational correlation time determination of nitroxide biradical aplication to solution studies. *J. Amer. Chem. Soc.* **96,** 335-337.

Milburn, M. V., Prive, G. G., Milligan, D. L., Scott, W. G., Yeh, J., Jancarik, J., Koshland, D. E., Jr. and Kim, S. H. (1991). Three-dimensional structures of the ligand-binding domain of the bacterial aspartate receptor with and without a ligand. *Science* **254,** 1342-7.

Montelione, G. T. and Anderson, S. (1999). Structural genomics: keystone for a Human Proteome Project. *Nature Structural Biology* **6,** 11-2.

Ogawa, S. and McConnell, H. M. (1967). Spin-label study of hemoglobin conformations in solution. *Proceedings of the National Academy of Sciences of the United States of America* **58,** 19-26.

Ogawa, S., McConnell, H. M. and Horwitz, A. (1968). Overlapping conformation changes in spin-labeled hemoglobin. *Proceedings of the National Academy of Sciences of the United States of America* **61,** 401-5.

Oh, K. J., Zhan, H., Cui, C., Hideg, K., Collier, R. J. and Hubbell, W. L. (1996). Organization of diphtheria toxin T domain in bilayers: a site-directed spin labeling study. *Science* **273,** 810-2.

Ottemann, K. M., Thorgeirsson, T. E., Kolodziej, A. F., Shin, Y. K. and Koshland, D. E. (1998). Direct Measurement of Small Ligand-Induced Conformational Changes in the Aspartate Chemoreceptor Using Epr. *Biochemistry* **37,** 7062-7069.

Ottemann, K. M., Xiao, W. Z., Shin, Y. K. and Koshland, D. E. (1999). A piston model for transmembrane signaling of the aspartate receptor. *Science* **285,** 1751-1754.

Pebay-Peyroula, E., Rummel, G., Rosenbusch, J. P. and Landau, E. M. (1997). X-ray structure of bacteriorhodopsin at 2.5 angstroms from microcrystals grown in lipidic cubic phases. *Science* **277,** 1676-81.

Perozo, E., Cortes, D. M. and Cuello, L. G. (1998). Three-dimensional architecture and gating mechanism of a K^+ channel studied by EPR spectroscopy. *Nature Structural Biology* **5,** 459-69.

Perozo, E., Cortes, D. M. and Cuello, L. G. (1999). Structural rearrangements underlying K^+-channel activation gating. *Science* **285,** 73-8.

Poirier, M. A., Xiao, W., Macosko, J. C., Chan, C., Shin, Y. K. and Bennett, M. K. (1998). The synaptic SNARE complex is a parallel four-stranded helical bundle. *Nature Structural Biology* **5,** 765-9.

Rabenstein, M. D. and Shin, Y. K. (1995). Determination of the distance between two spin labels attached to a macromolecule. *Proceedings of the National Academy of Sciences of the United States of America* **92,** 8239-43.

Rabenstein, M. D. and Shin, Y. K. (1996). HIV-1 gp41 tertiary structure studied by EPR spectroscopy. *Biochemistry* **35,** 13922-8.

Rees, D. C., DeAntonio, L. and Eisenberg, D. (1989). Hydrophobic organization of membrane proteins. *Science* **245,** 510-3.

Rink, T., Riesle, J., Oesterhelt, D., Gerwert, K. and Steinhoff, H. J. (1997). Spin-labeling studies of the conformational changes in the vicinity of

D36, D38, T46, and E161 of bacteriorhodopsin during the photocycle. *Biophysical Journal* **73,** 983-93.

Robinson, B. H., Haas, D. A. and Mailer, C. (1994). Molecular dynamics in liquids: spin-lattice relaxation of nitroxide spin labels. *Science* **263,** 490-3.

Salwinski, L. and Hubbell, W. L. (1999). Structure in the channel forming domain of colicin E1 bound to membranes: the 402-424 sequence. *Protein Science* **8,** 562-72.

Schrempf, H., Schmidt, O., Kummerlen, R., Hinnah, S., Muller, D., Betzler, M., Steinkamp, T. and Wagner, R. (1995). A prokaryotic potassium ion channel with two predicted transmembrane segments from Streptomyces lividans. *EMBO Journal* **14,** 5170-8.

Shin, Y. K., Levinthal, C., Levinthal, F. and Hubbell, W. L. (1993). Colicin E1 binding to membranes: time-resolved studies of spin-labeled mutants. *Science* **259,** 960-3.

Steinhoff, H. J. and Hubbell, W. L. (1996). Calculation of electron paramagnetic resonance spectra from Brownian dynamics trajectories: application to nitroxide side chains in proteins. *Biophysical Journal* **71,** 2201-12.

Steinhoff, H. J., Mollaaghababa, R., Altenbach, C., Hideg, K., Krebs, M., Khorana, H. G. and Hubbell, W. L. (1994). Time-resolved detection of structural changes during the photocycle of spin-labeled bacteriorhodopsin. *Science* **266,** 105-7.

Subramaniam, S. and Henderson, R. (1999). Electron crystallography of bacteriorhodopsin with millisecond time resolution. *Journal of Structural Biology* **128,** 19-25.

Sun, J., Voss, J., Hubbell, W. L. and Kaback, H. R. (1999). Proximity between periplasmic loops in the lactose permease of Escherichia coli as determined by site-directed spin labeling. *Biochemistry* **38,** 3100-5.

Sutton, R. B., Fasshauer, D., Jahn, R. and Brunger, A. T. (1998). Crystal structure of a SNARE complex involved in synaptic exocytosis at 2.4 A resolution. *Nature* **395,** 347-53.

Terwilliger, T. C., Waldo, G., Peat, T. S., Newman, J. M., Chu, K. and Berendzen, J. (1998). Class-directed structure determination: foundation for a protein structure initiative. *Protein Science* **7,** 1851-6.

Thorgeirsson, T. E., Xiao, W., Brown, L. S., Needleman, R., Lanyi, J. K. and Shin, Y. K. (1997). Transient channel-opening in bacteriorhodopsin: an EPR study. *Journal of Molecular Biology* **273,** 951-7.

Ujwal, M. L., Jung, H., Bibi, E., Manoil, C., Altenbach, C., Hubbell, W. L. and Kaback, H. R. (1995). Membrane topology of helices VII and XI in the lactose permease of Escherichia coli studied by lacY-phoA fusion analysis and site-directed spectroscopy. *Biochemistry* **34,** 14909-17.

Unger, V. M., Hargrave, P. A., Baldwin, J. M. and Schertler, G. F. (1997). Arrangement of rhodopsin transmembrane a-helices. *Nature* **389,** 203-6.

Voss, J., He, M. M., Hubbell, W. L. and Kaback, H. R. (1996). Site-directed spin labeling demonstrates that transmembrane domain XII in the lactose permease of Escherichia coli is an a-helix. *Biochemistry* **35,** 12915-8.

Voss, J., Hubbell, W. L., Hernandez-Borrell, J. and Kaback, H. R. (1997). Site-directed spin-labeling of transmembrane domain VII and the 4B1 antibody epitope in the lactose permease of Escherichia coli. *Biochemistry* **36,** 15055-61.

Voss, J., Hubbell, W. L. and Kaback, H. R. (1995a). Distance determination in proteins using designed metal ion binding sites and site-directed spin labeling: application to the lactose permease of Escherichia coli. *Proceedings of the National Academy of Sciences of the United States of America* **92,** 12300-3.

Voss, J., Hubbell, W. L. and Kaback, H. R. (1998). Helix packing in the lactose permease determined by metal-nitroxide interaction. *Biochemistry* **37,** 211-6.

Voss, J., Salwinski, L., Kaback, H. R. and Hubbell, W. L. (1995b). A method for distance determination in proteins using a designed metal ion binding site and site-directed spin labeling: evaluation with T4 lysozyme. *Proceedings of the National Academy of Sciences of the United States of America* **92,** 12295-9.

Wahl, P. and Weber, G. (1967). Fluorescence depolarization of rabbit gamma globulin conjugates. *Journal of Molecular Biology* **30,** 371-82.

Wang, Q., Voss, J., Hubbell, W. L. and Kaback, H. R. (1998). Proximity of helices VIII (Ala273) and IX (Met299) in the lactose permease of Escherichia coli. *Biochemistry* **37,** 4910-5.

Weber, T., Zemelman, B. V., McNew, J. A., Westermann, B., Gmachl, M., Parlati, F., Sollner, T. H. and Rothman, J. E. (1998). SNAREpins: minimal machinery for membrane fusion. *Cell* **92,** 759-72.

Weimbs, T., Low, S. H., Chapin, S. J., Mostov, K. E., Bucher, P. and Hofmann, K. (1997). A conserved domain is present in different families of vesicular fusion proteins: a new superfamily. *Proceedings of the National Academy of Sciences of the United States of America* **94,** 3046-51.

Weissenhorn, W., Dessen, A., Harrison, S. C., Skehel, J. J. and Wiley, D. C. (1997). Atomic structure of the ectodomain from HIV-1 gp41. *Nature* **387,** 426-30.

Wu, J., Voss, J., Hubbell, W. L. and Kaback, H. R. (1996). Site-directed spin labeling and chemical crosslinking demonstrate that helix V is close to helices VII and VIII in the lactose permease of Escherichia coli. *Proceedings of the National Academy of Sciences of the United States of America* **93,** 10123-7.

Yeh, J. I., Biemann, H. P., Prive, G. G., Pandit, J., Koshland, D. E., Jr. and Kim, S. H. (1996). High-resolution structures of the ligand binding domain of the wild-type bacterial aspartate receptor. *Journal of Molecular Biology* **262,** 186-201.

Zhao, M., Zen, K. C., Hubbell, W. L. and Kaback, H. R. (1999). Proximity between Glu126 and Arg144 in the lactose permease of Escherichia coli. *Biochemistry* **38,** 7407-12.

Chapter 5

EPR Spectroscopic Ruler: the Method and its Applications

WENZHONG XIAO[1] and YEON-KYUN SHIN[2]
[1]*Department of Chemistry, University of California, Berkeley, California 94720*
[2]*Department of Biochemistry and Biophysics, Iowa State University, Ames, Iowa 50011*

Abstract: The distance measurement method based on Fourier deconvolution of dipolar coupling in spin-labeled EPR spectra provides a new way of examining the structure and function of biological macromolecules. In this chapter, we describe a new approach that has been developed for effective and reasonably accurate data analysis, followed by discussions of several successful applications to interesting biological problems on membrane-associated proteins. This method of EPR spectroscopic ruler has emerged as a powerful tool to investigate the functions of membrane-associated proteins.

1. INTRODUCTION

X-ray crystallography and NMR have made important contributions to our understanding of the structure and function of many soluble proteins. Recent determination of crystal structures of ribosomes truly represents the success of protein crystallography (Ban et al., 1999; Clemons et al., 1999; Cate et al., 1999). Currently, there are more than 11,000 entries in the protein data bank (PDB), and the number is expected to increase rapidly in the future in pace with the swift progress of the structural genomics project.

In contrast, the progress in determining the structure of membrane proteins, which are as many as one third of the total encoded proteins in an organism (Gerstein, 1998), has been very slow. Structures of only a handful

Biological Magnetic Resonance, Volume 19: Distance Measurements in Biological Systems by EPR, edited by Berliner et al. Kluwer Academic/Plenum Publishers, New York, 2000

of membrane proteins have been determined. Although several promising crystallization procedures such as antibody conjugation (Iwata et al., 1995) and the cubic phase method (Pebay-Peyroula et al., 1997) have been developed, the generality of these methodologies has not been proven.

Furthermore, conformational changes triggered by ligands or membrane potentials are essential for the regulatory functions of membrane proteins. There is compelling evidence that membrane receptors, ion pumps, and ion channels are mechanically analogous to simple machines composed of basic components such as shafts, hinges and rotors. "Piston", "scissors", "see saw", "screw", "tilt" and "rotation" are the hypothetical terms describing movements in the membrane proteins. Despite extensive efforts, all of these models have not been experimentally tested, largely because high-resolution techniques remain ineffective for such dynamic problems of membrane proteins.

Could spin labeling EPR be an effective tool for these important, but difficult problems, although maybe with lower resolution? In this chapter we will try to provide some answers to this question. We will describe first the EPR dipolar distance measurement method (Rabenstein and Shin, 1995), followed by discussions of several successful applications to interesting biological problems on membrane-associated proteins (Poirier et al., 1998b; Ottemann et al., 1999; Thorgeirsson et al., 1997).

Modern molecular biology techniques are so advanced that native residues in a protein can be routinely replaced by cysteines, which provide specific labeling sites for thiol specific spin labels. This labeling method is often called site-directed spin labeling (SDSL) (Hubbell and Altenbach, 1994). We use SDSL to attach a pair of nitroxides to specific positions of a protein. The two nitroxides then experience dipolar interaction, which has an inverse cube dependence on the inter-spin distance (Figure 1). We are interested in the analysis of the dipolar broadened EPR spectra to determine the interspin distance.

The method discussed in this chapter is largely limited to 9 GHz continuous wave (CW) EPR spectroscopy. As an initial step towards the establishment of the methodology, an EPR spectroscopic ruler, a series of α-helical polyalanine peptides (Marqusee et al., 1989) containing the two nitroxides, was synthetically made in order to probe the accuracy, precision, and measurable distance range of the EPR method. Moreover, a new simplified dipolar interaction theory, called Fourier convolution-deconvolution (FCD), which requires no adjustable parameters, has been developed for an effective and reasonably accurate data analysis. In this chapter, we will start with the description of the theory. Next, we will explain the step-by-step data analysis using the FCD method. In particular, we will put some emphasis on the procedure by which we treat the problem of incomplete labeling. The problem of incomplete labeling is the major source of error existing in most double spin labeling EPR experiments. This method of distance measurements in protein molecules provides a new way of examining the structure and function of membrane-associated proteins.

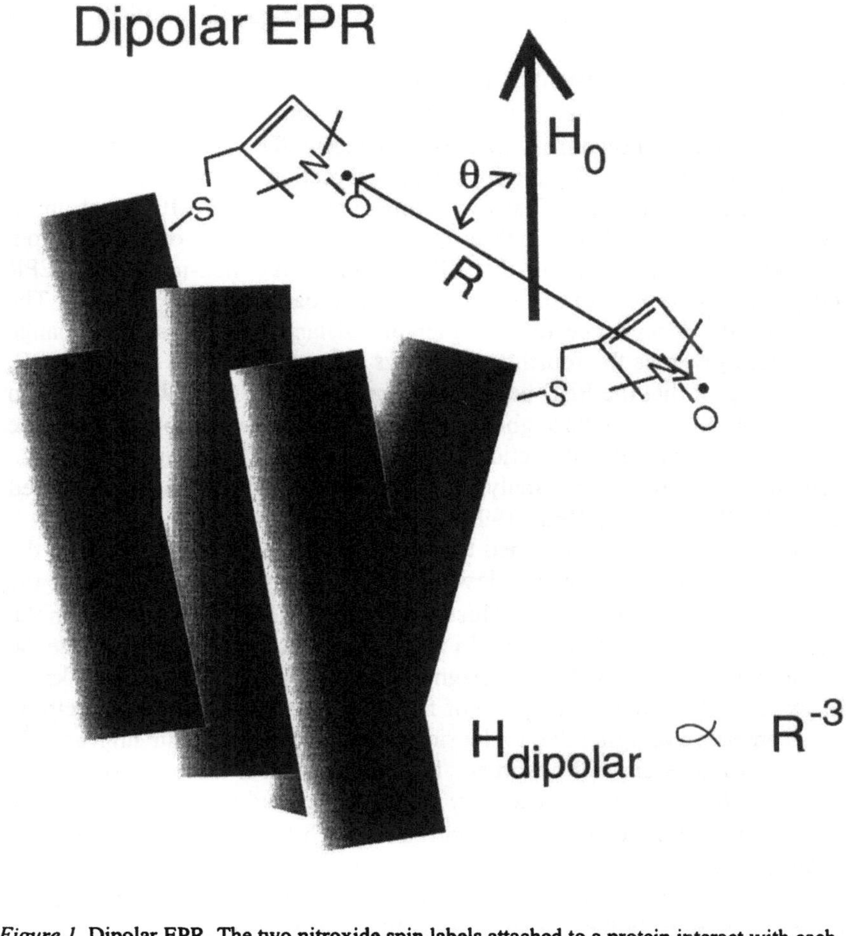

Figure 1. Dipolar EPR. The two nitroxide spin labels attached to a protein interact with each other through electron-electron dipolar interaction.

1.1 Review of the Theory for Dipolar Interactions between Two Nitroxides

The two nitroxide spin labels attached to a protein interact with each other through the electron-electron dipolar interaction (EED) (Figure 1). The biradical spin-Hamiltonian H is expressed as

$$H=h\Omega_1 S_{1z} + h\Omega_2 S_{2z} + h^2\gamma_e^2 (1-3\cos^2\theta)(3S_{1z}S_{2z}-S_1\bullet S_2)/2R^3 + JS_1\bullet S_2 \qquad [1]$$

where h is Planck's constant, γ_e is the gyromagnetic ratio of an electron, and J is the exchange integral. θ is the angle between the interspin vector and magnetic field z direction. Ω_1 and Ω_2 represent the resonance offsets for electron spins 1 and 2. S_{1z} and S_{2z} are the z direction components of spins S_1 and S_2, respectively. The first two terms give rise to the EPR spectrum in the absence of the EED, which is given as the third term of the equation. However, the existence of the EED perturbs the non-interacting EPR spectrum to give rise to an overall spectral broadening (see below). The EED has R^{-3}-dependence on the interspin distance R. The spin exchange interaction given as the fourth term decays exponentially as a function of R, and we expect that the EED becomes a dominant term when R is larger than 7 Å unless there is a through-bond interaction. We will not consider the short-range exchange interaction in this analysis. The spin hamiltonian including the EED can be analyzed rigorously for two geometrically fixed nitroxides (Hustedt and Beth, 1999).

When nitroxides are attached to a macromolecule such as a protein, the flexible nitroxide side chain allows significant degrees of freedom of the magnetic tensors at a given θ. This averaging effect must be included in the analysis of the spin hamiltonian. In reality, it is not trivial to incorporate the geometrical distribution of the magnetic tensors, because we do not know the extent of this motional degree of freedom. In most cases, we expect that this geometric averaging due to the side chain flexibility is quite high as long as the nitroxides are not absorbed, for example, on the protein surface or geometrically contained in a pocket. Under these conditions we may assume that the magnetic tensors are not correlated with the angle θ and the inter-dipole vector.

Under these conditions the EPR absorption lines at all spectral positions in the magnetic field for a given θ are split by 2B (in Gauss),

$$2B = \left(\frac{3}{2}\right) g_e \beta \, (3\cos^2\theta - 1)/R^3 \qquad [2]$$

where g_e is the isotropic g value of the electrons, β is the electron Bohr magneton($3/2\ g_e\beta = (30.3\ )^3 Gauss{\cdot}A^3$). Here we assume that the effect of g anisotropy is negligible at 9 GHz.

In most biological systems, except macroscopically oriented samples, the distribution of θ is isotropic. It is possible then to treat the spectrum of the two interacting spins as the convolution of the non-interacting powder pattern absorption spectrum with a dipolar broadening function D(R, B), known as a Pake pattern. The average splitting <2B> over the distribution D(R, B) is then:

$$< 2B > = (0.75)(\tfrac{3}{2})g_e\beta / R^3 \qquad [3]$$

Thus, the EPR spectrum Π(B) for the two nitroxides is described by:

$$\Pi(B) = \int_{-\infty}^{\infty} S(B')D(R, B'-B)dB' \ , \qquad [4]$$

where S(B) is the non-interacting EPR spectrum. However, in practice there will be a distribution of interspin distances due to the flexibility of the nitroxide side chain as well as the conformational variation of the macromolecules. In this situation the EPR spectrum is described by:

$$\Pi(B) = \int_{-\infty}^{\infty} S(B')M(B'-B)dB' , \qquad [5]$$

where M(B) is the weighted sum of the D(R,B) over the distribution of the distances P(R):

$$M(B) = \sum_R P(R)D(R,B) . \qquad [6]$$

1.2 Fourier Deconvolution

Since Π is approximated by a convolution of S and M in real space, Eq. 5 is simplified in Fourier space using the convolution theorem:

$$\Pi^*(\omega) = S^*(\omega) \bullet M^*(\omega), \qquad [7]$$

where the superscript * indicates the Fourier transformed functions and ω is the inverse variable of B in units of gauss^{-1}. M(B) is then obtained from the inverse Fourier transform of the division of Π * by S*:

$$M(B) = \frac{1}{\sqrt{2\pi}} \int_{-\infty}^{\infty} \exp(2\pi i \omega B) \frac{\Pi^*(\omega)}{S^*(\omega)} d\omega . \quad [8]$$

Thus, the average splitting <2B> for a given P(R), and subsequently the average separation of the two nitroxides is obtained from:

$$< 2B >= \int_{-\infty}^{\infty} | 2B | \bullet M(B) dB / \int_{-\infty}^{\infty} M(B) dB , \quad [9]$$

and

$$< R >= \left(0.75 \left(\frac{3}{2} \right) g_e \beta / < 2B > \right)^{\frac{1}{3}} \quad [10]$$

1.3 Spectral Analysis

To avoid the complication arising from an incomplete time averaging of θ in Eq. 2 the EPR spectra must be taken under a motionally frozen condition. This can be achieved by simply freezing the samples at liquid nitrogen temperature.

EPR spectra can be analyzed according to Eqs. 7-10 using fast Fourier transform routines (Figure 2), and the resulting spectra at each step of the analysis are shown in Figure 3. Both real and imaginary components of a dipolar spectrum in Fourier space [Π*(ω)/S*(ω)] contain high amplitude noise at high values of ω (Figure 3B). The noise can be suppressed by fitting the spectral decay in Fourier space with a sum of gaussian functions. In most cases, we have found that it is sufficient to use a sum of only two gaussians. The final dipolar spectrum is obtained by inverse Fourier transformation of the best fitting function. The distribution P(R) of interspin distance leads to broadening functions that are the sum of many Pake Patterns. This, in combination with the EPR line width, leads to broadening functions that do not look like Pake Patterns. The average distance is calculated using Eqs. 8-10.

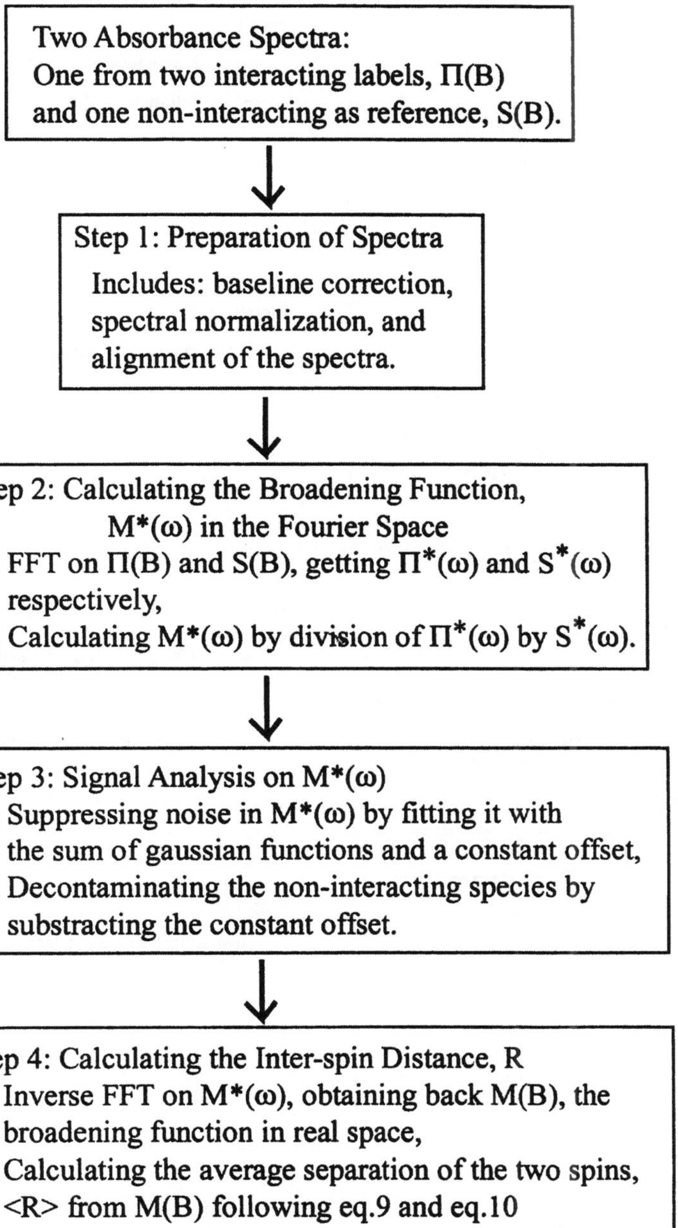

Figure 2. A flow chart of the algorithm for distance analysis from dipolar spin-spin interaction. The MATLAB program is available from the authors upon request.

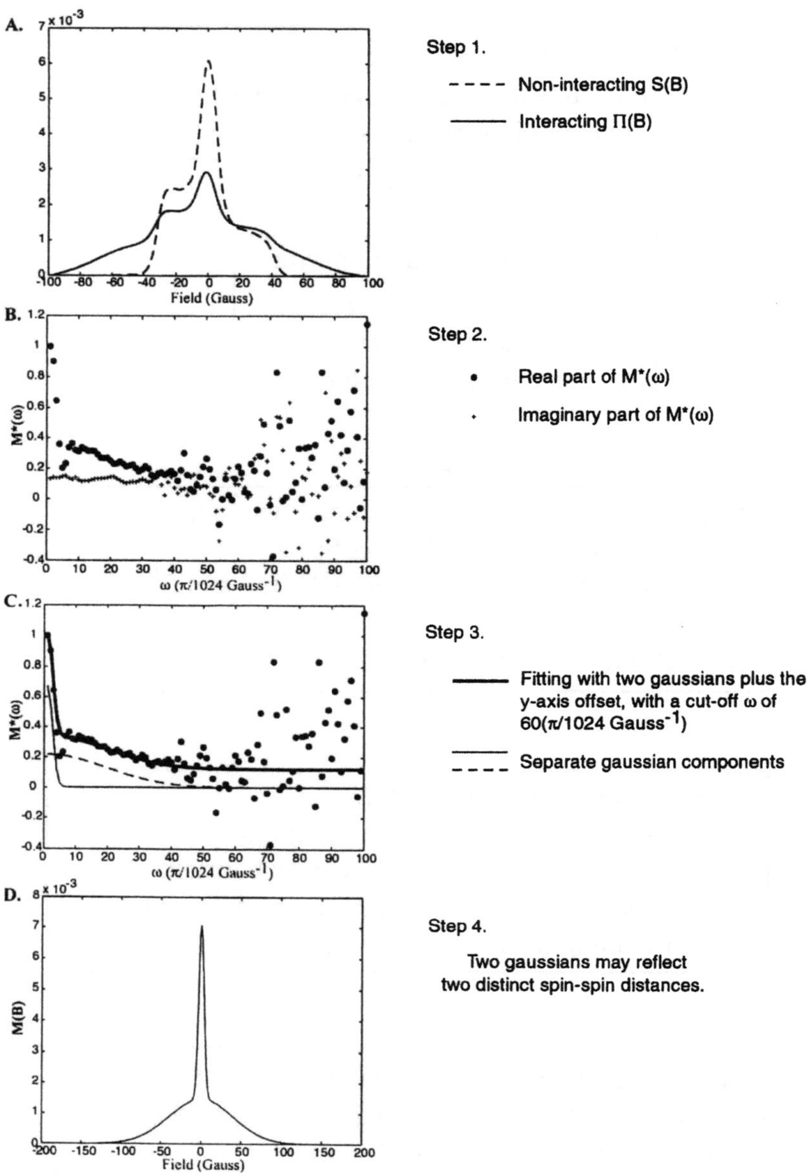

Figure 3. The resulting spectra at each step of the analysis outlined in Figure 2.

1.4 Model Systems: a Spectroscopic Ruler

To test the method, a series of short alanine-based α-helical peptides with two cysteines at various separations were synthesized and cysteine residues were labeled with thiol-specific methanethiosulfonate spin label. The distance obtained following the analysis discussed above are plotted vs. distances calculated based on the α-helical geometry in Figure 4.

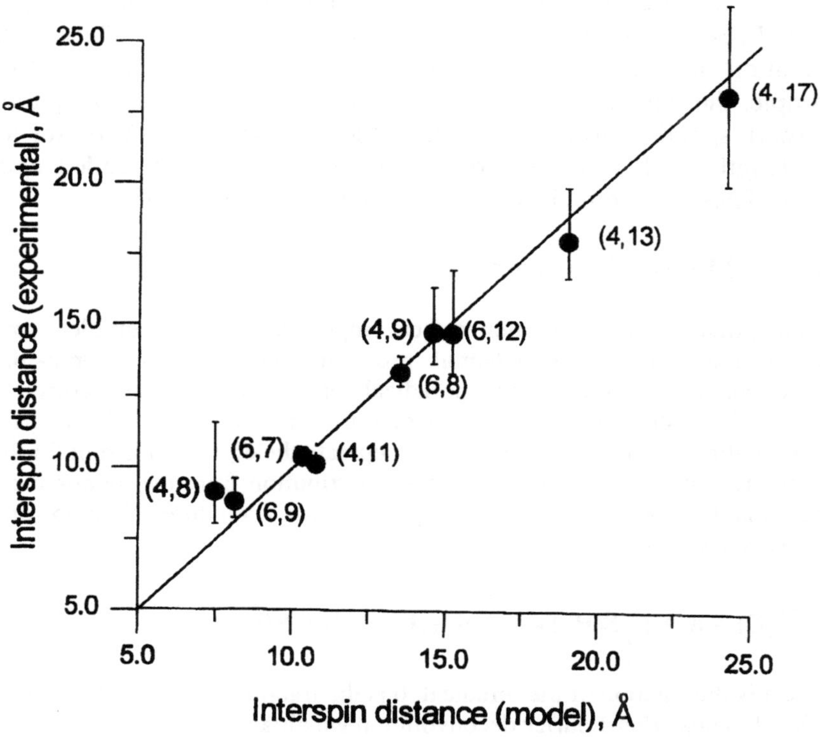

Figure 4. Inter-spin separations in a-helical peptides: measurement vs. model.The original peptide is Ac-(1)AAAAKAAAAKAAAAKAAAAKA(21)-NH2. At each data point, the numbers in parenthesis indicate the positions the two spin labels are attached to. (Figure reproduced from Rabenstein and Shin, 1995.)

The model nitroxide-nitroxide separation for each doubly-labeled variant was calculated assuming that the peptide is a rigid α-helix of a 5.41Å pitch and 1.50 Å translation per residue (Creighton, 1983).

Because the exact side chain conformations are unknown, the distance between the helix axis and the radical, known as the "arm length" is unknown. The "arm" is assumed to be the line perpendicular to the helix axis intersecting the cysteine ß-carbon. Spin label separations were calculated for various arm lengths and compared to the experimental data. An arm length of 6.7Å best fit the experimental data. There is good agreement between the experimental values and the theoretical values, with a standard deviation of 0.9Å for the entire range of 7.5 to 25Å. Furthermore, it should be noted that with spin labels 7 residues apart, the nitroxides are nearly eclipsed. Thus, their separation should be independent of the assumed arm length. For example, the separation is calculated to be 10.6Å for an arm-length of 4Å (corresponding to the position of the ß-carbon) and 11.6Å for an arm length of 13Å (corresponding to a completely extended spin label). These are both close to the measured value of 10.5Å.

1.5 Monoradical Impurities

For practical purposes, monoradical impurities due to incomplete spin labeling reactions are a potential problem when analyzing spin-spin interactions. EPR spectra with monoradical contamination can be incorrectly interpreted as showing larger nitroxide-nitroxide separation. The Fourier deconvolution method can, in principle, separate the dipolar spectrum for the interacting nitroxides from the spectral contribution from the monoradical contaminants. The composite EPR spectrum for a mixture of mono- and biradicals is given by:

$$\Pi(B) = a \bullet \int_{-\infty}^{\infty} S(B')M(B'-B)dB' + b \bullet S(B), \qquad [11]$$

where a is the fraction of the biradical, b is the fraction of monoradicals with a + b = 1. Thus, after Fourier deconvolution we obtain:

$$M'^*(\omega) = a\, M^*(\omega) + b, \qquad [12]$$

in the Fourier space and the non-interacting species contributes to the y-axis offset. This y-axis offset b may be readily subtracted from the dipolar spectrum in the Fourier space (Figure 5).

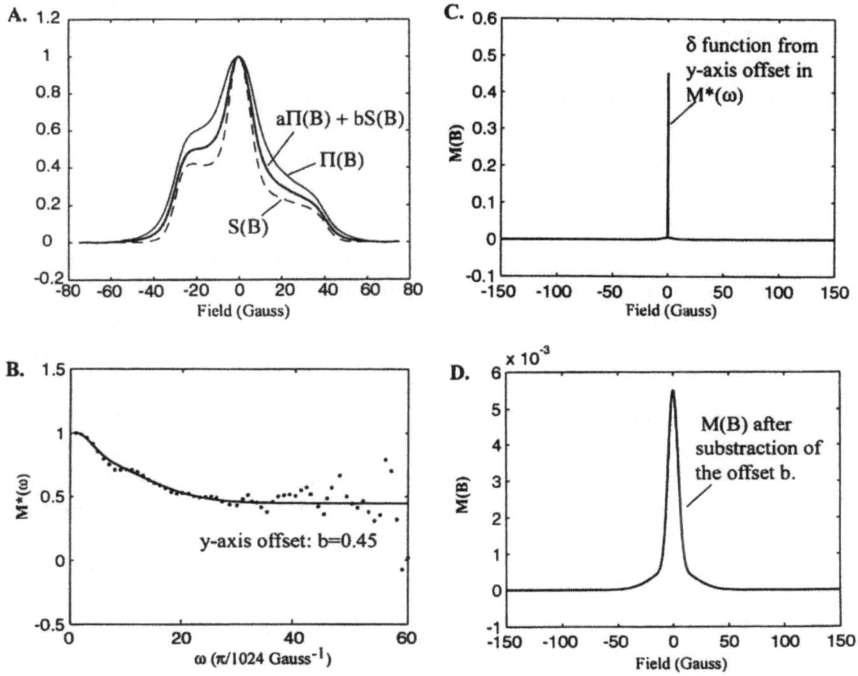

Figure 5. Analysis of monoradical impurities.

2. STRUCTURAL MODELING USING EPR-DETERMINED DISTANCES

Some proteins, including most membrane proteins, are not amenable to structural determination using NMR and X-ray crystallography. Using computational methods may be an alternative approach. However, it is still not possible to assign a unique fold of a protein using the prediction methods. The homology modeling has difficulties when the sequence is not closely related to the known structures (when similarity < 25%). On the other hand, the secondary structure prediction methods are relatively successful and they can predict the secondary structure with accuracy of as high as 75%. Spin labeling EPR can also help in identifying secondary structural elements in a protein using nitroxide-scanning strategies (Chapter 4 by Perozo and Mchaourab).

In this section we explore the approach of using the EPR distance measurement technique to provide multiple long-range distance constraints necessary to put the secondary structural elements together. For a protein sequence of a known secondary structure we intend to predict the 3-dimensional fold of the protein with the help of molecular modeling.

This combined modeling approach has been successfully tested for the core complex of the SNARE (soluble N-ethylmaleimide-sensitive factor attachment protein receptor) proteins (Poirier et al., 1998b).

What is the SNARE complex? Intracellular protein trafficking, as well as secretion of neurotransmitters and hormones, is accomplished by the recognition and the fusion of a cargo vesicle from the donor compartment to a target membrane, resulting in delivery of the cargo to the appropriate destination. This process requires highly conserved machinery, generically termed SNARE proteins. SNARE proteins include a protein on the surface of the cargo vesicle known as v-SNARE (i.e. the VAMP/Synaptobrevin family) and two proteins on the surface of the target membrane known as t-SNAREs (i.e. the syntaxin and SNAP-25 families) (for review, see Bennett & Scheller, 1994; Sudhof, 1995). The most thoroughly characterized SNARE proteins include the synaptic vesicle protein VAMP2 (a v-SNARE) and the presynaptic plasma membrane proteins syntaxin 1A and SNAP-25 (two t-SNAREs). These three proteins form a stable complex, called the SNARE complex, that is required for synaptic vesicle fusion with the plasma membrane and subsequent release of neurotransmitter (Broadie et al., 1995; Hayashi et al., 1994; Niemann et al., 1994).

Secondary structural elements of the core are known. Several biochemical and biophysical studies have identified the minimal interacting regions within the synaptic SNARE complex. They include the domains of syntaxin 1A and VAMP2 that precede their C-terminal transmembrane anchors and separate N- and C-terminal domains of SNAP-25 (Chapman et al., 1994; Hayashi et al., 1994; Kee et al., 1995; Poirier et al., 1998a). These elements are all 60-80 amino acids long and are predicted to be highly α-helical. Interestingly, a profile analysis based on the Hidden Markov Model suggested that these three proteins evolved from a common gene and belong to a single superfamily of membrane trafficking proteins (Weimbs et al., 1997, 1998).

It has been suspected that SNARE complex assembly may proceed through the formation of a coiled coil. There is compelling evidence for an extended α-helical coiled-coil structure for the SNARE complex. The complex formation induces a dramatic increase in the α-helicity of the SNARE components (Fasshauer et al., 1997a, b). The synaptic SNARE complex, as revealed by negative stain electron microscopy, is a 12-14 nm long rod, which is compatible with a structure of coiled coil (Hanson et al., 1997). Coiled coil analysis (Lupas, 1996; Berger et al., 1995) shows that each of the four regions exhibits at least one heptad repeat motif. The individual positions within this motif are designated a-g, with the a and d positions generating a hydrophobic α-helical contact surface. The heptad repeat motif is a hallmark of proteins that form α-helical coiled coils (Lupas, 1996).

Important questions regarding the structure of the synaptic SNARE complex include the orientation, position, and register of the participating helices. Recent studies have suggested that syntaxin 1A and VAMP2 are parallel (Hanson et al., 1997; Lin and Scheller, 1997; Otto et al., 1997). Little is known, however, about the organization of SNAP-25 in the synaptic SNARE complex.

Overall Strategy. We characterized the conformation of the SNAP-25 N-terminal domain experimentally, by detecting mobility of spin labels attached at a series of strategic positions of the region. We then utilized dipolar EPR spectroscopy to investigate the arrangement of individual SNARE components within the complex. Using the long-range distances measured with EPR in combination with the information about the secondary structure of the complex, we have modeled the three-dimensional structure of the complex with the help of computer simulation.

Spin labeling EPR. To incorporate a spin label at specific positions within the SNARE proteins, site-directed mutagenesis was first used to introduce single cysteine residues within the N- and C-terminal domains of SNAP-25 (SNAP-25[N] and SNAP-25[C]) and the C-terminal domain of syntaxin 1A (syn). In addition, two double cysteine mutants within SNAP-25[N] were generated in which the cysteines flank a proposed discontinuity between the H1 and H2 heptad repeat motifs (Fasshauer et al., 1997a). After protein expression and purification, cysteine residues were labeled with MTSSL, and SNARE complexes were formed by combining wild type or spin-labeled versions of SNAP-25[N], SNAP-25[C] and syntaxin 1A with wild type VAMP2 or spin-labeled VAMP1. The labeled positions within the SNARE protein are summarized in Figure 6.

EPR line shape helps the secondary structural assignment. The mobility of the nitroxide is a good indicator for the local environment. For example, spin labels attached to positions in a coiled coil are expected to be highly immobilized at core (a and d) positions, mobile at the surface f, b, and c positions, and the intermediate mobility is expected for e and g positions. We have used the EPR line shape to verify the secondary structural predictions for the SNARE core complex (Poirier et al., 1998b. Data are not shown here.). The detailed discussion of the EPR line shape is given in Chapter 4.

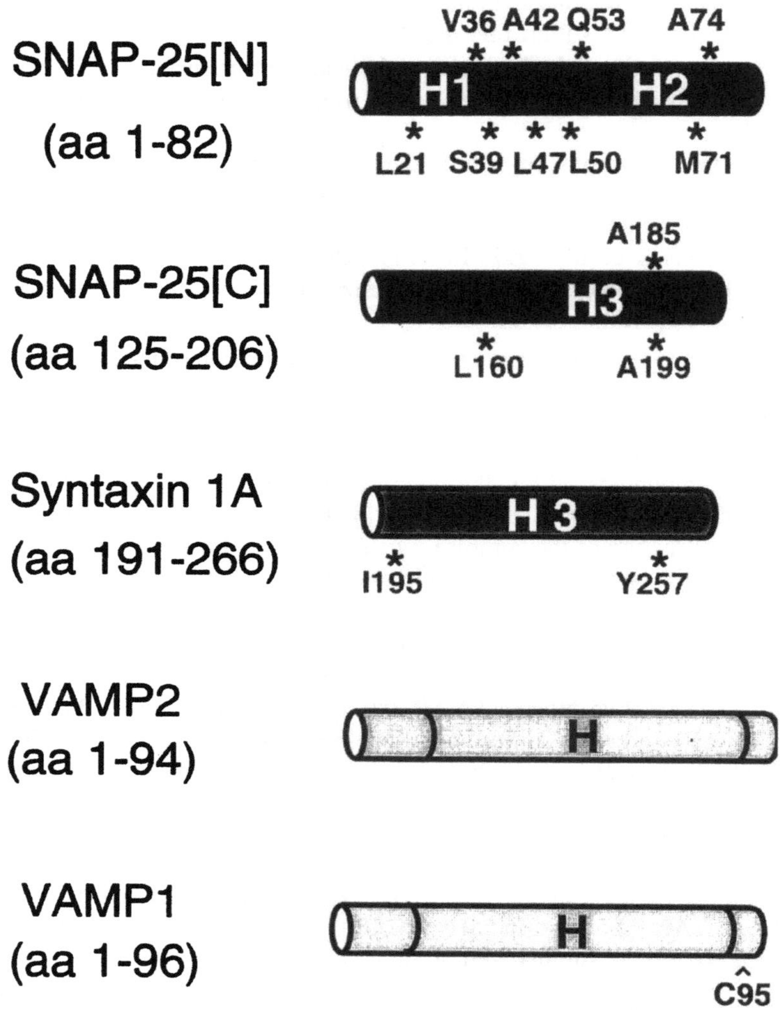

Figure 6. A schematic representation of recombinant SNARE proteins. Amino acid residues mutated to cysteine are denoted by an asterisk. The naturally occurring cysteine present at position 95 of VAMP1 is indicated by an arrowhead. This residue corresponds to position 93 of VAMP2, as judged by sequence alignment. Heptad repeat domains predicted by the COILS program (Lupas, 1996) within each protein are labeled (H1, 2, or 3). It should be noted that the COILS program, which is based on analysis of two- and three-stranded coiled coils, may not accurately predict four-stranded coiled-coil structures. A profile-based sequence analysis method (Weimbs et al., 1997, 1998) has led to the proposal of alternative registers for the H1 domain of SNAP-25[N] which is continuous with respect to H2 domain.

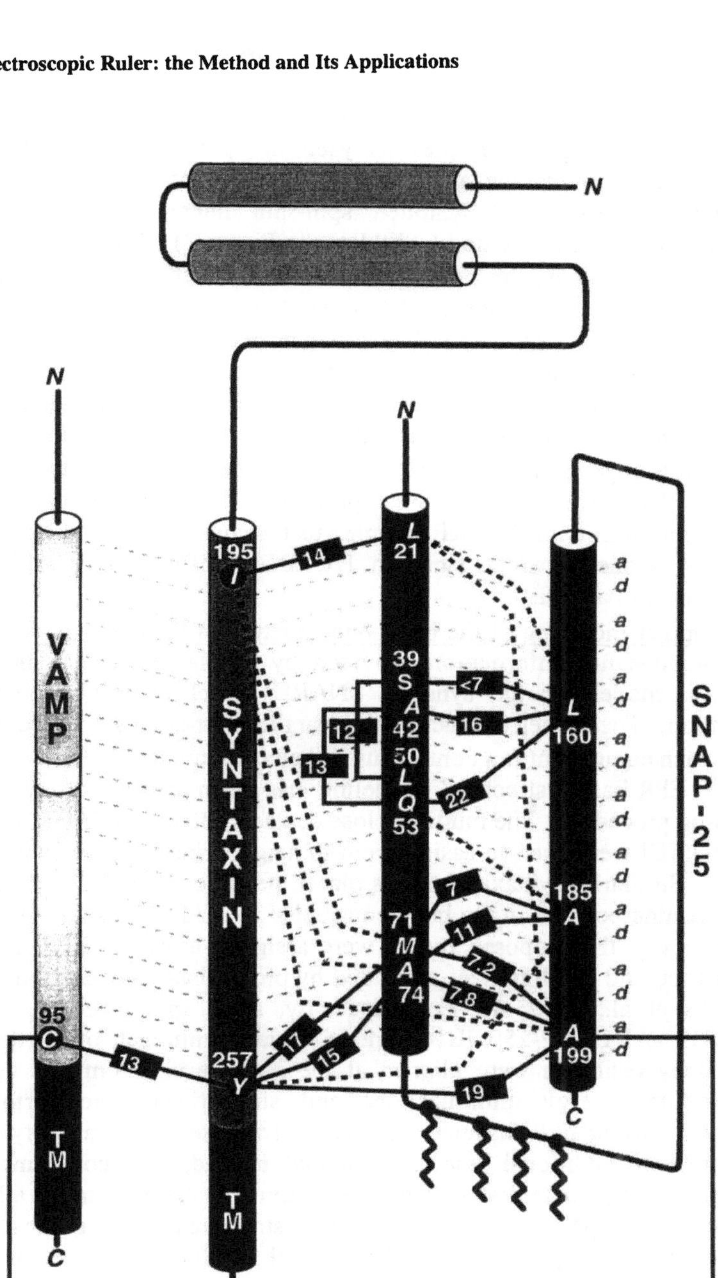

Figure 7. Schematic representation of SNARE complex components showing EPR-derived interspin distances. (Figure reproduced from Poirier et al., 1998b.)

Determination of the rough fold of the SNARE complex. For seven of the double spin-labeled samples analysed, no spin-spin interactions were detected indicating that the labels were >25Å apart (dotted lines in Figure 7). The remaining samples exhibited spin-spin interactions that provide interspin distance information (solid lines in Figure 7). For two spin labels separated by 7-25 Å, the Fourier deconvolution method described in the previous section was used to determine the distance from the dipolar broadening in each spectrum. The distance between spin labels that are closer than 7Å could not be accurately estimated due to complications generated by short-range Heisenberg spin-exchange (or J-coupling), as observed in the SNAP-25[N]S39C/SNAP-25[C]L160 spectrum (not shown). Such short-range interactions provide strong evidence of registering these two spin label sites at the closest proximity. The only arrangement of components within the SNARE complex that is consistent with the observed distances is one in which syntaxin 1A, SNAP-25[N], SNAP-25[C], and VAMP are all parallel.

Computer modeling of the backbone structure of the complex. We have used the distance information provided by dipolar EPR to generate a structural model for the synaptic SNARE complex using a computer simulation. The modeled structure is a parallel four-stranded coiled coil in which each component is a continuous α- helix (Figure 8).

The EPR-based structural modeling has been carried out with the following procedure. The multiple close contacts between SNAP-25[N] and SNAP-25[C] were used to assign them to neighboring positions within the bundle. Similarly, syntaxin 1A was placed neighboring SNAP-25[N] with the remaining position in the bundle occupied by VAMP2. The registers of the helices in the proposed model were assigned based on EPR distance information and an alignment generated by profile-based sequence analysis (Weimbs et al., 1997; 1998). Especially, since spin labels at SNAP-25[N]S39 and SNAP-25[C]L160 are in contact with each other, the two layers they reside in were aligned at the same level. From this general organization, energy minimizations and simulations were performed, without imposing EPR distance constraints. To preserve the integrity of the structure, two additional restraint functions instead, each consisting of a parabolic energy potential beyond the minimum and maximum allowed distances, were applied to confine the helical structure and tetrameric state of the complex during simulation (Porier et al., 1998b). The results were analyzed for consistency with the observed EPR distances. For the model presented in Figure 8, the averaged difference between the observed (nitroxide tip-to-tip) and modeled (Cα-to-Cα) distances was 2.8Å. This is a reasonable fit given the uncertainties about the conformation of spin-labeled side chains and possible local structural perturbations due to spin-labeling.

Reversing the positions of the helices within the bundle reduced the quality of the fit with the data.

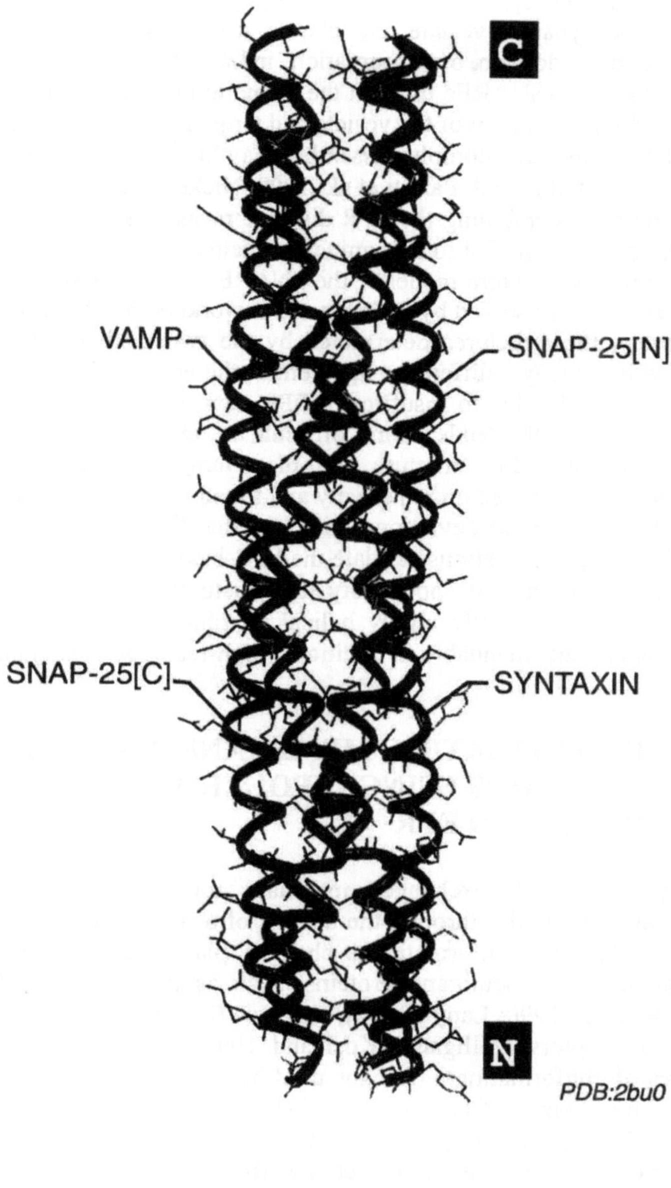

Figure 8. A structural model of the synaptic SNARE complex (PDB: 2bu0).

Possible mechanism inferred from the parallel four strand coiled coil structure. The fold of the complex structure determined here holds some surprises that hint at function. The complex is a parallel four-strand coiled coil. Thus the carboxyl termini of VAMP and syntaxin are tightened at the same end of the elongated rod. Recognizing that both VAMP and syntaxin anchor to the synaptic vesicle and plasma membrane by virtue of their carboxyl terminal domain, our observations indicate that the pairing of the v-SNAREs and the t-SNAREs through the formation of the SNARE complex drives the close apposition of the vesicle and target membranes. Some recent results suggest that this complex assembly may be sufficient for membrane fusion (Weber et al., 1998; Parlati et al., 1999; Nickel et al., 1999).

The approach combining the EPR distance measurements and computer modeling has been applied to determine a protein structure at the fold level. Shortly after this structure model of the SNARE complex was reported, the crystal structure was solved by Brünger and coworkers (Sutton, et al., 1998). Qualitatively, the structures determined by the two methods appear very similar, with the major difference apparent at the ends of the bundle. This difference could be due to insufficient EPR data within this region of the complex, making the ends more difficult to model. Still, the helix backbones of the EPR structure and the average of the three crystal structures were compared quantitatively and found to differ by an average of 1.2Å. Given this small deviation, it is clear that the structures are nearly identical. These observations validate the EPR-based approach to structural modeling, opening a new opportunity for structural investigation on other protein systems, especially highly helical membrane proteins and certain protein that are not amendable to traditional high-resolution techniques.

3. INVESTIGATING MEMBRANE PROTEIN DYNAMICS USING DIPOLAR AND TIME-RESOLVED EPR

Every living cell exchanges materials and information with the environment constantly, through the actions of a myriad of proteins on its membrane. Protein conformational changes play a central role in the functions of many membrane proteins, such as pumps (Tanford, 1983; Henderson et al., 1990; Lanyi, 1995), channels (Unwin, 1995; Perozo et al., 1999), and receptors (Milligan & Koshland, 1991; Spudich & Lanyi, 1996). The types of conformational changes used by these proteins, however, are generally not known. Characterizing these conformational changes has proven difficult since it requires not only a good structural model but also the detection of transient motion when the protein is in action.

Dipolar EPR spectroscopy has important advantages over other structural methods in studying such dynamic problems of membrane proteins. EPR spectra can be taken on spin-labeled samples either in the native cell membranes or in the reconstituted membranes. The conformational changes can be characterized directly by measuring distance changes between a pair of spin labels attached to two specific sites in the protein. Using time-resolved EPR, the protein movements in the μsec-msec time scale can be monitored in the real time. This combined use of dipolar and time-resolved EPR should be an effective method to answer essential mechanistic questions, such as the type, and the timing of conformational changes.

As an example, we applied this strategy to characterize the transient motion in bacteriorhodopsin (Thorgeirsson et al., 1997).

What is bacteriorhodopsin? Bacteriorhodopsin is a seven-α- helical transmembrane protein, with a retinal linked to K216 near the middle of helix G forming a protonated Schiff base (Henderson et al., 1990). The 26 kD protein uses the energy of photons absorbed by the retinal to actively pump out protons across the cell membrane. This is done through a series of steps known as a photocycle, and the kinetics at each step are well known (Lanyi, 1995). The proton gradient is then used for the flagella rotation and ATP synthesis. BR is the best-understood ion pump and has become a paradigm for active transporters as well as seven-α-helical transmembrane protein. Its studies have provided fundamental insights into the function of membrane pumps, and also into the structure and activation of structurally similar G-protein-coupled receptors.

Conformational switch in bR. Active translocation of protons across membranes requires alternating access between ion binding sites inside the pump and each of the membrane surfaces. In bR, this key event ensures the unidirectional proton transfer, proton release to the extracellular side and proton uptake from the cytoplasmic side. It has been a long-standing hypothesis that this switch involves a protein conformational change (Jardetzky, 1966; Lanyi, 1995). Indeed, several studies (Subramanian et al., 1993; Vonck, 1996) have detected some isolated movements at cytoplasmic side of helix F (for review, see Haupts et al., 1999).

Recently, high-resolution structures of the ground state bR as well as a major intermediate state (M) have been determined using x-ray crystallography. BR has been determined by electron and x-ray crystallography (for a recent review, see Haupts et al., 1999). These studies demonstrate that after initial photo-excitation, the redistribution of internal water molecules inside the protein appears to facilitate proton transport by providing hydrogen bonds and by stabilizing charges within the protein (Luecke et al., 1999 and Edman et al., 1999). However the conformational changes were not detected in contradiction to the previous electron microscopic studies (Subramanian et al., 1993; Vonck, 1996).

Experimental Design. Our study was aimed at understanding the mechanism of the reprotonation switch, which transiently opens up the cytoplasmic half of the channel for proton uptake, after proton release to the extracellular side. As the initial step, we characterized the transient conformational changes on the cytoplasmic interhelical loops of the protein. First, we used time-resolved EPR to determine the timing of the conformational change with respect to the photocycle. Then, we trap an intermediate state at low-temperature to determine distance changes.

Spin-labeling. Five cysteine mutants were constructed: two double mutants (S35C/M163C and A103C/M163C), and the corresponding single mutants (S35C, M163C, and A103C), all on the interhelical loops (Figure 9). Each mutant was then spin labeled in the native purple membrane. The protein concentration was accurately determined using UV spectrophotometry, and the spin concentration was also accurately determined by comparing the double integral of the spectrum with the Tempo standard solution. The spin labeling was quantitative (100(±5)%) except at position 35 where labeling was consistently 60 (± 3)%.

Both of the spin-labeled double mutants show EPR line broadening due to spin-spin interactions (Figure 11, left). These three positions triangulate the interhelical loops. Distance changes between interhelical loops can be monitored with these two doubly labeled mutants.

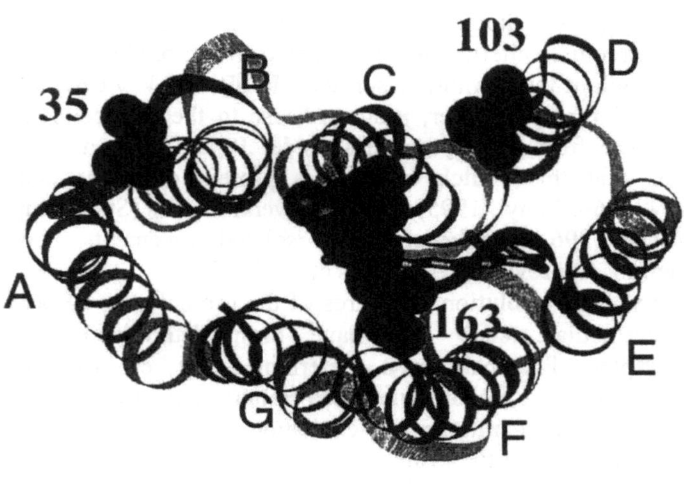

Figure 9. Bacteriorhodopsin structure in ribbon presentation (Henderson et al., 1990; Grigorieff et al., 1996), viewed from the cytoplasmic side. S35, A103, and M163 are shown in black, as well as the retinal and a key residue D96 in grey.

Time-resolved EPR. To monitor conformational changes in real time, we detected the time trace of the first derivative EPR spectrum after a light pulse using a Xe-arc lamp (Figure 10). The time-dependent spectral traces are shown at Figure 11. For both labeled A103C/M163C and labeled S35C/M163C a transient increase in the EPR amplitude during the photocycle is observed. No or small EPR spectral changes with singly labeled mutants implies no or small changes of tumbling motion of the nitroxide. Thus, the EPR change observed with the double mutants must be mainly due to a reduction in spin-spin interactions that result from an increase in the nitroxide-nitroxide separation. Using the same spin labeled sample, the kinetics of the photocycle was followed by the absorbance measurement of visible light in a separate experiment, and compared to the EPR change. The comparison of EPR and optical change shows that the time course of the distance change coincides very well with the life cycle of the N-intermediate. Thus the loop-opening movement occurs before the reprotonation of the key residue (D96) from the cytoplasmic side. This is also consistent with previous studies, which found movements of helix F upon formation of M_N or N states (Subramanian et al., 1993; Vonck, 1996).

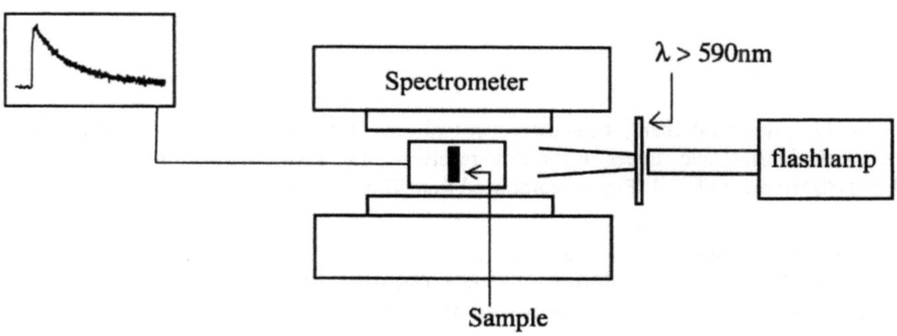

Figure 10. Diagram of the instrument set-up for time-resolved EPR spectroscopy. Room temperature EPR spectra were collected using a TM110 cavity (Bruker 4103TM). Samples were illuminated inside the cavity (l > 590 nm) using a Xenon flashlamp (350ms duration) for time-resolved measurements.

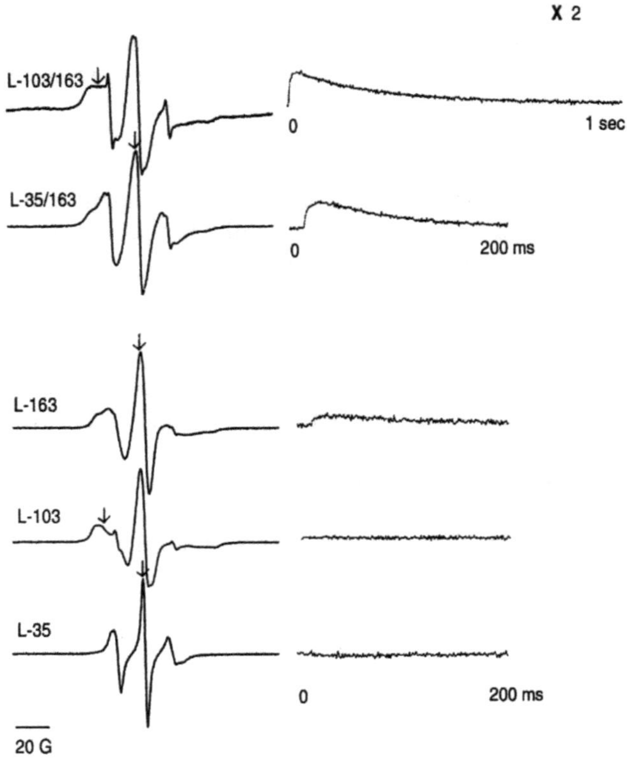

Figure 11. Structural changes monitored in real time. EPR spectra (left) for spin-labeled bacteriorhodopsins and traces from time-resolved EPR measurements following flash photolysis (right) at pH 7.0.(Figure reproduced from Thorgeirsson et al., 1997.)

Changes quantified by low temperature trapping experiment. Quantification of the distance changes requires low-temperature trapping. Low temperatures are also required to eliminate the effects of nitroxide motion on EPR spectra and allow the application of the Fourier deconvolution method to determine interspin distances. The time-resolved EPR measurements clearly show that the N-intermediates need to be trapped at low temperature, and the distances compared to those for the ground state.

To trap photo-stationary states, samples were illuminated, and continuous illumination produces photo-stationary states containing primarily N and the ground state bR. While illuminating the sample, the capillary is quickly inserted into liquid nitrogen and transferred to a loop-gap resonator that is pre-cooled to 140 K. In Figure 12 (A and B) the EPR absorption spectra for the dark-adapted doubly-labeled samples are compared with those of the photo-stationary states trapped at low temperature. The compositions of the photo-stationary states were determined from absorbance spectra of visible light, and the EPR absorption spectra for the pure N intermediate obtained using spectral subtraction based on the optically-determined composition (shown in Figure 12(C and D)).

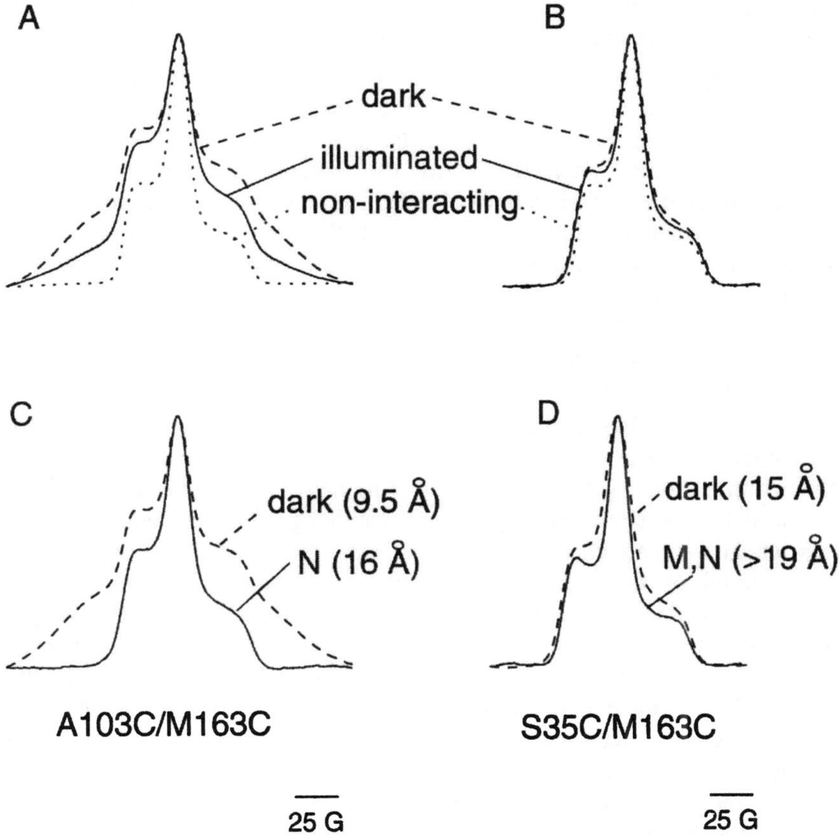

Figure 12. (A and B) Integrated low-temperature EPR spectra for doubly-labeled mutants in the dark and for photo stationary states trapped at 140 K; and (C and D), the corresponding intermediate spectra calculated by spectral subtraction using the compositions of the photo stationary states. The EPR spectrum for non-interacting label (L-163) is also shown in (A and B). (Figure reproduced from Thorgeirsson et al., 1997.)

The analysis of the spectra for A103C/M163C (as in Figure 3) reveals a distribution of distances in the dark, from ~ 8 to ~ 18 Å (as calculated from the two fitted Gaussians of the broadening function of the spectrum, respectively), with an average separation of 10 Å. This distribution perhaps indicates either some flexibility in the loop-structure or conformational heterogeneity. The average interspin distance is increased to 16 Å in the N intermediate. The same analysis of the spectra for S35C/M163C yields a change in average distance from 14.5 Å in the dark to >19 Å in the long-lived intermediate (Fig 12). The spectral contribution from the non-interacting component due to the incomplete labeling at position S35 was appropriately subtracted during the analysis. In conclusion, we observed a light-induced, ~5Å opening at the cytoplasmic end of bR.

Implications. The opening of the cytoplasmic loop region has been observed and followed in real time using EPR. The conformational change occurs before the proton uptake from the cytoplasm. This movement together with the outward tilt of the cytoplasmic end of helix F (Subramanian et al., 1993; Vonck, 1996), may constitute the conformational switch that changes the accessibility of the retinal Shiff base to the cytoplasmic region. Although the functional importance of this loop movement in bacteriorhodopsin has yet to be investigated further, the corresponding combination of F-helix (Farrens et al., 1996) and loop movements may occur in other heptahelical receptors where it could play a role in G-protein activation.

Perspectives. Recently, we have used the dipolar EPR method to study the transmembrane signaling motion in bacterial aspartate receptor (Ottemann et al., 1998, 1999). To distinguish among alternative models proposed for the signaling mechanism, we designed experiments in which spin labels were attached at a series of strategic positions of the receptor reconstituted in the membrane so that the distances between pairs of appropriately placed labels would give different results for different models. The ligand-induced movement was characterized as a small intrasubunit piston-type movement of one transmembrane helix downward relative to the other transmembrane helix. No spectral changes were detected when the samples were in detergent, suggesting that conformational changes are enhanced by protein-membrane interactions. It is therefore essential to conduct measurements on proteins in the membrane. This observation underscores the unique strength of the EPR method.

We also note that the EPR dipolar approach has been applied with good success, although qualitative, to the investigation of conformational changes in *Streptomyces* potassium channel, KcsA (Perozo et al., 1999) and rhodopsin (Farrens et al., 1996), where important mechanistic questions have been answered. Studies on the dynamics of membrane proteins would definitely help elucidate the mechanism of their functions. Information on the protein dynamics would be valuable for pharmaceutical developments, where membrane proteins are the major drug targets. EPR distance measurements have already made significant contributions and we expect that this method will solve many other important biological problems in the future.

4. REFERENCES

Ban, N., Nissen, P., Hansen, J., Capel, M., Moore, P., and Steitz, T. (1999). Placement of protein and RNA structures into a 5 A-resolution map of the 50S ribosomal subunit. *Nature* **400**, 841-847.

Bennett, M., and Scheller, R. (1994). A molecular description of synaptic vesicle membrane trafficking. *Ann. Rev. Biochem.* **63**, 63-100.

Berger, B., Wilson, D. B., Wolf, E., Tonchev, T., Milla, M., and Kim, P. S. (1995). Predicting coiled coils by use of pairwise residue correlations. *Proc. Natl. Acad. Sci. USA* **92**, 8259-8263.

Broadie, K., Prokop, A., Bellen, H., O'Kane, C., Schulze, K., and Sweeney, S. (1995). Syntaxin and synaptobrevin function downstream of vesicle docking in Drosophila. *Neuron* **15**, 663-673.

Cate, J., Yusupov, M., Yusupova, G., Earnest, T., and Noller, H. (1999). X-ray crystal structures of 70S ribosome functional complexes. *Science* **285**, 2095-2104.

Chapman, E., An, S., Barton, N., and Jahn, R. (1994). SNAP-25, a t-SNARE which binds to both syntaxin and synaptobrevin via domains that may form coiled coils. *J. Biol. Chem.* **269**, 27427-27432.

Clemons, W., May, J., Wimberly, B., McCutcheon, J., Capel, M., and Ramakrishnan, V. (1999). Structure of a bacterial 30S ribosomal subunit at 5.5 Å resolution. *Nature* **400**, 833-840.

Creighton, T. (1983). "*Proteins*", Freeman, New York, NY.

Edman, K., Nollert, P., Royant, A., Belrhali, H., Pebay-Peyroula, E., Hajdu, J., Neutze, R. and Landau, E. M. (1999). High-resolution X-ray structure of an early intermediate in the bacteriorhodopsin photocycle. *Nature* **401**, 822 – 826.

Farrens, D. L., Altenbach, C., Yang, K., Hubbell, W. L. and Khorana, H. G. (1996). Requirement of rigid-body motion of transmembrane helices for light activation of rhodopsin. *Science* **274**, 768-770.

Fasshauer, D., Bruns, D., Shen, B., Jahn, R., and Brünger, A. T. (1997a). A structural change occurs upon binding of syntaxin to SNAP-25. *J. Biol. Chem.* **272**, 4582-4590.

Fasshauer, D., Otto, H., Eliason, W., Jahn, R., and Brünger, A. T. (1997b). Structural changes are associated with soluble N-ethylmaleimide-sensitive fusion protein attachment protein receptor complex formation. *J. Biol. Chem.* **272**, 28036-28041.

Gerstein, M. (1998). Patterns of protein-fold usage in eight microbial genomes: a comprehensive structural census. *Proteins* **33**, 518-534.

Grigorieff, N., Ceska, T. A., Downing, K. H., Baldwin, J. M. and Henderson, R. (1996). Electron crystallographic refinement of the structure of bacteriorhodopsin. *J. Mol. Biol.* **259**, 393-421.

Hanson, P., Roth, R., Morisaki, H., Jahn, R., and Heuser, J. (1997). Structure and conformational changes in NSF and its membrane receptor complexes visualized by quick-freeze/deep-etch electron microscopy. *Cell* **90**, 523-535.

Haupts, U., Tittor, J., and Oesterhelt, D. (1997) Closing in on bacteriorhodopsin: progress in understanding the molecule. *Ann. Rev. Biophys. Biomol. Struct.* **28**, 367-399.

Hayashi, T., McMahon, H., Yamasaki, S., Binz, T., Hata, Y., Südhof, T., and Niemann, H. (1994). Synaptic vesicle membrane fusion complex: action of clostridial neurotoxins on assembly. *EMBO J.* **13**, 5051-5061.

Henderson, R., Baldwin, J. M., Ceska, T. A., Zemlin, R., Beckmann, E. and Downing, K. H. (1990). Model for the structure of bacteriorhodopsin based on high-resolution electron cryomicroscopy. *J. Mol. Biol.* **213**, 899-929.

Hubbell W. L., and Altenbach C. (1994). Investigation of structure and dynamics in membrane proteins using site-directed spin labelling. *Curr. Opin. Struct. Biol.* **4**, 566-573.

Hustedt, E., and Beth, A. (1999). Nitroxide spin-spin interactions: applications to protein structure and dynamics. *Ann. Rev. Biophys. Biomol. Struct.* **28**, 129-153.

Iwata, S., Ostermeier, C., Ludwig, B., and Michel, H. (1995). Structure at 2.8 Å resolution of cytochrome c oxidase from Paracoccus denitrificans. *Nature* **376**, 660-669.

Jardetzky, O. (1966). Simple allosteric model for membrane pumps. *Nature* **211**, 969-970.

Kee, Y., Lin, R., Hsu, S., and Scheller, R. (1995). Distinct domains of syntaxin are required for synaptic vesicle fusion complex formation and dissociation. *Neuron* **14**, 991-998.

Lanyi, J. K. (1995). Bacteriorhodopsin as a model for proton pumps. *Nature* **375**, 461-463.

Lin, R., and Scheller, R. (1997). Structural organization of the synaptic exocytosis core complex. *Neuron* **19**, 1087-1094.

Luecke, H., Schobert, B., Richter, H.T., Cartailler, J. P. and Lanyi, J. K. (1999) Structural changes in bacteriorhodopsin during ion transport at 2 Angstrom resolution. *Science* **286**, 255-260.

Lupas, A. (1996). Prediction and analysis of coiled-coil structures. *Methods Enzymol.* **266**, 513-525.

Marqusee S., Robbins V., Baldwin R. (1989). Unusually stable helix formation in short alanine-based peptides. *Proc. Natl. Acad. Sci.USA* **86**, 5286-5290.

Milligan, D. L., and Koshland, D. E. (1991). Intrasubunit signal transduction by the aspartate chemoreceptor. *Science* **254**, 1651-1654.

Nickel, W., Weber, T., McNew, J. A., Parlati, F., Sollner, T. H., and Rothman, J. E. (1999). Content mixing and membrane integrity during membrane fusion driven by pairing of isolated v-SNAREs and t-SNAREs. *Proc. Natl. Acad. Sci.USA* **96**, 12571-12576.

Niemann, H., Blasi, J., and Jahn. R. (1994). Clostridial neurotoxins: New tools for dissecting exocytosis. *Trends Cell Biol.* **4**,179-185.

Ottemann, K. M., Thorgeirsson, T., Kolodziej, A., Shin, Y.-K., and Koshland, D. E. (1998). Direct measurement of small ligand-induced conformational changes in the aspartate chemoreceptor using EPR. *Biochemistry* **37**, 7062-7069.

Ottemann, K. M., Xiao, W., Shin, Y.-K., and Koshland, D. E. (1999). A piston model for transmembrane signaling of the aspartate receptor. *Science* **285**, 1751-1754.

Otto, H., Hanson, P., and Jahn, R. (1997). Assembly and disassembly of a ternary complex of synaptobrevin, syntaxin, and SNAP-25 in the membrane of synaptic vesicles. *Proc. Natl. Acad. Sci. USA* **94**, 6197-6201.

Parlati, F., Weber, T., McNew, J., Westermann, B., Sollner, T., and Rothman, J. E. (1999) Rapid and efficient fusion of phospholipid vesicles by the alpha-helical core of a SNARE complex in the absence of an N-terminal regulatory domain. *Proc. Natl. Acad. Sci. USA* **96**, 12565-12570.

Pebay-Peyroula, E., Rummel, G., Rosenbusch, J., and Landau, E. (1997). X-ray structure of bacteriorhodopsin at 2.5 angstroms from microcrystals grown in lipidic cubic phases. *Science* **277**, 1676-1681.

Perozo, E., Cortes, D. M., and Cuello, L. G. (1999). Structural rearrangements underlying K+-channel activation gating. *Science* **285**:73-78.

Poirier, M., Hao, J., Malkus, P., Chan, C., Moore, M., King, D., and Bennett, M. (1998a) Protease resistance of syntaxin•SNAP-25•VAMP complexes. *J. Biol. Chem.* **273**, 11370-11377.

Poirier, M., Xiao, W., Macosko, J., Chan, C., Shin, Y.-K., and Bennett, M. (1998b). The synaptic SNARE complex is a parallel four-stranded helical bundle. *Nature Struct. Biol.* **5**, 765-769.

Rabenstein, M., and Shin, Y.-K. (1995). Determination of the distance between two spin labels attached to a macromolecule. *Proc. Natl. Acad. Sci.USA* **92**, 8239-8243.

Spudich, J. L. and Lanyi, J. K. (1996). Shuttling between two protein conformations: the common mechanism for sensory transduction and ion transport. *Curr. Opin. Cell. Biol.* **8**, 452-457.

Subramaniam S., Gerstein M., Oesterhelt D. and Henderson R. (1993). Electron diffraction analysis of structural changes in the photocycle of bacteriorhodopsin. *EMBOJ.* **12**, 1–8.

Südhof, T. (1995). The synaptic vesicle cycle: a cascade of protein-protein interactions. *Nature* **375**, 645-653.

Sutton, R. B., Fasshauer, D., Jahn, R., and Brünger, A. T. (1998). Crystal structure of a SNARE complex involved in synaptic exocytosis at 2.4 angstrom resolution. *Nature* **395**, 347-353.

Tanford, C. (1983). Translocation pathway in the catalysis of active transport. *Proc. Natl. Acad. Sci. U S A* **80**, 3701-3705.

Thorgeirsson, T. E., Xiao, W., Brown, L. S., Needleman, R., Lanyi, J. K., and Shin, Y.-K. (1997). Transient channel-opening in bacteriorhodopsin: an EPR study. *J. Mol. Biol.* **273**, 951-957.

Unwin, N. (1995). Acetylcholine receptor channel imaged in the open state. *Nature* **373**, 37-43.

Vonck, J. (1996). A three-dimensional difference map of the N intermediate in the bacteriorhodopsin photocycle: part of the F helix tilts in the M to N transition. *Biochemistry* **35**, 5870-5878.

Weber, T., Zemeiman, B., McNew, J., Westermann, B., Gmachi, M., Pariarti, F., Söllner, T., and Rothman, J. E. (1998). SNAREpins: Minimal machinery for membrane fusion. *Cell* **92**,759-772.

Weimbs, T., Low, S., Chapin, S., Mostov, K., Bucher, P., and Hofmann, K. (1997). A conserved domain is present in different families of vesicular fusion proteins: a new superfamily. *Proc. Natl. Acad. Sci. U S A* **94,** 3046-3051.

Weimbs, T., Mostov, K., Low, S., and Hofman, K. (1998). A model for structural similarity between different SNARE complexes based on sequence relationships. *Trends Cell Biol.* **8**, 260-262.

Chapter 6

TOAC

The Rigid Nitroxide Side Chain

Joseph C. McNulty and Glenn L. Millhauser
Department of Chemistry and Biochemistry, University of California, Santa Cruz, CA USA

Abstract: TOAC (2,2,6,6-tetramethylpiperidine-1-oxyl-4-amino-4-carboxylic acid) is an unnatural amino acid that contains a nitroxide ring rigidly attached at the backbone alpha carbon. Because the conformation of TOAC is highly constrained, distances determined between pairs of TOACs are not compromised by uncertainties in nitroxide side chain orientations. ESR detected interactions between TOACs have revealed accurate distances out to beyond 10 Å and, correspondingly, exciting new structural insights into the folds of both model and naturally occurring peptides. TOAC's history is recent, with the first report of its incorporation into a peptide via solid phase synthesis appearing in 1993 (Marchetto et al., 1993). While TOAC does offer significant advantages for biomolecular labeling, its incorporation presents unique challenges in peptide synthesis and design. This chapter will briefly describe synthesis of TOAC-containing peptides, placement within peptide secondary structure, spectroscopy and recent applications.

1. INTRODUCTION: INCORPORATING TOAC INTO POLYPEPTIDES

The TOAC amino acid, shown in Fig. 1, has been available for many years and it's synthesis via the TEMPONE intermediate was first described in 1967 (Rassat and Rey, 1967). There are a few early applications where TOAC was covalently linked to the end of a polypeptide but placing TOAC at an intermediate position is substantially more challenging. Typically, spin labels are added in a site-specific fashion to peptides and proteins via reagents that add a nitroxide to a

Biological Magnetic Resonance, Volume 19: Distance Measurements in Biological Systems by EPR, edited by Berliner et al. Kluwer Academic/Plenum Publishers, New York, 2000

unique side chain (e.g., MTSL modification of Cys) (Berliner et al., 1982; Altenbach et al., 1990; Millhauser, 1992). However, TOAC is an intact amino acid and site-specific placement requires either peptide synthesis or biosynthetic methods. Cornish et al. demonstrated an *in vitro* expression system derived from *Escherichia coli* for site selective placement of nitroxide amino acids (Cornish et al., 1994). While several unnatural amino acid probes were successfully incorporated into T4 lysozyme, there was no detectable expression of TOAC-containing mutants.

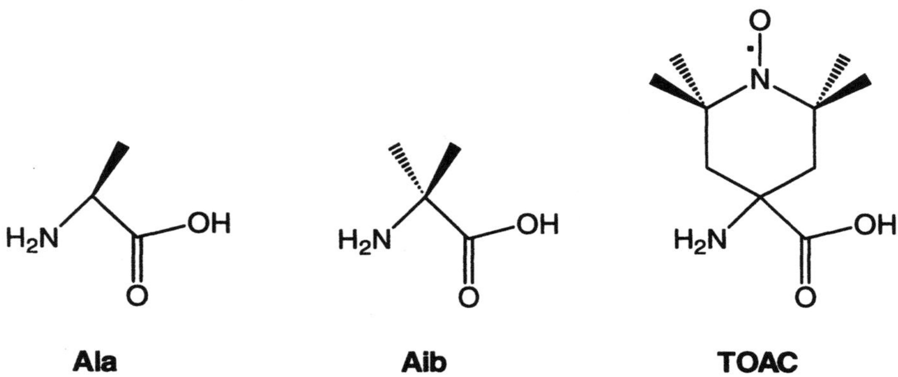

Figure 1. Chemical structures of the common amino acid alanine (Ala), the microbial amino acid α-aminoisobutyric acid (Aib) and the Aib analog TOAC.

Peptide synthesis offers an alternative method for spin label incorporation (Fields, 1997). Solid-phase peptide synthesis begins at the C-terminus and uses stepwise addition of N-terminally protected amino acids as shown schematically in Fig. 2. Such techniques are termed *solid phase* because the C-terminal amino acid is immobilized on a resin support to facilitate addition and removal of synthetic reagents. The final synthetic step is typically treatment with strong acid to deprotect side chain functional groups and cleave the peptide from the resin. Nitroxides often react in extreme chemical conditions and indeed the use of concentrated trifluoroacetic acid (TFA) for final deprotection usually results in the irreversible loss of the nitroxide spin. In 1993 Marchetto et al. showed that Fmoc-TOAC (where Fmoc is a common N-terminal protecting group) allowed for the incorporation of TOAC in solid phase synthesis (Marchetto et al., 1993). Use of HF instead of TFA in final deprotection resulted in simple N-O protonation, to give the hydroxyl amine, with full recovery of the paramagnetic group after mild treatment with base. Their approach resulted in the successful synthesis of the eight-residue peptide angiotensin II with a Pro-7 → TOAC substitution.

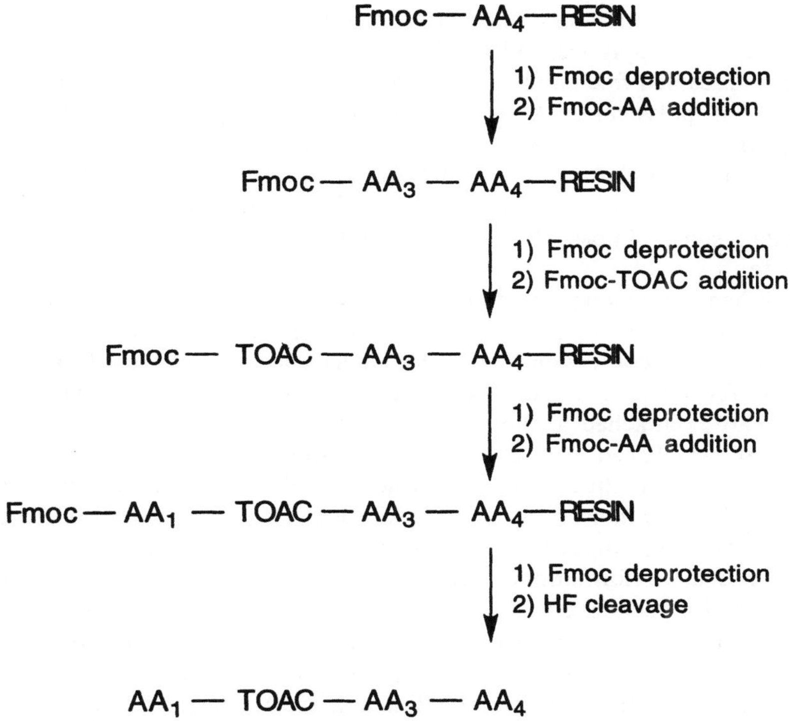

Figure 2. Schematic representation of solid phase synthesis of a four residue peptide. The resin is the solid support and synthesis starts at the C-terminus with each new amino acid added to the polypeptide N-terminus.

The work of Marchetto et al. established a powerful method for TOAC incorporation. Although this method has been widely exploited in solid phase and solution phase peptide synthesis, TOAC incorporation is nevertheless not trivial. The chief difficulty arises from TOAC's chemical structure. First, TOAC is bulky and steric hindrance contributes to low-yield couplings at the TOAC N-terminus (e.g., addition of AA_1 in Fig. 2). Second, TOAC has two side-chain carbons attached to the alpha carbon whereas common amino acids such as alanine (Ala, see Fig. 1) have a single side chain. According to Marchetto et al., this disubstitution at the alpha carbon may lower the TOAC amine nucleophilicity needed for efficient addition of the N-terminal amino acid. Thus, Fmoc-TOAC readily adds to elongate a polypeptide chain but the next amino acid is often difficult to couple and requires repeated attempts. Indeed, in the synthesis of angiotensin II, attachment of His-6 to TOAC-7 required three couplings. This is consistent with the experience of our lab and that of our collaborators. In addition, traditional methods for monitoring coupling don't apply well with TOAC. For example, the ninhydrin assay detects unreacted free amines and is widely used to look for unreacted N-termini of polypeptide chains. However, TOAC reacts poorly with ninhydrin and thus it can be difficult to determine whether one has successfully coupled to its N-terminus. Despite these challenges, however, repeated couplings and careful monitoring of detectable coupling steps almost always result in the desired TOAC containing peptide.

When designing peptide sequences, the choice of TOAC's placement within the polypeptide chain requires careful consideration. This is again a consequence of TOAC's unusual chemical configuration. Except for Gly, all mammalian protein amino acids have a single side chain carbon attached to the backbone alpha carbon (e.g., Ala in Fig. 1). It is well established that these normal amino acids readily incorporate into helical and β-sheet secondary structure. However, like the microbial amino acid α-aminoisobutyric acid (Aib, Fig. 1), TOAC is a disubstituted glycine with two carbons attached to the alpha carbon. Such amino acids have unusual conformational preferences. First, steric hindrance largely precludes the backbone torsion angles (ϕ, ψ) required for incorporation into β-strands or β-sheets. However, TOAC and Aib do readily incorporate into helices (α- and 3_{10}-, see below) and in fact work as strong helix stabilizers (Karle and Balaram, 1990). Second, TOAC is not chiral and thus can take on the conformation for either right or left handed helix. Thus, TOAC may be readily designed into helical domains or turns and contribute added stability to these domains.

Crystal structures of free TOAC and peptides containing TOAC have been characterized (Flippen-Anderson et al., 1996). Geometrical details have been carefully evaluated and demonstrate that, similar to Aib, TOAC readily adopts helical backbone torsion angles. Typical of unsaturated six-member rings, the TOAC nitroxide ring adopts a twist boat geometry as shown by the chemical structure in Fig. 3. The two conformations are mirror images and thus either twist boat conformation is allowed.

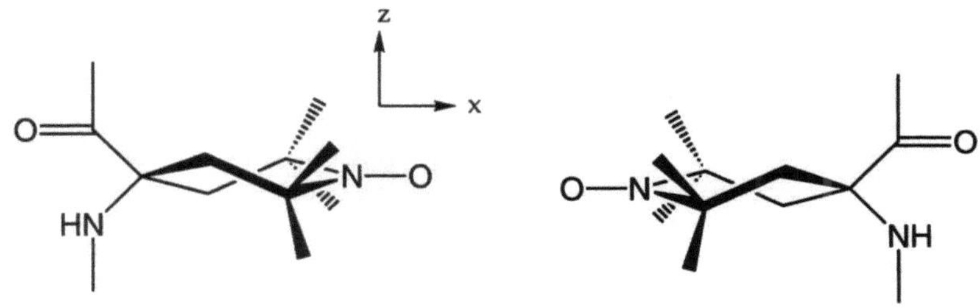

TOAC Twist-Boat Conformer $\mathbf{TB_0}$ TOAC Twist-Boat Conformer $\mathbf{TB_{180}}$

Figure 3. Twist boat (TB) conformations of TOAC. The subscripts 0 and 180 refer to ring puckering angles (Cremer and Pople, 1975; Hanson et al., 1998). The Cartesian axis system defining the nitroxide magnetic tensors is also shown.

When TOAC is incorporated into a helix, however, the twist boat conformation placing the nitroxide z-axis parallel to the helix axis (TB_0) is more common. Presumably, this orientation decreases steric interactions with neighboring side chains. This conformational preference is demonstrated, for example, by the crystal structure of the pentameric peptide

*p*BrBz-TOAC-Ala-Ala-TOAC-Ala-NH*t*Bu Pent-1,4

shown in Fig. 4 (where *p*BrBz is *para*-bromobenzoyl and NH*t*Bu is *tert*-butlyamino) (Toniolo et al., 1995). The Pent-1,4 peptide adopts the structure of a 3_{10}-helix (with i, i+3 CO-NH hydrogen bonding, see Toniolo and Benedetti, 1991) and Fig. 4 shows two TOAC rings extending away from the polypeptide backbone. The TOACs do not interfere with the helical structure and indeed the backbone conformation of this peptide is nearly identical to similar pentameric peptides composed solely of Ala and Aib. Also shown in Fig. 4 are enlargements of the TOAC portions of the crystal structure which clearly reveal the twist boat conformation.

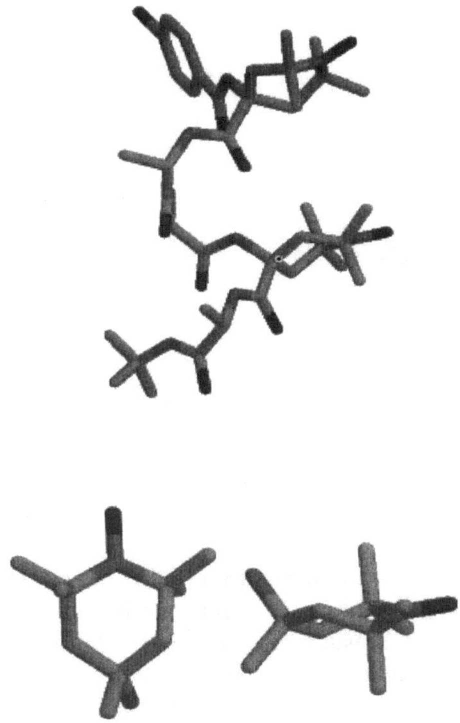

Figure 4. Crystal structure of the Pent-1,4 peptide (above) and TOAC (below) from two perspectives.

We summarize the properties and chemistry of TOAC with the following points:

- TOAC is a rigid nitroxide amino acid that may be incorporated into peptides and proteins via synthetic methods
- peptide synthesis with TOAC must use acids other than concentrated trifluoroacetic acid and coupling to the N-terminus of TOAC may require several attempts
- similar to Aib, TOAC is readily incorporated into helices and turns and stabilizes these backbone conformations
- TOAC adopts a twist boat conformation which, in a helix, places the nitroxide z-axis parallel to the helix axis

2. EVALUATING THE STRUCTURES OF SHORT HELICAL PEPTIDES

Short peptides composed of Ala and Aib are of interest because of their tendency to form stable secondary structure including turns, 3_{10}-helices and α-helices. However, most research until just a few years ago relied on X-ray crystallography to characterize conformations. It was not clear whether trends derived from crystal structure analyses would apply to peptides in solution. NMR is often a natural choice for such conformational investigations in solution, however, the lack of a proton at the Cα position in Aib can make spectrum assignment and structural characterization difficult. In addition, rapid interconversion among conformers can complicate peptide structure assignment by NMR. An alternative approach is to replace pairs of Aib or Ala with specifically placed TOACs and use ESR detected interactions between the nitroxide side chains as a probe of conformation.

The first peptide to be synthesized with two TOACs was the Pent-1,4 (sequence above) (Toniolo et al., 1995). As noted, the crystal structure of Pent-1,4 reveals the conformation of a nearly ideal 3_{10}-helix. Circular dichroism and infrared spectra suggest this structure persists in organic solvent. As shown in Fig. 4, the i, i+3 placement of the TOACs within a 3_{10}-helix places them in near proximity and separated by a single helical turn. ESR was performed at room temperature in various solvents and two representative spectra are shown in Fig. 5 (Toniolo et al., 1995).

Clearly, there are dramatic differences between the spectra. The spectrum obtained from the peptide in hexafluoroisopropanol (HFIP) follows a three line hyperfine pattern as one would expect from an uncoupled nitroxide. However, in MeOH the peptide gives a striking five-line pattern with the two new lines emerging between the expected hyperfine lines of a typical nitroxide spectrum. Similar five line patterns are well known in nitroxide biradicals and arise from J-coupling between the TOAC unpaired electrons (Luckhurst, 1976). This biradical spectrum is not influenced by variation in peptide concentration and thus the coupling arises from an intramolecular interaction between the TOAC nitroxides. Such couplings can arise from either through-bond or through-solvent interactions (Closs et al., 1992). However, there are fifteen covalent bonds separating the TOAC nitrogens and it is therefore unlikely that through-bond interactions contribute significantly (see below). The interaction must arise then from a through-solvent coupling and is therefore a sensitive probe of the spacing between the TOACs.

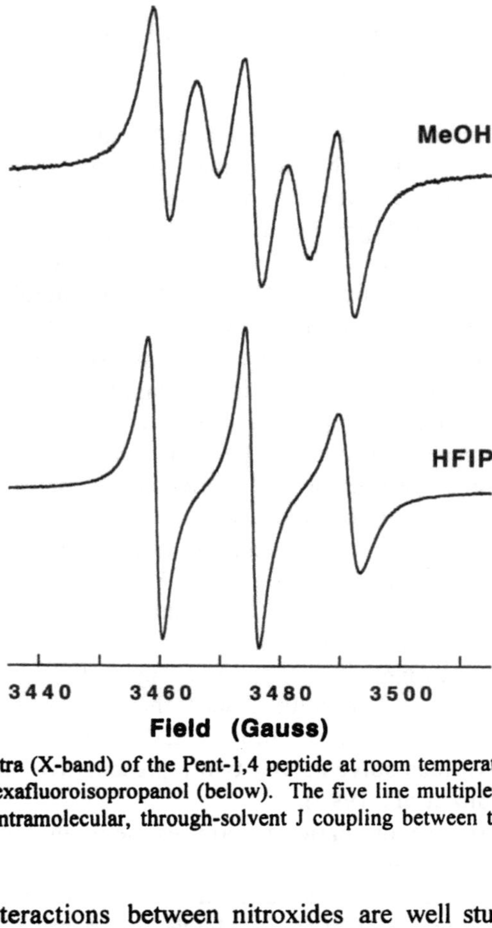

Figure 5. ESR spectra (X-band) of the Pent-1,4 peptide at room temperature in methanol (above) and hexafluoroisopropanol (below). The five line multiplet in MeOH demonstrates intramolecular, through-solvent J coupling between the TOACs.

Biradical interactions between nitroxides are well studied. In the motional narrowing regime, strong J coupling with $J >> a_N$, where a_N is the ^{14}N hyperfine coupling, gives a 1:2:3:2:1 five line pattern (Atherton, 1973; Luckhurst, 1976). Alternatively, weak coupling ($J << a_N$) gives slight broadening of the normal three hyperfine lines and intermediate cases ($J \approx a_N$) give rise to complicated multiplets. When folded as a 3_{10}-helix the distance between the TOAC N-O groups in Pent-1,4 is approximately 7.3 Å and this distance is similar to that in small synthetic biradicals that give strong J-coupling and corresponding five-line ESR multiplets. Thus, comparison of the two spectra suggests that in MeOH the peptide folds as a 3_{10}-helix but in HFIP the peptide is mainly unfolded. HFIP is a much more polar solvent than MeOH and, in general, we have found that high solvent polarity decreases helix content in short peptides.

Although the five line pattern for the Pent-1,4 in MeOH clearly demonstrates coupling between the TOACs, the multiplet pattern does not follow the intensity ratio of 1:2:3:2:1 expected for J >> a_N. Instead the second and fourth hyperfine lines appear to be less intense than the remaining three lines. Luckhurst has noted that time-dependent modulations of J-coupling can selectively broaden the second and fourth lines, which reduces their amplitudes and gives a characteristic spectrum of alternating linewidths (Luckhurst, 1966). For the Pent-1,4 dynamic modulations can arise from peptide backbone dynamics or rapid excursions to an unfolded state. Application of motional narrowing theory to describe the effect of J-coupling modulation gives the expression

$$T_2^{-1} = \frac{a_N^2}{4\,J^2}\, j(\bar{J}) \qquad (1)$$

for the width of the second and fourth lines where $j(\bar{J})$ is a spectral density determined at frequency $\bar{J}$ which, in turn, is the conformationally averaged J-coupling (Luckhurst, 1966). Eq. 1 suggests that if the time scale of the conformational fluctuations is temperature dependent, then the widths of the second and fourth lines should also be temperature dependent and indeed this is observed experimentally.

Encouraged by the strong coupling observed with 1,4 inter-TOAC spacing in the pentameric peptide above, the next goal was to determine the couplings around a full helical turn. The following hexameric peptides were examined:

Boc - TOAC- TOAC- Ala - Ala - Ala - Ala - O*t*Bu	Hex-1,2
Boc - TOAC- Ala - TOAC- Ala - Ala - Ala - O*t*Bu	Hex-1,3
Boc - TOAC- Ala - Ala - TOAC- Ala - Ala - O*t*Bu	Hex-1,4
Boc - TOAC- Ala - Ala - Ala - TOAC- Ala - O*t*Bu	Hex-1,5

where Boc and -O*t*Bu are *tert*-butyloxycarbonyl and *tert*-butoxy, respectively (Hanson et al., 1996). As with the Pent-1,4, these short sequences should fold as 3_{10}-helices according to trends from X-ray crystallography. Spectra obtained from the peptides in MeOH are shown in Fig. 6. The spectra exhibit sharp hyperfine lines and are therefore in the motionally narrowed regime. The nature of the multiplets and consequently the J-coupling vary significantly with the various inter-TOAC spacings. The Hex-1,2 gives a classical five-line strongly-coupled multiplet while the Hex-1,3 shows almost no evidence of J-coupling.

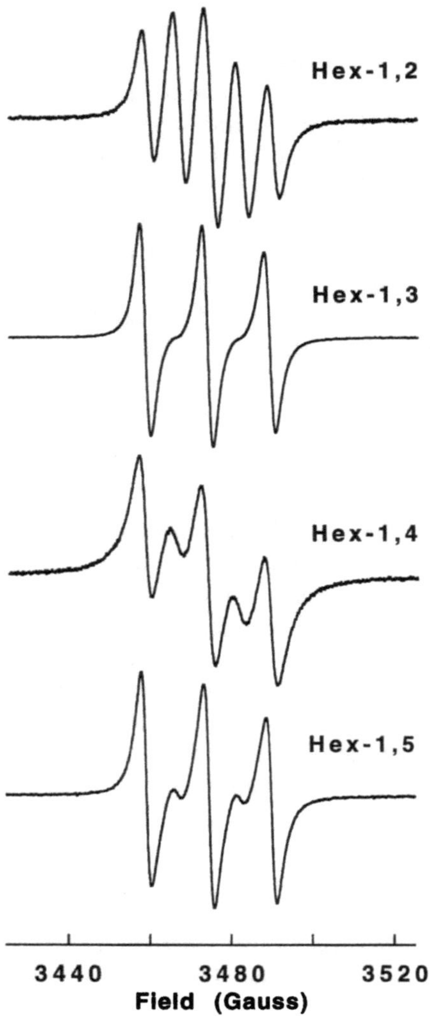

Figure 6. ESR spectra of the hexamers in MeOH demonstrating the effect of relative TOAC placement on observed J coupling. The pattern observed is consistent with the structure of a 3_{10}-helix.

To interpret the data we consider the expected distances between TOACs for the conformations of ideal 3_{10}- and α–helix characterized by i, i+3 and i, i+4 hydrogen bond acceptor-donor spacing, respectively (see below for a further discussion of α-helix). Both helix types are commonly found in peptides and proteins and there is current interest in determining the factors that control the equilibrium between these conformations. Molecular models of the peptides, in both the α- and 3_{10}-helical conformations, were built using crystallographically-determined backbone torsion ϕ,ψ angles (Toniolo and Benedetti, 1991). The resulting through-solvent distances between the nitroxide nitrogens ($D_{i,j}$) for the doubly-labeled peptides in each conformation are:

	$D_{i,i+1}$	$D_{i,i+2}$	$D_{i,i+3}$	$D_{i,i+4}$
3_{10}-helix	9.75 Å	9.94 Å	6.94 Å	10.82 Å
α-helix	9.54 Å	10.78 Å	7.70 Å	7.96 Å

Except for $D_{i,i+4}$, the distances between TOACs in the two conformations are comparable. 3_{10}-Helix is approximately 3.2 residues per turn and is more steeply pitched than α-helix at approximately 3.6 residues per turn. Consequently, $D_{i,i+4}$ is substantially shorter for α-helix than for 3_{10}-helix. The relative strengths of biradical interaction for the hexamers, as determined by the appearance of the second and fourth lines in the spectra, indicate an experimental distance ranking of $D_{1,2} < D_{1,4} < D_{1,5} \leq D_{1,3}$. According to the distances above, $D_{1,2}$ should be greater than $D_{1,4}$. Consequently, the strong J-coupling observed for Hex-1,2 suggests that another biradical mechanism contributes to its spectrum. One possible mechanism for the strong biradical signal is through-bond J-coupling. In contrast to through-solvent J-coupling, through-bond coupling arises from spin-polarization of connecting σ-bonds between the nitroxide unpaired electrons. Spin polarization decreases exponentially with the number of σ-bonds between labels. Work with randomly oriented acyl-alkyl biradicals, of seven to twelve carbons in length, has demonstrated such an exponential dependence of this through-bond contribution (Closs et al., 1992). In Hex-1,2, there are nine σ-bonds between the two TOAC nitroxide radicals. Using parameters from Closs et al., the estimated contribution from through-bond coupling in this peptide is approximately 230 gauss. This value is much greater than a_N (approximately 16 gauss) and is sufficient to give a 1:2:3:2:1 pattern.

For the Hex-1,3 -1,4 and -1,5 spectra, the strength of the biradical interaction is greatest in Hex-1,4 where the five line pattern is clearly observed, although the second and fourth lines are broadened (see above). The biradical interaction is weaker in Hex-1,5 , where only a hint of a five line pattern is observed, and nearly absent in Hex-1,3. The observation of little biradical interaction for Hex-1,3 indicates that the through-bond contribution is too weak to influence the spectral shape of an i, i+2 doubly TOAC labeled peptide. The experimental J-coupling hierarchy therefore indicates $D_{1,4} < D_{1,5}$. Comparison to the ideal distances tabulated above suggests that 3_{10}-helix is the dominant conformation, in agreement with crystallographic results for similar peptides. However, there is a detectable J-coupling in the Hex-1,5 and perhaps this arises from a small population of α-helical conformers.

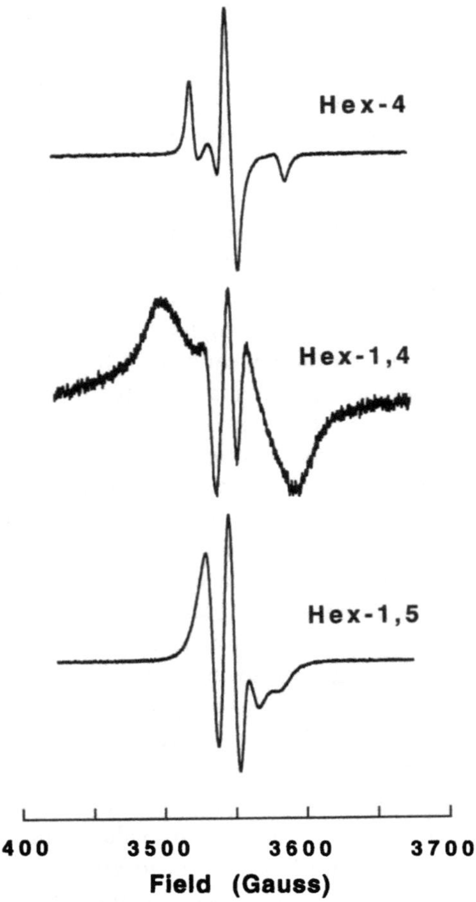

Figure 7. ESR spectra of Hex-1,4 and Hex-1,5 in MeOH at reduced temperature (200 K). The Hex-1,4 spectrum clearly shows dipolar coupling.

In addition to J-coupling, biradicals also exhibit electron-electron dipolar coupling. The strength of this interaction is determined by $1/D_{ij}^3$ (Luckhurst, 1976; Hustedt et al., 1997). For rapidly tumbling peptides (typically $\tau_R < 0.5$ ns), the dipole-dipole tensor averages to zero and does not contribute to the ESR spectrum. However, for slowly tumbling or immobile systems, the strength of the dipolar interaction serves as a measure of the distance between the coupled spins. To test the assignment of 3_{10}-helix as the dominant conformer, we examined Hex-1,4 and Hex-1,5 in fluid MeOH at a reduced temperature of 200 K (Hanson et al., 1996). As a control for peptide mobility, we also examined the singly spin labeled peptide Hex-4. The results are shown in Fig. 7. At low temperature, the peptide rotational correlation time increases and 200 K was found to be the highest temperature at which Hex-4 gave the expected spectrum for an immobile nitroxide. Hex-1,4 shows the broad pattern outside of the three line multiplet expected for a biradical exhibiting strong dipolar coupling (see below). Hex-1,5 also shows broadening as compared to the monoradical but significantly less than observed for Hex-1,4. Thus, the relative strength of the dipolar couplings agree with the J-couplings and provide strong support for the assignment of 3_{10}-helix as the dominant conformer of the hexapeptides.

Spectra shown in Figs. 5 - 7 demonstrate that pairs of TOACs, when in close proximity, give rise to profound distance-dependent couplings. In the motionally narrowed regime J-coupling gives rise to five line multiplets and slow tumbling species give clear dipolar couplings. Thus, double TOAC labeling allows one to address specific issues of local structure regardless of correlation time. Furthermore, the observed dipolar coupling for the Hex-1,5 at 200 K (Fig. 7) suggests that one can resolve distances to well beyond 10 Å in fluid solution at low temperature using continuous wave ESR.

3. STRUCTURE OF THE α-HELIX

The studies outlined above address peptide conformation in neat organic alcohols. Of course, for biological labeling it is essential that one can perform measurements in aqueous solution as well. For some time our laboratory has been interested in the detailed structures of α-helices in solution (e.g., see refs. Miick et al., 1992; Miick et al., 1993; Millhauser, 1995; Millhauser et al., 1997). Surveys of protein crystal structures have revealed a number of common distortions in α-helical structures including kinks and unusual winding pitches (Barlow and Thornton, 1988). Several years ago Marqusee and Baldwin demonstrated that alanine-rich peptides such as

$$\text{Ac-(Ala)}_4\text{-Lys-(Ala)}_4\text{-Lys-(Ala)}_4\text{-Lys-Ala-NH}_2 \qquad \text{3K}$$

(where Ac is an acetyl group) folded as helices in solution (Marqusee et al., 1989). For sufficiently long sequences the dominant conformation is α–helix. However, given the wide variety of geometries found for helices in proteins, it was not known whether these water soluble α–helical peptides would adopt the classical conformation of 3.6 residues per turn.

Given this unresolved but fundamental issue, and the newly emerging technology of using TOAC to rank distances, we decided to examine the detailed geometry of a water soluble α–helical peptide. Double TOAC labeling was introduced into 16 residue peptides to give the following sequences:

Ac-$(Ala)_4$-Lys-Ala-Ala-TOAC-Ala-Lys-$(Ala)_4$-Lys-Ala-NH_2	3KT-8
Ac-$(Ala)_3$-TOAC-Lys-TOAC-Ala-Ala-Ala-Lys-$(Ala)_4$-Lys-Ala-NH_2	3KT-4,6
Ac-$(Ala)_3$-TOAC-Lys-Ala-TOAC-Ala-Ala-Lys-$(Ala)_4$-Lys-Ala-NH_2	3KT-4,7
Ac-$(Ala)_3$-TOAC-Lys-Ala-Ala-TOAC-Ala-Lys-$(Ala)_4$-Lys-Ala-NH_2	3KT-4,8

which are based on the 3K sequence above. As with the hexameric peptides, the double TOACs are placed to map the geometry around a single helical turn.

Circular dichroism and ESR spectra for all the peptides are published elsewhere (Hanson et al., 1998) and unequivocally demonstrate that the 3KT sequences fold as α-helices with substantial stabilization coming from the TOACs. Analogous to the Hex-1,3 spectrum, the 3KT-4,6 spectrum shows little coupling between the TOACs. Thus, we focus here on the 3KT-4,7 and -4,8 ESR spectra. Derivative and absorption spectra from these peptides in fluid aqueous solution at 274 K are shown in Fig. 8 (Hanson et al., 1996; Hanson et al., 1998). Spectra for the 3KT-4,7 peptide shows a clear biradical interaction. Although there are sharp features present in the derivative representation, the absorption spectrum reveals substantial line broadening. There may also be additional transitions between the three hyperfine lines suggestive of a five-line pattern from strong J coupling. However, it is difficult to determine whether the line broadening arises from J coupling or emerging dipolar coupling.

The 3KT-4,8 spectrum reveals strong and unmistakable dipolar coupling. The large spectral "wings" lying outside of the three line pattern arise exclusively from dipolar interactions. J coupling may also be present but it is most likely masked by the broad dipolar pattern. The derivative spectrum shows a sharp three line spectrum but the absorption spectrum demonstrates that these sharp features represent a small amount of the total spin. The 3KT-4,8 is remarkable in that it closely approximates a classical rigid limit dipolar pattern yet the peptide is in fluid aqueous solution with correlation time of approximately 2 ns. Thus, the dipolar interaction is so strong that it is not significantly averaged by peptide tumbling.

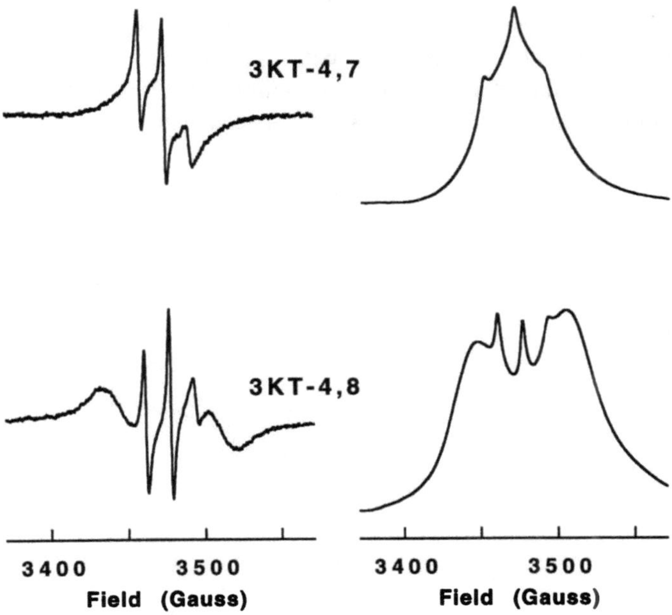

Figure 8. ESR spectra of 3KT-4,7 and 3KT-4,8 in aqueous solution at 274 K shown in derivative (left) and absorption (right) modes. Prominent dipolar coupling is observed in the 3KT-4,8 spectrum.

To see the benefit of working with TOAC labeling it is illustrative to compare the 3KT-4,7 and -4,8 spectra to spectra obtained with similar α-helical peptides but labeled with flexible side chains. Specifically, we investigated the longer 21 residue sequences:

Ac-$(Ala)_3$-Cys-Lys-Ala-Cys-Ala-Ala-Lys-$(Ala)_4$-Lys-$(Ala)_4$-Lys-Ala-NH_2 4K-4,7
Ac-$(Ala)_3$-Cys-Lys-Ala-Ala-Cys-Ala-Lys-$(Ala)_4$-Lys-$(Ala)_4$-Lys-Ala-NH_2 4K-4,8

where the Cys -SH groups were labeled with the MTSL reagent (Fiori et al., 1993; Millhauser, 1995). These peptides do not benefit from the extra stabilization afforded by the TOACs and thus the extra length relative to the 3K peptides insures that 4K peptides fold as primarily α-helix with little contribution from 3_{10}-helix.

Spectra for the 4K-4,7 and -4,8 are shown in Fig. 9. The hyperfine lines are substantially broader than found for monoradical spectra with the same correlation time and this is due to biradical couplings between the Cys-SL side chains. This broadening may arise from a number of effects including residual J-coupling, unaveraged dipolar coupling or dipole-dipole relaxation. As found with the 3KT sequences, the 4K-4,7 interaction is weaker than the -4,8 interaction but the difference is much less pronounced in these Cys-SL labeled 4K sequences.

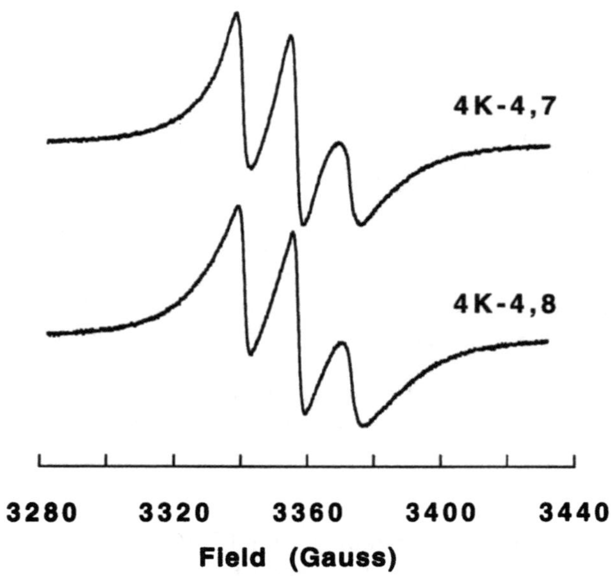

Figure 9. ESR spectra of the 4K peptides each doubly labeled with the MTSL reagent. The loss of spectral detail compared to that observed in Fig. 8 is due to Cys-SL side chain mobility.

As discussed above, the 3KT-4,8 spectrum in Fig. 8 shows clear dipolar coupling with near rigid limit splittings appearing outside the range of the normal three line nitroxide spectrum. This splitting is absent in the 4K-4,8 spectrum. Given the flexibility of the Cys-SL, there are most likely an ensemble of relative spin label side chain distances, the longer of which give spectra with little biradical coupling. In addition, spin label side chain dynamics may contribute to averaging of the dipolar tensor. Thus, TOAC double labeling with relatively fixed distances between side chains and limited side chain dynamics gives rise to pronounced dipolar couplings that are not observed for Cys-SL when on the surface of a peptide or protein. However, we do note that unaveraged dipolar couplings have been observed for Cys-SL pairs in proteins (e.g., Koteiche and McHaourab, 1999) and this presumably arises when the local environment restricts side chain mobility.

Returning to the spectra in Fig. 8, the dipolar interactions for the 3KT-4,8 peptide demonstrates a strong i, i+4 interaction not observed for the hexameric peptides. Thus, the 3KT sequences fold as α-helix (Hanson et al., 1996). Beyond confirmation of the α-helical conformation, though, a rather surprising result emerges from comparison of the 3KT-4,7 and -4,8 spectra. The tabulated distances in the previous section suggest that $D_{i,i+3} \approx D_{i,i+4}$ for α-helix, however, the broad dipolar wings observed specifically for the 3KT-4,8 instead suggests that $D_{i,i+3} > D_{i,i+4}$. To resolve this discrepancy, we obtained quantitative information about the i, i+3 and i, i+4 distances by employing spectral simulations from immobilized peptides (Hustedt et al., 1997). The 3KT-4,7 and -4,8 peptides were dissolved in MeOH and cooled to 200K and the spectra are shown in Fig. 10.

The 3KT-4,7 spectrum is complicated but nevertheless reveals evidence of a dipolar rigid limit pattern whereas the 3KT-4,8 spectrum is dominated by a classical dipolar pattern. This is particularly apparent in the absorption lineshape. The derivative spectra were fit using the procedure of Hustedt et al. to determine the distance and relative orientations between the nitroxide pairs. The best fits are shown superimposed on the experimental spectra and give the distances:

3KT-4,7	8.55 Å
3KT-4,8	6.59 Å

along with the relative orientations between the TOAC side chains (which are reported in Hanson et al., 1998). The simulations clearly capture the main features of each spectrum. However, none of the simulation trials were able to reproduce the small sharp details in the center of each spectrum. It is possible that these features arise from a small amount (< 1%) of unfolded peptide or monoradical contaminant.

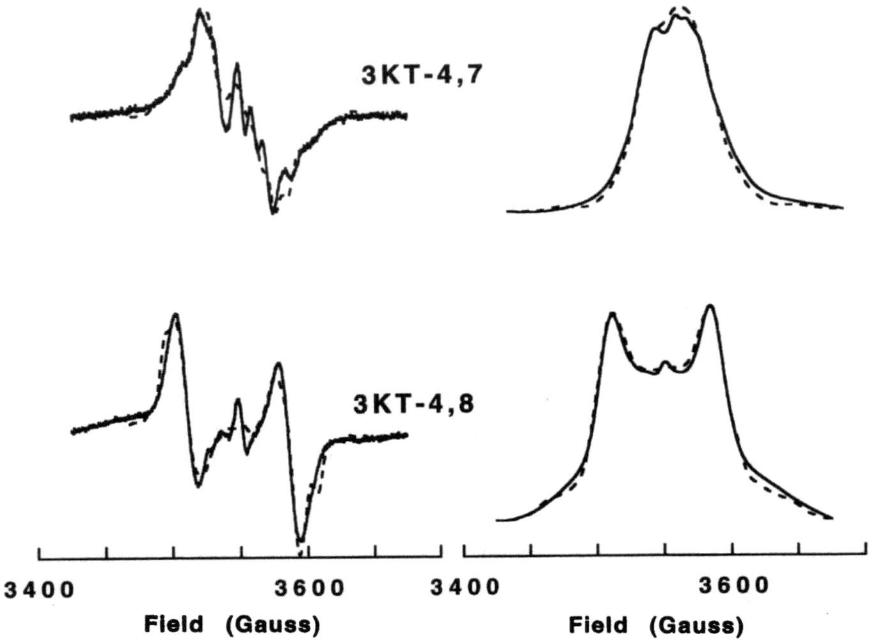

Figure 10. ESR spectra of the 3KT-4,7 and -4,8 peptides in MeOH at 200K (solid lines) in derivative (left) and absorption (right) mode. Simulations are shown as dashed lines.

The simulations confirm that indeed $D_{i,i+3} > D_{i,i+4}$. In order to assess the conformational implications of the distances obtained by simulation, computer models of TOAC-labeled helices were constructed. TOAC geometry was obtained from the published crystal structure and backbone torsion angles were assumed to be uniform along the polypeptide backbone. ϕ, ψ Backbone torsion angles were varied while monitoring the distances between the TOAC side chains. A good match to the experimentally-determined distances was obtained with the angles $\phi = -70^{o}$ and $\psi = -45^{o}$ (Hanson et al., 1998). Such angles are substantially more negative than published crystallographic values of -63°, -42°, respectively (Toniolo and Benedetti, 1991).

Results from the spectral fits are compared to the peptide model ($\phi = -70^{o}$, $\psi = -45^{o}$) for the 3KT-4,7 in Fig. 11. Given a fixed geometry for TOAC, spin label orientation is fully described by the positions of three atoms from the nitroxide ring. Thus, to simplify the representation, the position of each nitroxide as determined from the spectral fits is represented as spheres for the NO group and a single flanking nitroxide ring γ-carbon. All heavy atoms for the computer model are shown as sticks with N and O atoms in medium gray and black, respectively.

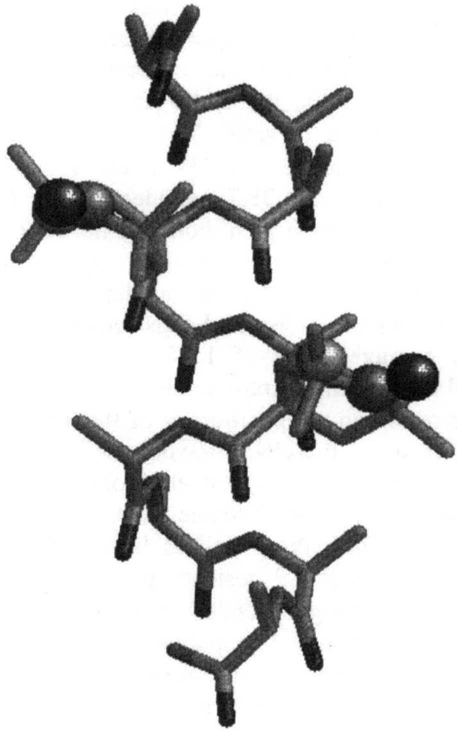

Figure 11. Comparison of the simulation results from Fig. 10 to a model of the 3KT-4,7 peptide containing two TOACs with uniform backbone torsion angles of ϕ = -70° and ψ = -45°. The model is shown as a stick representation while coordinates for the atoms determined from the ESR spectral simulations are shown as spheres (color code: C, light gray; N, medium; O, dark).

For the 3KT-4,7, the model provides an excellent match to both the distance and orientation determined from the spectral fit. The RMS deviation computed from comparing the six atoms shown from the model to the fit is approximately 0.20 Å. There is also good agreement between the model and fit from the 3KT-4,8 (not shown) but there is some discrepancy in the relative orientations between the labels. However, because strong dipolar coupling dominates the 3KT-(4,8) spectrum, we found that the spectral fits were not very sensitive to the orientations of the nitroxides.

The data reported suggest that the conformation of monomeric peptide α-helices in solution may be somewhat different from α-helices in peptide or protein crystals. α-Helices typically twist with 3.6 residues per turn (Creighton, 1993). However, the torsion angles $\phi = -70°$ and $\psi = -45°$ give a more open structure of approximately 3.8 - 3.9 residues per turn. It is interesting to consider the consequences of this backbone geometry. In a typical α-helix, the hydrogen-bond N···O distance (i.e., the distance from the N atom of amide NH donor to the O atom of the CO acceptor) is approximately 3.0 Å. With the $\phi = -70°$ and $\psi = -45°$ backbone geometry found for the 3KT peptides, the N···O distance is 2.7 Å which indicates that the hydrogen bonding network remains intact. However, it has been noted that the C'–O···N angle through a hydrogen bond can vary depending upon the surrounding medium (Barlow and Thornton, 1988). α-Helical hydrogen bonds formed in the interior of a protein exhibit an average angle of 157° whereas amides exposed to solvent are more tilted with an average angle of 148°. This additional tilt for exposed amides results in a repositioning of the amide carbonyl group away from the helix axis, thereby allowing the carbonyl to form an additional hydrogen bond to a nearby solvent molecule. For the $\phi = -70°$ and $\psi = -45°$ geometry found with double TOAC labeling, the C'–O···N angle is 140°, consistent with these helices being fully solvated. Thus, the open geometry of 3.8 - 3.9 residues per turn may be a favored structure because helical hydrogen bonds remain intact while allowing for solvation of the amide carbonyls. Unpublished NMR work from our lab has tentatively confirmed this unusual open α-helical structure. Thus, the precision offered from quantitative TOAC double labeling has revealed new geometrical details in the fundamental structure of the α-helix.

4. *THE TRICHOGIN GA IV PEPTIDE: RESOLVING MULTIPLE CONFORMATIONS*

Peptaibols are peptides of fungal origin, rich in Aib (Fig. 1), that possess a 1,2-aminoalcohol at the C-terminus (Benedetti et al., 1982). These unique peptides are of current interest because of their potent membrane modifying properties. As noted above, TOAC is an Aib analog and thus offers a new and exciting avenue for exploring the structure of these novel bioactive peptides. The longer peptaibols such as alamethicin form stable, membrane-spanning helices and are believed to assemble as barrel staves to form voltage-gated channels (Cafiso, 1994). The recently-isolated peptide trichogin GA IV brings new and unique characteristics to the family of peptaibols (Auvin-Guette et al., 1992). First, it has a lipophilic *n*-octanoyl group attached to the N-terminus and is therefore referred to as a *lipopeptaibol* (Auvin-Guette et al., 1992). The second distinguishing feature is its short length; trichogin GA IV is composed of only 10 amino acid residues and is therefore too short to span biological lipid bilayers. Nevertheless, trichogin GA IV exhibits remarkable membrane modifying properties similar to longer peptaibols, although the mechanism by which it causes membrane leakage is unknown.

The sequence of trichogin GA IV is

*n*Oct-Aib^{1}-Gly-Leu-Aib^{4}-Gly-Gly-Leu-Aib^{8}-Gly-Ile-Lol

where *n*Oct stands for *n*-octanoyl and Lol stands for leucinol. Preliminary circular dichroism and NMR studies in methanol suggest that the peptide forms a helix as expected for sequences containing Aib residues. The crystal structure was recently solved and confirmed a general helical character with residues one through four forming a distorted 3_{10}-helix and residues four through ten forming an α-helix. There is a kink at Gly 2 allowing for the switch from 3_{10}- to α-helix. Based on this crystal structure, it was proposed that a transmembrane pore could form via an alamethicin-like barrel stave mechanism but with each stave composed of a trichogin GA IV head-to-head dimer (Toniolo et al., 1994; Toniolo et al., 1996). This proposed mechanism of pore formation remains speculative.

We applied double TOAC labeling to map the solution structure of trichogin GA IV with the goal of determining its fold in a fluid environment (Anderson et al., 1999). The sequence has Aib residues at positions 1, 4 and 8 and thus the TOAC analogs trich-1,4, -4,8 and -1,8 represent all possible peptides with double Aib → TOAC substitutions. It has been demonstrated that these TOAC-labeled peptides retain the membrane-modifying properties of the parent trichogin GA IV (Monaco et al.,) and a crystal structure of the trich-4,8 showed that it adopts a mixed $3_{10}/\alpha$ fold similar to the parent peptide (Crisma et al., 1997; Monaco et al.,).

Conventional continuous wave spectra were obtained in MeOH at 200 K where rotational diffusion is too slow to average spectral rigid lattice features (see above). Derivative and absorption spectra for the three trichogin GA IV analogs are shown in Fig. 12. The trich-1,4 spectrum is clearly broader than a monoradical spectrum and gives a pattern that has been observed for other well-structured i, i+3 doubly-labeled helical peptides (see above). It is difficult to assign secondary structure without companion i, i+4 labeling in the same region of the peptide; however, the dipolar splitting is similar to that obtained from the α-helical 3KT-4,7 in MeOH at 200 K (Fig. 10).

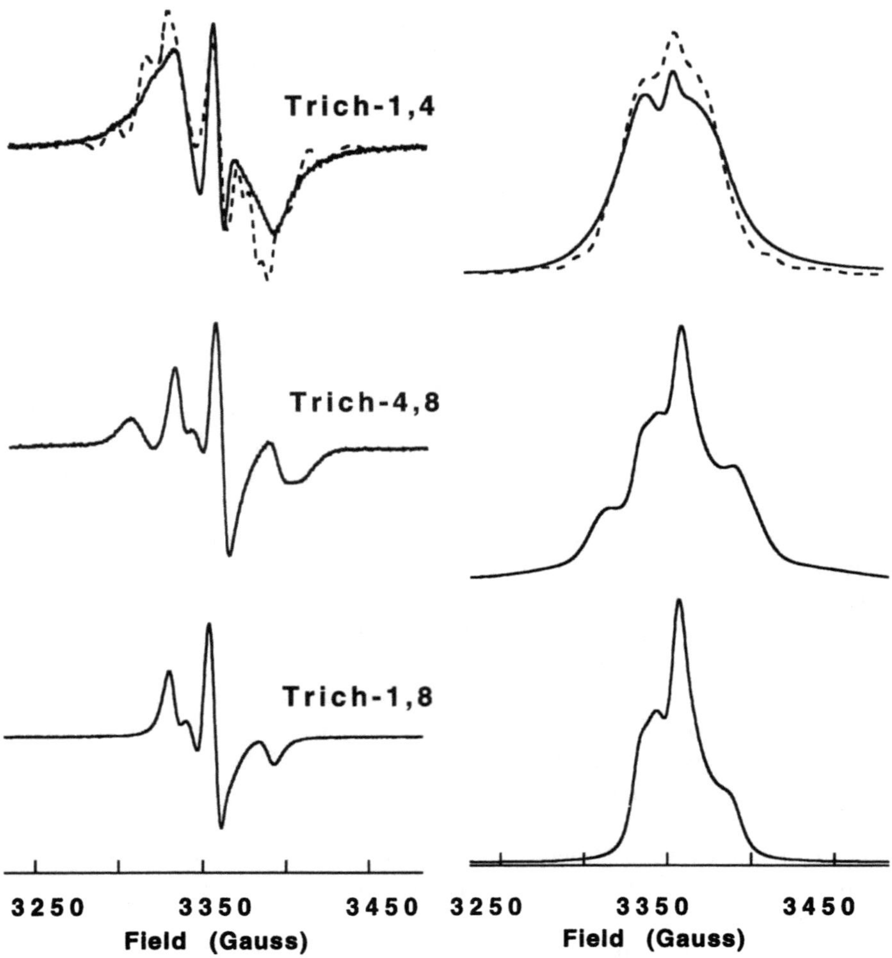

Figure 12. ESR spectra for the trichogin GA IV analogs in MeOH at 200 K in both derivative and absorption mode. Simulation of the trich-1,4 is also shown (dashed line).

We attempted to fit the trich-1,4 spectrum and the best simulation obtained is shown in Fig. 12. Fine features in the simulation are clearly absent in the experimental derivative spectrum. It is possible that some of these details are smoothed out by local conformation heterogeneity or small local motions. However, the absorption spectrum shows that the simulation captures the main spectral features and overall width but that the experimental spectrum is somewhat less sharp at the edges. The distance obtained from the simulation is 8.6 Å which is consistent with that obtained from the α-helical 3KT-4,7 (see previous section).

The trich-4,8 spectrum shows prominent outer wings consistent with strong dipolar coupling. However, the central region of the spectrum suggests there is substantial contribution from a peptide species devoid of dipolar coupling. Attempts at fitting the entire spectrum failed and the relative contributions of the dipolar coupled and uncoupled spectra exhibit a temperature dependence. Thus, it appears that the central and C-terminal regions of the peptide as probed by -4,8 double labeling are structurally heterogeneous with spectral contributions coming from both helical and unfolded peptide conformations. (Note: Rigorous purification was performed to insure the uncoupled spectrum did not arise from a monoradical contaminant.) Finally, the trich-1,8 spectrum exhibits very little dipolar coupling and is just slightly broader than that obtained from a monoradical species. Simulations were attempted but each fitting attempt converged to a different distance. Thus, we conclude that the broadening is too small to allow for determination of a well-defined distance.

The ESR spectra in Fig. 12 suggest that the N-terminal region is probably well structured although no single conformation gives a simulation that provides an ideal fit to the experimental data. The central and C-terminal regions are structurally heterogeneous with a contribution from an unfolded conformation. Given that the trichogin GA IV peptide exists in multiple conformations, it would greatly refine our knowledge of its structure if we had a method for determining an *average* distance between labels. For this we turn to measurement of half field intensities. In the presence of strong dipolar coupling, there is a mixing among the electron two-spin states resulting in a weak transition that appears at approximately half the field of the allowed transition. This transition is often referred to as "$\Delta M = 2$" although it actually arises from absorption of a single quantum (Atherton, 1973). The strength of this interaction is proportional to $1/r^6$ and becomes forbidden in the high field limit (or more specifically when the Zeeman interaction is much greater than the dipolar interaction). The Eaton laboratory has performed extensive studies of this signal and they have demonstrated that in the limit of low concentration the integral of the half field absorption is directly proportional to $1/r^6$ with negligible influence from relative nitroxide orientations (Eaton et al., 1983; Eaton and Eaton, 1989).

To measure distances using half-field transitions, we set up a half-field calibration curve using the hexameric peptides described above. Experiments were performed in 90:10 MeOH:EtOH at 120 K where the mixed alcohol solvent system allows for the formation of a solvent glass. The strength of the half field transition was determined for each peptide as a function of peptide concentration. Extrapolation to zero concentration (infinite dilution) gives the half field transition strength in the absence of intermolecular dipolar coupling (Eaton et al., 1983). The interspin distance was taken as the average of the internitroxide N···N and O···O distances for the TOAC residues as modeled in ideal 3_{10}-helices with each TOAC in the crystallographically-determined twist-boat conformation. Finally, the zero concentration integrals were plotted against these model distances thereby establishing a standard curve -- i.e., a half-field molecular ruler. The results shown in Fig. 13 are fit with a constant slope and zero intercept corresponding to the intensity$^{-1/6}$ $\propto$ distance relationship expected. The dashed lines in the plot represent the slope standard deviation which incorporates the scatter in this plot as well as propagated error associated with concentration and signal integral measurements. The data in Fig. 13 suggest that distances less than 10 Å may be reliably determined with error on the order of ± 1.0 Å.

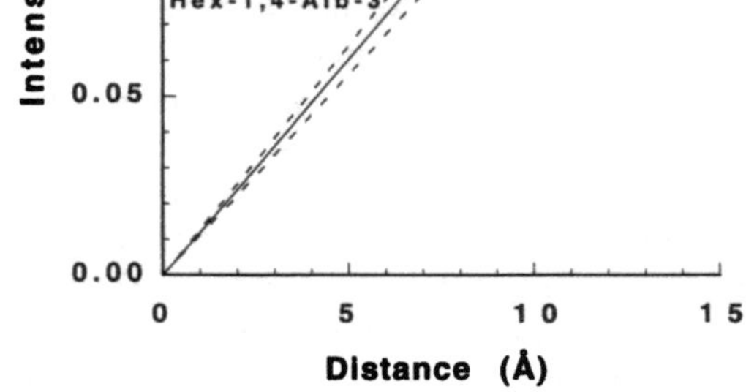

Figure 13. Half field signal calibration curve developed from the model hexameric peptides and two other peptides.

Concentration dependent half-field measurements were performed on the three trichogin GA IV analogs and the resulting distances reported below. Also reported below are hypothetical crystallographic distances determined from the trichogin GA IV crystal structure modeled with TOAC residues replacing the Aib's.

	Half Field (Å)	Crystallographic (Å)
Trich-1,4	7.9 ± 0.5	5.42
Trich-4,8	7.4 ± 0.5	7.28
Trich-1,8	12.0 ± 0.8	11.4

The trich-1,4 half field distance of 7.9 Å agrees reasonably well with 8.6 Å obtained from fitting the allowed ESR spectrum at 200 K. However, these ESR determined distances are substantially longer than that expected from our crystal structure based model. Nevertheless these distances are quite close to that expected for an α-helix or, to a lesser extent, a 3_{10}-helix. The trichogin GA IV crystal structure shows a kink at Gly 2 and computer models indicate that preservation of this backbone kink in trich-1,4 would bring the two TOACs in close proximity to give the 5.42 Å value above. Since the ESR data, from both fitting spectra and half field intensity, agree well with distances expected from conventional helical structures, we conclude that this kink is not preserved in solution and instead the N-terminal region of the peptide relaxes and forms a well structured helical turn.

The distance determined for the trich-4,8 matches well to that expected from the crystal structure and suggests that α-helix is the dominant folded structure in the C-terminal region of the peptide. There may be structural contributions from 3_{10}-helix or unfolded conformations, but in the presence of α-helix these may not contribute significantly to the half field intensity given the $1/r^6$ dependence. Allowed ESR spectra (see above) suggest that this region of the peptide is conformationally heterogeneous but clearly α-helix is a significant conformer. Finally, the trich-1,8 distance of 12.0 Å is only a little longer than the 11.4 Å expected from the crystal structure. Again, there may be unfolded conformers contributing to the overall structure but the approximate agreement between the half field and crystallographic values suggest that there is a significant population of peptide that is fully helical in solution.

The next issue is to resolve the apparent structural heterogeneity identified in the ESR spectra with the distances determined from the half-field measurements. The outermost features of the trich-4,8 spectrum match well to the wide splitting in the 3KT-4,8 spectrum (Fig. 10). This suggests that the distance for the folded conformer contributing to the trich-4,8 spectrum adopts approximately the same geometry as the 3KT-4,8 which is α-helix. As discussed above, the central region of the spectrum appears to be that from an uncoupled biradical which most likely arises from a population of unfolded conformers. Thus, the most conservative way to interpret the trich-4,8 spectrum is to model it as a superposition of only two spectra where one corresponds to that of an ideal -4,8 doubly labeled α-helix and the other to that of an uncoupled biradical. To perform this modeling we used the simulation of the 3KT-4,8 in Fig. 10 for the α-helical spectral component $\mathbf{S}_\alpha$ and a monoradical TOAC spectrum to represent the uncoupled biradical component $\mathbf{S}_\mathbf{u}$. Given these two normalized basis spectra, the normalized trich-4,8 absorption mode spectra were fit with the relation

$$\mathbf{S} = p_\alpha \mathbf{S}_\alpha + (1-p_\alpha)\, \mathbf{S}_\mathbf{u} \qquad (2)$$

where the single fitting parameter p_α represents the fractional population of α-helix. The resulting least-squares fits to trich-4,8 spectra obtained at 160 K and 180 K are shown in Fig. 14.

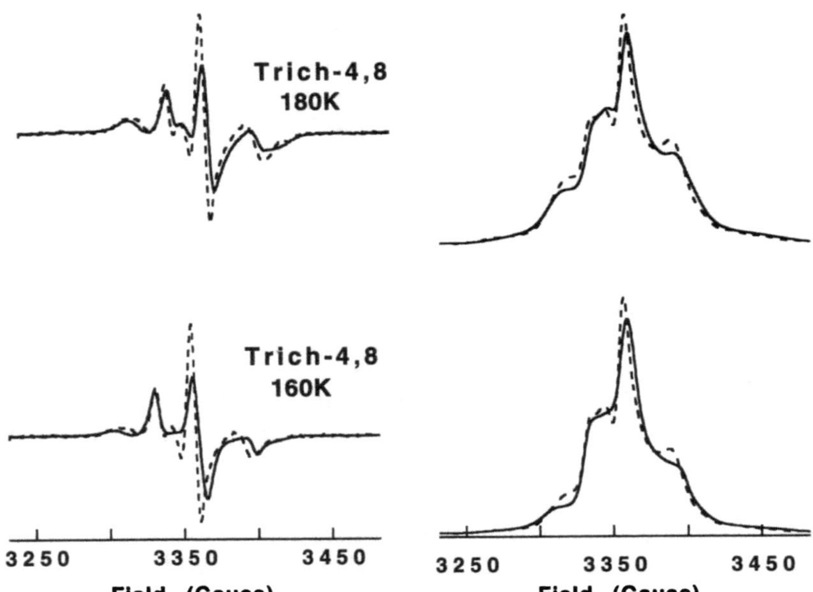

Figure 14. ESR spectra of the trich-4,8 at two temperatures below 200 K. The dashed lines overlaying the spectra are least squares fits to a superposition of two spectra where one corresponds to an uncoupled biradical and the other to the simulation of the 3KT-4,8 spectrum.

The absorption mode spectra show that this fitting procedure captures the main features of the trich-4,8 spectra. In contrast, the derivative spectra show that this fitting procedure does not reproduce the central line of the uncoupled component. The central line of the simulated spectrum is sharper by approximately 2 Gauss than the central line of the experimental trich-4,8 spectrum which, in turn, indicates that there is still residual biradical broadening in the central region of the trich-4,8 spectrum. It is interesting to note that the spectra in Fig. 14 reveal a temperature dependence where the population of the strongly coupled component decreases as the temperature is lowered.

We can now use these results to further interpret the half field result for the trich-4,8. Distances determined from half field intensities represent an average where each component is weighted by $1/r^6$. For the two species identified in the trich-4,8 spectra, the average half field-determined distance is given by

$$1/r^6_{avg} = p_\alpha \; 1/r^6_\alpha + (1-p_\alpha) \; 1/r^6_u \qquad (3)$$

where r_α and r_u correspond to the -4,8 distances for α-helix and unfolded conformers, respectively. Studies with the 3KT-4,8 suggest that r_α is approximately 6.6 Å. It is impossible to know r_u but we can estimate an upper bound on r_{avg} by setting $r_u = \infty$. Using $p_\alpha = 0.46$ determined from fitting the spectra obtained at 120K (where the half field measurements were obtained) we calculate $r_{avg} = 7.5$ Å which agrees well with the distance determined from the half field measurements. Thus, there is no discrepancy between the half field and allowed ESR measurements. To the contrary, reconciliation of these two measurements demonstrates that the central and C-terminal regions of trichogin GA IV are conformationally heterogeneous in solution and the half field measurement gives a slightly longer than expected distance because of the presence of unfolded conformers. These results are represented in Fig. 15 with molecular models of the two contributing conformers. On the left is regular α-helix and on the right is a partially unfolded peptide where the flexible residues Gly 5 and Gly 6 adopt a fully extended conformation ($\phi = 180°$, $\psi = 180°$). For this conformation the internitroxide distance in trich-4,8 would exceed 12 Å and give only weak dipolar coupling in the allowed ESR spectrum as observed.

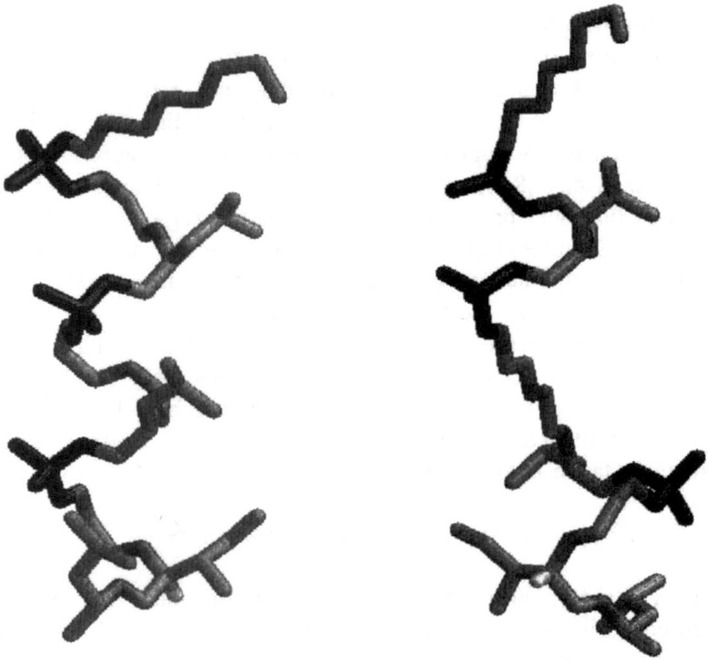

Figure 15. Schematic representation of the two relevant trichogin conformers suggested by the combined ESR results. The Aib's are shown in black and the remaining backbone and side chain atoms in gray. On the left is the α-helical conformer and on the right is the same structure but with fully extended backbone torsion angles ($\phi = 180°$, $\psi = 180°$) for Gly 5 and Gly 6.

5. SUMMARY AND OUTLOOK

From distinguishing polypeptide helix types, to revealing new folding geometries, to resolving multiple conformations in partially folded peptides, TOAC brings spin label ESR to a new level. The clear benefits arise from TOAC's rigidity which greatly eliminates uncertainties due to side chain conformations. Thus, one can map distances between TOAC N-Os back to quantitative distances between backbone attachment points. Given the long distance range available through double label ESR, one can now elucidate detailed folds and, if spectra exhibit a superposition of species, identify multiple conformations.

So far all applications have focused on peptide conformations. There is clear interest in performing similar experiments on proteins. Fortunately, new synthetic technologies are emerging which will allow for TOAC incorporation into polypeptides much larger than those described in this chapter. One such approach is chemical ligation whereby polypeptide segments are covalently linked (Wilken and Kent, 1998). In the spirit of this approach, we have recently reported synthesis and preliminary characterization of a two helix bundle where each helix contains a single TOAC and is tethered to a rigid cyclo-dipeptide template (Polese et al., 1999). In solution, five line multiplets and half field transitions were observed demonstrating that TOACs can resolve interactions *between* helical domains. Thus, there is little doubt that protein research in the near future will employ TOAC pairs for measuring critical intramolecular distances.

6. ACKNOWLEDGEMENTS

This work was supported by NIH Grant GM46870.

7. REFERENCES

Altenbach, C., Marti, T., Khorana, H. G. and Hubbell, W. L. (1990) Transmembrane Protein Structure: Spin Labeling of Bacteriorhodopsin Mutants. Science, **248**, 1088-1092.

Anderson, D. J., Hanson, P., McNulty, J., Millhauser, G., Monaco, V., Formaggio, F., Crisma, M. and Toniolo, C. (1999) Solution structures of TOAC-labeled trichogin GA IV peptides from allowed ($g \approx 2$) and half-field electron spin resonance. J. Am. Chem. Soc., **121**, 6919-6927.

Atherton, N. M. (1973) *Electron Spin Resonance: Theory and Applications* Wiley, New York.

Auvin-Guette, C., Rebuffat, S., Prigent, Y. and Bodo, B. (1992) Trichogin-GA-IV, an 11-Residue Lipopeptaibol From Trichoderma-Longibrachiatum. J. Am. Chem. Soc., **114**, 2170-2174.

Barlow, D. J. and Thornton, J. M. (1988) Helix Geometry in Proteins. J. Mol. Biol., **201**, 601-619.

Benedetti, E., Bavoso, A., Di Blasio, B., Pavone, V., Pedone, C., Toniolo, C. and Bonora, G. M. (1982) Peptaibol antibiotics: a study on the helical structure of the 2-9 sequence of emerimicins III and IV. Proc. Natl. Acad. Sci. USA, **79**, 7951-7954.

Berliner, L. J., Grunwald, J., Hankovszky, H. O. and Hideg, K. (1982) A Novel Reversible Thiol-Specific Spin Label: Papain Active Site Labeling and Inhibition. Anal. Biochem., **119**, 450-455.

Cafiso, D. S. (1994) Alamethicin: a peptide model for voltage gating and protein-membrane interactions. Ann. Rev. Biophys. Biomolec. Struct., **23**, 141-165.

Closs, G. L., Forbes, M. D. E. and Piotrowiak, P. (1992) Spin And Reaction Dynamics In Flexible Polymethylene Biradicals As Studied By EPR, NMR, And Optical Spectroscopy And Magnetic Field Effects - Measurements And Mechanisms Of Scalar Electron Spin Spin Coupling. J. Am. Chem. Soc., **114**, 3285-3294.

Cornish, V. W., Benson, D. R., Altenbach, C. A., Hideg, K., Hubbell, W. L. and Schultz, P. G. (1994) Site-specific incorporation of biophysical probes into proteins. Proc. Natl. Acad. Sci. USA, **91**, 2910-2914.

Creighton, T. E. (1993) *Proteins: Structures and Molecular Properties* Freeman, New York.

Cremer, D. and Pople, J. A. (1975) A general definition of ring puckering coordinates. J. Am. Chem. Soc., **97**, 1354-1358.

Crisma, M., Monaco, V., Formaggio, F., Toniolo, C., George, C. and Flippen-Anserson, J. L. (1997) Crystallographic Structure of a Helical Lipopeptaibol Antibiotic Analogue. Lett. Pept. Sci., **4**, 213-218.

Eaton, G. R. and Eaton, S. S. (1989) Resolved Electron-Electron Spin-Spin Splittings in EPR Spectra. In *Biological Magnetic Resonance: Spin Labeling Theory and Applications*, (L. J. Berliner and J. Reuben, Ed.) pp 339-397, Plenum Press, New York.

Eaton, S. S., More, K. M., Sawant, B. M. and Eaton, G. R. (1983) Use of the EPR Half-Field Transition to Determine the Interspin Distance and the Orientation of the Interspin Vector in Systems with Two Unpaired Electrons. J. Am. Chem. Soc., **105**, 6560-6567.

Fields, G. B., Ed. (1997) *Methods in Enzymology Vol. 289 Solid-Phase Peptide Synthesis*. San Diego, Academic Press.

Fiori, W. R., Miick, S. M. and Millhauser, G. L. (1993) Increasing Sequence Length Favors α-Helix Over 3_{10}-Helix in Alanine-Based Peptides: Evidence for a Length-Dependent Structural Transition. Biochemistry, **32**, 11957-11962.

Flippen-Anderson, J. L., George, C., Valle, G., Valente, E., Bianco, A., Formaggio, F., Crisma, M. and Toniolo, C. (1996) Crystallographic Characterization Of Geometry And Conformation Of TOAC, A Nitroxide Spin-Labelled C-Alpha,C-Alpha-Disubstituted Glycine, In Simple Derivatives And Model Peptides. Int. J. Pept. Protein Res., **47**, 231-238.

Hanson, P., Anderson, D. J., Martinez, G., Millhauser, G. L., Formaggio, F., Crisma, M., Toniolo, C. and Vita, C. (1998) Electron Spin Resonance and Structural Analysis of Water Soluble, Alanine-Rich Peptides Incorporating TOAC. Mol. Phys., **95**, 957-966.

Hanson, P., Martinez, G., Millhauser, G., Formaggio, F., Crisma, M., Toniolo, C. and Vita, C. (1996) Distinguishing Helix Conformations In Alanine-Rich Peptides Using The Unnatural Amino Acid TOAC and Electron Spin Resonance. J. Am. Chem. Soc., **118**, 271-272.

Hanson, P., Millhauser, G., Formaggio, F., Crisma, M. and Toniolo, C. (1996) ESR Characterization Of Hexameric, Helical Peptides Using Double Toac Spin Labeling. J. Am. Chem. Soc., **118**, 7618-7625.

Hustedt, E., Smirnov, A., Laub, C., Cobb, C. and Beth, A. (1997) Molecular distances from dipolar coupled spin-labels: The global analysis of multifrequency continuous wave electron paramagnetic resonance data. Biophys. J., **72**, 1861-1877.

Karle, I. L. and Balaram, P. (1990) Structural Characteristics of α-Helical Peptide Molecules Containing Aib Residues. Biochemistry, **29**, 6747-6756.

Koteiche, H. A. and McHaourab, H. S. (1999) Folding pattern of the alpha-crystallin domain in alphaA-crystallin determined by site-directed spin labeling. J. Mol. Biol., **294**, 561-77.

Luckhurst, G. R. (1966) Alternating Linewidths. A Novel Relaxation Process in the Electron Resonance of Biradicals. Mol. Phys., **10**, 543-550.

Luckhurst, G. R. (1976) Biradicals as Spin Probes. In *Spin Labeling Theory and Applications*, (L. J. Berliner, Ed.) pp 133-181, Academic Press, New York.

Marchetto, R., Schreier, S. and Nakaie, C. R. (1993) A Novel Spin-Labeled Amino Acid Derivative for Use in Peptide Synthesis: (9-Fluorenylmethyloxycarbonyl)-2,2,6,6-tetramethylpiperidine-N-oxyl-4-amino-4-carboxylic Acid. J. Am. Chem. Soc., **115**, 11042-11043.

Marqusee, S., Robbins, V. H. and Baldwin, R. L. (1989) Unusally stable helix formation in short alanine-based peptides. Proc. Natl. Acad. Sci. USA, **86**, 5286-5290.

Miick, S. M., Casteel, K. M. and Millhauser, G. L. (1993) Experimental Molecular Dynamics of an Alanine-Based Helical Peptide Determined by Spin Label Electron Spin Resonance. Biochemistry, **32**, 8014-8021.

Miick, S. M., Martinez, G. V., Fiori, W. R., Todd, A. P. and Millhauser, G. L. (1992) Short Alanine-Based Peptides May Form 3_{10}-Helices and Not α-Helices in Aqueous Solution [correction appears in Nature 1995, **377**, 257]. Nature, **359**, 653-655.

Millhauser, G. L. (1992) Selective Placement of Electron Spin Resonance Spin Labels: New Structural Methods for Peptides and Proteins. Trends Biochem. Sci., **17**, 448-452.

Millhauser, G. L. (1995) Views Of Helical Peptides - A Proposal For The Position Of 3(10)-Helix Along The Thermodynamic Folding Pathway. Biochemistry, **34**, 3873-3877.

Millhauser, G. L., Stenland, C. J., Hanson, P., Bolin, K. A. and van de Ven, F. J. (1997) Estimating the relative populations of 3(10)-helix and alpha-helix in Ala-rich peptides: a hydrogen exchange and high field NMR study. J. Mol. Biol., **267**, 963-974.

Monaco, V., Formaggio, F., Crisma, M., Toniolo, C., Hanson, P., Millhauser, G., George, C., Deschamps, J. R. and Flippen-Anderson, J. L. (1999) Determining the Occurrence of a 3_{10}-Helix and an α-Helix in Two Different Segments of a Lipopeptaibol Antibiotic Using TOAC, a Nitroxide Spin-Labeled C^{α}-Tetrasubstituted α-Aminoacid. Bioorg. Med. Chem., **7**, 119-131.

Polese, A., Anderson, D. J., Millhauser, G., Formaggio, F., Crisma, M., Marchiori, F. and Toniolo, C. (1999) First inter-chain peptide interaction detected by ESR in fully synthetic, template-assisted, two-helix bundles. J. Am. Chem. Soc., **121**, 11071-11078.

Rassat, A. and Rey, P. (1967) Nitroxides, XXIII. Préparation d'aminoacides radicalaires et de leurs sels complexes. Bull. Soc. Chim. Fr., 815-817.

Toniolo, C. and Benedetti, E. (1991) The Polypeptide 3_{10}-Helix. Trends Biochem. Sci., **16**, 350-353.

Toniolo, C., Crisma, M., Formaggio, F., Peggion, C., Monaco, V., Goulard, C., Rebuffat, S. and Bodo, B. (1996) Effect Of N-Alpha-Acyl Chain Length On the Membrane-Modifying Properties Of Synthetic Analogs Of the Lipopeptaibol Trichogin GA IV. J. Am. Chem. Soc., **118**, 4952-4958.

Toniolo, C., Peggion, C., Crisma, M., Formaggio, F., Shui, X. and Eggleston, D. S. (1994) Structure determination of racemic trichogin GA IV using centrosymmetric crystals. Nature: Struct. Biol., **1**, 908-914.

Toniolo, C., Valente, E., Formaggio, F., Crisma, M., Pilloni, G., Corvaja, C., Toffoletti, A., Martinez, G. V., Hanson, M. P., Millhauser, G. L., George, C. and Flippen-Anderson, J. L. (1995) Synthesis and conformational studies of peptides containing TOAC, a spin labeled $C^{\alpha,\alpha}$-disubstituted glycine. J. Pept. Sci., **1**, 45-57.

Wilken, J. and Kent, S. B. H. (1998) Chemical protein synthesis. Current Opinion in Biotechnology, **9**, 412-426.

Chapter 7

Depth of Immersion of Paramagnetic Centers in Biological Systems

Gertz I. Likhtenshtein
Department of Chemistry, Ben-Gurion University of the Negev, P.O.B. 653, Beer-Sheva 84105, Israel

Abstract: The theory of static and dynamic dipole-dipole and exchange spin-spin interactions between radicals and paramagnetic ions is surveyed. Methods of determination of depth of immersion and localization of a paramagnetic center, and the investigation of electrostatic potential around biologically important molecules are described in detail. Advantages and limitations of different methods are discussed. Applications of the methods to paramagnetic centers of enzymes (cytochrome P-450, the primary donor of the photosynthetic reaction center, and flavin-dependent alcohol oxidase) and to nitroxide spin probes in biological and model membranes are reviewed.

1. INTRODUCTION

Paramagnetic centers (PC) in biology play key roles in numerous biochemical and biophysical processes. These centers include Fe, Cu, Mo, Ni, and Co ions in active sites of metalloenzymes and metalloproteins, complexes of metals with proteins, nucleic acids and polysaccharides, drug metallocomplexes, flavin and semiquinone radicals, cation- and anion-radical intermediates in photosynthetic systems, spin labels and probes, and radicals in irradiated objects.

In solving problems of enzyme catalysis, molecular biophysics of proteins, biomembranes and molecular biology it is necessary to know the spatial disposition of individual parts. One must also know the depth of immersion of paramagnetic centers in a biological matrix, i.e. the availability of enzyme sites to substrates, distance of electron tunneling between a donor

Biological Magnetic Resonance, Volume 19: Distance Measurements in Biological Systems by EPR, edited by Berliner et al. Kluwer Academic/Plenum Publishers, New York, 2000

and an acceptor group, position of a spin-label in a membrane and in a protein globule, distribution of the electrostatic field around the PC, etc.

Structural studies of systems that cannot be obtained as pure single crystals are beset with serious difficulties. This problem can be solved with the use of ESR techniques if the lineshape or spin-relaxation parameters of the ESR signals from a paramagnetic center respond suitably to the approach of another paramagnetic center. Two types of spin-spin interactions can be distinguished: (1) dipole-dipole interaction arises because the magnetic dipole of one paramagnetic center induces a local magnetic field at the site of another paramagnetic group; (2) exchange interaction is caused by overlap of the orbitals occupied by unpaired electrons as the particles approach each other. In the overlap zone electrons lose their "magnetic memory" and spin exchange occurs. Because of differences in the dependence of spin-spin interactions on interspin distance r, exchange interaction may prevail over dipole-dipole interaction at $r < 8\text{-}10$ Å (Zamaraev et al., 1977)

The first applications of dipole-dipole spin-spin interactions to the investigation of protein surface topography were based on changes in the lineshape of nitroxide radical ESR spectra arising from interaction with a second nitroxide radical or a paramagnetic ion (Likhtenshtein, 1968; Likhtenshtein and Bobodzhanov, 1968; Taylor et al., 1969; Kulikov et al., 1972; Kokorin et al., 1972). Later the greater sensitivity of power saturation curves of a radical to interactions between the radical and paramagnetic ions was demonstrated (Kotel'nikov et al., 1975; Kulikov, 1976; Kulikov and Likhtenshtein, 1977; Case and Leigh, 1976). Dynamic spin exchange interaction during an encounter between radicals and paramagnetic ions diffusing freely in solution (spin label-spin probe technique) was employed to study the dynamics and microstructure of biological objects in the vicinity of added labels (Likhtenstein, 1976, 1988a,b, 1990, 1993; Likhtenshtein et al., 1970, 1972, 1986; Hyde et al., 1979; Zamaraev et al., 1977, 1981).

The effect of spin-spin interactions between paramagnetics is dependent on the distance of closest approach of the centers and, therefore, on the depth of immersion of the center in a matrix (Kotel'nikov et al., 1975; Kulikov and Likhtenshtein, 1977; Kulikov et al., 1979, 1981, 1987, 1989; Case and Leigh, 1976; Cherepanova et al., 1990; Bogatyrenko et al., 1991; Likhtenstein, 1976, 1988a,b, 1990, 1993; Averbakh et al., 1995). Under certain conditions a paramagnetic center can affect nuclear spin relaxation parameters monitored by NMR (Dwek, 1977; Hwang and Freed, 1975; Alexandrov, 1975; Berdnikov *et al.*, 1980; Borah and Bryant, 1981), which can provide information on the location of a nucleus.

Electrostatic interactions play a key role in the structure and function of biological molecules (Druzhinin et al., 1986; Bashford and Karplus, 1990;

Aquist et al., 1991; Loewenthal et al., 1993; Gilson, 1993; Anni et al., 1994; Hecht et al., 1995; Honig et al., 1995; Likhtenshtein et al., 1999). Association of proteins in solution and in membranes, protein-nucleic acid interactions, nucleic acid-nucleic acid interactions, enzyme-substrate complexation, chemical reactions in enzyme active sites, charge-transfer, voltage gating of membrane channels, folding and unfolding processes of biopolymers, etc. are all drastically affected by the strength and distribution of the electrostatic field around various regions in biological molecules. The frequency of encounters between two particles bearing electron or nuclear spins in solution depends on electrostatic interactions. Therefore, monitoring of effects of the electric charge of a paramagnetic probe (either nitroxide or metal ion) on the spin relaxation parameters of another particle by ESR or NMR allows one to investigate the distribution of electrostatic potential in the vicinity of the center of interest.

In the present chapter a general survey is made of theoretical bases and experimental methods of determination of depth of immersion of paramagnetic centers. A description of how each method is calibrated, details of how one actually uses the method, and examples of the method applied in biological systems will be emphasized.

2. SPIN-SPIN METAL-RADICAL INTERACTION

Depending upon the system to be studied, the spin-spin interaction may be dominated by exchange or dipolar interaction. In frozen or glassy solution the interaction is static, i.e. there is a fixed distance between the paramagnetic centers. In fluid solution the interaction may be dynamic or static. In this section the equations required to analyze data for these different cases are described.

2.1 Distances between isolated pairs

2.1.1 Analysis of CW power saturation curves in solids based on static dipolar interaction

Theoretical basis. According to the Solomon-Bloembergen (1956) theory, magnetic moments, and interspin distance are not the only factors affecting the rate of energy transfer or cross relaxation in a static magnetic field (H_0) as a result of the dipolar interaction. Other important factors are electron spin (S), the spin-phase (T_2) and spin-lattice (T_1) relaxation times, the ESR frequencies (ω_ι) of both interacting paramagnetics, and the angle (θ) between the interspin vector and the external field, H_0. An important factor is the product of relaxation time T_1 for the second center and the interaction energy express in frequency units ($\Delta\omega$). When $\Delta\omega^2 T_1^2 >> 1$, the effect of the second spin may be regarded as an interaction with a permanent dipole moment. When $\Delta\omega^2 T_1^2 << 1$, the interaction may be considered as a weak perturbation because of the fast relaxation.

Of particular interest for the solution of structural problems is the case involving one slower-relaxing paramagnetic center with T_{1s} and T_{2s} times that are long enough for the ESR line to be observed and for the ESR saturation curve to be obtained, and a second center characterized by short relaxation times, i.e., $T_{1f}\,\mu_f/r^3 << 1$. For such a pair (Kulikov, 1976; Kulikov and Likhtenshtein, 1974, 1977; Likhtenshtein, 1976, 1988a, 1993; Likhtenshtein et al., 1982, 1993; Bowman and Norris, 1982), the contribution of the second paramagnetic center to T_{1s} and T_{2s}.

$$\Delta\frac{1}{T_{1s}} = \frac{1}{T_{1s}} - \frac{1}{T_{1s}^0} = \frac{\mu_f^2\gamma^2}{6r^6 T_{1f}}\left[\frac{24}{5(\omega_f+\omega_s)^2} + \frac{12}{5\omega_s^2} + \frac{4}{5(\omega_f-\omega_s)^2}\right] \quad (1)$$

$$\Delta\left(\frac{1}{T_{2s}}\right) = \frac{1}{T_{2s}} - \frac{1}{T_{2s}^0} = \frac{4\mu_f^2\gamma^2}{15r^6} T_{1f} \quad (2)$$

In the derivation of Eq. (1) and (2) it is assumed that $T_{1f} = T_{2f}$, that $(\omega_f - \omega_s)^2 T_{1f}^2 >> 1$, that $(\omega_s)^2 T_{1f}^2 >> 1$, and the spherical average of orientation-dependent terms was used. Application of these formulas to distance estimation requires knowledge of the value of the T_{1f}. The T_{1f} factor cancels in the product $\Delta(1/T_{1s})\Delta(1/T_{2s})$, which facilitates calculation of r.

For estimation of distances between paramagnetic centers using Eqs. (1) and (2) it is necessary to know T_{1s} and T_{2s} for the radical. These values can be determined directly by pulse methods as described in Ch. 2. The relaxation times can be estimated by analysis of CW power saturation curves in solids from saturation curves recorded under conditions of rapid passage as proposed in (Safronov et al., 1969) and applied to nitroxide radicals in (Kulikov, 1976; Kulikov and Likhtenshtein, 1977). To apply this method it is important to specify conditions for recording the ESR signal. For example, the condition of fast passage for nitroxyl radicals with $(T_1T_2)^{1/2} \sim 10^{-5}$ s at T = 77 K are attained with a modulation frequency $\nu_m = 10^5$ Hz and the microwave amplitude close to the modulation amplitude H_m. At room temperature $(T_1T_2)^{1/2} \sim 10^{-7}$ s for nitroxide radicals, and fast passage conditions are not fulfilled. The power saturation curves are characterized by the microwave field intensities H_1* and H_1**, at which the amplitude of the ESR signal equals half of the curve maximum (Fig.1).

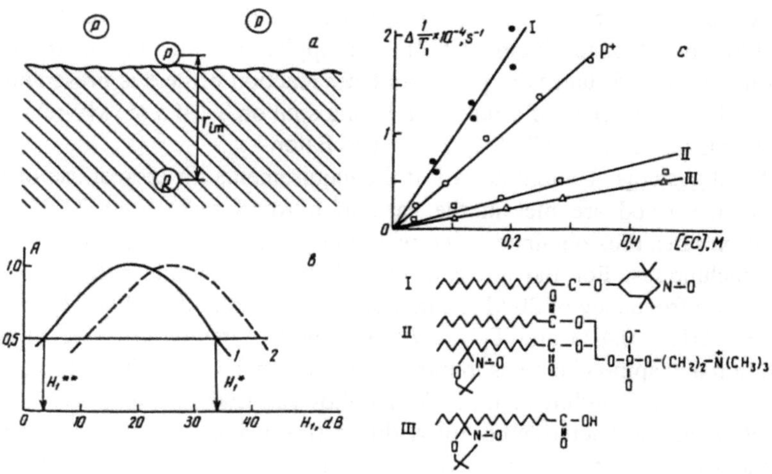

Figure 1. Estimation of depth of immersion (r_{im} = R_0) of the radical center (R) in a matrix from the effect of dipolar interaction with paramagnetic centers (P) in a solid solution; **a** diagram defining distances; **b** shape of the saturation curves without the interaction (1) and with interaction (2); **c** dependence of the dipole contribution on the concentration of ferricyanide [FC] for radicals in chromatophores of *R. rubrum*. I-III spin probes; P^+ cation of the bacteriochlorophyl dimer. (See Sections 2.2.1 and 4.1 for details). (Reproduced from Likhtenshtein, 1993).

The calculation of T_{1s} and T_{2s} involves the following steps: (1) The spin packet width $\Delta H_p = 1/\gamma\ T_{2s}$ (in G) is determined from the parameter $q = H_1^*/H_1^{**}$ using Table 1 on p. 56 of Kulikov and Liktenshtein (1977). (2) The ΔH_p value is used to calculate the parameter β using Table 2 on p. 56 of Kulikov and Likhtenshtein (1977). Tables 1 and 2 in this reference were calculated for a particular spectrometer and resonator and corresponding calculations would need to be performed to apply this method using a different system. (3) T_{1s} in sec. is calculated using $T_{1s} = 10^{-9}\beta/(H_1^*)^2$.

In solids T_{1f} usually decreases monotonically with increasing temperature (if no other dynamic process occurs that affects T_{1f}, such as a phase transition) so one can find experimentally the temperature at which the dipolar contribution is maximum. At this temperature $T_{1f} = \Delta\omega$ and the dipolar contribution is given by Eq. (3).

$$\Delta\left(\frac{1}{T_{1s}}\right) = \frac{\mu_f^2\gamma^2}{15r^6\Delta\omega} \tag{3}$$

Because μ_f depends on electron spin, the theoretical maximum distances (r_{max}) to a second paramagnetic center with spin S_f that could be determined were estimated as: r_{max} = 100 Å for S_f = 1/2 and r_{max} = 150 Å for S_f = 5/2 (Kulikov and Likhtenshtein, 1977). An application of this method to determination of distances in photosynthetic reaction centers is presented in Section 4.1. The effect of a rapidly-relaxing spin on T_1 of a slowly-relaxing spin also is discussed in Ch. 8 by Eaton and Eaton.

Model systems. Though not all of the requirements for correct use of the method described are met in the case of nitroxyl radicals at 77 K, the relaxation rates thus obtained agree reasonably with direct measurements by pulse techniques. For the nitroxide TEMPOL in ethanolic solution (at a modulation frequency of 20 kHz and modulation amplitude < 3 G) H_1^* = 7.5 10^{-3} G, ΔH_p = 0.4 G and T_1 = 2.7×10^{-4} s were obtained at 77 K. Pulse saturation and spin-echo measurements gave ΔH_p = 0.5 G and T_1 = 2.0×10^{-4} s (Kulikov, 1976; Salikhov et al., 1976; Kulikov and Likhtenshtein, 1977). For the cation of the bacteriochlorophyl dimer (the primary donor of reaction centers from *R. rubrum*) the CW saturation method give T_1 = 3.0×10^{-4} s (Kulikov et al., 1981) which is in a good agreement with the pulse-saturation methods (McElroy et al., 1974).

Taking into consideration that the estimated accuracy of the ΔH_p and T_1 measurements is about ±30% and there is a weak dependence of R_0 on spin relaxation rates (Eqs. 1, 2), one can estimate the accuracy of calculated r values for a pair with fixed distance as about ±5%. Nevertheless, the main problem that a researcher meets in investigation of spin-labeled biomolecules is an uncertainty in the label conformation. Either an average distance is measured or special analysis of experimental data is needed.

Application of this method is illustrated by estimation of distances between the heme group of human hemoglobin and the nitroxide fragment of three spin labels attached covalently to the β-93 SH-group (Kulikov, 1976). Saturation curves were recorded with the ESR-3 spectrometer with a magnetic field modulation frequency of 20 kHz. For deoxyhemoglobin (S_2 = 0) and fluorohemoglobin (S_2 = 5/2) spin-labeled with 4-acetamido-TEMPO the parameter q was found to be 38 and 20, respectively. Using these values the spin packet widths were calculated to be ΔH_p = 0.010 and 0.019 G, respectively, which corresponds to $\Delta(1/T_{2s}) = 4.1 \times 10^7$ s^{-1}. Using Eq. (2), to estimate the distance between the heme group and the label nitroxide fragment, r = 15.1 ± 1.2 Å. Independently, T_{1s}^0 = 1.8×10^{-4} and T_{1s} = 1.25×10^{-4} s were determined for the diamagnetic and paramagnetic samples of hemoglobin, respectively, by CW ESR method at 77 K. With the use of Eq. (3), these values give $\Delta(1/T_{1s}) = 2.4 \times 10^3$ s^{-1} and, according to Eq. (1) r = 17.0 ±1.0 Å. In the distance calculation the following values were used: magnetic moment of the high-spin heme, $\mu_2 = 5.8$ μ_B, where $\mu_B = 0.93 \times 10^{-20}$ $g^{1/2}cm^{5/2}s^{-1}$ is the Bohr magneton, the electron gyromagnetic ratio γ = 1.77×10^7 $cm^{1/2}g^{-1/2}$, ω_1 = 5.9×10^{10} rad/s, and ω_2 = 1.8×10^{11} rad/s. The distance calculated from the product $\Delta(1/T_{11})\ \Delta(1/T_{21})$ was found to be r = 15.9±0.8 Å. The values of r (in Å with an accuracy 5-6%) between heme and the nitroxide fragment are in reasonable agreement with each other (Table 1) and with the x-ray diffraction structure of hemoglobin (Perutz, 1989).

Table 1. Comparison of Distances (Å) between heme iron and spin labels on hemoglobin

Label/Methods	$\Delta(1/T_{1s})$	$\Delta(1/T_{2s})$	$\Delta(1/T_{1s})\Delta(1/T_{2s})$	A_F/A_C[b]
IA-TEMPO[a]	17.0	15.1	15.9	18.6
M-TEMPO[a]	15.2	15.2	15.2	16.3
PCMB-TEMPO[a]	16.5	15.1	15.8	17.3

[a]TEMPO, 2,2,6,6-tetramethylpiperidine-N-oxyl; IA, 4-iodoacetamido; M, 4-maleimido; PCMB, p-chloromercuribenzoate; [b]The A_F/A_{CN} method is based on measurement of the ratio of amplitudes of nitroxide ESR signals for labeled high-spin fluorohemoglobin and low-spin cyanohemoglobin based on theory developed in (Taylor et al., 1969).

2.1.2 Static exchange interactions in solids

The exchange interaction between pairs of spins leads to formation of two states: (1) a singlet state with antiparallel disposition of spins and (2) a triplet state with parallel disposition. The energy separation between the states equals 2J, where J is the exchange integral. The Hamiltonian for the spin exchange (SE) interaction between spins with operators S_1 and S_2 is Eq. (4).

$$H_{SE} = -2J_{SE}S_1S_2 \tag{4}$$

A negative sign of J_{SE} in Eq. (4) indicates antiferromagnetic interaction (Abragam and Bleaney, 1970). A vast literature is concerned with the quantitative investigation of exchange (see for example Plachy and Kivelson, 1967; Eastman et al., 1970; Likhtenshtein et al., 1970, 1972; Zamaraev et al., 1977, 1981; Eaton and Eaton, 1989; More et al., 1990; Likhtenshtein, 1976, 1993, 1995 and references therein).

Empirical data. The magnitude of the spin exchange integral J_{SE} characterizes the degree of overlap of molecular orbitals containing the unpaired electrons. J_{SE} is about 10^{14} s^{-1} at Van der Waals distance and decreases exponentially with increasing distance between spins in a vacuum or in homogeneous media (Zamaraev et al., 1977). The exchange interaction can be strengthened by "conductive" (for example, covalent bonds, conjugated structures, etc.) bridges via so called indirect exchange.

Experimental data on the dependence of exchange parameters (the rate of triplet-triplet energy transfer (k_{TT}) or the spin exchange integral (J_{SE})) on the distance (r) between interacting centers are approximated by the following equation (Likhtenshtein 1988a,b, 1993,1995; Likhtenshtein et al., 1982; Kotel'nikov et al., 1981).

$$k_{TT}, J_{SE} \sim 10^{14} \exp[-\beta(r-r_V)] \qquad (5)$$

where r_V is the distance at the Van der Waals contact. For a system in which centers that do not belong to a single molecule are separated by homogéneous “non-conducting” medium (for example, in solvents consisting of molecules with saturated chemical bonds), the dependence of k_{TT} on r is characterized approximately by β_{TT} = 2.6 Å. For a system in which the radical centers are linked by "conducting" conjugated bonds the dependence of J_{SE} on r followed Eq. (5) with β_{SE} = 0.3 Å^{-1} (Kotel’nikov et al., 1981; Likhtenshtein, 1988a,b, 1995, 1993). To first approximation spin exchange involves two orbitals with unpaired electrons and in the triplet energy transfer process four orbitals are involved (orbitals of the donor in the ground and exited states, and the acceptor in the ground and excited state). Taking this into consideration we can estimate β_{SE} for “non-conducting” media as β_{SE} = 1.3 Å^{-1}. In “non-conducting” media J_{SE} is damped by factor of about 4 for one Å increase in r (Eq. (5)). Since the smallest value of J_{SE} that can be determined experimentally is about 10^6 s^{-1}, the longest distance that can be derived by measuring J_{SE} is <10 Å for “non-conducting” systems. Examples of application of Eq. (5) to distances between pigments of photosynthetic reaction centers is given in Section 4.1.

The exchange phenomenon may be considered as an idealized model of electron transfer (ET) without a change in positions of nuclei of the electron donor (D) and electron acceptor (A) (Zamaraev et. al., 1977, 1981; Likhtenshtein 1993, 1995). The rate constant of outer sphere and long-distance electron transfer, $k_{ET} \sim V_{ET}^2 \sim S^2$, were V_{ET} is the resonance integral and S is the overlap integral that quantitatively characterizes the degree of overlap of the orbitals containing the unpaired electrons (Marcus and Sutin, 1985; Levich et al., 1970). Because the exchange integral J_{SE} is also proportional to S^2, a similarity of dependence of J_{SE} and k_{ET} on the distances between the centers may be expected. For “non-conducting” media such a dependence is characterized by the attenuation parameter $\gamma_{SE} = J_{SE}(0)/J_{SE}(r) = \exp[\beta_{SE}(r-r_v)]$ were $J_{SE}(0)$ and $J_{SE}(r)$ are the exchange integrals at the van der Waals distance and the distance r correspondingly. As one can see in Fig. 2, the slope of the plot of $\log\gamma_{SE}$ versus $(r-r_v)$ with β_{SE} = 1.3 Å corresponds to the slope of the dependence of $\log(k_{ET})$ on distance (r) for different stages of electron transfer in reaction centers of purple bacteria and photosystem II (PSII) of green plants (Likhtenshtein, 1995). A similar dependence of log γ_{SE} on $(r-r_v)$ was found for other biological and model systems (Gray et. al., 1994 and references therein).

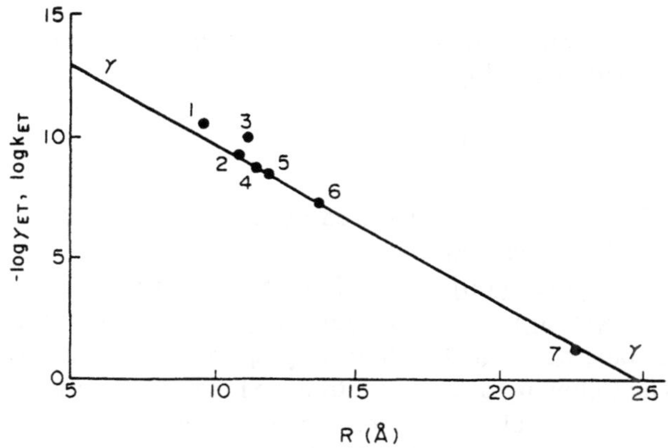

Figure 2. Dependence of the experimental rate constant of long-distance electron transfer (k_{ET}) in reaction centers of photosynthetic bacteria and plant Photosystem II on the edge-edge distance between the donor and acceptor center. The solid straight line is related to the dependence of the attenuation parameter for "non-conductive" media (γ_{SE}) (see text) (Likhtenshtein, 1995).

It is necessary to stress that exchange interactions within a molecule are markedly dependent on the nature of the chemical groups connecting the paramagnetic centers (Medzhidov et al., 1969; Parmon et al., 1980; Kotel'nikov et al., 1981; Eaton and Eaton, 1989; More et al., 1990; Likhtenshtein, 1993, 1995). Consideration of this important problem is beyond the scope of this chapter.

2.2 Distance of closest approach

2.2.1 Analysis of power saturation curves in solids (CW ESR)

The first attempts to employ CW ESR techniques for the investigation of location and depth of immersions of radicals were made in the early 1970's (Kotel'nikov et al., 1975; Kulikov and Likhtenshtein, 1974; Case and Leigh, 1976). The determination of the distance of closest approach of a "radical" and a paramagnetic species is performed by examining the effect of the dipole-dipole interaction between the radical and paramagnetics that are distributed uniformly in a vitrified matrix on the ESR saturation curves of the radicals (Kulikov et al., 1981). A similar approach examines the kinetics of spin-lattice relaxation for paramagnetics in solids (Bowman and Norris, 1982).

The observed relaxation rate enhancement for a radical, as a consequence of dipole-dipole interaction with a paramagnetic that is distributed uniformly in the vitrified matrix, consists of a combination of multiple individual curves as indicated by Eq. (6).

$$W = \Delta(1/T_{1s}) = 1/T_{1s} - 1/T_{1s}^0 = \Sigma\, w_{rp} \tag{6}$$

where w_{rp} is the relaxation rate induced by dipole-dipole interaction for an individual radical-paramagnetic ion pair.

Methods for estimation of a radical depth of immersion (Kulikov et al., 1981). Low concentrations of paramagnetic ions. At a low concentration of paramagnetic ions, $r_{av} > R_0$. The following approach was suggested for R_0 estimation. The following data were introduced into the computer to perform averaging of dipole-dipole interaction between radicals and paramagnetics: 1) starting saturation curve $A_0(H_1, H_1^{*0})$; 2) dimensionless parameter $\beta = T_{1s}^0 W_{max}$, were $W_{max} = 5\mu_f^2\gamma^2/3R_0^6$; and 3) dimensionless parameter $b = 0.5\, \alpha C_p(T_{1s}^0)^{1/2}$ where

$$\alpha = 1.5\text{x}10^8\, \mu_f\, (T_{1f})^{1/2}\ \text{cm}^3\text{s}^{-1/2} \tag{7}$$

To determine the β parameter and, therefore, W_{max}, the experimental dependence of the saturation curve width ($\Delta q = H_1^{**} - H_1^*$) on the H_1^* shift ($\Delta H_1^* = H_1^* - H_1^{*0}$) at various paramagnetic ion concentrations (Fig. 3) can be used. Having selected the β parameter (one of the curves 1-10 related to a given experiment), one can determine the b parameter from the calculated dependence of experimental ΔH_1^* on b at various β (Fig. 4) using the same curve number as selected from Fig. 3.

The distance of closest approach is calculated with Eq. (8).

$$R_0 = 7.6\, [(\tan\gamma)^2/\beta\,]^{1/6} \tag{8}$$

where $\tan\gamma = 0.5\alpha T_{1s}^0$ (in M^{-1}) is the slope of the plot of b versus C_p and R_0 is in Å. Eq. (8) is obtained by combination of Eq. (7) and the formulas for W_{max} and β. In this method, knowledge of the spin-lattice relaxation time of the radical, T_{1s}^0, is not needed.

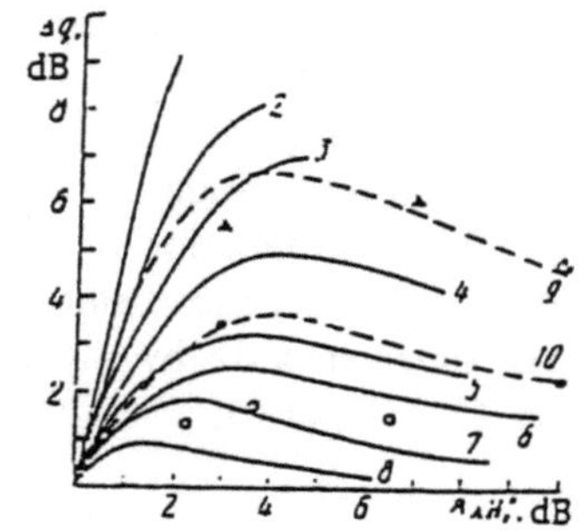

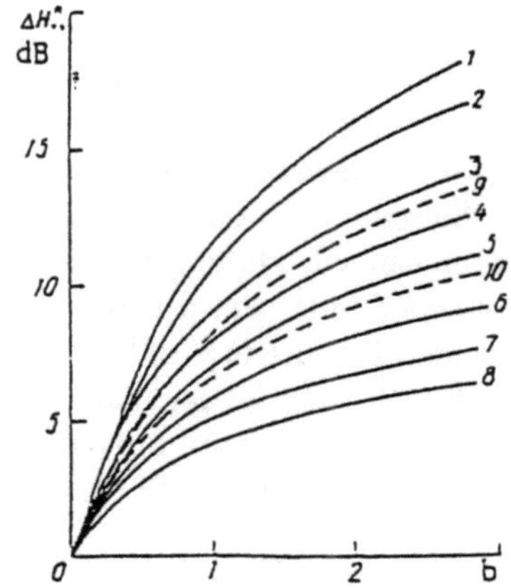

Figure 3. Relationship between the $\Delta q = H_1^{**} - H_1^{*}$ and $\Delta H_1^{*} = H_1^{*} - H_1^{*0}$ parameters of nitroxide saturation curves. Continuous lines 1-8 depict the dependences calculated for an aqueous glycerol solution at 77 K with b = 1100, 400, 200, 100, 45, 25, 11, and 4, respectively. Curve 9 pertains to the calculation for an aqueous glycerol solution at 35 K with b = 100, while curve 10 pertains to the calculation for an ethanol solution at 77 K with b = 25. Experimental points: $K_3Fe(CN)_6$, aqueous-glycerol solution, 77 K (O) and 35 K (▲); $NiCl_2$, ethanol solution, 77 K (•) (reproduced from Kulikov et al., 1981).

Figure 4. Calculated dependence of ΔH_1^* on parameter b (the designations of the curves are the same as in Fig. 3).

High concentration of paramagnetic ions. At high concentration of paramagnetic species, C_p, the average distance between paramagnetic ions $r_{av} = C_p^{-1/3}$ is less than the distance of closest approach to the buried radical, R_0. The overall relaxation rate, W, for this case is given by Eq. (9)

$$W = \Delta\left(\frac{1}{T_{1s}}\right) = \frac{A\pi\mu_f^2 T_{1f} C_p}{R_0^3} \tag{9}$$

where the numerical factor A depends on the geometry of the matrix that hinders approach of the radical and the paramagnetic ion. If the radical is located in the center of a sphere or is hidden under an infinite plane, A is equal to 16/9 or 2/9, respectively. If the radical is located at distance R_0 from the surface of a sphere with radius R, $A = (2 - R_0/R)^{-3} \times 16/9$.

The case where the paramagnetic ions have such a short relaxation time that $(\omega_s + \omega_f)T_{1f} \ll 1$ was considered. If the T_{2s} value and the condition of fast passage do not change when the paramagnetics are added, the power saturation curves are shifted without change of the curve width (Kulikov et al., 1981) and

$$W = \Delta \frac{1}{T_{1s}} = \frac{[(H_1^* / H_1^{*0})^2 - 1]}{T_{1s}^0} \qquad (10)$$

If the lineshape of the ESR signal in the presence of the paramagnetics was the same as in the absence of the paramagnetics it was assumed that T_{2s} was unchanged. Plotting experimental W versus C_p and using the values of the magnetic moment μ and T_{1f} of the paramagnetic ion one can derive the R_0 value. The product $\mu(T_{1f})^{1/2}$, that is necessary for determination of R_0 from Eq. (9), can be calculated from the experimental data for parameter α (Eq. 7). The value of $\mu(T_{1f})^{1/2}$ can be measured independently by pulse methods. Combination of Eqs. (7) and (9) gives Eq. (11) (R_0 in Å).

$$R_0 = 1.44 \times 10^{14}(A\alpha^2/\text{tg}\,\psi)^{1/3} \qquad (11)$$

where α is in $cm^3s^{-1/2}$ and tg ψ (in $M^{-1}s^{-1}$) is the slope of the plot of W vs C_p.

Application of CW saturation method. (Kulikov et al., 1981; Cherepanova et al., 1990). The first step of the procedure for determination of a depth of immersion of a radical in a matrix, R_0, is measurement of the distance of the closest approach of ions to free radical R_0^0. For the nitroxide radical TEMPOL in the presence of $Fe(CN)_6^{-3}$ in water-glycerol solution (1:1/ v:v) at 77 K the slope of the plot of W (Eq. (9)) for the radical as a function of the paramagnetic anion concentration (C_p) was found to be tg $\psi = 5.5 \times 10^5$ $M^{-1}s^{-1}$ and the slope of plot of parameter b as a function of C_p was $\alpha = 1.3 \times 10^{-17}$ $cm^3s^{-1/2}$. Assuming a spherical solvation shell around the anion (A=16/9), the use of Eq. (11), gives $R_0^0 = 21$ Å. Analogous systems containing this radical and $CoCl_2$ were characterized by the following values: tg $\psi = 1.8 \times 10^5$ $M^{-1}s^{-1}$, $\alpha = 1.3 \times 10^{-17}$ $cm^3s^{-1/2}$ and $R_0^0 = 14$ Å. The R_0^0 values for acetylacetone complexes of Ni^{2+}, Co^{2+} and aqua complexes of Ni^{2+} and Co^{2+} in ethanol at 20 – 77 K range from 9 to 12 Å or about 6-9 Å after subtraction of the nitroxyl radical radius. The values of corrected radii R_0^0 are in a good agreement with the radii of closest approach obtained with Stuart-Briegleb models. The unusually high value of $R_0^0 = 21 \pm 1.2$ Å ($R_0^0 \approx 18$ Å after correction for the nitroxyl radius) for the ferricyanide ion in glycerol-water mixture (1:1/v:v) at 20 – 77 K may be rationalized by formation of a thick shell of solvent molecules around the highly-charged ion.

To overcome difficulties that result from possible adsorption of paramagnetics on the surface of biomembranes, two approaches to determination of immersion depths in biological membranes were developed (Cherepanova et al., 1990). The first approach examined spin-labeled probes **IV** and **V** in a biomembrane. The paramagnetic ion, say Co^{2+} was introduced together with an ion antagonist, Pb^{2+} or Al^{3+}. These diamagnetic ions prevents adsorption of the paramagnetic ions on the membrane surface and the paramagnetic ions distribute only in the vitrified solvent phase.

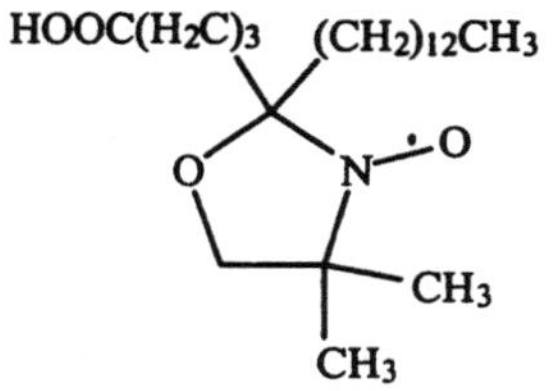

IV

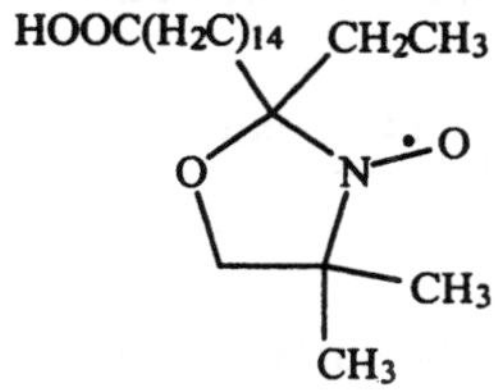

V

Table 2. Experimental values of R_0 for lipid spin probes **IV** and **V** in biological membranes, calculated with Eq. (11)[a]

Membrane	probe	paramagnetic	Diamagnetic agonist	tg $\psi 10^{-5}$ $M^{-1}s^{-1}$	$R_0 \pm 1$Å
M. capsulatus	**IV**	$CoCl_2$	$Pb(NO_3)_2$	0.6	11
"	**IV**	$CoCl_2$	$AlCl_3$	0.6	11
"	**V**	$CoCl_2$	$Pb(NO_3)_2$	0.25	18
"	**V**	$CoCl_2$	$AlCl_3$	0.25	18
Chromatophores	**V**	$CoCl_2$	$AlCl_3$	0.7	9
R. rubrum	**V**	$K_3Fe(CN)_6$	none	0.4	8
Chloroplasts (pea)	**IV**	$CoCl_2$	$AlCl_3$	0.9	8
"	**V**	$CoCl_2$	$AlCl_3$	0.45	13
Microsomes	**V**	$CoCl_2$	$AlCl_3$	0.45	13
"	$Co(AA)_2$				6[b]

[a]$T_{1s}^{0} = 2x10^{-4}$ s, values of parameter α are $1.3x10^{-17}$ $cm^3s^{-1/2}$ and $1.0x10^{-17}$ $cm^3s^{-1/2}$ for $K_3Fe(CN)_6$ and $CoCl_2$, respectively (Cherepanova et al., 1990). [b]See text, AA = acetylacetonate.

The investigation revealed a variety of spin-labeled lipid conformations in different biological membranes. The R_0 values for spin probes **IV-V** in *M. capsulatus* membranes are for more extended conformations then that in the chromatophore, chloroplast and microsome membranes (Table 2).

Spin probes **I-III** (Fig. 1) were introduced into membranes of chromatophores from ***Rodosphirillum rubrum*** and the effect of ferricyanide ion randomly distributed in the vitrified water-glycerol phase on the nitroxide saturation curves was investigated. The immersion depth values, R_0, of the nitroxide moieties determined with the use Eq. 9 were found to be 7, 20 and 22 Å for the probes **I, II** and **III** respectively. The R_0 values are in reasonable agreement with the values expected for the extended structure of the probes. Equation (9) was used because of the high concentration of ferricyanide.

The second method is based on the effect of paramagnetic ions adsorbed on the membrane surface (Cherepanova et.al., 1990). In such a case the spin-lattice relaxation rate of the radical

$$W = \Delta(\frac{1}{T_{1s}}) = \frac{(4/27)\pi\mu_f^2\gamma^2 T_{1f} C_p}{R_0^4} \tag{12}$$

Commonly, the surface concentration (C_p) is not known. Therefore the $R_0(X)$ value for radical X can be determined only by measuring the ratio of W_x in a sample with a previously-determined $R_0(St)$ for a standard radical and W_{St} in the sample with the same radical. According to Eq. (12)

$$W_x/W_{St} = R_0(X)^4/ R_0 (St)^4 \tag{13}$$

To study the localization of the hydrophobic probe, $Co(AA)_2$, the experimental values of $[(H_1^*/H_1^{*0})^2-1]$ for probes **IV** and **V** in *M. capsulatus* membranes were plotted versus the concentration of $CoCl_2$ and $Co(AA)_2$. Using Eq. (13) the average distances, R_0, from the probe location to the plane of the $Co(AA)_2$ localization was determined. Because the R_0 values for the probes were known from previous experiments, the immersion depth of $Co(AA)_2$ in the membrane (R_0 = 6 Å) could be calculated (Cherepanova et al., 1990).

Data on determination of depth of immersion of active centers of important biological objects, namely photosynthetic reaction centers, cytochrome P450 and flavin oxygenase are considered in section 4.

2.2.2 Dynamic exchange interaction in solution

In solutions of low viscosity electron spin-spin dipolar interactions between paramagnetics are averaged by fast rotation of molecules and the short lifetime of the encounter complex and, therefore, are very weak. In such a condition exchange interactions prevail over dipole-dipole ones.

When a "radical" in solution encounters another paramagnetic species the broadening of the Lorenzian line of the radical is related to the rate constant of the exchange relaxation (k_{ex}) as shown in Eq. (14) (Salikhov et al., 1971; Zamaraev et al., 1977; Likhtenshtein, 1976):

$$\Delta(1/T_{21}) \text{ is proportional to } \Delta H_L = 6.5\text{x}10^{-8}\, k_{ex}C \tag{14}$$

where ΔH_L is expressed in G, C is ion concentration in M, k_{ex} is in $M^{-1}s^{-1}$ and the factor $6.5\text{x}10^{-8}$ has units of G/s^{-1} and assumes a g value of 2.002. This equation allows one to determine the value of k_{ex}. In the case of inhomogeneous broadening of the initial line, adequate correction should be made for the value of the experimentally observed broadening (Marsh, 1989).

The value of k_{ex} also can be determined from the change in the rate of the spin-lattice relaxation of the radical in the presence of a paramagnetic species:

$$\Delta(1/T_{1s}) = k_{ex}C \tag{15}$$

Since $1/T_{1s}$ usually is more sensitive to spin-spin interaction than $1/T_{2s}$, especially in the region of slow rotation and in solids, this method widens the accessible range of k_{ex} values (Kulikov et al., 1983a; Hyde and Subczinski, 1989).

According to Salikhov et al. (1971)

$$k_{ex} = P_{ex} k_d = \frac{f_g f_{ns} k_d J^2 \tau_c^2}{(1 + J^2 \tau_c^2)} \tag{16}$$

where k_d is the rate constant of encounters in solution, P_{ex} is the probability of spin exchange in the course of life time (τ_c) of the encounter complex, f_g is the geometric steric factor, f_{ns} is the nuclear statistical factor and J is the exchange integral of interaction in the encounter complex at the direct contact between the particles. Eq. (16) is valid if $J >> \delta$ (the difference between the resonance frequencies of the spins), $\delta\tau_c >> 1$, and the ion spin $S_f = ½$.

If $J^2 \tau_c^2 >> 1$, k_{ex} is not dependent on J (strong exchange) and

$$k_{ex} = f_g f_{ns} k_d \tag{17}$$

In the case of weak exchange $J^2\tau_c^2 << 1$ and

$$k_{ex} = f_g f_{ns} k_d J^2\tau_c^2 \tag{18}$$

Substituting the definition of J from Eq. (5), with β=1.3 Å into Eq. (18) gives

$$k_{ex} = f_g f_{ns} k_d \tau_c^2 10^{28} \exp[2[-1.3(R_0 - r_V)]] \tag{19}$$

This equation can be used for estimation of the distance of closest approach (R_0) if other factors in the equation are known. For example, in water solution a typical value of f_g f_{ns} k_d = $2x10^9$ $M^{-1}s^{-1}$ for interaction between a nitroxide radical and a paramagnetic species and $\tau_d = 5x10^{-11}$ s. For a buried paramagnetic center, if the experimental $k_{ex} = 10^7$ M^{-1} s^{-1}, the value of (R_0-r_v) = 9 Å, based on Eq. (19).

A more general theory has been developed for exchange relaxation during encounters between paramagnetic particles in solution. According to the theory the probabilities of spin relaxation of paramagnetics during encounters depend in the general case on a number of parameters such as the exchange integral, the spin-lattice relaxation time, the duration of the encounters, and the differences between the resonance frequencies of the spins (δ) (Salikhov et al., 1971; Likhtenshtein, 1993).

3. INVESTIGATION OF ELECTROSTATIC EFFECTS IN MOLECULES IN SOLUTIONS

Methods of determination of the electrostatic potential and local charges in the vicinity of groups bearing magnetic moment are based on the sensitivity of spin-relaxation rates of radicals and nuclei to encounters with charged paramagnetic species in solution (Likhtenshtein et al., 1972; Likhtenshtein, 1976, 1993; Hecht et al., 1995). Measurements of the effect of paramagnetics with different charges on the spin relaxation rates of the nuclei or radical allow quantitative characterization of electrostatic effects.

3.1 Effect of charge on dipolar interactions between protons and a paramagnetic species

3.1.1 Theoretical grounds

Proton spin-lattice relaxation affected by paramagnetic ions and spin probes. According to a number of workers (Hwang and Freed, 1975; Alexandrov, 1975; Berdnikov *et al.*, 1980), the spin-lattice relaxation rate of proton nuclei, $1/T_{1p}$, upon encounter with a "radical" free electron, are dominated by dipolar interaction because of low electron spin density at the Van der Waals distance. The relaxation rate interaction can be described by Eq. (20) and (21):

$$1/T_{1p} = (\frac{4\pi}{9})(\frac{\gamma_p^2\gamma_e^2h^2}{R_0^6})S(S+1)T_{1e}[R\bullet]f(y) \qquad (20)$$

$$f(y) = \frac{(4+y)y^2}{9+9y+4y^2+y^3} \tag{21}$$

where γ_p and γ_e are the gyromagnetic ratios of protons and electrons, respectively; R_0 is the distance of the closest approach between proton and electron; $[R^\bullet]$ is the concentration of the paramagnetic species; $\tau_d = R^2/D$, R is the radius of the radical, $y = (\tau_d/T_{1e})^{1/2}$; T_{1e} is the electron spin-lattice relaxation time; D is the sum of diffusion coefficients of the proton-bearer and radical; and S is the electron spin. Due to the low value of γ_p, estimates have shown (Alexandrov, 1975; Berdnikov et al., 1980) that Eq. (20) is correct for any reasonable viscosity for the medium.

Experimental measurement of the rate constant of the proton spin-lattice relaxation, $k = \Delta(1/T_{1p})/dC$, and values of parameters in Eq. (20) (γ_p = $2.67\text{x}10^4$ $cm^{1/2}g^{-1/2}$, $\gamma_e = 1.76\text{x}10^7$ $cm^{1/2}$ $g^{-1/2}$) allow estimation of the distance of the closest approach of a proton in a matrix $R_{0(p)}$ and a paramagnetic freely diffused in solution. For example if k = 200 $s^{-1}M^{-1}$, $T_{1e} = 10^{-6}$ s and τ_d = $5\text{x}10^{-11}$ s (water solution) Eq. (20) and (21) gives $R_{0(p)}$ = 8 Å. This value was calculated taking into consideration the geometric sterical factor which is ~ 0.2 for a nitroxide.

Slopes, $k = d(1/T_{1p})/dC$, of the dependence of proton spin-lattice relaxation rate on the concentration of the nitroxide probes [R•] can characterize the dipole-dipole interaction that dominates in an encounter between a proton and paramagnetic species (Wien et al., 1972; Syrtsova et al., 1974; Hwang and Freed, 1975; Alexandrov, 1975; Dwek, 1977; Berdnikov et al., 1980; Borah and Bryant, 1981; Sletten et al., 1983; Niccolai et al., 1982, 1984). Variable k has the dimensions of a second-order rate constant ($M^{-1}s^{-1}$), and can be considered as an apparent relaxation rate constant. The superscripted index i = 0, +1 or –1 designates data acquired from radical probes with 0, +1 or –1 charge, respectively.

To explore the effect of electrostatic interaction on the dipole-dipole relaxation rate, the ratios of experimental values k^+/k^0 or k^-/k^0 were used to estimate the electrostatic potential, $U(R_0)$, in the vicinity of a proton (Hwang and Freed, 1975) as presented in Eq. (22):

$$\frac{k^{+-}}{k_0} = \beta_{el}\beta_I \tag{22}$$

where $$\beta_{el} = \frac{e^{U(R_0)/kT}}{\int_{R_0}^{\infty} \frac{e^{U(R)/kT}dR}{R^2}} \tag{23}$$

(if $U(R_0) < kT$ then $\beta_{el} = [U(R_0)/kT)\exp[U(R_0)/kT]/1-\exp[U(R_0)/kT])$

$$\beta_I = 10^{\left(\frac{\sqrt{I}}{1+\sqrt{I}}\right)} \tag{24}$$

and I is the ionic strength.

The electrostatic effect in the vicinity of a proton coming from charged functional groups (ammonium cation, carboxylate anion, etc.) in other regions of a target molecule X can be quantitatively characterized in an empirical manner by a relative apparent charge, α^X:

$$\alpha^x = \frac{Z^x_{app}}{Z^0} = \frac{U(R_0)^x_{exp}}{U(R_0)^0_{exp}} \tag{25}$$

where indices x or 0 are assigned to a generic target molecule or to a small charged model compound (ethylammonium cation, imidazolium cation, propionate anion, etc.), respectively. The α^X value depends on the position of the proton in the target molecule relative to the neighboring charged group. Taking $Z_0 = +1$ or -1, we can consider $Z^x{}_{app} = \alpha^X$ as a parameter indicating an electrostatic effect of the molecule on a charged particle placed in the vicinity of a given proton. Such a parameter can be used in the analysis of electrostatic factors effecting equilibrium and reaction rate constants for nuclei residing in particular local charge environments within the molecule.

3.1.2 Application to model small molecules

The simple organic molecules and amino acids in these investigations provide well-defined molecules to study the factors that influence electrostatic fields in more complex biologically important molecules. These model systems enable verification of experimental and theoretical approaches to this problem (Likhtenshtein et al., 1999). The only significant difference in the nitroxide spin-probes **VI-VIII** is the difference in the probe charges, Z_p.

The distances between these charges and charges Z_X in molecules under investigation are similar, thus the ratio of the apparent spin-lattice relaxation rate constants k^{+-}/k^0 reflects the effect of electrostatics upon the encounter interaction (Eq. 22). Measurement of these ratios allows one to calculate an NMR-based average electrostatic potential $U(R_0)_{exp}$ between the charged probe and a given target molecule proton and then to compare it with the theoretical one $U(R_0)_{calc}$ that is obtained by molecular modeling (Sundin,

1991). The rate constant k^0 is characteristic of the effect of a neutral nitroxide probe (VI) on the proton relaxation rate ($1/T_{1p}$) in the molecule under investigation.

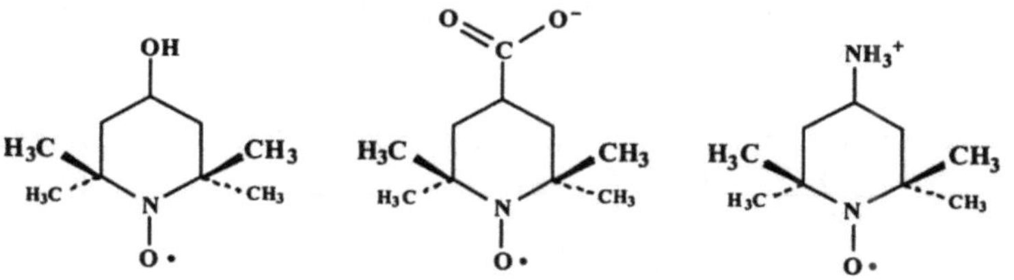

VI VII VIII

The spin-lattice relaxation rates ($1/T_{1p}$) of protons in small target molecules [aliphatic alcohols (ethanol, isopropanol, tert.-butanol), benzyl alcohol, propionate anion, ethylammonium cation, imidazolium cation) were measured as a function of concentration of spin-probes with different charges (Fig. 5). In each case, $1/T_{1p}$ was found to be proportional to the probe concentration. The rate constants, k, (slopes of $d(1/T_{1p})/dC$ where C is the concentration of the nitroxide radical) were then calculated (Table 3) and used to calculate the electrostatic potential $U(R_0)_{exp}$ (Eq. 22-24).

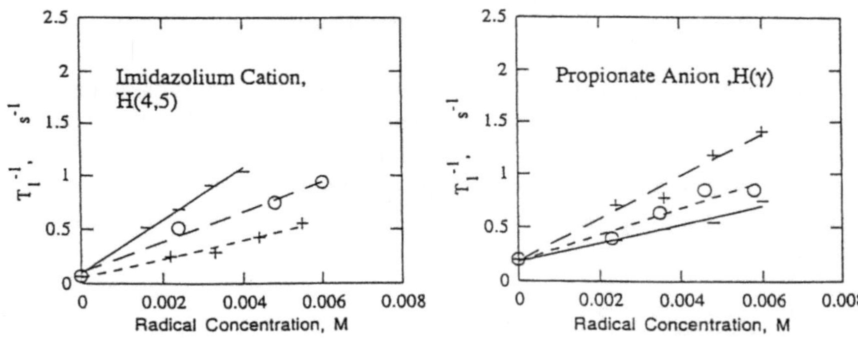

Figure 5. Spin lattice relaxation rate of individual protons of small charged molecules as a function of spin probe concentration. Neutral, negative and charged spin probes (VI-VIII) are respectively denoted as 0, -, and +. (Likhtenshtein et al., 1999)

A molecular modeling calculation of local electrostatic potential (Likhtenshtein et al., 1999) was applied to a number of small charged (ethylamine, propionate anion and imidazole) and uncharged (ethanol) molecules with an algorithm developed in (Depmeier et al., 1996). The sign and magnitudes of parameters $U(R_0)_{exp}$ and $U(R_0)_{calc}$ (Table 3) correspond to expected values from simple electrostatic considerations. The sign is positive for protonated ethylamine and imidazole, and negative for propionate anion. The absolute value of $U(R_0)_{calc}$ is very small for protons in neutral tert-butanol and is markedly larger for small charged molecules (propionate anion, ethylammonium cation, and imidazolium cation). Comparison of $U(R_0)_{exp}$ with $U(R_0)_{calc}$ values for protonated imidazole shows them to be in quantitative agreement (see Table 3) if the apparent dielectric constant is taken as $\varepsilon_{app} = 56$.

Table 3. Rate constants of spin-lattice relaxation of protons in small molecules under the effect of nitroxides **VI - VIII** of different charge.

Molecule	Proton	D_H[a]	k^0 ($M^{-1}s^{-1}$)	k^- ($M^{-1}s^{-1}$)	k^+ ($M^{-1}s^{-1}$)	$U(R_0)_{cal}$[b] (kcal/mol)	$U(R_0)_{exp}$[c] (kcal/mol)
Tert-Butanol			158±17	170±5	155±11		
Ethyl-amineH$^+$	H(α)	2.9	126±11	182±4	69±2	0.25	0.29
	H(β)	1.1	142±14	180±16	86±6	0.22	0.26
Propionate	H(α)	2.1	143±14	81±11	225±29	-0.16	-0.29
	H(β)	1.0	125±17	87±9	200±18	-0.14	-0.26
Imidazolium	H(2)	7.6	129±23	245±12	87±8	0.30	0.30
	H(4,5)	7.0	140±13	251±13	89±9	0.26	0.30

[a] D_H is the shift downfield from tetramethylsilane in ppm. [b] $\varepsilon_0 = 80$, $R_0 = 8$ Å, Ionic strength $I = 0.008$, [c]$U(R_0)_{exp}$ is the experimental electrostatic potential calculated using Eq. (22) and (23).

3.2 Impact of charge on spin exchange interactions between radicals and paramagnetic complexes

As discussed in section 2.2.2 exchange interaction is assumed to dominate over dipolar interaction in fluid solution. Calculations of local charge Z_X in the vicinity of a paramagnetic particle (such as the active site of metalloprotein or a spin label) colliding with a nitroxide or metallocomplex with known charge Z_p can be carried out with the use of the Debye equation (Eq. (26)).

$$\frac{k^+}{k^0} \text{ or } \frac{k^-}{k^0} = \frac{Z_p Z_x \alpha}{\exp(Z_p Z_x \alpha) - 1} \tag{26}$$

where k^+, k^- and k^0 are the rate constants of encounters for positively charged, negatively charged, and neutral uncharged particles, respectively; and $\alpha = e^2/k_B T \varepsilon_0 r$, where e is the charge of an electron; k_B is the Boltzmann constant; temperature T = 293 K, ε_0 is the dielectric constant of water, and r is the distance between the charges in the encounter complex (Debye, 1942). The values of (k^+/k^0) and (k^-/k^0) are determined by measurement of the rate constants of spin exchange using Eq. (14) and (15), which describe the effect of paramagnetic species on spin phase and spin-lattice relaxation rates of the radical.

Fig. 6 illustrates the effect predicted by Eq. (26) of the Z_0Z_X product upon the $\log(k^+/k^0)$ or $\log(k^-/k^0)$ at $\varepsilon_0 = 80$ [curves are drawn for a 1-11 Å range of inter-charge distances, r] (Grebenshchikov et al., 1972, Likhtenshtein, 1976, 1993).

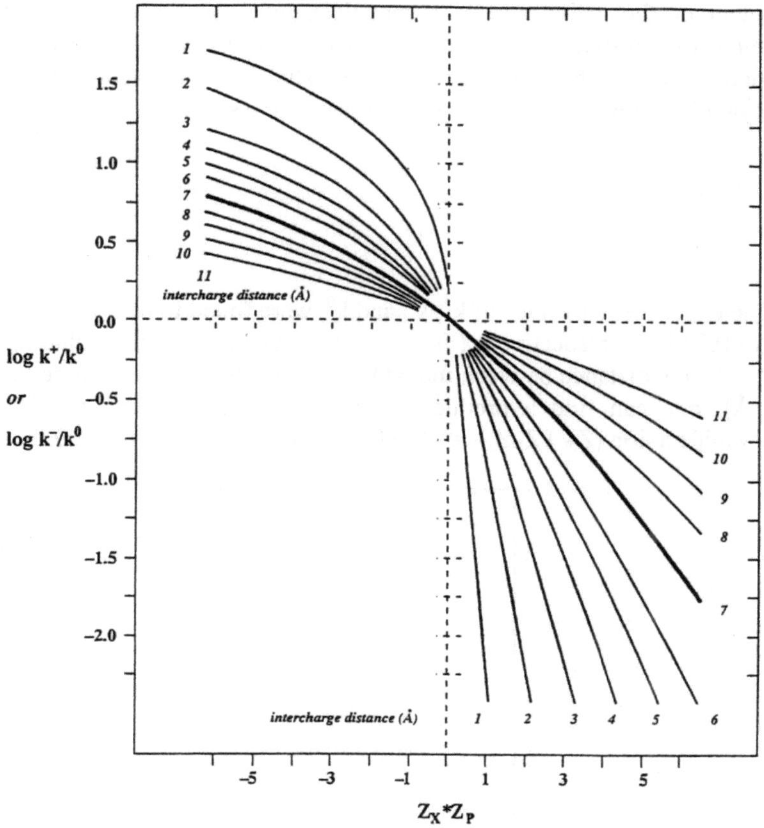

Figure 6. Variation in the relative rate-constants (k^+/k^0 or k^-/k^0) for nitroxide radicals in solution as a function of the Z_xZ_p product in the encounter between probe (p) and target (x) particles, calculated from the Debye equation (Eq. 26) for inter-charge distances r = 1-11 Å (Likhtenshtein, 1976).

This diagram allows one to calculate the Z_pZ_x product using the appropriate experimentally-measured $\log(k^+/k^0)$ or $\log(k^-/k^0)$ value and an r value estimated for the encounter complex. From the value of Z_xZ_p one can ascertain the Z_x charge in the vicinity of the region of the paramagnetic complex encountered by the charged nitroxide spin-probe since the Z_p charge of the nitroxide spin-probe is known.

This approach can be applied to two types of problems. It can be used to investigate the electrostatic fields in the vicinity of a spin label or spin probe using a second paramagnetic species with a different charge, ferricyanide anion or diphenylchromium cation, for example. (Likhtenshtein, 1976, 1993; Likhtenshtein et al., 1972). The second approach involves monitoring the effect of a paramagnetic species, such as a complex of paramagnetic ion with a protein or the active site of a metalloenzyme, on the spin relaxation parameters of nitroxide spin probes of different charges freely diffusing in solution. It can be illustrated by the study of interaction between ferricyanide anion and nitroxides of different charges in water solution.

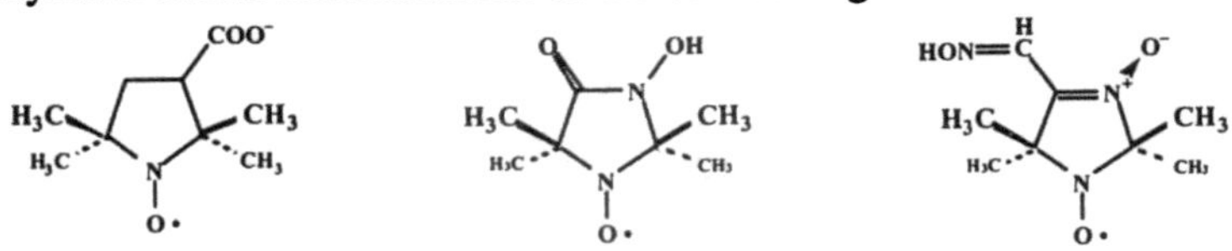

IX X XI

The experimental values of k^+, k^-, and k^0, were found to be $38x10^8$, $2x10^8$ and $6x10^8$ $M^{-1}s^{-1}$, respectively (Fig. 7). From the experimental value of log $k^+/k^- = 16$ the distance between the NO fragment and the ferricyanide ion (r = 6 Å), one can find a product of charges of radicals (Z_N = +1) and ferricyanide anion (Z_{FC}) $Z_NZ_{FC} = -2.9\pm0.1$ and therefore $Z_{FC} = -2.9\pm0.1$.

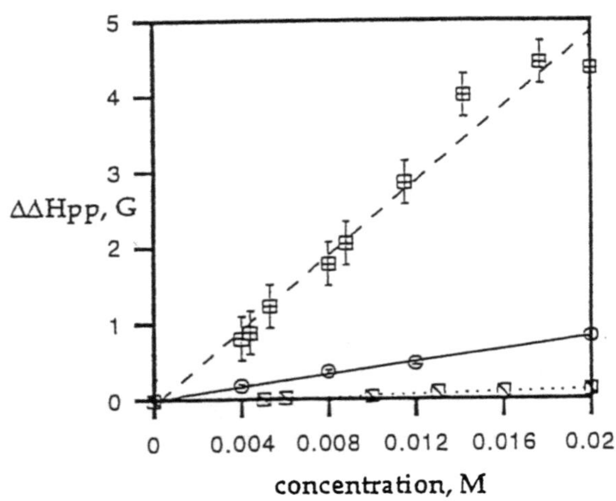

Figure 7. Dependence of peak-to-peak line-broadening, $\Delta\Delta H_{pp}$, of EPR signals for positively charged [VIII, dashed line, plus within square], negatively charged [IX, dotted line, diagonal within square] and neutral [X, solid line, symbol: circle] nitroxide spin-probes on the concentration of $Fe(CN)_6^{3-}$, where $\Delta\Delta H_{pp} = (\Delta H_{pp})_X - (\Delta H_{pp})_0$, and $(\Delta H_{pp})_X$ and $(\Delta H_{pp})_0$ are the linewidths in the presence and absence of $Fe(CN)_6^{3-}$; phosphate buffer, pH 6.9, ionic strength 0.09 M, 20 °C (Likhtenshtein et al., 1997).

A similar approach was employed in Hecht et al. (1995) for determination of the electrostatic potential near the surface of calf thymus DNA. Spin-spin interaction between an ^{14}N-nitroxide derivative of 9-

aminoacridine attached to DNA and free ^{15}N-labeled nitroxides of different charges was monitored by electron-electron double resonance (ELDOR). The electrostatic potential near the surface of DNA were calculated using a nonlinear Poisson-Boltzman equation. The calculated results were found to be in good agreement with the experimental potentials.

4. EXPERIMENTAL DATA ON BIOLOGICAL SYSTEMS

4.1 Photosynthetic reaction centers

The primary photophysical and photochemical processes of photosynthesis take place within small membrane-bound complexes of pigments and a protein of molecular mass about 102,000, which have been given the name "reaction center" (RC) (Michel and Deisenhofer, 1986; Feher, 1992). One mole of RC contains two moles of the bacteriochlorophyll dimer $(Bchl)_2$, two moles of the bacteriochlorophyll monomer (Bchl), two moles of bacteriopheophytin (Bph), two moles of ubiquinone (Q) and one mole of Fe. The RC converts the energy of light to the chemical energy of photoseparated charges, namely, the primary electron donor cation radical $(Bchl)_2^+$ and the primary electron acceptor FeQ_1^-. In the first photochemical step of the processes, during a time of the order of a picosecond after the $(Bchl)_2$ excitation, an electron is transferred to the (Bchl) molecule and then, in 5-10 ps, to Bph pigments and, in 200 ps, to the primary acceptor.

The mutual disposition of the pigments, distances between RC pigments, and the depth of immersion (R_0) of $(Bchl)_2^+$ in chromotophores from different photosynthetic bacteria and chloroplasts from green plants were determined by spectroscopic methods (Kulikov et al., 1979, 1981, 1983b; Likhtenshtein, 1988a, 1993; Likhtenshtein et al., 1982; Alakhverdiev et al., 1989; Cherepanova et al., 1990; Bogatyrenko et al., 1991).

In Kulikov et al. (1979) the distance between P^+ and FeQ_1^- (r_{PQ}) in chromatophores and reaction centers from *R. rubrum* was determined from the maximum value of $\Delta(1/T_{1s})$ for P^+ at 25 K. The measurements were carried out at 10-80 K in a mixture of buffer and glycerol (1:1 v/v) to make samples that were transparent to light. P^+ was produced by oxidation with ferricyanide or by freezing under illumination. The distance r_{PQ} was calculated with Eq. (3). It was found that $\Delta(1/T_{1s})_{max} = (1.0\text{- } 1.3)\times10^3\ s^{-1}$ for reaction centers and $(1.1\text{-}1.7)\times10^3\ s^{-1}$ for chromatophores, $\mu_f = 1.73$ Bohr magnetons (for low-spin FeQ^-), $\Delta\omega=6\times10^{10}\ s^{-1}$ (calculated from the g-factors of P^+ (2.00) and FeQ^- (1.82)). The distance is 30-32 Å for chromatophores and 31-32 Å for isolated reaction centers are in reasonable agreement with

r_{PQ1} obtained by X-ray analysis (Michel and Deisenhofer, 1986). From the x-ray data the distance between the center of the quinone ring and the point between two Mg atoms in $(BChl)_2$ is 28 Å.

The depth of immersion of P^+ into reaction centers of chromatophores from R. rubrum (R_P) was determined by the CW saturation method described in Section 2.1.1 (Kulikov et al. 1981, Cherepanova et.al., 1990). The contribution of $Fe(CN)_6^{3-}$ randomly distributed in the vitrified glycerol-water solution to the P^+ cation spin-lattice relaxation rate (W) was calculated with Eq. (10) using $T_{1s}^0 = 8x10^{-4}$ s. The W values depended linearly on the anion concentration with a slope tg $\psi = 5x10^5\ M^{-1}s^{-1}$. Eq. (11) with $\alpha = 1.3x10^{-17}$ $cm^3s^{-1/2}$ for $Fe(CN)_6^{3-}$ (Table 2) gives R_P = 8-10 Å. Independently, using paramagnetic $CoCl_2$ and antagonist $AlCl_3$ tgψ = $8x10^4\ M^{-1}s^{-1}$ and using α= $1.0x10^{-17}\ cm^3s^{-1/2}$ in Eq. (11) gives R_P = 8 Å. The R_P values are in reasonable agreement with the x-ray data (Michel and Deisenhofer, 1986). The distance of closest approach between P^+ and the hydrophobic probe $Co(AA)_2$, located in the chromatophores membranes at a depth of about 6 Å, was found to be 10 Å.

The effect of water-soluble $Fe(CN)_6^{3-}$ and lipid-soluble $Co(AA)_2$ on the spin-lattice relaxation rate of the primary electron donor cation radical $P890^+$ in reaction centers of chromatophors from *Chromatium minitissimum* was investigated. From the experimental parameters for this radical: $H_1^* = 2x10^{-2}$ G, q = 13, ΔH_p = 1.05 G, $T_1^0 = 8x10^{-5}$ s was calculated as described in Section 2.1.1. With the experimental value tg$\psi = 6x10^3\ M^{-1}s^{-1}$ for $Co(AA)_2$ and A = 2/9 (assuming that the radical is hidden under an infinite plane Section 2.2.1), Eq. (11) gives R_0 = 30 Å. This value is larger than that for RC from *R. rubrum*. Such a "discrepancy" appears because in the *C. minutissimum* chromatophores the RC is bound with cytochrome c_3 which serves as an electron donor for $(Bchl)_2^+$, whereas in the *R. rubrum* samples cytochrome c_3 was omitted. The R_0 value for $(Bchl)_2^+$ with the hydrophobic probe $Co(AA)_2$ was found to be about 10 Å for both *R. rubrum* and *C. minutissimum* reaction centers, which implies that the RC globule is "open" for interaction with the probe.

Eq. (5) with $\beta_{SE} = 1.3$ Å^{-1} allows determination of the distance between paramagnetic centers if the exchange integral J_{SE} is known or to estimate J_{SE} if distance is known. In bacterial reaction centers at temperatures < 10 K, a 60 G splitting of the ESR signal for an anion radical of bacteriopheophytin (Bph^-) is due to exchange interaction with Q-Fe^- (Tiede et al., 1976). From this splitting one can calculate the exchange integral $J = 5.4x10^8\ s^{-1}$. Based on the effect of magnetic field on the kinetics of electron transfer between Bph^- and the anion radical of the primary acceptor (the iron-quinone complex Q-Fe^-) it was concluded (Werner et al., 1978) that $J_{SE} = 1.5x10^8\ s^{-1}$ between the two centers. According to Eq. (5), the values of J correspond to r = 7.5-7.9 Å between BPh and $Q\text{-}Fe^{-}$ (Likhtenshtein et al., 1981, 1982). By measurement of the magnetic field effect on fluorescence of reaction centers from *R. spheroidas* a distance between $(Bchl)_2^+$ and BPh^- of about 7.0 Å was calculated from value $J_{SE} = 1.2 \times 10^9\ s^{-1}$ (Alfilov et al., 1980).

The depth of immersion of the primary donor $(Bchl)_2^+$ and the distances between other RC pigments determined by EPR spectrocopic methods are in good agreement with the X-ray model of RC reported later (Fig. 8) (Likhtenshtein, 1988b).

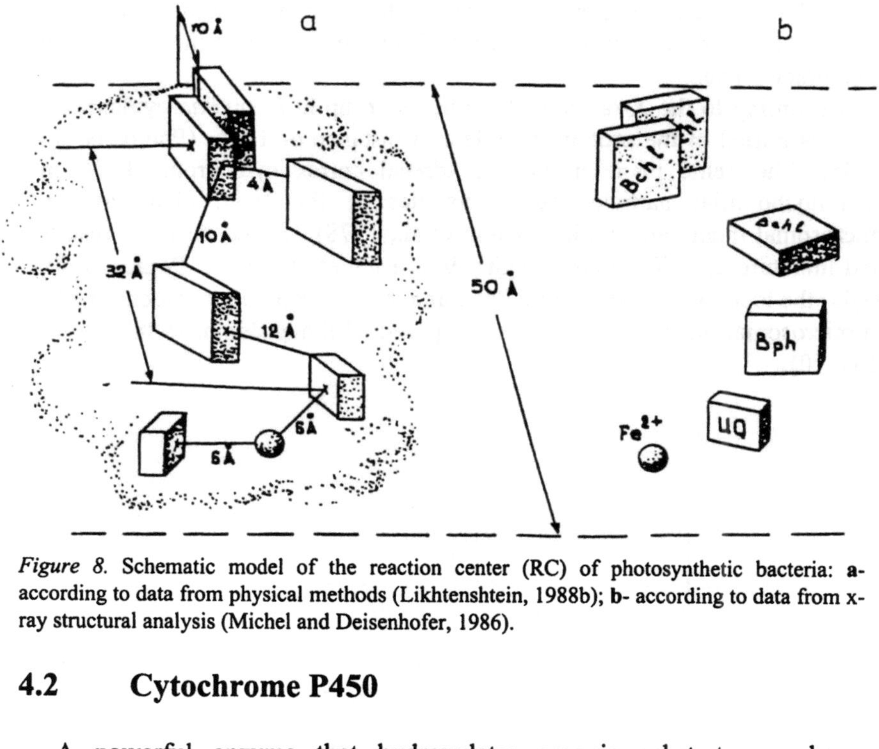

Figure 8. Schematic model of the reaction center (RC) of photosynthetic bacteria: **a**- according to data from physical methods (Likhtenshtein, 1988b); **b**- according to data from x-ray structural analysis (Michel and Deisenhofer, 1986).

4.2 Cytochrome P450

A powerful enzyme that hydroxylates organic substrates, such as hormones, hydrocarbons, organic acids, alcohols, amines is operative in animal and plant tissues, yeasts and bacteria. The hydroxylation is effected by an enzyme-hemoprotein containing protoheme IX. The reduction by the enzyme involves flavin reductases and electron carriers, such as adrenodoxin, rubredoxin and cytochrome b_5. Cytochrome P450 is so named because its reduced complex with carbon monoxide shows maximum absorption at 450 nm. The various forms of cytochrome P450 from liver microsomes and from *Pseudomonas putida* have a molecular mass about 49 kDa and one heme group per macromolecule.

The localization of the heme group in microsomes from rabbit liver was studied by measuring ESR saturation curves of the low-spin heme at 30 K in water-glycerol solution (1:1/v:v) (Kulikov et al., 1989). The central peak of the ESR spectrum (g = 2.25) was easily saturated. The effect of $Fe(CN)_6^{3-}$ ions or $Co(AA)_2$ located in the aqueous and lipid phases, respectively, on the H_1^* of the saturation curves was investigated (Fig. 9). Using Eq. (11) with $tg\psi = 6.1x10^4\ M^{-1}s^{-1}$ derived from the dependence of $[(H_1^*/H_1^{*0})^2 - 1]$ on the concentration of $Fe(CN)_6^{3-}$ and $\alpha = 1.3x10^{-17}\ cm^3s^{-1/2}$, A=2/9, and $T_{1s}^0 = 2.3x10^{-4}$ s for the heme group it was shown that the cytochrome P450 heme group is situated at a distance of 14 Å from the surface of the microsome membrane plane.

According to the X-ray model the heme group in *P. pudida* cytochrome P450 is buried in the protein globule at a distance of 10 Å. (Poulos et al., 1985). The heme plane in bovine adrenal cortex cytochrome P450 in submitochondrial particle multilayers lies in the same plane as the microsomal membrane surface (Blum et al., 1978). Based on these results and literature data (Case and Leigh, 1976; Rich et al., 1975; Poulos et al., 1985) the location of cytochrome P450 and the electron donor cytochrome b_5 in microsomal membranes has been proposed (Likhtenshtein, 1988b, 1993) (Fig. 10).

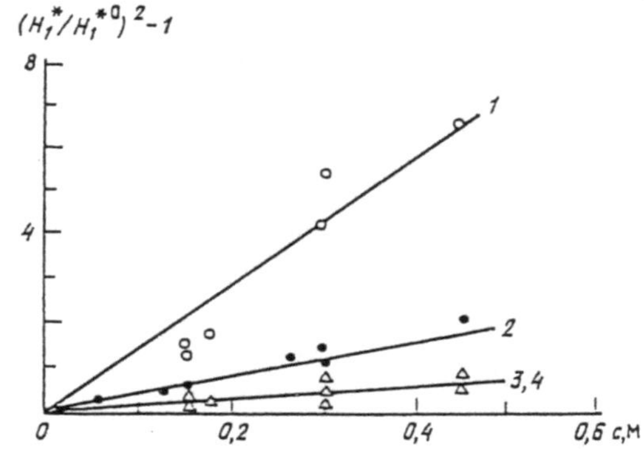

Figure 9. Dependence of $[(H_1^*/H_1^{*0})_2 - 1]$ on the concentration of $Fe(CN)_6^{3-}$ for cytochrome P-450 from rabbit liver microsomes in solid solution at 30K. 1- microsomes in glycerol-water (1:4/v:v); 2- isolated protein in a detergent in glycerol-water (1:4/v:v); 3 - microsomes in glycerol-water (1:1/v:v); 4 - in a detergent in glycerol-water (1:1/v:v). The second harmonic ESR signal was recorded with a modulation amplitude H_m = 18 G and modulation frequency = 20 kHz (Kulikov et al., 1989).

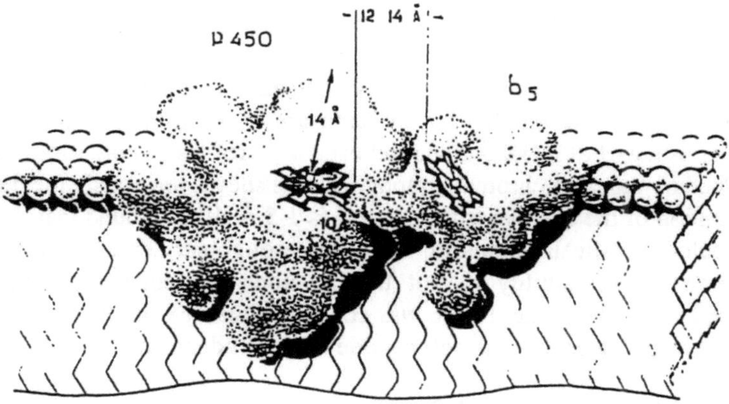

Figure 10. Schematic representation of localization of cytochromes P450 and b_5 in microsomal membranes during long distance electron transfer from b_5 to P450 (reproduced from Likhtenshtein, 1993).

4.3 Flavin-dependent alcohol oxidase

Flavin-dependent alcohol oxidase (AOX) from yeast catalyzes the conversion of methanol into formaldehyde and H_2O_2 and is a key enzyme in methanol metabolism. In an active state the enzyme is an oligomer of about 670 kDa, consisting of eight identical subunits and containing eight molecules of coenzyme, flavin adenine dinucleotide (FAD) prosthetic groups. Two or three FAD molecules are catalytically active in the oxidized state and approximately the same number of FAD molecules is distributed between the anionic flavin radical (AFR) and inactive oxidized flavin.

The spatial localization of the AFR in the AOX quaternary structure was investigated by two independent methods, CW ESR and tritium planigraphy (labeling by irradiation with hot tritium atoms followed by measuring the radioactivity of the surface moieties) (Averbakh et al., 1995).

The ESR experiments were performed in water-glycerol solution (1:1/ v:v) at 77 K. The effect of ferricyanide ions randomly distributed in the vitrified glassy medium on the shift of the AFR saturation curves was measured. For calculation of R_0 between the flavin radical and the paramagnetics, Eqs. (9) and (10) were used. $T_{1s}{}^0 = 2.8\text{x}10^{-4}$ s was determined by the algorithm described in Section 3.2.1 with $\Delta H_p = 1.38$ G, $q = H_1{}^{**}/H_1{}^*$ = 50. Values of $[(H_1{}^*/H_1{}^{*0})^2\text{-}1]$ obtained from the saturation curves were plotted versus the ferricyanide ion concentration, C_{FC}. Using the slope $W/C_{FC} = 1.8\text{x}10^5\ M^{-1}s^{-1}$, the value of $\mu^2T_{1f} = 1.0\text{x}10^{-50}$ s erg G^{-1} for ferricyanide (Likhtenshtein et al., 1993) and taking into consideration the value of R = 60 Å for the radius of the AOX octamer (A = 2/9) the immersion depth of paramagnetic AFR was evaluated as $R_0 = 4 \pm 2$ Å. It is suggested that FAD molecules are bound to the surface of the octamer, rather than to the subunit interfaces.

This conclusion was independently confirmed by the tritium planigraphy method. The FAD was labeled by exposure to hot tritium atoms in a free state or as a part of AOX and the radioactivity profiles of both samples were obtained by repeated chromatography. As was shown previously the depth of penetration of the reactive tritium atoms is 3-5 Å. It was found that the AFR molecules specifically bound to the enzyme oligomer demonstrate a 7-fold decrease in accessibility to hot tritium atoms, as compared with FAD labelled in the free state. It also was concluded that the riboflavin portion of FAD is buried deeper in the polypeptide then the adenosine.

4.4 Distribution of electrostatic potential around molecules of biological importance

4.4.1 Amino acids

The NMR spin probe method (Section 3.1) was used to study electrostatic effects in aminoacids. The effect of nitroxide spin-probe concentration on $1/T_{1p}$ for protons in basic and acidic amino acids was examined. The k^i values (i = 0, -, +) in Table 4 demonstrate that the slopes of $d(1/T_1)/d([R\bullet]) = k^i$, for specific protons in the charged molecules are sensitive to the charge of the nitroxide spin-probe (**VI-VIII**) relative to that in the charged target molecule.

The k^+/k^0 and k^-/k^0 ratios (Table 4) are markedly different for protons of positively-charged histidine and negatively-charged aspartate. The values of the local electrostatic potential $U(R_0)_{exp}$ calculated from the experimental data by Eq. (22)-(24) was found to be –0.30 and –0.31 kcal/mole for the H(α) and H(β) protons of aspartate, respectively.

Differences between k^0 and k^+ values for the protonated lysine side-chain protons are not significant since they fall within the error of their determinations. The average k^0/k^+ ratio of 1.46 for these nuclei indicates that the side-chain protons have a positively-charged environment. This average value is smaller than that measured for protons in ethylammonium cation and might be due to partial "quenching" by the neighboring negatively-charged carboxylate anion that can interact with the positive protonated amino groups. Similar to the case of propionate anion, the average 2.66, 2.83, and 2.46 k^+/k^- ratios calculated for the $H(\beta_I)$, $H(\beta_{II})$ and H(α) protons, respectively, in aspartate clearly show residence in negatively charged surroundings. Not surprisingly, the ratios for the zwitterion glycine H(α) protons were found to be essentially the same for nitroxides with different charges (Table 4).

Table 4. Rate constants of spin-lattice relaxation for protons of small molecules in solutions of nitroxides VI – VIII[a,b]

Molecule	Proton	D_H[c]	k^0 ($M^{-1}s^{-1}$)	k^- ($M^{-1}s^{-1}$)	k^+ ($M^{-1}s^{-1}$)
Lysine	H(α)	3.65	153±10	ND	105±1
	H(β)	1.80	131±15	ND	104±17
	H(γ)	1.34	162±19	ND	98±3
	H(δ)	1.61	135±15	ND	98±1
	H(ε)	2.91	147±18	ND	91±2
Aspartic acid	H(α)	3.80	148±8	94±15	ND
	H(β_I)	2.61	147±17	109±16	ND
	H(β)	2.71	187±11	97±10	ND
Glycine	H(α)	3.49	128±10	122±4	112±10
Histidine	H(α)	4.0	162±22	214±8	108±10
	H(β_I,β_{II})	3.1	161±36	208±6	108±14
	H(2)	8.5	180±8	272±6	99±12
	H(4)	7.5	183±5	258±1	96±11

[a]Data were analyzed with Eq. (20). pD = -log (concentration of D^+) = 4.5 for solutions of amino-TEMPO. [c]D_H is the shift downfield from tetramethylsilane in ppm.

4.4.2 Hemeproteins

Electron-carrier horse heat cytochrome c and dioxygen-carrier sperm-whale myoglobin served as models for determination of local electrostatic charges in the vicinity of paramagnetic active sites of metalloenzymes and metalloproteins (Likhtenshtein et al., 1997). Calculations of local charge Z_X in the vicinity of a paramagnetic particle can be performed using the diagram based on Eq. (26) for encounters between two charged particles in solution (Fig. 6). Neutral TEMPOL, positively-charged nitroxide **VIII** or negatively charged nitroxide **VII** were used to probe the local charge in the proteins' paramagnetic heme region. The experimental dependence of line-broadening, $\Delta\Delta H_{pp} = (\Delta H_{pp})_X - (\Delta H_{pp})_0$, for spin-probes with different charges **(VIII-XI)** on the concentration of $Fe(CN)_6^{3-}$ or heme protein (Fe^{3+}cyt c or Fe^{3+}Mb) are illustrated in Fig. 11. The values of k^i and the resulting charges, Z_H calculated with Eq. (26) are presented in Table 5. It was shown that cytochrome c causes a larger line-broadening ($\Delta\Delta H_{pp}$) for the negatively-charged nitroxide spin-probe (**IX**) compared to that observed for neutral **X**. Myoglobin results in a larger line-broadening for the neutral nitroxide spin-probe (**X**) than for the negatively charged species (**IX**). The spin-exchange rate constants in Table 5 show that the effect of the heme proteins on k^i is considerably weaker than that exerted by free hemin or by ferricyanide in solution.

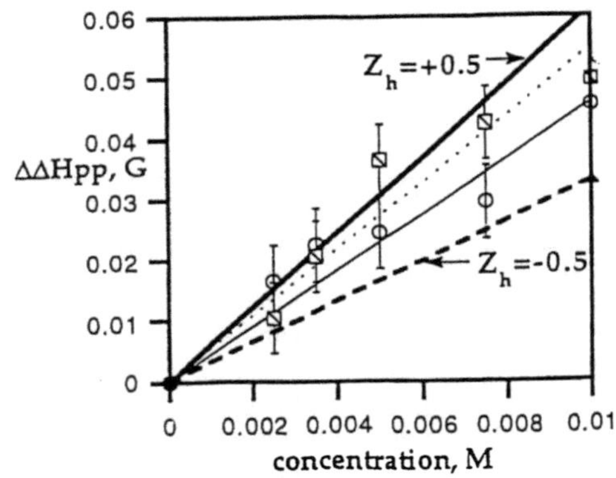

Figure 11. Dependence of line-broadening $\Delta\Delta H_{pp}$, of negatively charged [**IX**, dotted line, diagonal within square] and neutral [**X**, solid line, circle] nitroxide spin-probes on the concentration of cytochrome c; a) phosphate buffer, ionic strength 0.09 M, pH 6.9, 20 °C; the solid and dashed boldfaced lines are theoretical curves for Z_H = +0.5 or −0.5, respectively; b) distilled water, pH 10.7, 20 °C. (Likhtenshtein et al., 1997)

Table 5. Apparent spin-exchange rate-constants (k+, k-, and k0) of the positively charged (VIII), negatively charged (IX) and neutral (X) nitroxide spin-probes with hemin, ferricyanide anion, and heme proteins in aqueous solutions and apparent local charges (Z_H) in the vicinity of the heme groups.[a]

Paramagnetic species	pH	$k^+ \times 10^{-8}$ ($M^{-1}s^{-1}$)	$k^- \times 10^{-8}$ ($M^{-1}s^{-1}$)	$K^0 \times 10^{-8}$ ($M^{-1}s^{-1}$)	Z_H (±0.1)
Hemin[b]		c	c	24[d]	
$K_3Fe(CN)_6$[e]	6.9	38	2	6[f]	c
Cytochrome c (Fe^{3+})[g]	10.7	c	0.97±0.04	0.72±0.07[f]	+0.5
Cytochrome c (Fe^{3+})[e]	6.9	c	0.84±0.11	0.71±0.12[f]	+0.3
Cytochrome c (Fe^{2+})[e]	6.9			~0	
Myoglobin[g]	7.0	c	0.56±0.07	0.77±0.14[h]	-0.5

[a]0 °C. [b]Pyridine solvent (Grebenshchikov et al., 1972). [c]Not determined. [d]4-hydroxy-2,2,6,6-tetramethyl-1-piperidinyloxy spin-probe used, data taken from (Likhtenshtein, 1976). [e]Phosphate buffer, ionic strength 0.09 M. [f]Neutral probe **XI** used. [g]Distilled water. [h]Neutral probe **X** used.

According to the experimental rate constants of spin exchange for encounters between heme groups and nitroxides presented in Table 5, the accessibility of the heme group of cytochrome c to encounters with neutral spin-probe **X** is ~31-33 times lower than that observed for free hemin. At present, it is difficult to separate the effect of heme group immersion into the protein globular structure from that of association in the relatively concentrated solutions (2-10 mM) utilized in these studies.

The region of the cytochrome c active center is surrounded by a cluster of positively charged lysine residues. The existence of such a structural moiety is the basis for the specific affinity of this electron carrier to cytochrome c oxidase. However, in spite of this, our experiment did not locate significant local electrostatic charges on the protein interface in the vicinity of the heme group at pH 10.7 (isoelectric point) or pH 7 with an ionic strength of 0.09 M (Table 5). Therefore, while the charged lysine ε-amino groups of cytochrome c contribute significantly to the direct electrostatic interaction with the surface of cytochrome c oxidase, they apparently do not induce a markedly positive charge on the hemin. The existence of a large positive charge in this region could prevent electron transfer from Fe^{2+} cyt c.

The effect of dipole-dipole interaction between the Fe^{3+} heme group of myoglobin and water protons was used to study heme hydration and displacement in the pre-denaturational conformational transition of the molecule (Derzhansci et al., 1970).

While the differently charged spin-probes have been used so far on only a rather limited number of examples, the results show that these new methodologies show great promise for the investigation of local electrostatic fields in a wide range of biologically significant molecules.

5. CONCLUSIONS

Theoretical consideration of dynamic and static spin-spin interactions between paramagnetics and existing experimental data have shown that ESR spectroscopy is an effective tool for determination of the depth of immersion and location of paramagnetic centers of biological importance. The efficiency and reliability of the methods depend strongly on the characteristics of the object of interest and on the choice of an adequate method. In so doing, the advantages and limitations of a given method should be taking in consideration. The methods discussed in this chapter are based on CW measurements of electron spin relaxation rates ($1/T_1$, $1/T_2$). Pulse methods give accurate values of $1/T_1$ and $1/T_2$ if the values are in an accessible range. Nevertheless, under certain conditions, reasonable results can be obtained by CW ESR methods that are more familiar to chemists and biochemists then pulse techniques. For fast relaxing centers, when $1/T_1 > 10^8$ s^{-1}, only indirect methods of obtaining relaxation times can be employed.

In solutions of low viscosity electron spin-spin dipolar interactions between paramagnetics are averaged by fast rotation of molecules and the short lifetime of the encounter complex and therefore are very weak. In such a condition exchange interactions prevail over dipole-dipole ones. Taking into consideration the limitations in measurement of the rate constant of exchange interaction in solution we conclude that the maximum depth of immersion of a paramagnetic center in "nonconductive" media that can be determined by the dynamic spin exchange method is $R_0 \leq 9$ Å.

In contrast, in vitrified solids and highly-viscous solution the long-distance dipolar interactions more effective relative to the exchange ones, at least at r > 10 Å. Thus, parameters for saturation curves of nitroxide radicals, depending on $1/T_1$, are sensitive to the presence of paramagnetic Mn^{2+} up to an average distance between ions r < 90 Å (Kulikov and Likhtenshtein, 1974). During an encounter of a molecule bearing an aliphatic proton with a paramagnetic, say a nitroxide radical, in solution the main influence on the proton relaxation rates ($1/T_{1p}$, $1/T_{2p}$) is dipole-dipole interaction (Hwang and Fried, 1975). This interaction with paramagnetics has been used to study of the proton location. In these NMR experiments paramagnetics with relatively long spin relaxation times, say nitroxides or Cu^{2+} complexes, are more effective than other metal complexes (Fe^{3+}, Co^{2+}, Ni^{2+}, etc.) with short relaxation times (Berdnikov et al., 1980).

The distribution of local electrostatic potential around molecules of biological importance can be investigated in solution by measurement of the dipole-dipole interactions between nitroxide spin probes with different charges and protons of the molecule under study. Investigation of the effect of paramagnetic centers on spin-relaxation rates of nitroxide spin probes with different charges allows determination of the local electrostatic charge in the vicinity of active sites of metalloenzymes and metalloproteins.

A theoretical basis for determination of depth of immersion of protons and other nuclei bearing magnetic moments by the NMR–spin probe method has not yet been supported by experiments. The same conclusion may be made concerning determination of immersion depth of paramagnetic center, say metalloenzymes active sites, by measurement of the exchange interaction between the sites and spin probes freely diffusing in solution.

Most existing results on determination of depth of immersion of paramagnetic centers in biological matrices were obtained by the use of analysis of saturation curves in solids (CW ESR) and ESR line shape in solution. The powerful experimental potential of direct pulse methods has not yet realized.

6. ACKNOWLEDGMENT

The author is extremely grateful to Dr. Alexander Kulikov and to Professors Sandra and Gareth Eaton for valuable critical comments and fruitful discussions.

7. REFERENCES

Abragam, A. and Bleaney, B. (1970) *Electron Paramagnetic Resonance of Transition Ions,* Clarendon Press, Oxford.

Alakhverdiev, C. I., Kulikov, A.V., Klimov, V. V., Bogatyrenko, V. R., and Likhtenstein G. I. (1989) Determination of the depth of immersion of P680, pheophitin and secondary donor in photosystem 2 in pea subchloroplasts. *Biofizika,* **34**, 434-439 (In Russian).

Alexandrov, I. V. (1975) *Theory of Magnetic Relaxation*, Nauka, Moscow.

Alfilov, E. I., Vosnyak, V. M., and Kazanzev, A. P. (1980) Effect of magnetic field on fluorescence of *Rhodopseudomonas spheroidas*. *Biofizika* **25**, 498-505. (In Russian.)

Anni, H., Vanderkooi, J. M., Sharp, K. A., Yonetani,T., Hopkins, S. C., Herenyi, L. and Fidy, J. (1994) Electric field and conformational effects of cytochrome c and solvent on cytochrome c peroxidase studied by high resolution fluorescence spectroscopy. *Biochemistry*. **33**, 3475-3486.

Aquist, J., Luecke, H., Quiocho, F. A. and Warshel, A. (1991) Dipoles localized at helix termini of proteins stabilize charges. *Proc. Natl. Acad. Sci. USA* **88**, 2026-2030.

Averbakh, A.Z., Pekel, N.D., Seredenko V.I., Kulikov, A.V. and Gvozdev, R.I., (1995) Flavin-dependent alcohol oxidase from the yeast *Pichiapinus*. *Biochem. J.* **310**, 601-604.

Bashford, D., and Karplus, M. (1990) pK_as of ionizable groups in proteins: atomic detail from a continuum electrostatic model. *Biochemistry* **29**, 10219-10225.

Berdnikov B. M., Doktorov, A. B. and Makarshin, L. L. (1980) Dipole-dipole broadening of free radicals ESR spectra in the presence of paramagnetic ions. *Theor. Eksp. Chem.* (Kiev) **16**, 765-771. (In Russian)

Blum, H., Leigh, J. S., Salerno, J. C. and Ohnishi, T (1978) The orientation of bovin adrenal cortex cytochrome P-450 in submitochondrial particle multilayers. *Arch. Biochem. Biophys.* **187**, 153-157.

Bogatyrenko, V. R., Sabo, Ya. , Chamorovskii, S K., Zakharova, N.I., Kononenko, A. A. and Kulikov A. V. (1991) Study of localization of bacteriochlorophyl dimer and cytochrome c in reaction centers from *Chromatium minutissium* by ESR. *Biofizika* **36**, 289-290. (In Russian)

Borah, B., and Bryant, R. G. (1981) Nuclear magnetic resonance relaxation dispersion in an aqueous nitroxide system. *J. Chem. Phys.* **75**, 3297-3300.

Bowman, M. K. and Norris, J. R.(1982) Cross relaxation of free radicals in partially ordered solids. *J. Phys. Chem.* **86**, 3385-3390.

Case, G. D. and Leigh, J. S., Jr. (1976) Intramitochondrial position of cytochrome haem groups determined by dipolar interaction with paramagnetic cations. *Biochem. J.* **160**, 769-783,

Cherepanova, E.S., Kulikov, A.V. and Likhtenstein, G.I. (1990) Localization of paramagnetic centers relative to aqueous and lipid phases by ESR. *Biol. Membr.* (Moscow) **77**, 51-56. (In Russian).

Debye, P. (1942) Reaction rates in ionic solutions. *Trans.Electrochem. Soc.* **82**, 265-272.

Depmeier, B. J., Driessen, A. J., Hehre, W. J., Johnson, J., Peng, A. C., Lou, L. and Yu, J. (1996) *MacSpartan version 1.0.2*. Wavefunction, Irvine, CA.

Derzhansci, A., Georgieva, A., Kotev, K. and Atanasov, B. P. (1970) Heam hydration and displacement in the pre-denaturational conformational transition of the myoglobin molecule. *Biochim. Biophys. Acta* **214**, 83-93, .

Druzhinin, S. Yu., Fogel, V. R., Syrtsova, L. A., Likhtenstein, G. I. and Kotelnikov, A. I. (1986) Study on the cofactor center localization in nitrogenase by triplet labeling method. *Biofisika* **31**,16-21. (In Russian).

Dwek, R. A. (1977) *NMR in Biology*, Academic Press, New York.

Eastman, M. P., Bruno, G. V. and Freed, J. H. (1970) Studies of Heisenberg spin exchange. II Effect of radical charge and size. *J. Chem. Phys.* **52**, 2511-2522.

Eaton, G. R and Eaton, S. S. (1989) Resolved electron-electron spin-spin splittings in ESR spectra. *Biol. Magn. Reson.* **8**, 339-397.

Feher, G. (1992) Three-dimensional structure of the reaction center by X-diffraction from single crystals, *Is. J. Chem.* **32**, 375-378.

Gilson, M. K. (1993) Multiple-site titration and molecular modeling: two rapid methods for computing energies and forces for ionizable groups in proteins. *Proteins* **15**, 266-282.

Gray, H. B., and Ellis, W. R. Jr (1994) Electron Transfer. In: *Bioinorganic Chemistry.* (Bertini, I., Gray, G.B., Lippard S.J., and Valentine J.S. eds.) University Science Books, Mill Valley, California, 315-364.

Grebenshchikov, Yu. B., Ponomarev, G. V., Evstigneeva, R. P. and Likhtenshtein, G. I. (1972) Spin relaxation of a nitroxide radical upon interaction with porphyrin complexes of VO(II), Fe(III), NI(II), Cu(II) and hemoglobin. *Biofizika* **17**, 910-913. (In Russian).

Hecht, J. L., Honig, B., Shin, Y. and Hubbell, W. L. (1995) Electrostatic potential near the surface of DNA: comparing theory and experiment. *J. Phys. Chem.* **99**, 7782-7786.

Honig, B. and Nicholls, A. (1995) Classical electrostatics in biology and chemistry. *Science* **268**, 1144-1149.

Hyde, J. S. and Subczinski, W. K. (1989) Spin-label oximetry, *Biol. Magn. Reson.* **8**, 399-426.

Hyde, J. S., Swartz, H. M. and Antholine, W. E. (1976) The spin probe-spin label methods. In: *Spin Labeling. Theory and Application.*Vol. 2. (Berliner L. ed.) Academic Press, New York, pp. 72-113.

Hwang, L. P. and Freed, J. H.(1975) Dynamics effect of pair correlation function on spin-relaxation by translational diffusion in liquids. *J. Chem. Phys.* **63**, 4017- 4025.

Kokorin, A. I., Zamaraev, K. I., Grigoryan, G. L., Ivanov, V. P., and Rozantsev, E. G. (1972) Distance estimation between nitroxyl radicals. *Biofizika* **17**, 34-41.

Kotel'nikov, A. I., Likhtenshtein, G. I. and Gvozdev, R. I. (1975) The use of phenomenon of saturation of the ESR signals for study of relief of a macromolecule in the vicinity of paramagnetic center. *Studia Biophys.* **49**, 215-221.

Kotel'nikov, A.I., Fogel, V.R., Likhtenshtein, G.I., Postnikova, G.B., and Slyapnikova, E.A. (1981) The use of the exchange deactivation of triplet excited states for the investigation of the structure and electronic conductivity of proteins. *Mol. Biol.* (Moscow) **15**, 281-289. (In Russian)

Kulikov, A.V. (1976) Determination of distance between the nitroxide label and a paramagnetic center in spin-labeled proteins from the parameters of the saturation curve of the ESR spectrum of the label at 77K. *Mol. Biol.* (Moscow) **10**, 109-116. (In Russian)

Kulikov, A. V. and Likhtenstein, G. I. (1974) Application of saturation curves for evaluating distances in biological objects by the method of double spin-labels. *Biofisika* **19**, 420-424. (In Russian).

Kulikov, A. I. and Likhtenshtein, G. I. (1977) The use of spin-relaxation phenomena in the investigation of the structure of model and biological systems by method of spin labels. *Adv. Molecul. Relax. Proc.* **10**, 47-78.

Kulikov, A. I., Likhtenshtein, G. I., Rozantzev, E. G., Suskina, and Shapiro, A. V. (1972) Nitroxide bi- and polyradicals as standard models for distance estimation between the nitroxide moities. *Biofizika*, **17**, 42-49. (In Russian).

Kulikov A. V., Bogatyrenko, V. R., Melnikov, A. V., Syrtzova, L. A. and Likhtenshtein, G. I. (1979) Determination of distance between cation radical of bacteriochlorophyl dimer and anion of quinone in photosynthetic reaction center from *R. rubrum*. *Biofizika* **24**, 178-185. (In Russian).

Kulikov, A. V., Cherepanova, E. S, and Bogatyrenko V. R. (1981) Determination of the closest distance between a radical and a paramagnetic ion. *Theor. Exper. Chem.* **17,** 618-626.

Kulikov, A. I., Yudanova, E. I. and Likhtenshtein, G. I. (1983a) Investigation of the spin-exchange of nitroxide radicals using the continuous ESR spectrum saturation technique. *J. Phys. Chem.* (Moscow) **56,** 2982-2987. (In Russian)

Kulikov, A. V., Bogatyrenko, V. R., Likhtenstein, G. I., Allakhverdiev, S. I., Klimov, V. V., Shuvalov, V. A., Krasnovskii, A. A. (1983b) Magnetic interaction of Mn with anion radical of pheophytin and cation radical of chlorophyll in reaction centers of photosystem 2, *Biofizika* **28,** 357-363. (In Russian)

Kulikov, A.V., Cherepanova, E.S., Bogatyrenko. V.R., Nasonova, T.A,. Fisher, V.R. and Yakubov, H.M. (1987) Determination of the depth of immersion of radicals into biological matrices by ESR. *Bull. Acad. Sci. USSR, Div. Biol. Sci.* N 5, 7762-7769.

Kulikov, A. V., Cherepanova, E. S., Likhtenshtein, G. I., Uvarov, V. Yu. and Archakov, A. I. (1989) ESR Study of localization of cytochrome P450 in microsomes relative to aqueous and lipid phases. *Biologich. Membrany* **6,** 1085-1094. (In Russian)

Levitch, V. G., Dogonadze, R. R., German, E., Kuznetsov, A. N., and Kharakats, Yu. I. (1970) Theory of homogeneous reaction involving proton transfer. *Electrochem. Acta* **15,** 353-368.

Likhtenshtein, G. I. (1968) Determination of the topography of proteins group using specific paramagnetic labels. *Mol. Biol.* (Moscow) **2,** 234-240.

Likhtenstein, G. I. (1976) *Spin Labeling Methods in Molecular Biology*, pp. 66-88, John Wiley & Son, New York.

Likhtenstein, G. I. (1988a) *Chemical Physics of Redox Metalloenzymes Catalysis*. pp. 45-60, Springer Verlag, Heidelberg.

Likhtenshtein, G.I. (1988b) Structure and molecular dynamics of metalloenzymes studied by physical label methods. *J. Molec. Catal.* **47,** 128-129.

Likhtenshtein, G.I. (1990) Nitroxide in the solution of some problems of chemical biophysics. *Pure Appl. Chem.* **62,** 281-288.

Likhtenstein, G. I. (1993) *Biophysical Labeling Methods in Molecular Biology*, pp. 46-79, Cambridge University Press, New York, Cambridge.

Likhtenstein, G. I. (1995) Role of orbital overlap and local dynamics in long-distance electron transfer in photosynthetic centres and model systems. *J. Photochem. Photobiol. A: Chemistry* **96,** 79-92.

Likhtenshtein, G. I., and Bobodzhanov, P. Kh. (1968) Investigation of the structure and local conformational changes of proteins and enzymes using double paramagnetic labels. *Biofizika* **13,** 757-764.

Likhtenstein, G. I., Grebentchikov, Yu. B., Bobodzhanov, P. Kh. and Kokhanov, Yu. V. (1970) Study on the proteins microstructure by method of spin-label paramagnetic probe. *Mol. Biol.* (Moscow) **4,** 782-789.

Likhtenstein, G. I., Grebentchikov, Yu. B., Rosantev, E. G. and Ivanov, V. P. (1972) Study on the electrostatic charges in proteins by method of paramagnetic probes. *Mol. Biol.* (Moscow) **6,** 498-507.

Likhtenshtein, G. I., Kotelnikov, A. I., and Kulikov, A. V. (1981) Structure of reaction centers of photosynthetic bacteria. *Dokl. AN SSSR* **257,** 733-736. (In Russian)

Likhtenshtein, G. I., Kulikov, A. V., Kotelnikov, A. I. and Bogatyrenko, V. R. (1982) Structure and action mechanism of reaction centers of photosynthetic bacteria. *Photobiochem. Photobiol.* **3,** 178-182.

Likhtenshtein, G. I., Kulikov, A. V., Kotelnikov, A. I., and Levchenko, L. A. (1986) Methods of physical labels - a combined approach to the study of microstructure and dynamics of biological systems. *J. Biochem. Biophys. Meth.* **12**, 1-28.

Likhtenshtein, G. I., Kulikov, A. V. and Kotelnikov, A. V. (1993) Relaxation process involving nitroxyl radical in molecular biology. In: *Bioactive Spin Labels* (Zhdanov, R. I., ed.) pp. 125-151, Springer-Verlag, Heidelberg.

Likhtenshtein, G. I., Vaisbuch, I., Adin, I., Shames, A., and Glaser, R. (1997) Distribution of electrostatic-field around biological molecules studied by methods of spin-probes and NMR. *Biophys. J.* **72**, A129.

Likhtenshtein, G. I., Adin I., Krasnoselsky ,A., Vaisbuch, I. Shames, A., and Glaser, R. (1999) NMR and ESR studies of electric field distribution around biologically important molecules. *Biophys. J.* **77**, 443-454.

Marcus, R. A. and Sutin, N. (1985) Electron transfer in chemistry and biology. *Biochim. Biophys. Acta* **811**, 625- 632.

Marsh, D. (1989) Experimental methods in spin-label analysis. *Biol. Magn. Reson.* **8**, 255-285.

McElroy, J. D., Mauzerall, D. C., and Feher, G. (1974) Characterization of primary donor of bacterial photosynthesis.2. Kinetic studies of the light-induced ESR signal at g = 2.0026 and the optical absorbance at cryogenic temperatures. *Biochim. Biophys. Acta* **333**, 261-272.

Medzhidov, A. A., Likhtenshtein, G. I. and Kirichenko, L. A. (1969) Metallocomplexes with the paramagnetic ligands. *Bull. Acad. Sci. USSR (Chemistry)* N 3, 698-902.

Michel, H. and Deisenhofer J. (1986) X-ray diffraction studies on a crystalline bacterial photosynthetic center. A progress report and conclusions on the structure of the photosystem II reaction center. *In Encyclopedia of Plant Physiology, New Series*, v.19, (Stachelin, L.A. and Arntzen, C.J (eds) pp. 371-381, Berlin, Springer-Verlag.

More J. K., More K. M., Eaton G. R., and Eaton, S. S. (1990) Metal-nitroxyl interaction. 55: Manganase (III)-nitroxyl electron-electron spin-spin interaction. *Pure Appl. Chem.* **62**, 241-246.

Niccolai, N., Valensin, G., Rossi, C. and Gibbons, W. A. (1982) The stereochemistry and dynamics of natural products and biopolymers from proton relaxation spectroscopy: spin-label delineation of inner and outer protons of gramicidin S including hydrogen bonds. *J. Am. Chem. Soc.* **104**, 1534-1537.

Nicollai, N., Rossi, C., Valensin, G. Mascagni, P. and Gibbons, W. A. (1984) An investigation of the mechanisms of nitroxide-induced proton relaxation enhancements in biopolymers. *J. Phys. Chem.* **88**, 5689-5692.

Parmon, V. N., Kokorin, A. I., and Zhidomirov, G. M. (1980) *Stable Biradicals*, Nauka, Moscow. (in Russian)

Perutz, M.F. (1989) Mechanisms of cooperativity and allosteric regulation in proteins. *Quart. Rev. Biophys.* **22**, 139- 236.

Plachy, W. and Kivelson, D. (1967) Spin exchange in solution of di-tertiary-butyl nitroxide. *J. Chem. Phys.* **47**, 3312-3318.

Poulos, T. L., Finzel, B. C., Gunzalus, I. C., Wagner, G. C., and Kraut, J. (1985) The 2.6 Å crystal structure of Pseudomonas putida cytochrome P-450. *J. Biol. Chem.* **200**, 16122-16130.

Rich, P., Tiede, D.M, Bonner, W. D., Jr. (1979) Studies on the molecular organization of cytochromes P-450 and b_5 in the microsomal membrane. *Biochim. Biophys. Acta* **546**, 307-315.

Safronov S. N., Mstislavskii V. I., Safronova U. I. and Muromtsev, V. I. (1969) Method of determination of relaxation times from saturation curves of the signals in the fast transmission conditions. *Savodskaya Laboratoria* **35**, 1463-1465. (In Russian)

Salikhov, K.M., Doctorov, A.B., Molin, Yu.N., and Zamaraev, K.I. (1971) Spin relaxation of radicals and complexes upon encounters in solution *J. Magn. Reson.* **5**, 189-196.

Salikhov, K. M., Semenov, A. G., and Tsvetkov, Yu. D. (1976) *Electron Spin Echo and Its Application*. Nauka, Novosibirsk p. 342

Sletten, E., Jackson, J. I., Burns, P. D. and La Mar, G. N. (1983) Effects of cross relaxation on the analysis of T_1 data in paramagnetic proteins. *J. Magn. Reson.* **52**, 492-496.

Solomon, J. and Bloembergen, N. (1956) Nuclear magnetic interaction in HF molecule. *J. Chem. Phys*, **25**, 261-266.

Sundin, A. (1991) *MacMimic version 2.1*. InStar Software, Lund, Sweden.

Syrtsova, L. A., Likhtenstein, G. I., Pisarkaya, T. N., Berdinskii, V. L., Lezina, V. P. and Stepanyants, A. U. (1974) Estimation of the distance between the ATPase and substrate-binding sites in nitrogenase by the NMR-^{1}H method. *Mol. Biol.* (Moscow) **8**, 656-662. (In Russian)

Taylor, J. C., Leigh, J. S. and Cohn, M. (1969) The effect of dipole-dipole interaction between nitroxide radical and a paramagnetic ion on the line shape of the ESR spectra of radical. *Proc. Natl. Acad. Sci. USA* **64**, 219-206.

Tiede, D. M., Prince, R. C., Reed, G. H., and Dutton, P. L. (1976) EPR properties of the electron carrier intermediate between the reaction center bacteriochlorophyls and the primary acceptor. *FEBS Lett.* **65**, 301-304.

Wien, R. W., Morriset, J. D. and McConnell, H. M. (1972) Spin-label induced nuclear relaxation. Distances between bound saccharides, histidine-15, and tryptophan-123 on lysozyme in solution. *Biochemistry* **11**, 3707-3716.

Zamaraev K. I., Molin Yu. N., Salikhov K. M. (1977) *Spin Exchange. Theory and Physico-chemical Application*, Nauka, Novosibirsk pp. 267-278. (In Russian).

Zamaraev K. I., Molin Yu. N., Salikhov K. M. (1981) *Spin Exchange. Theory and Physico-chemical Application*. Springer-Verlag.

Chapter 8

Determination of Distances Based on T_1 and T_m Effects

Sandra S. Eaton and Gareth R. Eaton
Department of Chemistry and Biochemistry, University of Denver, Denver, Colorado 80208

Abstract: The distance between a rapidly-relaxing paramagnetic center and a slowly-relaxing center can be determined from the effect of the rapidly-relaxing center on the spin-lattice relaxation and on the spin echo dephasing of the slowly-relaxing center. The principles underlying these measurements are discussed and examples are given for high-spin and low-spin Fe(III).

In this chapter we address the interpretation of the interaction between one rapidly-relaxing metal and a single slowly-relaxing center at a fixed distance to determine the distance between the two centers.

During the 1970's and 80's extensive work was performed in Russia on spin echo decays, including analysis of the effects of electron-electron spin-spin interaction on the shapes and time constants for echo decays. The emphasis of that work was on the effect of the spatial distribution of slowly-relaxing paramagnetic centers on the shape of the spin echo decay. Some examples were included for rapidly-relaxing spins such as Cu(II) and Fe(II) (Dzuba et al., 1979, 1980). Because the emphasis of these studies was on spatial distributions of radicals, the effects on the echo decays were averaged over particular distributions of interspin distances. In many experiments the concentrations were sufficiently high that the effects of instantaneous diffusion had to be included in the data analysis. These experiments are reviewed by Raitsimring and Salikhov (1985). Hyde and Rao (1978) derived formulae for the effect of a rapidly-relaxing metal on T_1 of a slowly-relaxing species, such as a nitroxyl radical, under the assumptions that metal $T_1 = T_2$, interactions are purely dipolar, and all radial distances are allowed. Methods of analyzing relaxation enhancement to determine depths of immersion are

Biological Magnetic Resonance, Volume 19: Distance Measurements in Biological Systems by EPR, edited by Berliner et al. Kluwer Academic/Plenum Publishers, New York, 2000

described in ch. 7 by Likhtenshtein. The use of relaxation enhancement by Dy^{3+} to determine immersion depths of paramagnetic centers in photosystems is described in ch. 12 by Lakshmi and Brudvig.

1. EFFECT OF RAPIDLY-RELAXING METAL ON T_1 FOR SLOWLY-RELAXING SPIN

When a slowly-relaxing spin is coupled via dipolar or exchange interaction to a faster-relaxing spin the relaxation of the faster-relaxing spin provides a relaxation pathway for the slower-relaxing spin (Poole and Farach, 1971; Kulikov and Likhtenshtein, 1977). The stronger the dipolar and/or exchange interaction, the more effectively the relaxation of the fast relaxing spin is communicated to the more slowly-relaxing spin. In the limit of strong interaction the relaxation time for the slowly-relaxing spin becomes equal to that for the rapidly-relaxing spin. In the ensuing discussion we consider only the case of relatively weak interaction where the effect of the rapidly-relaxing spin can be viewed as a perturbation of the relaxation of the slowly-relaxing spin. A detailed discussion can be found in Poole and Farach (1971).

The Hamiltonian for the dipolar interaction can be expressed as

$$H_{dd} = g_1 g_2 \beta^2 [\frac{\mathbf{S}_1 \cdot \mathbf{S}_2}{r^3} - 3\frac{(\mathbf{S}_1 \cdot \mathbf{r})(\mathbf{S}_2 \cdot \mathbf{r})}{r^5}] \qquad (1)$$

In eq. (1) we have taken the general form of the dipolar interaction from ch. 1 that was expressed in terms of magnetic moments, μ, and substituted expressions in terms of g values. In eq. (1) it is assumed that g values are isotropic. A corresponding equation could be written for anisotropic g values, but eq. (1) makes it easier to follow the general arguments. The vector products in eq. (1) can be expanded in terms of S_x, S_y, and S_z. Since we are interested in spin transitions it is then convenient to replace S_x and S_y by the raising and lowering operators S_+ and S_- using the relationships $S_+ = ½ (S_x+iS_y)$ and $S_- = (1/2i)(S_x-iS_y)$. The dipolar Hamiltonian then becomes

$$H_{dd} = (g_1 g_2 \beta^2/r^3)(A + B + C + D + E + F) \qquad (2)$$

where

$$A = (1 - 3\cos^2 \theta)S_{1z}S_{2z}$$

$$B = -\frac{1}{4}(1-3\cos^2\theta)(S_{1+}S_{2-} + S_{1-}S_{2+})$$

$$C = -\frac{3}{2}\sin\theta\cos\theta e^{-i\phi}(S_{1z}S_{2+} + S_{1+}S_{2z})$$

$$D = -\frac{3}{2}\sin\theta\cos\theta e^{i\phi}(S_{1z}S_{2-} + S_{1-}S_{2z})$$

$$E = -\frac{3}{4}\sin^2\theta e^{-2i\phi}S_{1+}S_{2+}$$

$$F = -\frac{3}{4}\sin^2\theta e^{2i\phi}S_{1-}S_{2-}$$

θ is the angle between the interspin vector and the external magnetic field, and ϕ is the angle between the projection of the interspin vector on the external xy plane and the x axis (Poole and Farach, 1971). An isotropic exchange interaction between spins with isotropic g values adds a contribution to the A and B matrix elements.

Bloembergen et al. (1948, 1949, 1959) considered the perturbation of a slowly-relaxing nuclear spin by the relaxation of a faster-relaxing spin in terms of the matrix elements given in eq. (2). Their results can be expressed in a convenient form (eq. (3)) as shown by Kulikov and Likhtenshtein (1977).

$$\frac{1}{T_{1s}} = \frac{1}{T_{1s}^{o}} + S(S+1)[\underbrace{\frac{b^2T_{2f}}{1+(\omega_f-\omega_s)^2T_{2f}^{\,2}}}_{\text{B term}} + \underbrace{\frac{c^2T_{1f}}{1+\omega_s^{\,2}T_{1f}^{\,2}}}_{\text{C term}} + \underbrace{\frac{e^2T_{2f}}{1+(\omega_f+\omega_s)^2T_{2f}^{\,2}}}_{\text{E term}}]$$

$$b^2 = \frac{8}{3}[-\frac{J}{2} - \frac{1}{4}g_sg_f\beta^2\frac{(1-3\cos^2\theta)}{\hbar r^3}]^2 \qquad (3)$$

$$c^2 = 3g_s^{\,2}g_f^{\,2}\beta^4\frac{\sin^2\theta\cos^2\theta}{\hbar^2r^6}$$

$$e^2 = \frac{3}{2}g_s^{\,2}g_f^{\,2}\beta^4\frac{\sin^4\theta}{\hbar^2r^6}$$

where "f" and "s" denote the fast- and slow-relaxing spins, respectively, T_{1s}^{o} is T_1 for the slowly-relaxing spin in the absence of spin-spin interaction, T_{1s} is T_1 for the slowly-relaxing spin perturbed by the fast-relaxing spin, S is the electron spin on the faster-relaxing center, ω_f and ω_s are the resonant frequencies for the fast- and slow-relaxing spins, respectively, r is the interspin distance, J is the electron-electron exchange interaction for the

Hamiltonian written as $-JS_1 \cdot S_2$, and θ is the angle between the interspin vector and the external magnetic field. This equation is valid in the limit where T_{1f} is short enough that the dipolar splitting is partially collapsed, which requires that $1/T_{1f}$ > dipolar splitting in s^{-1} which means that $1/T_{1f} > g_s g_f \beta^2/\hbar r^3$. The A matrix elements do not contribute to the relaxation because these elements only contribute along the diagonal of the spin-spin interaction matrix. The "C term" in eq. (3) includes the contributions from matrix elements C and D in eq. (2) and the "E term" in eq. (3) includes the contributions from matrix elements E and F in eq. (2).

To emphasize the role of isotropic spin-spin coupling (exchange) the exchange and dipolar contributions to the enhancement of $1/T_1$ are sometimes written as separate terms (Abragam, 1961). However, both exchange and dipolar interaction contribute to the same matrix elements involving $S_{1+}S_{2-}$ and $S_{1-}S_{2+}$ (i.e. B in eq. (2)). Since the dipolar contribution to the interaction changes sign with orientation and since J has a sign, the sum of the squares is not equivalent to the square of the sum and therefore the two parts should be combined as shown in the b^2 coefficient of eq. (3) (Rakowsky et al., 1995) rather than adding a separate term for exchange. Combination of the two contributions prior to squaring the matrix element is particularly important when the exchange interaction is of the same order of magnitude as the dipolar interaction.

The maximum contribution for each term in eq. (3) occurs when the denominator is approximately one. For the B term this occurs when T_{2f}^{-1} is equal to the difference between the Larmor frequencies of the two paramagnetic centers. For the C term this occurs when T_{1f}^{-1} is equal to the Larmor frequency for the slow-relaxing spin. For the E term this occurs when T_{2f}^{-1} is equal to the sum of the Larmor frequencies for the slowly- and rapidly-relaxing spins. In a rigid lattice, when T_{1f}^{-1} or T_{2f}^{-1} are fast enough to be of the order of the Larmor frequency, it is likely that T_{1f} drives T_{2f} and that $T_{1f} = T_{2f}$. Since the relaxation rate also occurs in the numerators of the terms in Eq. (3), the faster relaxation rate required to maximize the contribution from the E term typically will result in smaller maximum contributions from the E term than from the C term. The relative importance of the B and C terms depends upon $\omega_f-\omega_s$ as well as on the metal relaxation rates.

To apply eq. (3) to NMR samples in fluid solution the average value of the orientation-dependent coefficients are used and the equation is then known as the Solomon-Bloembergen equation (Bertini and Luchinat, 1996). Application to EPR spectra in rigid lattice requires decisions on how to treat the orientation dependence of the dipolar interaction and how to treat T_{1f} and T_{1s}.

1.1 Application of Bloembergen Equation Based on an Average Value of T_{1s}

In early applications of eq. (3) to the determination of the distance between a spin label and a paramagnetic metal ion at 77 K i) the average values of the angular-dependent terms were used, ii) T_{1f} was determined from the CW linewidth of the EPR signal for the rapidly-relaxing center, and iii) T_{1s} and T_{1s}^{o} were determined by continuous wave (CW) power saturation (Kulikov and Likhtenshtein, 1977). At 77 K T_{1s} and T_{1s}^{o} for a nitroxyl spin label are so long that the "slow passage" criteria that typically are assumed for CW spectra are not satisfied and analysis of the power saturation curves requires equations based on rapid passage (Kulikov and Likhtenshtein, 1977; ch. 7 in this volume). The approximation of a single value for $1/T_{1s}$ and eq. (3) subsequently were used by other groups to determine interspin distances (Ohnishi et al., 1982; Brudvig et al., 1984; Scholes et al., 1984; Goodman and Leigh, 1985; Makinen and Wells, 1987). These applications have been reviewed in Eaton and Eaton (1988).

1.2 Application of Bloembergen Equation Based on Analysis of Saturation Recovery Curves without Knowledge of T_{1f} and T_{2f}

The anisotropy of the dipolar interaction means that T_{1s} for the slowly-relaxing partner is orientation dependent as shown in eq. (3). Even if the two spins in a spin-coupled pair have a well-defined relative orientation, when those pairs are randomly oriented in a powder or glassy sample, there is a distribution of values of T_{1s}. Application of eq. (3) based on a single value of T_{1s} is therefore a rather drastic approximation. The relaxation behavior can be analyzed more accurately by examination of the full shape of the saturation recovery (SR) curve. Due to the anisotropy of the dipolar interaction there will always be some probability, in a randomly oriented sample, of negligible dipolar contribution to T_{1s} and therefore some spins with $T_{1s} = T_{1s}^{o}$. As the value of r decreases, the distribution of values of T_{1s} widens, and therefore the shape of the saturation recovery curve depends upon r. Brudvig and co-workers proposed fitting the saturation recovery curve to limiting forms of the B or C terms of eq. (3) (Hirsh et al, 1992; Hirsh and Brudvig, 1993; Koulougliotis et al, 1994, 1995). Saturation recovery data were fitted to the sum of a dipolar contribution and an isotropic exchange contribution. The dipolar coefficient for a system with unknown interspin distance was compared with the dipolar coefficient for a slowly-relaxing spin at a known distance from the same rapidly-relaxing spin. This is an example of the utility of measurements of distances based on the relative effects of the same rapidly-relaxing paramagnetic center. Examples of this procedure are described in ch. 12 by Lakshmi and Brudvig.

1.3 Application of Bloembergen Equation Based on Analysis of the Saturation Recovery Curve with Knowledge of T_{1f} and T_{2f}

In the absence of information concerning T_{1f} and T_{2f} it may be difficult to determine whether the B or C term of eq. (3) is dominant. Also, several factors in addition to the angular dependence of the dipolar interaction may contribute to orientation dependence of T_{1s}. These factors could introduce errors in distances calculated based only on the shape of the saturation curve as described in section 1.2. i) We have observed that T_{1s}^{o} for nitroxyl spin labels depends upon the orientation of the spin label with respect to the external magnetic field (Du et al., 1995). Thus, even in the absence of interaction with a rapidly-relaxing metal, the saturation recovery curve for a spin label in an immobilized sample is not a single exponential. This orientation dependence of T_{1s}^{o} may have a negligible impact on T_{1s} when the interaction with the rapidly-relaxing center dominates T_{1s} but becomes increasingly important at longer interspin distances. ii) Anisotropy in g values contributes to the factor of ω_f - ω_s in the denominator of the B term and can make a significant contribution to the orientation dependence of T_{1s}. iii) For some metal ions T_{1f} and/or T_{2f} may be orientation dependent. For these reasons it is preferable to explicitly consider all sources of anisotropy rather than assuming that anisotropy is due solely to the dipolar interaction.

Eq. (3) was originally derived for NMR cases where $\Delta\omega$ is large relative to the spin-spin interaction, which is the first-order or AX approximation. In EPR spectra with overlapping signals, $\Delta\omega$ may be quite small for some orientations of the molecule in the external magnetic field. In the limit where $\Delta\omega$ is small enough that the denominator of the B term goes to 1, eq. (3) predicts that the effect on $1/T_{1s}$ is proportional to T_{2f}, which is not reasonable. To exactly describe the contribution to $1/T_{1s}$ as $\Delta\omega$ decreases would require equations comparable to those used in calculations of AB splitting patterns in CW spectra (Eaton et al., 1983). Based on saturation recovery data obtained for the effect of Cu(II) on T_1 of nitroxyl radicals (Rakowsky et al., 1995) we propose that a reasonable approximation can be obtained by replacing the B term in eq. (3) with eq. (4).

$$\text{B term} = \frac{b^2 T_{2f}}{1+(\omega_f-\omega_s)^2 T_{2f}^2 + b^2 T_{1f} T_{2f}}. \qquad (4)$$

The additional term in the denominator of eq. (4) is negligible when the product $\Delta\omega T_{2f}$ is large, however it ensures that in the limit of small $\Delta\omega$ and/or strong interaction (large b), the limiting contribution to $1/T_{1s}$ is $1/T_{1f}$. The physical model is as follows. For cases of relatively weak spin-spin interaction, the interaction can only enhance the spin-lattice relaxation rate of the slower-relaxing center to the extent that $1/T_{1s}$ becomes equal to the spin-lattice relaxation rate of the faster-relaxing center. This limiting behavior can also be important when T_{2f} is significantly shorter than the T_{1f}.

Examples of the application of the full form of eq. (3) are discussed in section 2 of this chapter. The goal of the studies is to determine how well the model works when information is available concerning T_{1f} and T_{2f}.

1.4 Application to Slowly-Tumbling Macromolecules in Fluid Solution

The discussion up to this point has assumed that the sample is immobilized, which requires that experiments be performed at low-temperature, in a glass with a glass-transition temperature above room temperature, or in a doped solid. However, if a macromolecule is tumbling at a rate that is slow compared with the magnitude of the dipolar coupling, the tumbling does not average the electron-electron dipolar coupling so one should be able to extract information concerning the dipolar interaction. Analysis of the dipolar interaction for a spin label attached to a slowly-tumbling macromolecule is complicated by local motion of the spin label. Flexibility of the linkage between the spin label and the protein typically results in partial averaging of the nitroxyl g and A anisotropy in fluid solution. However, if the lineshape can be monitored in the presence and absence of a paramagnetic metal, line broadening due to electron-electron spin-spin interaction could be used to determine interspin distances.

Voss et al. (1995a, 1995b) introduced Cu(II) binding sites into T4 lysozyme and into lactose permease of *Escherichia coli*. The effect of the Cu(II) on signal amplitudes for spin labels attached to the protein were monitored in solution at 22°C, in a solution at 22°C to which sucrose was added to decrease mobility, and in water:glycerol at –20°C. They demonstrated that the Cu(II) caused significant line broadening at both 22°C and –20°C. It was assumed that the Cu(II) T_1 was about 3 x 10^{-9} s, which is fast enough to justify the use of the Leigh model (Leigh, 1970) for distances greater than about 10 Å. Using the Leigh model Voss et al. calculated copper-nitroxyl distances between about 8 and 20 Å (Voss et al., 1995a, 1995b, 1998). Although there is no data available for room temperature Cu(II) relaxation rates in slowly-tumbling low-symmetry sites comparable to the sites in these macromolecules, the limited data in ch. 2 suggest that the

Cu(II) relaxation times in these samples are likely to be substantially longer than $3x10^{-9}$ s and that the Leigh model may not be valid. However, the experimental data clearly slow an observable effect, which can be useful for relative distance measurements even if the Leigh model is not applicable. A model appropriate for slower relaxing Cu(II) may be needed for quantitative absolute distance measurements.

1.5 Modification to Explicitly Treat S=5/2

As discussed below, we are interested in application of eq. (3) to high-spin Fe(III) with S = 5/2. As part of those studies we have derived an analog of eq. (3) explicitly for S = 5/2. The factor of S(S+1) in eq. (3) is based on the assumption that the Curie law is obeyed (Koenig, 1982). For systems with S>1/2 this assumption is valid at temperatures where kT is significantly greater than the ZFS. Thus, as a minimum, correction for the populations of energy levels is needed at lower temperatures. In addition, for systems with S>1/2 the factor of S(S+1) is an approximation. Derivation of the "B", "C", and "E" terms involves the use of the raising and lowering operators for S = ½, m_s= ±½ . The generalization of the equation to apply to other values of S ignores the differences between raising and lowering operators for transitions between levels with various values of m_s. The analog of eq. (3) for the case of S = 5/2 with zero-field splitting D much greater than the EPR quantum and explicit inclusion of the individual raising and lowering operators is eq. (5).

$$\frac{1}{T_{1s}} = \frac{1}{T_{1s}^{o}} + b^2[\frac{10(P_5+P_3)T_{2f}}{1+(4D-\omega_s)^2T_{2f}^2} + \frac{16(P_3+P_1)T_{2f}}{1+(2D-\omega_s)^2T_{1f}^2} + \frac{18P_1T_{2f}}{1+(\omega_f-\omega_s)^2T_{2f}^2}]$$
$$+c^2[\frac{(25P_5+9P_3+4P_1)T_{1f}}{1+\omega_s^2T_{1f}^2} + \frac{5(P_5+P_3)T_{1f}}{1+(4D)^2T_{1f}^2} + \frac{8(P_3+P_1)T_{1f}}{1+(2D)^2T_{1f}^2} + \frac{9P_1T_{1f}}{1+\omega_f^2T_{1f}^2}]$$
$$+e^2[\frac{10(P_5+P_3)T_{2f}}{1+(4D+\omega_s)^2T_{2f}^2} + \frac{16(P_3+P_1)T_{2f}}{1+(2D+\omega_s)^2T_{1f}^2} + \frac{18P_1T_{2f}}{1+(\omega_f+\omega_s)^2T_{2f}^2}] \qquad (5)$$

where P_1, P_3, and P_5 are the populations of the m_s = ±1/2, ±3/2, and ±5/2 energy levels, respectively, and other variables have the same meanings as in eq (3) and (4). The calculations of P_1, P_3, and P_5 assume that the Zeeman interaction is negligible compared to the zero-field splitting. The splittings between the ground state and excited states are 2D and 6D, respectively. In eq. (5) it is assumed that values of T_{1f} and T_{2f} are the same for transitions between any values of m_s. If this approximation is not valid, then it is necessary to explicitly consider separate values for each transition.

1.6 Predicted Frequency Dependence

As discussed in ch. 2 by Eaton and Eaton, there is relatively little information available concerning the frequency dependence of spin-lattice relaxation times, T_{1s} and T_{1f}. In the following discussion we assume that T_{1s} and T_{1f} are independent of frequency. These assumptions will have to be tested experimentally. For immobilized samples, when T_{1f} and T_{2f} are short enough that eq (3) or (5) are applicable, it seems reasonable to assume that $T_{1f} = T_{2f}$ because T_1 is driving T_2. Therefore, the following discussion is phrased in terms of T_{1f}.

The maximum effect of each of the terms in eq. (3) and (5) occurs when the second term in the denominator is ~ 1. For example, for the C term the maximum occurs when $\omega_s T_{1f} \sim 1$. As the microwave frequency, ω_s, is decreased a larger value of T_{1f} is required to make the $\omega_s T_{1f}$ product approximately 1. Since T_{1f} increases with decreasing temperature, the maximum effect shifts to lower temperature as the microwave frequency is decreased. Shifting the maximum to lower temperature causes the effect of the rapidly-relaxing metal on T_{1s} to be larger relative to T_{1s}^{o} because T_{1s} varies approximately as T^2 (see ch. 2). Thus the same magnitude of impact of the rapidly-relaxing spin becomes a larger fractional change in T_{1s} as the temperature is decreased. This effect should make it possible to determine longer distances at lower microwave frequencies. For example, for low-spin heme-cyanide the maximum effect of the iron on the average nitroxyl spin-lattice relaxation rate ($1/T_{1s}$-$1/T_{1s}^{o}$) at X-band occurs at about 75 K, where the iron $T_{1f} \sim T_{2f} \sim 2\times10^{-10}$ s (Rakowsky et al., 1995; Budker et al., 1995; Zhou et al., 2000b). Calculations based on eq. (3) and (5) used the g-values of methemoglobin-cyanide, T_{1f} for methemoglobin-cyanide, and T_{1s}^{o} for spin-labeled Zn-myoglobin in 1:1 water:glycerol, including the characteristic orientation dependence of nitroxyl T_1, distances ranging from 20 to 50 Å, and microwave frequencies of 9.25 and 2.5 GHz. At X-band the calculated impact of the low-spin iron on the nitroxyl saturation curve became too small to be significant at distances greater than about 30 Å. By contrast, at 2.5 GHz the calculated impact of the iron on the nitroxyl saturation curve was significant at distances up to about 40 Å. These calculations suggest that longer distances will be accessible at lower microwave frequencies.

2. DISTANCE DETERMINATION BASED ON T_1 EFFECTS – EXPERIMENTAL DATA

This section provides some examples of applications of the effect of a rapidly-relaxing spin on the spin lattice relaxation for a more slowly-relaxing spin to determine interspin distances in a discrete pair.

2.1 High-spin Fe(II)

The spin-lattice relaxation of the dark-stable tyrosyl radical in photosystem II is enhanced by interaction with a high-spin Fe(II) that has S = 2 (Hirsh et al., 1992). The non-exponential saturation recovery curve for the tyrosyl radical was analyzed assuming that the B term in eq. (3) dominates. Using a T_1 of the iron at 5 K of 5.6×10^{-8} s, a distance of 38 Å was obtained (Hirsh et al., 1992). The distance between the oxidized bacteriochlorophyll a dimer and Fe(II) in *Rhodobacter sphaeroides* is 28 Å. Since the Fe(II) center in photosystem II is believed to be very similar to that in *Rhodobacter sphaeroides,* the measured effect of the Fe(II) on a slowly-relaxing center at a known distance in *R. sphaeroides* could be used to relate shapes of saturation curves for slowly-relaxing centers in photosystem II to the distance between that center and the Fe(II), provided it could be shown that the relaxation enhancement was due to Fe(II) and not to some other paramagnetic center. Comparison of the effect of the high-spin Fe(II) on the relaxation rates for the bacteriochlorophyll a dimer and on the relaxation rate for the dark-stable tyrosyl radical in photosystem II using the C term of eq. (3) gave an iron-tyrosyl distance of 37 ± 5 Å (Hirsh and Brudvig, 1993). A similar comparison of the relaxation rates for chlorophyll_z and the oxidized bacteriochlorophyll a dimer gave an Fe(II) – chlorophyll_z distance of 39.5 ± 2.5 Å. In this case similar results were obtained with either the B or C terms of eq. (3) (Koulougliotis et al., 1994). The similarity of the effects of the Fe(II) on the relaxation rates for tyrosyls D and Z in manganese-depleted photosystem II demonstrated that the distance from the Fe(II) to each of the tyrosyls is the same (Koulougliotis et al., 1995). The analyses in these studies were based on the assumption that the exchange and dipolar contribution to the B term could be treated separately, despite the fact that they are derived from the same matrix elements (see section 1). This approximation may not have caused serious problems for these cases because the exchange contribution is likely to be very small at these long distances. Details of these experiments are given in ch. 12 by Lakshmi and Brudvig.

2.2 Low-spin Heme Fe(III)

We have examined the application of eq. (3) and (4) to spin-labeled complexes of low-spin Fe(III). Initial experiments were performed at X-band with methemoglobin spin-labeled at the naturally-occurring β-93 cysteine (Budker et al., 1995) and with a series of spin-labeled iron(III) porphyrins (Rakowsky et al., 1995). The metal-nitroxyl distances had previously been determined for the copper(II) complexes of the same spin-labeled porphyrins, using doped single crystals (Damoder et al., 1983). These copper-nitroxyl measurements provided a basis for evaluation of the iron-nitroxyl distances that were obtained by analysis of SR data for the spin-labeled Fe(III) complexes. SR data were recorded in the center of the nitroxyl EPR spectra at temperatures between about 10 and 140 K. Although the SR curves for the nitroxyl signal are not single exponentials, a qualitative sense of the effect of the iron on the nitroxyl can be represented by fitting the data to a single exponential. Figure 1 shows the values of T_1 obtained by single-exponential fits to the SR data for spin-labeled methemoglobin-cyanide and methemoglobin-imidazole as a function of temperature, along with values of T_1 obtained for spin label attached to oxyhemoglobin in which the iron is diamagnetic. The effect of the low-spin Fe(III) on the nitroxyl relaxation is greatest between about 60 and 100 K. The faster-relaxing iron in methemoglobin-cyanide has a greater effect on the nitroxyl relaxation than the slower-relaxing iron in methemoglobin-imidazole (Fig. 1).

In the center of the nitroxyl X-band spectrum there are contributions from all orientations of the molecule with respect to the external magnetic field, although the populations are not in the same statistical ratio as for a powder average because of the much larger nitrogen hyperfine splitting along the magnetic z axis than in the x,y plane. The calculations of the effect of the iron on the nitroxyl SR curves were performed with a locally-written program (MENOSR). The version of the program in use at the time that the methemoglobin data were analyzed assumed a random distribution of molecular orientations with respect to the external magnetic field and did not include orientation selection. At X-band, possibilities for orientation selection based on position in the nitroxyl CW spectrum are limited by the orientation dependence of the nitroxyl nitrogen hyperfine splitting (about 30 G). The selectivity is decreased by unresolved proton hyperfine coupling, which contributes about 7 G to the line widths in frozen solution. The interspin distances used in the calculations (about 15 Å) result in maximum dipolar splittings that are of the same order of magnitude as the nitroxyl hyperfine splitting, which largely eliminates possibilities for orientation selection at these interspin distances. Thus the approximations in the calculations are justified.

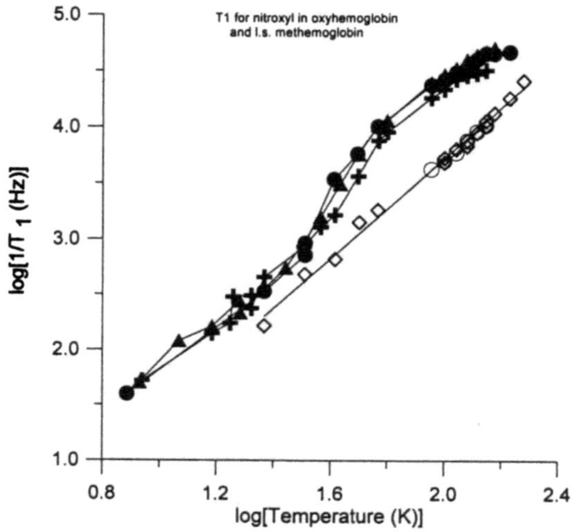

Figure 1. Temperature dependence of nitroxyl T_1 measured at the center of the spectrum in 1:1 buffer:glycerol in the absence and presence of low-spin Fe(III) at 9.2 GHz: (O, ◊) spin-labeled oxy-hemoglobin, two independent preparations; (▲,●) spin-labeled methemoglobin cyanide with two different spin labels; (+) spin-labeled methemoglobin imidazole. The solid line through the data for spin-labeled oxy-hemoglobin is the least squares fit to $\log(1/T_1) = 2.24\log(T)-0.767$. The other lines connect the data. (Reproduced from Seiter et al., 1995).

The g-values for the iron and nitroxyl in the absence of electron-electron interaction were obtained by simulation of CW spectra. Values of T_{1s}^{o} as a function of temperature were measured for spin-labeled oxy-methemoglobin (Budker et al., 1995) and for metal-free spin-labeled porphyrin (Rakowsky et al., 1995) in the same solvent as was used to study the corresponding spin-labeled iron complexes. The low-spin Fe(III) relaxation rates as a function of temperature were measured, as described in ch. 2. SR curves were recorded for the spin label attached to the heme proteins. Since the SR curves for spin label interacting with the Fe(III) include a distribution of relaxation rates, low observe powers were used to ensure that even the slower components of the distributions were çorrectly observed. SR curves were analyzed by computer simulations based on eq. (3) and (4). The simulations calculated the SR curve for each orientation of the spin-labeled heme in the external magnetic field and summed the contributions with weighting for the solid angles. The interspin distance was adjusted to give the best fit to the experimental SR curves at each temperature. The standard deviation of the average values of r obtained at a series of temperatures provides an estimate of the uncertainty in r. The shapes of the calculated curves are in good agreement with the experimental data (Fig. 2). The simulations shown in Fig. 2 as a function of temperature were obtained by changing only the input values of T_{1s}^o, T_{1f}, and T_{2f} as measured independently (Budker et al., 1995). The values of r for the spin-labeled iron porphyrins and spin-labeled methemoglobin are in the range of 10.5 to 15.5 Å. These values of r obtained by analysis of the SR curves are in good agreement with independent estimates of the interspin distances (Budker et al., 1995; Rakowsky et al, 1995). The calculated SR curves were weakly dependent upon the relative orientations of the g tensors for the iron and nitroxyl and it was estimated that uncertainty in the relative orientations corresponded to an uncertainty of about one Å in the interspin distance.

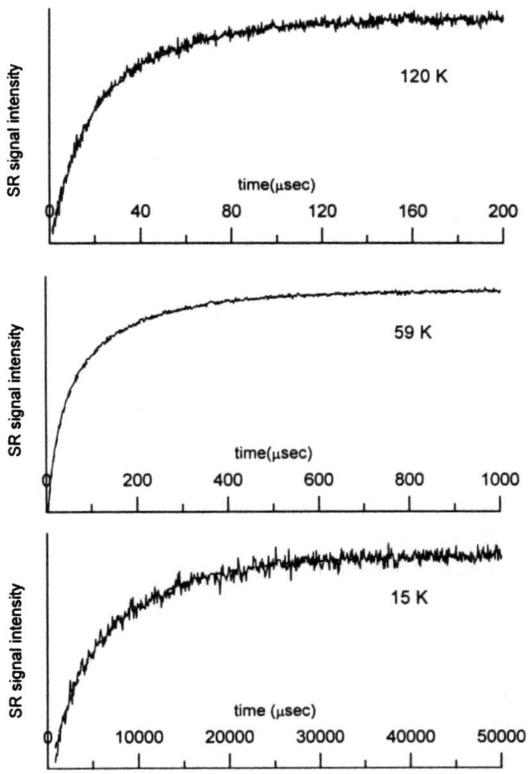

Figure 2. SR curves for the nitroxyl signal in spin-labeled methemoglobin cyanide in 1:1 buffer:glycerol at 9.2 GHz and 15, 59, and 120 K. The dashed lines were calculated for r = 15.5 Å as described in the text and in Seiter et al., 1995. (Reproduced from Seiter et al., 1995).

For distances in the range of 10 to 15 Å one might ask whether the iron-nitroxyl distance could have been determined by analyzing changes in the CW lineshape. Calculation of the distance based on the CW lineshape, by either simulation or deconvolution, requires an undistorted CW spectrum in the limit where the iron relaxation rate is slow compared with the dipolar interaction. For the low-spin heme complexes the iron relaxation rates become slow enough to satisfy this criterion below about 10 K. At these low temperatures the nitroxyl T_1, even in the presence of the iron, is so long that it is very difficult to obtain CW spectra that are not distorted by passage effects (Budker et al., 1995; Rakowsky et al., 1995). Since distortions due to passage effect would invalidate analysis of the CW spectra, distance determination based on the CW spectra are not likely to be reliable for these systems.

For iron-nitroxyl distances of the order of 10 to 15 Å the effect of the low-spin iron on the nitroxyl spin-lattice relaxation rate is so large that the SR curves calculated with eq. (3) are only weakly dependent on T_{1s}^{o}. However, at longer interspin distances, the effect of the iron becomes smaller and the calculated curves become increasingly dependent on T_{1s}^{o}. At X-band T_1 is about a factor of four longer when the magnetic field is along the nitroxyl z axis than when it is in the nitroxyl x,y plane (Du et al., 1995). SR curves measured in the middle of the nitroxyl spectrum do not fit well to a single exponential because of contributions for a range of orientations of the molecule with respect to the external field. The nitroxyl SR curves can be modeled well using the known orientation dependence of T_1 and inclusion in the simulated curve of only those orientations that contribute at the combination of microwave frequency and magnetic field that was used to record the experimental SR data, which is called "orientation selection" (Zhou et al., 2000b). The orientation dependence of T_{1s}^{o} and orientation selection have been incorporated into MENOSR and used to fit the SR curves for spin-labeled variants of metmyoglobin-cyanide. Distances ranged between about 19 and 30 Å. For comparison with these distances calculations were performed using crystallographic structures of metmyoglobin and the Insight II® software (Biosym/MSI, San Diego, CA). The iron-nitroxyl distances obtained from the modeling calculations are not without uncertainty. The spin labels on these variants are on the surface of the protein so structures are sensitive to solvation. Structures calculated with either water or glycerol solvation gave similar iron-spin label distances, but calculations without a solvent shell gave significantly different distances. The results shown in Table 1 are averages of the distances between the iron and the N and between the iron and the O of the nitroxyl radical. For each variant the iron-nitroxyl distance is significantly longer than the distance between the iron and β-carbon of the peptide because the mutations were introduced at surface residues and the spin label side-chain is constrained by the local protein structure to point out into the solution. For each myoglobin variant there is reasonable agreement between the interspin distances obtained by analysis of the SR data for the low-spin cyanide complexes and from the Insight calculations (Zhou et al., 2000b). These results indicate that iron-nitroxyl distances up to about 30 Å can be obtained at X-band for low-spin hemes with relaxation rates similar to those of metmyoglobin-cyanide. The slower-relaxing Fe(III) in imidazole-metMb-SL (ch. 2) has a smaller impact on the nitroxyl spin lattice relaxation than the faster relaxing Fe(III) in CN-metMb-SL so distances obtained for the imidazole adducts are less precise than for the cyanide complexes (Zhou et al., 2000b).

Table 1. Comparison of iron-nitroxyl distances (Å) obtained by analysis of SR data and calculations performed with Insight II for metmyoglobin variants

variant	Low-spin CN-MetMb[a]	High-spin formate-MetMb[a]	Insight II ® Fe-NO[b]	Fe- C_β[c]
K98C[d]	19.3 ± 1.3	(19)[e]	18.9 ±1.0	11.1
V66C	19.8 ± 0.8	21.6 ± 2.5	18.3 ± 0.9	12.3
H12C	27.9 ± 1.9	28.4 ± 2.2	31.8 ± 1.3	25.2
A53C	30.4 ± 0.8	27.8 ± 2.9	28.1 ± 1.9	23.4

[a]Uncertainties are calculated from standard deviations for distances obtained over a range of temperatures. [b]Uncertainties are standard deviations for distances obtained with different axial ligands and different solvation shells. [c]Distance from iron to β-carbon of naturally-occurring amino acid at this position, calculated from the X-ray crystal structure, Brookhaven pdb file pdb1mbw. [d]Variants are designated by the letter and position for the amino acid that was replaced by a cysteine. [e]Assumed distance used to determine high-spin Fe(III) relaxation rates that were then used in distance calculations for other variants with the same axial ligand on the Fe(III) based on eq. (5).

2.3 High-spin Heme Fe(III) with Zero-Field Splitting Greater than the X-band EPR Quantum

The determination of T_{1f} and T_{2f} for high-spin heme Fe(III) is complicated by the fact that the zero-field splittings (ZFS) are substantially larger than the X-band EPR quantum (Rakowsky et al., 1998; Seiter et al., 1998). For a Kramers ion such as high-spin Fe(III) with ZFS this large the only transitions that are observed at X-band are between $m_s = \pm 1/2$ energy levels. Thus measurements of the iron electron spin relaxation rates, either by SR or inversion recovery, or by the temperature-dependent contribution to the CW lineshape provide information only about this transition. In the following paragraphs we examine several approaches to this problem.

2.3.1 Assume that the $m_s = \pm 1/2$ transition dominates the B and E terms of eq. (3) and that T_{1f} for other transitions is approximately the same as for the $m_s = \pm 1/2$ transition

Our initial studies of the effect of high-spin Fe(III) on nitroxyl relaxation were performed for methemoglobin spin-labeled at the naturally-occurring β-93 cysteine (Seiter et al., 1998) and for the same series of spin-labeled porphyrins (Rakowsky et al., 1998) as were studied for low-spin Fe(III). The $m_s = \pm 1/2$ transitions have ω_f of the order of the EPR quantum, but other transitions have ω_f of the order of 2D or 4D where D is the zero-field splitting parameter. The large values of ω_f for transitions other than $m_s = \pm 1/2$ result in such large denominators for the B and E terms in eq. (3) and (4) that these other transitions do not contribute effectively to enhanced relaxation. To take account of the ineffectiveness of these transitions in enhancing the relaxation of the slowly-relaxing spin we proposed, as a first approximation, to add a weighting factor to the B and E terms. The weighting factor is the fraction of fast-relaxing spins with $m_s = \pm 1/2$ at a particular temperature and zero-field splitting calculated from the Boltzmann populations of the three Kramers' doublets. It was assumed that the other transitions make such a small contribution to the B and E terms that to first approximation they can be neglected. A similar factor does not apply to the C term because the denominator depends only on ω_s and not on ω_f. Since $m_s = \pm 1/2$ transitions dominate the contributions from the B and E terms, and the relaxation rates for other transitions are expected to be of the same order of magnitude as for the $m_s = \pm 1/2$ transitions, it was assumed that the relaxation rates obtained for the $m_s = \pm 1/2$ transitions could be used in eq. (3) and (4) to calculate the effect of the iron on the nitroxyl relaxation rate.

For the spin-labeled porphyrins the effect of the iron on the nitroxyl relaxation rate was studied for axial ligands = F^-, Cl^-, and Br^-. At the same temperature the iron relaxation rates for the three anions differ by almost an order of magnitude, but the interspin distances obtained from the nitroxyl SR curves were independent of axial ligand (Rakowsky et al., 1998) and the distances agreed well with distances obtained for the low-spin analogs (Rakowsky et al., 1995). Similarly, for spin-labeled methemoglobin the interspin distances obtained for the high-spin complexes with water or fluoride as the axial ligand (Seiter et al., 1998) were similar to values obtained for the low-spin analogs (Budker et al., 1995). For both sets of complexes the interspin distance for the high-spin complexes were about an Å longer than for the low-spin complexes, which suggested that the assumptions used in analyzing the SR curves might be causing systematic error.

2.3.2 Use of the more complete eq. (5) for S = 5/2 and assumption that T_{1f} for other transitions is approximately the same as for the $m_s = \pm 1/2$ transition

A series of spin-labeled variants of metmyoglobin have been examined with formate or fluoride as the axial ligand. These axial ligands put the Fe(III) in a high-spin form. When the SR curves for the interacting nitroxyl were analyzed, using the same values of the iron relaxation rates, with eq (3), including the correction for the fraction of iron in the $m_s = \pm 1/2$ state, and with eq. (5) the resulting distances agreed within about 1 Å for distances between about 18 and 30 Å (Zhou et al., 2000b). Thus, the approximations used in our original analyses do not appear to introduce a large error. However, it is preferable to use the more complete eq. (5) for analyzing SR curves for high-spin Fe(III).

2.3.3 Use of eq. (5) for S = 5/2 and a distribution of Fe(III) relaxation times

At a given temperature and within a series of related compounds, we have observed that the relaxation rate for the $m_s = \pm 1/2$ transitions of high-spin Fe(III) increases with increasing zero-field splitting. It is generally accepted that for metals with S > ½ there is a distribution of values of zero-field splitting (Fiamingo et al., 1989; Doctor et al., 1993). It is therefore expected that a distribution of ZFS values will result in a distribution of iron relaxation rates (Fiamingo et al., 1989). This expectation is consistent with our observations that the inversion recovery curves from which we calculate the iron relaxation rates at low temperature do not fit well to a single exponential and are better fit by a distribution of relaxation rates. Saturation recovery data for spin-labeled methemoglobin were analyzed assuming a single value of the iron relaxation rate at each temperature (Seiter et al., 1998). Preliminary results for variants of metmyoglobin indicate that incorporation of a distribution of values for the iron T_1 does not have a large impact on the calculated interspin distance.

2.3.4 Use of eq. (5) for S = 5/2 to estimate effective iron relaxation rates from the SR curves for an interacting nitroxyl at a known interspin distance

As noted previously, the X-band measurements of high-spin Fe(III) relaxation rates provide information only for the $m_s = \pm 1/2$ transition because transitions involving $m_s = \pm 3/2$ and $\pm 5/2$ are not observed. The temperature dependence of the relaxation rates obtained for the $m_s = \pm 1/2$ transitions in high-spin metMb is consistent with Orbach processes involving excited states at 2D and 6 D (Zhou et al., 2000a). When Eq. (5) was used to calculate iron-nitroxyl distances based on the relaxation times observed for the $m_s = \pm 1/2$ transitions, the calculated distances are shorter by 13% at shorter distances up to 33% at longer distances than the distances obtained for low-spin complexes of the same variants, which we attribute to systematic errors in the high-spin Fe(III) relaxation rates. It is expected that the relaxation rates will be faster for transitions involving $m_s = \pm 3/2$ and $\pm 5/2$ than for $m_s = \pm 1/2$ (Rubenstein et al., 1971). Lacking data for these unobserved transitions we used high-spin formate-metMbK98C-SL as a reference to estimate the effective iron relaxation rates. Based on results obtained for the low-spin analogs and on the results of the Insight II calculations (Table 1) it was assumed that the interspin distance was 19Å for metMbK98C-SL. The iron relaxation rates were then adjusted to fit the SR curves calculated using eq. (3) to the experimental data for the interacting nitroxyl radical. The resulting "effective" iron relaxation rates are faster than the rates determined from the observed transitions by a factor of about three for fluoride and six for formate (Zhou et al., 2000b). When these iron relaxation rates were used to analyze the saturation recovery curves for other spin-labeled metmyoglobin variants, the calculated interspin distances were in good agreement with distances obtained for the low-spin analogs (Table 1). In effect, this procedure for the high-spin Fe(III) complexes obtains distances for other variants, relative to the assumed distance for the K98C variant. For the same interspin distance, the faster relaxing iron in formate-metMb-SL (ch. 2) has greater impact on the nitroxyl SR curve than the slower relaxing iron in fluoride-metMb-SL, which permits more precise distance measurements in the formate complexes than in the fluoride complexes (Zhou et al., 2000b).

Because of the larger value of S for high-spin Fe(III) than for low-spin Fe(III) it might have been anticipated that the high-spin Fe(III) would have larger effects on the nitroxyl T_1 than low-spin Fe(III). However, at the same interspin distance, the maximal effect of the high-spin and low-spin Fe(III) on the nitroxyl relaxation are similar (Zhou et al., 2000b). The dominant reason why the high-spin Fe(III) is not more effective than low-spin Fe(III) in enhancing the nitroxyl relaxation rate is that in the relevant temperature range (~40 to 100 K) low-spin Fe(III) relaxes faster than high-spin Fe(III). The maximal % change in the nitroxyl relaxation rate due to interaction with the low-spin Fe(III) in CN-metMb-SL was observed at about 64 K, where the iron relaxation rate is about $6x10^9$ s^{-1}. The maximum % change in the nitroxyl relaxation rate due to interaction with high-spin Fe(III) in formate-metMb-SL occurs at about 57 K where the iron relaxation rate is about $3x10^8$ s^{-1}. The relaxation rate for the high-spin Fe(III) increases with increasing temperature more slowly than that of the non-interacting nitroxyl so when viewed as a fractional change in the nitroxyl relaxation rate, the high-spin Fe(III) becomes less effective at higher temperatures. In addition, for most orientations of the molecules with respect to the external magnetic field, the denominator of the "B" term in eq. (3) or the analogous terms in eq. (5) is substantially larger for high-spin Fe(III) than for low-spin Fe(III), which also diminishes the effectiveness of high-spin Fe(III) relative to low-spin Fe(III) (Zhou et al., 2000b).

3. EFFECT OF RAPIDLY-RELAXING METAL ON SPIN ECHO DEPHASING, T_M, FOR SLOWLY-RELAXING SPIN

The effect of the rapidly-relaxing metal on the slowly-relaxing spin is analogous to the effects of a physical motion, such as rotation of a methyl group, that averages nuclear spins to which the unpaired electron is coupled. Since the latter may be more intuitive than the former, we first consider the physical process then discuss the analogy.

In the temperature range where methyl group rotation is slow relative to electron-nuclear couplings, the unpaired electron is coupled to three inequivalent protons. As the rate of methyl group rotation increases, the three protons become equivalently coupled to the unpaired electron on the experimental timescale. The effects on CW EPR spectra of the dynamic averaging of electron-nuclear spin-spin splittings by methyl group rotation have been observed (Gamble et al., 1968). ESE measurements have been used to monitor this dynamic averaging process even when the splitting that is averaged is small relative to the overall line width (Dzuba et al., 1981; Tsvetkov and Dzuba, 1990; Nakagawa et al., 1992; Du et al., 1994). In the regime of intermediate rates of rotation, the averaging of inequivalent environments results in increased rates of two-pulse ESE decay.

Electron spin relaxation for the faster-relaxing partner of an electron-electron spin-spin coupled system is a dynamic process that averages inequivalent sites, analogous to a physical motion such as methyl rotation. When the relaxation rate of the more rapidly-relaxing spin is slow compared with the dipolar splitting, broadening or splitting of the CW spectrum is observed. As temperature is increased and the relaxation rate of the rapidly-relaxing center increases, the CW spectrum of the slowly-relaxing center broadens. Further increases in the relaxation rate cause collapse of the splitting due to the electron-electron coupling and then narrowing to an unbroadened spectrum. Examples of this phenomenon for iron-nitroxyl interaction that is large relative to the CW linewidths are given in Fielding et al. (1986). An example for smaller electron-electron interaction is shown in

Fig. 3 for a spin-labeled high-spin Fe(III) porphyrin with bromide as the axial ligand (Rakowski et al., 1998).

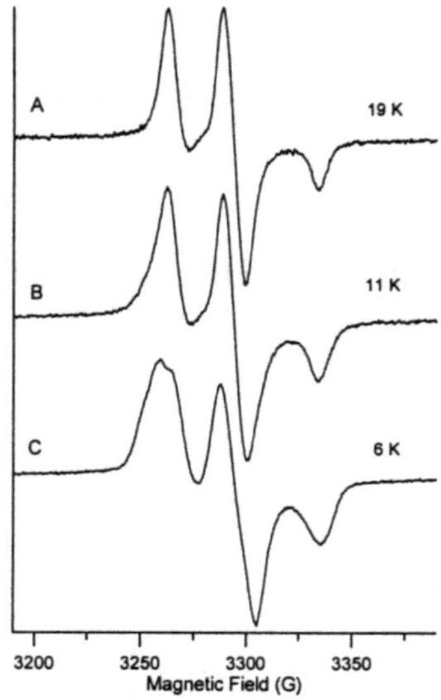

Figure 3. CW (9.24 GHz) spectra of the nitroxyl signal for a spin-labeled Fe(III) porphyrin with bromide as the axial ligand. Spectra were obtained with 1 G modulation amplitude at 100 kHz, 0.15 μW microwave power and temperatures of (A) 19 K, (B) 11 K, and (C) 6 K. (reproduced from Rakowsky et al., 1998).

Table 2. Impact of rapidly-relaxing spin on CW spectra and echo dephasing for slowly-relaxing partner in spin-coupled pair.

Temperature/ relaxation rate	low/slow	intermediate	high/fast
impact of dipolar interaction on CW spectra	potentially resolvable splitting	broadened	splittings averaged by the rapid relaxation
impact of dipolar interaction on echo dephasing	T_m determined by other factors	T_m decreases, goes through a minimum and then increases as T increases	T_m determined by other factors

The analogy between the effect of the rapidly-relaxing metal on the CW spectra and on the spin echo decay for a spin label to which it is dipolar coupled is summarized in Table 2. When the metal relaxation rate is slow relative to the dipolar interaction the T_m for the slowly-relaxing center typically is determined by interaction with neighboring solvent nuclei (see ch. 2). As temperature is increased and the metal relaxation rate increases, the rate of relaxation becomes comparable to the dipolar splitting, the dipolar splittings are averaged, and there are dramatic effects on T_m.

An example of this phenomenon is shown in Fig. 4 for nitroxyl interacting with low-spin Fe(III) in spin-labeled metmyoglobin. At low temperature the nitroxyl T_m in the low-spin Fe(III) complex is the same as in the absence of the low-spin Fe(III). As temperature increases there is increasing and then decreasing impact of the low-spin Fe(III) on the nitroxyl T_m. At higher temperatures T_m again becomes approximately equal to that observed in the absence of interaction with the Fe(III) (Fig. 4). For a spin label attached to oxidized cytochrome c Hilczer et al. (1995) observed a dramatic decrease in T_m in the vicinity of 25 K due to interaction with the low-spin heme Fe(III). This is the same temperature interval in which low-spin heme Fe(III) has a large impact on T_m for spin-labeled metmyoglobin (Fig. 4).

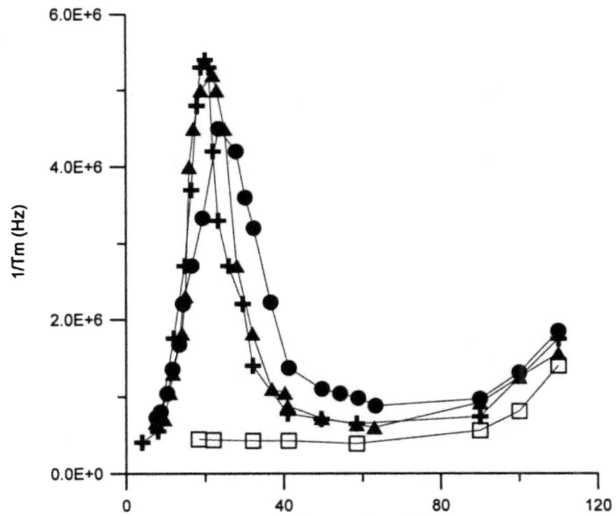

Figure 4. Temperature dependence of $1/T_m$ at the center of the nitroxyl signal in spin-labeled oxy-hemoglobin (□), two different spin labels attached to methemoglobin cyanide (+, ▲), and spin-labeled methemoglobin imidazole (●). Although the ESE curves are not single exponentials a single exponential fit (eq. (10)) was used to obtain a qualitative description of the temperature dependence of $1/T_m$. The lines connect the data points (reproduced from Seiter et al., 1995)

The effects on T_m in the "intermediate" region depend on distance and thus can be used to determine the interspin distance. In the following discussion we assume that the concentration of the sample is low enough and the pulse turning angles are small enough that there is negligible impact on the echo decay from instantaneous diffusion. Since the dipolar interaction is orientation dependent, the experimental spin echo decay curves are the superposition of the effects of collapsing the powder pattern of dipolar splittings. The following equations are for single orientations. Simulations for randomly oriented samples require summing a powder average of these echo decay curves.

Zhidomirov and Salikhov (1969) presented the following expression for the impact on the ESE curve of a process that they called "spectral diffusion" in a spin-coupled system.

$$E(2\tau) = R^{-2}\{\exp(-2\tau/\tau_c)[\tau_c^{-2}\sinh^2(R\tau) + R^2\cosh^2(R\tau) + R\tau_c^{-1}\sinh(2R\tau) + \Delta^2\sinh^2(R\tau)]\} \quad (6)$$

where

$E(2\tau)$ is the intensity of the echo as a function of τ

τ is the time between microwave pulses in the 2-pulse experiment

τ_c is the correlation time for the dynamic process

Δ is 1/2 the angular frequency difference between the two sites averaged by the dynamic process. For the spin-coupled pair Δ is the orientation-dependent spin-spin splitting, which includes the dipole and exchange contributions.

$$R^2 = \tau_c^{-2} - \Delta^2$$

When τ_c is significantly larger or smaller than $|\Delta|$, eq. 6 can be simplified as shown in eq. (7) and (8) (Kispert et al., 1982).

$$\text{for the slow averaging limit } (\tau_c^{-1} << \Delta) \quad E(2\tau) = \exp(-2\tau/\tau_c) \quad (7)$$

$$\text{for the fast averaging limit } (\tau_c^{-1} >> \Delta) \quad E(2\tau) = \exp[-2\tau(\frac{\Delta^2\tau_c}{2})] \quad (8)$$

The original expressions used T_{1f} as τ_c (Zhidomirov and Salikhov, 1969). It was subsequently recognized that this expression could be used to analyze the effects on spin echo decay due to rotation of a methyl group coupled to an unpaired electron (Kispert et al., 1982). In this case τ_c is the correlation time for the rotation. In the slow-averaging regime $T_m = \tau_c$ (eq. (7)) and is independent of interspin distance. In the fast-averaging regime $T_m = 2/(\Delta^2\tau_c)$ (eq. (8)), and therefore depends upon the interspin distance. These limiting expressions for T_m are the same as the approximate equations for rate constants that are used to analyze line widths in dynamic NMR experiments (Drago, 1977) because the same fundamental principles underlay the interpretation of dynamic NMR line shapes, the collapse of the spin-spin splittings in CW EPR spectra and the impact of electron spin relaxation of the fast-relaxing spin on the ESE data for the slowly-relaxing spin in a spin-coupled system (Nakagawa et al., 1992).

We propose that at some temperatures it also is necessary to consider T_{2f}. This can be understood as follows. In a discrete pair each slowly-relaxing spin is coupled to a particular rapidly-relaxing spin. A T_2 process for a particular rapidly-relaxing spin flips its magnetization and has the same effect on the slowly-relaxing partner as if the rapidly-relaxing spin had undergone a T_1 process. Through much of the temperature range of interest $T_{1f} = T_{2f}$, but that may not be the case at lower temperatures. In analyzing the spin echo data in our studies it was found that $\tau_c = \sqrt{T_{1f}T_{2f}}$ gives better agreement with the experimental data than $\tau_c = T_{1f}$ (Rakowsky et al., 1995). In analyzing a different problem, Wolf has shown that for inhomogeneously broadened lines, the statistics can give $\tau_c = \sqrt{T_{1f}T_{2f}}$ (Wolf, 1966).

In section 1 we discussed the model that Bloembergen developed for the effect of the rapidly-relaxing spin on T_1 for a slowly-relaxing spin. Bloembergen also considered the effect of the fast-relaxing spin on T_2 for the slower relaxing spin (Bloembergen et al., 1948, 1959; Bloembergen, 1949; Kulikov and Likhtenshtein, 1977). Most of the terms in the expression given by Kulikov and Likhtenshtein (1977) for the impact on T_2 are spectral density functions analogous to those in the expression for the impact on T_1 (eq. 3). However, in the expression for the impact of the faster-relaxing spin on T_2 of the slowly-relaxing spin there is an additional term as shown in eq. (9) (Kulikov and Likhtenshtein, 1977).

$$\frac{1}{T_{2s}} = \frac{1}{T_{2s}^0} + \ldots + \frac{\mu^2\gamma^2 T_{1f}}{3r^6}(1-3\cos^2\theta)^2 \tag{9}$$

where μ is the magnetic moment of the fast-relaxing spin, γ is the electron magnetogyric ratio, T_{1f} is T_1 for the fast-relaxing spin, θ is the angle between the magnetic field and the interspin vector, and r is the interspin distance.

The additional term in eq. (9) is the analog of eq. (8) with explicit statement of the dipolar splitting and with $\tau_c = T_{1f}$. Bloembergen included only this limiting expression (rather than the full form of eq. (6)) because his equations are valid only in the limit of short T_{1f}. At temperatures below about 150 K nitroxyl T_m values in proton-containing solvents or protein solutions are of the order of microseconds (see ch. 2), which is many orders of magnitude shorter than T_1. Therefore contributions to relaxation that are significant with respect to $1/T_1$ for nitroxyl are negligible with respect to

$1/T_m$. At temperatures where T_{1f}^{-1} is of the order of the electron-electron dipolar interaction, spectral density terms similar to those in eq. (3) have much smaller impact on T_{2s} and on spin echo decays than the term in eq. (9). Thus eq. (6) is consistent with the Bloembergen model, but permits analysis of the spin echo decays over a wider range of T_{1f} than would satisfy the criteria for application of the Bloembergen equations for T_{2s}.

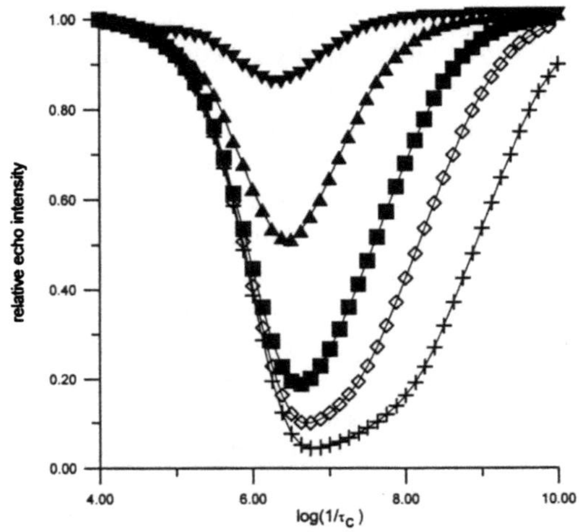

Figure 5. Intensity of spin echo calculated as a function of $1/\tau_c$ using Eqs. (6) – (8) with the time between pulses in the two-pulse sequence = 500 ns, with g values characteristic of low-spin metmyoglobin imidazole, and with the interspin vector along the iron z axis. The echo intensity is scaled relative to that expected for the same pulse timing in the limit of very long or short τ_c. Curves were calculated for interspin distances r = 15 (+), 20 (◇), 25(■), 35 (▲), and 50 (▼) Å.

An alternative to simulating the full shapes of the nitroxyl echo decay curves is to analyze the relative intensity of the echo, at a particular inter-pulse spacing, as a function of temperature. As temperature is increased and the metal relaxation rate passes through the regime in which it is comparable to the dipolar splitting, the echo intensity for the interacting spin label at constant inter-pulse timing decreases, goes through a minimum, and then increases again until it returns to the intensity expected in the absence of interaction. The magnitude of the intensity decrease depends on the interspin distance. Calculations of echo intensity as a function of $\log(1/\tau_c)$ are shown in Fig. 5. These particular curves were calculated for g values characteristic of low-spin metmyoglobin imidazole. The calculations show the dramatic effects expected as a function of the metal spin relaxation time. If there were no instrumental deadtime the echo decay rate would show corresponding trends. However, at shorter interspin distances the plots of $1/T_m$ are less "symmetrical" than at longer distances (Fig. 6). This distortion occurs because at the temperatures where the metal relaxation has the greatest impact on T_m, the spin-coupled pairs that have the orientations with respect to the external magnetic field that produce the largest dipolar couplings can have effective values of T_m that are short relative to the dead time of the spectrometer. These spins do not contribute to the echo decay curve, and therefore are not properly taken into account in the data analysis unless echo intensity is considered. Therefore, plots of effective $1/T_m$ as a function of temperature or $1/\tau_c$ (Fig. 6) are less useful in determining interspin distance than are plots of echo intensity. An additional advantage of analyzing echo intensities, rather than the full shape of the echo decay curve, is that distances can be determined without measuring the electron spin relaxation rates for the rapidly-relaxing center. One need only know that the metal relaxation rates are in the appropriate regime so as to be sure of the assignment of the relevant dynamic process.

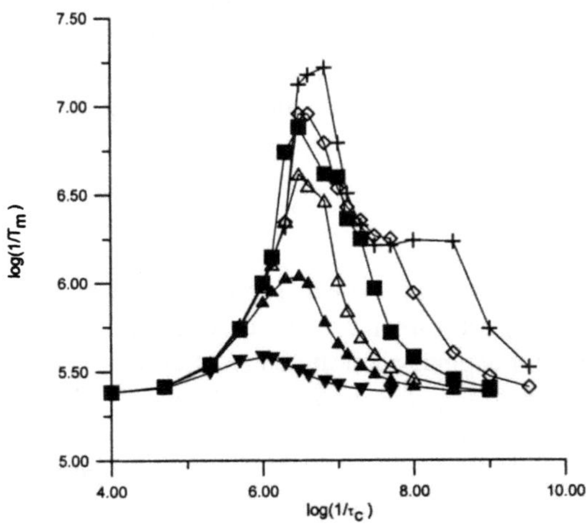

Figure 6. Values of spin echo dephasing rate constants $1/T_m$ obtained by fitting single exponentials to spin echo decays calculated using Eqs. (6) – (8) for a series of interspin distance and a deadtime of 120 ns. Curves were calculated for interspin distances r = 15 (+), 20(◊), 25(■), 30 (Δ), 35 (▲), and 50 (▼) Å.

4. DISTANCE DETERMINATION BASED ON T_M EFFECTS – EXPERIMENTAL DATA

This section provides some examples of applications of the effect of a rapidly-relaxing spin on the spin echo dephasing for a more slowly-relaxing spin to determine interspin distances.

4.1 Effect of Low-Spin Fe(III)

4.1.1 Analysis of full shape of echo decay curve

Spin echo decays as a function of temperature were obtained for spin labels attached to the β-93 cysteine of cyano-methemoglobin and imidazole-methemoglobin (Budker et al., 1995). Between about 5 and 50 K the interaction with the rapidly-relaxing low-spin Fe(III) dominates the spin echo dephasing. Because of the overlapping decay curves for the various orientations of the interspin vector relative to the external magnetic field, the decays are not simple exponential decays. However, a qualitative impression of the effects of the iron can be obtained by fitting the experimental spin echo curves to eq. (10).

$$Y(\tau) = Y(0)\exp[-(2\tau/T_m)^x] \qquad (10)$$

where $Y(\tau)$ is the intensity of the echo as a function of τ, the time between the two pulses. $Y(0)$, echo intensity extrapolated to time zero, and $Y(\tau)$ are in arbitrary units that depend upon the concentration of the sample, resonator Q, and instrument settings. The parameters x and T_m describe the shape of the echo decay and x depends upon the mechanism of dephasing (Brown, 1979; Salikhov and Tsvetkov, 1979). In the temperature range where interaction with the iron(III) dominates the echo dephasing the exponent x is about 1. Values of $1/T_m$ for the spin label in cyano-methemoglobin and imidazole-methemoglobin obtained by fitting the echo decays to eq. (10) are shown in Fig. 4 (Budker et al., 1995). The maximum effect of the iron on $1/T_m$ for the nitroxyl occurred at lower temperature (about 20 K) for cyano-methemoglobin than for imidazole-methemoglobin (about 30 K) because relaxation rates are faster for the low-spin Fe(III) in cyano-methemoglobin than in imidazole-methemoglobin (see ch. 2). As discussed in section 1, the Fe(III) has an impact on T_1 as well as T_m for the spin label. Therefore the spin echo data were obtained with long pulse-repetition times to ensure that all components of the distribution of relaxation rates were sampled appropriately. The time constants for the echo dephasing are easier to analyze when there is less echo envelope modulation, so decay curves were recorded with 40 and 80 ns 90° and 180° pulses rather than the shorter pulses that one would use when the goal is to maximize echo modulation.

The spin echo decay curves were simulated using a locally written program, TMDYNAM that is based on eq. (6) – (8). In this program Δ is calculated from the expressions that have been used to analyze the spin-spin splitting in CW EPR spectra (Eaton et al., 1983). The value of Δ was calculated for each orientation of the molecule in the magnetic field. Depending upon the relative values of $1/\tau_c$ and Δ, eq. (6), (7), or (8) was used to calculate the echo decay curve for that orientation. The calculated echo decay curve was multiplied by an addition decay characterized by T_m and x (eq. 10) observed for noninteracting nitroxyl, to account for other relaxation processes. The powder average decay curve for the frozen solution samples was obtained by summing the contributions from a spherical average of orientations.

The impact of orientation selection was tested in the calculations. Only those orientations of the molecule having resonance fields within a specified window around the observing field were included in the calculated summation. Most of the experimental ESE data were recorded at a magnetic field in the center of the nitroxyl spectrum with a microwave B_1 of about 5 gauss. In addition, unresolved proton hyperfine splitting spreads the signal from a particular orientation of the molecule over about 7 G (the inhomogenously broadened line width required to simulate the CW spectra at 100 K). Due to the magnitude of B_1, the proton hyperfine interaction, and the overlapping contributions from g anisotropy, decay curves calculated with orientation selection at the center of the spectrum (a window of 5 to 7 gauss) agreed within experimental noise levels with curves calculated without orientation selection. At a few temperatures echo decay curves were recorded as a function of position in the spectrum and indicated some orientation selection at the extremes of the spectra.

In the slow-averaging limit the ESE decay curve is independent of Δ (eq. (7)). For larger ratios of averaging rate to Δ, the decay curves depends on Δ (eq. (6) and (8)). Δ is orientation dependent, so for the powder-average ESE curves there is a superposition of curves with different values of Δ. For a given rate of metal relaxation, that rate might be fast relative to the dipolar interaction at some orientations, but intermediate or slow at other orientations in the powder average. Therefore the shape and intensity of the decay curve depends on r even for relatively slow metal relaxation rates. The value of r was adjusted to fit the full set of temperature dependent ESE decays (Budker et al., 1995). The resulting values of r for cyano-methemoglobin and imidazole-methemoglobin were 15 to 15.5 Å, which is in good agreement with values obtained by SR (Table 3).

Spin echo decay curves also have been analyzed for a series of spin-labeled low-spin Fe(III) porphyrins (Rakowsky et al., 1995). The values of r obtained for these complexes were between 11 and 15 Å, which are in good agreement with values obtained by SR for the same complexes. The through-bond pathways between the iron and nitroxyl are shorter for these complexes than for the methemoglobin samples, which raises the possibility of a through-bond exchange interaction. If isotropic exchange dominated the iron-nitroxyl interaction the decays would be single exponentials, which is not consistent with the experimental ESE decays. Inclusion of small exchange contributions did not improve the fit to the experimental data (Rakowsky et al., 1995).

Table 3. Iron-nitroxyl interspin distances (Å) for spin-labeled methemoglobin[a]

	CW lineshape	SR	ESE
Low-spin Fe(III)			
Im-metHb-ISL	14-17	15.5 ± 1	15 ± 1
CN-metHb-ISL	14-17	15 ± 1	15 ± 1
CN-metHb-MSL	14-17	15.5 ± 1	15.5 ± 1
High-spin Fe(III)			
F-metHb-ISL		15.8 ± 0.5	16 ± 2
H_2O-metHb-ISL		17.4 ± 0.3	17.5 ± 2
H_2O-metHb-MSL		17.1 ± 0.3	17 ± 2

[a]Spin label was attached to the β-93 cysteine. ISL is 4-(2-iodoacetamido)-TEMPO. MSL is 4-maleimido-TEMPO.

We have occasionally observed that the long echo decay times for small amounts of non-interacting (unbound) spin label can dominate the echo decay and give the impression that there is negligible iron-nitroxyl interaction. This problem can be recognized by the low intensity of the echoes for these samples, which demonstrates the importance of examining the intensity of the echoes as well as the decay time constants.

4.1.2 Analysis of echo intensity as a function of temperature

The intensity of the spin echo was measured as a function of temperature for four spin-labeled mutants of cyano-methemoglobin and imidazole-methemoglobin at an inter-pulse spacing of 230 and 526 ns. Echo amplitudes were measured with a Lecroy 9410 digital oscilloscope at pulse repetition times that are long relative to the nitroxyl T_1. The pulse repetition time was increased until further increase led to no change in the echo intensity. The echo intensities were corrected for Boltzmann factor by multiplying by T. Intensities were corrected for the temperature dependence of resonator Q by monitoring echo intensity as a function of temperature for a nitroxyl sample that was not interacting with a rapidly-relaxing metal. This temperature dependence was applied to the data in the presence of the iron. Corrected echo intensities were plotted as a function of temperature to determine the minimum intensity. The minimum echo intensity occurred at $1/\tau_c \sim 10^7$ s^{-1} for both CN^- and imidazole adducts (Y. Zhou et al, unpublished results). The minimum echo intensities were about 8% and about 40% of the noninteracting intensity for variants with an iron-nitroxyl distance about 19 Å and 28 Å, respectively. The substantial impact of the low-spin Fe(III) on the intensity of the nitroxyl spin echo at a distance of about 28 Å suggests that measurements of echo intensity will be useful to determine interspin distances that are substantially longer than 28 Å.

4.2 Effect of High-spin Fe(III) on Echo Dephasing

The dipole-dipole splitting is proportional to $\sqrt{S(S+1)}$, so for the same iron-nitroxyl distance larger splittings are observed for high-spin Fe(III) (S = 5/2) than for low-spin Fe(III) (S = ½). The maximum enhancement of $1/T_m$ increases as the dipolar splitting increases, so for the same interspin distance the effect on nitroxyl $1/T_m$ is larger for high-spin iron than for low-spin iron, which extends the range of distances at which this technique can be used.

For spin-labeled high-spin Fe(III) porphyrins with iron-nitroxyl distances less than about 15 Å, the effect of the Fe(III) on T_m for the spin label at temperatures between about 6 and 50 K is so great that echo intensity at accessible τ values was less than 10% of that expected for the known nitroxyl concentrations (Rakowsky et al., 1998). Thus, we conclude that these distances are too short to use the effect of the iron on the nitroxyl T_m to estimate distances. For a sample with an interspin distance of 15 Å the value of T_m became strongly temperature dependent above about 5 K, which is the temperature regime in which the iron relaxation rates are appropriate to average the dipolar couplings. Because of the S=5/2 spin of the iron, the averaging of the dipolar coupling is a six-site exchange. As a first approximation the spin echo decays were analyzed as a two-site exchange with a dipolar coupling that is larger than for S = ½ by a factor of $\sqrt{S(S+1)}$ (Rakowsky et al., 1998). The relaxation rates for the iron $m_s = \pm 1/2$ transitions were used in the simulations. The shapes of the spin echo decay curves were consistent with a distance of about 15 Å. Even at this distance the effect of the high-spin Fe(III) is so large that at some temperatures, for many orientations of the molecule with respect to the external magnetic field, $1/T_m$ is so fast that most of the contributions to the ESE curve occur during the instrumental deadtime and the primary effect on the observed echo is a decrease in echo intensity even at τ values as short as 100 ns.

ESE decay curves were obtained for methemoglobin spin-labeled at the β-93 cysteine and either fluoride or water as the axial ligand (Seiter et al.,

1998). The maximum effect of the high-spin iron on the nitroxyl T_m was observed at lower temperature (about 10 K) than for the low-spin hemes (see section 4.1.1) because relaxation rates in this temperature range are faster for high-spin Fe(III) than for low-spin Fe(III). The effect of the iron on the nitroxyl spin echo decay was approximated as a two-site exchange and the relaxation rates for the iron $m_s = \pm 1/2$ transitions (as discussed in section 2.3) were used in simulations based on eq. (6) - (8). The estimated distances were 16 to 17.5 Å (Table 3). These values are in good agreement with values obtained by SR for the same samples (Seiter et al., 1998). The effects of the high-spin iron on the echo decay curves were so dramatic at these distances that it is anticipated that significant effects will be observable at distances substantially longer than 17 Å.

A mutant of the ferric enterobactin receptor, FepA, containing a valine to cysteine (V338C) substitution was spin-labeled with MTSL. Binding of the ligand, ferric enterobactin (FeEnt), to the purified spin-labeled protein caused a significant decrease in the nitroxyl T_m at temperatures between about 30 and 70 K (Klug et al., 1998). This decrease in T_m is characteristic of the interaction between the iron and the nitroxyl in the temperature regime where the Fe(III) relaxation rate is comparable to the dipolar splitting. The largest effect of the iron on the spin label occurred at higher temperature (about 50 K) for this high-spin non-heme Fe(III) than for the high-spin heme Fe(III) examples discussed above because the iron relaxation rates are much slower for the non-heme Fe(III) in this protein than for the heme Fe(III) (see ch.2). The estimated interspin distance is in the range of 20 – 30 Å (Klug et al., 1998). Quantitative analysis of the iron-nitroxyl interaction in this case is complicated because the ZFS is of the order of the X-band EPR quantum. The ESE results, combined with the observation that in fluid solution the spin label mobility is greatly restricted by binding of FepA suggest that the V338C may occupy a hinge region connecting the ligand binding loop to the β-barrel embedded in the membrane (Klug et al., 1998).

5. SUMMARY AND GUIDE TO APPLICATIONS OF THESE TECHNIQUES

A wide variety of metalloproteins contain a rapidly-relaxing metal center such as heme or non-heme iron. In addition, a rapidly-relaxing paramagnetic metal can be substituted for a diamagnetic metal such as Co(II) for Zn(II) or Mn(II) for Mg(II). In such cases, the interaction of the metal spin with another spin can be used to estimate distances. In addition, metal binding sites can be engineered into proteins.

When the metal relaxation rate is comparable to the microwave frequency the contribution from the C term in Eq. (3) is maximized. For the same metal relaxation rate, the effect of the metal on the nitroxyl spin lattice relaxation rate ($1/T_{1s} - 1/T_{1s}^{o}$) is greater when the nitroxyl relaxation rate is slower. Since the nitroxyl relaxation rate decreases with decreasing temperature, it is possible to measure longer metal-radical distances for metals that relax quickly at lower temperatures than for more slowly-relaxing metals. For example, we predict that longer distances will be accessible for more rapidly-relaxing high-spin Co(II) than for slower relaxing high-spin Fe(III).

The factor of S(S+1) in Eq. (3) might give the impression that longer distances would be accessible for metals with larger values of S. This would be true if one were comparing two metals with the same relaxation rates and anisotropy of EPR transition energies. However, our experiments indicate that despite the larger value of S for high-spin Fe(III) comparable distances are accessible by measurements of SR data for low-spin Fe(III) and high-spin Fe(III) because for relaxation rates in typical protein environments the condition of $T_{1f}^{-1} \sim \omega_s$ is achieved at lower temperatures for low-spin Fe(III) than for high-spin Fe(III) (ch. 2, Seiter et al., 1998; Rakowsky et al., 1997). We estimate that for iron relaxation rates similar to those in aquo-methemoglobin or metmyoglobin-cyanide the upper limit for distance measurements by saturation recovery at X-band is about 30 – 35 Å (Seiter et al., 1998).

Dramatic effects on spin-echo dephasing are observed in the temperature interval in which the metal relaxation rate is comparable to the metal-radical dipolar interaction. These effects are expected to be observable out to longer distances than the effects on SR curves, especially for metals with $S>1/2$. In the spin echo experiments the larger moment for high-spin Fe(III) causes a larger dipolar splitting and therefore the effects on the echo decays are larger than for low-spin Fe(III). Calculations suggest an upper limit of about 50 Å for ESE measurements of distances between high-spin Fe(III) and a spin label by measurement of echo intensity as a function of temperature. There is a significant caveat – it will be important to be sure that the observed effects on the nitroxyl T_m are due to interaction with the rapidly-relaxing metal and not due to some other dynamic effect in the system.

Complete analysis of the saturation recovery curve or echo decays requires knowledge of the relaxation rates for the rapidly-relaxing metal ions. However comparison of relative effects for labels at different distances from the same paramagnetic center are likely to be powerful tools even when the metal relaxation rate is not known. It may also be possible to use these methods "in reverse" to determine the effective metal relaxation rates via the effect of the metal on the more slowly-relaxing spin. The reliability of relative methods depends upon how sensitively the effect of the metal on the more slowly-relaxing spin depends upon the relative orientation of the interspin vector and the g axes for anisotropic metal centers. Our preliminary results indicate that this is not the dominating source of uncertainty in the distance measurements.

Although dipolar splittings depend on r^{-3}, the impact of dipolar interaction on relaxation rates decrease as r^{-6}. The dependence of relaxation effects on r^{-6} has two impacts on the measurements of distances. First, the effects on relaxation drop off more rapidly with distance, so relaxation effects may not be useful in measuring distances as large as are accessible by DEER or the 2+1 sequence (see ch.1), which measure dipolar splittings and therefore depend on r^{-3}. Second, if there is a distribution of distances relaxation effects are more strongly selective for the shorter distances than are the splittings. However, when there are pairwise interactions, this 6th power dependence can be a great advantage, because the predominant effect is within the pair. Interaction with other pairs, even in the random distribution present in fluid or frozen solutions has much less impact.

6. ACKNOWLEDGMENT

The support of this work by NIH GM21156 is gratefully acknowledged.

7. REFERENCES

Abragam, A. (1961). *The Principles of Nuclear Magnetism*, Oxford University Press, Oxford, p. 289, 311.

Bertini, I. and Luchinat, C. (1996). *Relaxation in NMR of Paramagnetic Substances, Coord. Chem. Rev.* **150**, 77-110.

Bloembergen, N. (1949). On the Interaction of Nuclear Spins in a Crystalline Lattice. *Physica* **15**, 386-426.

Bloembergen, N., Purcell, E. M., and Pound, R. V. (1948). Relaxation Effects in Nuclear Magnetic Resonance Absorption. *Phys. Rev.* **73**, 679-712.

Bloembergen, N., Shapiro, S., Pershan, P. S., Artman, J. O. (1959). Cross-Relaxation in Spin Systems. *Phys. Rev.* **114**, 445-459.

Brown, I. M. (1979). Electron Spin-Echo Studies of Relaxation Processes in Molecular Solids, in *Time Domain Electron Spin Resonance*, Ed. by L. Kevan and R. N. Schwartz, Wiley, N. Y., ch. 6, 195-229.

Brudvig, G. W., Blair, D. F., and Chan, S. I. (1984). Electron Spin Relaxation of Cu_A and Cytochrome a in Cytochrome c Oxidase, *J. Biol. Chem.* **259**, 11001-11009.

Budker, V., Du, J.-L., Seiter, M., Eaton, G. R., and Eaton, S. S. (1995). Electron-Electron Spin-Spin Interaction in Spin-Labeled Methemoglobin, *Biophys. J.* **68**, 2531-2542.

Damoder, R., More, K. M., Eaton, G. R., and Eaton, S. S. (1983). Metal-Nitroxyl Interactions. 30. Single-Crystal EPR Spectra of Two Spin-Labeled Copper Porphyrins, *J. Amer. Chem. Soc.* **105**, 2147-2154.

Doctor, K. S., Gaffney, B. J., Alvarez, G., and Silverstone, H. J. (1993). EPR spectroscopy of interdoublet transitions in high-spin iron: application to transferrin oxalate, *J. Phys. Chem.* **97**, 3028-3033.

Drago, R. S. (1977). *Physical Methods in Chemistry*, Saunders, Philadelphia, p. 252.

Du., J.-L., Eaton, G. R., and Eaton, S. S. (1994). Effect of Molecular Motion on Electron Spin Phase Memory Times for Copper(II) Complexes in Doped Solids, *Appl. Magn. Reson.* **6**, 373- 378.

Du, J.-L., Eaton, G. R., and Eaton, S. S. (1995). Temperature, Orientation, and Solvent Dependence of Electron Spin-Lattice Relaxation Rates for Nitroxyl Radicals in Glassy Solvents and Doped Solids, *J. Magn. Reson. A* **115**, 213-221.

Dzuba, S. A., Raitsimring, A. M., and Tsvetkov, Yu. D. (1979). The distance distribution of radical-paramagnetic ion pairs studied by the electron spin echo method. Spatial irregularities of radical diffusion in glassy alcohols, *Chem. Phys.* **44**, 357-365.

Dzuba, S. A., Raitsimring, A. M., and Tsvetkov, Yu. D. (1980). Electron Spin-Echo Studies of Phase Relaxation Kinetics in Systems Containing two Types of Spins, *J. Magn. Reson.* **40**, 83-89.

Dzuba, S. A., Salikhov, K. M., Tsvetkov, Yu. D. (1981). Slow rotation ($\tau \geq 10^{-5}$ s) of methyl groups in radicals studied by pulsed ESR spectroscopy, *Chem. Phys. Lett.* **79**, 568-572.

Eaton, S. S. and Eaton, G. R. (1988). Interaction of Spin Labels with Transition Metals. Part 2., *Coord. Chem. Rev.* **83**, 29-72.

Eaton, S. S., More, K. M., Sawant, B. M., Boymel, P. M., and Eaton, G. R. (1983). Metal-Nitroxyl Interactions. 29. EPR Studies of Spin-Labeled Copper Complexes in Frozen Solution, *J. Magn. Reson.* **52**, 435-449.

Fiamingo, F. G., Brill, A. S., Hampton, D. A., and Thorkildsen (1989). Energy distribution at the high-spin ferric sites in myoglobin crystals, *Biophys. J.* **55**, 67-77.

Fielding, L., More, K. M., Eaton, G. R., and Eaton, S. S. (1986). Metal-Nitroxyl Interactions. 51. Collapse of Iron-Nitroxyl Electron-Electron Spin-Spin Splitting Due to an Increase in the Electron Spin Relaxation Rate for High-Spin Iron (III) When Temperature is Increased, *J. Am. Chem. Soc.* **108**, 8194-8196.

Gamble, W. L., Miyagawa, I., Hartman, R. L. (1968). E. S. R. study of stereospecific proton transfer in irradiated crystals of L-alanine, *Phys. Rev. Lett.* **20**, 415 - 418.

Goodman, G. and Leigh, J. S., Jr., (1985). Distance Between the Visible Copper and Cytochrome a in Bovine Heart Cytochrome Oxidase, *Biochemistry* **24**, 2310-2317.

Hilczer, W., Goslar, J., Gramza, M., Hoffmann, S. K., Blicharski, W., Osyczka, A., Turyna, B., and Froncisz, W. (1995). A resonance enhancement of the phase relaxation in the electron spin echo of nitroxide covalently attached to cytochrome c, *Chem. Phys. Lett.* **247**, 601-606.

Hirsh, D. J. and Brudvig, G. W. (1993). Long-Range Electron-Spin-Spin Interactions in the Bacterial Photosynthetic Reaction Center, *J. Phys. Chem.* **97**, 13216-13222.

Hirsh, D. J., Beck, W. F., Innes, J. B., and Brudvig, G. W. (1992). Using saturation recovery EPR to measure distance in proteins: Application to photosystem II, *Biochemistry* **31**, 523-541.

Hyde, J. S. and Rao, K. V. S. (1978) Dipolar-induced Electron Spin-Lattice Relaxation in Unordered Solids, *J. Magn. Reson.* **29**, 509-516.

Kispert, L. D., Bowman, M. K., Norris, J. R., and Brown, M. S. (1982). Electron spin echo studies of the internal motion of radicals in crystals: phase memory vs. correlation time, *J. Chem. Phys.* **76**, 26-30.

Klug, C. S., Eaton, S. S., Eaton, G. R., and Feix, J. B. (1998). Ligand-Induced Conformational Change in the Ferric Enterobactin Receptor FepA as Studied by Site-Directed Spin Labeling and Time-Domain ESR, *Biochemistry* **37**, 9016-9023.

Koenig, S. H. (1982). A classical description of the relaxation of interacting pairs of unlike spins: extension to $T_{1\rho}$, T_2, and $T_{1\rho\mathrm{off}}$ including contact interactions, *J. Magn. Reson.* **47**, 441-453.

Koulougliotis, D., Innes, J. B., and Brudvig, G. W. (1994). Location of chlorophyll$_z$ in photosystem II, *Biochemistry* **33**, 11814-11822.

Koulougliotis, D., Tang, X.-S., Diner, B. A., and Brudvig, G. W. (1995). Spectroscopic evidence for the symmetric location of tyrosines D and Z in Photosystem II, *Biochemistry* **34**, 2850-2856.

Kulikov, A. V. and Likhtenshtein, G. I. (1977). The use of spin relaxation phenomena in the investigation of the structure of model and biological systems by the method of spin labels, *Adv. Mol. Relax. and Interaction Proc.* **10**, 47-69.

Leigh, J. S., Jr., (1970). ESR Rigid-Limit Line Shape in a System of Two Interacting Spins, *J. Chem. Phys.* **52**, 2608-2612.

Makinen, M. W., and Wells, G. B. (1987). Application of EPR Saturation Methods to Paramagnetic Metal Ions in Proteins, *Metal Ions in Biological Systems* **22**, 129-206.

Nakagawa, K., Candelaria, M. B., Wilson, W. W. C., Eaton, S. S., and Eaton, G. R. (1992). Electron-Spin Relaxation Times of Chromium(V), *J. Magn. Reson.* **98**, 81-91.

Ohnishi, T., LoBrutto, R., Salerno, J. C., Bruckner, R. C., and Frey, T. G. (1982). Spatial Relationship between Cytochrome a and a_3, *J. Biol. Chem.* **257**, 14821-14825.

Poole, C. P., Jr., and Farach, H. (1971). *Relaxation in Magnetic Resonance*, Academic Press, New York, pp. 70-71, 196-200.

Raitsimring, A. M. and Salikhov, K. M. (1985). Electron spin echo method as used to analyze the spatial distribution of paramagnetic centers, *Bull. Magn. Reson* **7**, 184-217.

Rakowsky, M. H., More, K. M., Kulikov, A. V., Eaton, G. R., and Eaton, S. S. (1995). Time-Domain Electron Paramagnetic Resonance as a Probe of Electron-Electron Spin-Spin Interaction in Spin-Labeled Low-Spin Iron Porphyrins. *J. Amer. Chem. Soc.* **117**, 2049-2057.

Rakowsky, M. H., Eaton, G. R., and Eaton, S. S. (1997). Comparison of the effect of high-spin and low-spin Fe(III) on nitroxyl T_1 in a spin-labeled porphyrin, *Modern Applications of EPR/ESR from Biophysics to Materials Science*, C. Z. Rudowicz, K. N. Yu, and H. Hiraoka, eds., Springer, 19-24.

Rakowsky, M. H., Zecevic, A., Eaton, G. R., and Eaton, S. S. (1998). Determination of High-Spin Iron(III)-Nitroxyl Distances in Spin-Labeled Porphyrins by Time-Domain EPR, *J. Magn. Reson.* **131**, 97-110.

Rubenstein, M., Baram, A. and Luz, Z. (1971). Electronic and nuclear relaxation in solutions of transition metal ions with spin S = 3/2 and 5/2, *Mol. Phys.* **20**, 67-80.

Salikhov, K. M. and Tsvetkov, Yu. D. (1979). Electron Spin-Echo Studies of Spin-Spin Interactions in Solids, in *Time Domain Electron Spin Resonance*, Ed. by L. Kevan and R. N. Schwartz, Wiley, N. Y., ch. 7, 232-277.

Scholes, C. P., Janakiraman, R., Taylor, H., and King, T. E. (1984). Temperature dependence of the electron spin-lattice relaxation rate from pulsed epr of Cu_A and heme a in cytochrome c oxidase, *Biophys. J.* **45**, 1027-1030.

Seiter, M., Budker, V., Du, J.-L., Eaton, G. R., and Eaton, S. S. (1998). Interspin distances determined by time domain EPR of spin-labeled high-spin methemoglobin, *Inorg. Chim. Acta* **273**, 354-356.

Tsvetkov, Yu. D. and Dzuba, S. A. (1990). Pulsed ESR and molecular motions, *Appl. Magn. Reson.* **1**,179-194.

Voss, J., Salwinski, L., Kaback, H. R., and Hubbell, W. L. (1995a). A method for distance determination in proteins using a designed metal ion binding site and site-directed spin labeling: evaluation with T4 lysozyme, *Proc. Natl. Acad. Sci U.S.* **92**, 12295-12299.

Voss, J., Hubbell, W. L., and Kaback, H. R. (1995b). Distance determination in proteins using designed metal ion binding sites and site-directed spin labeling: Application to lactose permease of *Escherichia coli*, *Proc. Natl. Acad. Sci. U.S.* **92**, 12300-12303.

Voss, J., Hubbell, W. L., and Kaback, R. (1998). Helix packing in the lactose permease determined by metal-nitroxide interaction, *Biochemistry* **37**, 211-216.

Wolf, E. L. (1966). Diffusion effects in the inhomogeneously broadened case: high temperature saturation of the F-center electron spin resonance, *Phys. Rev.* **14**, 555-569.

Zhidomirov, G. M. and Salikhov, K. M. (1969). Contribution to the theory of spectral diffusion in magnetically diluted solids, *Sov. Phys. JETP* **29**, 1037-1040.

Zhou, Y., Bowler, B. E., Eaton, G. R. and Eaton, S. S. (2000a). Electron spin-lattice relaxation rates for high-spin Fe(III) in glass solvents at temperatures between 6 and 298 K, *J. Magn. Reson.*, accepted for publication.

Zhou, Y., Bowler, B. E., Lynch, K., Eaton, S. S., and Eaton, G. R. (2000b). Interspin distances in spin-labeled metmyoglobin variants determined by saturation recovery EPR, *Biophys. J.*, submitted for publication.

Chapter 9

Double-Quantum ESR and Distance Measurements

Petr. P. Borbat and Jack. H. Freed
Baker Laboratory of Chemistry and Chemical Biology Cornell University, Ithaca, NY 14853

Abstract: "Allowed" double quantum coherences (DQC) can now be routinely generated in disordered and oriented solids containing nitroxide biradicals and random distributions of stable radicals. The Pake doublets obtained from DQC pathways can be effectively used to determine a broad range of distances in the former case whereas decay constants yield concentrations in the latter. The DQC signals are strong and often comparable to standard single quantum signals. They are free of any large undesirable signals, so the DQ experiment is easy to perform. Their strong intensity permits the study of low concentrations of spins in samples typical of those ordinarily met in the case of doubly-labeled macromolecules such as proteins and polypeptides. The upper range of distances for systems labeled with nitroxides is estimated to be ca. 80 Å. In the limit of non-selective pulses the interpretation of DQC signals becomes independent of complicating geometric features which affect other ESR distance methods. The method is compared to other existing pulse distance measurement techniques and future improvements are also discussed.

1. INTRODUCTION: MOTIVATION

Multiple quantum coherence (MQC) is a well-known phenomenon in modern NMR, with many useful applications (Ernst *et al.*, 1987). On the other hand, it has been a considerable challenge to generate MQC in pulsed ESR. In recent work, Saxena and Freed (1997a) showed how "forbidden" coherence (FC) pathways can be utilized to generate double quantum coherence (DQC) spectra from a doubly labeled nitroxide (Saxena and Freed, 1996, 1997a). These FC pathways could be generated even under spectrometer conditions that were not favorable for producing DQC in nitroxides by allowed pathways, [e.g. insufficiently large microwave magnetic (B_1) fields in resonators with wide bandwidths]. A useful feature of the FC pathways is

Biological Magnetic Resonance, Volume 19: Distance Measurements in Biological Systems by EPR,
edited by Berliner et al. Kluwer Academic/Plenum Publishers, New York, 2000

that they are able to provide some orientational selectivity with respect to the angle between the interspin vector, $\boldsymbol{r}$, and the static magnetic field $\boldsymbol{B}_0$. This enhances their utility for measuring the distance between electron spins. A disadvantage is in weaker signals, and the precise adjustment of the B_1 field to its optimum magnitude is a complicating factor.

Given recent developments in pulsed ESR capabilities in our laboratory (Borbat and Freed, 1997) consisting of shorter dead-times, increased B_1's without a loss in resonator bandwidth, and increased signal-to-noise ratio (SNR) at higher working frequency, it proved possible to directly produce MQC by "allowed" pathways using several pulse sequences similar to those employed in NMR, but specially tailored for the ESR challenges (Borbat and Freed, 1999). These challenges include short relaxation times, T_2. They also include the typically broad spectral extent arising from electron-nuclear dipolar and g-tensor interactions that requires very strong B_1-fields in order to irradiate all the electron spins. Our successful results show, that by "allowed" coherence pathways, DQC signals can be achieved that are 10-100 times greater than for FC pathways. They also indicate some of the potentially rich uses to which MQC ESR can be employed. More recent results show that DQC signals can be obtained with smaller B_1 fields, especially in the case of large distances, but at the expense of SNR and resolution. Also, smaller B_1's considerably increase the minimum distance that can be reliably measured, (Borbat and Freed, 2000).

In the present chapter we emphasize the use of DQC for measurements of broad range of distances, which is important in solid-state studies and in biophysics. In the latter context one can use nitroxide bilabeled macromolecules to study aggregation and conformations in peptides and proteins as well as movement of their fragments (Farhbakhsh *et al.*, 1997; Thorgeirsson *et al.*, 1997). However, these biophysical applications were limited by an upper range of distance of ca. 20 Å (Rabenstein and Shin, 1995) since they are based upon cw-ESR. Also, whereas fluorescence energy transfer is capable of measuring distances greater than 20 Å, it requires two different types of fluorescent probes, and it is often limited by the uncertainty in the κ^2 parameter, which depends on the relative orientation of the transition-moments of the two chromophores. In the case of ESR, electron spins are oriented along $\boldsymbol{B}_0$.

Pulsed ESR methods have been used in the past by Milov *et al.* (1981, 1985), Ichikawa *et al.* (1985), Raitsimring and Salikhov (1985), Borbat *et al.*, (1977); to yield spatial distributions in solids. Several methods of practical importance have emerged. They are based on modifying the single quantum (SQ) echo produced by a 2 or 3 pulse sequence with a pulse at either the same frequency (Kurshev *et al.*, 1989; cf. Ch. 10) or at a different frequency (Milov *et al.*, 1998; Pfannebecker *et al.*, 1996; Pannier *et al.*, 1998; Larsen and Singel, 1993). The latter, based on selective pulses, which is known as DEER (double electron-electron resonance) or PELDOR, (Milov *et al.*, 1998; cf. Ch. 11), originated from a solid-state NMR concept (Emshwiller *et al.*, 1960). It has proven effective in a number of applications, including bi-labeled biomolecules. Unlike DQC, it is not based on coherent effects. While there are some similarities, we find that DQC based upon allowed pathways offers several advantages compared to DEER. For example, DQC with strong pulses can yield signals that are at least an order of magnitude greater, in fact comparable to single quantum (SQ) signals from non-selective pulses. This is a very desirable feature when working with small amounts of bilabeled biomolecule and/or when greater sensitivity is needed for measuring longer distances, up to ca. 80 Å.

Strictly speaking, the requirement of using strong (or nearly non-selective) pulses is not absolutely necessary for "allowed" DQC signals to be observed. However it is very beneficial to use pulses as strong as possible. Recent work has shown that some coherent SQ pulse sequences are also useful for distance measurements, but are not nearly as effective as the DQ signals produced by strong pulses (Borbat and Freed, 1999a, 2000).

2. THEORY OF DOUBLE QUANTUM COHERENCE

2.1 Multiple Quantum Coherences in ESR

"Allowed" multiple-quantum coherences in ESR were not observed until recently (Borbat, and Freed, 1999) although a form of "forbidden" DQC was reported by Saxena and Freed (1996) and a general theoretical approach for "forbidden" and "allowed" DQC was given by these authors (Saxena and Freed, 1997a). MQC were predicted by Tang, Norris *et al.*, (1995) and observed by Dzyuba *et al.*, (1996) in spin-correlated radical pairs formed by laser-induced dissociation (cf. Ch. 13). The relevant zero quantum (ZQ) coherence arises naturally from the initial polarized non-equilibrium state. This important special case is sufficiently different from the present study that we do not consider it further. The MQC experiment in ESR that we explore is based upon the same principles that were developed in pulsed NMR. In this section, we outline the basics of the double-quantum coherence ESR experiment using a simplified approach.

2.1.1 Double and Zero-Quantum Coherences

Multiple-quantum transitions and coherences have been treated in detail in numerous studies in the field of NMR. See for example the following references (Ernst *et al.*, 1987; Slichter, 1990; Cavanagh, 1996; Corio, 1966). Here, we outline the basic concepts and definitions in a way that is appropriate for the ESR experiment.

For a quantum system in thermal equilibrium, only the diagonal density matrix elements are non-zero and they represent the populations of the corresponding energy levels. All off-diagonal density matrix elements are zero. This is the fundamental property of random phases between any two states of the quantum system at thermal equilibrium. Non-zero off-diagonal density matrix elements correspond to coherent non-equilibrium superposition of states and therefore must vanish as the system approaches the equilibrium state. These elements are often referred to as coherences, and they can be produced by a coherent perturbation, for example by an *rf* pulse acting on the system. The action of the pulse yields such a coherent superposition between the states involved in the transition, which will persist until the system relaxes to thermal equilibrium.

In summary, coherences may be generated by coherently inducing transitions between states, and they are represented by the corresponding off-diagonal elements of the density matrix. Consider for example a system of two coupled spins of 1/2. In Fig. 1 we have four possible single-quantum transitions.

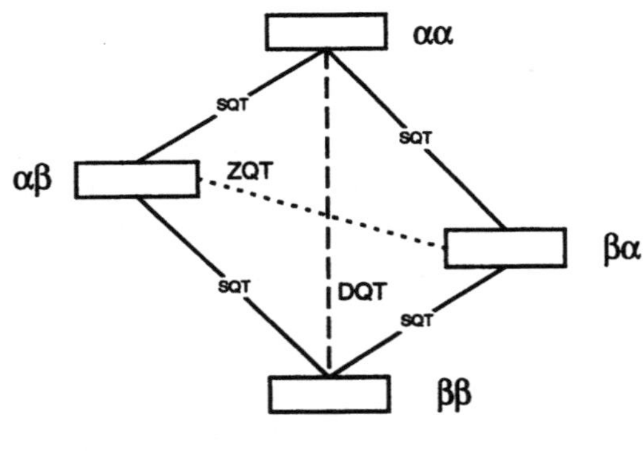

Figure 1. Energy level diagram for the two-spin system showing single (SQT), double (DQT), and zero (ZQT) quantum transitions.

These transitions are represented by the corresponding off-diagonal elements of the density matrix for the two spins. One can conveniently express the density matrix as the expansion in a complete set of the spin operators, (e.g. the density matrix for a single spin is given as a linear combination of the Pauli spin matrices and the unit operator; for many-spin systems one should take the direct product).

Then the coherences can be conveniently expressed by the spin operators corresponding to the elements of the density matrix for two spins defined in the product space $|m_1 m_2\rangle$. Single quantum transitions are represented by:

$$S_1^+ S_2^\alpha,\ S_1^+ S_2^\beta,\ S_1^- S_2^\alpha,\ S_1^- S_2^\beta \tag{1}$$

and those obtained by permuting the spins. Here,

$$S_i^\alpha \equiv \tfrac{1}{2} E_i + S_{iz},\quad S_i^\beta \equiv \tfrac{1}{2} E_i - S_{iz} \tag{2}$$

and E_i is the 2×2 unit operator for the ith spin. Associated with these transitions are the single-quantum coherences (SQC), of which four are "inphase" coherences, which have the form:

$$S_1^\pm (S_2^\alpha + S_2^\beta) = S_1^\pm E_2\,. \tag{3}$$

They correspond to the observable transverse magnetization. The other four are the antiphase coherences which cannot be directly detected but can evolve into observable SQC. They have the form:

$$S_1^\pm (S_2^\alpha - S_2^\beta) = 2S_1^\pm S_{2z}\,. \tag{4}$$

The double- and zero-quantum coherences (DQC and ZQC, respectively) are associated with the remaining two transitions. DQC and ZQC correspond to transitions with simultaneous flips or flip-flops of both spins, respectively, and can be represented as follows (Ernst *et al.*, 1987; Cavanagh *et al.*, 1996):

$$\begin{aligned} DQ_x &= \frac{1}{2}(S_1^+ S_2^+ + S_1^- S_2^-) = \frac{1}{2}(2S_{1x}S_{2x} - 2S_{1y}S_{2y}) \\ DQ_y &= \frac{1}{2i}(S_1^+ S_2^+ - S_1^- S_2^-) = \frac{1}{2}(2S_{1x}S_{2y} + 2S_{1y}S_{2x}) \end{aligned} \tag{5a}$$

$$\begin{aligned} ZQ_x &= \frac{1}{2}(S_1^+ S_2^- + S_1^- S_2^+) = \frac{1}{2}(2S_{1x}S_{2x} + 2S_{1y}S_{2y}) \\ ZQ_y &= \frac{1}{2i}(S_1^+ S_2^- - S_1^- S_2^+) = \frac{1}{2}(2S_{1y}S_{2x} - 2S_{1x}S_{2y})\,. \end{aligned} \tag{5b}$$

These coherences do not correspond to any experimental observables. They can only be detected indirectly in a pulse experiment. The multiple-quantum experiment is based on the fact that by a suitable pulse sequence, an antiphase SQC can be made to develop and can then be converted into higher orders of coherence and *vice versa*. The antiphase SQC also can be converted into an observable inphase SQC. These properties determine the strategy of the pulsed multiple-quantum experiment (Fig. 2). During the preparation period, the inphase SQC produced by the first pulse of the sequence evolves into an antiphase SQC due to the coupling between the spins. (More pulses can be applied during this period.) Then this coherence is converted by another pulse into ZQC, DQC or even higher orders of coherence (depending on the total spin of the system).

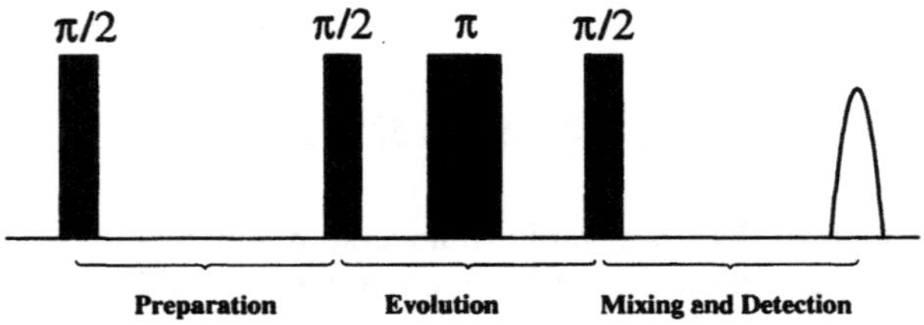

Figure 2. General scheme for the multiple-quantum experiment is exemplified by the 4-pulse sequence. The multiple-quantum pulsed experiment can be subdivided into the distinct periods:
1. During the preparation period single-quantum coherence evolves and becomes labeled by its interaction with other spins and then is converted into higher orders of coherence by the second pulse.
2. Evolution occurs when these newly created multiple-quantum coherences are allowed to evolve. (This period can, and normally will, include additional *rf* or field gradient pulses).
3. Mixing and detection starts with the last $\pi/2$ pulse, which transforms the higher orders of coherence into other orders of coherence including undetectable antiphase single-quantum coherence, which then evolves into observable inphase coherence.

At this stage, these higher orders of coherence evolve during the "evolution period" (with more pulses being applied during this period, as needed to complete the sequence or enhance performance). Finally, during the "mixing" and detection periods these coherences can be converted by an appropriate pulse into antiphase SQC, which evolves into the observable inphase SQC, that at last is detected as an FID or a spin-echo.

The incorporation of the evolution period for multiple-quantum coherence into the pulse sequence results in multiple-quantum filtering; i.e. the sequence transmits only those signals that directly involve the couplings between all of the spins involved in multiple-quantum transitions.

We illustrate these principles with specific examples that are relevant to the pulse schemes currently employed in multiple-quantum pulsed ESR experiments. We assume, for simplicity of presentation, that we have an isolated pair of weakly coupled spins of ½ and perfect, i.e. non-selective *rf* pulses. For a single pair of spins of ½ the highest order of coherence that can be produced is just two. Also we use a weak-coupling Hamiltonian (c.f. Sec. 2.2) in the rotating frame. We split the Hamiltonian into the two parts:

$$H = H_z + H_{zz} \tag{6a}$$

the first one contains the resonant offset terms

$$H_z = \Delta\Omega_1 S_{1z} + \Delta\Omega_2 S_{2z} \tag{6b}$$

while the last term is a dipolar coupling in the weak-coupling limit

$$H_{zz} = aS_{1z}S_{2z}\,. \tag{6c}$$

Here $\Delta\Omega_1$ and $\Delta\Omega_2$ represent the deviation of the resonant frequencies for the spins 1 and 2 respectively from the applied frequency, ω_{rf}. Also $aS_{1z}S_{2z}$ is the secular part of the spin-spin coupling. At this introductory stage, these assumptions are made to simplify the discussion.

2.1.2 Preparation

We describe the preparation of DQC by the use of 2 or 3 pulses. These pulse sequences are the standard preparation sequences in NMR (Ernst *et al.*, 1987) and were employed in the recent DQ ESR experiments by Borbat and Freed, (1999).

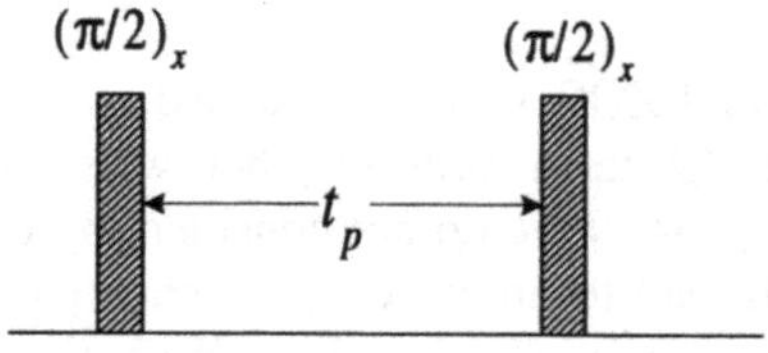

Figure 3. Preparation of multiple-quantum coherence with two pulses for the 4 and 5 pulse sequences.

For the two-pulse preparation sequence (cf. Fig. 3) the first pulse acts on both spins creating transverse magnetization that is proportional to inphase

SQC ($S_{1y}+S_{2y}$). Since S_{1y} and S_{2y} will yield equivalent results (but with subscripts permuted) we need only consider S_{1y} below. We shall employ, at this introductory stage, the standard Product Operator (PO) method (Sørensen *et al.*, (1983); Slichter, (1990); Ernst *et al.*, (1987); Cavanagh *et al.*, (1996); Shriver, 1992).

Here the pulses are taken as ideal $\pi/2$ or π pulses and according to Eqs. 6 the resonant offset terms in H_z commute with the dipolar coupling H_{zz} in the weak-coupling limit. For example, we find using the PO method that the effect of H_{zz} acting over a time, t on S_{1x} and S_{1y} is:

$$\begin{aligned} S_{1x} &\xrightarrow{H_{zz}t} S_{1x}\cos(at/2)+(2S_{1y}S_{2z})\sin(at/2) \\ S_{1y} &\xrightarrow{H_{zz}t} S_{1y}\cos(at/2)-(2S_{1x}S_{2z})\sin(at/2). \end{aligned} \tag{7}$$

One can see from Eq. 7 that the action of a finite spin-spin coupling term a during the time interval t will cause the inphase SQC's S_{1x}, S_{1y} to evolve as $\cos(at/2)$, whereas antiphase SQC's $S_{1x}S_{2z}$, $S_{1y}S_{2z}$ evolve as $\sin(at/2)$. That is, there are "coherent oscillations" between these two coherences. When we include the resonant offset terms as well, we find:

$$\begin{aligned} S_{1y} \xrightarrow{Ht} & (S_{1y}\cos\Delta\Omega_1 t+S_{1x}\sin\Delta\Omega_1 t)\cos(at/2) \\ & +2(S_{1x}S_{2z}\cos\Delta\Omega_1 t-S_{1y}S_{2z}\sin\Delta\Omega_1 t)\sin(at/2)\,. \end{aligned} \tag{8}$$

At the end of the time interval $t = t_p$ the second $\pi/2$-pulse converts the relevant antiphase SQ coherence into the sum of DQC and ZQC.

$$\begin{aligned} & 2S_{1x}S_{2z}\cos\Delta\Omega_1 t_p\sin(at_p/2)\xrightarrow{(\pi/2)_x}\cos\Delta\Omega_1 t_p\sin(at_p/2) \\ & \times[\underbrace{(2S_{1x}S_{2y}+2S_{1y}S_{2x})/2}_{DQ_y}-\underbrace{(2S_{1y}S_{2x}-2S_{1x}S_{2y})/2}_{ZQ_y}]. \end{aligned} \tag{9}$$

In this case the DQC and ZQC will be produced in equal amounts. (In the case of $\Delta\Omega_1=\Delta\Omega_2$ the ZQ_y term vanishes when we consider the similar expression derived for S_{2y}). In. 9 we do not consider any of the other terms on the rhs, since they will lead to unwanted coherence pathways, that must be canceled out by "phase cycling", cf. below. With the 3-pulse preparation sequence, which we consider next, we can obtain better conditions for generation of DQC. In this sequence, the first pulse again produces SQ coherence $S_{1y}+S_{2y}$.

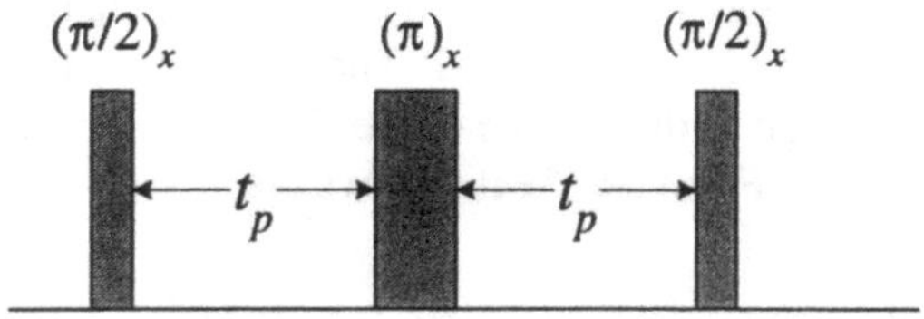

Figure 4. Preparation of DQC with three pulses used in the six-pulse sequence.

This coherence under the combined action of the spin-spin interaction, and the refocusing π-pulse, evolves into the sums of SQ inphase $S_{1y}+S_{2y}$ and antiphase $(S_{1x}S_{2z}+S_{1z}S_{2x})$ coherences:

$$(S_{1y}+S_{2y})\xrightarrow{Ht_p}\xrightarrow{(\pi)_x}\xrightarrow{Ht_p}(S_{1y}+S_{2y})\cos at_p \\ +2(S_{1x}S_{2z}+S_{1z}S_{2x})\sin at_p. \qquad (10)$$

Here, the refocusing π-pulse removes any dependence upon the frequency offsets $\Delta\Omega_1$ and $\Delta\Omega_2$. The last $(\pi/2)_x$ pulse in this three-pulse "sandwich" turns antiphase SQC into DQC:

$$2(S_{1x}S_{2z}+S_{1z}S_{2x})\sin at_p\xrightarrow{(\pi/2)_x}2(S_{1x}S_{2y}+S_{1y}S_{2x})\sin at_p. \qquad (11)$$

Thus all the SQC will be transformed into DQC when $\sin(at_p)=1$. For this case, there is no ZQC. Note, that if the last pulse in the sequence were applied along the y-axis, no DQC would be produced. Thus we see that in the case of non-selective pulses, and the preparation sequence of Fig. 4, only DQ_y is produced.

2.1.3 Evolution of DQC and ZQC due to the Frequency Offsets

During the evolution period the DQC and ZQC evolve due to resonant frequency offsets and are invariant with respect to the dipolar part of the Hamiltonian:

$$DQ_{x(y)}\xrightarrow{H_z t}DQ_{x(y)}\cos[(\Delta\Omega_1+\Delta\Omega_2)t]\pm DQ_{y(x)}\sin[(\Delta\Omega_1+\Delta\Omega_2)t] \\ ZQ_{x(y)}\xrightarrow{H_z t}ZQ_{x(y)}\cos[(\Delta\Omega_1-\Delta\Omega_2)t]\pm ZQ_{y(x)}\sin[(\Delta\Omega_1-\Delta\Omega_2)t]. \qquad (12a)$$

For the case of a typical ESR spectrum where there are wide spectral extents, signals are detected in the form of echoes, and the FID's decay too fast to be detected. To produce the maximum of the SQ echo at the final stage, the coherences in Eq. 12a should be refocused. The refocusing can be accomplished by a π-pulse placed at the middle of the DQC or ZQC pathway forming the sequence t_1 - π - t_1. From Eq. 5a it follows that

$$DQ_y \xrightarrow{\pi_x} -DQ_y \,, \qquad DQ_x \xrightarrow{\pi_x} DQ_x \,. \tag{12b}$$

Let us suppose that the second t_1 interval is changed by an amount δt. From Eqs. 12a and 12b one finds:

$$\begin{aligned} DQ_y(2t_1+\delta t) = DQ_y(0)\cos[(\Delta\Omega_1+\Delta\Omega_2)\delta t] \\ - DQ_x(0)\sin[(\Delta\Omega_1+\Delta\Omega_2)\delta t]. \end{aligned} \tag{12c}$$

Let us assume for simplicity that both spins have identical ESR spectra of the form $g(\Delta\Omega)=\pi^{-1/2}\Delta\Omega_0^{-1}\exp(-\Delta\Omega^2/\Delta\Omega_0^2)$ where $\Delta\Omega_0$ represents the spectrum width. Then we can integrate Eq. 12c over the distribution for both spins, to obtain the average signal for $DQ_y(2t_1+\delta t)$. We find:

$$DQ_y(2t_1+\delta t) = DQ_y(0)\exp(-\Delta\Omega_0^2\delta t^2/2)\,. \tag{12d}$$

The same result holds for ZQC. The reader can see that Eq. 12d describes an "echo" with maximum at $\delta t = 0$ and half width of $2(2\ln 2)^{1/2}\Delta\Omega_0^{-1}$. However, this DQ echo cannot be directly observed.

2.1.4 Mixing and Detection

At the end of the evolution period the final $\pi/2$ pulse (cf. Fig. 2) converts the refocused MQC (cf. previous Section) into SQ antiphase coherence. It will then evolve into observable inphase coherence. For the 6-pulse sequence described in Sec. 2.1.5 we have:

$$\begin{aligned} &\frac{1}{2i}(S_1^+S_2^+ - S_1^-S_2^-)\sin at_p \xrightarrow{(\pi/2)_x} \\ &-\frac{1}{2}(2S_1^+S_{2z} + 2S_2^+S_{1z} - 2S_1^-S_{2z} - 2S_2^-S_{1z})\sin at_p. \end{aligned} \tag{13}$$

This is essentially the reverse of Eq. 11. The antiphase coherences evolve during the mixing time into observable SQ coherences.

That is:

$$-\frac{1}{2}(2S_1^+S_{2z}+2S_2^+S_{1z}-2S_1^-S_{2z}-2S_2^-S_{1z})\sin at_p \xrightarrow{Ht}\xrightarrow{\pi}\xrightarrow{Ht_2}$$

$$-\frac{1}{2i}[S_1^+e^{-i\Delta\Omega_1\delta t_2}+S_2^+e^{-i\Delta\Omega_2\delta t_2}$$

$$-S_1^-e^{i\Delta\Omega_1\delta t_2}-S_2^-e^{i\Delta\Omega_2\delta t_2}]\sin at_p\sin(at+\delta t_2/2) \quad (14)$$

where $\delta t_2 \equiv t_2 - t$, ($t \equiv t_m - t_p$). The π-pulse refocuses the SQC at $t_2 = t$. One of the two counter rotating components in the square brackets of Eq. 14 is sufficient for calculation of the signal. This follows because the observed signal is proportional to the precessing magnetization:

$$M(t_2) = -\frac{2\,\mathrm{Im}\{Tr(\hat{\rho}(t_2)S^+)\}}{Tr(S^-S^+)}M_0 \quad (15a)$$

where M_0 is the static magnetization of the sample and $S \equiv S_1+S_2$. The DQ echo is singled out by appropriate phase cycling and/or by applying a sequence of pulsed field gradients. The echo signal becomes:

$$M(t_2) = M_0[\cos\Delta\Omega_1\delta t_2 + \cos\Delta\Omega_2\delta t_2]\sin at_p \sin[a(t+\delta t_2/2)]/2. \quad (15b)$$

The terms in the first set of square brackets, after averaging over all $\Delta\Omega_1$ and $\Delta\Omega_2$, produce an echo shape, with the maximum near $\delta t_2 = 0$. The echo envelope is modulated by $\sin(a\delta t_2/2)$. We see that the signal amplitude as a function of preparation and/or mixing times is modulated by $\sin(at_p)\sin(at)$. It is this modulation that is detected to obtain the modulation frequency a, from which the dipolar interaction, hence the distance, may be abstracted (cf. below), since for two coupled electron spins:

$$a = d(1-3\cos^2\theta)+2J \quad (15c)$$

where $d \equiv 2D/3 = \gamma_e^2\hbar/r^3$, with D the dipolar coupling constant (cf. Ch. 1), and J is the exchange integral. The angle θ is the angle between $\boldsymbol{r}$ and the direction of the static magnetic field, $\boldsymbol{B}_0$. Fourier transformation of $\cos[a(t_p-t)]$, that emerges from the above product of two sin-functions, produces the dipolar spectrum, which (in the case of $J = 0$) has the shape of a Pake doublet (Pake, 1948) with a width of 2/3 of the classical Pake doublet. In general, the dipolar spectrum has two distinct canonical points corresponding to the orientations of interspin vector parallel and perpendicular to $\boldsymbol{B}_0$. We denote these frequencies as $\omega_\parallel$ and $\omega_\perp$, respectively. The dipolar spectrum will be symmetric with respect to zero frequency, hence from Eq. 15d one finds:

$$\omega_\parallel = |-2d + 2J|, \quad \omega_\perp = |d + 2J|. \tag{15d}$$

The frequency $\omega_\perp$ is a singular point of the dipolar line shape, and therefore is the most prominent feature in the spectrum. Typically, the observed echo modulation oscillates with the frequency $\omega_\perp$.

The point of following a particular coherence pathway and "filtering out" the other "unwanted" coherence pathways is to select the one that best provides the desired information, in this case the coupling between the electron spins. In addition, the DQC pathways guarantee that the signal is just from interacting spins, *viz.* the doubly labeled species.

2.1.5 The Pulse Sequences Employed in the DQ ESR Experiment; "Allowed" Coherence Pathways: 4, 5, & 6 pulses

We considered a variety of pulse sequences, but we found the three illustrated in Figure 5 to be the most useful. These pulse sequences have been used in the DQC distance measurements. We also show their associated coherence pathways with coherence orders p at each stage. Only the coherence pathways shown survive the phase cycling. In the 4-pulse sequence of Fig. 5(a) the first two pulses in the sequence: $(\pi/2)_x$ - t_p - $(\pi/2)_x$ generate DQ_y and ZQ_y coherences, as we have discussed above. These coherences are refocused by the π-pulse in the evolution sequence (t_1 - π - t_1), and the fourth pulse converts them into antiphase coherences, $S_1^- S_{2z}$, $S_2^- S_{1z}$ which then evolve into observable SQ coherences, S_1^- and S_2^-. The DQ signal is obtained by monitoring the echo height at $t_2 = t_p$ as t_p is stepped out.

Phase cycling selects the DQC pathway. The ZQC cannot be separated from the SQ pathway other than by "detuning" the refocusing conditions on the MQC pathway, (cf. below). This experiment can also be referred to as double-quantum filtered COSY. The $(\pi/2) - t_1 - \pi - t_1 - (\pi/2)$ three pulse "sandwich" plays the role of producing the DQC, refocusing it, and then returning it to SQC. Thus it acts like a **filter**, which selects only those coherence pathways that include evolution of double-quantum coherence. As a result, it is only sensitive to the modulation in echo amplitude due to coupling between electron spins. The phase cycling associated with this three pulse "sandwich" preserves the DQC pathway, while suppressing the other coherence pathways.

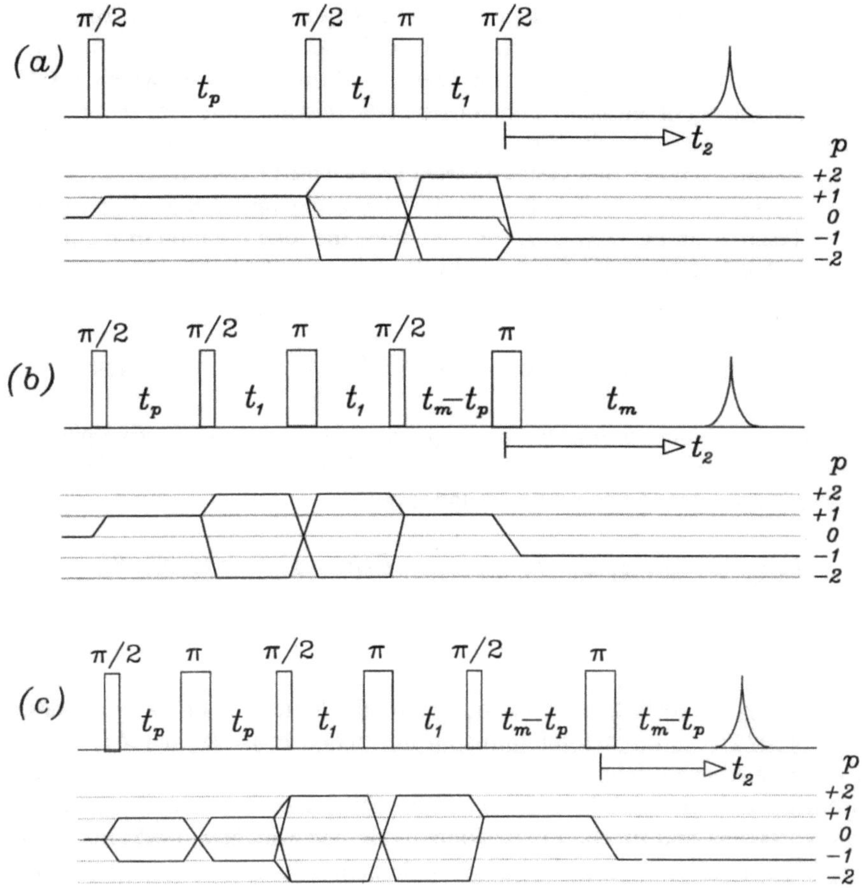

Figure 5. The 4-, 5- and 6-pulse schemes with the corresponding coherence pathways. (a)- the basic 4-pulse sequence leading to an echo at $t_2 = t_p$ from the coherence pathways illustrated. (b) - a 5 pulse variant wherein a π-pulse is added to the 4 pulse sequence to refocus the signal. This yields an echo at $t_2 = t_m$ for the coherence pathways shown. (c)- the 6 pulse sequence and associated relevant coherence pathways. It yields an echo for $t_2 = t_m - t_p$.

The 5-pulse sequence of Figure 5(b) employs an additional refocusing π-pulse. In this sequence, t_m is kept constant and t_p is varied. This eliminates the decay of the signal due to T_2 relaxation, since the echo is detected at a constant time: $2t_m+2t_1$ after the first pulse, independent of t_p.

The 6-pulse sequence of Figure 5(c) contains the 3-pulse preparation sequence $(\pi/2)_x - t_p - \pi_x - t_p - (\pi/2)_x$ which generates DQ coherence that is again refocused by the $t_1 - \pi - t_1$ sequence. The fifth pulse produces anti-phase coherences which evolve into observable SQ coherences, and the sixth pulse refocuses them to form an echo. This sequence is used with fixed t_m, and t_p is varied. The signal envelope is recorded as a function of $t_\xi \equiv t_m - 2t_p$.

The latter two cases allow for "zero dead-time" signals, which result from refocusing of antiphase coherences in the middle of the data acquisition interval. For example, for Fig. 5(c), when one steps out t_p in the range $0 \le t_p \le t_m$ keeping t_m fixed, this yields a signal vs. t_ξ which ranges from $-t_m \le t_\xi \le t_m$. Keeping t_m fixed also has the effect of canceling out the role of phase (or T_2) relaxation on the echo at $t_2 = t_m - t_p$, since the total duration of the pulse sequence is independent of t_p for both the five pulse (see above) and the six pulse sequences.

2.2 Product Operator Analysis for Arbitrary Pulses

In this section we drop the assumption of ideal or non-selective pulses, since this is rarely the case in ESR. The Hamiltonian for a radical pair in the case of nitroxides can be written in the form:

$$\begin{aligned} H/\hbar = &(\beta/\hbar)\mathbf{B}\cdot\mathbf{g}_1\cdot\mathbf{S}_1 + \mathbf{I}_1\cdot\mathbf{A}_1\cdot S_{1z} + (\beta/\hbar)\mathbf{B}\cdot\mathbf{g}_2\cdot\mathbf{S}_2 + \mathbf{I}_2\cdot\mathbf{A}_2\cdot S_{2z} \\ &+ D(3\cos^2\theta - 1)(S_{1z}S_{2z} - \frac{\mathbf{S}_1\cdot\mathbf{S}_2}{3}) + J(\frac{1}{2} - 2\mathbf{S}_1\cdot\mathbf{S}_2). \end{aligned} \tag{16}$$

Here, $\mathbf{g}_1$, $\mathbf{g}_2$ are the g-tensors and $\mathbf{A}_1, \mathbf{A}_2$ are the *hfi* tensors of the nitroxides, $D = 3\gamma_e^2\hbar/2r^3$ is the dipolar spin-spin interaction constant, and $J(r)$ is the electron exchange integral. The non-secular terms were dropped as is appropriate at high field. In addition, as is appropriate for the systems under study (Saxena and Freed, 1997a; Raitsimring and Salikhov, 1985; Milov *et al.*, 1998) nuclear Zeeman and quadrupole interactions were neglected. We shall use the Hamiltonian of Eq. 16 in the form:

$$H/\hbar = \Omega_1 S_{1z} + \Omega_2 S_{2z} + aS_{1z}S_{2z} + b_p(S_1^+S_2^- + S_1^-S_2^+)/2 \tag{17}$$

where $a = b + 2J$, $b_p = -b/2 + 2J$, with $b \equiv d(1 - 3\cos^2\theta)$, (cf. Sec. 2.1.4). Here Ω_1 and Ω_2 represent the Larmor frequencies of electron spins 1 and 2 due to their respective g and *hfi* tensors and their respective orientations, the details of which (Saxena and Freed, 1997a; Lee *et al.*, 1993) we suppress for compactness.

A rigorous treatment of DQ ESR for the nitroxide biradicals using the Hamiltonian of Eq. 16 was developed by (Saxena and Freed, 1997a). Detailed expressions, suitable for computer simulations, were provided for the cases of two and five-pulse "forbidden" echo sequences, as well as for the six-pulse "allowed" DQ sequence. This approach can be adapted to the other "allowed" pulse sequences used in the present study, and we have derived appropriate expressions. However, this again necessitates numerical solutions, and they do not provide much insight into the essence of the experiment until certain assumptions are made to render them more manageable. Instead, we employ an approximate approach (Borbat and Freed, 1999), suitable when dipolar and exchange couplings are weak in comparison with the electron spin Zeeman and hyperfine tensors. More precisely, the exited spectral extent should be much broader then dipolar and exchange couplings. This is the case for nitroxides with interspin distances exceeding 15 Å when B_1 is greater than ca. 30 G.

Given the reasonable approximation that for sufficiently broad spectral excitation $|\Delta\Omega_1 - \Delta\Omega_2| >> |a|$ as noted above, the pseudo-secular terms (i.e. terms in $S_1^{\pm}S_2^{\mp}$) can be ignored as is commonly done in analyses of weak dipolar and exchange interactions in pulsed ESR, (Salikhov *et al.*, 1976; Maryasov *et al.*, 1998), since they would only enter into the eigen-energies to order $a^2/(\Delta\Omega_1 - \Delta\Omega_2)$. Thus we end up with the weak-coupling Hamiltonian of Sec. 2.1.1, which we write in the rotating frame as:

$$H/\hbar \approx \Delta\Omega_1 S_{1z} + \Delta\Omega_2 S_{2z} + aS_{1z}S_{2z}. \tag{18}$$

The more complex case with the pseudo-secular term preserved (Borbat, Freed, 2000) can be treated as outlined in Appendix A.

We shall assume that the microwave pulses are intense enough that $|\gamma_e B_1/2| >> a$, so that during the pulse we have in the frame rotating at ω_{rf}, the applied microwave frequency:

$$H_r/\hbar \approx \Delta\Omega_1 S_{1z} + \Delta\Omega_2 S_{2z} + \frac{\gamma_e B_1}{2}[(S_1^+ + S_2^+)e^{-i\varphi} + (S_1^- + S_2^-)e^{i\varphi}] \tag{19}$$

where $\Delta\Omega_k \equiv \omega_{rf} - \Omega_k$. Here ϕ is the phase of the pulse such that for ϕ=0 (π/2), B_1 is along the rotating x (y) axis.

Whereas the standard PO method (Slichter, 1990) can now be applied to obtain the solution for the various pulse sequences for the case of non-selective pulses, as was done in Secs. 2.1.2-2.1.4, we must utilize Eq. 19 for ESR experiments, which is suitable for pulses of arbitrary strength, subject to the above condition $|\gamma_e B_1/2| >> |a|$. We must thus adapt the PO method to the case of arbitrary pulses.

To accomplish this, we first need to select a convenient set of basis operators with which to represent the density operator for the two interacting electron spins, and then to determine the transformation rules for the basis operators under the action of the arbitrary pulses and during the free evolution periods.

There are several convenient choices for the PO basis set:

$$\begin{gathered}(E_1, S_{1z}, S_{1x}, S_{1y}) \otimes (E_2, S_{2z}, S_{2x}, S_{2y}) \\ (S_1^{\alpha}, S_1^{\beta}, S_1^{+}, S_1^{-}) \otimes (S_2^{\alpha}, S_2^{\beta}, S_2^{+}, S_2^{-}) \\ (E_1/2, S_0^{(1)}, S_{+1}^{(1)}, S_{-1}^{(1)}) \otimes (E_2/2, S_0^{(2)}, S_{+1}^{(2)}, S_{-1}^{(2)}) \\ S_0^{(j)} = S_{jz}, \qquad S_{\pm 1}^{(j)} \equiv \mp S_j^{\pm}/\sqrt{2}\end{gathered} \tag{20}$$

which are the Cartesian, single-element, and spherical bases respectively. The symbol ⊗ implies the direct product. E_j, and S_{jz}, S_j^{+}, S_j^{-} are respectively the identity matrix and spin matrices of dimension 2 for the kth electron spin, as in Eqs. 1-3. Both Cartesian and spherical basis operators are very handy in describing coherences. In NMR the Cartesian basis is widely used due to its convenience in handling coherences under the action of non-selective *rf* pulses[1], whereas the spherical basis is more convenient for the description of the evolution of coherences. However, for the case of arbitrary pulses, which is typical for ESR, we found that the spherical basis is more convenient at all stages of the calculation especially for the complicated case of a full dipolar Hamiltonian. We actually used the modified spherical basis (Borbat and Freed, 1999) in the form of

$$(E_1/2, S_{1z}, S_1^{+}, S_1^{-}) \otimes (E_2/2, S_{2z}, S_2^{+}, S_2^{-}). \tag{21}$$

The pulse propagator superoperator $\tilde{R}$, is associated with H_r, as:

$$\tilde{R}\rho \Leftrightarrow e^{-i\frac{H_r t}{\hbar}} \rho e^{i\frac{H_r t}{\hbar}}. \tag{22a}$$

Here ρ is the density operator for two coupled spins and can be represented as:

$$\rho = \sum_i c_i B_i . \tag{22b}$$

Here, the complete basis set of 16 operators B_i is given by the following direct product operators:

[1] Only the Cartesian basis has the property $e^{-i\chi A}Be^{i\chi A}=B\cos\chi - i[AB]\sin\chi$ that makes it possible to express effect of multipulse sequences in terms of composite linear and bilinear rotations (Ernst *et al.*, 1987).

$$\begin{gathered} S_1^+E_2, S_1^-E_2, E_1S_2^+, E_1S_2^- \\ 2S_1^+S_{2z}, 2S_1^-S_{2z}, 2S_2^+S_{1z}, 2S_2^-S_{1z} \\ E_1E_2/2, S_{1z}E_2, E_1S_{2z}, 2S_{1z}S_{2z} \\ 2S_1^-S_2^+, 2S_1^+S_2^-, 2S_1^+S_2^+, 2S_1^-S_2^-. \end{gathered} \tag{22c}$$

The matrix multiplication in the 16 terms of Eq. 22c should be understood as the direct product. Consequently, these 16 direct product operators are represented by 4×4 matrices.

The basis sets given by Eqs. 20, 21, and 22c, after simple normalization are orthonormal in Frobenius trace metrics:

$$\mathrm{Tr}\{B_i^\dagger B_j\} = \delta_{ij} \tag{22d}$$

where $B_i^\dagger B_j$ implies simple matrix multiplication of the 4×4 matrices and Tr is the trace of this matrix product.

The pulse propagator superoperator $\tilde{R}$ becomes a direct product of the individual pulse propagator superoperators for the individual spins $\tilde{R} = \tilde{R}_1 \otimes \tilde{R}_2$ in the case when we can neglect the dipolar Hamiltonian during the pulse. In other words, we implicitly exclude "forbidden" coherence effects in this treatment. The next step is to determine the rules for transforming the basis operators. Rotational superoperators $\tilde{R}_1$ and $\tilde{R}_2$ can be represented in a matrix form using the Cayley-Klein parameters associated with a rotation about the *x*-axis by the kth *rf* pulse by a nominal rotation angle β_k (Bloom, 1955). The appropriate matrices $R_1(\beta_k)$ and $R_2(\beta_k)$ for the transformation of the operator basis set $(E_j/2, S_{jz}, S_j^+, S_j^-)$ are:

$$R_j(\beta_k) = \begin{pmatrix} 1 & 0 & 0 & 0 \\ 0 & C^{(j)}(\beta_k) & -\dfrac{S^{(j)}(\beta_k)^*}{2}e^{-i\varphi_k} & -\dfrac{S^{(j)}(\beta_k)}{2}e^{i\varphi_k} \\ 0 & -S^{(j)}(\beta_k)^* e^{i\varphi_k} & C_2^{(j)}(\beta_k)^* & S_2^{(j)}(\beta_k)e^{i2\varphi_k} \\ 0 & -S^{(j)}(\beta_k)e^{-i\varphi_k} & S_2^{(j)}(\beta_k)e^{-i2\varphi_k} & C_2^{(j)}(\beta_k) \end{pmatrix} \tag{23}$$

where the asterisk indicates complex conjugation. $\tilde{R}(\beta_k)$ is represented by a 16×16 matrix whose elements can be written out by expanding the direct product of $R_1(\beta_k)$ and $R_2(\beta_k)$. The elements of $R_j(\beta_k)$ are given by:

$$\begin{aligned} C_2^{(j)}(\beta_k) &= a_{jk}^2, \qquad & S_2^{(j)}(\beta_k) &= -b_{jk}^2 \\ S^{(j)}(\beta_k) &= 2a_{jk}b_{jk}, \qquad & C^{(j)}(\beta_k) &= |a_{jk}|^2 - |b_{jk}|^2. \end{aligned} \tag{24}$$

Here a_{jk} and b_{jk} are the Cayley-Klein parameters associated with a nominal rotation by angle β_k about the rotating x-axis, and they are:

$$a_k = \cos\beta_{ek}/2 - i\cos\psi_k \sin\beta_{ek}/2, \quad b_k = -i\sin\psi_k \sin(\beta_{ek}/2) \tag{25}$$

where

$$\beta_{ek} = \beta_k[1+(\Delta\Omega_k/\gamma_e B_1)^2]^{1/2} \text{ and } \tan\psi_k = \gamma_e B_1/\Delta\Omega_k\,. \tag{26}$$

For non-selective pulses Eq. 24 becomes:

$$C_2^{(j)}(\beta_k) = \cos^2\beta_k/2, \quad S_2^{(j)}(\beta_k) = \sin^2\beta_k/2$$
$$S^{(j)}(\beta_k)^* = i\sin\beta_k, \qquad C^{(j)}(\beta_k) = \cos\beta_k\,.$$

The evolution of the density operator between pulses is given by the free evolution propagator superoperator associated with Eq. 18, which includes dipolar and exchange effects. This superoperator can be represented by a matrix of dimension 16 by 16 in the PO basis. We need to know only a few of its elements here. Its relevant sub-block is given by Eqs. A5-A7. For example, for the transformation of $S_1^{\pm}$ and $2S_1^{\pm}S_{2z}$ we find from Eqs. A5-A7 for the case of zero pseudo-secular term (cf. App. A):

$$S_1^{\pm} \xrightarrow{Ht/\hbar} [S_1^{\pm}\cos\frac{at}{2} \mp i(2S_1^{\pm}S_{2z})\sin\frac{at}{2}]e^{\mp i\Delta\Omega_1 t}$$
$$(2S_1^{\pm}S_{2z}) \xrightarrow{Ht/\hbar} [(2S_1^{\pm}S_{2z})\cos\frac{at}{2} \mp iS_1^{\mp}\sin\frac{at}{2}]e^{\mp i\Delta\Omega_1 t}. \tag{27}$$

Eqs. 23 and 27 are sufficient for deriving expressions for the signals for each pulse sequence in the weak-coupling limit. The computational effort is greatly reduced by only tracking those coherence pathways of interest shown in Fig. 5, which are selected by the phase-cycling.

The general expressions for DQC and ZQC signals associated with the coherence pathways in Fig. 5 (Borbat and Freed, to be published) can now be written down.

One finds that all the signals may be written in the following form (or else as a sum of terms of this form):

$$V_{n,DQ(ZQ)}(\Delta\Omega_1, \Delta\Omega_2, \mathbf{r}_{12}, t_p) = G_n(\Delta\Omega_1)H_n(\Delta\Omega_2)F_n(\mathbf{r}_{12}, t_p)R_{n,DQ(ZQ)}\,. \tag{28}$$

The $F_n(\mathbf{r}_{12},t_p)$ represent the oscillations due to the dipolar interactions:

$$
\begin{aligned}
F_4(\mathbf{r}_{12},t) &= -(1/2)\sin^2(at_p/2) \\
F_5(\mathbf{r}_{12},t) &= -(S^{(2)}/2)\left(\sin^2(at_m/2)-\sin^2[a(t_m-t_p)]\right)+(C^{(2)}/2)\sin^2(at_p/2) \\
F_6(\mathbf{r}_{12},t) &= -\sin at_p \sin[a(t_m-t_p)] = (\cos at_m - \cos at_\xi)/2,\ (t_\xi \equiv t_m - 2t_p)
\end{aligned}
\tag{29}
$$

where for nominal $\pi/2$ and π pulses $S^{(2)} \equiv S_2^{(2)}(\pi)$, $C^{(2)} = 1 - S^{(2)}$. The $G_n(\Delta\Omega_1)$ and $H_n(\Delta\Omega_2)$ represent the excitation functions for spins 1 and 2, respectively. Also the $R_{n,DQ(ZQ)}$ represent the spin relaxation.

The detailed forms of the signal amplitudes, $V_{n,DQ(ZQ)}$ are given below. Since they are somewhat lengthy, we show here only the results when the nominal pulse rotation angles are set to $\pi/2$ and π as shown in Fig. 5. In the absence of relaxation they are for the 4, 5, and 6 pulse sequences (Borbat and Freed, 1999) :

$$
\begin{aligned}
V_{4DQ}(\Delta\Omega_1,\Delta\Omega_2,\mathbf{r},t_p) &= V_{4ZQ}(\Delta\Omega_1,\Delta\Omega_2,\mathbf{r},t_p) \\
&= -(1/2)\,\mathrm{Im}\{S^{(1)}(\pi/2)^*\}S_2^{(1)}(\pi)S_2^{(2)}(\pi) \\
&\times[|C_2^{(1)}(\pi/2)|^2 + S_2^{(1)}(\pi/2)^2]\,|S^{(2)}(\pi/2)|^2 \sin^2(at_p/2) \\
&\equiv \overline{V}_4(\Delta\Omega_1,\Delta\Omega_2)\sin^2(at_p/2)
\end{aligned}
\tag{30a}
$$

$$
\begin{aligned}
V_{5,DQ}(\Delta\Omega_1,\Delta\Omega_2,\mathbf{r},t_p) &= V_{5,ZQ}(\Delta\Omega_1,\Delta\Omega_2,\mathbf{r},t_p) \\
&= \overline{V}_4(\Delta\Omega_1,\Delta\Omega_2) \\
&\times S_2^{(1)}(\pi)[S_2^{(2)}(\pi)\sin^2 at_m/2 - (S_2^{(2)}(\pi)\sin^2[a(t_m-t_p)/2] \\
&\quad + (1-S_2^{(1)}(\pi))\sin^2 at_p/2]
\end{aligned}
\tag{30b}
$$

$$
\begin{aligned}
V_{6,DQ}(\Delta\Omega_1,\Delta\Omega_2,\mathbf{r},t_p) &= -\mathrm{Im}\{S^{(1)}(\pi/2)^*\}[S_2^{(1)}(\pi)S_2^{(2)}(\pi)]^3 \\
&\times|S^{(2)}(\pi/2)|^2 \sin at_p \sin a(t_m-t_p).
\end{aligned}
\tag{30c}
$$

Superscripts (j) for j = 1 or 2 are used in Eqs. 30 to label terms associated with the jth spin. We have set $\phi = 0$ for convenience. In the derivation of Eqs. 30 it was assumed that the spectrum is sufficiently broad, so that any unrefocused signals decay away rapidly, and any accidental refocusing is unlikely. [There are residual FID-like terms, which can be ignored, (cf. App. A), so they are not included in Eqs. 30]. Note that the dependence on $\Delta\Omega_1$ and $\Delta\Omega_2$

enters via the above definitions of β_{ek} and ψ_k. Note also that for Eqs. 30 there is an equivalent expression obtained from permuting the superscripts (1) and (2). The two give identical results, thus we write out only one term.

Amplitude factors $|C_2^{(1)}(\pi/2)|^2+S_2^{(1)}(\pi/2)^2|, |S^{(2)}(\pi/2)|^2$ in Eqs. 30 with good accuracy are nearly unity for the nitroxide experiments discussed in Sec. 4 and only $S_2^{(k)}(\pi)$ and $S^{(k)}(\pi/2)$ will determine the spectral excitation. We can account for the time evolution with respect to t_2 (Fig. 5) by simply multiplying the rhs of Eqs. 30 by:

$$\exp[i(\Delta\Omega_1+a/2)\delta t_2]e^{-\delta t_2/T_{2,SQ}} \tag{31}$$

where $\delta t_2 \equiv t_2 - t_p$ for Eq. 30a; $\delta t_2 \equiv t_2 - t_m$ for Eq. 30b, and $\delta t_2 \equiv t_2 - (t_m - t_p)$ for Eq. 30c. For the present, we are mainly interested in the case of $\delta t_2 = 0$. Note that in the limit of non-selective pulses (i.e. $B_1 \rightarrow \infty$) these equations simplify considerably and become:

$$\begin{aligned} V_{4,DQ} &= -(1/4)\sin^2(at_p/2) \\ V_{5,DQ} &= -(1/4)[\sin^2 at_m/2 - \sin^2 a(t_m - t_p)/2] \\ V_{6,DQ} &= -\sin at_p \sin a(t_m - t_p). \end{aligned} \tag{32}$$

These represent the ideal cases, where the signal, as a function of t_p, just shows the coherent oscillations due to dipolar (plus exchange) interactions, independent of $\Delta\Omega_k$. However, more generally, the $V_{n,DQ}$, $n = 4$, 5 or 6, signals will depend upon $\Delta\Omega_k$ via the spectral coverage factors associated with the finite pulses.

To account for relaxation we shall multiply Eqs. 30 by:

$$\begin{aligned} R_{4,DQ(ZQ)} &= \exp(-2(t_p T_{2,SQ}^{-1} + t_1 T_{2,DQ(ZQ)}^{-1})) \\ R_{5,DQ(ZQ)} &= \exp(-2(t_m T_{2,SQ}^{-1} + t_1 T_{2,DQ(ZQ)}^{-1})) \\ R_{6,DQ} &= \exp(-2(t_m T_{2,SQ}^{-1} + t_1 T_{2,DQ}^{-1})) \end{aligned} \tag{33}$$

respectively and Eq. 32 accordingly. Here $T_{2,SQ}$, $T_{2,DQ}$, and $T_{2,ZQ}$ refer to the T_2 decay constants characteristic of SQC, DQC, and ZQC, respectively (T_1 processes are neglected). This approach to the treatment of relaxation is a simplified one, (Saxena and Freed, 1997a). A rigorous approach is to include the relaxation superoperator in the Liouville-von Neuman equation for the density matrix as was done in the work of Lee *et al.*, (1993, 1994). This would help to account for any orientationally dependent relaxation effects and any correlation between magnetic tensors of the coupled radicals.

We shall choose Euler angles $\lambda_i \equiv (\alpha_i, \beta_i, \gamma_i)$, with γ_i set to zero, to represent the transformation from the dipolar frame to the magnetic frame of the ith nitroxide fragment (i = 1 or 2), and Euler angles $\eta \equiv (0, \theta, \zeta)$ to represent the transformation from the lab frame to the dipolar frame (Saxena and Freed, 1997a). Then the $G_n(\Delta\Omega_1)$ of Eq. 28 will depend upon λ_1 and η, while $H_n(\Delta\Omega_2)$ depends upon λ_2 and η, and $F(\mathbf{r}_{12}, t_p)$ just upon η as well as the magnitude $|\mathbf{r}_{12}| \equiv r$. It is then necessary to average the signal over an appropriate distribution in these variables, which we represent by angular brackets as:

$$D_n(t_p) = \left\langle G_n(\Delta\Omega_1(\lambda_1,\eta))H_n(\Delta\Omega_2(\lambda_2,\eta))F_n(r,\eta,t_p)\right\rangle_{r,\lambda_1,\lambda_1,\eta}. \quad (34)$$

The relaxation function $R_{n.DQ(ZQ)}(\Delta\Omega_1, \Delta\Omega_2, \lambda_1, \lambda_2, \eta)$ which has been dropped for convenience in Eq. 34 can be placed outside the angular brackets, since for nitroxides, which are the major application of our DQ methods, at the reduced temperatures used, T_2's are not substantially dependent upon λ_1, λ_2 nor upon η. In the nitroxide biradicals that we have studied so far, the contribution of the modulation of dipolar coupling to T_2 is very small, (Saxena and Freed, 1997b). Therefore there should not be noticeable dependence on η or r. Moreover, we have not seen a noticeable angular variation of the T_2's in the experiments with macroscopically aligned biradicals nor a significant variation of T_2 across a 2D-ELDOR spectrum (Saxena and Freed, 1997b), implying only a weak dependence on λ_1, and λ_2. We shall not include the factors of Eq. 32 in the sequel for brevity.

It is easy to see that when there is no correlation between the λ_i and η, then Eq. 34 yields the expected simpler form, Eq. 35:

$$\begin{aligned} D_n(t_p) &= \left\langle G_n(\Delta\Omega_1(\lambda_1,\eta))\right\rangle_{\lambda_1} \left\langle H_n(\Delta\Omega_2(\lambda_2,\eta))\right\rangle_{\lambda_2} \left\langle F_n(r,\eta,t_p)\right\rangle_{r,\eta} \\ &\equiv G_n^{AV} H_n^{AV} \left\langle F_n(r,\eta,t_p)\right\rangle_{r,\eta}. \end{aligned} \quad (35)$$

Thus one can simply integrate over the distribution of orientations and magnitudes of r in the sample. A FT with respect to t_p or t_ξ will then yield familiar Pake doublets. Another simple limiting case occurs when the pulses can be regarded as non-selective, so that

$$G_n(\Delta\Omega_1) \to G_n^{\infty} \text{ and } H_n(\Delta\Omega_2) \to H_n^{\infty} \quad (36)$$

independent of $\Delta\Omega_1$, $\Delta\Omega_2$, which follows immediately from the $B_1 \to \infty$ limiting forms of Eqs. 30 given in Eq. 29.

Then Eq. 34 becomes:

$$D_n(t_p) = G_n^{\infty} H_n^{\infty} \left\langle F_n(r,\eta,t_p) \right\rangle_{r,\eta} . \tag{37}$$

This is the case when all the spins in the sample contribute to the signal. We can thus write as inequalities:

$$G_n^{\infty} \geq G_n^{AV} \text{ and } H_n^{\infty} \geq H_n^{AV} . \tag{38}$$

2.3 Approach for General Analysis

When the condition $|\Omega_1 - \Omega_2| >> |a|$ no longer holds, the weak-coupling approach should be corrected. For nitroxides, this includes distances under 10-12 Å for non-selective pulses, corresponding to a large coupling, or exotic cases of mutual arrangement of the nitroxide moieties such that the magnetic axes substantially coincide, thereby leading to small frequency differences for a significant fraction of pairs. The full expressions for the signal in the 6-pulse sequence can be found in (Saxena and Freed, 1997a). They are obtained by tracking down the density matrix elements along the relevant coherence pathways using the complete spin-Hamiltonians instead of the approximate forms of Eqs. 18 and 19. As a result they become substantially more complex. In Appendix A we outline a suitable approach for obtaining expressions for the signals from "allowed" coherence pathways. The only assumption that is made in App. A is that the dipolar coupling during the pulse has only an insignificant effect on the signals. This is indeed the case for very intense pulses and longer distances, ≥12-15 Å.

2.4 Relation to Other ESR Techniques

A detailed review of cw ESR methods for determining distances using nitroxide bilabeling is given in (Hustedt and Beth, 1999). Cw ESR methods are very useful in the range of distances from 5 Å to 15 Å, and provide the opportunity to report on the mutual orientation of the nitroxide moieties under certain conditions. Even somewhat larger distances can be assessed by using perdeuterated ^{15}N spin-labels. A number of studies have emerged with use of cw-ESR due to its wide availability (Hustedt *et al.*, 1997; Rabenstein and Shin, 1996; Hanson *et al.*, 1996; Budker *et al.*, 1995). A summary of various cw and pulsed ESR distance measurement techniques is given in Ch.1 of this volume.

Here we note the relationships between pulsed ESR techniques which deal with the distance between the electron spins[2] and share common features.

The relevant techniques can be divided into the following categories:

1. *Incoherent techniques*:

These include double electron-electron resonance (DEER) (Milov *et al.*, 1998; Pannier, *et al.*, 2000; cf. Ch. 11). It usually makes use of selective pulses at two well-separated microwave frequencies, and hence it is applicable to systems with sufficiently broad ESR spectra, including nitroxides. It is practical in cases of broad spectra (e.g. ≥ ca. 50 G). It is the only sensible alternative in cases of widely separated spectra like the case of pairs of semiquinone radicals interacting with hydrogen atoms (Milov *et al.*, 1981). A further advantage of multiple-frequency techniques is the insignificant role played by the "pseudo-secular" part of the dipolar Hamiltonian. Also, nuclear modulation effects can be virtually absent if there is no spectral overlap from the selective pulses at the separated frequencies.

The "2+1" technique (Kurshev and Raitsimring, 1989; Ch. 10) can also be included in this group. It is closely related to the 3-pulse DEER technique, except for its use of only a single frequency (or else a large spectral excitation overlap in the 3-pulse DEER).

2. *Coherent techniques* which include:

i) Techniques that select a particular order of MQC. They include "allowed" and "forbidden" DQC and ZQC pulse sequences (Saxena and Freed, 1996; 1997a; Borbat and Freed, 1999).

ii) Techniques that manipulate SQ coherences and do not select MQC but instead detect indirectly the effect of generation of MQC's on the SQ echo amplitude. Thus, they do not display selectivity with respect to a given order of coherence. Examples include 5(6)-pulse sequences (cf. App. C; Borbat and Freed, 1999a, 2000) and 4-pulse sequences (Jeschke *at al.*, 2000).

3. INSTRUMENTATION:

3.1 2D-FT Pulse ESR Spectrometer

The multi-frequency pulsed two-dimensional FT-ESR spectrometer described elsewhere (Borbat and Freed, 1997) was used in all DQ experiments. Initial experiments were performed at both 9.2 and 17.3 GHz, but most results were obtained at the higher frequency, since it provided 4 times higher

[2] We do not consider here important distance measurement techniques which utilize relaxation phenomena (Ichikawa *et al.*, 1985; Dzyuba and Kawamori, 1996; Rakowsky *et al.*, 1998).

SNR, the maximum B_1~ 30 G and the shortest spectrometer dead-times $\tau_d \approx$ 30 ns. [The effective B_1 has been estimated using a technique based on a well-known method (Raitsimring and Salikhov, 1985)]. At X-band the maximum B_1 in the BLGR resonator with Q ≈ 50 was about 20 G and the dead-time was 5-10 ns longer. At 17.3 GHz the minimum $\pi/2$ (π) pulse widths were 3-3.5 ns (5-6 ns) in the low-Q dielectric resonator used. One more reason to prefer Ku-band was in the higher frequency and reduced magnitude of ESEEM. The signal acquisition unit includes two 200 Msps flash ADC converters with fast averaging capability. The repetition rate is 6-12 kHz depending on the record length. For sampling of the signals at time intervals shorter than 5 ns an interleaving technique was employed. The signal from the microwave bridge was detected in quadrature, and CYCLOPS phase cycling (Hoult and Richards, 1975; Gorcester and Freed, 1988; Ohba *et al.*, 1993) was always used to eliminate image signals in 2D collections.

For the successful implementation of DQ ESR the most important features are the intense pulses and the sensitivity, and to a somewhat lesser extent the dead time. The latter follows, because the most favorable DQ-ESR pulse sequences are based on the principle of "remote detection", i.e. they have "apparent" zero dead time.

3.1.1 Intense Pulses

Spectral excitation depends on the available B_1 and the type of signal. As we mentioned in Sec. 2.2, the expressions for the signal given by Eqs. 30 contain terms of the type $|S(\pi/2)|^2$ or $2(|C_2(\pi/2)|^2 + S_2(\pi/2)^2)$ which are close to unity for a broad range of $\Delta\Omega$'s, so they may be ignored in analyses of spectral excitation. The major factor limiting the spectral excitation arises from the products of terms as:

$$S_2(\pi) = \sin^2(\pi u/2)/u^2$$

where $u^2 \equiv 1 + \Delta\Omega^2/\gamma_e^2 B_1^2$. A small additional effect is from $S(\pi/2)$. In Fig. 6 we plot the amplitude factors G_6^{AV}, H_6^{AV} of Eq. 35, and their product M_6^{AV} calculated for an uncorrelated pair of ^{14}N nitroxides at X-band. At Ku-band the difference is marginal.

The product of three $S_2(\pi)$ functions arising from the π pulses in the 6-pulse sequence yields very reasonable values of about 0.5 for the amplitude factors given by G_6^{AV} or H_6^{AV} for a B_1 of 30 G, for which the overall amplitude factor, M_6^{AV}, is about 0.3. It is useful to note that the width of the spectral excitation predicted for a product of N $S_2(\pi)$ terms decreases slowly with N, as $N^{-1/2}$. For B_1 under 20 G the growth of M_6^{AV} is faster than linear, thus achieving $B_1 \geq 30$ G is essential for producing substantial DQ signals. We find that spectral distortions caused by correlations are determined by the H function of Eq. 35, whereas G only influences the overall signal strength.

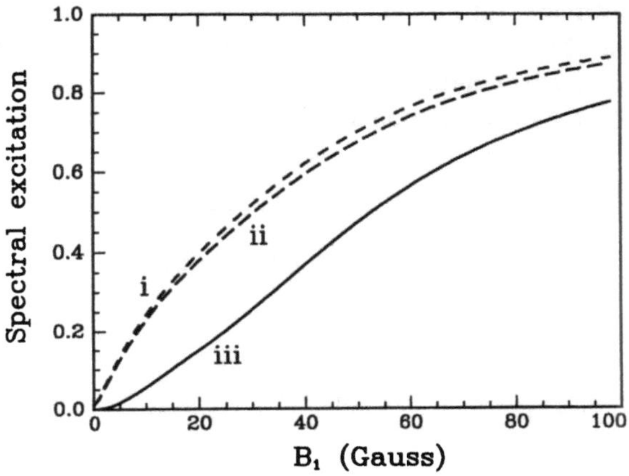

Figure 6. Spectral excitation in the 6-pulse DQ sequence as a function of B_1 calculated for an uncorrelated pair of ^{14}N nitroxides excited at the center of gravity of the ESR spectrum. (i) - G_6^{AV}, (ii) - H_6^{AV}, and (iii) - their product $H_6^{AV}G_6^{AV}$. Note that as $B_1 \rightarrow \infty$ one obtains $H_6^{\infty} = G_6^{\infty} = 1$. These results were calculated for 9.3 GHz. At 17.3 GHz differences are small, so the results are approximately valid at this frequency as well.

In the case of short distances (under 15 Å) the distortions of the Pake pattern caused by the pseudo-secular term (cf. App. A) are less significant at large B_1. For correlated nitroxides, we expect that by increasing B_1 to ca. 60 G and consequently H to 0.75 would greatly decrease the correlation effects (cf. Sec. 2.2). Further increases in B_1 only yield a steadily decreasing improvement as the asymptotic limiting value of unity is only slowly approached with increasing B_1.

We have studied the effects of weaker B_1's and shorter distances, (Borbat and Freed, 2000). With a B_1 of only 12 G (at 9.2 GHz using a TE_{012} resonator) we found that all the DQ and coherent SQ sequences yielded only weak signals for rigid biradical R-II, which displays a 16.2 Å distance (cf. Sec. 4.2). Strong signals are, however, obtained for longer distances under these conditions (e.g. with rigid biradical R-I, with a distance of 28.8 Å, cf. Sec. 4.2). At 17.3 GHz smaller B_1's still yield strong signals, but they are severely distorted by several factors (cf. App. A). Above about 25 Å there is no problem with using weaker pulses. Thus a large B_1 is very important for all these sequences to work reliably.

Simple estimates would indicate that for ^{15}N nitroxides only about 70% of the B_1, needed for ^{14}N, is required in order to achieve the same degree of spectral excitation.

The spectral excitation that has already been achieved corresponding to $B_1 = 30$ G is sufficient for producing strong DQ signals even in samples with a small number of spins, (cf. Sec. 4). Figure 7 shows an example of the SNR obtained under typical conditions (i.e. sample volume V_s of 15 μL, spin concentration C of 1 mM, data acquisition time, 9 s) for the biradical R-II (cf. Sec. 4.2).

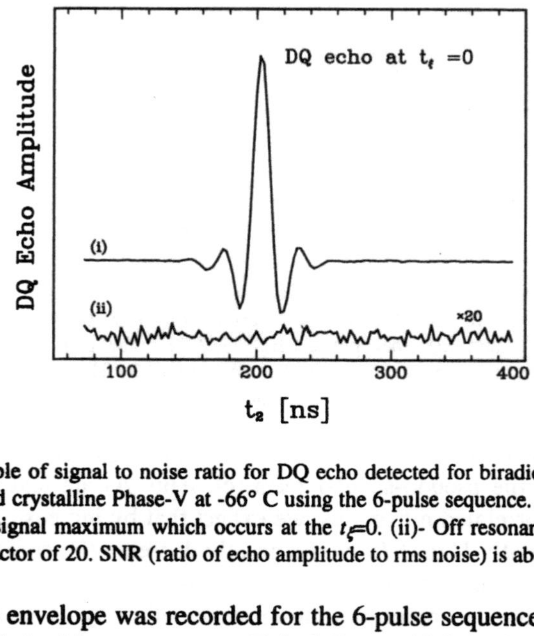

Figure 7. Example of signal to noise ratio for DQ echo detected for biradical R-II randomly oriented in liquid crystalline Phase-V at -66° C using the 6-pulse sequence. (i)- DQ echo was acquired at the signal maximum which occurs at the t_ξ=0. (ii)- Off resonance baseline vs. t_2 amplified by a factor of 20. SNR (ratio of echo amplitude to rms noise) is about 500.

The DQ echo envelope was recorded for the 6-pulse sequence at a temperature of -66° C. At this temperature, T_2 is 0.5 μs, which is about 4 to 8 times shorter than is typical for nitroxides at 77 K (Lindgren *et al.*,1997; Zecevic *et al.*, 1998).The amplitude of the DQ echo after correction for T_2-decay was about 0.15 of the COSY (Gorcester and Freed, 1988) signal, which is about 50% of the maximum expected for a B_1 of 30 G. The discrepancy is believed to be due to the effect of correlation of orientations of the nitroxide *hfi* tensors, so it should decrease with further increase of B_1 in the π-pulses

3.1.2 Short Dead-Times

The short dead times that were achieved are typically in the range of 30-35 ns at Ku-band (Borbat and Freed, 1997), although with careful adjustments we can obtain 22-25 ns. This is important for the 4-pulse DQ experiment and also for the COSY or 2D-ELDOR experiment. It is less important for 5 and 6 pulse DQ, where the DQ signal is detected when sufficient time has elapsed after the *rf* pulses. Of course, one must keep this elapsed time to a minimum to avoid signal loss due to T_2-type relaxation.

3.1.3 Signal Acquisition Schemes; 1D vs. 2D

It is sufficient to utilize only one-dimensional acquisition for most of the experiments. However, with our 2D FT spectrometer, we always acquire the whole echo shape, since we use a transient digitizer. The quadrature signals from the microwave bridge are fed into the digitizer, and they are accumulated for each step of the phase cycle. The data records are then combined according to the phase cycle table for selection of the desired DQ signal (cf. App. D). To obtain the 1D signal from the 2D data set we sample the signal amplitude as a function of t_p or t_ξ by selecting the necessary number of points from the data records. This "digital sampling window" serves the same function as a boxcar integrator.

In a 2D experiment all collected data are stored. Fourier transformation of the 2D time-domain signal with respect to t_p or t_ξ and t_2 yields the Pake pattern along the ESR spectrum (cf. Sec. 4.2.3). This information can be utilized to study correlations in the mutual orientation of the nitroxides. However the 2D experiment is more time consuming, because the sampling interval needed for 2D-FT ESR is larger than the sampling window in 1D data processing. The sampling window in the 1D experiment is about 0.5-1 times the full width at half height of the echo signal and typically is in the range of 5-15 ns. In the 2D experiment the sampling interval is ca. 100-150 ns. Thus, most of the echo is at a signal level well below its maximum, which is all that needed in 1D experiment. The larger window is essential for adequate spectral resolution in a 2D spectrum. As a result, the SNR in the 2D spectrum is several times worse than for the 1D Pake pattern. In order to compensate for this the acquisition time for the 2D experiment should be increased. In Sec. 4.2.3 we show results for 2D data collection with the nitroxide biradicals R-I and R-II.

3.1.4 Coherence Pathway Selection

3.1.4.1 Phase Cycling

Phase cycling is a powerful method for suppressing unwanted coherence pathways by alternating the phases of the *rf* pulses and then combining the detected signals appropriately. It was introduced in NMR and has become widely employed for the selection of desired coherence pathways in numerous types of 1D and 2D experiments (Ernst *et al.*, 1987). This method was brought into the ESR field (Fauth *et al.*, 1986; Gorcester *et al.*, 1988) in order to select desired signals in 1D and 2D pulsed electron spin-echo experiments.

The general rule for the selection of a particular coherence pathway is based on the fact that if a pulse propagator U, acting on the density matrix according to

$$\rho(t) = e^{iUt}\rho(0)e^{-iUt} \tag{39}$$

produces a change in the given coherence order p by Δp, then the same pulse propagator, but with the phase shifted by φ, i.e. $U(\varphi)$, will multiply $\rho(t)$ by $e^{-i\Delta p\varphi}$. Consequently, a sequence of pulse propagators ($U_1(\varphi_1)$, $U_2(\varphi_2)$ ... $U_N(\varphi_n)$)) will multiply the density matrix and hence the signal produced by a particular coherence pathway (Δp_1, Δp_2 .... Δp_{n},) by:

$$\exp(-i\sum_k \Delta p_k \varphi_k)\,. \tag{40}$$

Stepping the phase by $2\pi l/N$ ($l = 0, 1, \ldots, N$-1) and combining the detected signals with weighting factors $e^{i\Delta p_k \varphi_l}$ selects signals with a change in the coherence order equal to $\Delta p_k \pm Nm$ ($m = 0, 1, \ldots$) (Ernst *et al.*, 1987). For the $\pi/2$ phase increment one has N=4. Hence a phase cycle that selects $\Delta p = \pm 2$ also selects pathways $\Delta p = (\pm 6, \pm 10, \ldots)$ and suppresses the others.

For a phase increment of π, N is 2. Thus $\Delta p = (0, \pm 2, \pm 4, \ldots)$ are selected. We see that $\Delta p = 2$ and $\Delta p = 0$ are both selected. Therefore a phase cycle based on just the π increment cannot separate DQC from ZQC. The more pathways there are containing a given Δp_k that can interfere with the desired signal, the more pulse propagators should be included in the phase cycling scheme. It is well known that a series of pulse propagators (which represents unitary transformations) can be replaced by a single propagator representing the cumulative effect of the series. This also applies to a series of pulse propagators interspersed by free evolution propagators representing the effect of the spin Hamiltonian, Eq. 18. Such a cumulative propagator is referred to as a composite propagator or rotation in NMR (Sørensen, 1983).

For example, the 2 and 3 pulse sequences for preparation of DQC (cf. Sec. 2.1.2) can be replaced by a single composite propagator which produces changes in the order of coherence of $\Delta p = 0, \pm 1, \pm 2$. It is the last value that corresponds to DQC. Thus a phase cycle that selects $\Delta p = \pm 2$ is required. The virtue of using the composite propagator is that it is sufficient to consider just its phase shifts in constructing the phase cycling. Then for a given phase shift φ for the composite propagator each pulse in its sequence must be given the identical phase shift, φ.

Consider as an example the 6-pulse DQ sequence. The first three pulses may be considered as a composite propagator that produces DQC. The subsequent change in the coherence order Δp is ± 2. We should select all four pathways depicted in Fig. 5(c) with this change. Using $N = 4$, with the understanding that the phases of the three pulses should be incremented simultaneously, we arrive at a 4-step phase sequence with the phases of these 3 pulses changing as follows (0, 0, 0), ($\pi/2$, $\pi/2$, $\pi/2$), (π, π, π), ($3\pi/2$, $3\pi/2$, $3\pi/2$) and each of the 4 signals are weighed by factors (1, -1, 1, -1), when they are added. This phase cycle is, in principle, sufficient because it suppresses all of the pathways that yield SQ or ZQ coherence.

This phase cycle for the 6-pulse sequence is satisfactory if the phases are very accurate, which is hard to achieve in practice. Therefore, it is usually necessary to cycle the phases of the other pulses in order to attain an improved suppression of unwanted signals. Additionally, the phase cycle should be combined with the CYCLOPS sequence for suppression of signals at the image frequency. This increases the number of phase steps by a factor of four. The typical phase cycling that we employ for the 6-pulse sequence consists of 64 steps, which combined with CYCLOPS yields 256 steps (cf. App. D). The 4-pulse and 5-pulse sequences have phase cycles with the number of steps ranging from 64 to 128.

With these phase cycles we could suppress unwanted echoes by a factor of 300-3000 depending on the sharpness of the SQ echoes, (e.g. the smaller figure applies for nitroxides, which have sharper echoes, and the larger to E'-centers).

3.1.4.2 Pulsed Field Gradients

In the previous section we learned that to select coherence orders with $p \geq 3$, phase increments less than $\pi/2$ would be necessary. This is not a standard feature of existing pulsed ESR spectrometers. Such a phase increment would need to have a smaller absolute error, bringing an extra challenge to the task. It may be useful to consider an alternative solution.

One well-known approach for selection of high orders of coherence relies on using pulsed field gradients (PFG). This is routinely done in NMR for a variety of purposes including suppression of unwanted coherence orders (Bax *et al.*, 1985) The same technique could in principle be employed in the DQ ESR experiment. There is not yet an experimental demonstration in ESR, but we include this topic for its potential.

An underlying principle is based on a key property of multiple-quantum transitions. The frequency of a multiple-quantum transition is the sum of the frequencies of all the single quantum transitions that can be considered as making up the multiple-quantum transition. Thus a shift of resonant frequencies of all relevant SQ transitions by the same amount shifts the frequency of the multiple-quantum transition by p times that for the SQ transitions, with p the order of coherence of the multiple-quantum transition (Ernst *et al.*, 1987). When the field gradient pulse $G_z^{(i)}(\mathbf{r},t) = \nabla B_z^{(i)}(\mathbf{r},t)$, [with superscript i numbering the gradient pulse], is applied at a given point in a coherence pathway where the signal has coherence order p, the amplitude V of the observed signal becomes:

$$V = V_0 \int_{V_s} v_1(\mathbf{r}) \exp[-ip\Phi_i(r)] d\mathbf{r}\,. \qquad (41)$$

We introduced the phase:

$$\Phi_i(\mathbf{r}) \equiv \int_{\Delta t_i} \gamma_e B_z^{(i)}(\mathbf{r},t) dt \qquad (42)$$

where Δt_i is the duration of the ith gradient pulse. Also, $V_0 v_1(\mathbf{r})$ is the signal per unit volume in the absence of the gradient pulse. As a result, the signal amplitude decreases due to dephasing caused by a distribution of $\Phi_i(\mathbf{r})$ over the sample.

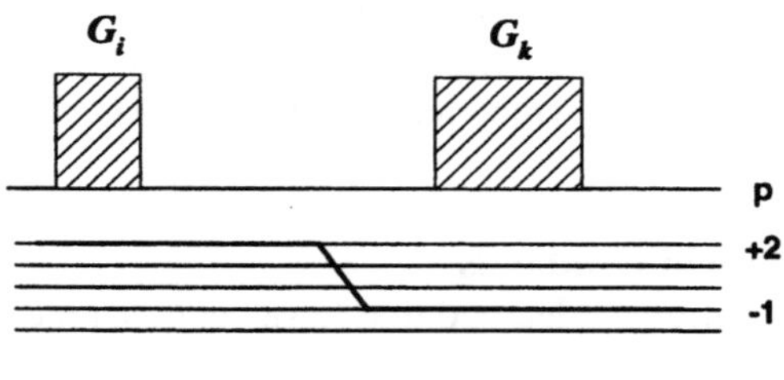

Figure 8. Selection of the coherence pathway using PFG's. Two field gradient pulses with equal field gradients but with widths of Δt and $2\Delta t$, respectively, are applied. The coherence pathway $2 \rightarrow -1$ is to be selected as is shown.

If a second gradient pulse with the same spatial distribution is applied, (cf. Fig. 8), then it is possible to refocus the phase introduced by the first gradient pulse. This occurs when the condition $p_i\Phi_i(\mathbf{r}) = -p_k\Phi_k(\mathbf{r})$ is satisfied. Here, p_i and p_k are the orders of coherence for the selected pathway when the ith and k^{th} gradient pulses are applied, and $\Phi_i(\mathbf{r})$ and $\Phi_k(\mathbf{r})$ are the phases introduced by PFG's i and k.

In the examples given in Sec. 4 for DQC a strong suppression of unwanted signals has been achieved using phase cycling. An estimate for suppression by PFG's can be made, assuming a sample with the dimensions of ($2X$, $2Y$, $2Z$), uniform square gradient pulses with components (G_X, G_Y, 0), and $v(\mathbf{r})$ = const.

The condition for refocusing $2\Phi_i(\mathbf{r}) = -\Phi_k(\mathbf{r})$ is met for the signal from the DQ coherence pathway 2 → -1 depicted in Fig. 8. Alternatively, for the SQ pathway 1 → -1, which is to be suppressed, we find from Eq. 41:

$$V = V_0 \frac{\sin(\gamma_e G_X X\Delta t)\sin(\gamma_e G_Y Y\Delta t)}{\gamma_e^2 G_X G_Y XY\Delta t^2} \qquad \text{(43a)}$$

For a gradient pulse with a width Δt of 20 ns, sample dimensions of 1.5×6×1.5 mm, and PFG of (70, 70, 0) G/cm, we find $V/V_0 \approx 0.063$. (The 0 → -1 pathway would be suppressed by a factor of 4 greater). Past work on PFG's for FT-ESR Imaging (Ewert *et al.*, 1991) suggests that such specs are not unreasonable. Suppression can be improved by fine tuning of gradients, such as to bring the sine functions in the numerator of Eq. 43a closer to zero. For selection of high orders of coherence this value would be adequate. Note also that signal selection with PFG's is highly specific with respect to the given order of coherence.

For such experimental conditions, PFG's would not provide as great a suppression as phase cycling does. Nevertheless PFG's could be very useful in the case of small spin clusters where, for spin-counting, it is necessary to separate higher orders of coherence. This would be a challenging task for phase cycling. PFG's could also be used in conjunction with phase cycling to improve the suppression of unwanted pathways by phase cycling in experiments with nitroxides.

4. EXAMPLES

4.1 Random Radicals (E' Centers in Fused Silica)

Paramagnetic centers with narrow lines are ideally suited as a model system to examine the features of DQC. The model system of E' centers in γ-irradiated silica has been studied by Borbat and Freed, (1999). E' centers have a narrow ESR spectral extent (ca. 5 G at Ku-band), thereby posing few instrumental challenges. This system provides a good illustration for all the pulse sequences. It was used for initial studies in order to work out the proper pulse sequences and to optimize phase cycling schemes. Because paramagnetic centers are randomly distributed, this system can also serve as an illustration for DQ experiment on random distributions. The weak-coupling approximation is adequate for magnetically dilute solids (Feldman and Lacelle, 1996).

Due to the small spectral extent Eq. 37 readily applies to the experiments on E'-centers in γ-irradiated fused silica shown in Figs. 9-11. To interpret such an experiment, we must sum over all the individual random radical-pair interactions. Formally, this averaging over the random distribution of pairs of spin is done by configurational averaging of

$$\left\langle \prod_j \cos a(\mathbf{r}_{1j})t \right\rangle_{\{\mathbf{r}_{1j}\}} . \tag{43b}$$

Such averaging has previously been performed by the Markov method (Chandrasekhar, 1943) for SQ coherence (Abragam, 1961), but it can be performed for DQC as well (Borbat and Freed, 1999). It yields, for example, for the envelope of the 4 and 6-pulse signals:

$$\begin{aligned} D_4(t_p) &= -\frac{1}{8}[1-\exp(-kt_p)] \\ D_6(t_\xi) &= \frac{1}{2}[\exp(-kt_m)-\exp(-k\,|\,t_\xi\,|)] \end{aligned} \tag{44}$$

where, $k = (4/3^{5/2})\pi^2\gamma^2\hbar C$ and C is the concentration (i.e. number density) of spins. In a 4-pulse sequence we observe exponential growth of the DQ signal that is faster at higher concentrations.

Note that for random distributions, all orders of coherence are produced by the preparation sequence. This results in the loss of amplitude of the DQ signal by a mechanism that is referred to as "instantaneous diffusion" in the literature (see for example, Salikhov, *et al.*, 1976; Raitsimring and Salikhov, 1985), and Eqs. 44 should be modified accordingly by multiplying them by $\exp(-k_1 t_m)$ where $k_I = (4/3^{5/2})\pi^2\gamma^2\hbar CH$ where in this case $H_n = H_n^\infty$ $(n = 4, 6)$

The simple time behavior expressed by Eq. 44 in conjunction with complete excitation was convenient for demonstrating some key features. We compared the DQ signals from different DQ sequences as well as with SQ signals.

The ratio 1:4 for the amplitude of the 4 (or 5) pulse vs. the 6-pulse DQ echoes predicted by Eq. 32 is experimentally demonstrated in Fig. 9. The result in this figure was obtained by the 6-pulse sequence of Fig. 5(c). It is a 2D acquisition in which the echo shape is obtained as a function of t_2; then the acquisition is repeated as the time interval t_{p1} in the preparation period is stepped out. Here we label the first t_p period in Fig. 5(c) as t_{p1} and the second as t_{p2}. The sum of t_{p1} and t_{p2} is kept constant. When $t_{p1}= t_{p2}$, the 6-pulse DQ echo is successfully refocused. This echo is in the center of the plot. When $t_{p1} \neq t_{p2}$ we detect two echoes from the upper and lower SQC pathways with $p = \pm 1$ in the t_{p1} portion of the preparation sequence. These two echoes are positioned symmetrically with respect to the 6-pulse echo, and their magnitudes are the same as expected for a 4(5)-pulse echo, i.e. ¼ of the 6-pulse DQ echo (cf. Fig. 9). The first and last π pulses of the 6-pulse sequence now only change the sign of these two echoes. Splitting of the 6-pulse echo into two 4(5) pulse type DQ echoes yields a factor of 2 reduction. Another factor of 2 emerges because half of the coherence pathways of the type

$$S_1^{\pm}S_{2z} \rightarrow S_2^{-}S_{1z},\ S_2^{\pm}S_{1z} \rightarrow S_1^{-}S_{2z}$$

are lost, since they cannot satisfy the refocusing conditions with such a timing arrangement.

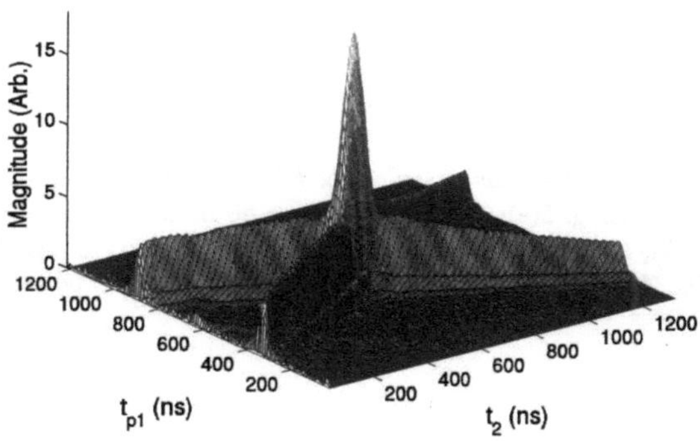

Figure 9. The relative amplitudes in the 4(5) and 6-pulse DQ echoes as exemplified by results on E'-centers. A modified 6-pulse acquisition scheme was used in 2D format. In the plot t_2 is the spectral coordinate and t_{p1} is the length of the first time interval in the preparation sequence.

In Fig. 10 we show results from samples of E' centers in γ-irradiated fused silica for the 4 pulse sequence. Again uniform coverage was obtained. By varying t_2 we record the actual shapes of the echoes. We show the 4-pulse SQ echo [signal (i)] and the 3-pulse stimulated echo [signal (v)] (arising from eliminating the π-pulse of Fig. 5(a)). They are equal but opposite in sign. Note, that both have magnitudes equal to COSY (cf. Sec. 4.2.5), if relaxation can be neglected. By phase cycling we separate the DQ signal from the sum of ZQ and SQ signals. Also, it was possible to suppress the ZQ signal from the latter sum yielding just the 4-pulse SQ by "detuning" the ZQ signal. This is achieved by setting the two t_1 periods in Fig. 5 to be unequal. The DQ signal (iv) is close to 25% of that for the stimulated echo at this t_p. By "retuning", (i.e. resetting the refocusing condition for DQC by equalizing the lengths of the two intervals t_1), we get the algebraic sum of the SQ and ZQ signals, (cf. ii). Finally with a modified phase cycle, we get the algebraic sum of SQ + ZQ + DQ, (cf. iii). One sees that the ZQ and DQ signals are equal, and they are opposite in sign to the SQ signal. These relationships are consistent with the theory given in the Sec. 2.2.

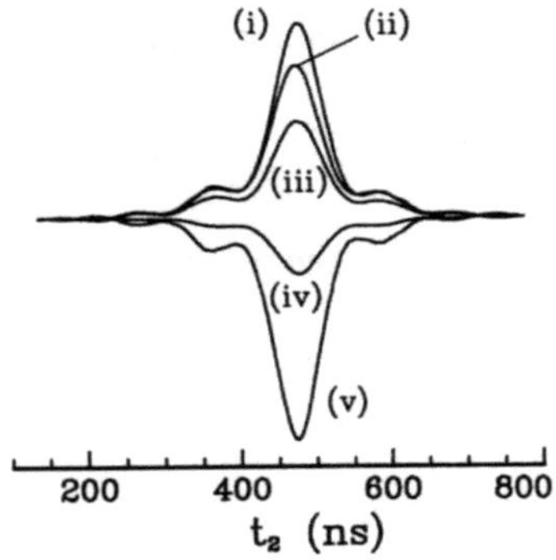

Figure 10. DQ, SQ, and ZQ signals observed for E' centers in γ-irradiated fused silica with the 4-pulse sequence of Fig. 5(a); t_p =475 ns and t_1= 50 ns with t_2 varied to give the echo shape: Case (i) the SQ signal; (ii) the algebraic sum of the SQ and ZQ signals; (iii) the algebraic sum of the SQ, ZQ, and DQ signals; (iv) the DQ signal; (v) the 3 pulse stimulated echo.

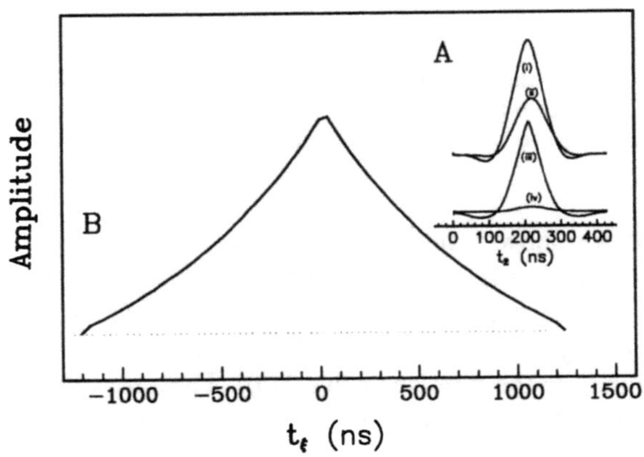

Figure 11. E' centers in γ-irradiated fused silica. The 6-pulse sequence of Fig. 5(c) was used. The $\pi/2$ pulses were set to 6.8 ns and the π-pulses to 13.2 ns. t_m = 1,200 ns, t_1 = 50 ns and the DQ signal is shown in B vs. t_ξ for a step size in t_p of 20 ns.A) echo shapes vs. t_2 for (i) the two-pulse COSY echo; (ii) the DQ echo at t_p = 200 ns. Cases (iii) and (iv) are respectively the COSY and DQ echoes for a more dilute sample.

Fig. 11 shows DQ echo shapes vs. t_2 for the 6-pulse sequence, as well as echoes from the two pulse COSY signal, for both a low and high concentration of E' centers. [Results from the latter sample were shown in Fig. 10] The DQ signal vs. t_ξ in Fig. 11(b) is also shown for the more concentrated sample. This figure illustrates refocusing of dipolar interactions for randomly distributed spins. An analysis of this signal in terms of a random distribution of interacting spins given by Eq. 44 yields a sample concentration of $1.7 \cdot 10^{18}$ spins/cm^3. This value is close to the value $2.0 \cdot 10^{18}$ spins/cm^3 that we obtained from a standard instantaneous diffusion measurement (Eaton and Eaton, 1993) for this sample.

One sees from Fig. 11(A) that for the high concentration sample, the DQ signal reached about 50% of its maximum value at this t_p (200 ns) by comparison with the COSY signal, (which is produced by two successive $\pi/2$ pulses), whereas it is relatively much smaller for the low concentration sample due to its considerably slower signal growth (cf. Eq. 44).

4.2 Bilabeled Organic Molecules

As an example of a system that exhibits a large distance, ca. 30 Å, the nearly linear rigid nitroxide biradical: piperidinyl-CO_2-(Phenyl)$_4$-O_2C-piperidinyl, (R-I), (Borbat and Freed, 1999) was studied. It could be aligned in the liquid-crystal mixture: Phase V (Meirovitch *et al.*, 1982), by slowly cooling the 0.5 mM solution into the nematic phase in an orienting magnetic field of 13.5 kG. The nematic phase persists to low temperatures, and the sample remained fully aligned in the temperature range from -57°C to -66°C, where the experiments were performed. For comparison, the sample was also quickly frozen in zero magnetic field to yield an isotropic powder (glass). As a model system with a shorter distance, ca. 16 Å, the biradical (R-II) piperidinyl-CO_2-Phenyl-O_2C-piperidinyl was used. It was studied in the LC phase V.

4.2.1 4, 5, and 6 Pulse Sequences: Oriented Samples

We show in Figs. 12 and 13 a series of DQ signals obtained with the pulse sequences in Fig. 5. In Fig. 12A (curves (i) - (ii)) we show the DQ signal from the 4-pulse sequence of Fig. 5(a) for the biradical R-I for the cases of 0° and 90° alignment respectively. The t_p minimum here was 35 ns and one observes that for 0° alignment a significant build-up of signal has occurred during this "dead-time". The experiments are affected by decay due to a T_2 of about 0.8 μs. In addition, there is ESEEM, due to matrix and nitroxide protons, which does not pose a problem, since it occurs at 26.2 MHz. This is still much greater than the dipolar modulation we discuss next. In addition, the relative intensity of the ESEEM is low. The modulation of the DQ signals due to dipolar interaction is strong for both orientations. Fourier transformation yields frequencies of 2.13±0.15 MHz and 4.2±0.15 MHz respectively for 90° and 0° tilt angles for both the 4 and the 5-pulse sequences. The decay of the oscillations, which is more pronounced at 0° tilt, is presumably due to a small distribution of interspin distances, as well as to an order parameter close to, but not exactly unity. In Fig. 12(B) (iii, iv) we show signals obtained with the 5-pulse sequence of Fig. 5(b). These signals have the characteristic features of no decay due to T_2, a baseline offset, and they are a sum of two components, (cf. Sec. 2.2). The second component is considerably weaker and can be noticed in (iii) for large t_m-t_p.

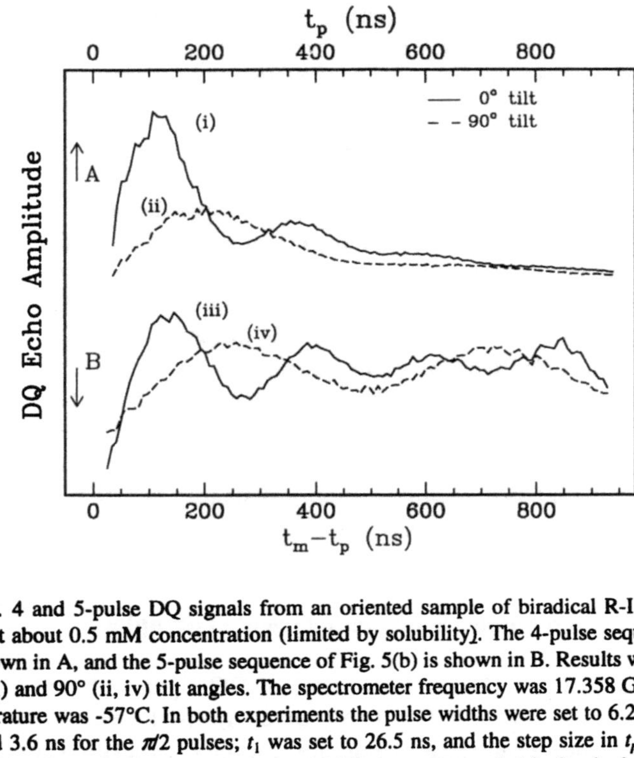

Figure 12. 4 and 5-pulse DQ signals from an oriented sample of biradical R-I dissolved in Phase V at about 0.5 mM concentration (limited by solubility). The 4-pulse sequence of Fig. 5(a) is shown in A, and the 5-pulse sequence of Fig. 5(b) is shown in B. Results were obtained at 0° (i, iii) and 90° (ii, iv) tilt angles. The spectrometer frequency was 17.358 GHz; the sample temperature was -57°C. In both experiments the pulse widths were set to 6.2 ns for the π pulses and 3.6 ns for the $\pi/2$ pulses; t_1 was set to 26.5 ns, and the step size in t_p was 8 ns. A total of 200 points yielding a t_p maximum of 1.6 μs were sampled in 9 min for the 4-pulse sequence (only the first 114 points are shown), and 114 points were sampled in 49 min for the 5-pulse sequence. The signal for the 5-pulse sequence is plotted as a function of t_m-t_p, so as to correspond to the graph of the 4-pulse sequence.

The characteristic dipolar modulation frequencies of 2.13 and 4.2 MHz for 90° and 0° tilt angles respectively, readily yield the distances from Eq. 15c and Eqs. 30a and 30b. For the 4-pulse sequence the modulation obeys $\sin^2(at_p/2) = (1-\cos at_p)/2$. From Eq. 15d, $a(90°) = \gamma_e^2\hbar/r^3$ and

$a(0°) = -2\gamma_e^2\hbar / r^3$ corresponding to the 1:2 ratio observed. These results are readily converted into r. We obtain r = 28.8 Å using the conversion r[Å] = $10(51.9/f\,[\text{MHz}])^{1/3}$.

Fig. 13 shows the results obtained with the 6-pulse sequence for the same aligned sample of R-I.

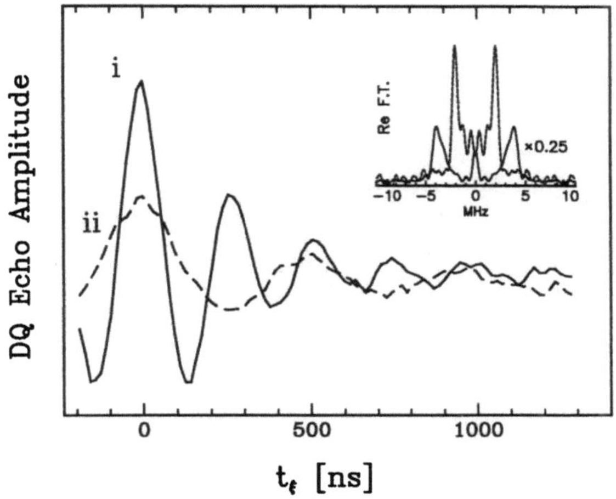

Figure 13. Rigid biradical R-I aligned in LC Phase-V. Data collected with 6-pulse sequence are plotted for two orientations (i) 0° and (ii) 90°. Only part of the signal was collected, given its symmetry vs. t_ξ. Inset shows Fourier transforms of the corresponding envelopes. Solid lines: 0°; dashed lines: 90°.

For the oriented samples DQ echoes are quite strong and the signal can be collected in a matter of minutes even over an interval as long as 1 μs. We obtained the same characteristic frequencies in the 6-pulse scheme as with the 4 and 5 pulse schemes, since its signal oscillates as $\cos(at_\xi)$. The DQ echo is stronger for 0° tilt angle, which is qualitatively consistent with the proposed geometrical structure of the biradical (cf. below), where the z-axes of the respective *hfi* tensors of nitroxides are nearly perpendicular to the interspin vector. For such a case at the 0° tilt angle $\Delta\Omega \approx A_\perp$ for the $m = \pm 1$ nuclear manifolds. Hence with limited spectral excitation stronger signals are produced. This observation is supported by comparison of ESR spectra obtained by Fourier transformations of the COSY signal at both tilt angles.

The above results indicate how macroscopically aligned samples provide the greatest opportunity for accurate distance measurements: the dipolar line shape from the FT (cf. Inset to Fig. 13) is just a narrow line. This also leads to increased sensitivity. Macroscopically aligned samples are ideally suited for extracting distance distributions, order parameters, and structural parameters.

Also, the range of measurable distances would be broader, since at 0° tilt the dipolar splitting is larger (cf. App. B). In addition, the relative angles between interspin vectors of different types of radical pairs could be determined from the angular dependence of their dipolar splitting (Astashkin *et al.*, 1997).

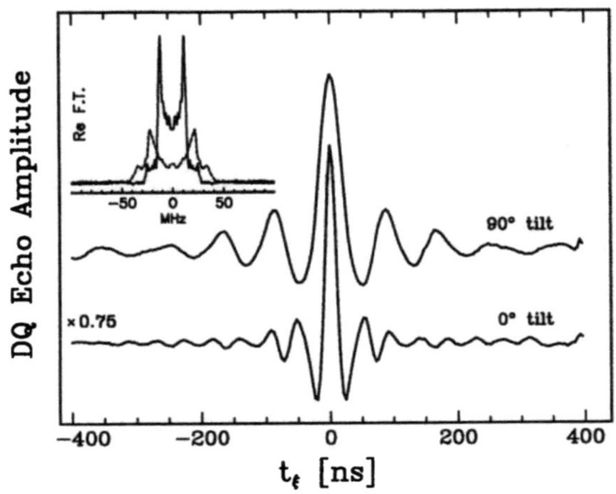

Figure 14. Biradical R-II aligned in Phase-V at -69°C. Signal envelopes in the 6-pulse sequence are shown for 0° and 90° tilt angles. Inset: Real FT of the left halves of the signal envelopes for 0° tilt (dashed line) and 90° tilt (solid line). Concentration is about 1 mM. The collection time was 15 min, and we show unprocessed signals.

Next, we show the results for the shorter biradical R-II, also aligned in Phase-V in Fig. 14. The signal is detected with high SNR. The frequency of 11.9 MHz at 90° tilt angle can be readily read from the FT spectra (Inset) and it translates into a distance of 16.2 Å. We note that at this relatively short distance, effects caused by the pseudo-secular term of the dipolar Hamiltonian can be noticed (cf. App. A). While they are small for 90° tilt angle, for 0° tilt angle the frequency $3\nu_\parallel/2 = 36$ MHz ($\nu_\parallel \equiv \omega_\parallel/2\pi$) is clearly seen. These pseudo-secular effects plus small deviations from perfect alignment and also the weak ESEEM line at 13 MHz (cf. Sec. 4.2.4) lead to the Pake-like appearance in Fig. 14.

4.2.2 Random vs. Oriented Samples

We have shown in Figs. 13 and 14 the DQ signals obtained from the 6-pulse sequence of Fig. 5(c) for aligned samples. The same samples can be converted to disordered ones by rapid freezing. Figs. 15 and 16 show the disordered cases, which are the ones more typically encountered. We see the

DQ modulation pattern plotted vs. t_ξ. The signals, of course, have zero dead time as a function of t_ξ, and they are symmetric about $t_\xi = 0$. They represent the "dipolar echo" which occurs at $t_\xi = 0$, when refocusing conditions for dipolar interaction are met. We also show the real part of the FT of the time domain signals.

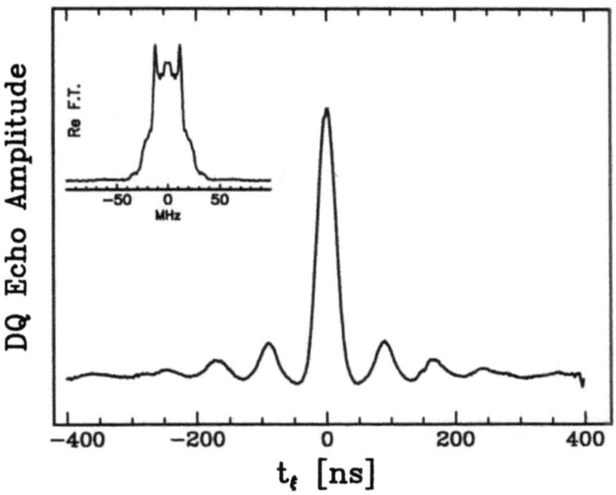

Figure 15. Results from the 6-pulse DQ sequence. Biradical R-II was studied in disordered LC Phase V at -69° C. Inset shows Fourier transform of the left half of the signal envelope. The two characteristic frequencies of the Pake doublet (cf. Insert) 11.9 and 25 MHz translate into a distance between NO groups of 16.2 Å, and the exchange integral, J is -0.25 MHz. The biradical concentration is about 1 mM. (A signal at 0 MHz corresponds to a residual baseline vs. t_ξ). Data collection took 1 hour for 400 points in t_ξ, but with improved time sweep capabilities it can be performed in about a minute. We show the unprocessed signal.

The DQ signal for R-II is strong with a SNR of about 300 at the signal maximum. The concentration could be reduced to 100 μM and would still permit an accurate distance determination. The Pake doublet is well-resolved and both canonical frequencies $\nu_\parallel$ and $\nu_\perp$ can be accurately determined. From Eq. 15c,d using the experimental values for $\nu_\parallel$ and $\nu_\perp$ equal to 25 MHz and 11.9 MHz, respectively, we can find that d is 12.3 MHz and J is -0.25 MHz. The value of 12.3 MHz for d yields the distance of 16.2 Å between the NO groups. The shape of the Pake doublet shows characteristics due to non-negligible effects of the pseudo-secular terms (cf. App. A), but at the distance ca. 16 Å these effect can be noticed only at 0° tilt angle where d is ca. 8.2 G.

The DQ signal for R-I (cf. Fig. 16) was acquired over a long enough sampling interval of 1.56 μs to produce a nearly ideal Pake doublet after subtraction of a very weak linear slope, that is caused by the DQ signal from the interaction of nitroxide spins on different molecules (cf. Sec. 4.3). The intermolecular DQ signal from the randomly distributed spins, is quite small in comparison to the DQ signal caused by the intramolecular coupling. While the DQ signal in Fig. 16 is a very good one, it is not yet optimal, because full spectral excitation has not yet been achieved. The Pake doublet can readily be interpreted (cf. above) to yield a distance between nitroxides of 28.8±0.5 angstroms.

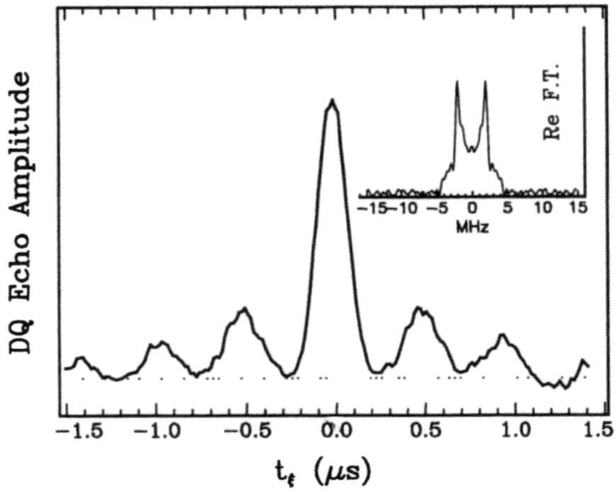

Figure 16. DQ signal using 6-pulse sequence at -62° C for disordered sample of rigid biradical R-I in LC Phase-V. Data were collected over a sampling interval of 1.56 μs with 6 hours of acquisition. No data smoothing or filtering was applied. Inset shows the Pake doublet obtained by the real Fourier transformation of the sum of the left and right halves of the signal envelope after folding the spectrum over the point $t_\xi = 0$ and subtracting a linear baseline. The concentration was 0.5 mM.

Molecular modeling by means of the PCMODEL v.7 software package yielded a value of 28.06 Å (27.81 Å) for the H-H (N-N) distance between the diamagnetic NH groups, which had to be used in the modeling instead of $\dot{\mathrm{NO}}$.

The data collection for Fig. 16 took a substantial but reasonable time, for this not very concentrated sample, (≈0.5 mM), but it could be made shorter since the S/N in Fig. 16 is greater than necessary. A reasonable upper limit for the sampling interval for this sample with its T_2 of 0.8 μs is about 2 μs. Note that the 1.56 μs sampling interval used in Fig. 16 is already sufficient to measure a 43 Å distance with ±5 Å accuracy, and a 2 μs interval increases it to 47 Å. Such estimates are based on the simple assumption that one needs at least one full period of oscillation of the frequency $\nu_\perp = d$, the singular component of the Pake doublet. Larger distances can be assessed with longer T_2's. With a T_2 of 2 μs a distance of 70 Å could be determined with the same relative error, whereas for T_2 in the range of 1 - 1.5 μs such a distance could be estimated with tolerable accuracy.

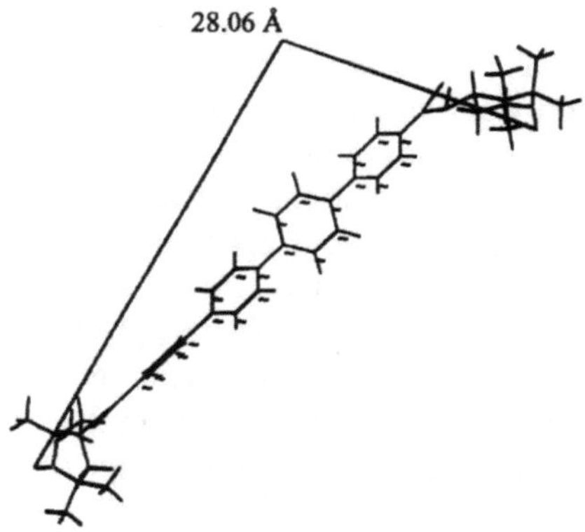

Figure 17. Geometrical model of the rigid biradical R-I, generated by PCMODEL v7.0 software package

For aligned samples the upper range of distances extracted from $\nu_{\parallel} = 2d$ would be about 80 Å. Further discussion relating to the upper range of distances, can be found in App. B.

4.2.3 Correlation of Orientations (2D Experiments)

We found (cf. Sec. 3.1.1) that even with a B_1 as large as 30 G, π-pulses did not provide uniform excitation for the ^{14}N-labeled nitroxides. FT's of our data for 4, 5, and 6-pulse sequences for the aligned sample of R-I show that the amplitude ratios of the spectral peaks for the 90° orientation relative to that for the 0° orientation are about 1.5, 1, and 0.65, respectively. This indicates that the nitroxide moieties are correlated and indicates the steady decline in spectral excitation with increasing number of pulses in the sequence. When there is spatial correlation between *hfi* tensors of the nitroxides and ***r***, then Eq. 34 needs to be used for simulations. It has been shown in DEER (Maryasov *et al.*, 1998) that selective excitation gave rise to modulation of the Pake doublet by a function $\xi(\cos\theta)$ which they called the geometrical form-factor. It was also shown there that in the case of inhomogeneous broadening typical for nitroxides, $\xi(\cos\theta)$ is a fairly smooth function of $\cos\theta$ and typically varies within a factor of 2. It can badly distort Pake doublets, but does not change characteristic frequencies. Maryasov *et al.* (1998) suggested that by varying B_0 it could be possible to extract some structural information related to the mutual orientations of the nitroxides *hfi* tensor principal axes by modeling variations of $\xi(\cos\theta)$ with magnetic field. It is reasonable to expect that a 2D DQ experiment can be utilized for obtaining this kind of information. Such an experiment provides the shape of a Pake pattern as a function of position in the ESR spectrum. That is, a 2D Fourier transformation of the echo shapes $F(\omega_1,\omega_2)=FT\{D_n(T,t_2)\}_{T,t_2}$ collected as a function of the time T (which is t_ξ for the 6-pulse sequence and t_p for 4-pulse sequence) would yield variations of the Pake doublet displayed along $f_1 = \omega_1/2\pi$ as a function of position in the ntroxide spectrum, (i.e. along $f_2 = \omega_2/2\pi$) due to orientational correlations. Let us consider how this can happen. Let us first ignore any variation in spectral coverage of the second spin for simplicity. Excitation of the first spin at a certain frequency offset $\Delta\Omega_1(\lambda_1,\eta)$ selects a range of orientations (λ_1,η) corresponding to this frequency offset, if the orientations of the magnetic tensors of this nitroxide is correlated with the orientation of interradical vector, $\boldsymbol{r}$, (cf. discussion of Eqs. 34 and 35). Only biradicals with a certain range of angles $\eta \equiv (\theta,\varphi)$ will be excited corresponding to a non-uniform distribution for $(3\cos^2\theta -1)/2$. This will result in a variation of the Pake pattern over the ESR spectrum. [Clearly, this type of argument holds equally well when there is correlation for both nitroxides with $\boldsymbol{r}$].

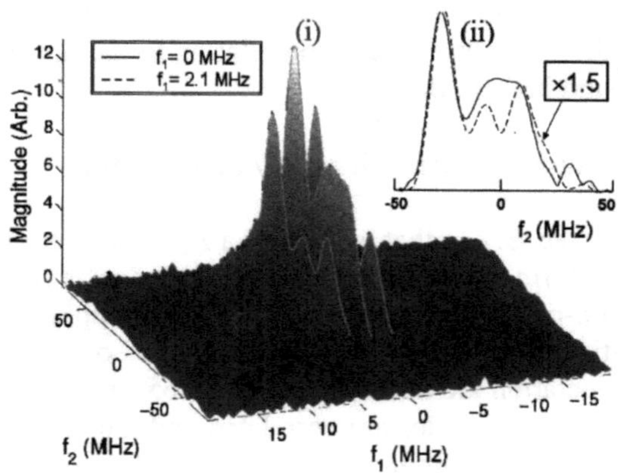

Figure 18. The 2D 6-pulse DQ experiment with biradical R-I disordered in the LC Phase-V at -62° C. (i) - The 2D data were collected in this experiment, which also provided the data in the 1D format depicted in Fig. 16. The spectrum is the magnitude FT vs. f_1 and f_2, which are the respective FT variables of t_ξ and t_2. No baseline correction was made. The inset (ii) displays the $f_1 = 0$ MHz slice (solid) and the $f_1 = 2.1$ MHz slice (dashed) superimposed. The latter is at the maximum of the DQC Pake doublet whereas the former is mainly a background that depends only on $\langle G_6(\Delta\Omega_1)\rangle_{\Delta\Omega_1}\langle H_6(\Delta\Omega_2)\rangle_{\Delta\Omega_2}$. Hence the difference between them is a result of correlation between $\mathbf{r}_{12}$ and the nitroxide magnetic tensors.

We show in Fig. 18 such a 2D spectrum, which clearly shows variations in the Pake pattern along f_1 with spectral resonant frequency, f_2. Fig. 18 is from a disordered sample of R-I. A detailed analysis of these 2D spectra should enable important structural information to be obtained. In Fig. 18 we can clearly see substantial variations of the intensity of the $\nu_\perp$ component of the Pake doublet. This component is visible as a sideband in the stack plot. The line at zero frequency has its origin from the FT of the slope, which is caused by the DQ signal from interactions with the other spins in the sample, and therefore corresponds to the uncorrelated case. The sidebands vary quite substantially, with a lower intensity of $\nu_\perp$ in the center of the spectrum. This indicates that the z-axes of the *hfi* tensors are oriented at an angle that is substantially different from the interspin vector. This conclusion is consistent with the molecular model of the biradical (cf. Fig. 17).

4.2.4 ESEEM Effects

So far, we have not discussed ESEEM effects in this chapter. As long as oscillations from ESEEM occur at significantly different frequencies from those resulting from DQ modulation, simple Fourier transformation methods can easily separate them. However, ESEEM can interfere with the relevant signal in some cases. At this point it seems appropriate to give an aproximate picture of ESEEM in the DQ experiment. Nuclear modulation is known to be a potential problem not only in other coherent distance measurement techniques such as “2+1” (Kurshev *et al.*, 1988) but also in 4-pulse DEER, when overlap of spectral excitations at the two applied frequencies is not negligible, (Pannier *et al.*, 2000).

A rigorous treatment of ESEEM for radical pairs could yield very cumbersome expressions for DQ pulse sequences even for the case of non-selective pulses, for which it should be possible to arrive at a closed form expression. This is why we give a more approximate picture of ESEEM effects with the example of the 6-pulse sequence. We assume that ESEEM is determined predominantly by the SQ pathways, i.e. by preparation and detection periods where it attains the well-known form for a 2-pulse echo (Dikanov and Tsvetkov, 1992):

$$M(t)=\prod_{n=1}^{N}(1-2k_n^2\sin^2(\omega_{\alpha,n}t/2)\sin^2(\omega_{\beta,n}t/2)) \tag{45a}$$

where $\omega_{\alpha,n}$, $\omega_{\beta,n}$, k_n are given by

$$k_n^2=\frac{\omega_{I,n}^2(T_{zx,n}^2+T_{zy,n}^2)}{\omega_{\alpha,n}^2\omega_{\beta,n}^2} \tag{45b}$$

$$\omega_{\alpha,n},\omega_{\beta,n}=[(\omega_{I,n}\pm T_{zz,n})^2+(T_{zx,n}^2+T_{zy,n}^2)/4]^{1/2}.$$

Here t is either t_p or $t_m - t_p$, N is the number of interacting nuclei with their respective *hfi* tensors $\mathbf{T}_n$ and Zeeman frequencies $\omega_{I,n}$. We shall ignore the action of the DQC refocusing pulse sandwich $(\pi/2) - t_1 - \pi - t_1 - (\pi/2)$, which introduces constant phase and amplitude factors. The signal envelope in the 6-pulse sequence is then modulated by $M(t_p)M(t_m - t_p)$ and can be rewritten as:

$$M(t_m, t_\xi) \approx 1 - 4\sum_n k_n^2 (1 - \cos\frac{\omega_{\alpha,n} t_m}{2}\cos\frac{\omega_{\alpha,n} t_\xi}{2} - \cos\frac{\omega_{\beta,n} t_m}{2}\cos\frac{\omega_{\beta,n} t_\xi}{2}$$
$$+ \frac{1}{2}\cos\frac{\omega_{n+} t_m}{2}\cos\frac{\omega_{n+} t_\xi}{2} + \frac{1}{2}\cos\frac{\omega_{n-} t_m}{2}\cos\frac{\omega_{n-} t_\xi}{2}). \tag{46a}$$

Here $\omega_{n\pm} = \omega_{a,n} + \omega_{\beta,n}$ and the terms in k_n^4 are neglected. For matrix nuclei and weakly-coupled nuclei of nitroxides, $\omega_{a,n}, \omega_{\beta,n} \approx \omega_I$ and Eq. 46a becomes:

$$M(t_m, t_\xi) \approx 1 - k(\frac{3}{4} + \frac{1}{4}\cos\omega_I t_m \cos\omega_I t_\xi - \cos\frac{\omega_I t_m}{2}\cos\frac{\omega_I t_\xi}{2}) \tag{46b}$$

with k is a spatial average of the k_n^2 over all interacting nuclei. One immediately finds that the modulation frequencies are scaled down by a factor of 2, and that "blind spots" will emerge, where the ESEEM at a given frequency is suppressed. (Such blind spots occur in ESEEM for sequences with more than two pulses (Schweiger, 1990).

With respect to ESEEM, we note that the 4-pulse DQ sequence resembles HYSCORE (Höfer *et al.*, 1986; Gemperle *et al.*, 1990). Using an expression for HYSCORE from Dikanov and Tsvetkov, (1992) for weakly-coupled protons we derived:

$$M(t, t_1) \approx 1 - 2k\sin^2(\omega_I t/2)\sin^2[(\omega_I (t + 2t_1)/2].$$

The line at ω_I will be suppressed by the effect of "blind spots" when $2\omega_I t_1 = \pi(2n+1)$, ($n = 0, 1, \ldots$), i.e. for $t_1 = 9.5$ ns, 28.5 ns, etc. at 17.3 GHz. We typically set $t_1 \approx 28$-30 ns and have successfully suppressed the line at ω_I. The same type of "blind spots" effect can take place in the 6-pulse sequence as well, (cf. Eq. 46b).

Experiments we have performed have confirmed that Eq. 46b gives a qualitatively correct picture. In some of our experiments, we made use of the blind spots in order to suppress ESEEM at the proton Zeeman frequency, by adjusting t_m. Finally, we remind the reader that ESEEM of the DQ signal from electron spin pairs in diluted systems does not constitute a problem for two reasons. First, at Ku-band the modulation frequencies from matrix nuclei, which are typically protons, occur well outside the Pake doublet. [In the 4 or 5 pulse sequences at 17.3 GHz, the proton ESEEM at 26.3 MHz is higher in frequency than the dipolar modulations for $r > 15$ Å. In the 6-pulse sequence the proton ESEEM frequency is 13.1 MHz and is outside the doublet for $r > 20$ Å]. Second, the modulation is shallow in the DQ experiment, and we often do not see it. At X-band, the modulation due to solvent nuclei is about 4 times stronger and nuclear modulation frequencies are about 2 times lower than at 17.3 GHz.

A problem arises when a weak DQ signal from electron spin pairs is imposed on a strong DQ signal from randomly distributed radicals or from the radicals grouped in small clusters, (e.g. in cases of very high concentration). In the case of randomly distributed spins at high concentration, the DQ signal from pairs decreases due to generation of higher orders of coherence, and the intermolecular DQ signal from radicals in the bulk becomes larger. With increase of the sampling interval t_m, the DQ signal from the randomly distributed spins grows according to Eqs. 44. This signal appears as a nearly linear slope modulated by ESEEM. At sufficiently high concentration, this modulation can become comparable to the DQ signal from pairs. Prior to Fourier transformation this signal should be removed. For reliable subtraction, it is advantageous to reduce the modulation depth, which is accomplished by using the suppression effect of the "blind spots".

4.2.5 Comparison with COSY, etc.

It is clear that the simplest pulse experiment from which it is possible to extract distance information is just the two pulse Hahn echo sequence. In the weak coupling limit the expression for the primary echo signal amplitude (under the same conditions that Eqs. 30 for DQ signals were derived) is[3] :

$$V_{2,SQ}(\Delta\Omega_1,\Delta\Omega_2,\mathbf{r},t_p)=-\mathrm{Im}\{S^{(1)}(\beta_1)^*\}S_2^{(1)}(\beta_2) \times[1-2S^{(2)}(\beta_2)\sin^2(at_p/2)]. \tag{47}$$

In COSY $\beta_1=\beta_2=\pi/2$. In the limit of non-selective $\pi/2$ pulses, Eq. 47 becomes:

$$-V_{2,SQ}(\Delta\Omega_1,\Delta\Omega_2,\mathbf{r},t_p)\approx(1+\cos at_p)/4 \tag{48a}$$

[3] A somewhat lengthy expression for the 2-pulse echo for coupled spins was derived by Yudanov et al. (1969) in the limit of non-selective pulses. In the weak coupling limit it simplifies to Eq. 47 if FID-like terms (cf. App. A) are ignored.

which is to be compared with Eq. 32. Note that

$$-V_{4,DQ}(\Delta\Omega_1,\Delta\Omega_2,\mathbf{r},t_p)\approx(1-\cos at_p)/8 \tag{48b}$$

from Eq. 32 as $S^{(2)}(\pi/2)\approx 0.5$ for sufficiently non-selective pulses. (The actual value is 0.454 for a B_1 of 30 G for ^{14}N at X-band), Therefore Eq. 48a holds even for realistic pulses. However for B_1=30 G at X-band the factor $\mathrm{Im}\{S^{(1)}(\pi/2)^*\}S_2^{(1)}(\pi/2)$ for COSY is 0.75, whereas the factor

$$2\,\mathrm{Im}\{S^{(1)}(\pi/2)^*\}S_2^{(1)}(\pi)S_2^{(1)}(\pi)[|C_2^{(1)}(\pi/2)|^2+S_2^{(1)}(\pi/2)^2]|S^{(2)}(\pi/2)|^2$$

for $V_{4,DQ}$ (cf. Eq. 30a and Sec. 3.1.1) is 0.44. This means that $V_{4,DQ}$ will be about 4 times less than $V_{2,DQ}$ under realistic conditions.

Thus, it would appear that in the case of ^{14}N nitroxides, the COSY sequence can be effectively used, given sufficient spectral excitation and short dead-times. We examine this with examples using biradicals R-I(II).

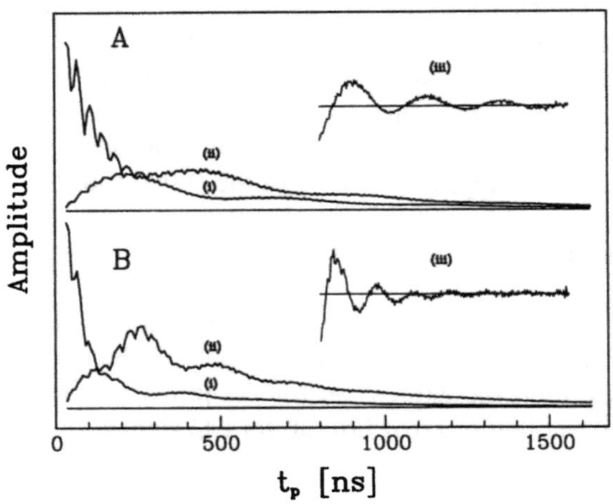

Figure 19. Comparison of the COSY and 4-pulse DQM experiments with the rigid biradical R-I aligned in LC phase V at -57° C. Results are shown for 90° (A) and 0° (B) tilt angles, respectively. (i) – DQM, [scaled up by a factor of 4 in (A)]; (ii) – COSY; (iii) - DQM with exponential decay removed. Pulse widths are 3.5/3.5 ns for COSY and 3.5/3.5/6/3.5 for the DQ experiment.

In Fig. 19 we show COSY and 4-pulse DQ signals for macroscopically aligned biradical R-I for two orientations. Note that both show the low-frequency oscillations due to dipolar coupling (DQM), but COSY shows relatively larger ESEEM effects (which are seen as high frequency noise in insets), and both show effects from T_2 decay.

The situation is totally different in the case of disordered samples. Fig. 20 shows the envelope of the primary echo [$\beta_1 = \pi/2$, $\beta_2 = \pi$] for R-II. It follows from Eq. 47 that for the primary echo sequence the relative effect of DQ modulation must be stronger than for COSY, since $S_2(\pi) \approx 1$ for strong pulses. Despite this, it is very difficult to obtain the dipolar modulation from the primary echo. After removing the decaying background the residual signal is Fourier transformed and shown in the inset of Fig. 20. In addition to the strong ESEEM pattern that is typical for weakly coupled protons, a weak solitary peak at 11.9 MHz is clearly visible. Even if we believe that it is the singular component of the Pake doublet (i.e. $\nu_\perp$), we cannot reliably translate it into a distance without a prior knowledge of J. That is we need both $\nu_\perp$ and $\nu_\parallel$ to extract J (cf. Eq. 15d), but the latter frequency is buried under the ESEEM line at the proton Zeeman frequency.

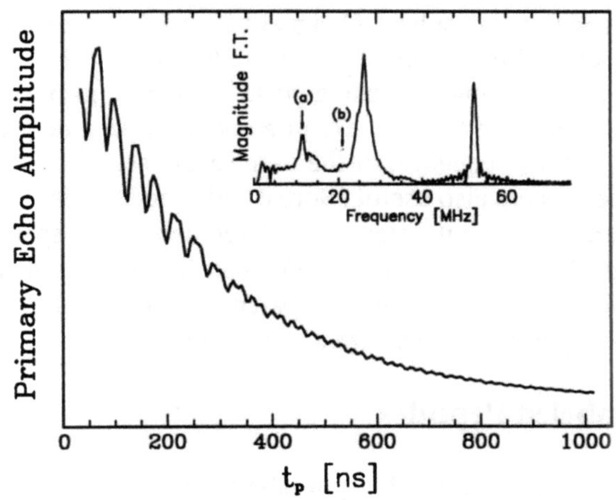

Figure 20. Primary echo ($\pi/2$-τ-π-τ) for the R-II in disordered LC Phase V at -66° C. Pulse widths were 3.5 and 6 ns. Inset: Fourier transformation of the primary echo after removing the decay. The weak solitary peak at 11.9 MHz (a) is likely the singular component of the Pake doublet and the small feature (b) at 22 MHz could be the $\nu_\parallel$.

The reader can compare this result with that obtained using the 6-pulse sequence for the same radical (cf. Fig. 15). In the 6-pulse sequence we obtained a Pake doublet with high SNR and extraction of the distance is a routine task. The electron exchange integral, (small in this case), was also obtained. In the 2-pulse echo (cf. Fig. 20) as we have pointed out above, matters are much more problematic. Even for this Ku-band experiment, this peak is weaker than the ESEEM component at ν_I (26.2 MHz). The second characteristic frequency of the Pake doublet (25 MHz for R-II) is hidden under the line at ν_I. Note that at X-band both lines will be buried under the ν_I (ca. 13.5 MHz) and $2\nu_I$ lines, respectively. [These lines will be about 4 time more intense at X-band]. Furthermore, the subtraction of the decaying component of the signal practically always produces spurious low-frequency components in the spectrum.

A range of low frequencies of instrumental origin are typically in the signal envelope, as well. All these frequencies yield a low-frequency band which could overlap with the dipolar spectrum in the case of long distances, leaving little hope for extraction of useful data. The situation becomes especially poor in the case of weak pulses. Thus in favorable cases, e.g. aligned systems, one can utilize COSY for distance measurements. When dipolar modulation is shallow, it becomes progressively more difficult to extract distance information from the COSY experiment, even with a dead-time of 35 ns, in cases where distances are sufficiently short (<20 Å) or else distributions in distance damp the oscillations quickly (cf. Sec. 4.3.1). The same problem exists with the 4-pulse DQ signal but to a lesser extent, because in this pulse sequence ESEEM effects can be reduced (cf. 4.2.4). The situation can be improved by adding refocusing pulses to both sequences, and thereby eliminating T_2 decay and to some extent reducing dead time problems. This produces a "2+1" SQ sequence, or the 5-pulse DQ sequence. Coherent 5(6) pulse SQ sequences are also useful (Borbat and Freed, 1999a; 2000), but we found them less convenient unless B_1 is large and/or the spectrometer frequency is high. However, greater benefits can be found in the 6-pulse DQ sequence, which entirely eliminates "apparent" dead-time, deeply suppresses background, and yields stronger signals, given adequate B_1.

4.3 Bilabeled Peptides

In addition to rigid biradicals, spin-labeled synthetic peptides have been studied, (Borbat and Freed, 1999). They are described in (Saxena and Freed, 1996) and have the structure: MTSL-CPPPC-MTSL, (R-III), MTSL-CPPPPC-MTSL, (R-IV), where MTSL is the (1-oxyl-2,2,5,5,-tetramethyl-pyrroline-3-methyl) methanethiosulphonate spin label, and C and P are amino acids cysteine and proline, respectively. The concentration was about 1.0 mM for R-III and 0.75 mM for R-IV in glycerol/water/trifluoroethanol mixture, which was buffered with MOPS (3-[N-morphilino] propane sulphonic acid). They were studied at about -82°C. All samples were about 15-20 μL in volume. In Fig. 21 we show the DQ signal, arising from the 6-pulse sequence, from the polyproline biradical R-IV, where T_2 is 340 ns. The Pake pattern generated by Fourier transforming the left half of the original signal is also shown. There are no well-defined Pake components of the type seen in Figs. 15-16 for R-I(II), which indicates that there may be a range of distances between the two ends in this synthetic peptide.

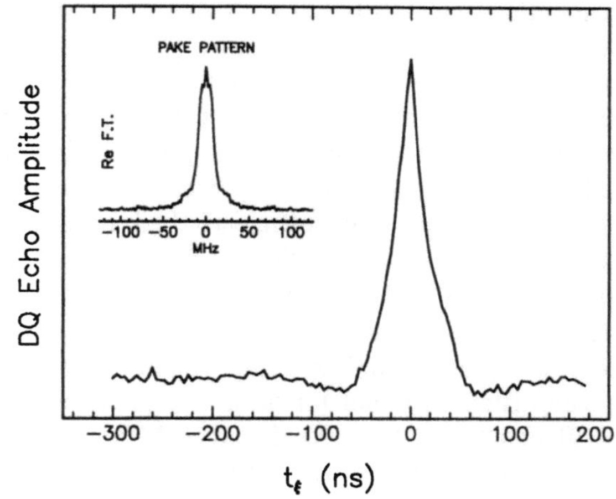

Figure 21. DQ signal, obtained at -82°C for the biradical R-IV dissolved in a glycerol-H_2O-trifluoroethanol mixture. The 6-pulse sequence of Fig. 1(c) was used at 17.35 GHz. Pulse widths were 3.2 ns for $\pi/2$ pulse and 6.0 ns for π-pulses. The Pake doublet from the real part of the FT of the signal is shown in the inset. No components are well resolved, suggesting a distribution of distances.

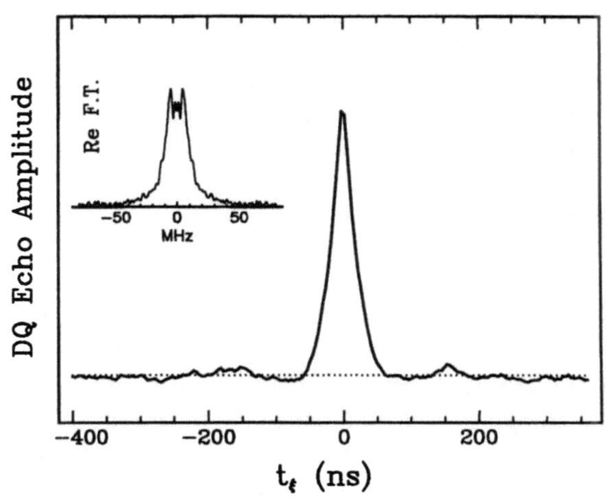

Figure 22. DQ signal, obtained at -82°C for the biradical R-III dissolved in a glycerol-H_2O-trifluoroethanol mixture. Concentration was 0.75 mM. The 6-pulse sequence of Fig. 1(c) was used at 17.35 GHz at the conditions of Fig. 21. The Pake pattern generated by taking the real part of the FT of left half of the signal is shown in the inset.

One can interpret the weak splitting frequency of 4.5 MHz and the half-width of the doublet (9 MHz) to yield an average distance of 22.6 Å with a distribution of distances characterized by a half width of 2 Å by simulating the line shape using distance distributions of different width. This is consistent with the value previously found for R-IV (Saxena and Freed, 1996, 1997). We also found that a 4-fold increase in the concentration of this peptide to 3 mM leads to peptide aggregation, such that the DQ signal from the intramolecular pairs is imposed on the larger signal from intermolecular interactions. This larger signal yields the nearly linear slope that is characteristic of random spin distributions with sufficiently high local concentrations[4]. We did not observe this kind of aggregation for R-III at 3 mM, but could not rule out other possibilities, e.g. dimerization. The Pake pattern for R-III also does not exhibit pronounced singularities of the classical Pake doublet and a small distribution of distances similar to that found for R-IV can well account for this. The distances found for these two peptides are very close, with that for R-III about 1 Å less. We found it difficult to reproduce the shape of the signal envelope in the time domain for R-III(IV) using simple considerations. Also, the DQ signal was not as strong as in the case of R-I or R-II. Two possible explanations are as follows. 1) There are two (or more) conformers one with (or more) separated by a shorter distance (especially for R-IV). This can yield a sharp featureless signal for the latter. 2) The presence of dimers with one of the distances short could also yield a similar result. In this case detection of higher orders of coherence can help to clarify the matter. These are matters for future study.

As an additional note, by analogy to other methods of distance measurements (Milov *et al.*, 1999; Rabenstein and Shin, 1995), one can attempt to subtract out the DQ signal arising from intermolecular dipolar interactions by obtaining the DQ signal from singly-labeled peptides under identical conditions. We have had some preliminary success in this way.

4.3.1 Distribution of Distances

Our examples above, as we have interpreted them, indicate the challenges when there is a distribution of distances. For this case one has to sum over the dipolar line shapes using the proper distance distribution in addition to consideration of possible correlations between nitroxide *hfi* tensors. The net result is a rapid damping of oscillations in the time-domain and a featureless line in the Pake pattern from which it may be very difficult to extract the radial distribution. An estimate of the half width of the distribution can certainly be made from the width of the Pake doublet. The situation can be even more challenging if electron exchange is not negligible. Obviously, our two

[4] Precisely, the slope is exponential in the case of uniform spatial distributions (cf. Eq. 44).

peptide examples cannot be fit reliably with many parameters. In the case of macroscopically aligned systems the situation is predicted to be more encouraging, because distance distributions will show up directly in the form of the observed line shape especially in the limit of non-selective pulses. A study of the angular dependence of the line shape would yield data sufficient to address more complex cases of spatial distributions.

The 6-pulse sequence with sufficiently non-selective pulses seems to be the most suitable technique in the case of distance distributions: it yields the least distorted signals in the time-domain due to virtually zero apparent dead-time, its high time resolution, and its remarkable filtering properties (Borbat and Freed, 2000). It is relatively insensitive to the correlation of *hf* tensors, given adequate spectral excitation. Thus, only the distribution of distances and the dipolar line shape need to be considered. The role of the latter can be minimized by using an oriented sample as we already noted. Current spectral excitation should be improved, and we see several reliable ways of achieving this, (cf. Sec. 5.2).

4.4 "Forbidden" Coherences

4.4.1 "Forbidden" Coherence Pathways

The coherence pathways discussed in Sec. 2 are not the only ones that can give rise to DQC. They are based upon "allowed" coherence pathways, i.e. those that yield DQC which are then converted to detectable SQC by means of hard or non-selective pulses. Even when arbitrary pulses were considered in Sec. 2.2 the coherence pathways remained the same "allowed" ones. The arbitrary pulses referred to the fact that the B_1 was not necessarily large enough to cover the whole nitroxide spectrum; it still ignored the dipolar interaction during the pulse.

In fact, in the case of arbitrary pulses, the dipolar interaction can, in conjunction with the *rf* pulse, produce DQC, as was shown in the extensive numerical analysis of Saxena and Freed (SF), (1997a). In particular, a two-pulse sequence is sufficient to first produce the DQC and then convert and detect it as SQC. In this case the first pulse becomes the full preparation period, i.e. the equivalent of Fig. 3 for non-selective pulses. How can that be? SF found that the first pulse should be a selective one; i.e. a long weak pulse. In particular, of the cases that they considered, the best was in the region of $\tau = 80$ ns and $B_1 = 1.8$ G for an $r \approx 18.5$ Å (D=4.4 G). Thus the dipolar interaction can act during the pulse.

A relatively simple way of analyzing this pulse is by a method due to Salikhov *et al.*, (1996), based on the Suzuki-Trotter (S-T) formula (Suzuki, 1985). They show that the arbitrary pulse can be split up into a series of *n* small $\Delta\tau$ segments (where $n\Delta\tau = \tau_{pulse}$). Within each segment, the role of the pulse and of H (Eq.18) can be treated as separate. Then Fig. 3 becomes the $n = 2$ approximant to this pulse, (except that if the selective pulse is nominally $\pi/2$, then each sub-pulse for $n = 2$ is just $\pi/4$). As a result, the forbidden coherence (FC) becomes weakly allowed. SF estimate that FC will range from 14 to 60 times weaker than the allowed pathways from their simulations, which assumed both pulses to be identical. Actually, it would be better to use a strong second $\pi/2$ pulse to more effectively convert the DQC into antiphase SQC. In addition, preliminary experiments have indicated that this FC approach would be better achieved by adding refocusing π pulses after the first and second pulses, by analogy to what is done for the "allowed" pulse sequence of Fig. 5(b). In fact, it would yield the 4-pulse FC pathway analogue of the 5-pulse allowed coherence pathway of Fig. 5(b), where the first two pulses of Fig 5(b) are replaced by a single selective pulse.

The low order analysis based on the S-T formula (n=2) outlined above shows that the "mixing in" of DQC depends on the ratio:

$$|D/\gamma_e B_1| = \frac{3\gamma_e \hbar}{2r^3 B_1} |1 - 3\cos^2\theta| \tag{50}$$

It thus favors those biradicals with interspin vector parallel to $\boldsymbol{B}_0$, for which the magnitude is maximized. This can provide some orientational selectivity by analogy to that provided by aligning the sample (cf. Sec. 4.2.1), but it corresponds to a weak partial alignment. However, higher order S-T theory (i.e. $n > 2$) can predict FC pathways where $|D/\gamma_e B_1|$ appears to a power greater than unity, implying stronger orientational selectivity.

In their original work, SF (1996) found that they could obtain a FC from a 5-pulse sequence, which is a modified form of Fig. 5(b), from R-IV which shows a more substantial modulation pattern than what is seen in Fig. 21, but exhibits much weaker overall S/N. It was fit by numerical simulations by the rigorous theory of SF (1997a), but the effects of distributions in angles λ_1 and λ_2 and distance, r were not considered due to extensive computation times. Clearly, further exploratory work is required before FC can be usefully applied to distance measurements.

5. DISCUSSION

The studies described in the previous section have shown that DQC experiments on bilabeled nitroxide molecules yield the Pake doublets from which one can readily extract distances. This is most successful if there is no correlation between the orientations of the nitroxide moieties relative to $\mathbf{r}_{12}$, (i.e. Eq. 35). In general this need not be true, so that one has, in principle, to perform a complex analysis based upon Eq. 34 that is analogous to rigorous analyses of any other cw or pulsed ESR method. However, in the DQC experiment it is possible, by using a large enough B_1 field relative to the spectral extent, to almost completely suppress any effects of the nitroxide moieties, (i.e. Eq. 37). Current experimental conditions should be somewhat closer to this limit for ^{15}N labeled nitroxides. Additional improvements (cf. Sec. 5.2) should enable approaching this limit for ^{14}N nitroxides. This "decoupling" is a virtue of the DQC experiment not shared by DEER or "2+1" based upon selective pulses, or by cw-ESR methods (Hustedt and Beth, 1999). Fortunately, such matters have negligible effects in our DQC experiments on the characteristic frequencies, $\nu_{\perp}$ and $\nu_{\parallel}$ but do affect the shape of the dipolar spectra. Such matters become of greater concern in the case of biological systems that in our experience show broad distributions of r, so the Pake patterns become poorly resolved or completely unresolved. In such cases, any and all distortions due to correlations and other effects cannot be ignored. *The best way to deal with them is to supply nonselective excitation and double-quantum filtering*. Our experience has indicated that selective pulses should be used with great caution.

Correlation can probably be best addressed by 2D experiments such as illustrated in Fig. 18. Alternatively, the first pulse of the sequences of Fig. 5 could be made selective, and its field position stepped out in a series of experiments. The 2D method, however, provides minimal unwanted effects due to broad spectral excitation with the multiplex advantage. It promises better frequency resolution, that is only limited by the sampling window, and consequently would be expected to result in a more reliable analysis.

Another important advantage of the DQC experiment is that the (nearly) non-selective pulses lead to (nearly) complete excitation of all the spins thereby enhancing signal strength, which is not the case for techniques based upon selective pulses. This should be of considerable value for spin-labeled biomolecules available only in small amounts and typically at low concentrations. In addition, we note that, unlike other ESR methods, the DQC experiment is unaffected by any singly labeled molecules (unless they are present at high concentrations, which in turn will produce more substantial problems to other ESR techniques).

We have seen how the 6-pulse sequence effectively provides zero deadtime. This is achieved by refocusing dipolar coupling at the (virtual) zero time[5], $t_{\xi} = 0$. In addition, the extremely narrow pulses employed in the method means that they do not contribute significant uncertainty to the time periods for spin evolution. The 4-pulse DQ sequence lacks the zero deadtime feature, (but can yield stronger signals in some cases when there is re-

[5] This is analogous to a 4-pulse DEER sequence (Pannier *et al.*, 2000)

duced coverage from the finite B_1). We do observe ESEEM in the DQC experiments due to surrounding nuclei. It has a small or no effect on the signal in dilute systems (cf. Sec. 4.3.4). In the case of concentrated systems, as we have shown, it is possible to use the suppression effect that results in "blind spots" in the proton modulation in the 4 and 6-pulse sequences.

All in all, we believe these results show how the development of multiple quantum coherences in ESR can lead to a range of new and useful techniques, especially in applications to distance measurements.

5.1 Comparison with Other Pulsed ESR Techniques

When dipolar coupling between two electron spins is present, virtually any SQ signal carries distance information in some form, because its amplitude becomes modulated by the dipolar coupling as antiphase coherences are created (cf. Eq. 27). The simplest technique is of course the Hahn echo, which, in the case of a short dead time, can yield at least one of the two canonical frequencies of the Pake doublet, as we demonstrated with the example of the biradical R-II (cf. Fig. 20). ESEEM is not an unavoidable problem for nitroxides, especially at higher spectrometer frequencies. With very short dead-times a 2-pulse echo technique could be successful (cf. Sec. 4.2.5). Nevertheless, it was instructive to compare the results obtained on the same sample using a Hahn echo and the 6-pulse DQ sequence, (cf. Sec. 4.2.5) to show that the latter is superior.

An improvement to the 2-pulse SQ technique can be made by adding one or more pulses. Several SQ pulse sequences for distance measurement emerge, including DEER and "2+1" sequences (cf. Sec. 2.2).

In the 3-pulse SQ technique known as the "2+1" pulse sequence (Kurshev *et al.*, 1989; cf. Ch. 10), the added pulse has in the past been mainly a selective one, (but this is not necessary). An advantage of this sequence in comparison with the 2-pulse echo is that it eliminates decay due to T_2 and any artifacts that emerge from incrementing the interpulse distance. However, the informative part of the signal consists of two overlapping signals depending on different time variables. These signals are difficult to separate, and there is a large background. As a result, an FT of the signal does not result in a Pake pattern. ESEEM could be strong in "2+1" and it has received special consideration, (Kurshev *et al.*, 1988, Raitsimring *et al.*, 1995). However, in the limit of hard pulses, the nuclear modulation will disappear (Raitsimring, Freed *et al.*, 1995), albeit this is hard to achieve due to the spectral excitation challenges in this sequence.

Coherent SQ sequences (Borbat and Freed, 1999a, 2000; cf. App. C) have also been proposed. They essentially use the refocusing of dipolar coupling in the same fashion as was done in the DQ 6-pulse sequence. Another

SQ sequence has been reported recently by Jeschke *et al.*, 2000. The deficiency of these SQ sequences is that they cannot provide DQ filtering, thus they are useful mainly with very high B_1's and/or well-defined distances (as opposed to distributions in distance).

DEER is a viable SQ technique with many advantages. First of all, ESEEM is virtually absent in 3-pulse DEER, which is an incoherent experiment. The apparent dead-time in 3-pulse DEER is typically limited by the pulse widths, which for a B_1 of 2-4 G are in the range of 40-20 ns, making it difficult to detect $\nu_{\parallel}$ or to obtain the Pake doublet when there is a distribution of distances, because this leads to a rapid damping or even an absence of oscillations (cf. Sec. 4.3).

The 4-pulse version of DEER largely eliminates the dead-time problem of 3-pulse DEER, at the expense of using only part of the pulse sequence length available for evolution of coherences due to dipolar interactions. This should reduce the maximum measurable distance (R_{max}) up to a factor of $2^{1/3}$ as compared to other techniques, (cf. App. B). Nevertheless, the benefits gained due to shortening of the dead-time will be more important. It was clearly demonstrated that this method produces reliable distance information over a broad range of distances and for a variety of systems, (Pannier *et al.*, 2000). The same advantage of zero dead-time is shared by the 6-pulse DQ sequence, but with a much finer time resolution enabling measurements of distances as short as ca. 10 Å.

In most typical cases that we have encountered, the strong dipolar oscillations seen for bilabeled molecules with very rigid structures are rarely observed. For broad distance distributions the oscillations are shallow or completely smoothed out. With weak pulses it becomes problematic, if at all possible, to separate this small smooth signal from the large baseline, which has a steep slope that can be modulated by low ESEEM frequencies, as we have observed in some solvents.

DEER utilizes selective pulses and the signal amplitude is proportional to the product of the excitation profiles at both frequencies (Maryasov *et al.*, 1999). The desirability of increasing B_1 is limited by the rapidly increasing overlap in the spectral excitations at the two irradiating frequencies. Our estimates show that even for ^{14}N nitroxides, B_1 is limited by ca. 4.0 G even when pulses are placed at opposite ends of the spectral bandwidth (the most favorable case). (Here we used a criterion that the spectral overlap should be no more than 10% of the useful signal). The corresponding amplitude factor for the DEER signal was estimated to be about 0.005 of the ideal primary echo signal from all of the spins. The signal amplitude of DEER could be improved by a factor of 5-10 if a high Q resonator were used. However, Pannier *et al.* (2000) found it necessary to use a low-Q LGR to enable excitation at both frequencies. The reduced receiver bandwidth in DEER should yield an improvement by a factor of 2 or 3 (to about 0.01). In making this estimate, we assumed that in DEER the sampling window, pulse widths, and Q are all set to their optimum values, matching the maximal permissible B_1, (cf. above). More considerations about sensitivity can be found in App. B.

Note that in DEER a pulse that is too wide would ultimately lead to a distortion of the Pake pattern when $0.5\gamma_e B_1 \sim d$. This corresponds to ~20 Å at a B_1 of 4 G. As we noted above, the DQ technique is virtually free from this limitation.

A small amplitude factor in DEER necessitates using higher concentrations and/or large samples, in order to acquire the signal in a reasonable time, especially for larger distances. This may be a reason why, in DEER, the small dipolar modulation is usually imposed on the steep slope of a larger signal (Milov *et al.*, 1999; Pannier *et al.*, 2000) which is caused by intermolecular interactions amongst the randomly distributed spins in the sample. The slope can be removed by conducting experiments at different concentrations (Milov *et al.*, 1999). However this introduces noise and one cannot be fully confident that the separation of useful signal from the large offset is reliably obtained.

By contrast, the DQ technique, due to its sensitivity, is most beneficial for low-concentration samples, which can be of small size. Along with cw ESR methods it can become the technique of choice in distance measurements for many biological systems, which normally are neither concentrated nor available in large quantities.

Finally, there should be large distortions and a reduction of signal with selective pulses, when the *hfi* tensors of the nitroxides are correlated (Maryasov *et al.*, 1998). The DQ sequences are potentially free from this problem, which would virtually disappear at sufficiently large B_1's.

It is appropriate to compare the key equipment used in the pulsed ESR techniques. Although they are different, the equipment is similar in many aspects. In 3-pulse DEER it is better to use 2 high-power pulse sources, [usually TWTA's], to avoid interference between pulses that occurs in the source when the pulses are close to each other. Nevertheless, single high-power sources, have been successfully used (Larsen and Singel, 1993; Pfannebecker *et al.*, 1996). In 4-pulse DEER or in the DQ technique one source is adequate, because the pulses do not need to be very close. "2+1" typically uses the standard ESE equipment "as is" or with minor modifications, but an optimal performance, as that in DEER, can be achieved with use of two TWTA's. The source in the DQ experiment should have an output of at least 2 kW, which is in fact readily available. TWT's with 0.5-1 GHz bandwidth and even greater power are found on the market.

One important difference concerns the probes. Different types of resonators are in use. Bimodal cavity resonators with the modes tuned to different frequencies have the advantage of high Q in the modes and an opportunity of tuning the modes independently in DEER. The samples can be conveniently studied at low-temperatures using standard fingertip dewars or gas-flow systems (Milov *et al.*, 1998). However, the filling factor is rather poor leading to a loss of SNR. In the DQ experiment low-Q structures including dielectric, loop-gap, and other types of small resonating and delay structures are the most appropriate. Thus low-Q single-mode resonators are in routine use, and they are simpler devices than bimodal structures. However cavity resonators could also be used, and they are more convenient for cryogenic temperatures. We have successfully employed a standard Varian rectangular resonator at X-band for DQC and SQC distance measurements (Borbat and Freed, 2000). It was possible to achieve $\pi/2$ pulses as short as 7.0 ns with our TWT for this cavity. Nevertheless, for best results one should use a high-power source and a small low-Q resonator.

The unavoidable reduction of Q, needed to provide the wide spectral bandwidths to accommodate the short pulses in the DQ experiment (or the different frequencies in DEER), decreases the SNR. This is more than offset by the rapid growth of the amplitude of DQ signals (as the pulse width is decreased) which is a product of the spectral excitations of the two spins as well as by the suppression of destructive effects of correlation and the pseudo-secular term on the signal intensity. This growth is limited only by reaching the condition of full spectral excitation (B_1 of 10 G for E'-centers) or by the available B_1, whichever comes first.

Experimentally DQC is less laborious than other distance measurements by pulsed ESR, because the relevant signal is not overwhelmed by a large unwanted signal.

It is not surprising that we conducted most of our experiments at Ku-band due to its advantages for this technique. Other techniques, including DEER, could also benefit from higher frequency. Commercial spectrometers working at X-band can be used 'as is' for the DQ experiment, but for best results with nitroxides a somewhat smaller loop-gap or dielectric resonator would be highly desirable in order to enhance B_1.

Finally we wish to point out a unique advantage of DQ, or more precisely MQ techniques. More complex structures, such as aggregates of spin-labeled proteins, would involve several interacting electron spins. One should be able to generate and detect the higher orders of coherence that emanate, by analogy to NMR (Baum *et al.*, 1985). This would provide additional insight into such systems.

5.2 Improvements in the Future

As we have seen, given sufficient spectral excitation, the DQ distance measurement technique offers the best sensitivity in comparison to other pulsed methods. It is fully capable of providing accurate distance information over a broad range of distances from 10-15 Å to about 80 Å (cf. Appendix B).

It has been demonstrated that we can generate strong DQ signals at conditions (190-230 K) that are far more challenging than those at cryogenic temperatures (i.e. below ca. 100 K), and can extract distances using small samples (ca. 10 μL) with concentrations of spin even as low as 0.1 mM. This is equivalent to 50 μM concentrations and 500 pmoles of doubly-labeled species. A worst case situation of short T_2's (ca. 0.3-0.5 μs) and distances larger than 25 Å would require about 5-10 nmoles of doubly-labeled molecules. It is also important that solutions of bilabeled proteins are at low enough concentrations, especially, for large molecular weight proteins. This is especially true for measurement of large distances, when concentrations should be low. The DQ experiment can meet this challenge. There is much potential for future improvements of the technique, and we expect that the current limits can be reduced with the additional improvements, that we now describe.

The most important improvement would be the generation of more intense pulses. It greatly improves not only DQC experiments but coherent SQ techniques as well. Using as an example the 6-pulse sequence, we found that there would be a significant increase in the signal amplitude if the B_1 were increased from 30 to 60 G. This increase would be even greater for correlated nitroxides.

The coverage is mainly determined by the π-pulses. It would be sufficient to increase B_1 just for the π-pulses by a factor of 2 and make them of the same length as the $\pi/2$-pulses. As we described, we employ a TWTA with rather typical parameters. It delivers about 2 kW at X and Ku bands and can be boosted up to 3 kW at X-band. For the two-fold increase of B_1 in the resonators that are currently used, the required power would need to be increased by a factor of 4 to the level of about 8 kW. One can use either a more powerful traveling-wave tube or else utilize output pulse compression (Farkas *et al.*, 1989) in order to complete the task.

Another possibility is to explore the applicability of composite pulses, (Levitt, 1979; Crepeau *et al.*, 1989) in DQ sequences. Another improvement, which is being implemented, is to use a higher working frequency, i.e. Ka-band corresponding to 35 GHz (which would also further improve the sensitivity). At this frequency, B_1's can be created as large as 60 G with 1 kW pulses in a dielectric resonator of small size, This would be sufficient for both ^{15}N and ^{14}N nitroxides at this frequency. The amount of sample could be reduced by a factor of 4 to 5, with a small gain in concentration sensitivity being likely.

Additionally, ESEEM will provide practically no problem at this frequency even in those special cases (small clusters of spin-labeled molecules) that might be expected to pose difficulties, such as when there are long distances (35-80 Å) and when perdeuterated spin-labels and solvents may be desirable or even necessary. At 35 GHz the deuterium ESEEM frequencies are about 4 and 8 MHz in the 6-pulse DQ experiment, and therefore outside the Pake doublet for distances exceeding 30 Å. The ESEEM modulation depth also would decrease as the frequency squared.

Additional improvements should include better phase cycling with a smaller increment to enable one to separate higher orders of coherences in cases of clusters of spin-labeled molecules. Also, pulsed field gradients could be employed to ease the task of separation of high orders of coherence in this case.

ACKNOWLEDGMENTS. We are grateful to Professor A. Rassat for the gift the sample of nitroxide biradical, R-I, and to Professor I. A. Grigorev who kindly supplied the biradical R-II. We thank Dr. M. Ge for help in preparing samples of spin-labeled polypeptides R-III(IV), Professor C. F. Wilcox for help with the molecular modeling of R-I, Drs. S. Saxena and Bosong Xiang for useful discussions, and C. R. Dunnam who brought to our attention the binary pulse compression technique.

The study was supported by NIH Grants RR07126 and GM25862 and NSF Grant CHE9615910.

APPENDICES

A. Effects of the Pseudo-Secular Terms.

In Sec 2.2 we described the theory for DQC pulse sequences, where (1) the pseudo-secular dipolar terms in H were neglected, (cf. Eq. 18), and (2) the dipolar interaction was completely neglected during the pulse, (cf. Eq.19). If both of these approximations are removed, then the solution to the resulting general problem requires numerical simulations (Saxena and Freed, 1997a). Since we are primarily interested in allowed coherence pathways and strong pulses, assumption (2) remains a good approximation (with the pulses that we used ($B_1 \cong 30$ G) for $r \geq 11.5$ Å using estimates from perturbation theory). Thus, in this Appendix we shall remove assumption (1) so that Eq. 17 transformed to the rotating frame will need to be used for H:

$$H/\hbar = \Delta\Omega_1 S_{1z} + \Delta\Omega_2 S_{2z} + aS_{1z}S_{2z} + b_p(S_1^+S_2^- + S_1^-S_2^+)/2\,. \qquad \text{(A1)}$$

This admits of analytic, though complicated solution. The approach is to utilize the basis set of Eqs. 21 and 22c to describe the free evolution periods when H, given by Eq. A1, is acting. Then by analogy to Eq. 22a:

$$\tilde{Q}\rho \Leftrightarrow e^{-i\frac{H}{\hbar}t}\rho e^{i\frac{H}{\hbar}t} \equiv P\rho P^\dagger\,. \qquad \text{(A2)}$$

The superoperator $\tilde{Q}$ is represented by a matrix with dimension 16×16 in the operator space with the basis vectors of Eq. 22c. The Hamiltonian H can be diagonalized by a suitable unitary transformation U: $U^\dagger HU = \Lambda$. The propagators on the rhs of Eq. A2 become $P = Ue^{-i\Lambda t}U^\dagger$.

The matrix elements of $\tilde{Q}$ can be determined by applying $\tilde{Q}$ to the basis vectors $\tilde{Q}B_i = Q_{ij}B_j$. Using the scalar product as described by Eq. 22d we can write:

$$Q_{ij} = \mathrm{Tr}\{B_j^\dagger PB_iP^\dagger\}\,. \qquad \text{(A3)}$$

For evolution of SQC coherences one may reduce the dimension of the problem by considering only the relevant sub-block of Q. We also use only the relevant sub-blocks of the pulse propagator superoperator $\tilde{R}(\beta_k)$ (cf. Eqs. 22a-23) in the derivations. To account for evolution of SQC we introduce the relevant super vectors **I**, **O** of inphase and antiphase SQC's respectively:

$$\begin{aligned} \mathbf{I} &\equiv (\mathbf{I}^+, \mathbf{I}^-), & \mathbf{O}^\pm &\equiv (\mathbf{O}^+, \mathbf{O}^-) \\ \mathbf{I}^\pm &\equiv (S_1^\pm, S_2^\pm), & \mathbf{O}^\pm &\equiv (2S_1^\pm S_{2z}, 2S_2^\pm S_{1z}). \end{aligned} \qquad \text{(A4)}$$

With these definitions, we can write for the free evolution of SQC during a time interval t using the relevant sub-block of Q:

$$\begin{pmatrix} \mathbf{I}' \\ \mathbf{O}' \end{pmatrix} = \begin{pmatrix} \mathbf{V}_t & \mathbf{W}_t \\ \mathbf{W}_t & \mathbf{V}_t \end{pmatrix} \begin{pmatrix} \mathbf{I} \\ \mathbf{O} \end{pmatrix} \qquad \text{(A5)}$$

where the prime denotes the coherences at the end of the evolution period.

The sub-blocks $\mathbf{V}_t$ and $\mathbf{W}_t$ are :

$$\mathbf{V}_t = \begin{pmatrix} \mathbf{V}e^{-i\Delta\Omega_s t} & 0 \\ 0 & \mathbf{V}^* e^{i\Delta\Omega_s t} \end{pmatrix}, \quad \mathbf{W}_t = \begin{pmatrix} \mathbf{W}e^{-i\Delta\Omega_s t} & 0 \\ 0 & \mathbf{W}^* e^{i\Delta\Omega_s t} \end{pmatrix} \tag{A6}$$

which in turn contains the 2×2 matrices:

$$\mathbf{V} = \begin{pmatrix} \alpha c & -i\beta s \\ -i\beta s & \alpha^* c \end{pmatrix}, \quad \mathbf{W} = \begin{pmatrix} -i\alpha s & \beta c \\ \beta c & -i\alpha^* s \end{pmatrix}. \tag{A7}$$

Here

$$\begin{aligned} &\Delta\Omega_s = (\Delta\Omega_1 + \Delta\Omega_2)/2, \quad s = \sin(at/2), \quad c = \cos(at/2) \\ &\alpha = \cos(Rt/2) - ic_{2\Phi}\sin(Rt/2), \quad \beta = -is_{2\Phi}\sin(Rt/2) \\ &c_{2\Phi} = \Delta\Omega/R, \; s_{2\Phi} = b_p/R, \; R^2 = b_p^2 + \Delta\Omega^2, \; \Delta\Omega = \Delta\Omega_1 - \Delta\Omega_2. \end{aligned} \tag{A8}$$

We assumed that $\Delta\Omega_1 > \Delta\Omega_2$ for a positive sign of R. When the pseudo-secular term in b_p of Eq. A1 can be neglected, $\mathbf{V}_t$ and $\mathbf{W}_t$ become diagonal, since $\beta \to 0$; also $\alpha \to \exp(-i\Delta\Omega t/2)$. Then the matrix form is no longer needed and we arrive at Eq. 27 for the evolution. The pulse propagators have a similar block form; details may be found in (Borbat and Freed, 2000), where we also provide the super vector forms of DQ and ZQ coherences and the matrix form for the double quantum filter. This matrix form of the PO formalism allows implementation in "Mathematica" (Wolfram Research) although many of the derivations were easily performed by hand.

After straightforward matrix multiplication one obtains the signal as the scalar product :

$$V = Tr\{\rho^{\dagger} I^{-}\}. \tag{A9}$$

The calculations are generally lengthy and somewhat complicated, but they simplify considerably when we use the 6 pulse sequence of Fig. 5(c), i.e. nominal rotation angles $\pi/2$ and π. In this important case the expansion is straightforward, given the assumption that for the terms caused by pseudo-secular part of the dipolar coupling we should take $\Delta\Omega_1 \approx \Delta\Omega_2$. This greatly simplifies calculations and results in more compact expressions. We found that we should modify the $F_n(\mathbf{r}_{12}, t_p)$ of Eqs. 29. For example, the signal envelope in the 6-pulse sequence is now given by $F_6(\mathbf{r}_{12},t_p) = F(t_p)F(t_m\text{-}t_p)$, where now $F(t)$ has the form :

$$F(t) = [c_{2\Phi}^2 + s_{2\Phi}^2 \cos(Rt)]\sin(at) - s_{2\Phi}\sin(Rt)\cos(at)\,. \qquad \text{(A10)}$$

The expressions for the other DQ and SQ signals that are used in distance measurements have similar forms, but we do not give them here. Eq. A10 (and Eqs. 30) do not include "FID-like" terms that occur at $t_\xi \approx 0$. They are caused by the transfer of antiphase coherences between coupled spins that is possible with non-ideal first and last refocusing π pulses and they have the form:

$$\left\langle K(\Delta\Omega_1,\Delta\Omega_2)e^{i(\Delta\Omega_1-\Delta\Omega_2)t_\xi}\right\rangle_{\lambda_1,\lambda_2} \qquad \text{(A11)}$$

where $K(\Delta\Omega_1,\Delta\Omega_2)$ is an unwieldy function of the $\Delta\Omega_i$. They do not carry useful information on the dipolar interaction. Such terms are minimal for nominally π refocusing pulses. Moreover, we found that for the 6-pulse sequence there is a factor of 1/64 in this "FID-like signal", As a result, it is typically two orders of magnitude smaller than the "useful" signal.

Numerical simulations, which we have performed for uncorrelated nitroxides, have shown that, as the distance between the spins becomes shorter, the shape of the Pake doublet changes in an expected way due to the pseudo-secular term in Eq A10. The general tendency is to reduce the intensity of the singular part of the spectrum and ultimately to produce a featureless Pake pattern, as the distance becomes very short (<10 Å). The characteristic frequencies $3d/2$ emerge from Eq. A10. One finds from Eq. A10 that when $\Delta\Omega_1$ is equal to $\Delta\Omega_2$, $F(t) = \sin(3dt/2)$. This is expected for dipolar coupled like spins (Slichter, 1990). These results explain the slightly smoothed shape of the Pake doublet for R-II and the appearance of weak peaks at $\pm 3d/2$ on the Pake doublet (cf. Fig. 16). In the simulations for the distance of 16.2 Å, the effect was distinct, but relatively small for B_1 ca. 30 G. The situation can be different for correlated *hfi* tensors, but this has yet to be explored in detail. This effect becomes pronounced for distances under 15 Å, where the dipolar broadening is very distinct in the ESR spectra of coupled nitroxides. This range of distances is readily studied by cw ESR, (cf. Ch. 3-Ch. 7). For longer distances, where cw methods are difficult to apply, the effects of the pseudo-secular term are not very significant, and it becomes appropriate to utilize the simple treatment given in the Sec. 2.2. With the example of biradical R-II we have seen that we should account for its effects when the distances are under 20 Å. But what is perhaps most significant, these simulations have also shown that an increase in B_1 to 60 G reduces the scale of distortions substantially. This is one more reason for developing ways to increase B_1 (cf Sec. 5.2). As we have already observed in a number of experiments (Borbat and Freed, 2000), great caution should be exercised in using weak B_1 pulses in coherent experiments, especially in the case of sufficiently broad distance distributions. The dipolar signals from spins at insufficiently large separations (e.g. $r \leq$ 20-25 Å) may decrease by orders of magnitude and/or the dipolar spectrum could be too distorted to reflect distances correctly. We observed this for both SQ and DQ methods, i.e. irrespective of the presence of a DQ filter. Signals in most practical cases, when there are distributions of distances, are usually non-oscillating, and it is practically impossible to reliably separate them from the background in SQ coherent sequences and "2+1" if the pulses are weak.

B. Sensitivity and the Upper Range of Measurable Distances

The important issue of sensitivity in a pulsed ESR distance measurement experiment has not been discussed so far in this chapter. Therefore, it seems appropriate to make this estimate and in addition to evaluate an upper range of distances that can be reliably determined in the DQC experiment. We also wish to compare the DQ and DEER experiments with respect to concentration sensitivity. We consider the relevant case of coupled nitroxides, which we assume have uncorrelated *hfi* tensors. For the SNR we can write as an estimate:

$$\mathrm{SNR} = S_1\sqrt{ft}K\exp[-(\frac{2t_m}{T_2})^{\kappa} - 2kCH^{AV}t_m], \qquad K = \mu G^{AV}H^{AV}. \tag{B1}$$

Here, μ is a constant, S_1 is the SNR in a single-shot data acquisition for a standard SQ sequence, such as COSY, for the condition where the signal intensity is at its maximum [i.e. $t_m \approx 0$]. Other parameters are as follows: f is the repetition rate of the pulse sequence, t is the duration of the experiment, $2t_m$ is the time between the first pulse in the pulse sequence and the echo. This time is assumed to be fixed, and it is measured by the duration of the SQC pathways, (since the DQ pathway is relatively short). K is the ratio of the signal of interest (the DQ echo amplitude or the decrease of the echo intensity in the case of DEER) to a standard signal which for example can be COSY at $t_m = 0$ in the limit of hard pulses. Note that $K \propto G^{AV}H^{AV}$ where G^{AV} and H^{AV} are the spectral excitations for the first and the second spins, respectively, in both methods. The first term in the exponent is due to T_2 relaxation, which for nitroxide radicals at low temperatures is mainly determined by nuclear spin diffusion and methyl group rotations (Milov *et al.*, 1973; Zecevic *et al.*, 1998; Saxena and Freed, 1997b). The second term in the exponent is due to instantaneous diffusion, with C is the concentration of electron spins, and k is a constant (cf. Sec. 4.1). For S_1 we use the estimate from Borbat and Freed, (1997)

$$S_1 \propto \omega^{3/2} C V_s (Q / V_c \Delta\omega)^{1/2} . \tag{B2}$$

Here Q is the Q-value for the working resonator, ω is the applied *mw* frequency, $\Delta\omega$ is the receiver bandwidth (bw), V_c is the effective volume of the resonator, $V_s = V_c \eta$ is the volume of the sample, with η is the filling factor. We assume that the following conditions $\omega Q^{-1} \approx \Delta\omega \approx G^{AV} \Delta\Omega_s$ should result in an optimal SNR. Here $\Delta\Omega_s$ is the total spectral extent. The first condition matches the resonator bw to the receiver bw, and the second matches them to the extent of the spectrum that is excited. Thus we have :

$$S_1 \propto \omega Q (\eta V_s)^{1/2} C \cong \omega^2 (\eta V_s)^{1/2} C / (G^{AV} \Delta\Omega_s) . \tag{B3a}$$

From Eqs. B1 and B3a we find that the SNR $\propto \omega^2 (\eta V_s)^{1/2} C H^{AV}$.

It is appropriate to define and discuss the concentration sensitivity at this point. We assume for the moment that resonators with the same V_c and η, and Q-values are used in both methods. For simplicity, we define the concentration sensitivity as the value of C for a short enough τ (such that instantaneous diffusion does not matter in Eq. B1), that yields a SNR = 1. From Eqs. B1 and B3a we find that

$$C_{\min} \propto [H^{AV} \omega^2 (\eta V_s)^{1/2}]^{-1} . \tag{B3b}$$

Eq. B3b leads to two observations. First, the concentration sensitivity improves as ω^2, making it desirable to increase the working frequency at least up to the point where the spectral width starts to be mainly determined by the g-tensor anisotropy, (i.e. for frequencies substantially greater than 35 GHz for nitroxides). Second, in the DQC, H^{AV} could be close to unity (cf. Sec. 3.1.1), whereas in DEER it is about an order of magnitude smaller. Consequently, the concentration sensitivity of the DQ technique should be about an order of magnitude greater.

In DEER the volume of the resonator can be larger than in DQC in some cases (at X-band), thus narrowing the gap in concentration sensitivity for the samples that are not volume limited, due to the factor $V_s^{1/2}$. Cavity resonators can admit the largest samples but, due to the degradation of η, they cannot compete favorably with DR or (B)LG resonators.

At Ku-band, DR or (B)LG resonators with $\eta \approx 1$ and maximal volumes V_c are used in DQC. Thus DEER, at this band, would not benefit from a larger sample. However, at X-band V_s has to be reduced by a factor of ca. 8 from the maximal possible volume ($V_s \approx V_c \approx 160$ μL) at this frequency range to attain sufficient B_1. Thus, this reduces the gain in sensitivity in comparison to DEER at X-band by a factor of 2.8. Of course, at Ku-band we have an additional increase in sensitivity, compared to X-band, of ~3.4[6].

The upper range of measurable distances is ultimately limited by the phase memory time which is determined by both T_2 and instantaneous diffusion. In measurements of very large distances in the range of 40-80 Å it is necessary to lower the concentration until the instantaneous diffusion becomes relatively unimportant in comparison with T_2 relaxation, and then to maximize T_2. In the case of nitroxides at cryogenic temperatures, nuclear spin diffusion of the solvent protons is the major process leading to a rapid phase relaxation with a time dependence $\exp[-(2t_m/T_2)^{\kappa}]$ (Milov *et al.*, 1973; Zecevic *et al.*, 1998; Lindgren *et al.*, 1997). For nitroxides, T_2's of 2-4 μs and κ ca. 1.5-2.5 are typical at temperatures under ca. 120 K, whereas for the temperature range in our studies, ($T \geq 190$ K) T_2's are in the range of 0.2-1 μs and $\kappa \approx 1$ (Saxena and Freed, 1997b). For biological systems, the use of D_2O as a solvent can considerably slow down phase relaxation, and it leads to a phase relaxation as $\exp(-2t_m/T_2)$ with T_2's of order tens of microseconds. It should be noted that our experiments were performed at a local maximum of T_2 for $T > 150$ K. For distances under 30 Å, these T_2's were sufficient, nevertheless, for longer distances than we have explored, temperatures under 100K would be advisable. At temperatures ca. 190 K, T_1 (ca. 25-30 μs) is short enough that the repetition rates of 10 kHz, that we used, allow for virtually complete spin relaxation between pulse sequences.

For extracting distance information with reasonable accuracy, it is necessary to determine, at the very least, the most prominent frequency in the Pake pattern, i.e. $\nu_\perp$. A minimum signal sampling interval could be $1/2\nu_\perp$ (Larsen and Singel, 1993). However, we do not consider this as sufficient, so we take $1/\nu_\perp$ as a more reasonable minimum (i.e. we set $t_m = 1/\nu_\perp$).

With these considerations in mind, we can deduce from Eq. B1 an estimate for an upper range of measurable distances R_{max} in our studies, (where $\kappa \approx 1$):

$$R^3_{\max} = \frac{\gamma_e^2 \hbar T_2}{4\pi(1 + kCH^{AV}T_2)} \ln[\frac{S_1\sqrt{ftK}}{\mathrm{SNR}_{acc}}]. \tag{B4}$$

[6] Of course, one must take cognizance of the wall thickness of the sample tube relative to its diameter at higher frequencies.

We introduced into Eq. B4 an acceptable SNR denoted by SNR_{acc}. In our spectrometer at Ku-band at 200 K, $S_1 \approx 1$ occurs in the COSY with 30 G pulses at about 1 nmole of nitroxides in 20 μL. Thus, $S_1 \approx 10^9 CV_s$. Substituting into Eq. B4, $C = 100$ μM, $V_s = 20$ μL, $T_2 = 0.5$ μs, $S_{acc} = 10$, $f = 10$ kHz, $t = 10^4$s, $K = 0.4$, we find from Eq. B4 an $R_{max} \approx 44$ Å for DQC. At temperatures below 80 K, T_2's are typically in range 2-4 μs, leading to an increase in R_{max}, (but one must consider κ as well).

In measurements of large distances, C should be sufficiently small. In fact, to avoid a significant contribution from instantaneous diffusion, we require $kCH^{AV}T_2 << 1$. However, if we just assume the weaker inequality:

$$q \equiv \ln[\frac{S_1\sqrt{ft}K}{SNR_{acc}}] >> kCH^{AV}T_2 \tag{B5}$$

then to first order in $kCH^{AV}T_2/q$, we obtain

$$R^3_{\max} = \frac{\gamma_e^2 \hbar T_2 q^{\mu}}{4\pi(1+\mu k CH^{AV}T_2 q^{\mu-1})} \tag{B6}$$

where $\mu \equiv 1/\kappa$. For $T = 80$ K, $C = 100$ μM, $V_s = 20$ μL, $SNR_{acc} = 10$, $f = 2$ kHz, $t = 10^4$ s, $K = 0.4$, $k = 5 \cdot 10^8\ M^{-1}$ we find that for $T_2 = 3$ μs and $\kappa = 2$ (nuclear spin diffusion), q is 6.8 and $R_{max} = 58$ Å and similarly for $T_2 = 1$ μs and $\kappa = 1$ (methyl group rotation) we have $R_{max} = 55$ Å. Thus we find that the upper range of distances by DQC is strongly limited to about 60 Å for a protonated or fluorinated solvent; (we remind the reader that Larsen and Singel's relaxed criterion for t_m (see above) yields 72 Å).

A larger range of distances can only be measured with acceptable accuracy if the solvent is perdeuterated. In this case T_2's may be in the range of tens of microseconds and $\kappa = 1$. For $T_2 = 10$ μs and $C = 100$ μM, one estimates an $R_{max} = 110$ Å, although a more conservative estimate might be somewhere in the range of 80 Å.

Another issue arises regarding the possibility of working with small samples at low concentrations. If the distances are not very large, relaxation is not a major problem. In this case the DQ technique, for a smaller sample, lowers the concentration requirement by at least an order of magnitude (cf. Sec. 5.1) compared to DEER, (in addition to providing less distorted spectra). At Ku-band, a sample with a C ≈ 200 μM and a volume of 10 μL has just 2 nmoles of spins and could easily be studied at T=100 K, for distances under 25 Å.

C. 5-PULSE SQ SEQUENCE

In addition to the 6-pulse DQ sequence (DQ6) which features zero deadtime with respect to the dipolar signal (as is the case in 4-pulse DEER), we also introduced at about the same time a 5-pulse SQ (SQ5) sequence, which was the first SQ sequence with such a property (Borbat and Freed, 1999a).

Here, we would like to briefly describe this sequence. The SQ5 sequence resembles the DQ6 sequence of Fig. 5c, except that the evolution period (consisting of the two t_1 intervals with a π pulse between them) is just replaced by a sufficiently long time interval T so that all coherences which contain frequency shift operators relax by T_2 - type relaxation processes.

The pulses are arbitrary in intensity and, as in the DQ sequences, the nominal rotation angles of the pulses, and their relative phases do matter. The only coherences which survive after T correspond to longitudinal magnetization and 2-spin longitudinal dipolar order which, in turn, are formed from corresponding in-phase and antiphase coherences at the end of the preparation period. Generally, we can describe the sequence as follows:

$$\begin{aligned} &S_{1z}+S_{2z}\xrightarrow{R(\alpha)Q(t_p)R(\beta_1)Q(t_p)}-(S_{1y}+S_{2y}),\,2S_{1x}S_{2z}+2S_{2x}S_{1z} \\ &\xrightarrow{R(\beta)} S_{1z}+S_{2z},\,2S_{1z}S_{2z},\,\ldots \xrightarrow{Q(T)} S_{1z}+S_{2z},\,2S_{1z}S_{2z} \\ &\xrightarrow{R(\beta)} S_{1y}+S_{2y},\,2S_{1x}S_{2z}+2S_{2x}S_{1z}\xrightarrow{R(\beta)Q(t_2)R(\beta_1)Q(t_2)}-(S_{1y}+S_{2y}) \end{aligned} \tag{C1}$$

Here α is the rotation angle of the 1st pulse, β_1 is that for the 2nd and 5th pulses, and finally β is for the 3rd and 4th pulses. (We use these arbitrary angles because different specific choices favor different pathways). In general, in this 5-pulse sequence with arbitrary pulses, there are 6 coherence pathways given in Eqs. C2 which cannot all be separated by setting proper phases and/or other means:

$$\begin{aligned} &(1)\ S_y\to -S_y\to S_z\to S_y\to -S_y \\ &(2)\ S_y\to -S_y\to S_z\to S_y\to -S_y\cos at_2 \\ &(3)\ S_y\to -S_y\cos at_p\to S_z\to S_y\cos at_p\to -S_y\cos at_p \\ &(4)\ S_y\to -S_y\cos at_p\to S_z\to S_y\cos at_p\to -S_y\cos at_p\cos at_2 \\ &(5)\ S_{1y}\to 2S_{1x}S_{2z}\sin at_p\to 2S_{1z}S_{2z}\to 2S_{1x}S_{2z}\sin at_p\to -S_{1y}\sin at_p\sin at_2 \\ &(6)\ S_{1y}\to 2S_{1x}S_{2z}\sin at_p\to 2S_{1z}S_{2z}\to 2S_{2x}S_{1z}\sin at_p\to -S_{2y}\sin at_p\sin at_2 \end{aligned} \tag{C2}$$

Here, $t_2 \equiv t_m - t_p$. The first three pathways correspond to those cases where one or both of the π pulses do not flip the coupled spins. The first pathway is dominant in the case of selective pulses and forms a "baseline". Pathways (2) and (3) are unwanted. Pathway (4) was our original interest in this sequence. All these pathways are suppressed in the DQC experiment. The pathways (5) and (6) resemble those in DQC as we can deduce from the following comparison.

$$\begin{aligned}&\mathrm{SQ5}:\ 2S_{kx}S_{lz}\rightarrow 2S_{kz}S_{kz}\rightarrow 2S_{kx}S_{lz},\ 2S_{lx}S_{kz}\\&\mathrm{DQ6}:\ 2S_{kx}S_{lz}\rightarrow S_{kx}S_{ly}+S_{ky}S_{lx}\rightarrow 2S_{kx}S_{lz},2S_{lx}S_{kz}\end{aligned}\tag{C3}$$

In the general case, the signal can be expressed as a sum of two distinct signals V_1 and V_2 corresponding to pathways (1-4) and (5, 6) respectively. For arbitrary pulses, using our PO technique (cf. App. A) we find that

$$\begin{aligned}V_1&=\frac{1}{2}\mathrm{Im}\{S(\alpha)^*\,|S(\beta)|^2\,e^{i\Phi}+S(\alpha)S(\beta)^{*2}e^{-i\Phi}\}[S_{\beta 1}^{(1)}]^2\\&\quad\times[C_{\beta 1}^{(2)}+S_{\beta 1}^{(2)}\cos(at_p)][C_{\beta 1}^{(2)}+S_{\beta 1}^{(2)}\cos(at_2)]\\V_2&=\frac{1}{2}\mathrm{Im}\{(S^{(1)}(\alpha)^*S^{(1)}(\beta)e^{i\Phi}-S^{(1)}(\alpha)S^{(1)}(\beta)^*e^{-i\Phi})(S^{(1)}(\beta)^*C^{(2)}(\beta)\\&\quad+S^{(2)}(\beta)^*C^{(1)}(\beta))\}C^{(2)}(\beta)[S_{\beta 1}^{(1)}S_{\beta 1}^{(2)}]^2\sin(at_p)\sin(at_2)\end{aligned}\tag{C4}$$

Here, $S_{\beta 1}^{(k)}\equiv S_2^{(k)}(\beta_1)$, $C_{\beta 1}^{(k)}\equiv 1-S_{\beta 1}^{(k)}$, Φ is the relative phase of the 3rd and 1st pulses. Originally, we were primarily concerned with the signal V_1 (Pathway 4 in Eq. C2; $\Phi = 0$; $\alpha, \beta = \pi/2$, $\beta_1 = \pi$ in Eq. C4), which in the limit of nonselective pulses is just $\frac{1}{2}(\cos at_m + \cos at_\xi)$, and for selective pulses is about 1.4 times stronger than the 6-pulse DQC. Signal V_2 resembles the DQ6 signal but it is about a factor of 1.5 times weaker, as follows from simulations.

Why should one prefer one signal over the other? Signal (4) is typically about 2 times stronger and does not have any unwanted relaxation (cf. below). Thus, despite the presence of unwanted signals (1-3) it is useful. Signals (5 and 6) are similar to that in DQ6 but about 1.5-2 times weaker, and one of them (6) is subject to unwanted relaxation caused by nuclear spin diffusion, which is a major T_2 mechanism (cf. App. B and Eq. C5). Thus, in the case of very large distances, one may think of signal (6) as problematic, since the T_2 relaxation would usually be dominated by nuclear spin diffusion (cf. App. B). This signal can be removed by adjustment of the pulse sequence or by adding sufficiently nonselective inverting π pulses during T.

The advantage of the V_1 signal is that there is no extra relaxation with respect to t_ξ. This is the same useful property as in DEER. To explain how relaxation can enter into a constant time experiment, we consider the case which arises when nuclear spin diffusion is causing T_2 decay given by (cf. App. B):

$$\exp[-(2t_m/T_2)^\kappa] \tag{C5}$$

for the decay of Pathway (5), whereas for Pathway (6) the decay is :

$$\exp\left[-\left(\frac{t_m - t_\xi}{T_2}\right)^\kappa - \left(\frac{t_m + t_\xi}{T_2}\right)^\kappa\right] \tag{C6}$$

This expression yields approximately a Gaussian signal envelope in t_ξ. For t_ξ=0 the exponent is $2(t_m/T_2)^\kappa$ in Eq. C6 while in Eq. C5 it is $2^\kappa(t_m/T_2)^\kappa$. For a typical case of $\kappa = 2$, the useful interval for sampling the dipolar interaction is reduced by a factor of 2 in the former case due to T_2 decay. Thus in the case (Eq. C6) the latter decay rate of the signal is substantially reduced. This special feature may be useful in order to detect the spin-echo signal at long times. At first glance, an increase of the interval allotted for the dipolar evolution leads to a corresponding increase in resolution of the dipolar spectrum. In fact, it may well be an illusion. Any appreciable increase in resolution does not occur, as is clear from Eq. C6. There is an additional (approximately) Gaussian broadening due to very rapid decay in t_ξ, which effectively shortens the useful interval in t_ξ to ca. $T_2/2^{\frac{1}{\kappa}}$. There is no such broadening in the basic SQ5 sequence (4).

Despite the flexibility of the 5-pulse SQ sequence it does not produce better results than the 6-pulse DQC sequence. With the intense pulses that we employ, it is more difficult to tune this sequence for a clean V_2 signal, as compared with DQ6. We have invariably ended up with better results using the 6-pulse DQC than with the 5-pulse SQC. Maybe in future work it will be possible to find cases where the SQ sequence would be more appropriate.

D. PHASE CYCLES

According to Fig. 5, the signal in the 4(5) pulse sequence is a sum of two signals from the pathways shown. In the 6 pulse sequence of Fig. 5c there are 4 pathways. We denote the pathways as *a*, *b* for 4(5) pulse sequence and *a*, *b*, *c*, *d* for the 6-pulse sequence. The phases for the signals from corresponding pathways are shown under phase tables; (recall ϕ_k is the phase of the kth pulse). The conditions for generation of DQ signal impose the restriction (cf. Sec. 3.1.4.1): $\phi_1 = \phi_2 = \phi_3$.

Phase cycling combines the signals *a*, *b* (*a*, *b*, *c*, *d*) constructively and suppresses unwanted pathways that interfere with the desired signal.

PHASE TABLE FOR 4-PULSE DQ SEQUENCE[a]

Step number	Phase				Memory address[b]	
	ϕ_1	ϕ_2	ϕ_3	ϕ_4	I	II
1	x	x	x	x	1	2
2	y	y	x	x	-1	-2
3	-x	-x	x	x	1	2
4	-y	-y	x	x	-1	-2
5	x	x	y	x	1	2
6	y	y	y	x	-1	-2
7	-x	-x	y	x	1	2
8	-y	-y	y	x	-1	-2
9	x	x	-x	x	1	2
10	x	y	-x	x	-1	-2
11	-x	-x	-x	x	1	2
12	-y	-y	-x	x	-1	-2
13	x	x	-y	x	1	2
14	y	y	-y	x	-1	-2
15	-x	-x	-y	x	1	2
16	-y	-y	-y	x	-1	-2
17	x	x	x	y	2	-1
18	y	y	x	y	-2	1
19	-x	-x	x	y	2	-1
20	-y	-y	x	y	-2	1
21	x	x	y	y	2	-1
22	y	y	y	y	-2	1
23	-x	-x	y	y	2	-1
24	-y	-y	y	y	-2	1
25	x	x	-x	y	2	-1
26	y	y	-x	y	-2	1
27	-x	-x	-x	y	2	-1
28	-y	-y	-x	y	-2	1
29	x	x	-y	y	2	-1
30	y	y	-y	y	-2	1
31	-x	-x	-y	y	2	-1
32	-y	-y	-y	y	-2	1

Phases ϕ_a, ϕ_b for selected pathways in 4 and 5-pulse sequences:

4–pulse: $\phi_a = \phi_1 + \phi_2 - 4\phi_3 + \phi_4$ $\quad \phi_b = \phi_1 - 3\phi_2 + 4\phi_3 - 3\phi_4$

5–pulse: $\phi_a = \phi_1 + \phi_2 - 4\phi_3 + 3\phi_4 - 2\phi_5$ $\quad \phi_b = \phi_1 - 3\phi_2 + 4\phi_3 - \phi_4 - 2\phi_5$

(a) This phase table is expanded by the addition of 2 step CYCLOPS to 128 steps.

(b) The memory addresses I and II are used to store the two quadrature signals. The factors in these columns indicate which quadrature output is stored (1 or 2) and the (+ or -) sign indicates whether they are added to or subtracted from I and II.

Step number	Phase				Memory address[b]	
	ϕ_1	ϕ_2	ϕ_3	ϕ_4	I	II
33	x	x	x	-x	-1	-2
34	y	y	x	-x	1	2
35	-x	-x	x	-x	-1	- 2
36	-y	-y	x	-x	1	2
37	x	x	y	-x	-1	-2
38	y	y	y	-x	1	2
39	-x	-x	y	-x	-1	-2
40	-y	-y	y	-x	1	2
41	x	x	-x	-x	-1	-2
42	x	y	-x	-x	1	2
43	-x	-x	-x	-x	-1	-2
44	-y	-y	-x	-x	1	2
45	x	x	-y	-x	-1	-2
46	y	y	-y	-x	1	2
47	-x	-x	-y	-x	-1	-2
48	-y	-y	-y	-x	1	2
49	x	x	x	-y	-2	1
50	y	y	x	-y	2	-1
51	-x	-x	x	-y	-2	1
52	-y	-y	x	-y	2	-1
53	x	x	y	-y	-2	1
54	y	y	y	-y	2	-1
55	-x	-x	y	-y	-2	1
56	-y	-y	y	-y	2	-1
57	x	x	-x	-y	-2	1
58	y	y	-x	-y	2	-1
59	-x	-x	-x	-y	-2	1
60	-y	-y	-x	-y	2	-1
61	x	x	-y	-y	-2	1
62	y	y	-y	-y	2	-1
63	-x	-x	-y	-y	-2	1
64	-y	-y	-y	-y	2	-1

PHASE TABLE FOR 5-PULSE DQ SEQUENCE:

This phase table is based on the first 16 steps of the 4-pulse DQ sequence given in Table 1. A 64 step sequence is generated as follows. For the first 16 steps, add $\phi_5 \rightarrow x$; for the 2nd 16 steps, just add $\phi_5 \rightarrow -x$; for the 3d 16 steps, just add $\phi_5 \rightarrow$ y; for the 4th 16 steps, just add $\phi_5 \rightarrow$ -y. The memory addresses for the first 16 steps are the same as in Table I; for the second 16 steps

[b] The memory addresses I and II are used to store the two quadrature signals. The factors in these columns indicate which quadrature output is stored (1 or 2) and the (+ or -) sign indicates whether they are added to or subtracted from I and II.

change signs of original entries in columns I, II; for the third 16 steps, exchange entries for I and II and then change signs of entries in column II; for the last 16 steps, again swap entries in I and II, but now change signs in column I. This 64 step phase table is also expanded to 128 steps by adding the 2 step CYCLOPS.

PHASE TABLE FOR 6-PULSE DQ SEQUENCE[c]

Step Number	Phase						Memory Array[d]	
	ϕ_1	ϕ_2	ϕ_3	ϕ_4	ϕ_5	ϕ_6	I	II
1	x	x	x	x	X	x	1	2
2	y	y	y	x	x	x	-1	-2
3	-x	-x	-x	x	x	x	1	2
4	-y	-y	-y	x	x	x	-1	-2
5	x	x	x	y	x	x	1	2
6	y	y	y	y	x	x	-1	-2
7	-x	-x	-x	y	x	x	1	2
8	-y	-y	-y	y	x	x	-1	-2
9	x	x	x	-x	x	x	1	2
10	y	y	y	-x	x	x	-1	-2
11	-x	-x	-x	-x	x	x	1	2
12	-y	-y	-y	-x	x	x	-1	-2
13	x	x	x	-y	x	x	1	2
14	y	y	y	-y	x	x	-1	-2
15	-x	-x	-x	-y	x	x	1	2
16	-y	-y	-y	-y	x	x	-1	-2
17	x	x	x	x	-x	x	-1	-2
18	y	y	y	x	-x	x	1	2
19	-x	-x	-x	x	-x	x	-1	-2
20	-y	-y	-y	x	-x	x	1	2
21	x	x	x	y	-x	x	-1	-2
22	y	y	y	y	-x	x	1	2
23	-x	-x	-x	y	-x	x	-1	-2
24	-y	-y	-y	y	-x	x	1	2
25	x	x	x	-x	-x	x	-1	-2
26	y	y	y	-x	-x	x	1	2
27	-x	-x	-x	-x	-x	x	-1	-2
28	-y	-y	-y	-x	-x	x	1	2
29	x	x	x	-y	-x	x	-1	-2
30	y	y	y	-y	-x	x	1	2
31	-x	-x	-x	-y	-x	x	-1	-2
32	-y	-y	-y	-y	-x	x	1	2

[c] This phase table is combined with 4 step CYCLOPS yielding 256 steps.

[d] See the Table for 4-pulse sequence, footnote b.

	ϕ_1	ϕ_2	ϕ_3	ϕ_4	ϕ_5	ϕ_6	I	II
33	x	x	x	x	y	x	2	-1
34	y	y	y	x	y	x	-2	1
35	-x	-x	-x	x	y	x	2	- 1
36	-y	-y	-y	x	y	x	-2	1
37	x	x	x	y	y	x	2	-1
38	y	y	y	y	y	x	-2	1
39	-x	-x	-x	y	y	x	2	-1
40	-y	-y	-y	y	y	x	-2	1
41	x	x	x	-x	y	x	2	-1
42	x	y	y	-x	y	x	-2	1
43	-x	-x	-x	-x	y	x	2	-1
44	-y	-y	-y	-x	y	x	-2	1
45	x	x	x	-y	y	x	2	-1
46	y	y	y	-y	y	x	-2	1
47	-x	-x	-x	-y	y	x	2	-1
48	-y	-y	-y	-y	y	x	-2	1
49	x	x	x	x	-y	x	-2	1
50	y	y	y	x	-y	x	2	-1
51	-x	-x	-x	x	-y	x	-2	1
52	-y	-y	-y	x	-y	x	2	-1
53	x	x	x	y	-y	x	-2	1
54	y	y	y	y	-y	x	2	-1
55	-x	-x	-x	y	-y	x	-2	1
56	-y	-y	-y	y	-y	x	2	-1
57	x	x	x	-x	-y	x	-2	1
58	y	y	y	-x	-y	x	2	-1
59	-x	-x	-x	-x	-y	x	-2	1
60	-y	-y	-y	-x	-y	x	2	-1
61	x	x	x	-y	-y	x	-2	1
62	y	y	y	-y	-y	x	2	-1
63	-x	-x	-x	-y	-y	x	-2	1
64	-y	-y	-y	-y	-y	x	2	-1

Phases ϕ_a, ϕ_b, ϕ_c, ϕ_d for selected pathways in 6-pulse sequence:

$$\phi_a = \phi_1 - 2\phi_2 - \phi_3 + 4\phi_4 - \phi_5 - 2\phi_6 \qquad \phi_c = -\phi_1 + 2\phi_2 + \phi_3 - 4\phi_4 + 3\phi_5 - 2\phi_6$$

$$\phi_b = \phi_1 - 2\phi_2 + 3\phi_3 - 4\phi_4 + 3\phi_5 - 2\phi_6 \qquad \phi_d = -\phi_1 + 2\phi_2 - 3\phi_3 + 4\phi_4 - \phi_5 - 2\phi_6$$

REFERENCES

Abragam, A. (1961) The principles of nuclear magnetism. (Oxford Univ. Press, NY), Ch. 4.

Astashkin, A.V., Hara, H. and Kawamori A. (1997) The pulsed electron-electron double resonance and "2+1" electron spin echo study of the oriented oxygen-evolving and Mn-depleted preparations of photosystem II. *J. Chem. Phys.*, **108**, 3805-3812.

Bloom, A.L. (1955) Nuclear induction in inhomogeneous fields. *Phys. Rev.* **98,** 1105-1112.

Baum, J., Munowitz, M., Garroway. A.N. and Pines, A. (1985) Multiple-quantum dynamics in solid state NMR. *J. Chem. Phys.* **83**, 2015-2025.

Bax, A., de Jong, P.G., Mehlkopf, A.F. and Smidt, J. (1980) Separation of the different orders of NMR multiple-quantum transitions by the use of pulsed field gradients. *Chem. Phys. Lett.*, **69**, 567-570.

Borbat, P.P., Berdnikov, V.M., Milov, A.D. and Tsvetkov, Yu.D. (1977) Spatial distribution and diffusion of hydrogen atoms formed on photolysis of Fe(II) in frozen sulfuric acid water solutions. *Sov. Phys. Solid State*, **19**, 628-663.

Borbat, P.P. Crepeau, R.H. and Freed, J.H. (1997) Multifrequency two-dimensional Fourier transform ESR: X/Ku-band spectrometer. *J. Magn. Reson.*, **127**, 155-167.

Borbat, P.P. and Freed, J.H. (1999) Multiple-quantum ESR and distance measurements. *Chem. Phys. Lett.*, **313**, 145-154.

Borbat, P.P. and Freed J.H. (1999a) Progress in multiple-quantum ESR Spectroscopy: powerful tool for distance measurements, at *"41st Rocky Mountain Conference on Analytical Chemistry"*, Aug. 1-5, Denver, Colorado.

Borbat, P.P. and Freed, J.H. (2000) Double and single quantum coherence ESR for distance measurements: effects of microwave B_1 and frequency, (To be published).

Budker, V., Du, J.-L., Seiter, M., Eaton, G.,R., Eaton, S.S. (1995) Electron-electron spin-spin interaction in spin-labeled low-spin methemoglobin, *Biophys. J.*, **68**, 2531-2542.

Cavanagh, J., Fairbrother, W.J., Palmer III, A.G., Skelton, N.J. (1996) Protein NMR Spectroscopy. Academic Press, San Diego.

Chandrasekhar, S. (1943) Stochastic problems in physics and astronomy. Revs Modern. Phys. **15**, 165-191

Corio, P.L. (1966) Structure of high-resolution NMR spectra. Academic Press, NY.

Crepeau., R.H., Dulcic, A., Gorcester, J., Saarinen, T. and Freed, J.H. (1989) Composite pulses in time-domain ESR. *J. Magn. Reson.*, **84**, 184-190.

Dikanov, S.A., Tsvetkov, Yu.D. (1992) Electron spin echo envelope modulation (ESEEM) spectroscopy. CRC Press, 21.

Dzuba, S.A., Bosh, M.K., Hoff, A.J. (1996) Electron spin-echo detection of quantum beats and double-quantum coherence in spin-correlated radical pairs of protonated photosynthetic reaction centers. *Chem. Phys. Lett.*, **248**, 427-433.

Dzuba, S.A. and Kawamori, A. (1996) Selective hole burning in EPR: spectral diffusion and dipolar broadening. *Concepts in Magnetic Resonance* **8,** 49-61.

Eaton, S.S. and Eaton, G.R. (1993) Irradiated fused-quartz standard sample for time-domain EPR. *J. Magn. Reson.* **A. 102**, 354-356.

Emshwiller, M., Hahn, E.I., Kaplan, D. (1960) Pulsed nuclear resonance spectroscopy. *Phys. Rev.*, **118**, 414-424.

Ernst, R.R., Bodenhausen, G., Wokaun, A. (1987) Principles of nuclear magnetic resonance in one and two dimensions. Clarendon Press, Oxford.

Ewert, U, Crepeau, R.H., Lee, S., Dunnam C.R., Xu, D. and Freed, J.H. (1991) Fourier transform electron spin resonance imaging. *Chem. Phys. Lett.*, **184**, 25-33.

Farhbakhsh, Z.T., Huang, Q.-L., Ding, L.-L., Altenbach, C., Steinhoff, H.J., Horwitz, J. and Hubbell, W.L. (1995) Interaction of α-crystallin with spin-labeled peptides. *Biochemistry*, **34**, 509-516.

Farkas, Z.D., Spalek, G. and Wilson P.B. (1989) RF pulse compression experiment at SLAC, *Proceedings of the 1989 IEEE particle accelerator conference*. Chicago, March 20-23, vol. 1, 132-134.

Fauth, J.-M., Schweiger, A., Braunschweiler, L., Forrer, J. and Ernst, R.R. (1986) Elimination of unwanted echoes and reduction of dead time in three-pulse electron spin-echo spectroscopy. *J. Magn. Reson.*, **66**, 74-85.

Feldman, E.B., Lacelle, S. (1996) Configurational averaging of dipolar interactions in magnetically dilute spin networks. *J. Chem. Phys.* **104**, 2000-2009.

Gemperle , C., Aebli, G., Schweiger, A., and Ernst, R. R., (1990) Phase cycling in pulse EPR. *J. Magn. Reson.*, **88**, 241-256.

Gorcester, J. and Freed, J.H. (1988) Two-dimensional Fourier transform ESR correlation spectroscopy. *J. Chem. Phys.*, **88**, 4678-4693.

Hanson, P., Anderson, D.J., Martinez, G., Millhauser, G., Formaggio, F., Crisma, M., Toniolo, C. and Vita, C. (1999) Electron spin resonance and structural analysis of water soluble, alanine-rich peptides incorporating TOAC. *Mol. Phys.*, **95**, 957-966.

Hanson, P., Millhauser, G., Formaggio, F., Crisma, M., Toniolo, C. (1996) ESR characterization of hexameric, helical peptides using double TOAC spin-labeling. *J. Am. Chem. Soc.*, **118**, 7618-7625.

Hara, H., and Kawamori, A. (1997) A selective hole burning method applied to determine distances between paramagnetic species in Photosystems. *Appl. Magn. Reson.*, **13**, 241-257.

Höfer, P., Grupp, A., Nebenfuhr, H. and Mehring, M. (1986) Hyperfine sublevel correlation (HYSCORE) spectroscopy: A 2D ESR investigation of the squaric acid radical. *Chem. Phys. Lett.*, **132**, 279-282.

Hoult, D.I. and Richards (1975) Critical factors in the design of sensitive high resolution nuclear magnetic resonance spectrometers. *Proc. Roy. Soc.*, (*Lond*), **344**, 311-340.

Hustedt, E.J., Smirnov, A.I., Laub, C.F., Beth, A.H. (1997). Molecular distances from dipolar coupled spin-labels: the global analysis of multifrequency continuous wave electron paramagnetic resonance data. *Biophys. J.* **72**, 1861-1877.

Hustedt, E.J. and Beth, A.H. (1999) Nitroxide spin-spin interactions: applications to protein structure and dynamics. *Annu. Rev. Biophys. Biomed. Struct.*, **28**, 129-153.

Ichikawa, T., Wakasugi, S.Ichi and Yoshida, H. (1985) Structure of spurs in γ-irradiated alcohol matrices by electron spin-echo method. *J. Phys. Chem.*, **89**, 3583-3586.

Jeschke, G., Bauer C., Pannier M., Leporini D. and Spiess, H.W. (2000) Pulse high-field EPR on nitroxide spin probes in polymers, at "*High-frequency electron paramagnetic resonance: technology and applications*", Amsterdam, April 12-14.

Kurshev, V.V., Astashkin, A.V., and Raitsimring, A.M (1988) Modulation Effects in 2+1 Electron spin echo pulse sequence. *J. Struct. Chem.*, **29**, 62-68.

Kurshev, V.V., Raitsimring, A.M. and Tsvetkov, Yu.D. (1989) Selection of dipolar interaction by the "2+1" pulse train ESE. *J. Magn. Reson.*, **81**, 441-454.

Larsen R.G. and Singel, D.J. (1993) Double electron-electron resonance spin-echo modulation. Spectroscopic measurement of electron spin pair separation in orientationally disordered solids. *J. Chem. Phys.*, **98**, 5134-5146.

Lee, S., Patyal, B.R. and Freed, J.H. (1993) A two-dimensional Fourier transform electron spin resonance (ESR) study of nuclear modulation and spin relaxation in irradiated malonic acid. *J. Chem. Phys.*, **98**, 3665-3689.

Lee, S., Budil, D.E. and Freed, J.H. (1994) Theory of two-dimensional electron spin-resonance for ordered and viscous fluids. *J. Chem. Phys.*, **101**, 5529-5558.

Levitt, M.H. (1979) Composite pulses. *Progr. NMR. Spectr.*, **18**, 61-122.

Lindgren, M., Eaton, G.R., Eaton, S.S., Johnsson, B-H, Hamarström, P., Svensson, M. and Carlsson, U. (1997) Electron spin echo decay as a probe of aminoxyl environment in spin-labeled mutants of human carbonic anhydrase II. *J. Chem. Soc., Perkin Trans.* **2**, 2549-2554.

Maryasov, A.G., Tsvetkov, Yu.D., Raap, J. (1998) Weakly-coupled radical pairs in solids: ELDOR in ESE structure studies. *Appl. Magn. Reson.*, **14**, 101.

Meirovitch, E., Igner, D., Igner, E., Moro, G. and J.H. Freed, J.H. (1982) Electron-spin relaxation and ordering in smectic and supercooled nematic liquid crystals. *J. Chem. Phys.*, **77**, 3915-3937.

Milov, A.D., Salikhov, K.M. and Tsvetkov, Yu.D. (1973) Phase relaxation of hydrogen atoms stabilized in an amorphous matrix. *Sov. Phys. Solid State*. 15, 802-806.

Milov, A.D., Salikhov, K.M., Shirov, M.D. (1981) Application of the double resonance method to electron spin echo in a study of the spatial distribution of paramagnetic centers in solids. *Sov. Phys. Solid State*, **23**, 565-569.

Milov, A.D., Maryasov, A.G. and Tsvetkov, Yu.D. (1998) Pulsed electron double resonance (PELDOR) and its applications in free-radical research. *Appl. Magn. Reson.*, **15**, 107-143.

Milov, A.D., Maryasov, A.G., Tsvetkov, Yu.D., Raap, J. (1999) Pulsed ELDOR in spin-labeled polypeptides. *Chem. Phys. Lett.*, **303**, 135-143.

Milov, A.D., Ponomarev, A.B. and Tsvetkov, Yu.D. (1985) Modulation beats of double electron-electron resonance in spin echo for biradical systems. J. Chem. Struct., **25**, 710-713.

Ohba, Y., Satoh, R., Kikuchi, T., Yamauchi. S. and Iwazumi. M. (1993) A new technique for accurate measurement and adjustment of the phase of the microwave pulse in pulsed EPR spectroscopy. *J. Magn. Reson.*, **A103**, 282-287.

Pake, G.E. (1948) Nuclear resonance absorption in hydrated crystals: fine structure of the proton line. *J.Chem.Phys.*, **16**, 327-336.

Pannier, M., Veit, S., Godt, A., Jeshke, G. and Spiess, H.W. (2000) Dead time free measurement of dipole-dipole interactions between electron spins. *J. Magn. Reson.*, **142**, 331-340.

Pfannebecker, V., Klos, H., Hubrich, M., Volkmer, T., Heuer, A., Wiesner, U. and Spiess, H.W. (1996) Determination of end-to-end distances in oligomers by pulsed EPR. *J. Phys. Chem.*, **100**, 13428-13432.

Rabenstein, M.D. and Shin, Y.-K. (1995) Determination of the distance between two spin labels attached to a macromolecule. *Proc. Natl. Acad. Sci., USA*, **92**, 8239-8243.

Rabenstein, M.D., Shin, Y.-K. (1996), HIV-1 gp41 tertiary structure studied by EPR spectroscopy. *Biochemistry*, **35**, 13922-13928.

Raitsimring, A.M. and Salikhov, K.M. (1985) Electron spin echo method as used to analyze the spatial distribution of paramagnetic centers *Bulletin of Magn. Reson.*, **7**, 184-217.

Raitsimring, A., Crepeau, R.H. and Freed, J.H. (1995) Nuclear modulation effects in "2+1" electron spin-echo correlation spectroscopy. *J. Chem. Phys.*, **102**, 8746-8762.

Rakowsky, M.H., Zecevic, A., Eaton, G.R. and Eaton, S.S. (1998) Determination of high – spin iron(II)-nitroxyl distances in spin-labeled porphyrins by time-domain EPR. *J. Magn. Reson.*, 11, 97-110.

Salikhov, K.M., Semenov, A.G. and Tsvetkov, Yu.D. (1976) Electron spin echo and its applications. Nauka, Novosibirsk, (in Russian).

Salikhov, K.M., Schneider, D.J., Saxena, S., Freed, J.H. (1996) A theoretical approach to the analysis of arbitrary pulses in magnetic resonance. *Chem. Phys. Lett.*, **262**, 17-26.

Saxena, S., Freed, J.H. (1996) Double quantum two-dimensional Fourier transform spin resonance: distance measurements. *Chem. Phys. Lett.*, **251**, 102-110.

Saxena, S., Freed, J.H. (1997a) Theory of double quantum two-dimensional electron spin resonance with applications to distance measurements. *J. Chem. Phys.*, **107**, 1317-1340.

Saxena, S. and Freed, J.H. (1997b) Two-dimensional electron spin resonance and slow motions. *J. Phys. Chem.*, **A101**, 7998-8008.

Schweiger, A. (1990) New trends in pulsed spin-resonance methodology. In *Modern Pulsed and Continuous-wave Electron Spin Resonance*, (Kevan, L., Bowman, M.K., eds), pp 43-118, Wiley, NY.

Shriver, J. (1992) Product operators and coherence transfer in multiple-pulse NMR experiments. *Concepts in Magn. Reson.*, **4**, 1-33.

Slichter, C.P. (1990) Principles of magnetic resonance, 3rd enl. and updated edn., Springer-Verlag, Berlin-Heidelberg-New-York.

Sørensen, O.W., Eich, G.W., Levitt, M.H., Bodenhausen, G. and Ernst, R.R. (1983) Product operator formalism for the description of NMR pulse experiments. *Progr. NMR Spectr.*, **16**, 163-192.

Suzuki, M. (1985) Decomposition formulas of exponential operators and Lie exponentials with some applications to quantum mechanics and statistical physics. *J. Math. Phys.*, **26**, 601-612.

Tang, J., Norris, J.R. (1995) Multiple-quantum EPR coherence in a spin-correlated radical pair system. *Chem. Phys. Lett.*, **233**, 192-200.

Thorgeirsson, T.E., Xiao, W., Brown, L.S., Needleman, R., Lanyi, J.K. and Shin, Y.-K. (1997) Transient channel-opening in bacteriorhodopsin: an EPR study. *J. Mol. Biol.*, **273**, 951-957.

Yudanov, V.F., Salikhov, K.M., Zhidomirov, G.M. and Tsvetkov, Yu.D. (1969) Modulation effects in the electron spin echo of biradical systems. *Theor. and Experim. Chem.*, *5*, 663-668.

Zecevic, A., Eaton, G.R., Eaton, S.S. and Lindgren, M. (1998) Dephasing of electron spin echoes for nitroxyl radicals in glassy solvents by non-methyl protons. *Mol. Phys.*, **95**, 1225-1263.

Chapter 10

"2+1" Pulse Sequence as Apllied for Distance and Spatial Distribution Measurements of Paramagnetic Centers.

A. Raitsimring
Department of Chemistry, University of Arizona, Tucson, AZ 85721 USA

Abstract Applications of one of the varieties of electron spin-echo method, the "2+1" pulse sequence used for the measurement of distances in radical pairs with fixed and distributed distances, and for the determination of the parameters of spatial distribution in more complicated cases (fractals, clusters, non-random distributions) are discussed. Special attention is given to features of the "2+1" technique such as suppression of dipolar interactions of randomly distributed spins, which allows the selection of a dipolar interaction between radical labels in proteins, even if protein molecules aggregate. Experimental conditions that allow the separation of nuclear modulation from electron dipolar related modulation, or minimize the appearance of ESEEM in a "2+1" experiment are also discussed.

1. INTRODUCTUION.

A simple way to determine the distance between paramagnetic centers and/or their spatial distribution is by making measurements of their static dipolar interactions. When the static dipolar interactions are small, and are covered by non-homogeneous line broadening, one may use electron spin echo (ESE) based techniques for their detection. In ESE, static dipolar interactions show themselves as an additional modulation or decay/recovery. This is caused, as originally suggested by Klauder and Anderson (Klauder *et al.*, 1962), by an

Biological Magnetic Resonance, Volume 19: Distance Measurements in Biological Systems by EPR, edited by Berliner et al. Kluwer Academic/Plenum Publishers, New York, 2000

instantaneous variation in the local dipolar field at the position of the observed spin (A), by flipping the interacting spin (B) with a microwave (mw) pulse. While considering this phenomenon for randomly distributed spins in 3D space, Klauder and Anderson called it "instantaneous diffusion" (ID), to differentiate it from phase relaxation, which is caused by spectral diffusion. The phenomenon of ID in ESE can be destroyed by random spin-flips, which may be caused, for example, by spin-lattice relaxation or spin diffusion. Qualitatively, all of the features of ID (experimental and theoretical), such as the dependence of dipolar related decay/modulation on pulse parameters, the shape of EPR spectrum, and the rate of spin flips induced by spin-lattice relaxation and/or spin diffusion were investigated, described and explained in the numerous publications (Raitsimring *et al.*, 1974, 1975, 1985; Salikhov *et al.*, 1981; Dzuba *et al.*, 1980; Popov *et al.*, 1978; Tregub *et al.*, 1983). It was understood that in general, a simple two-pulse technique is well suited for ID measurements. Initially, measurements of dipolar interactions were made just by using a two pulse primary echo (Tregub *et al.*, 1983; Raitsimring *et al.*, 1985; Raitsimring *et al.*, 1980). The need, however, to eliminate substantial decay due to electron-nuclear interactions (Salikhov *et al.*, 1981) that mask the kinetics caused by ID resulted in the invention and implementation of two additional advanced pulse sequences which employed three pulses. These pulse sequences were the "2+1" pulse sequence (Kurshev *et al.*, 1988, 1988, 1989), and the pulsed double electron-electron resonance (DEER), configured for probing dipolar interactions (Milov *et al.*, 1981, 1998). Both techniques are structured in a similar way. Two pulses, the first and third, produce a primary spin echo signal, the variation of which is detected while the position of the second pulse, placed between the first and third is varied. In the "2+1" technique, all three pulses have the same carrier frequency, ω_1. The second pulse can be generated using either one mw oscillator, or two independent mw oscillators which operate at the same mw frequency. The latter, although slightly more complicated in instrumental implementation, is better suited for measurements, because only the desired echo remains coherent (substantially facilitating phase cycling). Both options have been used in various experiments. In the pulsed DEER technique, the primary echo is generated at ω_1 and the additional pulse is applied at $\omega_2 \neq \omega_1$. As a result, the "2+1" sequence and a pulsed DEER are complementary techniques for the measurement of dipolar interactions. The "2+1" technique is useful for measuring dipolar interactions between similar paramagnetic centers that have the characteristic width of an EPR spectra comparable with the pulse amplitude, and may be simultaneously excited by a single pulse. For instance, nitroxide radicals (often used as labels) have a characteristic width (in the S-X (2-12GHz) bands) of about 60 Gs. These radicals fall into this category because pulses of comparable amplitude (~30 Gs, (Borbat *et al.*, 1997, 1999, this book), are available in pulsed EPR spectroscopy. Conversely, the application of pulsed DEER is more useful for the determination of distances of dissimilar paramagnetic centers with well separated EPR spectra, where one type of spin serves as the observed spin (A), and the other one is used as the pumped spin (B). Albeit both techniques appear to be very similar, they manipulate dipolar interactions in different ways. In the pulsed DEER, the observed spin always acquires phase in addition to the phase that was already acquired by the primary echo. In the "2+1" technique, the observed spin can either recover the phase acquired in the primary echo, or (as in pulsed DEER) gain an additional phase, depending on the pulse parameters. This feature of the "2+1" technique allows the rigorous manipulation of the dipolar interactions. It may be "switched on" or "switched off", and the unwanted dipolar interactions can be suppressed. The "2+1" technique may be also used in the DEER mode, despite the fact that all of the pulses are applied at the same frequency. Below, we consider the features of the "2+1" technique, and the means by which dipolar interactions can be manipulated.

2. THEORY

Resonance spins, static dipolar interaction. A diagram of the pulse settings in the "2+1" pulse sequence is presented in Fig. 1. The events that occur under pulse actions in this technique are easy to understand when considering the simplest model, which is a pair of spins, weakly coupled by a static dipolar interaction. The term " static" here means that during the echo formation, which usually occurs in time interval that does not exceed 2-3 μsec, spins flip only by means of a pulse. The term "weakly" means that the dipolar interactions are not only much less than the Zeeman frequency, but in general, the characteristic width of an EPR spectrum of a given paramagnetic species. The external magnetic field, B_0, is chosen to be directed along the Z-axis of the rotating coordinate frame (RCF), while the microwave field, B_1, is directed along the X-axis. We designate the first spin (A) in a pair as the probe that is used solely for observation, and is responsible for echo generation. The second spin (B) does not contribute to the spin echo signal. It flips by a mw-pulse, which provides a variation in the local dipolar field *D* at the location of the first spin. We introduce a simple and well-known expression (Landau *et al.*, 1989) that describes the probabilities of spin flips after a resonant microwave pulse of duration t_p and nominal flip angle $\theta = \gamma B_1 t_p$ is applied. Namely, we introduce $\check{s} = \sin^2\theta/2$, the probability for a spin to be flipped by a mw-pulse, $\check{c} = \cos^2\theta/2$, the probability for the spin to *not* be flipped, and $\check{S}$=(sinθ/2)*(cosθ/2), which is the average projection of the spin on the Y-axis. To generate a spin echo signal, the first spin should be rotated by the first pulse in the Y direction, resulting in the projection

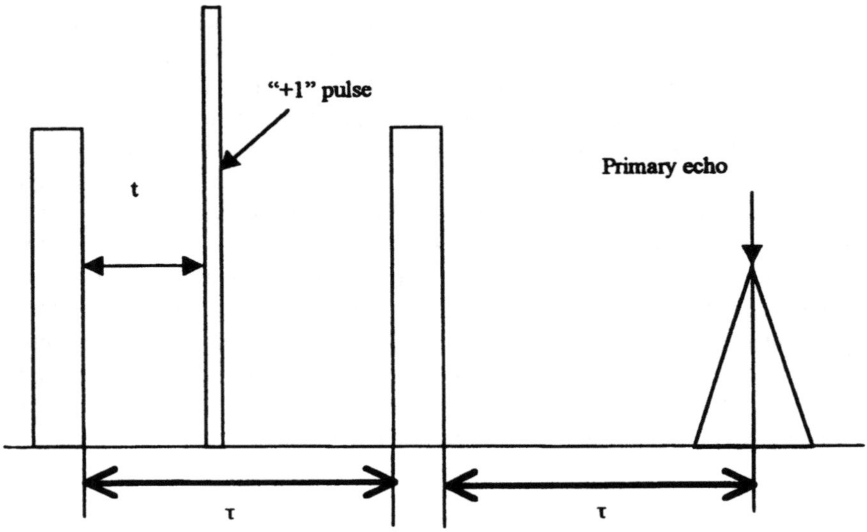

Figure 1. Diagram of the "2+1" pulse sequence. The primary echo is generated by the first and third pulses and separated by time interval τ (held constant). Variable time interval is t, the distance between "+1" and the first pulse.

S, and by the third pulse in the (-Y) direction with a probability š. Also, the first spin should not be rotated by the second pulse at all (with a probability č). The spin echo, therefore, with an amplitude of *V* after the action of these three pulses, is the product of the corresponding projection and probabilities:

$$V=\check{S}^{(1)}\times\check{c}^{(2)}\times\check{s}^{(3)} \qquad (1)$$

where the upper indices in brackets relate to the pulse number. The change in the ESE signal amplitude results from uncompensated phase that the first spin acquires while precessing in XY plane due to the flipping of the second spin by the second and third pulses. Since the second spin may be either flipped or not flipped by these pulses, four situations, denoted by (++), (+-), (- +), and (- -) are possible, where (+) means flip and (-) means no flip. For instance, (-+) implies that the second spin is flipped by third pulse and is not flipped by the second one. The probability of this situation is $\check{c}^{(2)}\check{s}^{(3)}$, etc. Fig. 2 shows the uncompensated phase of the first spin for the above cases. It can be seen that the additional phase acquired by the time 2τ depends on the second pulse position, *t*, in two cases (+-) and (++). In the case of (+-), the phase ***increases*** with *t*, and for (++) it ***decreases*** with *t*. Obviously, the case (+-) describes pulsed DEER. In pulsed DEER, (+-) is the only possible situation, because when the frequency separati-on between pumping and observing pulse is large, spin B has no chance of being flipped by the third pulse.

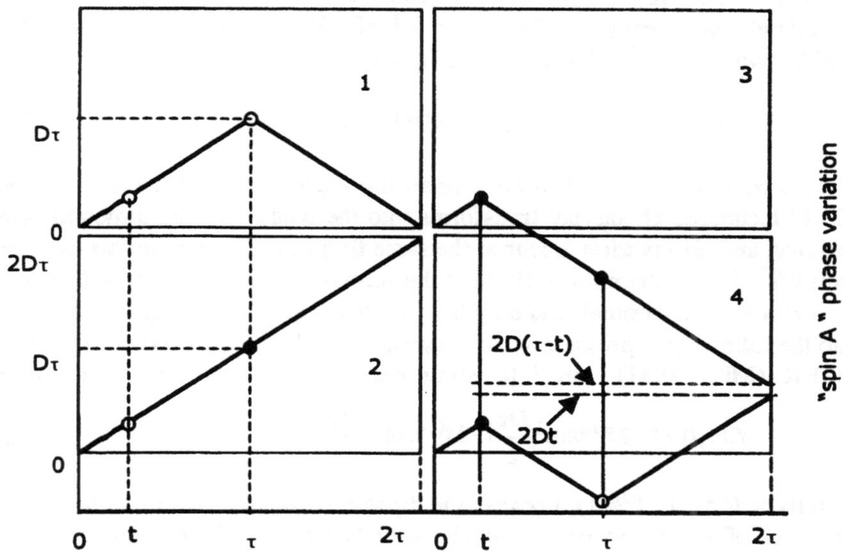

Figure 2. Schematic representation of four possible situations that occur with a phase of spin A, depending on flipping (black circle) or non flipping (open circle) of spin B by the respective pulse of the "2+1" sequence. D is the value of the dipolar interactions between spins A and B in a pair. The "2+1" sequence creates a similar situation to the standard primary echo (diagrams 1 and 2) and pulsed DEER (diagram 4). Diagram 3 shows the unique feature of the "2+1" sequence that occurs due to flipping spin B by two pulses, resulting in phase recovery of spin A.

Situations (--) and (-+), where the second pulse does not turn spin B, represent the conventional primary echo. Finally, the case of (++) is intrinsic only for the "2+1" method, and makes the "2+1" method distinct from pulsed DEER and primary echo. The second pulse ***eliminates*** the phase that first spin has already acquired while forming the primary spin echo signal. Depending, therefore, on the pulse parameters when the "2+1" technique is used, one may obtain either a pulsed DEER effect (additional phase acquisition), or a phase recovery, or a combina- tion of both, which allows a variety of means to manipulate *d-d* interactions. Summing up all four situations with the appropriate probabilities, we can immediately derive that the spin echo signal of the first spin V_1 is modulated by *d-d* interactions with the second spin as

$$V(\tau,t) \propto V_1 \times f_2(\tau,t)$$

$$f_2(\tau,t) = c^{(2)}c^{(3)} + c^{(2)}s^{(3)}\langle\cos D\tau\rangle + s^{(2)}s^{(3)}\langle\cos D(\tau - t)\rangle + s^{(2)}c^{(3)}\langle\cos Dt\rangle \qquad (2)$$

where the subscripts and superscripts in brackets relate to the spin and pulse number, respectively. Let us now consider various pulse settings. First, if a second pulse is absent (i.e., $\check{c}^{(2)}=1$ and $\check{s}^{(2)}=0$), Eq.2 immediately reduces to a well known expression that describes dipolar modulation in primary echo:

$$f_2(\tau,t) \;=\; \check{c}^{(3)}+\check{s}^{(3)}\{\cos D\tau\} \;=1-2\check{s}^{(3)}\{\sin^2\frac{D\tau}{2}\} \tag{3}$$

If the second and third pulses have small flip angles, i.e. $\check{s}^{(2)}$ and $\check{s}^{(3)}<<1$, the "2+1" technique effectively transforms into the pulsed DEER, although both pumping and observation occur at the same frequency. Indeed, in this case the term $\check{s}^{(2)}\check{s}^{(3)}$ (the probability of flipping the second spin by both pulses) becomes second order, infinitesimal, and may be neglected. In this case, Eq. 2 transforms into the following expression:which describes dipolar modulation of the pulsed DEER (Milov,1981). An interesting case is when the second pulse is

$$f_2(\tau,t)\approx 1-2\check{s}^{(3)}\{\sin^2\frac{D\tau}{2}\}-2\check{s}^{(2)}\{\sin^2\frac{Dt}{2}\} \tag{3a}$$

exactly π ($\check{c}^{(2)}=0$; $\check{s}^{(2)}=1$), because the dipolar modulation acquired during the formation of the primary echo should be entirely recovered by the second pulse. This is not feasible, however, because such a pulse setting gives zero amplitude for the observed spins. An exact π third pulse ($\check{c}^{(3)}=0$; $\check{s}^{(3)}=1$) completely eliminates the DEER effect in the "2+1" pulse sequence,

$$f_2(\tau,t)\propto \;\check{c}^{(2)}\{\cos D\tau\}+\check{s}^{(2)}\{\cos D(\tau-t)\} \tag{4}$$

In this situation, a $\pi/2$ second pulse is optimal for the phase recovery and amplitude. It gives half the amplitude of the initial spin-echo signal, which is modulated as:

$$f_2(\tau,t)\propto\; \cos\frac{Dt}{2}\times\cos D(\tau-\frac{t}{2}) \tag{5}$$

Eqs. 2-5 give an idea of how it is possible to manipulate dipolar interactions by means of adjusting the mw-pulse when the spins are resonant (i.e., the difference between the Zeeman frequency and the carrier frequency of the mw-pulse is zero). In a real experiment, the Zeeman frequencies are distributed, forming an EPR line shape, g($\Delta\omega$). As a result, mw-pulses are not able to rotate all of the spins at the same angle. Such a frequency distribution is easy to represent using the above expressions, as shown below. Also, the spatial distribution of radicals or ions is not limited to a distribution of pairs that have a fixed distance and angle. We therefore show below the expressions that describe the "2+1" kinetics for various types of spatial distributions for paramagnetic centers.

Real EPR spectra and various spatial distributions. In accordance with (Bloom, 1955), to take into account the non-resonant excitation of spins by mw-pulses, one just has to replace $\sin^2\theta/2$ in Eqs. 2-5 with $\beta(\beta^*)$, and $\cos^2\theta/2$ with α (α^*) respectively, where

$$\alpha_i=\cos\frac{\omega_{efi}t_{pi}}{2}-i\cos\phi_i\sin\frac{\omega_{efi}t_{pi}}{2};\beta_i=-i\sin\phi_i\sin\frac{\omega_{efi}t_{pi}}{2}$$
$$\phi_i=atan\frac{\omega_{Ni}}{\Delta\omega};\qquad \omega_{efi}=\sqrt{(\omega_{Ni}^2+\Delta\omega^2)} \tag{6}$$

In Eq.6, ω_{Ni} is the amplitude of the i -th pulse, t_{pi} is the duration, and $\Delta\omega$ is the difference between the carrier and the Zeeman frequencies. The shape of the "2+1" spin-echo signal in accordance with (Bloom, 1955) is calculated as:

$$V_i(t')=\Im m\left\{\int\exp(i\Delta\omega t')(\alpha_1^*\beta_1)(\alpha_2^*\alpha_2^*)\,(\beta_3^2)\;g_i(\Delta\omega)d(\Delta\omega)\right\} \tag{7}$$

for any spatial distribution. In Eq.7, t' is the spin echo coordinate. The dipolar modulation for *pairwise distribution of radicals* is expressed as:

$$f_k(\tau,t)=1-\int 2\beta_3\beta_3^*\sin^2\frac{D\tau}{2}+2\beta_2\beta_2^*\times$$
$$\left[\sin^2\frac{Dt}{2}-\beta_3\beta_3^*\left(\sin^2\frac{Dt}{2}-\sin^2\frac{D(\tau-t)}{2}+\sin^2\frac{D\tau}{2}\right)\right]g_k(\Delta\omega) \tag{8}$$

where k and i=1 or 2. If the EPR spectra of both partners in a coupled system are the same, $k\equiv i$. For dissimilar systems, $g_1(\Delta\omega)$ and $g_2(\Delta\omega)$ are different, and the signal in the "2+1" sequence is the sum of the signals resulting from the first and second spin:

$$V(t,t',\tau)=V_1(t')\times f_2(\tau,t)+V_2(t')\times f_1(\tau,t) \tag{9}$$

To calculate the kinetics for various spatial radical distributions, one should be more specific in describing the dipolar interaction. We introduce this in simple form as:

$$D(r)=\gamma^2\hbar(1-3\cos^2\vartheta)/r^3 \tag{10}$$

where r is the distance between radicals, and ϑ is the angle between the direction of the external magnetic field, B_0, and r. *If pairs do not have a unique orientation* (i.e., they are not in a single crystal nor in a highly anisotropic g-frame) but do have fixed distance, ϑ is arbitrary, and the resulting dipolar modulation is an average over ϑ:

$$<f(\tau,t|r)>=\frac{1}{2}\int f(\tau,t|r,\vartheta)\sin\vartheta d\vartheta \tag{11}$$

An example of the kinetics for such a spatial distribution is shown in Fig. 3.

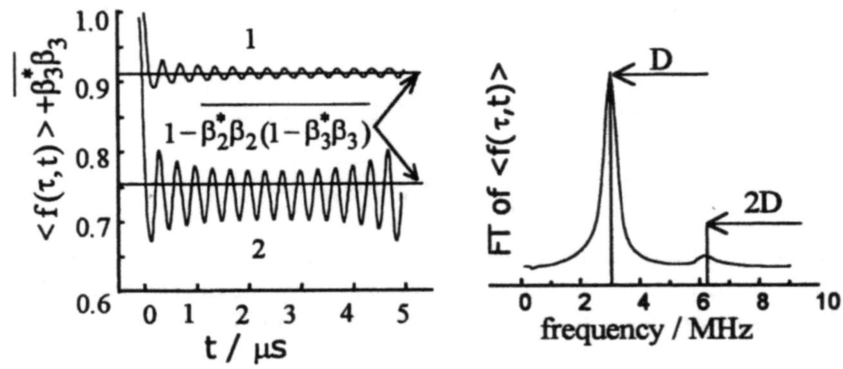

Figure 3. (left). "2+1" kinetics at various $\beta\beta^*$ calculated for D=3 MHz (distance between radicals ~25Å .Trace 1: $\beta_2\beta_2^* = \beta_3\beta_3^* = 0.1$ ("quasi" pulsed DEER pulse settings); trace 2: $\beta_2\beta_2^* = \beta_3\beta_3^* = 0.5$; Both traces level off at ~300-500 ns (e.g., Dt ~0.9-1.5). (right). FT of any trace from top representing dipolar spectrum of radical pair with fixed distance. Peak at 2D is an order of magnitude less than peak at D and might be easily missed in experiment.

It can be seen that $\langle f(\tau, t|r)\rangle$ is the modulation that decays only to a certain level. This level depends only on pulse parameters and shape of the EPR spectrum. Indeed, if all time delays ($x \equiv t, \tau, \tau - t$) are large enough (i.e. $D^*x \geq$ 1-1.5), then:

$$2\int \sin^2\frac{Dx}{2} d\cos\vartheta \to 1 \qquad (12)$$

and

$$f(\tau,t|r)_{x\to\infty} \to \int (1-\beta_3\beta_3^*)(1-\beta_2\beta_2^*) g(\Delta\omega) d(\Delta\omega) \equiv \overline{(1-\beta_3\beta_3^*)(1-\beta_2\beta_2^*)} \qquad (13)$$

The leveling off of the modulation is found to be useful for determining the number of radicals in clusters, and will be considered later. Fourier transform of $\langle f(\tau,t|r)\rangle$ over t is shown in Fig. 3, right. It consists of two peaks at frequencies D and $2D$, with the first peak having a much greater intensity. The intra-pair distance is readily calculated from position of the more intense peak. Note that the position of the intense peak is D, in contrast with $3D/2$ splitting that would be observed in a "Pake pattern" of a CW EPR spectrum (Abragam, 1961). This difference occurs because only the secular part of the dipolar Hamiltonian is involved in the formation of the line shape in this particular spin-echo technique, and for this particular case of isotropic g-values and weak dipolar interaction, as it was defined above. The reason why the pseudo secular term (Carrington *et al.*, 1967)

$$B(S_{1x}S_{2x}+S_{1y}S_{2y}) \qquad B=-D/2 \tag{13a}$$

of a dipolar interaction does not show itself in dipolar modulation in this particular case is easy to understand. Let us take, for simplicity, a primary echo. As shown in (Salikhov *et al.*, 1969), for primary echo, the modulation caused by both secular and pseudo secular terms is

$$f_2(\tau)\sim\cos^2\theta/2\times$$
$$\left\{\frac{3}{4}+\frac{R^2+B^2}{4R^2}+\frac{\Delta\omega^2}{4R^2}\cos R\tau-\left[\frac{(R-B)}{8R}\cos(B+R)\tau+\frac{(R+B)}{8R}\cos(B-R)\tau\right]\right\}$$
$$+\sin^2\theta/2\left\{\frac{\Delta\omega^2}{R^2}\cos D\tau+\frac{B^2}{R^2}\cos D\tau\cos R\tau+\frac{B}{R}\sin D\tau\sin R\tau\right\} \tag{13b}$$

where $R=(B^2+\Delta\omega^2)^{1/2}$. If the spins are resonant, $\Delta\omega\equiv0$, and

$$f_2(2\tau)\sim\cos^2\theta/2+\sin^2\theta/2\ \times\cos(D-B)\tau \tag{13c}$$

After averaging over space, the FT of this signal gives the conventional "Pake pattern" because D-B= 3/2 D. If, however, $\Delta\omega \gg B$ (***case of weak dipolar interaction***) and $\Delta\omega$ has some distribution, then averaging over $\Delta\omega$, (i.e. over EPR line shape, $g(\Delta\omega)$) reduces all terms that include $\Delta\omega$ to zero. In this case the resulting modulation is

$$f_2(2\tau)\sim\cos^2\theta/2+\sin^2\theta/2\ \times\cos D\tau \tag{13d}$$

and reproduces Eq.3. Therefore, when the EPR line width substantially exceeds the characteristic magnitude of the dipolar interaction one can use a Hamiltonian without including pseudo secular term. An example of such a system is nitroxide radical pairs. For these, the characteristic line width is about 200 MHz, and B term does exceed 10 MHz even for the spin labels which had the shortest interpair distance (~15 Å), of those used in the ESE experiments. When using the "2+1" technique, the characteristic time for a modulation to level off, t^*, allows us to immediately estimate the limits of the distances as $r\propto(1\text{-}1.5)\times(Dt^*)^{1/3}$ (or $r\propto(50\text{-}60)\times(t^*)^{1/3}$ (Å) if t^* is given in microseconds). To obtain a spectrum without large distortions, one should start the acquisition at time x, when $Dx\leq0.1$. The minimal distance is determined by the pulse amplitude and dead time, t_d. The pulse amplitude should be sufficient to excite dipolar interactions, and therefore should be ~2*D*. Dead time is estimated for the "2+1" technique to be ~(2-3)t_p. With recently reported 3 ns pulses, (Borbat *et al.*, 1997) that provide pulse amplitude of 80 MHz for $\pi/2$ pulse and t_d ~6-10 ns, this minimal distance may be evaluated as 12-14 Å. Maximal distance is determined by phase relaxation time and/or sensitivity of the instrument. The distance between radicals in a pair, which would make them distinguishable from radicals in the bulk, should be at least half of the average distance between bulk radicals (Raitsimring *et al.*, 1985). There should also be a sufficient time separation between the first and third pulses. For organic radicals like nitroxide, one can expect to measure a maximal separation of 100 Å, based, for evaluation, on the lowest detectable average concentration of $\sim 10^{17}$ cm^{-3} (corresponding to an average distance between radicals ~200Å) and a phase relaxation time of ~2 μsec, which allows one to have a pulse separation of ~4 μsec. Finally, when a coupled system does not have *a fixed intra-pair distance*, the resulting kinetics are obtained by averaging $<f(\tau,t|r)>$ over the distribution $\phi(r)$:

$$f(\tau,t) = \langle\langle f(\tau,t|r)\rangle\rangle = \int <f(\tau,t|r)>\phi(r)dr \tag{14}$$

In this case, the kinetics are smooth traces without pronounced (or only slightly pronounced) modulation, as shown in Fig.4.

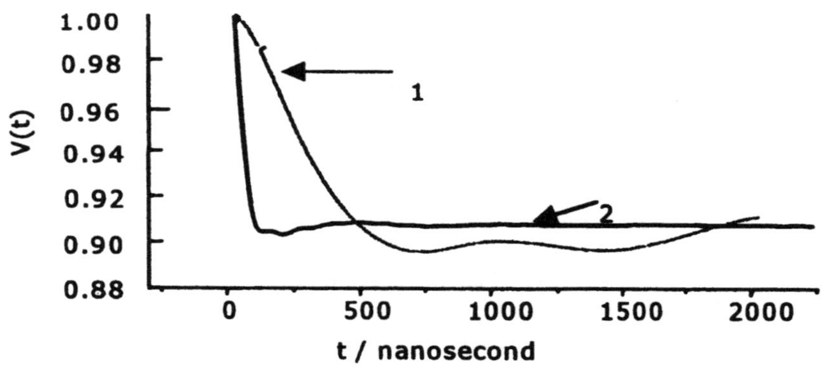

Figure.4. Trace 1:"2+1" kinetics calculated for pair with a rectangular distribution of D, within the limits of 2-10 MHz (distances from 17 to 30 Å) and trace 2: within the limits of 0.25-1.25 MHz (distances from 34 to 60 Å). In all calculations, $\beta_2\beta_2^* = \beta_3\beta_3^* = 0.1$. Note the variation in time where the kinetics leveled off reflects the average distance for a pair.

Cluster distribution is the general case of radical generation under ionizing radiation. Often, in limited volumes (called spurs), more than two radicals are generated by secondary electrons. It is likely that if a protein has more than two close binding sites for radical labels, the resulting distribution will also be a cluster distribution. Since the phase relaxation of each partner in a cluster is considered as independent, the resulting kinetics of the radicals in clusters can be represented as a generalization of Eq. 9:

$$\overline{V(t,t',\tau) = < \sum_{i=1..n} V_i(t') \times \prod_{i,l \neq i} f_{i,l}(\tau,t|r) >} \qquad (15)$$

where n is the number of radicals in a cluster, $f_{ij}(\tau,t|r)$ is the modulation caused by interaction of the i th-spin in the cluster with j th-spin. In Eq. 15, <...> means averaging over space, and $\overline{}$ is averaging over $g(\Delta\omega)$. As in the case of a pair, the modulation decays down to a certain level, which depends, in addition to the pulse parameters and shape of the EPR spectrum, on the number of spins in the cluster. If the radicals in a cluster are the same, then:

$$<f(\tau,t|r)>_{x\to\infty} \to \left[\overline{(1-\beta_3\beta_3^*)(1-\beta_2\beta_2^*)}\right]^{(n-1)} \qquad (16)$$

For "weak" pulses (i.e. $\beta\beta^* \ll 1$), Eq. 16 simplifies and reduces to the expression:

$$<f(\tau,t|r)>_{x\to\infty} \to 1-(n-1)(\overline{\beta_3\beta_3^*}+\overline{\beta_2\beta_2^*}) \qquad (17)$$

The number of radicals in a cluster, n, as this expression shows, can be obtained directly from the extent of the leveling off of the experimental kinetics, since the $(\overline{\beta_3\beta_3^*}+\overline{\beta_2\beta_2^*})$ is calculated easily from known EPR shape and pulse parameters, in accordance with Eqs. 6-8. In addition, the characteristic distance between radicals may be found from the characteristic time where the kinetics start to flatten.

Arbitrary random radical distribution in D-dimensional space. Generally, Eq. 15 describes the kinetics for any particular kind of distribution. In some cases, however, it is possible to derive more manageable analytical expressions. For example, in the case of randomly distributed radicals, averaging over space can be performed directly by using the Markov's method (Abragam, 1962), transforming Eq. 15 into the following expression:

$$V_D(t,\tau,t') \propto V(t')\exp\left(-C_D\int_{V_D} dr\,(1-f(t,\tau|r))\right) \qquad (18)$$

where $f(\tau,t|r)$ is determined by Eq. 8, D is the dimension of space (D=1-3) and C_D is the D-dimensional concentration of radicals. Evaluation of Eq. 18 gives:

$$V_D(t,\tau,t') \propto V(t')\exp\left(-C_D\left(\frac{\gamma^2\hbar t}{2}\right)^{D/3} I\,\eta\right) \qquad (19)$$

where I and η are:

$$I=\int_{-1}^{1}|1-3x^2|^{D/3}\int_{0}^{\infty}\sin^2\left(\frac{1}{z^3}\right)d(z^D);\qquad \eta=\overline{(\beta_2\beta_2^*(1-2*\beta_3\beta_3^*))} \tag{20}$$

For D = 2-3, a good approximation for I is 1.68. For instance, in the case of the random radical distribution in *3-D* space, one can obtain a familiar expression

$$V_{D=3}(t,\tau,t')\propto$$
$$V(t')\exp\left(-C\times 8.1*10^{-13}\,2*t*\eta\right)=V(t')\exp\left(-2*\Delta*t*\eta\right) \tag{21}$$

(Raitsimring *et al.*, 1974) for spin echo kinetics caused by dipolar interactions: where the volume concentration, C, is given in cm^{-3}. As follows from Eq. 21, the inter-radical decay can be completely suppressed by an appropriate choice of pulse parameters, which will set η equal to zero. In the case of complete excitation, $\eta \sim \cos\theta_3$ and equals zero when the third pulse has a rotational angle of $\pi/2$. This feature of the "2+1" method becomes useful when investigating radical pairs with distributed distances, because instead of signal modulation, smooth kinetics are produced. Indeed, the observed kinetics, $V(t,\tau,t')$ are the products of inter- and intra-pair kinetics:

$$V(t,\tau,t')\propto V_D(t,\tau,t')V_p(t,\tau,t') \tag{22}$$

Depending on the average concentration used in the measurements, and for the particular case of distance distribution, V_p may be completely obscured by V_D. The "2+1" method allows us to simply set V_D constant and detect directly only the intra-pair kinetics. It also follows from Eq. 19 that the kinetics are sensitive to the type of spatial distribution. For instance, it can be used for differentiating between structural conformations of labeled bio-polymers (rod/clew and intermediate situations) using the time behavior of the kinetics. For a linear distribution, the logarithm of the signal decay is $\sim t^{1/3}$. For a clew distribution, this time dependence varies between $t^{2/3}$ and t. The kinetics /modulation, therefore, detected by the "2+1" pulse technique allows the determination of the characteristic distance in similar radical pairs, the number of radicals in clusters, the type of spatial distribution, and the suppression of unwanted dipolar interactions. Finally, the measurement of dipolar interactions is not limited to radicals or species with an isotropic g-value. Paramagnetic ions can have a highly anisotropic g-tensor. In this case, the dipolar interactions (e.g., in *3-D* space, Δ) for the axial g-tensor are represented in the following form (Kurshev *et al.*, 1988):

$$\Delta=\frac{4\pi^2}{9\sqrt{3}}\bar{g}^2\beta^2\hbar^{-1}C \tag{23}$$

where

$$\tilde{g}^2 = [g_\perp^4 \sin^2\theta + g_\parallel^4 \cos^2\theta]/g^2_{eff}; \qquad g_{eff}^2 = [g_\perp^2 \sin^2\theta + g_\parallel^2 \cos^2\theta];$$

Distortion of the static dipolar interaction by random spin flips. As we have stated, all of these conclusions are valid as long as intrinsic spin flips do not interfere with the spin flips induced by pulses. Any departure from static conditions results in an alternative path to the ID process (which causes, as mentioned, a modulation of the signal, or reversible relaxation) which is spectral diffusion. This results in irreversible phase relaxation of the signal, and can not be suppressed by adjustment of the pulse parameters. As we mentioned above, in the case of a pair, the FT of the kinetics gives two peaks at frequencies equal to D and $2D$. Random flips decrease the intensity of these peaks. Using a simple model for the two-position "sudden jump" random flips, we show below how the intensities of the peaks vary as the rate of random spin flips increases. A simple Hamiltonian, which includes the dipolar interactions of a pair, and the independent interaction of each spin with the lattice is:

$$\mathcal{H} = \hbar\omega_1 \hat{S}_{1z} + \hbar\omega_2 \hat{S}_{2z} + \hbar D \hat{S}_{1z}\hat{S}_{2z} + V_1(t) + V_2(t) \tag{24}$$

The operator $V_k(t)$ in this Hamiltonian describes non-adiabatic spin interactions with the lattice, and induces random spin flips. The calculation of the spin-echo signal for this Hamiltonian can be done in the standard manner, utilizing a density matrix formalism:

$$V(t,\tau|r) = Tr[\hat{S}_y \rho(\tau,t)] \tag{25}$$

where we assume that the spin rotations are around the X-axis of the rotating coordinate frame and the detector is placed along Y axes. The kinetic equation for the density matrix is:

$$\frac{d\rho}{dt} = -i[\hbar\omega_1 \hat{S}_{1z} + \hbar\omega_2 \hat{S}_{2z} + \hbar D \hat{S}_{1z}\hat{S}_{2z}] + R(t) \tag{26}$$

where $R(t)$ is the relaxation operator describing the density matrix variations due to random spin flips. The relaxation operators can be obtained using the Redfield theory (Slichter, 1996), assuming that the correlation times of lattice variables, τ_c, are sufficiently short. In this case, R(t) is expressed by only two parameters: $W_1 = 2\,|\langle\alpha|V_1(0)|\beta\rangle|^2\,\tau_c$ and $W_2 = 2\,|\langle\alpha|V_2(0)|\beta\rangle|^2\,\tau_c$, where W_1 and W_2 are the average frequencies of the spin flips, and are related to the spin lattice relaxation time T_1 by $W = 1/2T_1$. Assuming, for simplicity, that a pair has two similar radicals, and omitting a simple but cumbersome calculation, one can readily obtain the resulting expression for the spin-echo signal:

$$V(\tau,t,t'|r) = V(t')\exp(-2W_1\tau)\times\Bigg(\exp(-2W\tau)\times\Big\{\overline{(chR\tau+\frac{W}{R}shR\tau)^2}$$
$$\overline{+(1-\beta_2\beta_2^*)(1-2\beta_3\beta_3^*)}\frac{D^2}{4R^2}sh^2R\tau+$$
$$+\overline{\beta_2\beta_2^*}\Big[\frac{D^2}{2R^2}shRt\,shR(\tau-t)\times(chR\tau+\frac{W}{R}shR\tau)$$
$$\overline{+(1-2\beta_3\beta_3^*)}\frac{D}{2R}shR\tau(\frac{WD}{2R^2}shR\tau\,shR(\tau-t)+\frac{D}{2R}shR(\tau-2t)\Big]\Big\}\Bigg) \tag{27}$$

where *sh* and *ch* are the hyperbolic sine and cosine, and $R^2 = W^2 - D^2/4$. In spite of the fact that the spins are similar, we specified $W_1 = W$, which relates the flip frequencies of the observed spin. As W→0, this expression approaches Eq.8, derived at the "static limit". The FTs of $V(\tau,t,t'|r)$ are shown in Fig.5 at various rates of random spin flips.

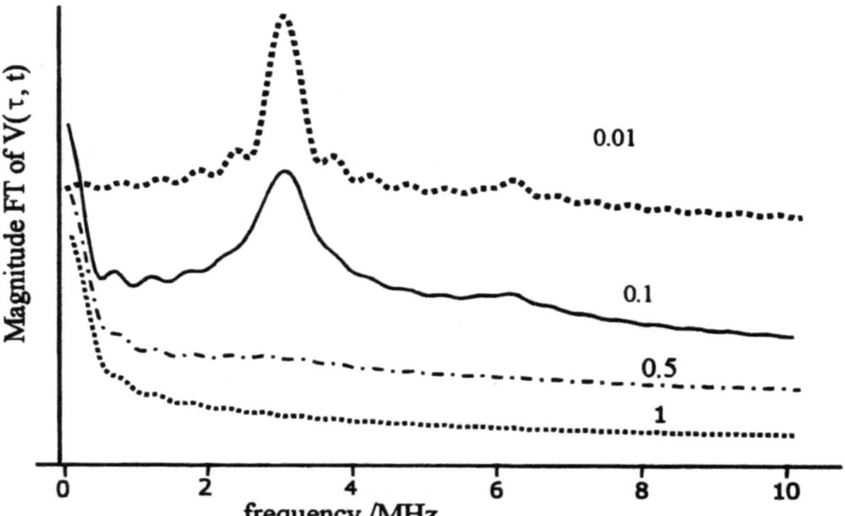

Fiure 5. Variation of a spectrum of a radical pair with rate of random spin flips. These rates (MHz) are shown at the top of every trace. For the calculations, the interval t was taken as 2 μsec; D=3 MHz. The spectrum for the "static" dipolar interaction is not detectable at Wt≥1. Maximal magnitude of every trace is normalized to unity.

As follows from this figure, at Wx≥1, the static dipolar spectrum is already barely visible. This fact was observed in (Salikhov *et al.*, 1981), where it has been shown that just one random flip during echo formation practically destroys the instantaneous diffusion mechanism, and replaces it with the mechanism of spectral diffusion. When the rate of random flips increases, the kinetics for randomly distributed spins are altered, and for Wx<<1, the logarithm for the kinetics becomes $\sim x^{(D/6)}$ (Tregub *et al.*, 1983). For distance or distribution measurements, therefore, it is essential to prove that the system is at a "static" limit. Although spin-lattice relaxation can be suppressed by decreasing the temperature, the spin diffusion (which consequently generates spectral diffusion), originated by the mutual flips of spins due to dipolar interactions, does not depend on temperature (Raitsimring *et al.*, 1975). Experiments, therefore, have to be performed at as low an average concentration as possible, and the characteristic dipolar interactions must be substantially less than the characteristic EPR line width. For example, for nitroxide radicals, the average concentration must not exceed 2-3 mM.

Masking the static dipolar interaction with electron-nuclear modulation. Simply by a judicious selection of pulse settings, the "2+1" technique allows an immediate attainment of one of the initial goals: avoiding the decay caused by nuclear spectral diffusion. This is because the primary echo in this experiment does not move, and is used only for detection. Electron-nuclear interactions, however, cause not only spectral diffusion, but also nuclear modulation (ESEEM). The initial idea was that the ESEEM would be nonexistent in the "2+1" method, but it was observed in the very first experiments. This phenomenon gave rise to new varieties of Pulse Adjustable (PA) ESEEMS (Kurshev *et al.*, 1988; Raitsimring *et al.*, 1995,1996; Borbat *et al.*, 1995). When measuring electron-electron dipolar interactions the presence of ESEEM is a disadvantage. The physical nature, however, of ESEEM and electron -electron dipolar modulation is different, and because ESEEM in the "2+1" technique is sensitive to pulse parameters in a different way, these two types of modulations can be separated. An adequate theory that describes the nuclear modulation in PA techniques (and in the "2+1" technique as well) was recently developed, and may be found elsewhere (Raitsimring *et al.*, 1995,1996). The mathematical description of this phenomenon still remains too cumbersome to present here, therefore we will just briefly comment on the peculiarities of the "2+1" modulation. The modulation frequencies in the "2+1" technique are the same as in primary echo. The ESEEM, for example, of the system S=1/2, I=1/2, contains four frequencies: two fundamental, ω_α and ω_β, and two combinational, $\omega_\alpha - \omega_\beta$, $\omega_\alpha + \omega_\beta$. The differences in the intensities of these lines can be substantially higher or lower than those observed in conventional primary echo. Generally, substantial nuclear modulation appears if the inverse pulse duration is of the same order of magnitude as either the nuclear Zeeman frequency (ω_I), or a non-secular part of hyperfine interaction (hfi), and the pulse amplitude is of the same order of magnitude as either ω_I, or the secular part of the hfi. Such pulse settings have to be avoided as much as possible. Even short pulses (5-10 ns), however, having standard rotational angles (~π) can lead to a substantial ESEEM if the hfi is large enough. In the case, therefore, of measurements of short intra-pair distances between radicals (~15 Å), the dipolar modulation may be in the range of the proton ESEEM in the X-P bands (8-18 GHz). Such peculiarities may result in ambiguities when interpreting data. An approach for avoiding possible misinterpretation (at least if the pair consists of radicals with a low g-value anisotropy), and for separating modulations of a different nature, is to perform the experiment at different operational frequencies (e.g., through the S-X bands (2-12GHz)), while keeping the same pulse settings. In such experiments, nuclear frequencies are varied while dipolar frequencies remain practically constant, and can be easily separated. If such an approach is not available, one can use short pulses with small turning angles (30^0-60^0) in the experiment, which would probably not generate an ESEEM. Preliminary standard ESEEM experiments (primary and stimulated echo, and HYSCORE) would also help, giving an idea about ESEEM frequencies. Other methods that were used are discussed in the experimental section.

3. EXPERIMENT.

In this section, we briefly consider the experimental data that show the application of the "2+1" technique for manipulating dipolar interactions, the measurement of distances and distribution functions in radical pairs, and the determination of the parameters of spatial distributions in more complicated cases.

Random distribution of radicals and ions in 3D-space. The most simple model of the system, where we are confident that *3D* random radical distributions and "static" conditions have been reached, is H or D atoms stabilized in a γ-irradiated, 8 M solution of sulfuric acid (Kurshev *et al.*, 1989). This system is characterized by rather narrow individual lines (6-14 MHz) with a separation of ~1400 MHz between the lines of H, or ~200 MHz between the lines of the triplet (D). Because the pulse amplitude is much less than the separation between the lines, each line can be considered as independent. Small line widths allow the complete (or close to complete) excitation of the spins at rather mild pulse parameters. Fig. 6 represents the experimental kinetics of H atoms for θ_3 equal to $\pi/3$ and $2\pi/3$, while θ_2 was kept constant at ~0.9π. For individual lines of the H atom, such pulse parameters allow η to vary from approximately -0.4 to 0.5 (see Eqs.20-21). This produces either suppression or enhancement of the dipolar interactions in the kinetics through adjustments in the pulse parameters, which

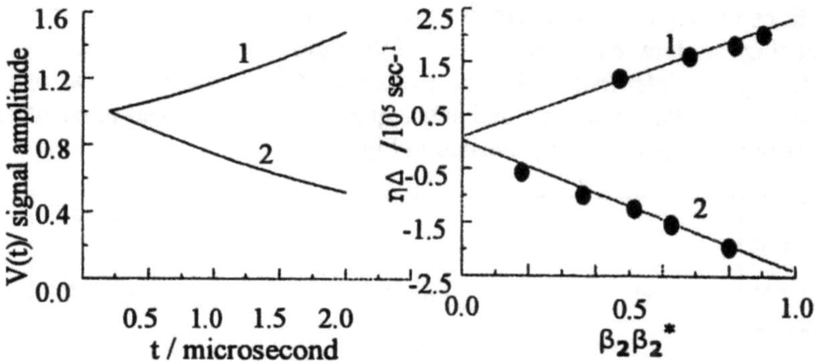

Figure 6. (left). Normalized kinetics of H atom (concentration 7.6 x 10^{17} cm^{-3}) at nominal rotational angles of θ_2 =160° and θ_3 =120° (curve 1); θ_2 =160° and θ_3 =60° (curve 2); (right). Dependence of $\eta\Delta$ (see Eq. 21) for D atoms on $\beta\beta^*$ at two values of θ_3: 60° and 120° for curves 1 and 2, respectively.

is a distinguishing feature of the "2+1" technique. For D atoms, where complete excitation of individual spins was achieved, $\eta\Delta$ is represented in Fig.6, where η was varied over wide limits. As one can see from this figure, similar to the previous case, $\eta\Delta$ was either positive or negative, depending on the parameters of the third pulse. The same condition can be reached for the nitroxide radical by adjusting the pulse parameters, as is shown in Fig. 7.

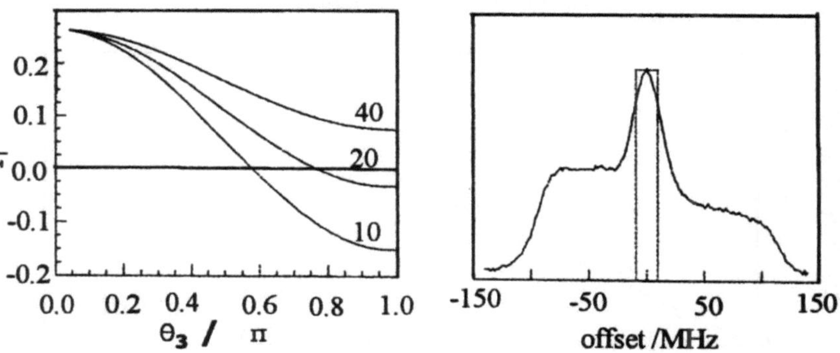

Figure 7. (left) Calculated dependence of an η for TEMPO on the flip angle of the third pulse when the duration of the third pulse (shown on the top of each trace in ns) is varied. The flip angle and duration of the second pulse were held constant at $\pi/2$ and 10 ns, respectively. (right) ESE detected spectrum of TEMPO. The rectangle shows the field position where the pulses were applied.

With a nominal $\pi/2$ second pulse, which has a duration of 10 ns, the dipolar interactions between randomly distributed spins can be completely suppressed by adjusting the duration and amplitude of the third pulse. A different way to regulate dipolar interactions for the nitroxide radical which can be more convenient during the experiment, and only requires the simultaneous attenuation of preset pulse amplitudes, is presented in Fig. 8 a,b.

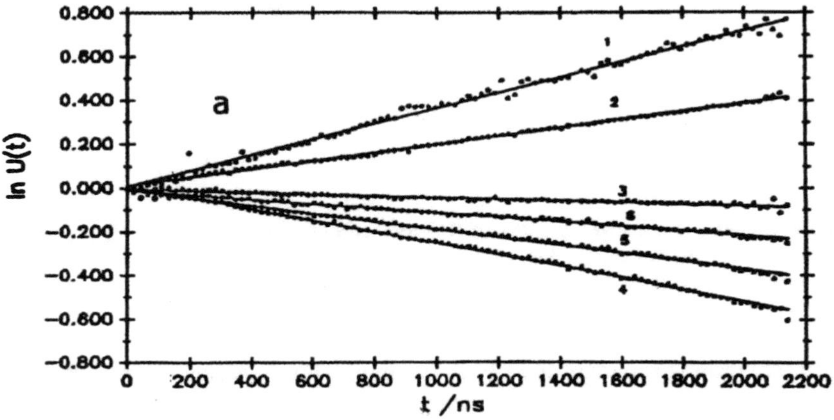

Figure 8a. The "2+1" signal kinetics of the TEMPO in a glassy water /ethylene glycol solution recorded at various pulse amplitudes. Durations (16 ns) and amplitudes of all pulses are equal. Pulse amplitudes are −0.5, -1, -3, -6, and −7.5 dB for curves 2-6, respectively. For trace 1, the flip angle is equal to 0.8π, and corresponds to an initial attenuation of 0 dB. The radical concentration is equal to 0.4 mM.

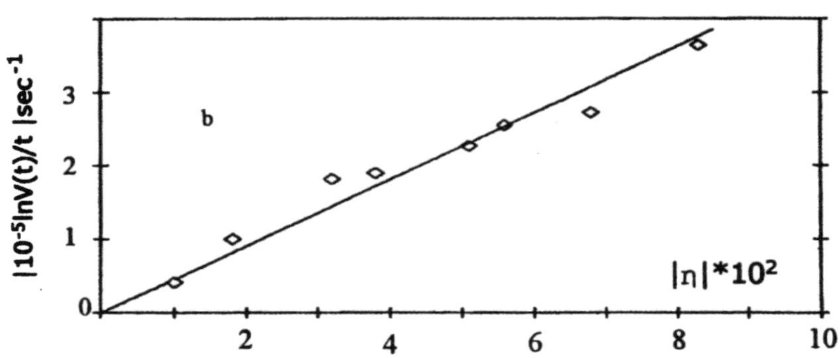

Figure 8b. Moduli of slopes of the traces 1-6, shown in (8a) vs η

In this case, all three pulses were given equal durations and amplitudes, and their amplitudes were varied simultaneously. This results in a variation of η from negative to positive values, and does not require separate adjustments to each pulse. Such a pulse adjustment has been used to suppress the strong background from dipolar interactions from "pseudo" randomly distributed spins, so that dipolar interactions in a pair of nitroxide labels in hemoglobin could be investigated. To test the more complicated case of ions with a highly anisotropic g-tensor, Er^{3+} in a single crystal of $CaWO_4$ have been investigated (Kurshev *et al.*, 1988). For these ions, g-values vary over very wide limits, from 8.38 to 1.247. All measurements were performed at sub-liquid He temperatures to provide a "static" limit for dipolar interactions. The results of these measurements, which are shown in Fig.9, confirm that the expression for dipolar interactions includes $\overline{g^2}$ instead of g_{ef}^2 (see Eq. 23).

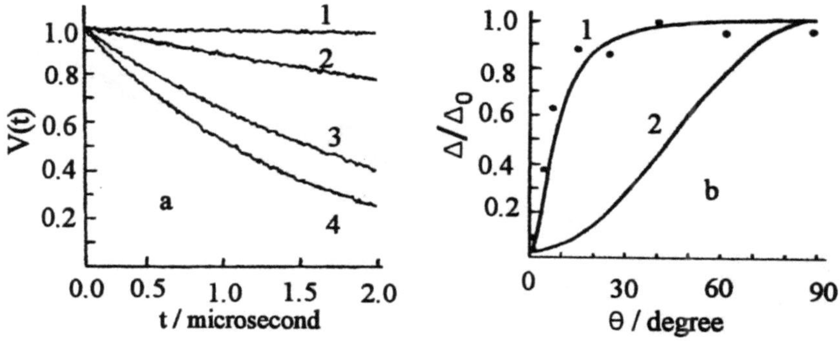

Figure 9. (a) The kinetics of the "2+1" ESE signal for Er^{3+} at different orientations of a single crystal of $CaWO_4$ relative to an external magnetic field, $\mathbf{B_0}$. The angle between the **c** axis of the crystal and $\mathbf{B_0}$ was 0°, 5°, 15°, and 90° for traces 1-4, respectively. Measurements were taken at 3° K. The ion concentration was 4 x $10^{16}cm^{-3}$. (b). Data points (dots) - experimental dependence of dipolar interactions, Δ, on angle Θ between main axis of **g**-tensor and $\mathbf{B_0}$. Curves - dependence of (1) $\overline{g^2}$ and (2) g^2_{eff} on Θ. $\Delta_0 = 8.5 \times 10^5\ s^{-1}$.

Random distribution of radicals in non-integer space and determination of the dimension of fractal objects. In 1980-90, a new hypotheses which assumed that a catalyst could be treated as a *D*-dimensional fractal object (2<*D*<3) was introduced (Pfeifer *et al.*, 1983; Klafter *et al.*, 1984). Some physical methods, like small angle X-ray scattering (SAXS), and quenching of luminescence, confirmed this hypotheses. Indeed, dimension *D* of such a fractal can be determined by means of investigating the kinetics caused by dipolar interactions between paramagnetic centers placed on the surface of such an object (Kutsovcii *et al.*, 1991). Magnesium oxide having a specific surface area, S^B, from 22 to 219 m^2/g, was chosen as a model. The specific surface area was determined from the absorption of nitrogen. The relationship between S^B and the *D*-dimensional surface, $S^D = N/C_D$, determined from ESE measurements in accordance with Eq. 19, is given by $S^B = S^D\{(D/2)\,\Gamma(D/2)\sigma^{(1-D/2)}\}$, (Mandelbrot, 1982), where Γ(x) is the gamma-function, and σ is the surface area of molecule used for specific surface area measurements. The so called F_s^+ centers (holes trapped in the surface defects of the catalyst during γ-irradiation) were used as a paramagnetic probe. Simultaneously, γ-irradiation also generates volume- trapped holes, called F_v^+ centers. These centers have a slightly different g-factor, and by this means can be separated from F_s^+. Because MgO has a small fraction of magnetic nuclei, the EPR line width of both centers is approximately 10 MHz, with a separation between the line maxima of 20 MHz. These centers can therefore be investigated separately. Fig. 10 shows the "2+1" kinetics that were measured for two samples having different specific surface areas.

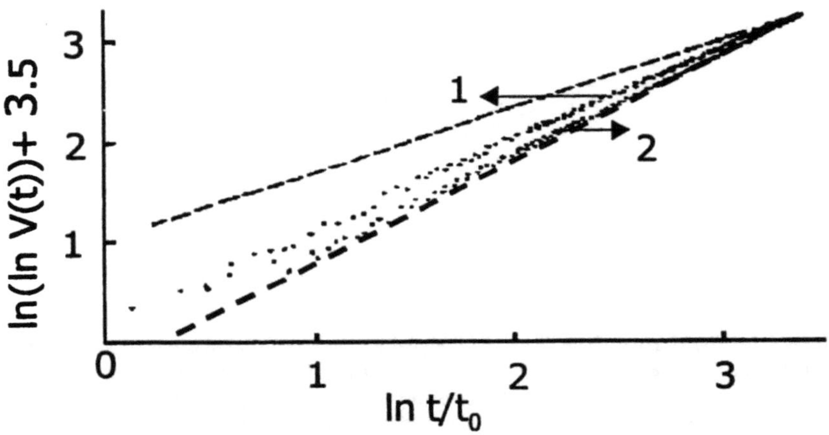

Figure 10. Kinetics of the "2+1" ESE signal represented in double logarithmic coordinates for the determination of the fractal dimension D. Dashed lines: expected kinetics for D=2 or 3 (upper and lower lines); dotted lines 1 and 2 are experimental data collected for MgO samples with specific surface areas of 74 and 219 m^2/g, respectively. ESE signals are normalized to the initial signal at t_0 =30 nsec. Measurements were performed in the "quasi" DEER mode; the primary echo was generated by equal pulses with a duration of 70 ns, nominal rotational angle of π/3 and separation of 6-8 μsec; the "+1" pulse has a duration and rotational angle of 35 ns and π/2, respectively.

The results immediately show that (i) these particular catalysts can be considered as fractal objects, and (ii) the fractal dimension of the sample depends on the specific surface area, and varies from 2.65 ± 0.05 to 2.85±0.05 when the specific surface area varies from 22 to 219 m^2/g. These results, together with SAXS measurements, are presented in Table I. Similar measurements performed for F_v^+ and for deuterium atoms randomly distributed in glassy water/acid solutions (see above) result in *D*= 3±0.05.

Table I. Fractal dimension D of MgO samples determined from the analysis of "2+1" ESE and SAXS data.

Specific surface area/ m^2/g	D_{ESE}	D_{SAXS}
22	2.65±0.05	2.6±0.1
74	2.70±0.05	2.6±0.1
100	2.75±0.05	2.7±0.1
219	2.85±0.05	

Non-random distribution of paramagnetic centers. Transition metal ions in low concentrations are often used to modify the properties of molecular sieves, such as aluminophosphate (ALPO) and silico- aluminophosphates (SAPO). It was not clear, however, whether the incorporation procedure resulted in inclusion of these ions into the catalyst network, or if the ions just precipitated onto the surface of the microcrystals. Obviously, in the former case, one would expect a $3D$-random distribution of ions, as was just demonstrated for F_v^+ centers in highly dispersed magnesium oxide. In the later case, one would expect a random spatial ion distribution, from fractals to clusters. The "2+1" technique has been applied to investigate the distribution of Mn^{2+} in AlPO-5 samples (Levi *et al.*, 1991), prepared in various ways, and having different average concentrations of Mn^{2+} (as determined by CW EPR). As a standard, a frozen glassy water-glycerol solution of Mn^{2+} (where one can expect a random $3D$ distribution of ions) was used. All measurements were performed at 4 K to reach the "static" limit for dipolar interactions. Fig.11 shows the "2+1" kinetics obtained for a standard solution at various values of η. As one can see from Fig.11, these kinetics, as expected for a $3D$ random distribution, are exponential, and their slopes were found to be linearly dependent on η, as well as on the ion concentration. Different results were obtained for MnAlPO-5 samples, as is shown in Fig. 12. The observed signal was very weak, and therefore the interval t, where measurements were performed, was insufficient to allow a conclusion to be made regarding the deviation of the logarithm of the kinetics from the linear dependence on t. The kinetics, therefore, were treated as an exponential. For such an approximation, the slope of the kinetics on a semi-logarithm co-ordinate allows estimate of the concentration of randomly distributed ions. It was found that the variation of the average concentration by an order of magnitude (from 4×10^{19} cm^3 to 4×10^{18} cm^{-3}) did not reflect on the slopes of the kinetics, and that these slopes were what was expected for randomly distributed ions with a concentration of (3-4) x 10^{18} cm^{-3}. In differently prepared samples (using the

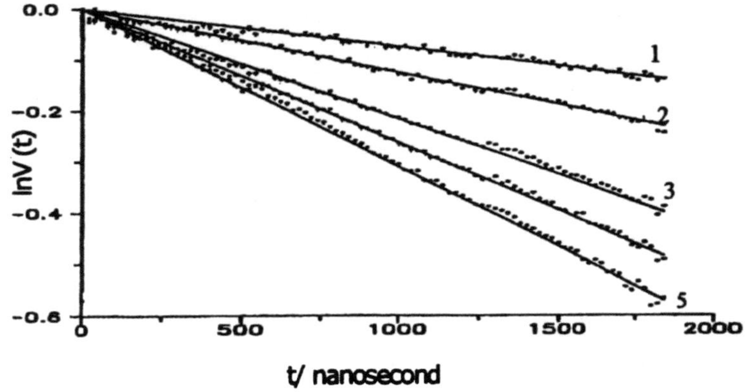

Figure 11. The "2+1" signal decay of Mn^{2+} in a glassy water/glycerol solution (concentration= 6×10^{18} cm^{-3}) at various η: 1) 2.4 x10^{-3}; 2) 3.43 x 10^{-3}; 3) 5.85 x 10^{-3} 4) 8.25 x 10^{-3}; 5) 2 x 10^{-2}

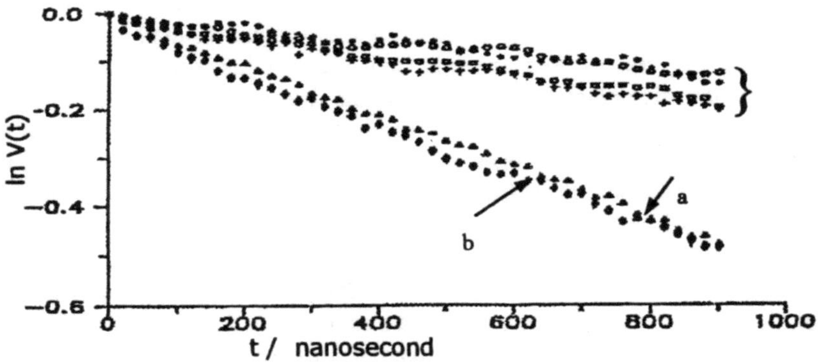

Figure 12. The "2+1" signal decay of Mn^{2+} in various ALPO-5 samples. The curves marked as (}) were collected for "as-synthesized" samples in which the average concentration of Mn^{2+} was varied within the limits of (3.7 to 38.5) x 10^{18} cm^{3}. Curve a: an "impregnated" sample with an average concentration of 5.1 x 10^{19} cm^{-3}. For comparison, curve b shows the data collected from sample with a concentration of 10^{19} cm^{-3}. The pulse setting for all kinetics gives $\eta = 10^{-2}$

so-called impregnation procedure), the average concentration and the concentration found from ESE kinetics differed by five fold, and were 5 x 10^{19} cm^{-3} and 1 x 10^{19} cm^{-3}, respectively. All of these measurements allow the conclusion to be drawn that ions are stabilized on the surface of the microcrystals (which are actually long needles having micron and sub-micron dimensions). As long as the surface has enough room to accommodate ions (at very low average concentrations), the CW and ESE measurements will give similar results. Increasing the average concentration results in the generation of "islands" or a "quasi-phase" of ions, with a high local concentration, and the possible exchange interaction makes the relaxation time too short to be detected in spin-echo measurements. When an "impregnation" procedure is used (stirring the ALPO and a salt of the ion together), the surface area of the microcrystals increases, allowing it to accommodate more randomly distributed ions (in agreement with the experimental observations).

Radical pairs with fixed distances. Plant photosystem II (PSII) (Astashkin *et al.* 1994, 1998) . In PSII, light irradiation generates two types of tyrosine radicals, Y^+_D and Y^+_z, that have practically identical EPR spectra, with a characteristic width of ~60 MHz. These tyrosine radicals belong to different proteins D_1 (Y_z) and D_2 (Y_D), and are thought to be located symmetrically with respect to a primary electron donor, P680, but with different functions in PSII. Because the line width of these radicals is narrow, and since they have identical g-values, a pulsed double resonance application at different pumping and observation frequencies was not feasible. For this reason, the "2+1" technique was applied for measuring the distances between these paramagnetic centers. In the first experiment, the durations of the second and third pulses were set at 24 and 16 ns, and their flip angles were set to $\pi/3$ and $\pi/6$, respectively. With such small flip angles, the "2+1" technique represents a "quasi"pulsed-DEER, since the probability of a spin rotation by both pulses (term $(\beta_3\beta_3^*)(\beta_2\beta_2^*)$ in Eq. 8) is about an order of magnitude less than $\beta_2\beta_2^*$, the probability of rotation by only the second pulse. For this reason, the spin echo kinetics can be simplified, and may be represented by:

$$V(\tau,t)\propto 1-2\overline{\beta_3\beta_3^*}\sin^2\frac{D\tau}{2}-2\overline{\beta_2\beta_2^*}\sin^2\frac{Dt}{2} \qquad (28)$$

The experimental kinetics are shown in Fig. 13 (a,b,c and d), demonstrating the low-frequency modulation which results from the dipolar interactions between Y^+_D and Y^+_z with $D\approx 2$ MHz. The distance evaluated from this D value was approximately 30 Å. To prove that modulation comes from dipolar interactions, the measurements were repeated on the same sample after dark adaptation, where only the Y^+_D radical remains. This sample did not show any modulation resulting from dipolar interactions. Also, the experimental kinetics at the applied pulse setting (amplitu- de of 5 MHz, and a rather long pulse duration in comparison with ω_H), did not show any observable proton related modulation, as was expected from the known properties of the "2+1" ESEEM. Further investigations of the same system were performed in membrane-oriented PS II, at two orientations of the membrane with the magnetic field directed along the membrane plane, and then perpendicular to membrane plane. The kinetics for the former orientation are shown in Fig.13b, along with those for dark-adapted samples, which, as in the previous case, did not show any dipolar-related modulation. Both illuminated and dark-adapted samples, however, showed considerable proton related modulation, since rather large pulse amplitudes were used. Proton modulation in this case was completely eliminated by subtraction of the kinetics of the dark-adapted sample from the illuminated sample. The resulting kinetics are shown in Fig. 13 c and d, and represent the modulation related only to dipolar interactions between Y^{+}_{D} and Y^{+}_{z}. Since the sample was specifically oriented, these kinetics show modulations with frequencies of D and $2D$ as well. Simulations, shown in the same Figs. 13 c and d, confirm that $D \approx$ 2 MHz, and that the distance between Y^{+}_{D} and Y^{+}_{z} is in the limits of 29 to 30Å.

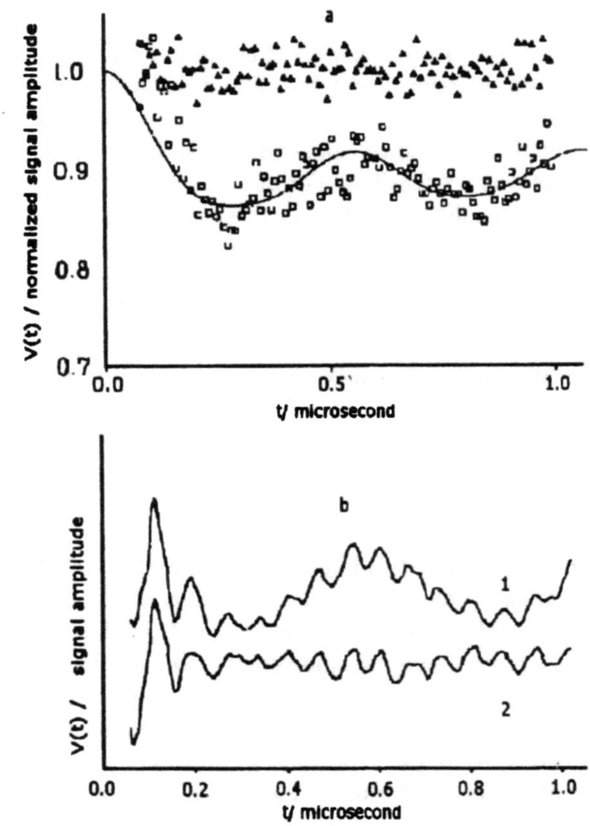

Figure 13 (a). The kinetics of the "2+1" ESE signal of tyrosine radicals. Open squares: obtained from the Mn-depleted PSII illuminated sample. Filled triangles: same sample after dark adaptation. Solid line: calculated kinetics in accordance with Eq. 8 having dipolar interaction constant of 2 MHz. All traces are normalized to unity at t=0. The mw- amplitude of pulses in this experiment was about 5 MHz (much less than the ω_H ~15 MHz). (b). The kinetics of the "2+1" ESE signal collected for magnetic field directed along the normal to the membrane plane. Traces 1 and 2 correspond to illuminated and dark-adapted samples. All pulses in this experiment have the same rotational angle, $2\pi/3$, and pulse duration, 16 ns. The amplitude of the mw-field was ~20 MHz, therefore the proton modulation is very pronounced.

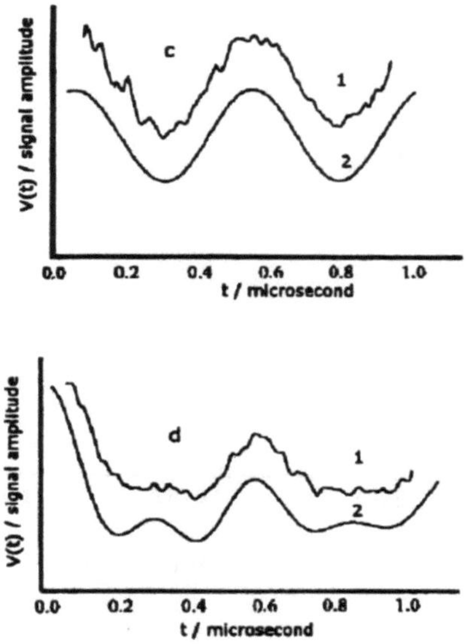

Figure 13. (c). Curve 1: experimental results for "light minus dark" samples with the magnetic field directed along normal to membrane plane; Curve 2: simulated kinetics for D=2 MHz; the angle between the radius-vector which connects Y^{+}_{D} and Y^{+}_{z} to normal of membrane plane is 80° (d). same as above, with magnetic field in membrane plane.

Similar results were obtained for biradical 3B (see Pannier *et al.*, 2000). This biradical consists of two nitroxide radicals, separated by nominal distance of 28Å . This sample was sent to AR by Dr. G. Jeschke . The experimental kinetics and FT's are shown in Fig 13 e,f.. The spectrum shows proton related lines, centered at the proton Larmor frequency (~9.7 MHz) and two peaks, located at ~2.1MHz ***(D)*** and ~4.2 MHz ***(2D)***. The distance between labels as calculated from these data is 29Å, which is in a good agreement with nominal distance.

Radical pairs with distributed distances. Spin-labeled tetrameric hemoglobin (Raitsimring *et al.*, 1992). The "2+1" technique has been applied to determine the distance between nitroxide spin labels tethered to each β-chain of a tetrameric hemoglobin. This was just a single example of where all the features of the "2+1" technique were used. This form of hemoglobin is stable in an aqueous buffer. When frozen for measurements, the solution does not make glass, and proteins aggregate in high local concentrations. It was found by analyzing the "2+1" kinetics, that local concentrations of the labels (about 7 mM) exceed the average concentrations in these samples by 4-6 fold. High local concentrations cause a very fast decay if pulse parameters are chosen to maximize the effect from intra- pair dipolar interactions, and this decay completely masks the desired intra-pair kinetics. In this experiment, therefore,

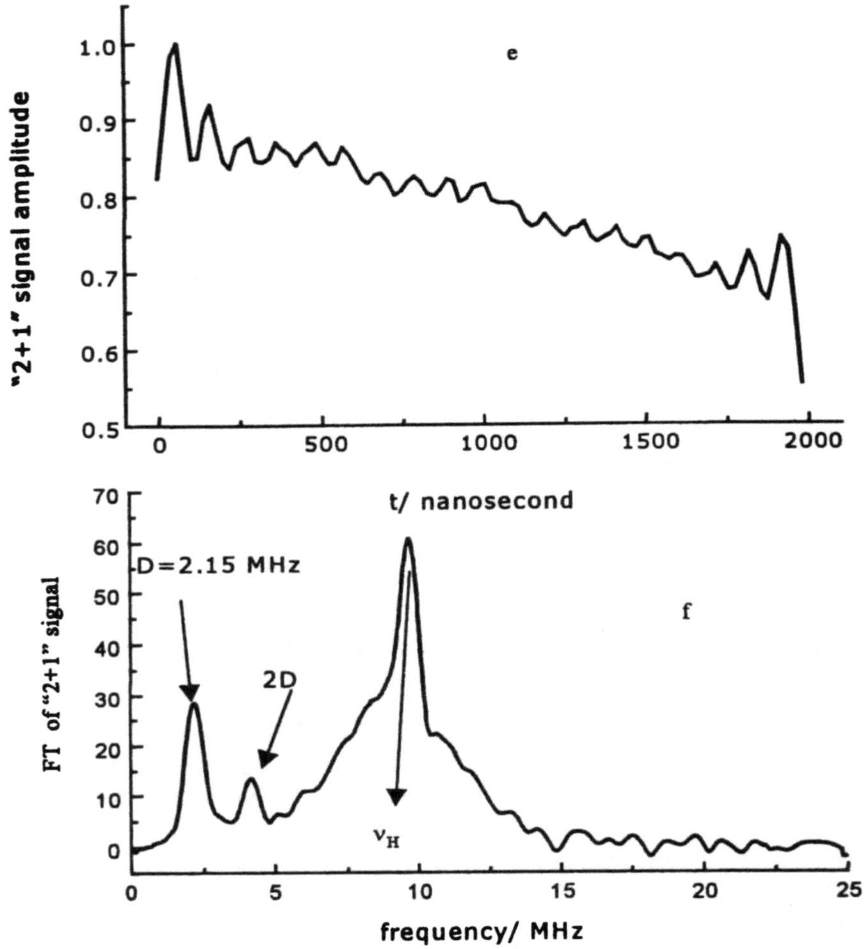

Figure 13. The kinetics (e) and FT (f) of the "2+1" ESE signal of biradical 3b (Pannier et al., 2000)in a glassy toluene/ methanol solution. Pulse setting are as (3) of Figure 8a. The radical concentration is equal to ~ 1.5 mM. Temperature of measurements -25 K; B_0 = 2282 Gs.

pulses were adjusted so that background bulk dipolar interactions in aggregates were completely suppressed, leaving only the kinetics caused by intra-pair interactions (albeit non-optimized). Such pulse adjustments also resulted in substantial proton modulation which, however, does less harm than decay from aggregates. The results of these measurements are presented in Fig. 14. As shown in Fig. 14, these unwanted effects (aggregation and proton modulation), although suppressed as much as possible, caused rather large relative uncertainties in the resulting intra-pair kinetics. Probably, due to these uncertainties

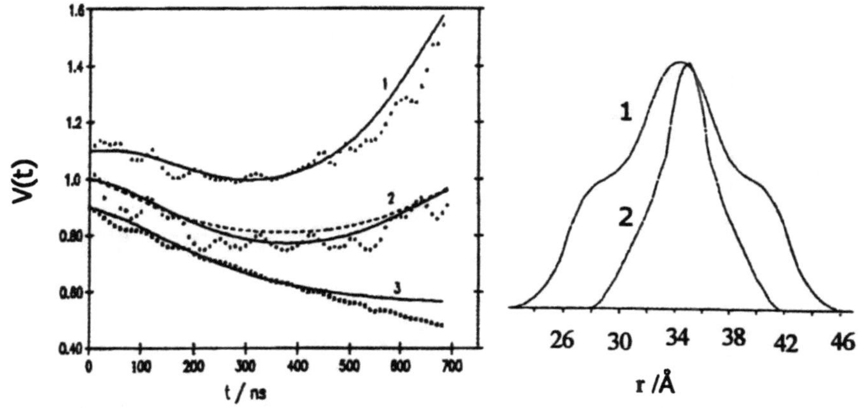

Figure 14. (left) The kinetics of the "2+1" ESE signal for spin-labeled hemoglobin in buffer solution. Durations (16 ns) and amplitudes of all pulses were set equal, with nominal rotational angles of 1) 0.82π; 2) 0.65π; 3) 0.4π. Experimental results are represented by dots; simulations for distribution functions 1 and 2, respectively (shown on right) are represented by dashed and solid lines. The local concentration for simulations was evaluated as 4×10^{18} cm^3, exceeding the average concentration by six fold. (right) The spin label pair distribution functions within tetrameric hemoglobin. 1) random positioning of each label on a sphere with radius of 6 Å and distance between sphere centers of 34 Å.(2) same as (1) but with the deviation restricted to ~3 Å along the radius vector connecting the sphere centers.

the analysis of the kinetics resulted in the conclusion that spin labels may have distances distributed over a broad range, from 28 to 41 Å (see Fig. 14). Such a distribution was allowed by the protein structure.

Radical clusters in γ-irradiated alcohols. (Ichikawa *et al.*, 1991). The γ-irradiation of a frozen glassy solution of alcohols generates so-called spurs (or clusters) - the entities that contain mostly two to four radicals. The distances between radicals in clusters are intrinsically distributed, because radical formation is preceded by thermalizing the sub-excitation electrons, which have a distribution in thermalization lengths. Radical formation is also preceded by the migration of "dry" holes and electrons, and their recombination before they are trapped and transformed into radicals. An example of intra-cluster kinetics for irradiated methanol are shown in Fig. 15. First, by utilizing an essentially infinite time limit (see Eq. 16), it was possible to determine that spurs in this particular case contain only two radicals. Then, from the characteristic time where the kinetics are leveling off (t~2 μsec), the characteristic distance between radicals can be evaluated (60-70 Å). Finally, simulation of the kinetics resulted in pair distribution function:

$$f(r)dV = \frac{3}{2\pi\sqrt{\pi}r_0^3}\exp(-\frac{r^6}{r_0^6})dV \tag{29}$$

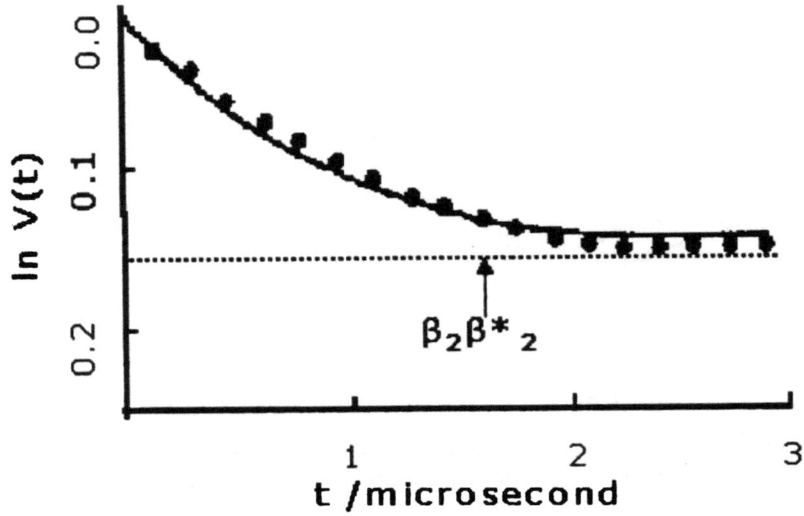

Figure 15. Logarithmic plot of "2+1" ESE signal decay for hydroxyalkyl radicals stabilized in spurs of γ-irradiated D_4-methanol (dots). The average radical concentration=1.6 x 10^{17} cm^{-3}. The curve was obtained by subtracting the ln of the kinetics corresponding to randomly distributed radicals at a given average concentration from the ln of the experimental data. The experiment was performed under "quasi " DEER conditions. The limiting value of ln V(t) for the pair is shown in the figure by dashed line. Simulated kinetics are represented by the solid line. Simulations were performed using the distance distribution function (see Eq. 29) with r_0 =76Å

Table 2 shows the parameters of this function as found for different alcohols.

Table 2. Intra-pair distance distribution of hydroxyalkyl radicals in γ-irradiated alcohol matrices found from analysis of "2+1" kinetics.

Alcohol	r_0/Å	<r> / Å
CD_3OD	72	56
C_2D_5OD	72	56
$(CD_3)_2CDOD$	70	53
n-C_4D_9OD	65	50

where <r> is the distance where 50% of partners are found, i.e.

$$\int_0^{<r>} f(r)dV = 0.5 \qquad (30)$$

4. CONCLUSION.

Development of methods for the determination of radical spatial distributions was prompted by problems in radiation and photochemistry. Whilst many of the phenomenological radiation-chemical characteristics were successfully investigated, a specific challenge was to learn about the transformation of energy from the ionizing radiation to radical formation. It has been understood that this energy transformation is controlled exclusively by spatial characteristics such as the thermalization length of subexcitation electrons, dry hole and dry electron migration until trapping, etc. The interest in investigating spatial distributions of paramagnetic centers declined after this understanding was reached. It seems that now one can witness the rebirth of interest in these kinds of measurements, stemming from biological challenges such as the determination of protein structures, chains of electron transfer, etc. This means that the time invested in the development of these techniques was not wasted, and that the "2+1" technique as shown above can be successfully applied to solve these types of problems. Moreover, the progress achieved in the last decade in hardware (fast mw switches and modulators, LNAs, fast DAQs, new types of probes, etc.) has made possible the recent implementation of multipulse techniques which are more complicated than the "2+1" and DEER. These techniques utilize single quantum (Pannier *et al.*, 2000) as well as double quantum coherences (Borbat *et al.*, 1999; this book, chapter 9). Any of them can now be used for distance/distribution measurements, and the particular choice depends on the instrumental capabilities, the particulars of the problem and the taste of the investigator.

Acknowledgment. The idea of the "2+1" technique came to author in 1982 in Novosibirsk. The initial experiments and theoretical calculations were performed by his student (at the time), Dr. V.Kurshev, who currently works at the University of Houston. The author enjoyed working with Dr.Kurshev, and is extremely grateful to him for numerous fruitful discussions on the properties this technique. Theoretical problems concerning ID, spectral diffusion, radical distribution, etc., were discussed and solved during a close, long term collaboration with Professor K.Salikhov (Kazan Physical Technical Institute RAS). The author is deeply thankful to Prof. Salikhov. Initial understanding of the nature of ESEEM in the "2+1" technique has been achieved in very helpful discussions with Dr. A. Astashkin (now at the University of Arizona). The author is thankful to Dr. Astashkin for these discussions. Further understanding of the nature of ESEEM in PAS was reached in collaboration with Drs. J.H. Freed and R. Crepeau (Cornell University). The author is extremely grateful to Prof. Freed for his exhaustive discussions on the theory of "1+2" SECSY, and to Dr. R . Crepeau who, for the first time taught author how to use 2D-spin echo methods. Prof. D. Goldfarb (Weizmann Institute of Science) gave the author an opportunity to apply the "2+1" technique to catalysts. The author is highly appreciative of the delightful collaboration with Prof. Goldfarb during his visit to the Weizmann Institute of Science, and thanks her for her enthusiastic support. The author highly values the scientific and moral encouragement that he received from Profs. J. Peisach (Albert

Einstein College of Medicine), L. Kevan (University of Houston), Dr. M. Bowman (Pacific Northwest National Laboratory), J.H. Enemark and F. A. Walker (University of Arizona) and is grateful to all of them. Translation from author's English to conventional language was performed by Dr. M. Raisglid, whose help is gratefully appreciated The author is extremely thankful to Dr. G. Jeschke for biradical sample with fixed distance between radicals.

REFERENCES

Abragam, A. The principles of nuclear magnetism, (1961) Clarendon: Oxford, pp.126-128.

Astashkin, A.V., Kodera, Y. and Kawamori, A. (1994) Distance between tyrosines Z+ and D+ in plant photosystem II as determined by pulsed EPR, *Biochim. Biophys. Acta,* **1187**, 89.

Astashkin, A.V., Hara,H. and Kawamori, A. (1998) The Pulsed electron-electron double resonance and "2+1" electron spin echo study of the oriented oxygen-evolving and Mn-depleted pre parations of photosystem II, *J.Chem.Phys.,* **108**, 3805.

Bloom, A. (1955) Nuclear induction in inhomogenious fields. *Phys.Rev.*, **98**, 1105.

Borbat, P. and Raitsimring, A. (1995) "1+2" pulse train -a new variety of pulse adjustable ESEEM spectroscopy. *J. Magn. Res.*, **A114**, 261.

Borbat, P.P., Crepeau, R.H. and Freed, J.H. (1997) Multifrequency two-dimensional Fourier transform ESR: An X/Ku-band spectrometer. *J.Magn. Reson.*, **127**, 155.

Borba, P.P. and Freed, J.H. Double Quantum ESR and Distance Measurements. This book, Chapter 9.

Borbat P.P. and Freed J.H. (1999) Multiquantum ESR and distance measurements. *Chem. Phys.Lett.* , **313**, 145.

Carrington, A.. and McLachlan, A.D. (1967) Introduction to Magnetic Resonance. Harper and Row , p.31.

Dzuba, S.A., Raitsimring, A.M., and Tsvetkov, Yu.D. (1980) ESE studies of the phase relaxation kinetics in system containing two types of spins. *J. Magn.Res.*, **40**, 83.

Ichikawa, T., Raitsimring, A.M. and Kurshev, V.V. (1991) The spatial distribution of radicals in gamma-irradiated alcohols. *J.Phys. Chem.*, **95**, 3565.

Klafter, J., and Blumen, A. (1984) Fractal behavior in trapping and reactions. *J.Chem.Phys.*, **80**, 875.

Klauder, J., and Anderson, P. (1962) Spectral diffusion decay in spin resonance experiments. *Phys.Rev.*, **125**, 912 .

Kurshev, V.V., Raitsimring, A.M., and Salikhov, K.M. (1988) Angular dependence of the dipolar broadening of the ESR line of the single crystal containing paramagnetic centers with an anisotropic g-tensor : Er^3 in $CaWO_4$. *Sov. Phys. Solid State*, **30**, 239.

Kurshev, V.V., Astashkin, A.A., and Raitsimring, A.M. (1988) Modulation effects in a "2+1" electron spin echo pulse sequence . *J. Struct. Chem.*, **29**, 62. (translation of *Zhurnal Strukturnoi Khimii*, **29**, 73, (1988)).

Kurshev,V.V., Raitsimring,A.M., and Tsvetkov,Yu.D. (1989) Selection of the dipolar interaction by the "2+1" pulse train. *J. Magn. Reson*,, **81**, 441.

Kutsovscii, Ya.E., Kurshev, V.V., Raitsimring, A., Aristov, Yu.I., and Parmon, V.M. (1991) Fractal nature of the surface of highly dispersed MgO using electron spin echo. *Doklady Physical Chemistry (A translation of the physical chemistry section of Doklady Akademii Nauk USSR)*, **316**, 133.

Landau, L.D., and Livshitz, E.M. (1989) *Quantum Mechanics*, III. Nauka, Moskva p.231.

Levi, Z., Raitsimring, A.M., and Goldfarb, D. (1991) ESR and electron spin echo study of MnAlPO5 catalyst. *J.Phys.Chem.*, **95**, 7830.

Mandelbrot, B. (1982) The Fractal Geometry of the Nature. Freeman, San Francisco, ch.3.

Milov, A.D., Salikhov, K.M., and Schirov, M.D. (1981) Application of double resonance method to electron spin echo in a study of the spatial distribution of paramagnetic centers in solids . *Sov. Phys. Solid State*, **23**, 565.

Milov, A.D., Mariasov, A.G., and Tsvetkov, Yu.D. (1998) Pulsed electron double resonance (PELDOR) and its application in free radical research. *Appl. Magn. Res.*, **15**, 107.

Pannier, M., Veit, S., Godt, A., Jeschke, G., and Spiess H.W. (2000) Dead-Time free measurement of dipole-dipole interactions between electron spins. *J.Magn.Res.*, **142** , 331.

Pfeifer, P. and Avnir, D. (1983) Chemistry in non-integer dimensions between two-and three. I. Fractal theory of heterogeneous surfaces. *J.Chem.Phys.*, **79**,3558.

Popov, V.I., Raitsimring, A.M., Salikhov, K.M. , and Tsvetkov, Yu.D. (1978) Electron spin echo investigation of mutual flip-flop electron spin reversals involving forbidden transitions. *Sov. Phys. Solid State* , **20**, 985.

Raitsimring, A., Salikhov, K.M., Umanskii, B.A., and Tsvetkov, Yu.D. (1974) Instantaneous diffusion in the electron spin echo of paramagnetic centers stabilized in solid host. *Sov. Phys. Solid State* **16**, 492.

Raitsimring, A.M., Salikhov, K.M., Bychkov, S.F., and Tsvetkov, Yu.D. (1975) Spin and spectral diffusion at 4.2.K due to dipole-dipole interaction between paramagnetic centers. *Sov. Phys. Solid State* , **17**, 303.

Raitsimring, A.M., Samoilova, R.I., and Tsvetkov Yu.D. (1980) The structure of the radical track in the methanol irradiated by T-beta particles. *Radiat.Phys.Chem.*, **15**, 553.

Raitsimring, A.M., and Salikhov, K.M. (1985) ESE method as used to analyze the spatial distribution of the paramagnetic centers. *Bull. Magn. Res.*, **7**, 184.

Raitsimring, A., Peisach, J., Lee, H.C. and Chen, X. (1992) Measurement of distance distribution between spin labels in spin-labelled hemoglobin using an electron spin echo method. *J.Phys.Chem.*, **96**, 3526.

Raitsimring, A., Crepeau, D.H., and Freed,J.H. (1995) Nuclear Modulation Effects in "2+1" electron spin echo correlation spectroscopy. *J.Chem.Phys.*, **102**, 8746.

Raitsimring, A., and Borbat, P. (1996) Electron Spin Echo Modulation for "1+2" pulse train in a weak interaction limits. *Chem.Phys.Let.*, **262**, 8.

Salikhov, K.M , Yudanov, V.F., Raitsimring, A.M., Zhidomirov, G.M., and Tsvetkov, Yu.D. (1969) Modulation Phenomena in Electron Spin Echo. In: Colloque AMPERE XV, Amsterdam, North-Holland, p. 278-284.

Salikhov, K.M., Raitsimring, A.M., and Dzuba, S.A. (1981) The theory of the ESE signal decay resulting from dipole-dipole interaction between paramagnetic centers in solids. *J. Magn. Res.*, **42**, 255.

Slichter, C.P. Principles of Magnetic Resonance. (1996) Springer -Verlag New-York-Berlin-Heidelberg, 3-d edition, ch. 5.

Tregub, V.V. and Raitsimring, A.M. (1983) ESE decay of ionic track in beta irradiated frozen solutions of sulfuric acid. Numerical calculations and experiment. *Chem.Phys.*,**77**, 123.

Chapter 11

Double Electron-Electron Resonance

Methodical Advances and Application to Disordered Systems

G. Jeschke, M. Pannier, and H.W. Spiess

Max-Planck-Institute for Polymer Research, Postfach 3148, D-55021 Mainz, Germany

Abstract: Determination of distances larger than about 2 nm by EPR with high precision requires pulse EPR experiments that can separate electron-electron couplings from all other contributions to the Hamiltonian and from relaxational broadening. The double electron-electron resonance (DEER) experiment as one of these methods is described and guidelines for its use are given. It is concluded that the experiment requires only minor modifications of commercial hardware and can be used for measurements in the distance range between 1.5 and 8 nm. For distances up to 5 nm, precision in non-crystalline systems is usually limited by conformational distributions rather than by the method. Four-pulse DEER is demonstrated to allow for the determination of broad distance distributions even at the lower end of this distance range by overcoming dead time problems. The measurement of ion cluster sizes and intercluster distances in ionomers is described to exemplify application of this method to a problem for which other methods for distance determination fail.

1. INTRODUCTION

Distance measurement by EPR is based on the dependence of dipole-dipole couplings on the inverse cube of the distance. As this relation follows from first principles, a very high precision is expected if the dipole-dipole coupling can be separated from other interactions. For the most important case of two electron spins, the only interaction that cannot be separated from the dipole-dipole coupling is the exchange coupling. If the two electron spins whose distance is to be measured are both well localized on the length scale of their distance and if this distance is large enough (about 1 nm), it is usually a good approximation to assume that the dipolar coupling acts

Biological Magnetic Resonance, Volume 19: Distance Measurements in Biological Systems by EPR, edited by Berliner et al. Kluwer Academic/Plenum Publishers, New York, 2000

between two point dipoles (point-dipole approximation) and that the exchange coupling is isotropic. For even larger distances (above 2 nm), the exchange coupling can often be neglected altogether. The problem of a precise distance measurement with high resolution and for distances as large as possible thus boils down to a precise measurement of the electron-electron coupling.

While being technically simple and requiring short measurement times, continuous-wave (CW) EPR is not well suited for such a precise measurement in solid samples or frozen solutions, since the resolution of the electron-electron couplings obtained from EPR spectrum analysis is limited by line broadening due to unresolved hyperfine couplings, *g* anisotropy, and relaxation. The domain of CW EPR distance measurements therefore lies in the range below 1.5 or perhaps 2 nm, where couplings are large enough to be resolved in the EPR spectra or at least influence lineshapes significantly. Precision is limited anyway in this range due to possible failure of the point-dipole approximation, so that a complete separation of electron-electron couplings from other interactions is not feasible. Note however that the semi-quantitative information on short distances obtained from CW EPR studies in solution is in many cases sufficient to establish fairly detailed structural models. Relaxation measurements can access larger distances than CW EPR, but precision may still be limited by additional assumptions made in deriving dipole-dipole couplings from relaxation times. Ultimate resolution and precise determination of large distances is achieved by separating the electron-electron coupling from all other contributions in the spin Hamiltonian as well as from relaxational broadening. Pulse EPR is well suited for such separation of interactions. There are two general ways for isolating the electron-electron coupling. First, one can make use of the time-dependence of the generation of double-quantum coherence involving both electron spins as described by Borbat and Freed in Chapter 9 (Saxena and Freed, 1996, Saxena and Fredd 1997, Borbat and Freed, 1999). Second, one can use an electron-electron double-resonance (ELDOR) technique (Milov et al. 1981, Milov et al. 1984) as discussed in the present Chapter. A limiting case of the ELDOR technique is the „2+1“ experiment where the two microwave (m.w.) frequencies coincide (Kurshev et al., 1989). Separation of the electron-electron coupling from other interactions is not complete in this case, but often still sufficient to obtain meaningful distance information. This experiment is discussed by Raitsimring in Chapter 10. The present Chapter explains both the basic pulse sequence for three-pulse double electron-electron resonance (DEER) in section 2 and an extended four-pulse DEER experiment providing dead-time free time-domain data in section 3. This extension is valuable for characterizing broad distributions of distances, whose signal contributions otherwise decay within the dead time.

The determination of distance distributions in disordered systems is an important advantage of magnetic resonance techniques as compared to most other techniques for distance measurements. The technical requirements of DEER experiments are discussed in section 4 with particular emphasis on our own setup that is based on only slight modifications of a commercial spectrometer.

Assuming complete separation of interactions, the theoretical limit of resolution and maximum measurable distance is determined by the lifetime of transverse magnetization of electron spins. This puts the upper limit of the distance range at about 8 nm and indicates a precision of about 0.01 to 0.4 nm for distances between 2 and 5 nm. The lower limit is determined by the requirement of an excitation bandwidth exceeding at least the electron-electron coupling. This can be achieved down to about 1.5 nm with the experimental setup described in this Chapter. Smaller distances can be accessed more easily by CW EPR. A more detailed discussion of limitations of the technique and of requirements on the spin probe is given in section 5. A critical assessment of the performance of DEER as compared to other EPR methods for distance determination is also provided.

Some aspects of the application of DEER to the measurement of distances between nitroxide spin probes or spin labels are discussed in section 6 before providing an application example in section 7. This example is concerned with the measurement of ion cluster sizes and intercluster distances in homopolymers and blockcopolymers that carry ionic chain endgroups (ionomers). It is demonstrated that both the cluster sizes of about 2 nm and the intercluster distances between 4 and 6 nm can be determined with adequate precision.

2. THREE-PULSE DEER

The three-pulse DEER sequence displayed in Fig. 1(a) consists of a two-pulse echo sequence $\pi/2-\tau-\pi$ on a fixed observer m.w. frequency and an additional m.w. π pulse on a fixed pump frequency whose position is varied between the positions of the two observer pulses (Milov *et al.*, 1981; Milov *et al.*, 1984). The principle is best understood by considering a pair of two coupled electron spins A and B. The A spins are influenced only by the observer sequence while the B spins are influenced only by the pump pulse and do not contribute to the signal. A and B spins may be located in like paramagnetic centers or in unlike centers. In the absence of the pump pulse, the echo sequence would lead to complete refocusing of all interactions that cause inhomogeneous line broadening, including electron-electron coupling.

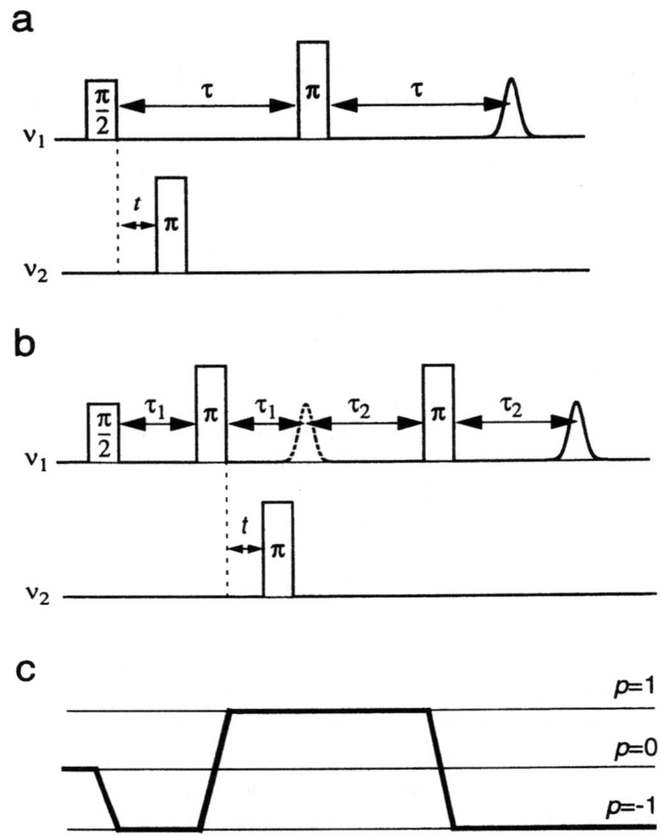

Figure 1. DEER pulse sequences. a) Three-pulse sequence consisting of a Hahn echo on the observer frequency ν_1 and a π pulse on the pump-frequency ν_2. Time t is varied, τ is fixed. b) Four-pulse sequence consisting of a refocused echo subsequence on the observer frequency ν_1 and of a π pulse on the pump frequency ν_2. Time t is varied, τ_1 and τ_2 are fixed. c) Coherence transfer pathway for the A spins in the four-pulse sequence assuming quadrature detection (coherence order p=-1 at the time of detection). For single-channel detection, the mirror image pathway is also observed.

Only homogeneous broadening, mainly due to relaxation processes is not refocused. Consider now the effect of the pump pulse inserted at time t after the first pulse of the observer echo sequence with interpulse delay τ ($t < \tau$). For time t, an A spin experiences a local field from the coupled B spin that depends on the spin state of the B spin. At time t, the B spin is flipped which inverts this local field. The A spin now evolves for time τ-t in the changed local field, before it is flipped itself by the observer π pulse. Evolution for time τ after the π pulse no longer aligns the magnetization vector along the same axis where it started after the initial observer pulse; rather the A spin gains a phase $\phi = \omega_{ee}t$, where ω_{ee} is the coupling between the A and B spin. Accordingly, the echo amplitude is modulated with $\cos(\omega_{ee}t)$. A spectrum featuring a line at ω_{ee} is thus obtained by a cosine Fourier transformation. Note that there is no relaxational decay of the echo unless longitudinal relaxation of B spins is significant. This is because the duration of the observer echo sequence is constant. Longitudinal relaxation of B spins can often be suppressed to a sufficient extent by lowering the measurement temperature. The achievable precision of the measurement of ω_{ee} still depends on the phase memory time T_m of the A spins. This is because t is limited by τ and τ itself cannot be chosen arbitrarily long as some echo intensity must be left for observation. The trade-off between resolution and sensitivity is, however, straightforward.

These considerations on a single spin pair can be extended easily to macroscopic systems. For an ensemble of isolated spin pairs with different ω_{ee}, the time-domain signal is just the sum of the time-domain signals of all the single pairs. If one A spin interacts with several B spins, as is usually the case, the time-domain signal is the product of the time-domain signals of the A-B pairs. In the following we first consider the case of isolated spin pairs.

In a disordered system for which the point-dipole approximation is valid and for which the exchange coupling J can be assumed to be isotropic, the electron-electron coupling is given by

$$\omega_{ee} = \omega_{dd}(3\cos^2\theta - 1) + J\,, \tag{1}$$

with

$$\omega_{dd} = \frac{g_A g_B \mu_B^2 \mu_0}{4\pi\hbar}\frac{1}{r^3}\,, \tag{2}$$

where g_A and g_B are the g values of the two spins, μ_B is the Bohr magneton, r is the distance of the two spins and θ is the angle between the static field axis and the axis connecting the loci of the two electron spins. For an isotropic distribution of angles θ, the spectrum has the shape shown in Fig. 2 from which both J and r can be inferred, provided g_A and g_B are known.

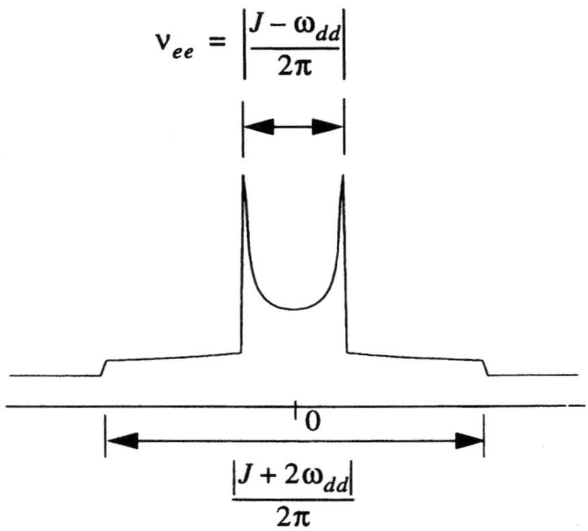

Figure 2. Schematic powder pattern due to dipolar and exchange couplings between two electron spins (simulation for $\omega_{dd}/2\pi$= 4.5 MHz, $J/2\pi$= 1.7 MHz).

If the exchange coupling can be neglected, the powder pattern degenerates to a Pake pattern and r can be directly calculated from the position of the singularities $\nu_{ee}(90°)= \omega_{ee}(90°)/(2\pi)$ corresponding to $\theta = 90°$ by the formula

$$r = \left(\frac{52.041 MHz}{\nu_{ee}(90°)}\right)^{1/3} \text{nm} \tag{3}$$

The assumption of an isolated spin pair is of course an idealization that cannot be realized in practice: There is always a statistical distribution of spins in a sample with average couplings that depend on the volume concentration C of the spin probe. Furthermore there is usually a part of A spins whose coupling partners B are not excited by the pump pulse. Considering this and a result by Klauder and Anderson (1962), one can write the normalized three-pulse DEER signal as

$$V(t) = \exp(-kCF_B t)[1 - \int_0^{\pi/2} \lambda(1-\cos(\omega_{ee}t))\sin\theta d\theta] \tag{4}$$

with

$$k = \frac{2\pi g_A g_B \mu_B^2 \mu_0}{9\sqrt{3}h} \tag{5}$$

where F_B is the fraction of the B spins excited by the pump pulse and λ is a modulation depth factor that is related to but not identical with F_B. In general, λ depends on θ as it includes effects of orientation selection. In a determination of only one well-defined distance, λ can be treated as a fitting parameter that does not strongly influence the result for r. To be able to obtain meaningful distances from a fit, modulation depths λ should not be smaller than 3-5%. In principle, both r and C can be obtained by fitting the signal, but note that the determination of C requires a good estimate of F_B and is therefore less precise. Calibration of the method with samples of known bulk concentration is strongly advised. It is also apparent that the exponential damping of the cosine function corresponds to broadening of the powder pattern displayed in Fig. 2, thus limiting the resolution and hence the precision of r. For long distances r, the corresponding modulation is overdamped. It is therefore important to use small bulk concentrations of the spin probe to obtain utmost resolution in such cases. Theory and applications of three-pulse DEER have been reviewed recently in some detail by Milov et al. (1998). Biological applications include the measurement of distances between the Mn-cluster and the redox active tyrosine residue in photosystem II with a precision of 0.02 nm (Hara et al., 1996), determination of the angle included by the vector between the Mn-cluster and the tyrosine and the membrane normal in an oriented sample of the same system (Astashkin et al., 1998), and a distance determination between an Mo(V) and Fe(III) center in chicken liver sulfite oxidase (Enemark et al., 1999).

3. FOUR-PULSE DEER

The signal contributions due to spin pairs with a broad distribution of small (<1.5 nm) and intermediate (1.5-2.5 nm) distances decay almost completely within a few tens of nanoseconds. Unfortunately, such small values of t are not accessible with the three-pulse DEER experiment, since the m.w. pulse durations are also of this order of magnitude and pulse overlap leads to signal distortions. In practice, a dead time of at least three times the m.w. pulse length can hardly be avoided. This problem can be solved by extending the pulse sequence by one more observer pulse as shown in Fig. 1(b) (Pannier et al., 1999).

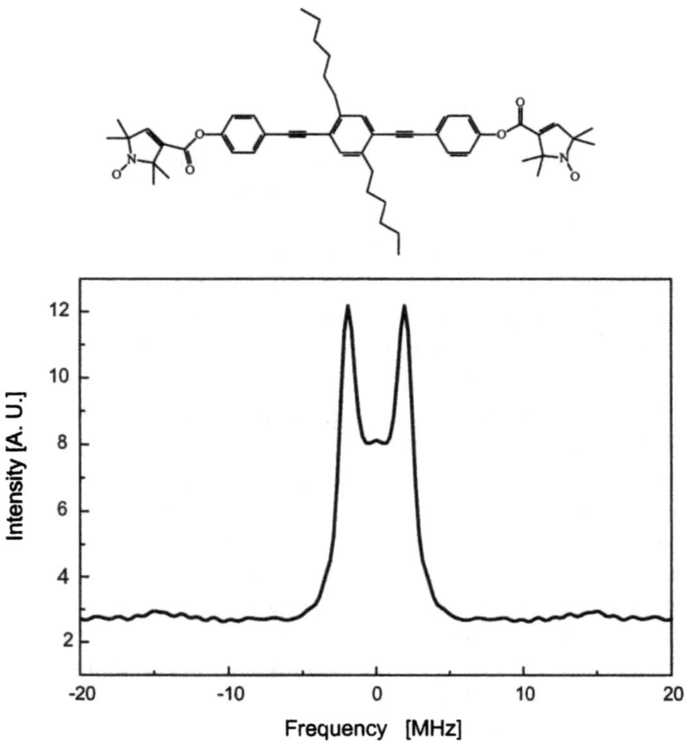

Figure 3. Experimental powder pattern obtained with the four-pulse DEER experiment for a rigid-rod biradical with a length of 2.83 nm.

The first two pulses of the observer sequence lead to echo formation at time $2\tau_1$. The magnetization defocuses again and is once more refocused by the third observer pulse, leading to echo formation at time $2(\tau_1 + \tau_2)$. The pump pulse is now inserted between the second and third pulses of the observer sequence. It can thus be applied at the time of complete refocusing of the A spin magnetization ($t = \tau_1$) without any interference between the m.w. pulses if τ_1 is chosen sufficiently long. The basic principle of the experiment remains the same as for three-pulse DEER and the equations above still apply, if t is now substituted by $t' = |\tau_1 - t|$. The length of the whole pulse sequence of $2(\tau_1 + \tau_2)$ instead of 2τ for the three-pulse sequence leads to a somewhat smaller echo amplitude at comparable resolution, since the upper limits for t' and t are given by τ_2 and τ, respectively. Additional signal loss occurs by introducing one more pulse. Nevertheless, the sensitivity of four-pulse and three-pulse DEER is similar, for disordered sys-

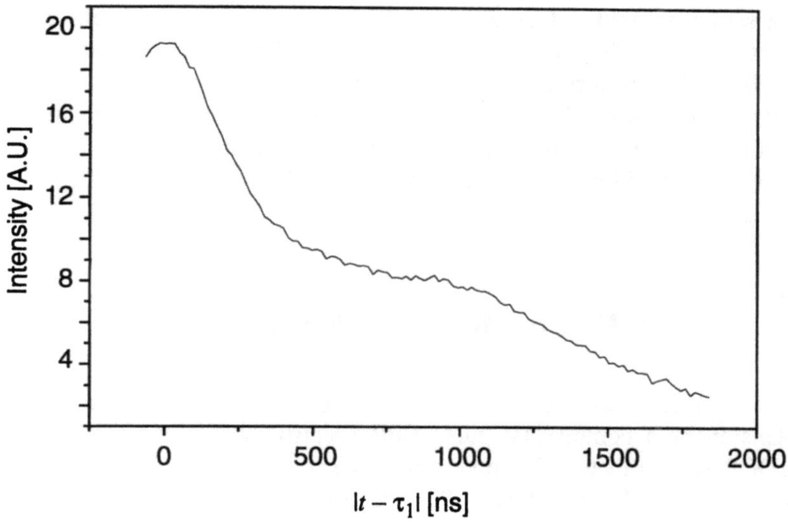

Figure 4. Experimental time-domain data obtained with the four-pulse DEER experiment for a rigid-rod biradical with a length of 3.8 nm.

tems, since modulation depths are maximum for the small t or t' that are accessible only with the four-pulse version. Sensitivity can be further improved by detecting a series of stimulated observer echoes instead of only one observer echo by applying a train of $\pi/2$ observer pulses after the final π observer pulse (Pannier et al., 1999).

In contrast to three-pulse DEER, four-pulse DEER features a symmetric coherence transfer pathway concerning excitation and detection as can be seen in Fig. 1(c). It thus conforms to a principle for designing pulse sequences that is well established in NMR spectroscopy (Ernst et al. 1987). For $\tau_1 = \tau_2 = \tau$, the whole sequence is symmetric. The variation of the echo amplitude on varying t' should then also be symmetric, except for a effects of longitudinal relaxation of B spins. Our experiments with nitroxide biradicals and spin probes in polymer matrices at a temperature of 15 K have shown no evidence for an asymmetry of the signal about t'=0; hence longitudinal relaxation of B spins can be safely neglected. Observation of the dead-time free signal allows one to obtain a whole powder pattern rather than only the positions of the θ=90° singularities by Fourier transformation of the time-domain data (Fig. 3). Note that the feet of the pattern are still somewhat suppressed due to orientation selection. If exchange coupling is expected to be significant and ω_{dd} and J need to be separated, this problem has to be overcome (section 6).

Four-pulse DEER has been shown to yield reliable distances in a study of a series of rigid-rod biradicals with lengths between 1.65 and 2.77 nm (Martin et al., 1998). Fig. 4 shows that more than a full oscillation is still clearly visible for a rigid-rod biradical as long as 3.8 nm. The experimental distance of 3.63 nm agrees well with the N-N distance of 3.79 nm found from a force-field conformer search for the structure of the corresponding keton using the Merck Molecular Force Field (Jeschke et al., 2000).

4. TECHNICAL REQUIREMENTS

Both three-pulse and four-pulse DEER require either an ELDOR extension of a commercial pulse EPR spectrometer or a homebuilt pulse ELDOR spectrometer. Our own setup uses minimum modifications that are made on the Bruker ESP 380 spectrometer to accomodate ELDOR experiments. Microwave power from an external tunable source is fed into one m.w. pulse forming unit (MPFU) of the spectrometer thus providing two pump pulse channels. A second MPFU is left unchanged to provide two observer pulse channels. As an external m.w. source, we tested both an HP8350B sweep oscillator with HP-86290B RF plug-in module and a YIG oscillator (Avantek AV 78012) customized by Magnettech GmbH Berlin.

With the sweep oscillator, a Miteq m.w. amplifier (AMF-5S-8012-18) was required to obtain sufficient m.w. power. All recent measurements were done with the cheaper YIG oscillator. Note that no phase coherence between the pump and observer channels is required, since both types of excitation interact with distinct subsets of spins. Indeed it is advantageous to use non-coherent sources, as this eliminates part of the artifacts that can arise from an overlap of the excitation bands (Kurshev et al., 1989).

Necessary excitation bandwidths and resonator properties are determined by the expected spectrum of the spin pair that is shown schematically in Fig. 5. In consists of two doublets with equal splitting ω_{ee} and with an EPR frequency difference $\Delta\Omega$ that depends on the resonance offsets of the A and B spins. The observer and pump pulses must be able to simultaneously excite both lines of the A and B doublet, respectively. For distances above 1.5 nm, pulse lengths of 32 ns are sufficient to provide this kind of spectral coverage. At the same time, exclusive excitation of A and B spins by observer and pump pulses, respectively, requires a sufficient separation $\Delta\Omega$ of the doublets. For most probes, the range of possible $\Delta\Omega$ is much larger

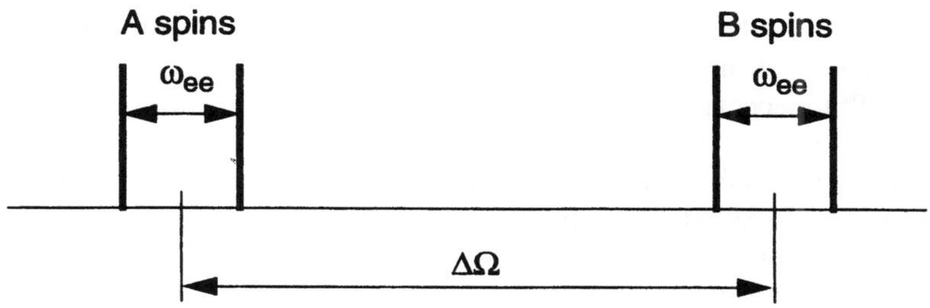

Figure 5. Stick spectrum of a pair of coupled electron spins consisting of two doublets with equal splitting. The case $\Delta\Omega >> \omega_{ee}$ applies to the DEER experiment. Pulses at the observer and pump frequency must be able to excite both lines of the doublet of A and B spins, respectively, and should not excite transitions in the other doublet.

than the excitation bandwidth, so that a convenient difference of the two m.w. frequencies can be chosen. We found that a difference of 60 MHz for nitroxides provides good spectra with little distortion from overlap of the excitation ranges. This is particularly true for distances above 2 nm where pulse lengths of 40 ns are sufficient. As there is an optimum excitation bandwidth, $\pi/2$ and π pulses should feature the same bandwidth and thus the same length to obtain best results.

Excitation of A spins by the pump pulse can also take place via forbidden ($\Delta m_S=1$, $\Delta m_I=1$) transitions. Such transitions are closer to the pump frequency than the allowed transitions by approximately $\nu_I = 14$ MHz for protons at X band frequencies. This effect leads to nuclear modulations at the proton Zeeman frequency ν_I for small *t*. Such modulations can largely be suppressed by choosing a τ_1 value for which $\nu_I\tau_1$ is approximately an integer number (Pannier et al., 1999). Nevertheless, some residual oscillations may still be observed if pulse lengths of 32 ns or shorter are needed to access also distances below 2 nm (see Fig. 7).

Bimodal resonators were used in the first DEER experiments (Milov et al., 1981, Milov et al., 1999). This approach has the advantage that the bandwidths of the two modes can be of the order of ω_{ee}, while a single-mode resonator must cover the whole spectrum with a width of the order of $\Delta\Omega + \omega_{ee}$. Bimodal resonators can therefore feature higher *Q* values, thus minimizing power requirements and optimizing detection efficiency. Effects of observer and pump pulse overlap are also less dramatic, which may be an advantage in the three-pulse DEER experiment. Note however that this advantage is spoiled by pulse overlap in the amplifier, unless two separate high-power m.w. amplifiers for pump and observer pulses are used. Most bimodal resonator designs also have the disadvantage of a low filling factor.

A broad-band slow-wave structure gave better results than a bimodal loop-gap resonator in the work of Larsen and Singel (1993). Combination of a commercial spectrometer with a home-built bimodal resonator also proved feasible (Pfannebecker et al., 1996). However, for routine applications we prefer a strongly overcoupled commercial dielectric ring resonator probehead (Q_L~100). The electric nuclear double resonance (ENDOR) resonator EN4118X-MD4 from Bruker is particularly well suited, because of its already intrinsically lower Q value, but the standard dielectric ring resonator probehead without ENDOR coils performs almost as well. Sufficient bandwidth is obtained by overcoupling. This setup proved to be more stable and reliable than the one with a home-built bimodal resonator and is easier to realize.

5. LIMITATIONS AND COMPARISON WITH OTHER METHODS

As mentioned above, limitations of excitation bandwidth and failure of the point-dipole approximation set a lower limit of r = 1.5 nm for DEER distance measurements. At such short distances, CW EPR measurements can compete with DEER in terms of precision while being easier to perform. The upper limit of the distance range is given by the minimum frequency $\nu_{ee}(90°)$ that still leads to modulations which can be safely recognized in the time-domain signal. Our experience suggests that frequencies as low as 0.1 MHz may be measured, if bulk concentration and measurement temperature are optimized. Bulk concentration should be small enough to just avoid overdamping of the modulation by the factor $\exp(-kCF_Bt)$ in eq (4). Measurement temperatures of 15 K or even lower may be required to ensure slow transverse relaxation and to improve the Boltzmann factor to obtain sufficient signal at such low concentrations. A minimum detectable frequency of 0.1 MHz would put the upper distance limit to 8 nm according to eq (2). Even larger distances may still lead to time-domain signal shapes that can be distinguished from a case of homogeneously distributed radicals, but we do not think at the moment that more than qualitative information can be obtained in this range. The resonance shift of the A spins by the pump irradiation does not limit application of the method, because the actual observation is made by changing the durations of two periods of *free evolution* of the A spins. Assuming an error of the frequency measurement of 0.1 MHz for the larger frequencies, we obtain a precision of about 0.03 nm at r_{AB}= 2 nm and of 0.12 nm at r_{AB}= 4 nm. These values should be understood as rough approximations. They indicate that below 5 nm the error of DEER distance measurements is smaller than the width of typical distance distributions in most disordered systems and even in many biradicals.

There are also some limitations to the type of EPR probe molecules to which DEER can be applied. If the EPR spectrum of both probes is so narrow that observer and pump excitation cannot be separated properly, it is advisable to use the "2+1" experiment described by Raitsimring in Chapter 10 (Kurshev et al., 1989). On the other hand, if the spectra of both probes are very broad, only a small fraction of the B spins can be excited by the pump pulses and λ and F_B in eq (4) become very small. This leads to a sensitivity problem that has so far prevented us from extending our approach for nitroxide probes in completely disordered systems to a commercial high-field EPR spectrometer operating at W-band frequencies. Note however, that the combination of two transition metal probes (Mo(V) and Fe(III)) at X band frequencies seems to be feasible (Enemark et al., 1999). We were also successful in observing four-pulse DEER modulations for a rigid-rod biradical of 2.8 nm length at W band with partial orientation of the biradical in a liquid crystal.

With respect to resolution, precision, and accessibility of large distances, DEER is among the best EPR methods for distance determination. At the moment, its performance is met only by the multiple-quantum methods discussed by Borbat and Freed in Chapter 9 (Borbat and Freed, 1999). Multiple-quantum methods require excitation of the whole spectrum of the spin pair rather than separate excitation of the two doublets. This has the advantage over DEER that the powder patterns are not distorted by orientation selection, but imposes technical requirements that are much harder to satisfy. The technical requirements of DEER in turn are higher than for the "2+1" experiment discussed in Chapter 10, with the advantage of DEER being that data analysis is more straightforward and separation of the electron-electron couplings is complete. If one spin has a much faster relaxation rate than the other, relaxation measurements are also an alternative for distance measurements with low technical requirements. Note however that a larger number of spectra must be acquired and data analysis is more complicated.

The advantage of simple data analysis based on first principles combined with the dead-time free data acquisition of four-pulse DEER is crucial if distance distributions with several maxima are to be characterized in systems lacking long-range order (see section 7). Among all physical methods for distance determination in the nanometer range, only multiple-quantum EPR seems to be able to match the performance of DEER for such applications.

6. DEER ON NITROXIDES

Nitroxide spin probes and spin labels comprise the most important class of EPR probes for distance determination as they are stable, commercially available at affordable prices, and above all rather easy to attach to systems of interest via covalent bonds or non-covalent interactions. EPR properties, molecular dynamics, and chemistry of nitroxide radicals are well documented (Berliner 1976, Berliner 1979, Berliner and Reuben, 1989). In a detailed study of the three-pulse DEER experiment applied to a nitroxide biradical it has been pointed out that some pecularities must be considered in the interpretation of the results (Larsen and Singel, 1993). Excitation of all orientations of a nitroxide radical moiety by m.w. pulses with the usually available power is possible only for those transitions where the ^{14}N nucleus is in its $m_I = 0$ state, so that the hyperfine coupling vanishes. Only one of the two m.w. frequencies can be set to this transition, which means that for either the A or B spins only part of the spectrum is excited. This corresponds to the selection of certain orientations of the nitroxide moiety with respect to the static magnetic field direction. This set of orientations is correlated by the molecular geometry to a certain set of angles between the axis connecting the loci of the two electron spins and the static field direction. In other words, the angle θ in eq (2) does not exhibit an isotropic distribution and a weighting factor $G(\theta)$ has to be added.

Such orientation selection does not usually impede observation of the $\theta = 90°$ singularities but may well suppress the $\theta = 0°$ feet of the powder pattern as is apparent from Fig. 3. This spectrum was acquired by setting the pump frequency to the $m_I = 0$ transition and the observer position to the high-field shoulder as schematically depicted in Fig. 6 (Pannier et al. 1999). Similar results with respect to orientation selection are obtained for the opposite choice of pump and observer frequency as it was used in earlier work (Larsen and Singel 1993, Pfannebecker et al. 1996). The choice of Singel et al. has the advantage of leading to a slower exponential decay due to the homogeneously distributed radicals and hence potentially higher resolution by decreasing F_B in eq (3), while the choice in Fig. 6 maximizes modulation depth. This is advantageous, if resolution is anyway limited by the distribution of distances, as it is often the case even in biradicals due to conformational disorder. If orientation selection is to be overcome, one needs to repeat the experiment for a number of different observer positions while keeping the pump position fixed on the m_I=0 transition.

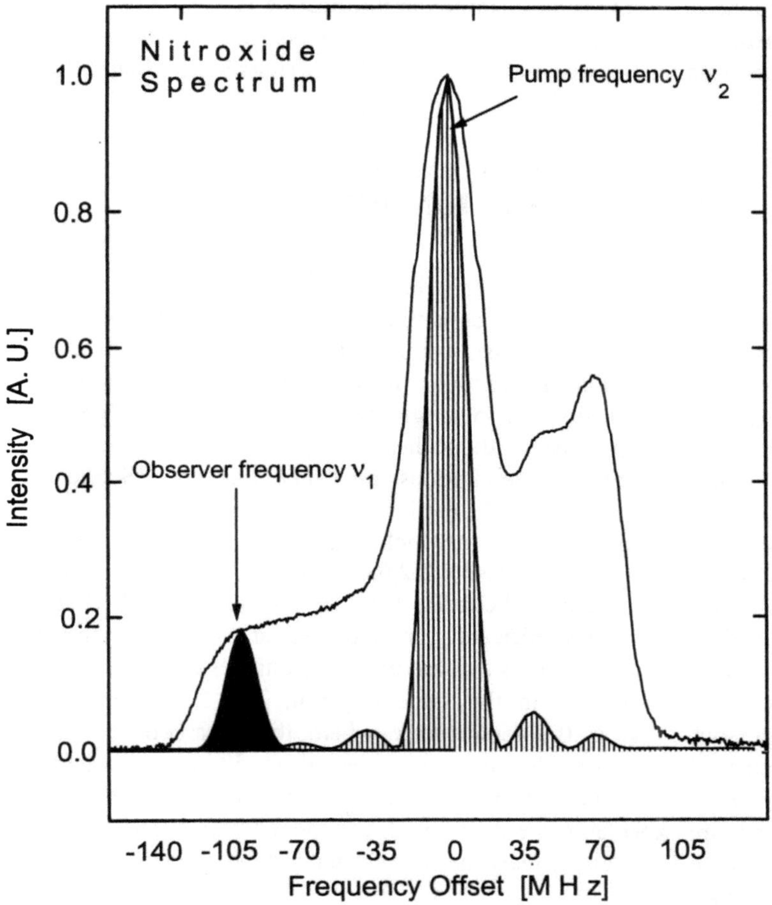

Figure 6. Schematic drawing of the excitation positions and approximate excitation profiles for pump and observer pulses in a nitroxide spectrum. Pump and observer position can be exchanged to obtain somewhat higher resolution at given concentration, albeit at the expense of lower modulation depths.

In principle, orientation selection effects contain information on the relative orientation of the two nitroxide moities as has been examined in both experimental (Larsen and Singel, 1993) and theoretical work (Maryasov et al. 1998). In practice, these effects are rather subtle for systems lacking macroscopic order at least at X-band frequencies. Determining dihedral angles with some confidence by this method may require going to high frequencies where orientation selection becomes more pronounced. If only semi-quantitative information is needed or if one wants to distinguish between simple geometrical models, orientation selection at X-band frequencies may however be sufficient (Larsen and Singel, 1993).

7. CLUSTER SIZES AND INTERCLUSTER DISTANCES IN IONOMERS

One of the recent trends in polymer research is to generate supramolecular structures by combining different order principles that may either act cooperatively or may counteract each other (Zentel et al., 1997). One of the possible interactions leading to regular structures is the electrostatic interaction of ionic groups attached to the polymer chain, especially if the polymer without these groups is nonpolar. Such polymers with ionic sidegroups or endgroups are called ionomers (Schlick, 1996, Eisenberg and Kim, 1998). In earlier work we have found that endgroup ionomers form ionic clusters to which ionic nitroxide spin probes like the potassium salt of 4-carboxy-2,2,6,6-tetramethyl-piperidin-1-oxyl (K-TEMPO) can be attached (Schädler et al. 1997, Schädler et al. 1998). Such attached probes are significantly immobilized and exhibit apparent EPR glass transition temperatures that are up to 50 K higher than the ones of non-ionic nitroxide probes in the same material. The EPR glass transition temperature T_{50G} is the temperature where the separation of the outer extrema of the nitroxide spectrum is equal to 50 G, i.e., intermediate between the rigid and fast motion case. The sizes and distances of the ionic clusters are expected to be distributed, and indeed it has been possible only in one case to obtain at least the intercluster distance by small angle X-ray scattering (SAXS).

A larger variety of supramolecular structures is obtained if the ionic endgroups are attached to one or both chain ends of a diblock copolymer. For such samples, SAXS measurements have provided information on the morphology of the diblock copolymer but not on the size and distance of the clusters. To measure both cluster sizes and intercluster distances by EPR techniques, approximately two K-TEMPO probe molecules must be attached to an ionic cluster. This leads to a broad distribution of distances of probe pairs attached to the same cluster even if the cluster sizes would be uniform. Dead-time free detection of the signal is therefore mandatory to determine cluster sizes.

Our measurements on a series of ionomers with different molecular weight derived from both homopolymers and diblock copolymers revealed a

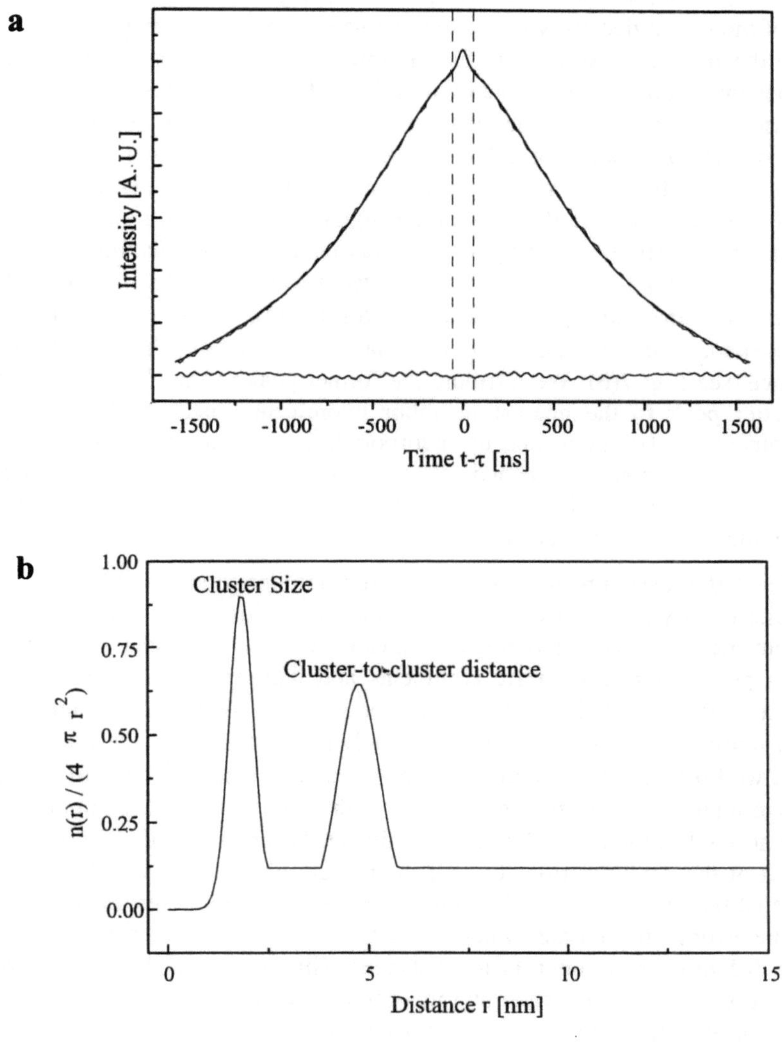

Figure 7. Four-pulse DEER on a zwitterionic diblock copolymer a) Experimental data (dotted line), fit function, and residual. The oscillations in the residual are proton modulations. The part of the signal between the two dashed lines can only be accessed with the dead-time free four-pulse sequence b) Distance distribution corresponding to the fit.

characteristic signature of such samples in the DEER signal that can be recognized in Fig. 7(a) where the signal of a zwitterionic diblock copolymer with a molecular weight of 11 kDalton is displayed (Pannier et al., 2000). Part of the signal decays within a few ten nanoseconds with respect to $t' = 0$ while the remaining signal exhibits a slow modulation. The fast decaying component could not be detected with any confidence by three-pulse DEER because of the dead time, see area between the dashed lines in Fig. 7(a). It corresponds to a fairly broad distribution of small distances. The slow modulation indicates a second maximum in the distance distribution at larger distances. Indeed all time-domain signals can be fitted by assuming a distance distribution consisting of two Gaussian peaks with maxima at about 1.8-2.2 nm and 4.5-7 nm. A uniformly distributed background is assumed to account for the exponential decay corresponding to the more or less homogeneous distribution of pair distances exceeding the intercluster distance, see Fig. 7(b). We attribute the former peak to the cluster size and the latter peak to the nearest neighbor intercluster distance. The former assignment has been checked by incorporating surfactant spin probes with the nitroxide moiety attached at different distances from the ionic headgroup. As expected, the apparent cluster size increases with the distance of the nitroxide from the headgroup.

The latter assignment is supported for homopolymer samples by the successful SAXS measurement of the intercluster distances. Both assignments are also in agreement with estimates of the number of polymer chains per cluster from steric arguments and they are selfconsistent with respect to the correlation between cluster size and intercluster distance at given number of ionic endgroups and polymer density. For the displayed signal we find a cluster size of 1.77 nm with a peak width of 0.87 nm and an average intercluster distance of 4.74 nm with a variation of 2.31 nm. Model calculations have been performed to estimate the precision of these values. A shift of the peak positions by half their width already leads to deviations of the fit that are clearly visible with bare eyes. We are therefore confident that the error of the average values is below 0.3 nm for the cluster sizes and below 0.5 nm for the intercluster distances. This quality of the data allows one to discuss trends on changing the polymer, kind of ionic endgroups, or molecular weight. Details will be published elsewhere (Pannier et al., 2000).

8. CONCLUSION

The DEER method is well suited for distance determination in the range between 1.5 and 8 nm. A precision of about 0.03 nm at the lower end of this range and of about 0.2 nm in the middle of this range can be expected. DEER can be realized with minor modifications of a commercial spectrometer and without modifying commercial probeheads. The modification of the spectrometer can be achieved at a small fraction of the cost of the spectrometer itself. DEER works well with nitroxide spin probes/spin labels, but can also be applied to measure distances between organic radicals and transition metal ions or between two transition metal ions. By overcoming dead-time problems, the four-pulse DEER experiment allows for characterizing broad distance distributions even at the lower end of the applicable distance range. By the same token, this experiment is capable of recording the shape of powder spectra quite faithfully. With these opportunities, DEER distance measurements have become a valuable tool for the characterization of ionomers based on both homopolymers and diblock copolymers.

9. ACKNOWLEDGEMENT

The authors wish to thank A. Godt and U. Wiesner for providing the samples discussed in this contribution. We are also grateful to A. Raitsimring, M. Bowman, P. Borbat, and J. Freed for helpful discussions. Financial support from the Deutsche Forschungsgemeinschaft and the Bundesministerium für Bildung und Forschung (BMBF) is gratefully acknowledged.

10. REFERENCES

Astashkin A.V., Hara H., and Kawamori A. (1998) The pulsed electron-electron double resonance and "2+1" electron spin echo study of the oriented oxygen-evolving and Mn-depleted preparations of photosystem II ,J. Chem. Phys. 108, 3805-3812.

Berliner L.J. (Ed.) (1976) Spin Labeling: Theory and Applications, Vol. 1, Academic Press, New York

Berliner L.J. (Ed.) (1979), Spin Labeling: Theory and Applications, Vol. 2, Academic Press, New York

Berliner L.J. and Reuben J. (Eds.) (1989) Biological Magnetic Resonance, Vol. 8, Plenum, New York

Borbat P.P. and Freed J.H. (1999) Multiple-quantum ESR and distance measurements, Chem. Phys. Lett., 313, 145-154.

Eisenberg A., Kim J.-S. (1998) Introduction to Ionomers, Wiley-Interscience, New York.

Enemark J.H., Astashkin A., Kedia R., Kozliouk V., McMaster J., Pacheco A, Raitsimring A.M., and Valek M. (1999) Pulsed electron-electron double resonance (ELDOR) studies of Mo(V)/Fe(III) centers, J. Inorg. Biochem. 74, 123-123.

Ernst R.R., Bodenhausen, G., and Wokaun A. (1987) Principles of Nuclear Magnetic Resonance in One and Two Dimensions, Clarendon, Oxford.

Hara H., Kawamori A., Astashkin A.V., and Ono T. (1996) The distances from tyrosine D to redox-active components on the donor side of Photosystem II determined by pulsed electron-electron double resonance, Biochim. Biophys. Acta Bioenergetics 1276, 140-146.

Jeschke G., Pannier M., Godt A., and Spiess H.W. (2000), Dipolar Spectroscopy and Spin Alignment in Electron Paramagnetic Resonance, Chem. Phys. Lett., *submitted.*

Klauder J.R. and Anderson P.W. (1962) Spectral Diffusion Decay in Spin Resonance Experiments, Phys. Rev. 125, 912-932.

Kurshev V.V., Raitsimring, A.M., Tsvetkov Yu.D. (1989) Selection of Dipolar Interaction by the "2+1" Pulse Train ESE, J. Magn. Reson. 81, 441-454.

Larsen R.G. and Singel, D.J. (1993) Double Electron-Electron Resonance Spin-Echo Modulation – Spectroscopic Measurement of Electron-Spin Pair Separations in Orientationally Disordered Solids, J. Chem. Phys. 98, 5134-5146.

Martin R.E., Pannier M., Diederich F., Gramlich V., Hubrich M., and Spiess H.W. (1998) Determination of end-to-end distances in a series of TEMPO diradicals of up to 2.8 nm length with a new four-pulse double electron electron resonance experiment, Angew. Chem. Int. Ed. 37, 2834-2837.

Maryasov A.G., Tsvetkov Y.D., and Raap J. (1998) Weakly coupled radical pairs in solids: ELDOR in ESE structure studies, Appl. Magn. Reson. 14, 101-113.

Milov A.D., Salikhov K.M., and Shirov, M.D. (1981) Application of ELDOR in electron-spin echo for paramagnetic center space distributions in solids, Fiz. Tverd. Tela (Leningrad) 23, 975-982.

Milov, A.D., Ponomarev A.B., and Tsvetkov Y.D. (1984) Electron electron double-resonance in electron-spin echo- model biradical systems and the sensitized photolysis of decalin, Chem. Phys. Lett. 110, 67-72.

Milov A.D.; Maryasov, A.G.; Tsvetkov, Y.D. (1998) Pulsed electron double resonance (PELDOR) and its applications in free-radicals research, Appl. Magn. Reson. 15, 107-143.

Pannier M., Veit S., Godt A., Jeschke G., Spiess H.W. (2000) Dead-time free measurement of dipole-dipole interactions between electron spins, J. Magn. Reson. 142, 331-340.

Pannier M., Schädler V., Schöps M., Wiesner U., Jeschke G., and Spiess H.W. (2000), Determination of Ion Cluster Sizes and Cluster-to-Cluster Distances in Ionomers by For-Pulse Double Electron Electron Resonance Spectroscopy, Macromolecules, *submitted.*

Pfannebecker V., Klos M., Hubrich M., Volkmer T., Heuer A, Wiesner U., and Spiess H.W., (1996) Determination of end-to-end distances in oligomers by pulsed EPR, J. Phys. Chem. 100, 13428-13432.

Saxena S. and Freed J.H. (1996) Double quantum two-dimensional Fourier transform electron spin resonance: Distance measurements, Chem. Phys. Lett. 251, 102-110.

Saxena S. and Freed J.H. (1997) Theory of double quantum two-dimensional electron spin resonance with application to distance measurements, J. Chem. Phys. 107, 1317-1340.

Schädler V., Franck A., Wiesner U., and Spiess H.W. (1997) EPR studies on telechelic polymers: Characterization of ion multiplets, Macromolecules, 30, 3832-3838.

Schädler V., Kniese V., Thurn-Albrecht T., Wiesner U., and Spiess H.W. (1998) Self-assembly of ionically end-capped diblock copolymers, Macromolecules 31, 4828-4837.

Schlick S. (Ed.) (1996) Ionomers: Characterization, Theory, and Applications, CRC Press, New York.

Zentel R., Galli G., Ober C.K. (Eds.) (1997) Manipulation of Organization in Polymers Using Tandem Molecular Interactions, Macromol. Symp. 117, 1-304.

Chapter 12

Electron Paramagnetic Resonance Distance Measurements in Photosystems

K. V. Lakshmi and Gary W. Brudvig
Department of Chemistry, Yale University, P.O. Box 208107, New Haven, CT 06520-8107

Abstract: EPR distance measurements have been extensively employed to characterize the structure and function of the photosynthetic reaction centers from plants, bacteria and algae. Such measurements have also been pivotal in establishing the feasibility of EPR distance measurements in biological systems. In this chapter, we present an overview of distance measurements in the purple non-sulfur bacterial, Photosystem II and Photosystem I reaction centers. The distance measurement techniques described involve both line shape analyses and spin-lattice relaxation measurements by cw and pulsed methods. Also described are EPR studies of oriented membranes and single crystals that facilitate the measurement of interspin distances and relative orientations of redox cofactors.

1. INTRODUCTION

Photosynthetic reaction centers (RC) are energy-transducing pigment-protein complexes present in bacteria, algae and higher plants that efficiently harness light energy to drive cell functions. In the case of plants and algae, a useful byproduct, dioxygen, is generated and released into the atmosphere. In addition to serving a vital function in nature, RCs are important structural models for membrane bound protein-cofactor complexes and electron-transfer studies. As is the case with most membrane protein complexes, structural information by x-ray and electron crystallography has been slow in coming due to the complexity of the systems and the difficulty in preparing

Biological Magnetic Resonance, Volume 19: Distance Measurements in Biological Systems by EPR, edited by Berliner et al. Kluwer Academic/Plenum Publishers, New York, 2000

diffraction-quality crystals. Application of NMR structure determination methods to RCs has been limited by the large size of the protein complexes (~ 600 kDa) and the paramagnetism of the cofactors. Fortunately, most of the electron-transfer intermediates in the reaction centers are paramagnetic species and are accessible by electron paramagnetic resonance (EPR) techniques. In particular, EPR distance measurements have been pivotal in establishing the sequence, rates and mechanism of key electron-transfer events in the RCs and are uniquely suited to the study of these systems as the larger magnetic moment of the electrons provides access to longer distances (up to about 50 Å between redox centers). The focus of this chapter is on *cw* and pulsed EPR methods to measure distances in photosynthetic reaction centers. Other EPR studies on correlated spin pairs in photosynthetic reaction centers are described in Chapter 13 by Dzuba and Hoff.

Exchange and dipolar couplings between a pair of slowly relaxing electron spins can be manifested as line shape effects in EPR spectra. Typical splitting patterns observed in powder, oriented membrane and single crystal EPR spectral line shapes can be used to determine interspin distances, relative orientations and the extent of orbital overlap between nearest neighbor cofactors in RCs. For a slow-relaxing spin interacting with a fast relaxer, dipolar and exchange couplings can be manifested as spin-relaxation enhancements of the slow-relaxing spin. Spin-relaxation measurements of isolated spin pairs can be used to measure directly long-range interspin distances in RCs. Distance determination due to line shape effects and spin-lattice relaxation effects are complementary techniques. Line shape effects are manifested in the static limit while spin-relaxation effects are observed in the fast-relaxing limit when line shape effects are averaged out. Spin-relaxation measurements have an advantage over line shape effects in that they can access longer distances and weaker couplings.

1.1 Photosynthetic Reaction Centers

Photosynthetic reactions involve an initial light-driven charge separation followed by rapid electron transfer in the RCs. Typically, activation of reaction centers begins by light capture within an antenna pigment-protein complex. Transfer of the excitation to a donor chlorophyll in the RC leads to the formation of a short-lived charge-separated state, that is stabilized by electron transfer along a chain of electron donors and acceptors. The reactions lead to longer lived charge-separated states in which a diffusible electron carrier is reduced. There exist two types of photosynthetic organisms, oxygen-evolving plants, algae and cyanobacteria, and non-oxygen-evolving green or purple photosynthetic bacteria. Anoxygenic purple non-sulfur photosynthetic bacteria contain only one photosystem, the bacterial reaction center (bRC), and undergo cyclic electron transfer, whereas plants and cyanobacteria contain two photosystems, photosystems II and I (PS II and PS I), that participate in unidirectional electron transfer. The electron-transfer chain in purple photosynthetic bacteria generates a transmembrane pH gradient that is used to drive cell functions while that in oxygenic organisms couples the breakdown of water to dioxygen in PS II with the reduction of $NADP^+$ to NADPH via PS I yielding reducing equivalents for use in the carbon-fixation reactions.

2.3-2.65 Å resolution x-ray crystal structures of bRCs (Allen *et al.*, 1987, Deisenhofer *et al.*, 1984, Deisenhofer *et al.*, 1985, Ermler *et al.*, 1994, Lancaster & Michel, 1999) are available and have provided a structural basis for analysis of the primary electron-transfer processes of photosynthesis. More recently a high-resolution x-ray crystal structure of PS I has been determined (Krauß *et al.*, 1996), but the structure of PS II is known only at an 8 Å resolution (Rhee *et al.*, 1998). A variety of structural and mechanistic aspects of RC function remain unclear, even in the bRC. For example, the cofactors in bRCs are arranged along two branches A and B which are related by a pseudo-C_2-axis of symmetry relative to the non-heme Fe(II) (Figure 1). The rate of transfer is known to be at least 20 times larger along the Q_A branch although the reason for this preference is not well understood. Electron-transfer rates in photosynthesis depend on the distance and orientation between the cofactors and on the electron densities of the cofactors and intervening media. In addition to the spatial arrangement, knowledge of dynamic electronic structures and overlap of electron orbitals is necessary to elucidate the mechanistic details of the electron-transfer events.

Fortunately, photosynthetic RCs contain natural paramagnetic spin probes that are a part of the electron-transfer chain making EPR spectroscopy a powerful tool to probe structure-function relationships in the RCs. Magnetic interactions between pairs of spin probes in charge-separated states of the photosynthetic reaction yield critical structural data such as the distance and relative orientation of the redox cofactors. Magnetic interactions also aid characterization of integer-spin species, such as the non-heme Fe(II), that cannot be directly observed using EPR spectroscopy. In this section, we review the properties of the bRC, PS II and PS I complexes and present an overview of EPR distance measurements in photosynthetic reaction centers.

1.1.1 Bacterial Reaction Centers

bRCs were first isolated from *Rb. sphaeroides* (Feher *et al.*, 1989) and are composed of three subunits (L, M and H), nine cofactors that include four bacteriochlorophylls (a special pair, P_{870}, and two auxiliary 'voyeur' chlorophylls), two bacteriopheophytins ($bPheo_A$, $bPheo_B$), two quinones (Q_A, Q_B) and a non-heme Fe(II). The bRC core consists of a heterodimer of L and M subunits that are related by a pseudo-C_2 symmetry axis along the non-heme Fe(II)-P_{870} vector (Figure 1). The absorption of a photon excites the primary electron donor (P_{870}) from its ground state, P_{870}, to its excited state P_{870}^*. This excitation is followed by rapid transfer of the electron from P_{870}^* to $bPheo_A$ generating a $P_{870}^+bPheo_A^-$ charge-separated state with a quantum yield that is close to unity. The $P_{870}^+bPheo_A^-$ charge-separated state is further stabilized by electron transfer to the acceptors Q_A and Q_B and the missing electron on P_{870}^+ is replaced by a secondary donor (a cytochrome). The electron is ultimately transferred to a pool of diffusible quinones that carry the reducing equivalents on to the cytochrome bc_1 complex.

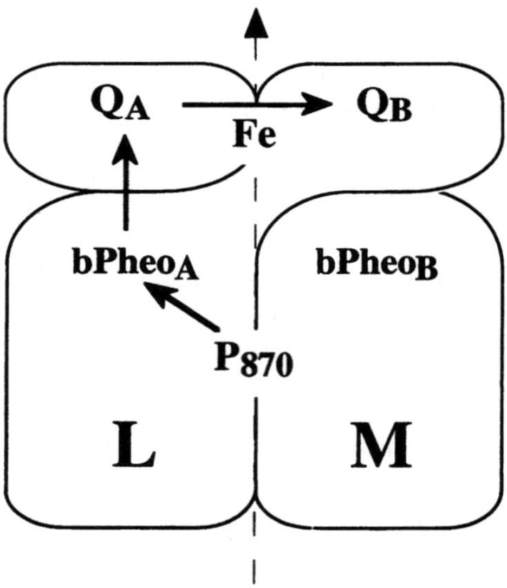

Figure 1. Schematic depicting the L and M subunits and cofactors in the bacterial reaction center core from *Rb. sphaeroides*. The solid arrows indicate the sequence of electron transfer through the protein complex. Also shown by a dashed arrow is the pseudo two-fold axis of symmetry that bisects the non-heme Fe(II) and the bacteriochlorophyll special pair, P_{870}.

The two terminal electron acceptors, Q_A and Q_B, in the bRC are known to be associated with the non-heme Fe(II) although the precise role of the Fe(II) center is not clear. It was initially thought that the Fe(II) may directly be involved in the electron-transfer chain (the 'iron-wire' hypothesis). However, it has been shown that Fe(II) can be replaced with other divalent metals with the same electrostatic charge and ligand-binding properties but with vastly different redox properties indicating that the Fe(II) may stabilize the structural features necessary in the quinone electron-transfer chemistry (Feher *et al.*, 1974). Removal of the non-heme Fe(II) or replacement with other divalent diamagnetic cations leads to very different EPR signals for the quinones, Q_A and Q_B, indicating the presence of strong magnetic interactions between the iron-quinone pairs. Powder and oriented line shape simulations of the experimental EPR signals of the Fe(II)-Q_A^- complex have established the non-heme Fe(II)-Q_AQ_B distances and the orientation of the Fe(II) relative to the membrane plane (Butler *et al.*, 1984, Dismukes *et al.*, 1984).

Historically, some of the earliest EPR applications were conducted on paramagnetic charge-separated states in the bRCs, thus establishing the feasibility of EPR techniques to study biological macromolecules. This is in part due to the simplicity of the system and due to the relative ease of detecting light-induced radicals. Later work benefited from the development of methods for isolation and purification of the bRCs. In addition to providing information on the bRC itself, EPR studies of the bRC have also been used to calibrate distance measurements in PS I and PS II and have served as a means to assess the accuracy of distance estimates obtained. For example, the P_{870}^+-Fe(II) interaction in the bRC from *Rb. sphaeroides* has served as a model system to determine accurate parameters for saturation-recovery and progressive microwave power-saturation measurements in PS II.

1.1.2 Photosystem II

Photosystem II (PS II) contains the catalytic water-oxidation site which is the source of electrons in the electron-transfer chain in oxygenic systems. The reaction center core of PS II is comprised of a heterodimer of polypeptides (D1 and D2) that are structurally homologous to the L and M subunits in the bRC. The D1/D2 complex contains a collection of redox cofactors that are also homologous to those in the bRC and are arranged with a pseudo-C_2 symmetry. These include a chlorophyll electron donor (P_{680}), two pheophytins ($Pheo_A$ and $Pheo_B$), two quinones (Q_A and Q_B) and a non-heme Fe(II). However, PS II is more complex than the bRC and contains a number of additional polypeptides and a tetranuclear manganese (Mn_4) cluster that is the active site for water oxidation (Figure 2).

The initial absorption of a photon of light by the antenna complexes in PS II results in the formation of an excited singlet state of a specialized chlorophyll *a* (P_{680}). The excited P_{680}^* state rapidly transfers an electron to a pheophytin electron acceptor forming the charge-separated $P_{680}^+Pheo_A^-$ state. The oxidation of $Pheo_A^-$ by Q_A and the reduction of P_{680}^+ by a redox-active tyrosine called Y_Z stabilizes the charge-separated state by spatially delocalizing the electronic charges. Q_B is a diffusible secondary plastoquinone that oxidizes Q_A^- and delivers the electron to the cytochrome b_6f complex. The electrons are then passed on to PS I and ultimately used for $NADP^+$ reduction via a *ferredoxin:NADP⁺ oxidoreductase*.

The tetranuclear manganese cluster (Mn_4) in the active water-splitting site in PS II re-reduces $Y_Z^\bullet$ and prepares the protein for the next photochemical turnover. As shown in Figure 3, the Mn_4 oxidation states are described by the Kok S-state (or charge-storage state) cycle (Kok *et al.*, 1970) and a molecule of dioxygen is released during the fourth turnover via the S_4 state. The x-ray structures of the bRC have provided a basic working model for the PS II RCs in plants and cyanobacteria. The primary sequences of the D1 and D2 polypeptides are homologous and exhibit 5% sequence identity to the L and M subunits of the bRCs. However, PS II contains a water-splitting active site that is not present in the bRC. A recent 8 Å resolution electron crystallography structure of manganese-depleted plant photosystem II by Rhee *et al.* (1998) reveals the positions of the transmembrane helices in the D1, D2 and CP47 polypeptides but lacks information regarding the Mn_4 cluster and the position and orientation of the redox-active cofactors in the catalytic water-oxidation site. On the basis of manganese chemistry and structure prediction, several models have been proposed for oxygen evolution in the active water-splitting site of PS II (Limburg *et al.*, 1999, Yachandra *et al.*, 1996).

Y_Z is the Tyr-161 residue of the D1 polypeptide in PS II (cyanobacterial numbering) and is known to be the redox link between the Mn_4 cluster and P_{680}. The stabilization of charge-separated species in the electron transfer chain is efficiently achieved by the reduction of P_{680}^+ by Y_Z. In addition to its role as a redox cofactor, Y_Z is also thought to be involved in proton-coupled electron transfer with the Mn_4 cluster (Force *et al.*, 1997, Gilchrist *et al.*, 1995, Hoganson & Babcock, 1997, Limburg *et al.*, 1999, Peloquin *et al.*, 1998, Szalai *et al.*, 1998a, Szalai *et al.*, 1998b). Determination of the distance and orientation between the Mn_4 cluster and Y_Z is crucial to elucidating the mechanism of proton- and electron-transfer events between

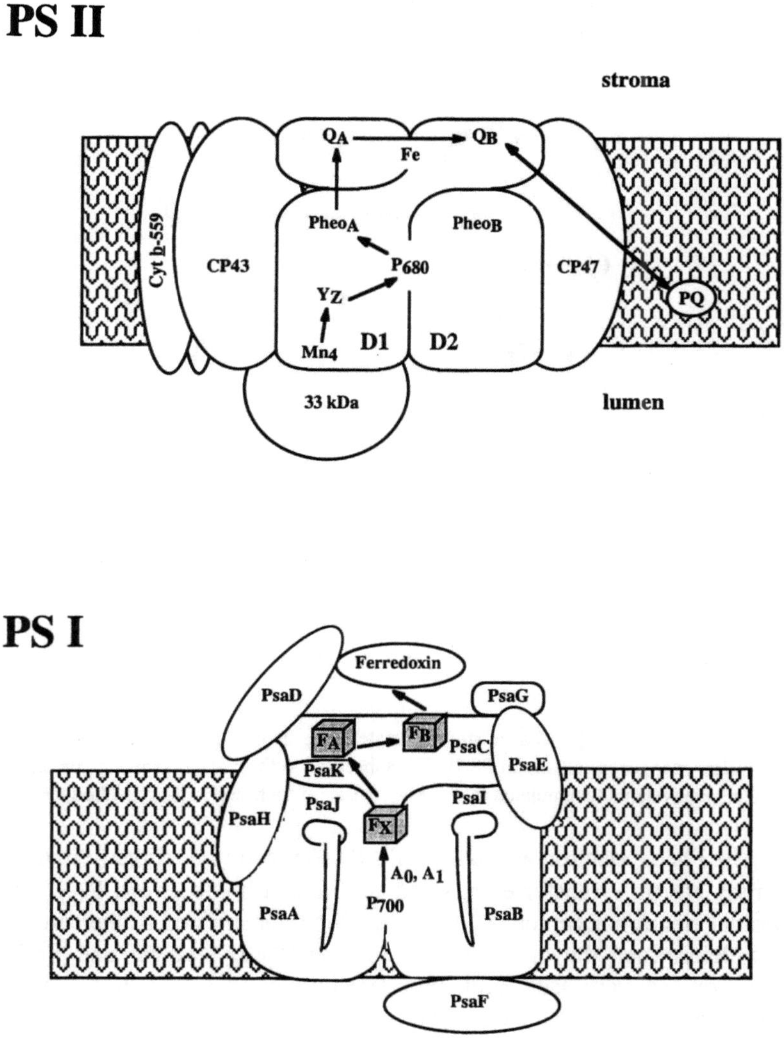

Figure 2. Schematic depicting key subunits and cofactors in PS II and PS I. The arrows indicate the sequence of electron transfer. D1 and D2 are the reaction center core subunits of PS II, CP43/CP47 are tightly associated light-harvesting proteins, the extrinsic 33 kDa protein is the Mn-stabilizing protein, and Cyt b-559 is a heme protein that participates in a secondary path of electron transfer. For PS I, PsaA-PsaK denote polypeptide subunits. The PsaA/PsaB subunits form the reaction center core, and PsaC binds the [4Fe4S] clusters, F_A and F_B.

the two cofactors. Line shape analyses of powder and oriented PS II membranes have successfully yielded the distance and relative orientation of $Y_Z^{\bullet}$ and the Mn_4 cluster in the active water-splitting site (Dorlet *et al.*, 1998, Lakshmi *et al.*, 1999a, Lakshmi *et al.*, 1998, Peloquin *et al.*, 1998).

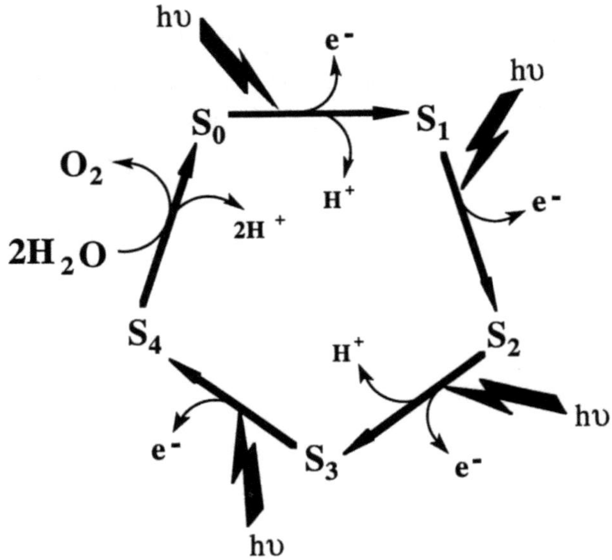

Figure 3. The Kok S- (or charge-storage) state cycle depicting the cyclic oxidation of the tetranuclear manganese cluster in PS II. The subscripts to the S states represent the number of oxidizing equivalents accumulated by the Mn_4 cluster at each step in the oxidation cycle.

The $Pheo_A$ electron acceptor in PS II is known to be directly involved in primary electron transfer events and serves as a link between P_{680}^+ and the terminal electron acceptors Q_A and Q_B. Deligiannakis and Rutherford (1996) have determined the distance between the $Pheo_A^-$ radical and Fe(II) by saturation-recovery EPR and have unequivocally established the spatial arrangement of $Pheo_A$ in the protein complex.

Y_D is a redox-active residue located in the D2 polypeptide of PS II in a symmetry-related position as Y_Z (relative to the pseudo-C_2-axis of symmetry through the non-heme Fe(II)-P_{680} vector). Y_D does not have a well-established function in PS II, but over the years, there have been proposals of indirect and direct involvement of Y_D in the electron-transfer events in PS II. The oxidation of the S_0 to S_1 state of the Mn_4 cluster by $Y_D^{\bullet}$ has been thought to stabilize the manganese complex during prolonged periods of darkness (Debus, 1992). It was initially believed that Y_D and Y_Z are symmetrically placed in the PS II reaction center because Y_D and Y_Z are located in homologous positions in the D2 and D1 polypeptides, respectively, and also form neutral radicals with hydrogen bonds to homologous histidine residues in the D1 and D2 polypeptides. Probing the depth of Y_D within the thylakoid membranes by progressive microwave power saturation measurements and estimating the Y_D-Fe(II) distance by saturation-recovery EPR studies have established that Y_D and Y_Z are symmetrically placed with respect to the non-heme Fe(II) (Hirsh *et al.*, 1992a, Innes & Brudvig, 1989, Koulougliotis *et al.*, 1995) in PS II. Analysis of the magnetic interaction of $Y_D^{\bullet}$ and the Mn_4 cluster in the S_2 state of the tetranuclear manganese cluster has revealed that Y_D is distant from the Mn_4 cluster (Hara *et al.*, 1996, Koulougliotis *et al.*, 1997). Based on these results, Y_D is considered unlikely to be directly involved in water-splitting chemistry.

Chl_Z in PS II has been proposed to be involved in the photooxidation of Cyt b_{559} in an alternate electron-transfer pathway to reduce P_{680}^{+}. Chl_Z photooxidation is thought to be involved in protection of PS II against photoinactivation (Stewart & Brudvig, 1998). Progressive microwave power-saturation and direct saturation-recovery measurements have been used to determine the depth of Chl_Z within PS II membranes and the Chl_Z-Fe(II) distance in PS II (Koulougliotis *et al.*, 1994), respectively. The Chl_Z-Fe(II) distance measurements identified the location of possible histidine ligands to Chl_Z in the D1 and D2 polypeptides in PS II which led to site-directed mutagenesis experiments that confirmed D1-His118 as the ligand to Chl_Z (Stewart *et al.*, 1998).

Application of the '2+1' pulse sequence to distance measurements in PS II are described in Chapter 10 by Raitsimring. The use of out-of-phase echo to obtain distances in photosystems is described in Chapter 13 by Dzuba and Hoff.

1.1.3 Photosystem I

PS I is a light-driven *plastocyanin:ferredoxin oxidoreductase* located in the thylakoid membranes and is the site for the second photochemical electron-transfer reaction of plant photosynthesis (Golbeck & Bryant, 1991). PS I contains three [4Fe-4S] clusters ($F_{X/A/B}$), a photochemically active chlorophyll dimer (P_{700}), a primary monomeric chlorophyll electron acceptor (A_0) and a secondary phylloquinone electron acceptor (A_1) that participate in the electron-transfer reactions (Figure 2). The absorption of a photon by the antenna chlorophyll initiates a chain of photochemical events starting with the formation of the excited state, P_{700}^{*}, of the P_{700} chlorophyll dimer that results in a charge separation between P_{700} and a series of acceptor molecules, A_0, A_1 and F_X.

A recently published 4.0 Å resolution x-ray crystallography structure by Krauß *et al.* (Fromme *et al.*, 1996, Krauß *et al.*, 1996) unambiguously assigns the location of the P_{700} chlorophylls, the accessory chlorophylls, the electron acceptor (A_0) and the iron-sulfur cluster F_X in PS I. Although it is possible to determine the separation between the terminal iron-sulfur electron-acceptors F_A and F_B (12 Å) at this resolution, the two centers are indistinguishable in this structure due to a pseudo two-fold symmetry in the PsaC protein to which they are bound. An unambiguous determination of the spatial arrangement of the iron-sulfur clusters F_X, F_A and F_B within the protein is key to understanding the chain of terminal electron-transport events in PS I (Golbeck & Bryant, 1991). Saturation-recovery EPR techniques have been used to determine the topography of the P_{700}, A_1 and F_X cofactors in PS I (Berry *et al.*, 1997). Dipolar spin-spin interactions in the EPR spectra of oriented PS I membranes (Guigliarelli *et al.*, 1993) and relative orientations of the **g** tensors of the iron-sulfur clusters in PS I single crystals (Kamlowski *et al.*, 1997a, Kamlowski *et al.*, 1997b) have been examined to determine the orientation of the F_A and F_B clusters within the protein complex. More recently, a qualitative study using progressive microwave power-saturation techniques has been used to unambiguously assign the relative positions of the F_A and F_B clusters in PS I (Lakshmi *et al.*, 1999b).

2. EPR DISTANCE MEASUREMENTS

2.1 Theory

2.1.1 Exchange and Dipolar Couplings

As reviewed in Chapter 1, spin-spin interactions are comprised of isotropic and anisotropic exchange and dipolar couplings. The interactions encountered can be in a strong, weak or moderate coupling regime. The experimental techniques and theoretical treatments to determine spin-spin couplings in each of the regimes are vastly different. In this chapter, discussion of exchange and dipolar couplings in photosynthetic reaction centers is confined to weakly coupled systems.

In a weak-coupling limit, there are two types of interactions possible between a pair of spins separated by a distance r, namely a dipolar interaction (D) and an isotropic exchange interaction (J). The total spin Hamiltonian for the pair of interacting spins is represented as,

$$\mathcal{H} = \mathcal{H}_{Z,1} + \mathcal{H}_{A,1} + \mathcal{H}_{Z,2} + \mathcal{H}_{A,2} + \mathcal{H}_{J} + \mathcal{H}_{D} \qquad (1)$$

where $\mathcal{H}$ is the total Hamiltonian and $\mathcal{H}_{Z,1}$, $\mathcal{H}_{Z,2}$, $\mathcal{H}_{A,1}$, $\mathcal{H}_{A,2}$, $\mathcal{H}_{J}$ and $\mathcal{H}_{D}$ are the Zeeman (Z) and hyperfine (A) terms for electron 1 and electron 2 and, the exchange and dipolar terms, respectively. The dipolar interaction (D) is a through-space anisotropic interaction that depends on the interspin point-dipole distance (r) and the relative orientation of the interspin vector in the applied magnetic field (θ). The exchange interaction (J) is an electrostatic repulsion between the spins arising from a weak bonding interaction between the pair of interacting spins (see Chapter 1 for a detailed discussion).

2.1.2 Line Shape Effects

In the static limit, exchange and dipolar couplings between a pair of interacting spins can be manifested as line shape changes in the EPR signals of the two spins. As reviewed in Chapter 1, well-defined splittings are observed in the spectrum when $\mathcal{H}_{D} >> \mathcal{H}_{J}$ and the dipolar and exchange couplings ($D \pm J$) are much larger than the inhomogeneous line widths of EPR resonances of the non-interacting spins. Exchange and dipolar couplings depend on the interspin distance and numerical simulations of line shapes of the interacting species in powder samples can provide a precise determination of the dipolar and exchange couplings between the two interacting spins. The Hamiltonian represented by eqn 1 can either be solved by numerical diagonalization or by perturbation theory (Smith & Pilbrow, 1974). The nature of the individual spin systems and their interaction often determines the choice of the numerical technique used to solve the total spin Hamiltonian. Single-crystal EPR analysis or that in oriented membrane samples can also yield the orientation of the anisotropic contributions (**g**, **A** anisotropies and dipolar interactions) to the spin Hamiltonian. Membrane-bound proteins that are oriented upon planar quartz or other substrates have been used for a variety of oriented EPR measurements. As will be discussed in section 3.1.1, the extent of disorder (or the mosaic spread) has to be suitably accounted for in a single-crystal or oriented-membrane study.

2.1.3 Spin-Relaxation Effects

In a coupled spin system, the exchange and dipolar terms in the interaction Hamiltonian can be modulated by dynamic processes arising from spin relaxation. If the spin-lattice relaxation rate of one of the spins is rapid in comparison to the spin-spin coupling between the two spins, an averaging of the spin-spin interactions due to rapid relaxation results in the collapse of the splitting patterns observed in the *cw* EPR spectra ((Fielding *et al.*, 1986, Szalai *et al.*, 1998a) and Chapter 1). In the rapid-relaxation limit when line shape effects are averaged, spin-spin interactions can be manifested as spin-relaxation effects on the slow-relaxing spin in the presence of the fast-relaxing spin.

While non-interacting isolated radicals exhibit single-exponential spin-lattice relaxation transients, non-single-exponential spin-lattice relaxation transients are observed for a pair of interacting spins that are randomly oriented in a rigid lattice (Hirsh *et al.*, 1992a). A 'dipolar model' has been used to treat such dipole-dipole interactions of a pair of spins in a rigid lattice, at a fixed distance and random orientation relative to the applied magnetic field (Figure 4) (Hirsh *et al.*, 1992a, Hirsh & Brudvig, 1993). Spin-relaxation rate constants determined from the dipolar model have been used to probe the distance, orientation and magnetic properties between pairs of interacting spins in RCs (Berry *et al.*, 1997, Deligiannakis & Rutherford, 1996, Hirsh *et al.*, 1992a, Hirsh & Brudvig, 1993, Koulougliotis *et al.*, 1994, Koulougliotis *et al.*, 1997, Koulougliotis *et al.*, 1995).

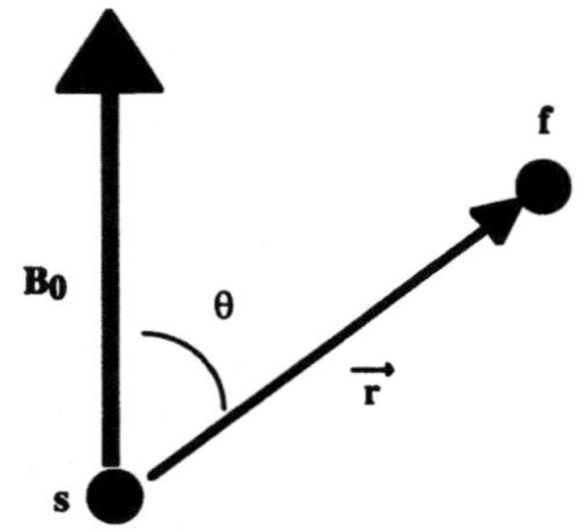

Figure 4. Schematic representation of the pairwise interaction between a slow-relaxing (s) and fast-relaxing (f) spin separated by a distance r and at a fixed angle θ relative to the applied magnetic field, B_0.

In the presence of an applied magnetic field, B_0, the field experienced by a slow-relaxing species s is a combination of the applied field and a local fluctuation caused by spin flips of the fast-relaxing species f. If species f has a much more rapid spin-relaxation rate, $1/T_{1f}$, this will enhance the spin-lattice relaxation rate of the species s. The observed spin-lattice relaxation rate for the species s ($1/T_{1s,observed}$) is then the sum of the intrinsic ($1/T_{1s,intrinsic}$) and dipolar enhanced ($1/T_{1s,dipolar}$) contributions to the relaxation rates for the species s and the spin-lattice and spin-spin relaxation rates are represented as k_{1i} or k_{2i} and $k_{1\theta}$ and $k_{2\theta}$, respectively, where the subscripts i and θ refer to the intrinsic and dipolar-enhanced relaxation rates. Based on the theory of dipolar enhanced spin-lattice relaxation by Bloembergen (1949) and Abragam (1961), the dipolar-enhanced spin-relaxation rates of a slow-relaxing free radical spin in the presence of fast-relaxing paramagnetic centers are given by (Kulikov & Likhtenshtein, 1977),

$$k_{1\theta} = \left(\frac{1}{T_{1\theta}}\right) = \frac{\gamma_s^2 \mu_f^2}{r^6}\left\{\frac{1}{6}B + 3C + \frac{3}{2}E\right\} \tag{2}$$

$$k_{2\theta} = \left(\frac{1}{T_{2\theta}}\right) = \frac{\gamma_s^2 \mu_f^2}{r^6}\left\{\frac{1}{3}A + \frac{1}{12}B + \frac{3}{2}C + 3D + \frac{3}{4}E\right\} \tag{3}$$

where γ_s is the gyromagnetic ratio for the slow spin, μ_f is the magnetic dipole moment for the fast-relaxing spin, r is the interspin distance and θ is the angle between the interspin vector and the applied magnetic field. The terms A-E are as defined by the dipolar alphabet (see Chapter 8). The observed rates of spin-lattice and spin-spin relaxation are described as,

$$k_{1obs}(\theta) = k_{1i} + k_{1\theta} \tag{4}$$

$$k_{2obs}(\theta) = k_{2i} + k_{2\theta} \tag{5}$$

A knowledge of the observed and intrinsic relaxation rates can lead to an estimation of the dipolar enhancement in relaxation and accurate distance measurements can be arrived upon from eqns 2 and 3. However, an exchange interaction between the two spins may produce an additional contribution to the relaxation rates if the interspin distance is less than ~ 10 Å, and the appropriate equations are given in Rakowsky *et al.* (1995) and in Chapter 8.

Methods available to determine spin-lattice relaxation enhancements include indirect *cw* progressive microwave power-saturation and direct pulsed saturation-recovery techniques (see Chapter 2). Spin-lattice relaxation parameters have also been estimated from line shapes of EPR resonances (Un *et al.*, 1994) at high microwave frequencies.

3. DISTANCE MEASUREMENTS IN PHOTOSYNTHETIC REACTION CENTERS

3.1 Line Shape Analyses in the Static Limit

3.1.1 Spectral Simulations of the $S_2Y_Z^{\bullet}$ Interaction in Powder and Oriented PS II Membranes

In this section, we consider an example of the numerical simulations of experimental $S_2Y_Z^{\bullet}$ spectra from powder and oriented acetate-treated PS II membranes. These studies have provided information on the distance between the Mn_4 cluster and Y_Z, and also the orientation of these centers within the membrane. Inhibitory treatments on the donor side of PS II, followed by illumination and rapid freezing, are known to trap the protein in the $S_2Y_Z^{\bullet}$ state (Boussac & Rutherford, 1988, Gilchrist *et al.*, 1995, McLachlan & Nugent, 1993, Szalai & Brudvig, 1996, Szalai *et al.*, 1998a, Tang *et al.*, 1996). The broad 'split signal' EPR spectrum from the $S_2Y_Z^{\bullet}$ state in acetate-treated PS II, containing both the split $Y_Z^{\bullet}$ EPR signal and the S_2 multiline signal from the Mn_4 cluster, has enabled the analysis of the magnetic interactions giving rise to this spectrum (Dorlet *et al.*, 1998, Lakshmi *et al.*, 1998, Peloquin *et al.*, 1998). The splittings in the $S_2Y_Z^{\bullet}$ interaction spectra depend on J and r, and the relative signal intensities depend on the energy differences between the transitions of the interacting S_2 and $Y_Z^{\bullet}$ species. Hence, simulation of the $S_2Y_Z^{\bullet}$ EPR signals requires that the extensive ^{55}Mn (I = 5/2) hyperfine interactions of the Mn_4 cluster in the S_2 state (S = 1/2) be accurately reproduced prior to modeling the interaction spectra.

The mixed-valence Mn_4 cluster in the S_2 state consists of an S = 1/2 electron spin interacting with four ^{55}Mn I = 5/2 nuclear spins. In light of the large number of hyperfine interactions present within the metal cluster, the S_2 multiline spectrum has been simulated by using a perturbation theory approach (Bonvoisin *et al.*, 1992, Lakshmi *et al.*, 1998, Peloquin *et al.*, 1998, Zheng & Dismukes, 1996). The simulation approaches used by Bonvoisin *et al.* (1992) and Lakshmi *et al.* (1998) to model the Mn_4 S_2 multiline spectrum use effective hyperfine couplings for the four Mn nuclei. A different approach has been adopted by Dismukes and coworkers (Zheng & Dismukes, 1996) where a dimer-of-dimers and a distorted trigonal geometry have been used to model the Heisenberg exchange couplings within the Mn_4 cluster. As previously mentioned by Zheng and Dismukes (1996) and Lakshmi *et al.* (1998), due to the small number of resolved features in the S_2 multiline spectrum and the large number of adjustable parameters, it is difficult to assign unique exchange-coupling models and fixed geometries to the Mn_4 cluster. However, the hyperfine couplings constants obtained for the Mn_4 cluster in the S_2 state in all three studies are within 15-20% of each other and compare well with those measured by ENDOR spectroscopy (Peloquin *et al.*, 1998).

The parameters obtained in the S_2 multiline simulations (Lakshmi *et al.*, 1998) have been used to model the interaction spectrum between the S_2 state of the Mn_4 cluster (S = 1/2) and the $Y_Z^•$ radical (S = 1/2) in the $S_2Y_Z^•$ state of acetate-inhibited PS II. The spin Hamiltonian used is represented by,

$$\mathcal{H} = \mathcal{H}_{Z,Y_Z} + \mathcal{H}_{Z,Mn_4} + \sum_{i=1}^{4} \mathcal{H}_{A,Mn_4} + \mathcal{H}_D + \mathcal{H}_J \tag{6}$$

where $\mathcal{H}_Z >> \mathcal{H}_A \sim \mathcal{H}_J \sim \mathcal{H}_D$. The spin Hamiltonian includes Zeeman terms for the S = 1/2 Mn_4 cluster ($\mathcal{H}_{Z,Mn_4}$) and the tyrosyl radical ($\mathcal{H}_{Z,Y_Z}$), hyperfine coupling to the Mn_4 cluster ($\mathcal{H}_{A,Mn_4}$), and dipolar ($\mathcal{H}_D$) and isotropic exchange ($\mathcal{H}_J$) interactions between the S_2 state and $Y_Z^•$. The Zeeman interaction was taken to be the zero-order interaction in a perturbation theory calculation and the hyperfine coupling terms were summed over all four equivalents of ^{55}Mn. The spin-spin interaction between the S_2 state and $Y_Z^•$ is weak in comparison to the couplings within the Mn_4 cluster in the S_2 state and was viewed as a second-order perturbation. It is significant to note that only for acetate-inhibited PS II are the spectral features from the S_2 state well-resolved in the $S_2Y_Z^•$ spectrum. Analysis of signals of both spins contributing to the interaction have imposed greater constraints on the line shape simulations.

Shown in Figure 5 are experimental and simulated $S_2Y_Z^•$ spectra at X- and Q-band EPR frequencies, respectively. The X-band EPR spectrum (Figure 5a) consists of split features in the central g = 2 region and flanking hyperfine peaks arising from the interacting S_2 state that are spread over ~ 2000 G. The peak-to-peak splitting in the central features is ~ 240 G. Previous studies have established that the interacting signal is in the static limit at 7 K and there is no line shape averaging observed in the $S_2Y_Z^•$ line shape at temperatures ≤ 20 K (Szalai *et al.*, 1998a). Simulations have indicated that both the peak positions as well as the relative intensities are sensitive to the magnitude of the dipolar and exchange couplings. The experimental and simulated powder spectra at both X- and Q-band EPR frequencies have been reproduced by an interspin distance of 7.7 ± 0.5 Å and an exchange coupling of -280 x 10^{-4} cm^{-1}.

The $S_2Y_Z^{\bullet}$ 'split' signal in acetate-treated PS II membranes has also been modeled by Britt and coworkers (Peloquin *et al.*, 1998) and McCracken and coworkers (Dorlet *et al.*, 1998). The exchange and dipolar couplings obtained by line shape simulations of EPR and pulsed ENDOR spectra by Britt and coworkers were -850 MHz and 150 MHz (r = 8.1 Å), respectively (Peloquin *et al.*, 1998). An interspin distance of 8.1 Å obtained in this study is significantly longer (but closer to other current estimates) than a previous distance estimate of 4.5 Å obtained by treating the split signal as a purely dipolar interaction arising from a tyrosyl radical and an S = 1 Mn_4 cluster in the $S_2Y_Z^{\bullet}$ state in Ca^{2+}-depleted PS II membranes (Gilchrist *et al.*, 1995).

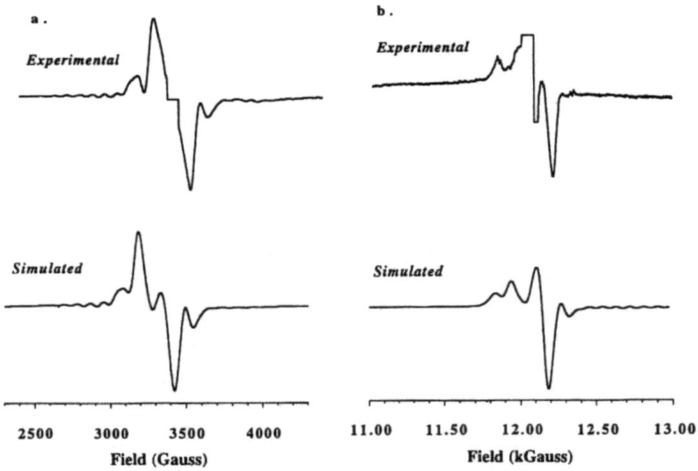

Figure 5. Experimental and simulated $S_2Y_Z^{\bullet}$ signals in acetate-treated PS II at (a) X-band, and (b) Q-band EPR frequencies. (Modified from (Lakshmi *et al.*, 1998).)

The calculations by Dorlet *et al.* (1998) have also assumed the presence of both dipolar and exchange couplings between the S_2 and $Y_Z^{\bullet}$ spins. The total spin Hamiltonian in this study was solved by a diagonalization approach and the exchange and dipolar couplings were found to correspond to an interspin distance of 8.2 Å which is in good agreement with estimates obtained by other groups.

The estimated magnitude of the exchange coupling between $Y_Z^{\bullet}$ and the S_2 state of the Mn_4 cluster obtained in the $S_2Y_Z^{\bullet}$ studies described is in the range of 270-300 x 10^{-4} cm^{-1} which is lower than the J value observed in some systems at this interspin distance (Coffman & Buettner, 1979). The $S_2Y_Z^{\bullet}$ samples in these investigations were trapped by inhibiting the donor-side of PS II by acetate binding. It is thought that acetate inhibits the donor side of PS II by disrupting the hydrogen bond between $Y_Z^{\bullet}$ and a ligand on the Mn_4 cluster in the S_2 state as suggested by an extremely short distance between the acetate methyl group and the Mn_4 cluster found by pulsed ENDOR spectroscopy (Force *et al.*, 1997). This could alter orbital overlap between the two interacting spins decreasing the magnitude of exchange coupling observed.

The effects of under or over estimating the dipolar and exchange couplings have been examined (Dorlet *et al.*, 1998, Lakshmi *et al.*, 1998). It has been observed that a spectral simulation with a predominant exchange coupling is closer to the experimental spectrum than a purely dipolar interaction but does not entirely reproduce the peak positions and relative intensities of the experimental $S_2Y_Z^{\bullet}$ spectrum suggesting the presence of both dipolar and exchange components (Dorlet *et al.*, 1998, Lakshmi *et al.*, 1998).

In the case of the $S_2Y_Z^{\bullet}$ interaction, the dominant contribution to the spin-spin interaction is the exchange coupling which makes evaluation of the dipolar coupling difficult. The orientation dependence of the dipolar interaction in the external magnetic field can be used to distinguish the dipolar contribution from the isotropic exchange contribution thus yielding both the magnitude and the orientation of the dipolar vector relative to the membrane geometry. The oriented sample EPR technique has been used to determine the relative orientation of the Mn_4 cluster and $Y_Z^{\bullet}$ in the water-splitting site in acetate-treated PS II membranes (Lakshmi *et al.*, 1999a).

The treatment of mosaic spread is important in the analysis of oriented membranes. For example, it can be assumed that the Z1 axis for the Mn_4 cluster has a well-defined orientation relative to the normal to the membrane (Figure 6). However, even in relatively well-oriented samples, there is variation (mosaic spread) in the orientation of the membrane normal relative to the normal to the substrate. The analysis of the mosaic spread in the 1-D ordering of the membranes used in the oriented $S_2Y_Z^{\bullet}$ study (Lakshmi *et al.*, 1999a) is analogous to that used by Schiller *et al.* (1998) to analyze EXAFS data for PS II samples oriented on Mylar films. As noted by Schiller *et al.* (1998), a distribution in α (the angle between the manganese Z1 axis and the substrate normal) is not equivalent to a distribution in θ_D (the angle between the interspin vector and the substrate normal) because of the nonlinear relationship between α and θ_D shown in eqn 7. This is a more rigorous way to analyze the orientation dependence of EPR spectra than previously applied treatments of the disorder as a simple Gaussian distribution in orientations of the magnetic field with respect to the axes of the paramagnetic center or the interspin vector.

$$\cos\theta_D = -\sin\alpha\sin\varepsilon\sin\eta + \cos\alpha\cos\varepsilon \qquad (7)$$

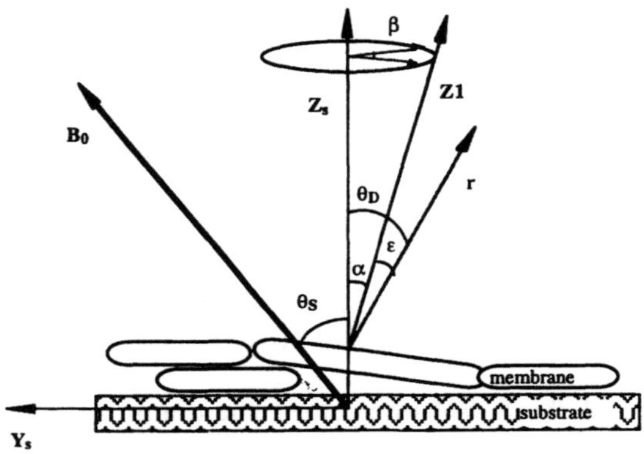

Figure 6. Schematic representing the axes definitions used in the simulations to model the orientation dependence of the $S_2Y_Z^{\bullet}$ spin-spin interaction for acetate-treated PS II membranes oriented on Mylar films. The angle θ_S describes the orientation of the substrate normal, Z_S, in the magnetic field, B_0; Euler angles α and β describe the orientation of the Mn_4 Z-axis (Z1) relative to the substrate normal, Z_S; ε and η define the spatial relationships within the spin-coupled pair and the Mn_4 axes. (Modified from (Lakshmi *et al.*, 1999a).)

Shown in Figure 7a are experimental and simulated oriented $S_2Y_Z^{\bullet}$ spectra at X-band. Similar to the powder spectra shown in Figure 5a, oriented $S_2Y_Z^{\bullet}$ spectra contain both the split features in the g ~ 2.0 region and the Mn_4 multiline features spanning ~ 2000 G. Both the relative intensities as well as the peak positions of the $Y_Z^{\bullet}$ and S_2 multiline features are sensitive to the orientation of the membrane normal relative to the applied magnetic field. As can be seen in Figure 7a-c, good agreement between the experimental and simulated peak positions and relative intensities was obtained with an interspin distance of 7.9 ± 0.2 Å, and θ_D of 75°.

3.1.2 Angular Dependence of the Mn_4-$Y_D^{\bullet}$ Interaction in PS II Membranes

The angular dependence of spin pairs has also been estimated by line shape analysis of high-field (245 GHz) EPR spectra of PS II (Un *et al.*, 1994). Observing the spectrum at high magnetic fields affords the possibility of orientation selection in the $Y_D^{\bullet}$ spectrum. At lower magnetic fields, the $Y_D^{\bullet}$ EPR line shape is dominated by hyperfine features whereas at higher fields, the **g** anisotropy is more pronounced and the orientation dependence of the anisotropic components has been used to determine the orientation of the $Y_D^{\bullet}$ **g** tensor within the PS II membrane (Un *et al.*, 1994).

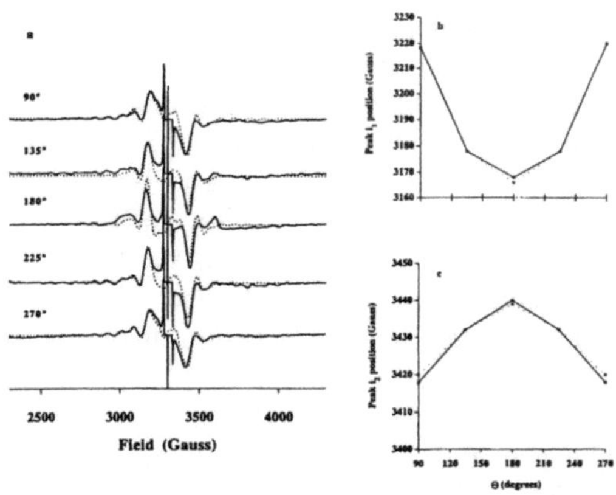

Figure 7. (a) Experimental and simulated $S_2Y_Z^{\bullet}$ spectra at different orientations, and (b-c) rotation plots showing experimental and simulated peak positions at different orientations. The experimental data are shown by solid lines and the simulations are shown by dotted lines. (Modified from (Lakshmi *et al.*, 1999a).)

Un *et al.* (1994) have studied high-field $Y_D^{\bullet}$ spectra at 4 K in the presence of the S = 1/2 S_2 state of the Mn_4 cluster. A large variation of the $Y_D^{\bullet}$ spectral intensity was observed that depended on the g value and on the presence of the Mn_4 cluster. This effect was attributed to saturation effects together with relief of saturation due to the interaction of $Y_D^{\bullet}$ with the paramagnetic Mn_4 cluster in PS II. The relative orientation of $Y_D^{\bullet}$ and other paramagnetic centers was determined by line shape analysis of the $Y_D^{\bullet}$ EPR signal including the effects of saturation based on eqns 8 and 9, assuming that scalar exchange interactions due to the Mn_4 cluster and the non-heme Fe(II) are negligible,

$$\frac{1}{T_{1I}} = \frac{1}{T_{1Ii}} + \frac{1}{6} D \frac{T_{2S}}{1 + (\omega_I - \omega_S)^2 T_{2S}^2} (1 - 3\cos^2\theta)^2 \quad (8)$$

$$\frac{1}{T_{2I}} = \frac{1}{T_{2Ii}} + \frac{1}{3} D T_{1S} (1 - 3\cos^2\theta)^2 \quad (9)$$

where T_{1Ii} and T_{2Ii} are the intrinsic relaxation times of spin I ($Y_D^{\bullet}$) and $D = \gamma_I^2 \gamma_S^2 h^2 S(S+1) / 4\pi^2 r^6$.

The relative orientation of $Y_D^{\bullet}$ and the non-heme Fe(II) has been measured in manganese-depleted PS II membranes and the two effects (from the Mn_4 cluster and the non-heme Fe(II)) in manganese-containing PS II samples were deconvolved (Un *et al.*, 1994). Using the manganese T_2 values obtained from spin-echo modulation and the T_1 values calculated from saturation-recovery curves, the Mn_4-$Y_D^{\bullet}$ distance was found to be 25-35 Å. A structural model was proposed by taking a Fe(II)-$Y_D^{\bullet}$ distance of 38 Å determined by Hirsh *et al.* (1992a) and placing the Mn_4 and Fe(II) on the C_2 axis of symmetry resulting in a Mn_4-$Y_Z^{\bullet}$ distance of 25 Å (Un *et al.*, 1994). This model contradicts emerging structural information obtained from recent electron-crystallography studies and other EPR distance measurements (Lakshmi *et al.*, 1999a, Rhee *et al.*, 1998). It is difficult to assess the reliability of distance and orientation determinations from orientation selection of high-field EPR spectra as the method has not previously been applied to a known system as a benchmark prior to its application on a complex system such as PS II.

3.1.3 Spectral Simulations of Powder and Oriented Iron-Quinone EPR Signals in bRCs

The reduced electron acceptor quinone cofactors, Q_A^- and Q_B^-, were first proposed to be magnetically coupled to the non-heme Fe(II) in reaction centers by Okamura *et al.* (1975) and it was thought that the electron may be localized on the quinone center in the complex. In a study by Feher and coworkers, the reaction centers from *Rb. sphaeroides* were Q_B depleted to isolate the Fe(II)-Q_A^- interaction. In the bRC iron-quinone complex, the magnetic interactions were taken to be a combination of zero-field splitting of Fe(II), exchange and dipolar interactions. The effective spin Hamiltonian was modeled as a sum of the crystal-field terms for the high spin Fe(II), an anisotropic Zeeman term for Fe, a Zeeman term for Q_A^-, and an anisotropic **J** term for the coupling between the Fe and Q_A^- (Butler *et al.*, 1984).

$$\mathcal{H} = D[S_{Fe,z}^2 - \frac{1}{3}S_{Fe}(S_{Fe}+1)] + (\frac{E}{D})(S_{Fe,x}^2 - S_{Fe,y}^2) + \mu_B \mathbf{B_0} \cdot \mathbf{g_{Fe}} \cdot \mathbf{S_{Fe}}$$
$$+ g_Q \mu_B \mathbf{B_0} \cdot \mathbf{S_Q} - \mathbf{S_{Fe}} \cdot \mathbf{J} \cdot \mathbf{S_Q} \qquad (10)$$

where D and E/D are the zero-field splitting parameters for Fe(II), $\mathbf{g_{Fe}}$ is the Fe **g** tensor, g_Q is the isotropic g value for Q_A^-, **J** is the magnetic interaction tensor, $\mathbf{S_{Fe}}$ and $\mathbf{S_Q}$ are the spin operators for Fe(II) and Q_A^-, respectively, and $\mathbf{B_0}$ is the applied magnetic field. The Fe(II)-Q_A^- system has ten states $(2S_{Fe} + 1)(2S_Q + 1)$. The asymmetric crystal field of Fe(II) (S = 2) splits the degenerate Fe manifold into five energy levels. Further coupling with Q_A^- splits each level into a doublet where the spacings within the doublet correspond to the spin up and spin down of the coupled Q_A^- (S = 1/2). At X-band, spin transitions occur only within the Kramer's doublets. It has been proposed that the Fe(II)-Q_A^- spectrum is a superposition of signals from the ground and first excited doublets and that the external magnetic field causes an anisotropic Fe(II) magnetic moment that in turn causes a broadening of the coupled Q_A^- EPR signals. This was experimentally verified by obtaining the Fe(II)-Q_A^- spectrum at two different temperatures (1.4 and 4.2 K) where either the ground doublet or the first excited doublet dominates the high- or low-field regions of the spectrum, respectively. It is known that the temperature dependence for the reduced quinone signals in the presence of Fe(II) is governed by the transitions between the Fe(II) levels.

Feher and coworkers have studied the Fe(II)-Q_A^- interaction in *Rb. sphaeroides* at 1, 9 and 35 GHz EPR frequencies and have found that the EPR spectrum for a reduced quinone electron acceptor consists of a broad signal centered at g = 1.8 with wings ranging from g ~ 5 to g < 0.8. Butler *et al.* (1980) previously optimized the parameters that characterize the electronic state of the Fe(II) by magnetic susceptibility measurements. The zero-field splitting parameters, D and E, the isotropic g value, g_{iso}, and an isotropic exchange interaction have been determined for the Fe(II)-Q_A^- complex. Subsequently in a later study, an anisotropic magnetic interaction between the quinone and Fe(II) was introduced to model accurately the interaction Hamiltonian (Butler *et al.*, 1984). For a perturbation analysis, the interaction Hamiltonian in eqn 10 was divided into a zeroth-order Hamiltonian ($\mathcal{H}_0$) and a perturbation ($\mathcal{H}'$), where,

$$\mathcal{H}_0 = D[S_{Fe,z}^2 - \frac{1}{3}S_{Fe}(S_{Fe}+1)] + (\frac{E}{D})(S_{Fe,x}^2 - S_{Fe,y}^2) \qquad (11)$$

$$\mathscr{H}' = \mu_B \mathbf{B_0} \cdot \mathbf{g}_{Fe} \cdot \mathbf{S}_{Fe} + g_Q \mu_B \mathbf{B_0} \cdot \mathbf{S}_Q - \mathbf{S}_{Fe} \cdot \mathbf{J} \cdot \mathbf{S}_Q \tag{12}$$

and the effective Hamiltonian was computed to the third order. Both a perturbation analysis and an exact numerical diagonalization solution to the total spin Hamiltonian have indicated that the observed EPR spectrum arises from the two lowest doublets in the ground manifold of the Fe(II)-Q_A^- system that are separated by 3 K. A best spectral fit was found for J_x = -0.13 K, J_y = -0.58 K and J_z = -0.58 K and a Gaussian spread of 0.14 and 0.06 K was introduced in J_y and J_z, respectively, to account for the loss of well-defined splittings in the g = 1.8 peak of the Fe(II)-Q_A^- signal. Assumption of a pure anisotropic dipolar or a combination of anisotropic exchange and dipolar couplings have yielded distances of 2.6 Å and > 2.6 Å, respectively, while the isotropic exchange coupling has predicted a distance of approximately < 10 Å (Butler *et al.*, 1984). This has been found to be consistent with the nearest atom distance of ~ 7 Å between the non-heme Fe(II) and Q_A^- observed in the x-ray crystallography structure (Deisenhofer *et al.*, 1984, Deisenhofer *et al.*, 1985). Thus, EPR and x-ray analyses have established that there is no direct coordination between the quinone and Fe(II). Most likely, the orbital overlap observed is via Fe(II) ligands intervening between the two centers. Also, given that the spin or oxidation state of the non-heme Fe(II) does not change during the electron-transfer reaction, it is not directly involved in the Q_A^- to Q_B electron-transfer mechanism.

The orientation dependence of the Q_A-Fe(II)-Q_B^- and Q_A^--Fe(II)-Q_B^{2-} EPR line shapes has also been studied in oriented cells of *Rps. viridis* (Dismukes *et al.*, 1984). Magnetic fields were used to orient the prolate ellipsoid shaped cells of *Rps. viridis* and a distinct orientation dependence was observed for the iron-quinone complex in an applied magnetic field which was indicative of a precise orientation of the non-heme iron within the protein complex. EPR spectra either perpendicular or parallel to the plane defined by the long axis of the bacterium permitted observation of the Zeeman field preferentially out of plane or preferentially in plane relative to the surface of the membranes. The line shapes were simulated using a spin interaction Hamiltonian similar to that shown in eqn 10 and both an exact numerical solution and a perturbation calculation were performed. It was found that for a model which is disordered in the plane of the membrane, the orientation of the Fe(II) axis of largest zero-field splitting is tilted 64 ± 6° relative to the membrane normal. The ligand field for Fe(II) has a low symmetry with D of 7.0 ± 1.3 cm^{-1}, E of 1.7 ± 0.5 cm^{-1} and E/D = 0.26. The rhombic character of the ligand field of the Fe(II) was seen to increase in a sample with just Q_A^- relative to a sample in the Q_A^-Fe(II)Q_B^{2-} state, indicating that Q_B influenced the ligand-field symmetry and splitting of the non-heme Fe(II). The exchange couplings between Fe(II)-Q_A^- and Fe(II)-Q_B^- were found to be 0.12 ± 0.03 cm^{-1} and 0.06 cm^{-1}, respectively. Line-broadening of the Q^- coupling to the magnetic sublevels of Fe(II) were inferred to be a result of a dipolar coupling between them and a line-broadening analysis yielded an interspin distance of 6.2 - 7.8 Å (Dismukes *et al.*, 1984).

The spectral assignments of the Fe(II)-Q_A^- EPR signals by Dismukes *et al.* (1984) are, however, in conflict with the results obtained by Feher and coworkers (Butler *et al.*, 1984). Dismukes *et al.* (1984) have attributed the high-field portion of the EPR spectrum to the excited doublet transitions and the low-field part of the spectrum to ground state doublet transitions. In addition, the sign of J was found not to affect calculations by Dismukes *et al.*, but the transitions in the Feher calculations are switched upon changing the sign of J. Judging from the temperature dependence, the low-field part of the spectrum is from the excited state and not the ground state (Butler *et al.*, 1984). In view of this result, the numerical values of the zero-field and exchange couplings obtained by Dismukes and coworkers are approximate and not so much a function of the different bacterial system used in these studies (Hoff, 1986).

3.1.4 Spectral Simulations of EPR Signals from Iron-Sulfur Clusters in Oriented PS I Membranes and PS I Single Crystals

The iron-sulfurs clusters in PS I exhibit distinct EPR spectra. The F_X EPR signal has characteristic principal g turning points at g_x = 1.79, g_y = 1.86 and g_z = 2.04 and non-interacting F_A and F_B clusters have g values of g_x = 1.86, g_y = 1.94, and g_z = 2.05 and g_x = 1.89, g_y = 1.94, and g_z = 2.05, respectively. When both the F_A and F_B centers are reduced, the g_x peak for the F_A^- - F_B^- interaction spectrum is found to shift to g = 1.89. The shift in g values in the $F_A^-F_B^-$ state indicates the presence of a magnetic interaction between the two clusters.

More recently, the structural organization of the [4Fe-4S] clusters F_A and F_B have been investigated by EPR line shape analyses of oriented PS I membranes (Guigliarelli *et al.*, 1993) and PS I single crystals (Kamlowski *et al.*, 1997a, Kamlowski *et al.*, 1997b). The two studies have employed different strategies to determine the relative orientation of the two [4Fe4S] clusters F_A and F_B. The former investigates the orientation dependence of the interacting $F_A^-F_B^-$ EPR signals in oriented PS I membranes (Guigliarelli *et al.*, 1993) while the latter investigates non-interacting F_A^- or F_B^- EPR signals in PS I single crystals to determine their relative orientation by analogy to a bacterial ferredoxin (Kamlowski *et al.*, 1997a, Kamlowski *et al.*, 1997b).

Guigliarelli *et al.* (1993) investigated the relative arrangement of F_A and F_B in the protein complex by studying magnetic interactions in the $P_{700}^{+}F_A^{-}F_B^{-}$ ($F_A^{-}F_B^{-}$) state in PS I membranes oriented on Mylar films. The spin Hamiltonian describing the Zeeman, dipolar and exchange interactions between the [4Fe-4S] was solved by numerical diagonalization. In the oriented spectra, when the mosaic spread was not too large, the amplitude of each of the peaks in the interacting $F_A^{-}F_B^{-}$ spectrum was dependent on the orientation relative to the applied magnetic field and reached a maximum when the field was aligned very close to the magnetic axis associated with the peak.

As discussed in section 3.1.1, accurate treatment of the mosaic spread leading to line broadening of observed resonances in oriented EPR spectra is of importance. In the case of non-interacting centers in proteins, the line broadening observed is predominantly from **g** or hyperfine strain and a Gaussian distribution is usually employed to model such strain. However, in the case of interacting spins, the broadening could be due to conformational effects on the anisotropic spin-spin interactions. Guigliarelli *et al.* (1993) assumed that because the spin Hamiltonians are in the high-field limit i.e. $\mathcal{H}_Z >> \mathcal{H}_D$ and $\mathcal{H}_J$, the distribution of the Zeeman term is the dominant contribution to the line broadening observed. But it was observed that broadening in the experimental spectra was too much to be accounted for by **g** strain and further analysis indicated that additional sources of broadening were the exchange and dipolar interactions. A J and **r** strain was incorporated into the simulation to obtain a better fit.

Simulation of oriented $F_A^{-}F_B^{-}$ spectra yielded the orientations of each of the magnetic axes relative to the membrane normal. However, knowledge of these six angles was not sufficient to determine the Euler angles that define the orientation of the interspin vector. Triplets of Euler angles were calculated that accounted for the experimental oriented spectra and only one set of values for the crystal axes, a, b and c, corresponded to one of these triplets. The angle between the F_A-F_B vector and the membrane normal was deduced to be 30° and the interspin distance was determined to be 10 Å (Guigliarelli *et al.*, 1993).

In a separate EPR study, Kamlowski *et al.* (1997a and 1997b) have examined the magnitude and orientation of the **g** tensors of the F_A^{-} and F_B^{-} clusters in PS I single crystals to determine the orientation of the PsaC protein (Figure 2). A PS I single crystal was illuminated under controlled conditions to generate predominantly either the F_A^{-} or F_B^{-} state. A unit cell of PS I contains twelve iron-sulfur centers and the orientation dependence of each of these was mapped as a function of the angle relative to the applied magnetic field. Significant overlap of resonances at some orientations caused the EPR lines to be inhomogeneously broadened leading to an angular dependence of both the peak intensities and line widths. In principle, the mosaic spread of a single crystal could account for such an orientation dependence as each crystal had a non-negligible mosaic spread which leads to a distribution of g values resulting in angular dependence of the EPR line width. The broadening was determined by the variation of the line position with molecular orientation. As this variation approached zero near the principal axes of the **g** tensor, broadening due to the mosaic spread was smallest when a principal axes was parallel to the applied magnetic field. Another important contribution to the angular dependence of the peak intensities and line widths was found to be the relaxation times, T_1 and T_2, and it was concluded that both the homogeneous line width and the saturation behavior of an EPR line are governed by the relaxation times and can thus be orientation dependent.

The **g** tensors of F_A^- and F_B^- were seen to exhibit vastly different orientation dependences. Peak positions as a function of the rotation angle α yielded two sets of six rotation curves that were numerically fit for F_A^- or F_B^- to determine the orientation of the **g** tensors relative to the c-axis. Kamlowski *et al.* (1997a) estimated the relative orientation of the F_A^- or F_B^- **g** tensors by finding a unique transformation matrix relating the two iron-sulfur centers. The transformation matrices were calculated based on the structure of the bacterial ferredoxin of *Peptococcus aerogenes* as it contains a pseudo-C_2^{Fd} axis that relates two iron-sulfur clusters as well as their protein environments. Only two calculated matrices were found to be compatible with experiment and calculation. The relative orientations of F_A^- or F_B^- have been used to predict the orientation of the PsaC protein within the PS I complex. However, a two-fold symmetry axis inherent to the ferredoxin model leaves a two-fold ambiguity in the structural conclusion regarding the orientation of the PsaC protein.

3.2 Spin-Lattice Relaxation Enhancement Measurements

3.2.1 Progressive Microwave Power-Saturation Measurements of Spin-Lattice Relaxation Effects

In an EPR experiment, as long as spin-lattice relaxation can return the sample to its equilibrium spin distribution faster than the absorption of radiation perturbs it, increasing the applied microwave power yields EPR signals with higher intensities. At sufficiently high applied microwave power, the relaxation pathways available to the spin are saturated and further increase in the applied microwave power does not result in an increase of the observed EPR signals. In a *cw* progressive microwave power-saturation experiment, the spins are progressively subjected to higher and higher microwave power. The experimentally determined value of the microwave power at half-saturation, $P_{1/2}$, coupled with a theoretical prediction of relaxation processes of the spin, can provide structural and magnetic information on the spin system.

The expressions for progressive microwave power saturation have been derived by Portis (1953) and Castner (1959) and have been extended to the limit of inhomogeneously broadened resonances in biological systems by Beinert and Orme-Johnson (1967). The EPR signal amplitude for an inhomogeneously broadened line is given by the empirical equation,

$$S = K\frac{\sqrt{P}}{(1+\frac{P}{P_{1/2}})^{b/2}} \tag{13}$$

where P is the observe microwave power, $P_{1/2}$ is the microwave power at half-saturation, b is the inhomogeneity factor (b = 1 in the inhomogeneous limit and b = 3 in the homogeneous limit) and K is a proportionality constant. A plot of the EPR signal intensity as a function of microwave power level can be used to determine the $P_{1/2}$ value for the system.

Distance measurements using progressive microwave power-saturation techniques to study enhanced spin-lattice relaxation rates of protein bound radicals have been applied to a wide variety of systems. Relaxation enhancements from pairwise interactions of paramagnetic centers within a single protein provide information on the location of redox sites in the protein (Brudvig *et al.*, 1984, Goodman & Leigh Jr., 1985, Makinen & Wells, 1987, Sahlin *et al.*, 1987, Stevenson *et al.*, 1986). The effect of exogenous relaxation enhancers, such as Dy^{3+} or Gd^{3+} complexes, provide information on the distance from redox centers to the protein surface (Blum *et al.*, 1983, Blum *et al.*, 1981, Blum *et al.*, 1980, Hales, 1993, Hyde & Rao, 1978, Innes & Brudvig, 1989, Oliver & Hales, 1993). For example, Ohnishi and coworkers have studied the effect of Dy^{3+} complexes on the spin-lattice relaxation behavior of heme, iron-sulfur and copper complexes in a number of proteins to probe the location of the redox centers (Blum *et al.*, 1980). Brudvig and coworkers have extended this dipolar analysis to include a distribution model for exogenous Dy^{3+} ions to analyze the spin-lattice relaxation enhancement of $Y_D^{\bullet}$ and Chl_Z^+ radicals in PS II (Innes & Brudvig, 1989, Koulougliotis *et al.*, 1994).

As is the case for fast-relaxing Dy^{3+} ions interacting with an organic radical, $(\omega_s + \omega_f)^2 T_{2f}^2$, $\omega_s^2 T_{1f}^2$, $(\omega_s - \omega_f)^2 T_{2f}^2 << 1$ and this leads to a simple approximation of eqn 2 where the denominators of the B, C and E terms tend to unity. When $T_{1f} >> T_{2f}$, the B and E terms can be ignored and the C term is dominant (eqn 14).

$$\frac{1}{T_{1s,\text{dipolar}}} = \frac{3\gamma_s^2\mu_f^2}{r^6\omega_s^2} T_{1f} \sin^2\theta \cos^2\theta \tag{14}$$

The progressive microwave power at half-saturation, $P_{1/2}$, is a direct reflection of the spin-lattice relaxation rate $(1/T_{1s})$ of an EPR-active species as shown in eqn 15. $P_{1/2}$ also depends on T_{2s} which complicates the use of $P_{1/2}$ values for distance determinations. However in many cases, T_{2s} is independent of the fast relaxer.

$$P_{1/2} \propto \frac{1}{(\gamma^2 T_{1s} T_{2s})} \tag{15}$$

$P_{1/2,dipolar}$ values can be determined experimentally by measuring the spectral intensities of the slow-relaxing species s with $(P_{1/2,observed})$ and without $(P_{1/2,intrinsic})$ the presence of the fast-relaxing species f (eqn 16). These $P_{1/2,dipolar}$ values can then be used to derive the separation between the slow- and fast-relaxing spins.

$$P_{1/2,\text{observed}} = P_{1/2,\text{intrinsic}} + P_{1/2,\text{dipolar}} \tag{16}$$

The change in $P_{1/2}$ caused by a given concentration of fast relaxers $(\Delta P_{1/2})$ is the difference of $P_{1/2,observed}$ and $P_{1/2,intrinsic}$ and,

$$\Delta P_{1/2} = P_{1/2,\text{dipolar}} \propto \frac{1}{T_{1s,\text{dipolar}}} \tag{17}$$

Eqns 14 and 17 predict that the change in $P_{1/2}$ $(\Delta P_{1/2})$ will have a r^{-6} distance dependence.

In the case of a slow-relaxing spin within the protein interacting with a distribution of fast-relaxing spins in a frozen solution, a sum over all inter-spin distances between the slow-relaxing spin and the ensemble of fast-relaxing spins has to be considered and this requires the assumption of a model to evaluate the distribution of fast relaxers. Because all orientations of the radial vector with respect to the applied magnetic field are equally probable, the Dy^{3+}-enhanced spin-lattice relaxation rate of the slow-relaxing spin in an average environment can be expressed by the equation,

$$\Delta P_{1/2} = C[Dy^{3+}]\Sigma_i r^{-6} \tag{18}$$

where C is a constant, $[Dy^{3+}]$ is the concentration of the added Dy^{3+}-chelate complex and the sum runs over all volume elements in the sample (Innes & Brudvig, 1989).

If $\Sigma_i r^{-6}$ were expressed as a simple function of the distance from the free radical to the surface of the protein, the distance of closest approach to the protein embedded spin could be determined. Previously, this distance was either approximated as the distance to a closest Dy^{3+} or the ensemble of perturbing fast relaxers was considered to be a single average site (Oliver & Hales, 1993). Blum and coworkers have also used a simple r^{-6} distance dependence to model the effect of Dy^{3+} ions on the relaxation properties of several slow-relaxing sites in proteins.

A detailed analysis has been conducted by Innes and Brudvig (1989) by performing an integration of the $1/r^6$ term over the nonexcluded volume of a protein assuming either surface binding or a random distribution of Dy^{3+} ions in solution. In order to perform such an integration, several models based on the overall shape of the protein were tested, *i.e.* an ellipsoidal shape for water-soluble myoglobin nitroxide (MbNO), cylindrical for detergent solubilized bRCs and an infinite planar sheet for a membranous PS II protein complex. Only the model assuming no specific-binding sites with a random distribution of Dy^{3+} in solution yielded distance estimates in agreement with known structures of the bRC from *Rb. sphaeroides* and myoglobin nitroxide. Modifications to the distribution model proposed by Innes and Brudvig have been made by Oliver and Hales (1993) where it is assumed that the protein (nitrogenase) is spherical with either a random distribution of Dy^{3+} on the surface of the protein or in solution. This method has the advantage that only the radius of the spherically-approximated protein is required for the analysis.

Using the equations to approximate the distance dependence for a given structural model, the proportionality constant C (eqn 18) can be determined by plotting the change in $P_{1/2}$ ($\Delta P_{1/2}$) as a function of [Dy-EDTA]. The constant C should be the same for any free radical interacting with the Dy^{3+} ions in solution at the same temperature and instrumental conditions. The constant factor C can be determined by conducting $P_{1/2}$ measurements on model systems. Because the constant C is dependent on the structural model assumed for the protein and the instrumental conditions, it is best calibrated on a known protein prior to its application on an unknown system as demonstrated in the cases described below (Innes & Brudvig, 1989, Koulougliotis *et al.*, 1994).

Progressive microwave power saturation has been successfully used to determine the depth of $Y_D^{\bullet}$ in PS II from the outer and inner surfaces of the thylakoid membranes (Innes & Brudvig, 1989). Figure 8 shows the change in $P_{1/2}$ of the NO-ligated form of myoglobin nitroxide (MbNO), P_{870}^{+} in the bRC and $Y_D^{\bullet}$ in PS II as a function of increasing concentrations of Dy^{3+}-EDTA. The analysis of the distance dependence of spin-lattice relaxation enhancement caused by multiple randomly distributed exogenous spins in MbNO and the bRC was used to characterize the location of the $Y_D^{\bullet}$ and Chl_Z^{+} radicals in PS II membranes. The depth of $Y_D^{\bullet}$ from the outer and inner membrane surfaces in PS II was found to be 26 Å and 27 Å, respectively (Innes & Brudvig, 1989).

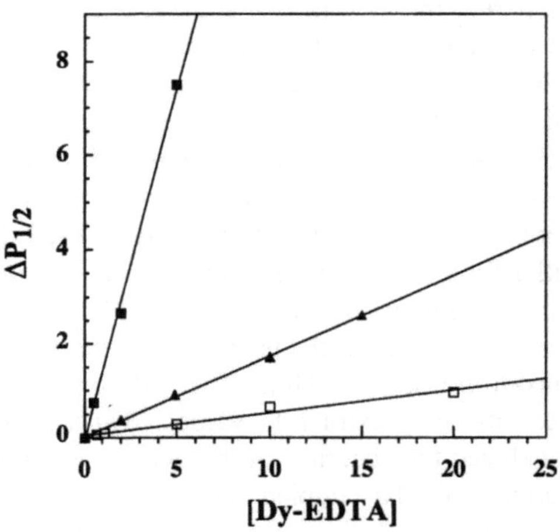

Figure 8. The change in $P_{1/2}$ values of (a) MbNO (■), (b) P_{870}^{+} in the bRC (▲) and (c) $Y_D^{\bullet}$ in PS II (□). $\Delta P_{1/2}$ is the difference between $P_{1/2}$ in samples with and without Dy^{3+}-EDTA present. (Modified from (Innes & Brudvig, 1989).)

Using progressive power-saturation techniques, the depth of Chl_Z^{+} has also been found to be approximately equidistant from the stromal and luminal surfaces of PS II (Koulougliotis *et al.*, 1994). This result in conjunction with the non-heme Fe(II)-Chl_Z saturation-recovery EPR distance measurement (described in section 3.2.2) has been used to assign Chl_Z to an accessory chlorophyll monomer bound to either D1-H118 or D2-H117. The location of D1-H118 as the axial ligand to Chl_Z was later confirmed by a site-directed mutagenesis study (Stewart *et al.*, 1998).

Previously, Ni^{2+} and Gd^{3+} were employed as fast relaxers to determine the location of redox components within mitochondrial membranes using progressive microwave power-saturation techniques (Case and Leigh, 1976, Case *et al.*, 1976). However, in recent studies, Dy-EDTA or Dy-HEDTA are chosen as exogenous relaxation agents because Dy^{3+} is a potent relaxer and bulky EDTA or HEDTA complexes reduce the possibility of specific-binding effects on the membrane surface. Also in the case of PS II studies, the chelate complexes do not damage the Mn_4 cluster in the active water-oxidation site, whereas the free metal ions do (Ghanotakis *et al.*, 1985).

The microwave power saturation behavior of $Y_D^{\bullet}$ has also been examined by Styring and Rutherford (1988) in PS II samples illuminated by a series of flashes where each flash advanced the S states by one oxidizing equivalent. The $P_{1/2}$ values of $Y_D^{\bullet}$ were found to depend on the S state of the Mn_4 cluster and an apparent increase in $P_{1/2}$ at 20 K was observed in the S_2, S_3 and S_0 states. An unusual temperature dependence of the $P_{1/2}$ values of $Y_D^{\bullet}$ around 20-30 K in the S_1 and S_2 state was taken to be a reflection of the oxidation states and magnetic properties of the Mn_4 cluster. However, detailed studies by Galli *et al.* (1996) have shown that analysis of power-saturation measurements at low temperatures is complicated and can yield anomalous temperature dependencies (Innes & Brudvig, 1989, Kodera *et al.*, 1992, Styring & Rutherford, 1988) as organic radicals are in the extreme saturation limit at such low temperatures (Galli *et al.*, 1996).

$P_{1/2}$ values are also dependent on the properties of the microwave cavity and the spin-spin relaxation time, T_2. Thus, it is difficult to make adequate approximations to quantitatively extract T_1 values. Brudvig and coworkers have systematically analyzed the accuracy of the determination of $P_{1/2}$ values of the slow-relaxing spin in dipolar-coupled systems (Beck *et al.*, 1991, Galli *et al.*, 1996). It was found that the conventional expressions used for determining $P_{1/2}$ values are not appropriate for moderate to strongly dipolar-coupled spins, as eqn 13 describing the EPR signal intensity as a function of $P_{1/2}$ does not directly apply to a collection of dipolar-interacting spins. In deriving eqn 13, it was assumed that all spins on-resonance saturate equivalently. In the case of dipolar-coupled spins, however, the spin-lattice and spin-spin relaxation of each species depend on the orientation of the dipolar vector relative to the applied magnetic field which will lead to an inhomogeneous behavior. The distribution of T_1 values encountered in a saturation-recovery measurement (described in section 3.2.2) is also manifested as a distribution of $P_{1/2}$ values for the interacting spin. This is effectively demonstrated by the experimental and simulated power-saturation curves for $Y_D^{\bullet}$ in manganese-depleted PS II shown in Figure 9

(Galli *et al.*, 1996). $Y_D^\bullet$ in manganese-depleted PS II exhibits an enhanced spin-lattice relaxation rate due to a dipolar coupling with the non-heme Fe(II). The dashed line shows a numerical fit to the data with b (the inhomogeneity factor) set to its lower limit of unity. The curve fits poorly and caution should be exercised in the use of eqn 13 for analysis of the power-saturation curves as the dipolar interaction results in a distortion of the curves such that eqn 13 is not valid. Treating $P_{1/2}$ and b as variables in the fit yields the solid curve which is a better match to the experimental power-saturation curve and a best fit is obtained with $b = 0.71$. However, a value of $b < 1.0$ is not physically meaningful, because $b = 1$ is the lower limit, but is diagnostic of the presence of a dipolar interaction. It has been suggested that the value of b obtained from the best fit could provide a qualitative estimate of the extent of dipolar coupling in a pairwise spin-spin interaction. Because an exchange interaction does not distort power-saturation curves, it should also be possible to separate the exchange and dipolar contributions to spin-lattice relaxation enhancement by analyses of power-saturation data.

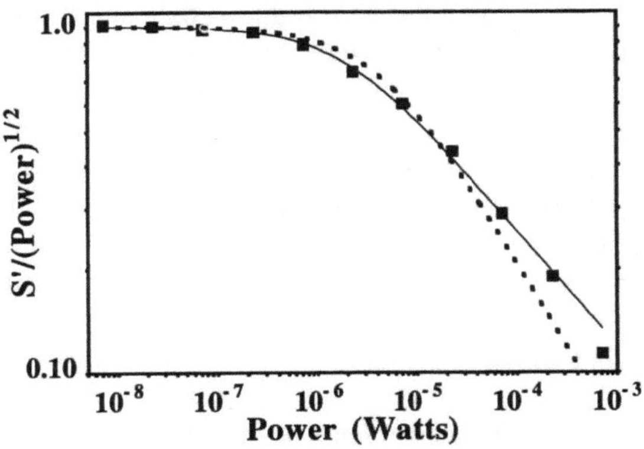

Figure 9. Power-saturation curves for $Y_D^\bullet$ in manganese-depleted PS II membranes at 8 K. Solid squares are experimental data points and the solid and dashed lines are simulated curve fits. The dashed line is a best fit obtained with the parameter b fixed at unity while the solid line is the best fit obtained by floating b as a variable and converging at $b = 0.71$. (Modified from (Galli *et al.*, 1996).)

Rapid-passage effects (see Chapter 2) can also contribute to the deviation of power-saturation data from predicted behavior as eqn 13 assumes that the change in the applied magnetic field is slow compared to the response of the spin system. Qualitatively, increasing the acquisition temperature, decreasing the modulation frequency and addition of exogenous fast relaxers should decrease contributions from rapid-passage effects (Innes & Brudvig, 1989).

The examples described so far have been for a slow-relaxing $S = 1/2$ spin interacting with a distribution of Dy^{3+}. The treatment is different in the case of a high-spin slow relaxer. Oliver and Hales (1993) have analyzed the case of a $S = 3/2$ spin state in the MoFe-protein (M centers of component 1 of nitrogenase) interacting with a distribution of Dy^{3+} on the surface.

Progressive microwave power saturation has also been used to make a qualitative determination of relative distances between different spin pairs. For example, using power-saturation techniques the relative locations of the [4Fe4S] clusters, F_A and F_B, with respect to P_{700} have been determined in PS I (Lakshmi *et al.*, 1999b). The power-saturation behavior of P_{700}^+ was investigated in the $P_{700}^+F_A^-$ and $P_{700}^+F_A^-F_B^-$ charge-separated pairs in mercury-treated and wild-type PS I, respectively. Mercury treatment depletes the F_B cluster in PS I enabling the examination of spin-lattice relaxation enhancement effects solely due to F_A. The $P_{1/2}$ of P_{700}^+ was found to be greater when both F_A and F_B were reduced than when only F_A was reduced. The experimental $P_{1/2}$ values were compared with values calculated by using crystallographic distances obtained by Krauß *et al.* (1996) and by assuming that either F_A or F_B was closer to P_{700}. It was found that the $P_{1/2}$ values calculated by assuming F_A in closer proximity to P_{700} agreed well with the experimental $P_{1/2}$ determined in the $P_{700}^+F_A^-F_B^-$ pair. This study unambiguously assigned the relative location of the two terminal iron-sulfur clusters in PS I. Comparison of relative effects due to the presence of F_A^- and $F_A^-F_B^-$ and calibration using the $P_{1/2}$ of P_{700}^+ in the presence of F_A^- eliminated the need for additional parameters in the data analysis.

3.2.2 Saturation-Recovery Measurements of Spin-Lattice Relaxation Effects

As demonstrated by progressive microwave power-saturation measurements described in section 3.2.1, long range spin-spin interactions between a pair of fast- and slow-relaxing spins results in spin-lattice relaxation enhancement of the slow-relaxing spin. The technique of saturation recovery can be used to obtain a direct measure of the spin-lattice relaxation rate. A dipolar model that accounts for the non-single exponential decay transients observed for interacting spin pairs can be applied to obtain long-range distances between spin pairs (also described in Chapter 8).

Because $k_{1\theta}$, the dipolar spin-lattice relaxation rate, is a function of the orientation of the dipolar vector relative to the applied magnetic field (Figure 4) and a powder EPR sample has a random distribution of orientations between 0 and π, the normalized intensity of the saturation-recovery transient at time t, I(t), is given by the eqn,

$$I(t) = 1 - N\int_0^{\pi} \sin\theta \exp(-(k_{1scalar} + k_{1\theta})t)d\theta \qquad (19)$$

where N is a scaling factor and $k_{1scalar}$ and $k_{1\theta}$ are the isotropic and orientation-dependent dipolar contributions to the relaxation rate (see eqns 2-5). For large interspin distances, it is assumed that $k_{1scalar} = k_{1in}$, where k_{1in} is the intrinsic spin-lattice relaxation rate of the species. A quantitative analysis of I(t) in order to determine the interspin distance, r, requires a model for the angular dependence of $k_{1\theta}$.

Equations 2 and 3 give the expressions for $k_{1\theta}$ and $k_{2\theta}$, respectively, and these equations are valid in the Redfield limit and require that T_{1f}, $T_{2f} << T_{1s}$ and T_{1f}, $T_{2f} << T_{2s}$. The choice of the appropriate term of the dipolar alphabet [B, C or E] that has the most contribution to $k_{1\theta}$ requires prior knowledge of the spin-lattice and spin-spin relaxation times of the fast-relaxing spin (T_{1f} and T_{2f}) and the Larmor frequencies of the slow and fast relaxer, ω_s and ω_f, respectively (see Chapter 8). However, there are two possible limits in which these equations can be simplified:

(a) For a spin with a rapid spin-lattice relaxation rate, $(\omega_s + \omega_f)^2 T_{2f}^2$, $\omega_s^2 T_{1f}^2$, $(\omega_s - \omega_f)^2 T_{2f}^2 << 1$, leads to an approximation of eqn 2 where the denominators of the B, C and E terms tend to unity. When, $T_{1f} >> T_{2f}$, T_{2f} can be ignored and in this limit, the dipolar relaxation rate depends only on the C term and is given by,

$$k_{1\theta} = k_{1d}^{C} \sin^2\theta \cos^2\theta \qquad (20)$$

where

$$k_{1d}^{C} = \frac{3\gamma_s^2\mu_f^2}{r^6\omega_s^2 T_{1f}} \qquad (21)$$

For eqn 21 to apply at X-band EPR frequency requires that T_{1f} and $T_{2f} << 10^{-11}$ s.
(b) In the opposite limit, when $\omega_f^2 T_{1f}^2$, $(\omega_s + \omega_f)^2 T_{2f}^2$, and $(\omega_s - \omega_f)^2 T_{2f}^2 >> 1$, and in addition, $(\omega_s - \omega_f)^2 << \omega_s^2$ and $(\omega_s + \omega_f)^2$, the C and E terms are negligible and the B term dominates. $k_{1\theta}$ is given by,

$$k_{1\theta} = k_{1d}^{B}(1 - 3\cos^2\theta)^2 \tag{22}$$

where

$$k_{1d}^{B} = \frac{\gamma_s^2 \mu_f^2}{6r^6 \omega_s^2 (1 - g_f / g_s)^2 T_{2f}} \tag{23}$$

For eqn 23 to apply at X-band EPR frequency requires that T_{1f} and $T_{2f} >> 10^{-11}$ s.

Using the approximations described above, saturation-recovery EPR has been used to characterize the spin-lattice relaxation enhancement of P_{870}^+ by Fe(II) in the bRC from *Rb. sphaeroides*. Norris and coworkers (1980) compared the spin-lattice relaxation rates of P_{870}^+ in Fe(II)-containing and Fe(II)-depleted bacterial reaction centers. An enhanced spin-lattice relaxation of P_{870}^+ in the Fe(II)-containing sample was ascribed to a dipolar interaction with the S = 2 Fe(II). From these data, the distance between P_{870}^+ and Fe(II) was estimated to be 12 Å, which is significantly shorter than the value of 28 Å determined by x-ray crystallography. The error arose for two reasons. First, the saturation-recovery EPR data were fit to a single exponential, even though the relaxation transients were non-single-exponential as now understood to arise from a dipolar relaxation mechanism (eqn 19). Second, and more importantly, incorrect relaxation times for Fe(II) were used. It is often not possible to measure the relaxation times for the fast-relaxing spin and this is a problem for accurate distance estimates using eqns 2 and 3. More recently, Hirsh and Brudvig (1993) have studied the spin-lattice relaxation of P_{870}^+ in Fe(II)-containing bRCs in comparison to model cation radicals. The dipolar model predicts a non-single-exponential spin-lattice relaxation behavior for the P_{870}^+ radical in the presence of Fe(II) and this was observed. In a saturation-recovery analysis, the C-term was assumed to be the dominant contribution, r was fixed at 28 Å and the parameters for Fe(II) were obtained by fitting the saturation-recovery data for P_{870}^+ interacting with Fe(II) (Figure 10).

The geometric and magnetic parameters of the non-heme Fe(II) in the bRCs have subsequently been used to model the $Y_D^\bullet$-Fe(II) interaction in manganese-depleted PS II (Hirsh *et al.*, 1992a). In intact manganese-containing PS II samples, $Y_D^\bullet$ exhibits a spin-lattice relaxation enhancement due to a dipolar coupling with the non-heme Fe(II) as well as the Mn_4 cluster. However, manganese depletion of PS II isolates the effect of the Fe(II) on $Y_D^\bullet$. Analysis of the saturation-recovery transients in this case has yielded a $Y_D^\bullet$-Fe(II) distance of 38 Å.

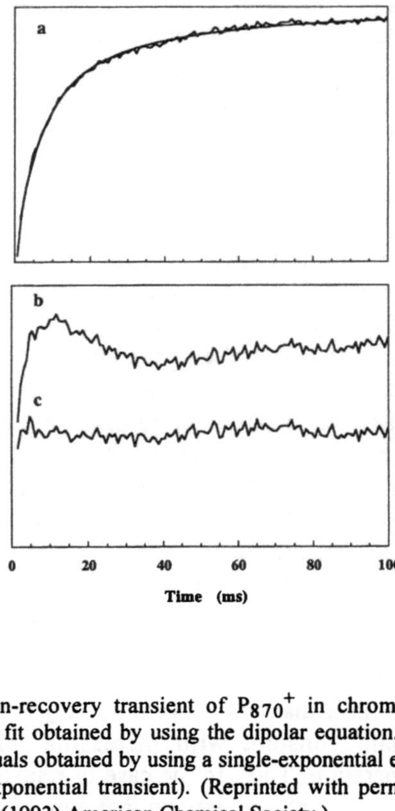

Figure 10. (a) Saturation-recovery transient of P_{870}^+ in chromatophore-bound bRCs superimposed with a best fit obtained by using the dipolar equation. Also shown (b and c, respectively) are the residuals obtained by using a single-exponential equation and the dipolar equation (a non-single-exponential transient). (Reprinted with permission from (Hirsh & Brudvig, 1993). Copyright (1993) American Chemical Society.)

The P_{870}^+-Fe(II) interaction in bRCs has also been used to deduce the $Y_Z^\bullet$-non-heme Fe(II) distance in *Synechocystis* (cyanobacterial) PS II (Koulougliotis *et al.*, 1995). The $Y_Z^\bullet$-Fe(II) distance measurement was conducted in a Y_D -less PS II mutant allowing for the observation of $Y_Z^\bullet$ without interfering effects from the dark-stable $Y_D^\bullet$ radical. The dipolar rate constants were extracted from saturation-recovery data acquired between 4 -

70 K. The temperature dependence exhibited by the $Y_Z^{\bullet}$ and $Y_D^{\bullet}$ signals in manganese-depleted PS II were identical indicating that the spin-lattice relaxation observed in both cases is due to the same fast-relaxing spin. In a saturation-recovery analysis based on the Fe(II) parameters from the bRC, the distances from $Y_D^{\bullet}$ and $Y_Z^{\bullet}$ to the non-heme Fe(II) have been found to be 37 ± 5 Å. This was the earliest spectroscopic demonstration that the two redox-active tyrosine residues, Y_D and Y_Z, in PS II are in symmetric locations relative to the Fe(II), consistent with the presence of a C_2 axis of symmetry in PS II reaction centers. Subsequently, a distance of 42 ± 2 Å between $Y_D^{\bullet}$ and non-heme Fe(II) was measured by selective hole burning which is a spin-echo analog of the saturation-recovery measurements of relaxation enhancement (Hara & Kawamori, 1997).

Saturation recovery has also been used to probe the location of Chl_Z^+ in manganese-depleted PS II. The spin-lattice relaxation transient of Chl_Z^+ has been observed to exhibit non-single-exponential behavior due to a dipolar coupling with the non-heme Fe(II). Manganese depletion of PS II ensured that Chl_Z^+ is dipolar coupled solely to the non-heme Fe(II) (Koulougliotis *et al.*, 1994). By using the Fe(II)-P_{870}^+ distance in bRCs for calibration, the Chl_Z^+-Fe(II) distance in PS II has been estimated to be 39.5 ± 2.5 Å.

The choice of the approximation leading to the use of the B or C term can also be judged by observing the temperature dependence of the dipolar rate constant, k_{1d}. If k_{1d} exhibits a temperature dependence similar to T_{1f}, then the choice of the C term can be taken as a valid approximation. This is well illustrated by an example of the temperature dependence of k_{1d} and $k_{1scalar}$ for a $Y_D^{\bullet}$ tyrosine radical interacting with a non-heme Fe(II) moiety in manganese-depleted PS II (Figure 11) (Hirsh *et al.*, 1992a). As can be seen in Figure 11, interaction of $Y_D^{\bullet}$ with the non-heme Fe(II) results in temperature dependent dipolar spin-lattice relaxation rate, k_{1d}, with k_{1d} (and $k_{1\theta}$) increasing with temperature. In the limit that eqn 20 is valid, k_{1d} would be proportional to $1/T_{1f}$, indicating that k_{1d} should increase with increasing temperature. This is indeed the experimental trend that is observed in Figure 11. However, in the limit that eqn 22 is valid, k_{1d} is proportional to $1/T_{2f}$, which should remain constant with increasing temperature if T_{2f} is independent of temperature. Thus, in the case of relaxation enhancement by the non-heme Fe(II), it is valid to assume that eqn 20 represents the correct limit for the evaluation of the dipolar relaxation rate. In the temperature regime where $T_{1f} = T_{2f}$, the temperature dependence of k_{1d} does not provide a basis for distinguishing between the B and C contributions.

Also, as can be observed in Figure 11, the temperature dependence of the scalar rate constant ($k_{1scalar}$) of $Y_D^{\bullet}$ interacting with a non-heme Fe(II) in manganese-depleted PS II is similar to that of UV-generated tyrosyl radicals in a model system. In both cases, the temperature dependence is $T^{2.4}$. The temperature dependence for k_{1d} is $T^{1.7}$ indicating that the primary contribution to $k_{1scalar}$ is the intrinsic spin-lattice relaxation rate of the tyrosine radical and not a contribution from an exchange interaction between $Y_D^{\bullet}$ and the non-heme Fe(II) in PS II. In the presence of a significant exchange contribution, the temperature dependence exhibited by $k_{1scalar}$ and k_{1d} would have been similar (Hirsh *et al.*, 1992b).

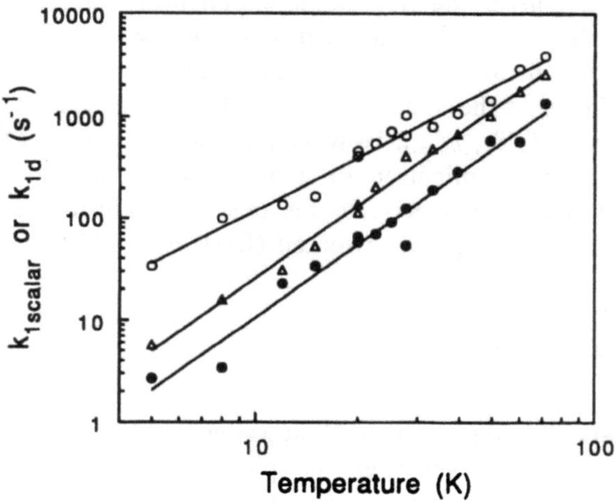

Figure 11. Temperature dependence of the scalar ($k_{1scalar}$, ●) and dipolar (k_{1d}, O) spin-lattice relaxation rates of the $Y_D^{\bullet}$ radical interacting with the non-heme Fe(II). Also shown are the single-exponential spin-lattice relaxation rates of UV-generated tyrosine radicals (△). The solid lines indicate the best fits to the data yielding $k_{1scalar} = 0.0459 \times T^{2.36}$ and $k_{1d} = 2.21 \times T^{1.73}$. (Reprinted with permission from (Hirsh *et al.*, 1992a). Copyright (1992) American Chemical Society.)

An estimate of T_{1f} and/or T_{2f} is required for the evaluation of the C or B terms, respectively. These can either be obtained by direct EPR measurements or from Mössbauer spectroscopy. A direct measurement of T_{2f} is often tricky as T_{2f} is usually too fast to measure within the given time resolution of an EPR spectrometer. T_{2f} measurements are also often complicated by interference from spin-diffusion and cross-relaxation processes. However, T_{2f} can sometimes be indirectly estimated from temperature-dependent measurements. In the case of dipolar-enhanced spin-lattice relaxation measurement of the phyllosemiquinone radical in PS I, A_1^-, by an iron-sulfur cluster, F_X^-, it has been shown that T_{2f} is independent of temperature below 10 K. Above 10 K, $T_{2f} = T_{1f}$ as the line width of the F_X^- EPR signal is determined by its T_1 relaxation time. Thus T_{2f}, above

10 K, is approximated as being equal to the experimentally measured value of T_{1f} (Deligiannakis *et al.*, 1998). T_{2f} values have also been approximated by estimating the line widths of the fast relaxer in the regime where line widths are temperature dependent. However, unless measured directly or calibrated by comparison with a similar spin system with known magnetic parameters, simple estimation of T_{1f} and T_{2f} has often been a source of error in distance measurements by relaxation effects (Hirsh *et al.*, 1992a).

As described above, the non-single-exponential saturation-recovery transient of the slow relaxer s combined with a knowledge of the intrinsic spin-lattice relaxation rate of the non-interacting slow spin could yield a quantitative estimate of the spin-lattice relaxation enhancement due to the fast-relaxing species f, $T_{1s,dipolar}$. However, a potential complication of this technique is that the application of a microwave pulse to measure a saturation-recovery transient perturbs the observed spin-lattice relaxation rates. As reported by Brudvig and coworkers (Beck *et al.*, 1991),

$$\frac{1}{T_1(\text{observed})} = \frac{1}{T_1(\text{actual})} + CP \tag{24}$$

where $1/T_1$(observed) is the observed T_1 at power P, $1/T_1$(actual) is the T_1 at the limit of zero observe power where the spin is unperturbed by the observe microwave power and C is a proportionality constant. The strategy used to overcome this problem is to measure the spin-lattice relaxation rate, $1/T_1$(observed), at several power levels and extrapolate its value to zero observe power to obtain an accurate measure of the relaxation rate for the spin without a perturbation due to the observe microwaves. As discussed in section 3.2.3, theoretical extrapolation to zero observe power is not necessary for T_1 measurements conducted with a spin-echo-detected pulse sequence.

Direct measurements of spin-lattice relaxation rates by EPR spectroscopy are also complicated by the inhomogeneous line width of EPR resonances that are usually broader than the excitation profile of the microwave pulse. Under these conditions, saturation-recovery and electron spin-echo measurements are at best only hole-burning experiments and spectral diffusion and cross relaxation are interfering relaxation pathways that are encountered. Spectral diffusion results in the spreading of saturation over an inhomogeneously broadened line from spins that are on resonance to other spins that are off-resonance due to their different nuclear hyperfine, dipolar or quadrupolar coupled states. All three phenomena take place during the T_1 evolution of longitudinal magnetization. Cross relaxation is a result of flip-flop interactions between the electron spins. In a frozen solution of UV-generated radicals, where spin pairs are in close proximity, both spectral diffusion and cross relaxation can occur and the dominant contribution is on the order of the spin-spin relaxation time, T_2. Whereas in protein-embedded radicals, the spin pairs are isolated from each other via a protein matrix making cross-relaxation processes due to spin-spin interactions unlikely and the dominant contributions are mainly on the order of the spin-lattice relaxation time, T_1. Contributions from spectral diffusion and cross relaxation interfere with a saturation-recovery measurement by enhancing the apparent relaxation rate of the slow relaxer. This results in larger observed spin-lattice relaxation rates leading to the calculation of shorter interspin distances.

The recovery of the Z-magnetization in time, when the inhomogeneous broadening of the EPR resonance is larger than the pump pulse microwave field, is described by $M_Y^{rec}(t)$. It contains three contributions to the time-dependent evolution, the actual recovery of the Z-magnetization, the free induction decay of magnetization along X- or Y- and the response of the Z-magnetization remaining after the incident observe microwave pulse.

$$M_Y^{rec}(t) \propto [\frac{\gamma_e B_1 T_2}{1+(\delta\omega T_2)^2}(Z_0 - 1)\exp(\frac{-t}{T_1})] \tag{25}$$

where ω = observe frequency; $\delta\omega$ = difference in frequency (ω - ω_0) between the observe microwave field and the resonant frequency of the spins; Z_0 = residual magnetization remaining on-resonance after the pulse and $2B_1$ = amplitude of the observe microwave magnetic field.

Spin packets which are slightly off-resonance also make a contribution (spin diffusion and cross relaxation) to the apparent recovery observed in a saturation-recovery measurement. $M_Y^{FID}(t)$ represents the coherent magnetization in the X-Y plane of the rotating frame. In solid samples and proteins within the slow tumbling domain, $T_{2s} << T_{1s}$, and, therefore, the FID would not interfere with the saturation-recovery measurement. However, in liquids, $T_{2s} \sim T_{1s}$ which leads to a significant contribution from the FID. $M_Y^{resid}(t)$ represents the evolution of the residual Z magnetization that is remaining after the pulse.

$$M_Y^{FID}(t) \propto \frac{\pi}{(T_1 T_2)^{1/2}}\exp[-\gamma_e B_2 (\frac{T_1}{T_2})^{1/2} t] \tag{26}$$

$$M_Y^{resid}(t) \propto \frac{\pi}{(T_1 T_2)^{1/2}}\frac{B_1}{B_2}\exp[-\gamma_e B_2 (\frac{T_1}{T_2})^{1/2} t] \tag{27}$$

where B_2 is the field produced by the pump pulse.

The goal in a saturation-recovery experiment is to enhance M_Y^{rec} and minimize M_Y^{FID} and M_Y^{resid} by choosing the right conditions for observing the EPR signal. Using high power levels for the pump pulses maximizes the field B_2 and in doing so minimizes M_Y^{resid}. Furthermore, because the time constant for M_Y^{FID} and M_Y^{resid} is a function of B_2, increasing the effective field will minimize these components. M_Y^{FID} is usually reduced by taking advantage of its phase dependence. By alternating the phase of the pump pulse between 0° and 180° while adding the signals obtained in half-cycles (known as the pump phase-modulation technique), the contribution of M_Y^{FID} can be nulled. Complete recovery of M_Y^{rec} can also be monitored by acquiring a pre-trigger period of the T_1 transient for calibration purposes. However, if $T_1 >> T_2$, these contributions do not interfere with the measurement of saturation-recovery transients.

It has been shown that in the case of T_2 processes, application of short saturating pulses (~ 50 μs) prior to a saturation-recovery measurement suppresses these contributions. Application of long saturating pulses (~ 3-5 ms) (Beck *et al.*, 1991, Dalton *et al.*, 1972a, Dalton *et al.*, 1972b) or a train of saturating pulses (Percival & Hyde, 1975, Smigel *et al.*, 1974) have been shown to suppress spin diffusion on the order of the T_1 relaxation time. The length of the pulse applied to suppress spin-diffusion processes on the order of T_1 depends on the rates of the processes and typically Pulse Length ~ $1/k_X$ where k_X = rate of spectral diffusion or cross-relaxation (Beck *et al.*, 1991).

Another complication encountered in the analysis of a non-single-exponential saturation-recovery transient is the presence of **g** and **A** anisotropies in the slow- and fast-relaxing spins. The dipolar model described in 2.1.3 and 3.2.2 considers isotropic spins and thus eqns 20-23 contain only g_{iso} terms. However, **g** anisotropy in the slow relaxer would lead to the saturation of only a part of the total EPR resonance in a saturation-recovery measurement. Only the spins that are on-resonance will contribute to the observed spin-lattice relaxation transient. The interspin vector is fixed relative to the **g** tensor of the slow relaxer, which restricts sampling the possible orientations to a subset of values. At the extremes of the powder pattern (at g_x and g_z) there is only a single orientation of the interspin vector. More orientations are accessible as the g value approaches the average g value. The values of θ at a certain field position can be predicted if the **g** tensor and the orientation of the interspin vector relative to the principal axis system are known, and the observed saturation-recovery transient can be summed over these orientations to correct for this problem. However, most slow relaxers described in this chapter and studied to date are nearly isotropic organic radicals with small **g** and **A** anisotropies, in which case the saturation-recovery transient will sample all orientations.

On the other hand, **g** anisotropy in the fast relaxer is a case that is often encountered in saturation-recovery measurements. The qualitative effects of **g** anisotropy in a fast relaxer have been examined by Rutherford and coworkers (Deligiannakis *et al.*, 1998) by determining the distance between the phyllosemiquinone (A_1^-) and the [4Fe-4S] cluster F_X^- in PS I. The reduced F_X center in PS I is known to possess significant **g** anisotropy (Golbeck, 1992) while the phyllosemiquinone radical A_1^- is isotropic. The saturation-recovery transients at 4.7 K were modeled using the B term (eqns 22-23) and representative sets of angles. With the exception of a small set of angles, most F_X-A_1 distances obtained fell between 10-20 Å (Figure 12). The dipolar contribution was maximized for 0° and 90° and minimized at the magic angle. Therefore, the polar angles 47-51° match the optimal θ angles for only a limited range of Euler angles. This limits the probability that the interspin vector will be oriented relative to F_X in such a way that its effective g value is 2.0. Thus, taking $g^{effective}$ to be the average g^{av} in this case appeared to be a valid assumption.

The dipolar model (eqns 19-23) describes a pairwise interaction between slow- and fast-relaxing S = 1/2 spins. However, analysis of the non-single-exponential saturation-recovery transient can be considerably different in the case of fast-relaxing spins with S > 1/2 (detailed discussion by Eaton and Eaton in Chapter 8). For a fast-relaxing spin with S > 1/2 there is more than one Δm_S transition and each of these transitions can have different potentials for inducing transitions in the slow-relaxing spin. Also, each Δm_S transition may have different values of T_{1f} and T_{2f}. This affects the approximations leading to the choice of the B or C term having a larger influence on k_{1d}. For the B term, there is no longer a single g_f or a single $\mathbf{g}_f$ tensor. In this case, the resonant frequencies for the $\Delta m_S = 1$ transitions become a complicated function of the **g** anisotropy, the resonant frequency of the slow relaxer and the zero-field splittings. However, these complications can be simplified when a single $\Delta m_S = 1$ transition dominates the relaxation enhancement. This was demonstrated by the saturation-recovery analysis conducted on P_{870}^+ in bRCs (Hirsh & Brudvig, 1993). P_{870}^+ in this system displays non-single-exponential spin lattice relaxation behavior due to its interaction with the non-heme Fe(II) (S = 2). Despite making the assumption of a single $\Delta m_S = 1$ transition for the high-spin S = 2 Fe(II) and applying the S = 1/2 dipolar model to analyze the non-single-exponential relaxation transients, an interspin distance of 28 Å is obtained that is in good agreement with previously published crystallographic data. Although the analysis used by Hirsh *et al.* (1992) to investigate the $Y_D^\bullet$-Fe(II) interaction ignored the complications arising from a S = 2 spin, it nevertheless resulted in an accurate distance estimate because the study considered a ratio of parameters between PS II and the bRCs. By using the known distance in the bRC and the homologous non-heme Fe(II) center to calibrate the relaxation enhancement, it was not necessary to rely on specific values for the relaxation times T_{1f} and T_{2f}.

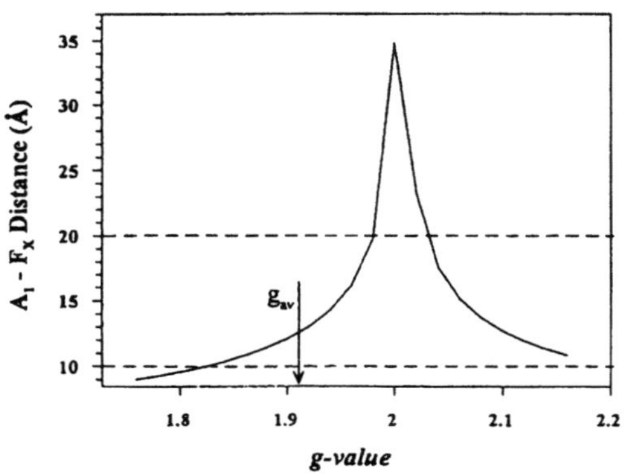

Figure 12. Plot describing the effect of the **g** tensor on the calculated F_X-A_1 distance in PS I. The experimental saturation-recovery transients are fit to the B term (eqns 22-23). (Reprinted with permission from (Deligiannakis *et al.*, 1998). Copyright (1998) American Chemical Society.)

The effects of a large line width which gives a distribution of Larmor frequencies, the presence of excited spin manifolds and spin diffusion through hyperfine pathways on a saturation-recovery experiment have also been analyzed by examining the $Y_D^{\bullet}$-Mn_4 interaction in PS II (Koulougliotis *et al.*, 1997). Previous studies had estimated the $Y_D^{\bullet}$-Mn_4 distance to be 28-43 Å but the non-single-exponential character of the spin-lattice relaxation transient had been incorrectly assumed to be the result of $Y_D^{\bullet}$ interacting with two separate Mn centers at different distances (Evelo *et al.*, 1989). Koulougliotis *et al.* (1997) examined the effect of the different spin states of the Mn_4 cluster at different temperatures. The temperature dependence of the spin-lattice relaxation enhancement experienced by $Y_D^{\bullet}$ has been analyzed by accounting for the temperature variation of the effective magnetic moment of the S_2 state of the Mn_4 cluster. Below 10 K, the ground state (S = 1/2) was found to dominate whereas at T > 30 K, the first excited state dominates the relaxation. Because the S_2 multiline signal is dominated by ^{55}Mn nuclear hyperfine couplings, nuclear relaxation processes allow frequencies near resonance to be accessed resulting in greater T_1 enhancement. In the dipolar model, the two spins are described as isotropic with fixed values for the Larmor frequency. However, the two spins may have a distribution of Larmor frequencies depending on the EPR line shape. The S_2 multiline EPR signal from the Mn_4 cluster in PS II is centered at ~ g = 2 and contains approximately 18-22 lines with a spectral width of about 2000 Gauss. Consequently, when analyzing the $Y_D^{\bullet}$-Mn_4 interaction in the S_2 state of PS II, a fixed Larmor frequency approximation is invalid. The study used a dipolar model that included the line shapes of both $Y_D^{\bullet}$ and the S_2 multiline to analyze the $Y_D^{\bullet}$-Mn_4 interaction. If the Larmor frequency ω_f remains fixed on the timescale of spin-lattice relaxation, the values of ω_s and ω_f for the pairwise interaction would vary statistically throughout the range of the inhomogeneous line shape. When the magnetic moments of both spins are isotropic, the equation for I(t) can be extended by integrating over the spectral line shapes of both interaction signals.

$$I(t) = 1 - N\int\int\int g_s(\omega_s) g_f(\omega_f) \sin\theta [\exp - (k_{1scalar} + k_{1\theta})] d\omega_s d\omega_f d\theta \quad (28)$$

where ω_s and ω_f are the Larmor frequencies of the slow- and fast-relaxing spins, respectively, and $g_s(\omega_s)$ and $g_f(\omega_f)$ are line shape functions.

Detailed analysis of the effect of the line shape of the fast relaxer demonstrates that as the line width of the fast relaxing spin increases, it becomes less effective as a relaxation enhancer (if the B term dominates) because more of the pairwise interactions involve large values of $(\omega_s - \omega_f)^2$. With a 2000 G line width for the S_2 multiline signal, the line width of ~ 10 G for $Y_D^{\bullet}$ was negligible and was approximated as a fixed Larmor frequency. Simply including the line width for the S_2 multiline signal was seen to yield a physically unreasonable short $Y_D^{\bullet}$-Mn_4 distance of 13 Å. In order to account for relaxation enhancement of $Y_D^{\bullet}$ by Mn_4 at a longer distance, Koulougliotis *et al.* (1997) have proposed a new dynamic dipolar model. The line width of the S_2 EPR signal is from ^{55}Mn hyperfine interactions and the ^{55}Mn nuclear spin-lattice relaxation could modulate the effective Larmor frequency of the S_2 state multiline EPR signal when occurring on the same timescale as the T_1 measurements. Indeed, nuclear transitions do occur on this timescale and are the basis for pulse field-sweep EPR measurements of ^{55}Mn complexes (Falkowski *et al.*, 1986, Tan *et al.*, 1991). Thus, the Larmor frequency of the fast relaxer changes during the course of the experiment and causes the effective $(\omega_s - \omega_f)^2$ to be lower. Examination of the relaxation behavior of the Mn_4 cluster in the S_2 state indicated an unusual decrease in these rates below 3 K and this decrease was attributed to the slowing of electron-nuclear dynamic processes such that they no longer contributed to the spin-lattice relaxation rates. The electron-nuclear interactions are manifested as a stronger T_1 enhancement of the slow relaxer, $Y_D^{\bullet}$ and lead to significant effects over a longer $Y_D^{\bullet}$-Mn_4 distance. Saturation-recovery data at 1.8 K have been used to obtain a conservative $Y_D^{\bullet}$-Mn_4 distance estimate of ≥ 22 Å.

3.2.3 Spin-Echo Measurements of Spin-Lattice Relaxation Effects

As mentioned in section 3.2.2, saturation-recovery techniques require the measurement of T_1 relaxation times at several power levels that can be extrapolated to zero power to minimize the perturbation due to the observing microwaves. However, this can be avoided by the use of electron spin-echo (ESE) techniques, as the time evolution of the recovering longitudinal magnetization in a spin-echo-detected saturation-recovery experiment proceeds under the influence of microwave pulses. The most commonly used echo sequence is a two-pulse Hahn echo. A three-pulse stimulated echo is also sometimes used to determine T_1 relaxation times. Stimulated echoes can preserve magnetization for longer echo intervals than two-pulse echoes. In addition, an echo-detected saturation-recovery sequence includes either a short pulse, a long saturation pulse or train of pulses to create non-equilibrium longitudinal magnetization and also eliminate interference from spin-diffusion and cross-relaxation processes.

Brudvig and coworkers have systematically compared the results of *cw* and ESE-detected saturation-recovery techniques to characterize the spin-diffusion effects that contribute to the spin-lattice relaxation of the tyrosyl radicals in PS II ($Y_D^{\bullet}$) and ribonucleotide reductase (Beck *et al.*, 1991). Two-pulse spin-echo-detected spectra are potentially complicated by spin-diffusion and cross-relaxation effects. The phase-memory times allow for evaluation of contributions from nuclear and electron-electron spin-spin modulation to the observed magnetization.

The relative positions of the electron-transfer components in PS I have been probed by ESE saturation-recovery techniques (Berry *et al.*, 1997). The relaxation transients for A_0^-, A_1^- and P_{700}^+ were found to be non-single-exponential in the presence of the [4Fe-4S]$^-$ centers. The P_{700}-F_X distance was determined to be 25-35 Å and the distance between A_1-F_X and A_0-F_X to be 20-30 Å. The progress of bulk magnetization in these experiments was detected by using a Hahn echo sequence. A three-pulse stimulated echo sequence was used in the case of F_X. In this study, the analysis of the spin-lattice relaxation transients of P_{700}^+, A_0^- and A_1^- in the presence of F_X^- was based on the dipolar model given in eqn 19. However, as mentioned in the previous section, the presence of **g** anisotropy in the [4Fe-4S] clusters in PS I can lead to a more complicated analysis. Also, as has been noted in a recent progressive microwave power-saturation study conducted by Brudvig and coworkers, the presence of triplet oxygen in PS I samples contributes to the observed spin-lattice relaxation enhancement of P_{700}^+ (Lakshmi *et al.*, 1999b). The P_{700}^+ center in PS I is fairly solvent exposed leading to increased accessibility to paramagnetic triplet oxygen. It is thus necessary to evaluate the contribution to the spin-lattice relaxation enhancement from dissolved oxygen prior to conducting quantitative spin-lattice relaxation measurements.

Deligiannakis and Rutherford (1996) have examined the spin-lattice relaxation behavior of the pheophytin radical ($Pheo_A^-$) in the presence of the non-heme Fe(II) in PS II. The $Pheo_A^-$ radical was studied in manganese-depleted PS II membranes with doubly-reduced Q_A (Q_A^{2-}) to eliminate magnetic interactions with Q_A^- and the Mn_4 cluster, thus permitting the investigation of the pairwise interaction between the $Pheo_A^-$ and high-spin non-heme Fe(II). The $Pheo_A^-$-Fe(II) coupling was calibrated by comparison to the $Y_D^•$-Fe(II) coupling reported previously yielding a distance of 20 ± 4 Å between the $Pheo_A^-$ radical and the iron center. The error in the distance estimate was determined by taking into account both the experimental error and estimated error from the presence of a distributed dipole over the pheophytin cofactor.

Kodera *et al.* (1995) have estimated the $Y_Z^•$-Mn_4 distance in PS II by ESE-detected saturation-recovery techniques. This study used a previously determined $Y_D^•$-Mn_4 distance as a calibration and the $Y_Z^•$-Mn_4 distance was estimated to be 15 Å based on experiments conducted at 80 K. However, a $Y_Z^•$-Mn_4 distance of 15 Å is longer than the ~ 8 Å distance determined by EPR line shape simulations (see section 3.1.1). As discussed by Koulougliotis *et al.* (1997), saturation-recovery transients measured at higher temperatures may contain contributions from higher excited states of the Mn_4 fast relaxer that could result in inaccurate interspin distances. The apparent error in the $Y_Z^•$-Mn_4 distance in this study could also be due to heterogeneity in preparing the states of the protein.

3.3 Pulsed Electron-Electron Double-Resonance Distance Measurements

Dipolar interactions between paramagnetic species can also be probed by the pulsed electron-electron double-resonance (PELDOR) EPR technique. Consider a pairwise interaction of two paramagnetic species, 1 and 2, respectively. In the PELDOR technique, the spins are excited by three microwave pulses. The first and third pulses constitute a Hahn echo sequence with a time interval τ and have a carrier frequency ω_1 that is resonant with the EPR transitions of the paramagnetic species 1. The echo sequence forms the primary ESE signal of these spins. The second pulse is a 'pump' 180° pulse that is executed during the first τ interval of the echo sequence. It is separated from the first pulse by a time interval τ' ($\tau' \neq \tau$) and has a carrier frequency ω_2, resonant with the species 2. If the two spins are spin coupled, excitation of the spins of species 2 can change the local magnetic field of the spins of species 1 resulting in incomplete refocusing of magnetization after the echo sequence. Thus, the amplitude of the primary ESE signal would exhibit an oscillating dependence on the second pulse position. In a PELDOR experiment, the frequency of oscillations of the ELDOR trace are determined by the dipolar interaction (Hara *et al.*, 1996). A detailed discussion of the PELDOR technique and its applications has previously been provided by Tsvetkov and coworkers (Milov *et al.*, 1998).

Kawamori and coworkers (Hara *et al.*, 1996) have used PELDOR to determine the distance from $Y_D^\bullet$ to the Mn_4 cluster in PS II. The ESE ELDOR experiment was performed with a fixed echo interval (τ) of 1000 ns and varying τ' between 8 and 960 ns. The microwave pump pulse was set to 9.68 GHz, the resonant frequency for $Y_D^\bullet$, and the first and third pulses were set to 9.79 GHz, the resonant frequency of the S_2 multiline signal from the Mn_4 cluster. The ELDOR traces for $Y_D^\bullet$ interacting with the Mn_4 cluster exhibited oscillations due to the dipolar interaction between the two centers. Calculations yielded a dipolar coupling of 2.6 ± 0.03 MHz that corresponds to a $Y_D^\bullet$-Mn_4 interspin distance of 27.1 ± 0.15 Å, which is in good agreement with distances previously determined by selective hole burning (Kodera *et al.*, 1994).

Typically, incomplete excitation was observed in $Y_D^\bullet$ as the spectral width of the $Y_D^\bullet$ signals (~2 mT) exceeded the excitation width of the 180 degree pump pulse (~1.4 mT). Also, $Y_D^\bullet$ in PS II was thought to interact with both the Mn_4 cluster in the S_2 state and the Fe in Cyt b_{559}. The effects from the two paramagnets on $Y_D^\bullet$ were deconvolved by using the difference in their T_2 relaxation times at 12 K. The interaction of $Y_D^\bullet$ with Cyt b_{559} did not exhibit an oscillatory ELDOR trace suggesting a very weak dipolar interaction between the two spins.

Similar PELDOR experiments were also performed on Ca^{2+}-depleted PS II to determine the dipolar coupling between $Y_D^\bullet$ and the $S_2Y_Z^\bullet$ state of the OEC (Hara *et al.*, 1996). However, analysis of the ELDOR trace was complicated by interference from unconverted centers with the Mn_4 cluster in the S_2 state. The percentage of the centers in the S_2 state was estimated to be ~ 15 % based on the premise that the T_2 relaxation time for the S_2 state is smaller than that of the $S_2Y_Z^\bullet$ state. As ELDOR oscillations are additive, a difference between the ELDOR trace from the mixture of the S_2 and $S_2Y_Z^\bullet$ states and the trace from the $Y_D^\bullet$-S_2 interaction was used to isolate the effects solely due to the $Y_D^\bullet$-($S_2Y_Z^\bullet$) interaction. Other EPR evidence at the time indicating that the species interacting with $Y_D^•$ was a dipolar-coupled system in itself required treatment of the orientation dependence of its dipolar vector. Detailed analysis by Hara *et al.* (1996) has yielded a $Y_D^•$-($S_2Y_Z^•$) coupling of 1.99 ± 0.03 MHz corresponding to a point dipole interspin distance of 29.7 ± 0.15 Å.

Application of the '2+1' pulse sequence, described in Chapter 10, has provided the following information concerning PS II. The distance between $Y_Z^•$ and $Y_D^•$ is ~ 30 Å (Astashkin *et al.*, 1994). In oriented Mn-depleted PS II samples the angle between the membrane normal and the vector connecting $Y_D^•$ to $Y_Z^•$ is 100 ± 2° and the angle between the membrane normal and the vector connecting $Y_D^•$ to the Mn_4 cluster is 110 ± 2° (Astashkin *et al.*, 1998). These estimates have yielded a $Y_Z^•$-Mn_4 distance of ~ 14 Å and a $Y_Z^•$-Mn_4 orientation of ~ 100° relative to the membrane normal. However, the $Y_Z^•$-Mn_4 distance and orientation obtained in this study are larger than the current estimates of an ~ 8 Å distance and 75° orientation relative to the membrane normal (see section 3.1.1). In a study by Shigemori *et al.* (1998), the distance between $Y_D^•$ and Q_A^- was found to be 38.8 ± 1.1 Å and the distance between $Y_D^•$ and Chl_Z^+ was found to be 29.4 ± 0.5 Å. In Fe^{2+}-depleted, oriented preparations, the angle between the membrane normal and the vector connecting P_{680} to Y_D was determined to be 42° (Yoshii *et al.*, 1999).

4. SUMMARY

As has been demonstrated in this chapter, when used appropriately EPR distance measurements are a powerful tool to elucidate structural and electronic properties in photosynthetic reaction centers. Both spectral line shapes and relaxation properties of a pair of spins can be a used to probe exchange and dipolar couplings between interacting paramagnetic cofactors in the RCs. Due to their relative simplicity and ease of purification, bacterial reaction centers have served as suitable models for progressive power-saturation and saturation-recovery EPR techniques. Oriented and powder line shape simulations led to some of the earliest distance estimates in RCs and have yielded useful information about the structure and function of the Fe(II)-quinone complex.

The structure and function of the OEC in PS II has been extensively probed using EPR distance measurement techniques (Figure 13). X-ray and electron diffraction data on PS II have been limited due to the difficulty in preparing well-ordered manganese-containing crystals. EPR distance and orientation measurements have been instrumental in establishing the distance and relative orientation of almost all the primary cofactors in PS II. The proximity of $Y_Z^{\bullet}$ to the Mn_4 cluster and determination of the orientation of the interspin vector relative to the membrane normal has made for detailed investigations into the mechanistic aspects of proton-coupled electron transfer between $Y_Z^{\bullet}$ and the Mn_4 cluster. The center-to-center point-dipole distance between Y_Z and the Mn_4 cluster has been found to be 7.5 - 8.5 Å suggesting a hydrogen bond between Y_Z and a ligand on the metal cluster (possibly a hydroxyl group) that facilitates proton-coupled electron transfer to the tyrosyl residue, Y_Z.

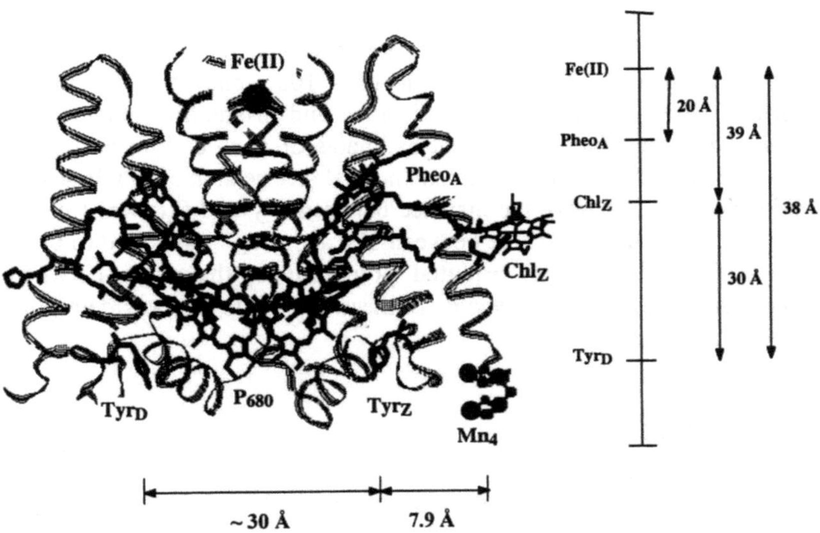

Figure 13. A theoretical model of PS II (Svensson *et al.*, 1996) combined with a summary of distances and relative orientation of cofactors determined by EPR distance measurements. The distances are summarized on a ruler, although the interspin vectors are not collinear.

Determination of the distance between the dark-stable redox-active tyrosine, $Y_D^{\bullet}$, and the non-heme iron and the Mn_4 cluster supported the presence of a C_2-axis of symmetry in PS II similar to that in the bRCs. Furthermore, the distance of ≥ 22 Å between $Y_D^{\bullet}$ and the Mn_4 cluster has unequivocally established that Y_Z and Y_D are not symmetrically placed relative to the metal cluster indicating that Y_D is not involved in the primary electron-transfer pathways. Measurement of the pheophytin-Fe(II) distance has indicated its proximity to the quinones, thus establishing that $Pheo_A$ acts as a bridge between the primary donor P_{680} and the acceptor quinone Q_A.

The mechanism of photoinactivation in PS II has long been of interest to the photosynthesis community. Measurement of the Chl_Z-Fe(II) distance in

PS II provided evidence for the presence of an accessory redox-active Chl, distinct from the pigments present in the bRCs. The estimated location of Chl_Z has enabled the search of potential ligands in the D1- and D2- subunits. Chl_Z is now a well-established player in a secondary electron-transfer pathway in PS II that may play a role in protection of PS II from photoinactivation.

A 4 Å resolution x-ray crystallography structure has only recently been available for PS I and lacks the precise location of the A_1 phylloquinone and the terminal iron-sulfur clusters, F_A and F_B (Krauß *et al.*, 1996). Critical distance estimates between the key cofactors in PS I have been determined by using EPR distance measurements (Figure 14). Saturation-recovery measurements have determined a distance of 25-35 Å between P_{700}^+ and the F_X iron-sulfur cluster and a distance of 20-30 Å has been estimated between A_0, A_1 and F_X, respectively.

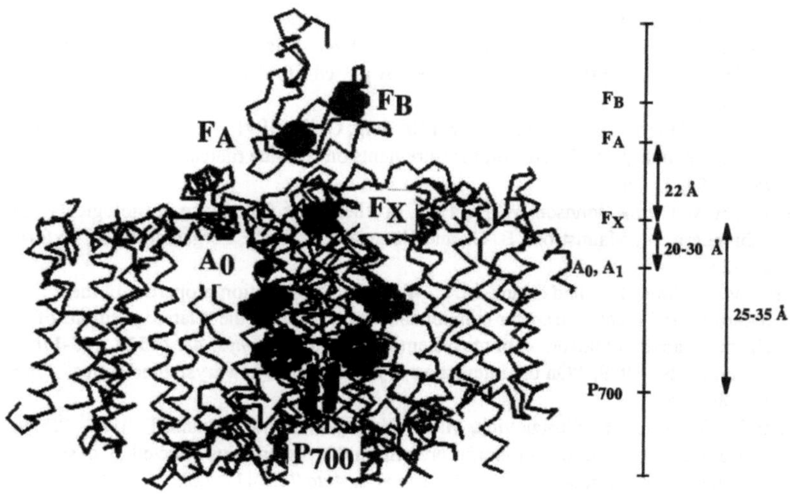

Figure 14. PS I protein structure (Krauß *et al.*, 1996) combined with a summary of distance and orientation information determined by EPR distance measurements. The distances are summarized on a ruler, although the interspin vectors are not collinear.

The x-ray structure of PS I indicates a 22 Å distance between F_X and F_A or F_B, but there is a two-fold ambiguity in the orientation of the PsaC protein to which they are linked (Krauß *et al.*, 1996). EPR of oriented membranes, single crystals of PS I and progressive power-saturation measurements of P_{700}^+ interacting with the terminal iron-sulfur clusters has unambiguously assigned the positions of the F_A and F_B clusters. It is now known that F_A is proximal to F_X while F_B is more distant.

5. REFERENCES

Abragam, A., 1961, *The Principles of Nuclear Magnetism*, Clarendon Press, Oxford.

Allen, J. P., Feher, G., Yeates, T. O., Komiya, H. and Rees, D. C., 1987, "Structure of the reaction center from *Rhodobacter-Sphaeroides* R-26 - The cofactors .1" *Proc. Natl. Acad. Sci. USA* **84**, 5730-5734.

Astashkin, A. V., Hara, H. and Kawamori, A., 1998, "The pulsed electron-electron double resonance and '2+1' electron spin echo study of the oriented oxygen-evolving and Mn-depleted preparations of photosystem II" *J. Chem. Phys.* **108**, 3805-3812.

Astashkin, A. V., Kodera, Y. and Kawamori, A., 1994, "Distance between tyrosines Z^+ and D^+ in plant photosystem-II as determined by pulsed EPR" *Biochim. Biophys. Acta* **1187**, 89-93.

Beck, W. F., Innes, J. B., Lynch, J. B. and Brudvig, G. W., 1991, "Electron spin-lattice relaxation and spectral diffusion measurements on tyrosine radicals in proteins" *J. Magn. Reson.* **91**, 12-29.

Beinert, H. and Orme-Johnson, W. H., 1967, in "Magnetic Resonance in Biological Systems" (Ehrenberg, A., Malmström, B. G., and Vänngård, T., Eds.), Pergamon Press, Oxford, pp 221-247.

Berry, M. C., Bratt, P. J. and Evans, M. C. W., 1997, "Relaxation properties of the photosystem I electron transfer components: Indications of the relative positions of the electron transfer cofactors in photosystem I" *Biochim. Biophys. Acta* **1319**, 163-176.

Bloembergen, N., 1949, "On the interaction of nuclear spins in a crystalline lattice" *Physica* **15**, 386-426.

Blum, H., Bowyer, J. R., Cusanovich, M. A., Waring, A. J. and Ohnishi, T., 1983, "Spin-lattice relaxation rates of iron-sulfur proteins and heme proteins affected by dysprosium complexes and temperature" *Biochim. Biophys. Acta* **748**, 418-428.

Blum, H., Cusanovich, M. A., Sweeney, W. V. and Ohnishi, T., 1981, "Magnetic interactions between dysprosium complexes and soluble iron-sulfur proteins" *J. Biol. Chem.* **256**, 2199-2206.

Blum, H., Leigh, J. S. and Ohnishi, T., 1980, "Effect of dysprosium on the spin-lattice relaxation time of cytochrome *c* and cytochrome *a*" *Biochim. Biophys. Acta* **626**, 31-40.

Bonvoisin, J., Blondin, G., Girerd, J.-J. and Zimmerman, J.-L., 1992, "Theoretical study of the multiline EPR signal from the S_2 state of the oxygen evolving complex of photosystem II" *Biophys. J.* **61**, 1076-1086.

Boussac, A. and Rutherford, A. W., 1988, "Nature of the inhibition of the oxygen-evolving enzyme of photosystem II induced by NaCl washing and reversed by the addition of Ca^{2+} or Sr^{2+}" *Biochemistry* **27**, 3476-3483.

Brudvig, G. W., Blair, D. F. and Chan, S. I., 1984, "Electron spin relaxation of Cu_A and cytochrome *a* in cytochrome *c* oxidase comparison to heme, copper, and sulfur radical complexes" *J. Biol. Chem.* **259**, 11001-11009.

Butler, W. F., Calvo, R., Fredkin, D. R., Isaacson, R. A., Okamura, M. Y. and Feher, G., 1984, "The electronic-structure of Fe^{2+} in reaction-centers from *Rhodopseudomonas-Sphaeroides* .3. Electron-paramagnetic resonance measurements of the reduced acceptor complex" *Biophys. J.* **45**, 947-973.

Butler, W. F., Johnston, D. C., Shore, H. B., Fredkin, D. R., Okamura, M. Y. and Feher, G., 1980, "The electronic structure of Fe^{2+} in reaction centers from *Rhodopseudomonas Sphaeroides* 1. Static magnetization measurements" *Biophys. J.* **32**, 967-992.

Castner, T. G., 1959, "Saturation of the paramagnetic resonance of a V center" *Phys. Rev.* **115**, 1506-1515.

Coffman, R. E. and Buettner, G. R., 1979, "Limit function for long-range ferromagnetic and anti-ferromagnetic superexchange" *J. Phys. Chem.* **83**, 2387-2392.

Case, G. D. and Leigh, J. S., 1976, "Intramitochondrial positions of cytochrome heme groups determined by dipolar interactions with paramagnetic cations" *Biochem. J.* **160**, 769-783.

Case, G. D., Ohnishi, T. and Leigh, J. S., 1976, "Intramitochondrial positions of ubiquinone and iron-sulfur centers determined by dipolar interactions with paramagnetic ions" *Biochem. J.* **160**, 785-795.

Dalton, L. R., Kwiram, A. L. and Cowen, J. A., 1972a, "Electron spin-lattice and cross relaxation in irradiated malonic acid" *Chem. Phys. Lett.* **14**, 77-81.

Dalton, L. R., Kwiram, A. L. and Cowen, J. A., 1972b, "Electron spin-lattice relaxation in molecular crystals-S = 1/2" *Chem. Phys. Lett.* **17**, 495-499.

Debus, R. J., 1992, "The manganese and calcium-ions of photosynthetic oxygen evolution" *Biochim. Biophys. Acta* **1102**, 269-352.

Deisenhofer, J., Epp, O., Miki, K., Huber, R. and Michel, H., 1984, "X-ray structure analysis of a membrane protein complex - Electron density map at 3 Å resolution and a model of the chromophores of the photosynthetic reaction center from Rhodopseudomonas Viridis" *J. Mol. Biol.* **180**, 385-398.

Deisenhofer, J., Epp, O., Miki, K., Huber, R. and Michel, H., 1985, "Structure of the protein subunits in the photosynthetic reaction center of Rhodopseudomonas Viridis at 3 Å resolution" *Nature* **318**, 618-624.

Deligiannakis, Y., Hanley, J. and Rutherford, A. W., 1998, "Spin-lattice relaxation of the phyllosemiquinone radical of photosystem I" *Biochemistry* **37**, 3329-3336.

Deligiannakis, Y. and Rutherford, A. W., 1996, "Spin-lattice relaxation of the pheophytin, $pheo^-$, radical of photosystem II" *Biochemistry* **35**, 11239-11246.

Dismukes, G. C., Frank, H. A., Friesner, R. and Sauer, K., 1984, "Electronic interactions between iron and bound semiquinones in bacterial photosynthesis - Electron paramagnetic resonance spectroscopy of oriented cells of *Rhodopseudomonas-Viridis*" *Biochim. Biophys. Acta* **764**, 253-271.

Dorlet, P., Di Valentin, M., Babcock, G. T. and McCracken, J. L., 1998, "Interaction of Y_Z with its environment in acetate-treated photosystem II membranes and reaction center cores" *J. Phys. Chem. B* **102**, 8239-8247.

Ermler, U., Fritzsch, G., Buchanan, S. K. and Michel, H., 1994, "Structure of the photosynthetic reaction center from Rhodobacter-Sphaeroides at 2.65-angstrom resolution - Cofactors and protein-cofactor interactions" *Structure* **2**, 925-936.

Evelo, R. G., Styring, S., Rutherford, A. W. and Hoff, A. J., 1989, "EPR relaxation measurements of photosystem II reaction centers: Influence of S-state oxidation and temperature" *Biochim. Biophys. Acta* **973**, 428-442.

Falkowski, K. M., Scholes, C. P. and Tayleo, H., 1986, "Pulse field-sweep electron-paramagnetic-resonance - A method of extracting hyperfine information from inhomogeneously broadened electron-paramagnetic-resonance lines of bioinorganic systems" *J. Magn. Reson.* **68**, 453-468.

Feher, G., Allen, J. P., Okamura, M. Y. and Rees, D. C., 1989, "Structure and function of bacterial photosynthetic reaction centers" *Nature* **339**, 111-116.

Feher, G., Isaacson, R. A., McElroy, J. D., Ackerson, L. C. and Okamura, M. Y., 1974, "On the question of the primary acceptor in bacterial photosynthesis:Manganese substituting for iron in reaction centers of *Rhodopseudomonas Sphaeroides* R-26" *Biochim. Biophys. Acta* **368**, 135-139.

Fielding, L., More, K. M., Eaton, G. R. and Eaton, S. S., 1986, "Metal-nitroxyl interactions. 51. Collapse of iron-nitroxyl electron-electron spin-spin splitting due to the increase in the electron spin relaxation rate for high-spin iron(III) when temperature is increased" *J. Amer. Chem. Soc.* **108**, 8194-8196.

Force, D. A., Randall, D. W. and Britt, R. D., 1997, "Proximity of acetate, manganese and exchangeable deuterons to tyrosine $Y_Z^\bullet$ in acetate inhibited photosystem II membranes: Implications for the direct involvement of $Y_Z^\bullet$ in water-splitting" *Biochemistry* **36**, 12062-12070.

Fromme, P., Witt, H. T., Schubert, W.-D., Klukas, O., Saenger, W. and Krauß, N., 1996, "Structure of photosystem I at 4.5 Å resolution: A short review including evolutionary aspects" *Biochim. Biophys. Acta* **1275**, 76-83.

Galli, C., Innes, J. B., Hirsh, D. J. and Brudvig, G. W., 1996, "Effects of dipole-dipole interactions on microwave progressive power saturation of radicals in proteins" *J. Magn. Reson. B* **110**, 284-287.

Ghanotakis, D. F., Babcock, G. T. and Yocum, C. F., 1985, "Structure of the oxygen-evolving complex of photosystem II - calcium and lanthanum compete for sites on the oxidizing side of photosystem II which control the binding of water-soluble polypeptides and regulate the activity of the manganese complex" *Biochim. Biophys. Acta* **809**, 173-180.

Gilchrist, M. L., Ball, J. A., Randall, D. W. and Britt, R. D., 1995, "Proximity of the manganese cluster of photosystem II to the redox-active tyrosine Y_Z" *Proc. Natl. Acad. Sci. USA* **92**, 9545-9549.

Golbeck, J. H., 1992, in "Structure and function of photosystem I" *Ann. Rev. Plant Phys. Plant Molec. Biol.* **43**, 293-324.

Golbeck, J. H., and Bryant, D. A., 1991, "Photosystem I" *Current Topics in Bioenergetics* **16**, 83-177.

Goodman, G. and Leigh Jr., J. S., 1985, "Distance between the visible copper and cytochrome *a* in bovine heart cytochrome oxidase" *Biochemistry* **24**, 2310-2317.

Guigliarelli, B., Guillaussier, J., More, C., Sétif, P., Bottin, H. and Bertrand, P., 1993, "Structural organization of the iron-sulfur clusters in Synechocystis 6803 photosystem I. EPR study of oriented thylakoid membranes and analysis of the magnetic interactions" *J. Biol. Chem.* **268**, 900-908.

Hales, B. J., 1993, "Intrinsic and extrinsic paramagnets as probes of metal-clusters" *Metallobiochem., PT D* **227**, 384-395.

Hara, H. and Kawamori, A., 1997, "A selective hole burning method applied to determine distances between paramagnetic species in photosystems" *Appl. Magn. Reson.* **13**, 241-257.

Hara, H., Kawamori, A., Astashkin, A. V. and Ono, T., 1996, "The distances from tyrosine D to redox-active components on the donor side of photosystem II determined by pulsed electron-electron double resonance" *Biochim. Biophys. Acta* **1276**, 140-146.

Hirsh, D. J., Beck, W. F., Innes, J. B. and Brudvig, G. W., 1992a, "Using saturation recovery EPR to measure distances in proteins - Applications to photosystem II" *Biochemistry* **31**, 532-541.

Hirsh, D. J., Beck, W. F., Lynch, J. B., Que, L. J. and Brudvig, G. W., 1992b, "Using saturation-recovery EPR to measure exchange couplings in proteins: Applications to ribonucleotide reductase" *J. Amer. Chem. Soc.* **114**, 7475-7481.

Hirsh, D. J. and Brudvig, G. W., 1993, "Long-range electron spin-spin interactions in the bacterial photosynthetic reaction center" *J. Phys. Chem.* **97**, 13216-13222.

Hoff, A. J., 1986, "Magnetic-interactions between photosynthetic reactants" *Photochem. Photobiol.* **43**, 727-745.

Hoganson, C. W. and Babcock, G. T., 1997, "A metalloradical mechanism for the generation of oxygen from water in photosynthesis" *Science* **277**, 1953-1956.

Hyde, J. S. and Rao, K. V. S., 1978, "Dipolar-induced electron spin-lattice relaxation in unordered solids" *J. Magn. Reson.* **29**, 509-516.

Innes, J. B. and Brudvig, G. W., 1989, "Location and magnetic-relaxation properties of the stable tyrosine radical in photosystem II" *Biochemistry* **28**, 1116-1125.

Kamlowski, A., van der Est, A., Fromme, P., Krauß, N., Schubert, W.-D., Klukas, O. and Stehlik, D., 1997a, "The structural organization of the PsaC protein in photosystem I from single crystal EPR and x-ray crystallographic studies" *Biochim. Biophys. Acta* **1319**, 199-213.

Kamlowski, A., van der Est, A., Fromme, P. and Stehlik, D., 1997b, "Low temperature EPR on photosystem I single crystals: Orientation of the iron-sulfur centers F-A and F-B" *Biochim. Biophys. Acta* **1319**, 185-198.

Kodera, Y., Dzuba, S. A., Hara, H. and Kawamori, A., 1994, "Distances from tyrosine D^+ to the manganese cluster and the acceptor iron in photosystem II as determined by selective hole-burning in EPR spectra" *Biochim. Biophys. Acta* **1186**, 91-99.

Kodera, Y., Hara, H., Astashkin, A. V., Kawamori, A. and Ono, T. A., 1995, "EPR study of trapped tyrosine Z^+ in Ca-depleted photosystem II" *Biochim. Biophys. Acta* **1232**, 43-51.

Kodera, Y., Takura, K. and Kawamori, A., 1992, "Distance P_{680} from the manganese complex in photosystem II studies by time-resolved EPR" *Biochim. Biophys. Acta* **1101**, 23-32.

Kok, B., Forbush, B. and McGloin, M., 1970, "Cooperation of charges in photosynthetic O_2 evolution .1. A linear 4 step mechanism" *Photochem. Photobiol.* **11**, 457-475.

Koulougliotis, D., Innes, J. B. and Brudvig, G. W., 1994, "Location of $chlorophyll_Z$ in photosystem II" *Biophys. J.* **66**, A114-A114.

Koulougliotis, D., Schweitzer, R. H. and Brudvig, G. W., 1997, "The tetranuclear manganese cluster in photosystem II: Location and magnetic properties of the S_2 state as determined by saturation-recovery EPR spectroscopy" *Biochemistry* **36**, 9735-9746.

Koulougliotis, D., Tang, X.-S., Diner, B. A. and Brudvig, G. W., 1995, "Spectroscopic evidence for the symmetrical location of tyrosines D and Z in photosystem II" *Biochemistry* **34**, 2850-2856.

Krauß, N., Schubert, W.-D., Klukas, O., Fromme, P., Witt, H. T. and Saenger, W., 1996, "Photosystem I at 4 angstrom resolution represents the first structural model of a joint photosynthetic reaction centre and core antenna system" *Nature Struct. Biol.* **3**, 965-969.

Kulikov, A. V. and Likhtenshtein, G. I., 1977, "Use of spin relaxation phenomena in investigation of structure of model and biological systems by method of spin labels" *Adv. Mol. Relax. Interact. Processes* **10**, 47-79.

Lakshmi, K. V., Eaton, S. S., Eaton, G. R. and Brudvig, G. W., 1999a, "Orientation of the tetranuclear manganese cluster and tyrosine Z in the O_2-evolving complex of photosystem

II: An EPR study of the $S_2Y_Z^{\bullet}$ state in oriented acetate-inhibited photosystem II membranes" *Biochemistry* **38**, 12758-12767.

Lakshmi, K. V., Eaton, S. S., Eaton, G. R., Frank, H. A. and Brudvig, G. W., 1998, "Analysis of dipolar and exchange interactions between manganese and tyrosine Z in the $S_2Y_Z^{\bullet}$ state of acetate-inhibited photosystem II via EPR spectral simulations at X- and Q-Bands" *J. Phys. Chem. B* **102**, 8327-8335.

Lakshmi, K. V., Jung, Y.-S., Golbeck, J. H. and Brudvig, G. W., 1999b, "Location of the iron-sulfur clusters F_A and F_B in photosystem I: An electron paramagnetic resonance study of spin relaxation enhancement of P_{700}^{+}" *Biochemistry* **38**, 13210-13215.

Lancaster, C. R. D. and Michel, H., 1999, "Refined crystal structures of reaction centers from Rhodopseudomonas Viridis in complexes with the herbicide atrazine and two chiral atrazine derivatives also lead to a new model of the bound carotenoid" *J. Mol. Biol.* **286**, 883-898.

Limburg, J., Szalai, V. A. and Brudvig, G. W., 1999, "A mechanistic and structural model for the formation and reactivity of a Mn-V=O species in photosynthetic water oxidation" *J. Chem Soc., Dalton Trans.*, 1353-1361.

Makinen, M. W. and Wells, G. B., 1987, "Application of electron-paramagnetic-resonance saturation methods to paramagnetic metal ions in proteins" *Metal Ions in Biological Systems* **22**, 171-206.

McLachlan, D. J. and Nugent, J. H. A., 1993, "Investigation of the S3 electron paramagnetic resonance signal from the oxygen-evolving complex of photosystem 2: Effect of inhibition of oxygen evolution by acetate" *Biochemistry* **32**, 9772-9780.

Milov, A. D., Maryasov, A. G. and Tsvetkov, Y. D., 1998, "Pulsed electron double resonance (PELDOR) and its applications in free radicals research" *Appl. Magn. Reson.* **15**, 107-143.

Norris, J. R., Thurnauer, M. C. and Bowman, M. K., 1980, "Electron spin echo spectroscopy and the study of biological structure and function" *Adv. Biol. Med. Phys.* **17**, 365-416.

Okamura, M. Y., Isaacson, R. A. and Feher, G., 1975, "Primary acceptor in photosynthesis: Obligatory role of ubiquinone in photoactive reaction centers of *Rhodopseudomonas Sphaeroides*" *Proc. Natl. Acad. Sci. USA* **72**, 3491-3495.

Oliver, M. E. and Hales, B. J., 1993, "Using dysprosium complexes to probe the nitrogenase paramagnetic centers" *Biochemistry* **32**, 6058-6064.

Peloquin, J. M., Campbell, K. A. and Britt, R. D., 1998, "Mn-55 pulsed ENDOR demonstrates that the photosystem II "split" EPR signal arises From a magnetically-coupled mangano-tyrosyl complex" *J. Amer. Chem. Soc.* **120**, 6840-6841.

Percival, P. W. and Hyde, J. S., 1975, "Pulsed EPR spectrometer 2." *Rev. Sci. Instrum.* **46**, 1522-1529.

Portis, M., 1953, "Electronic structure of F centers: Saturation of the electron spin resonance" *Phys. Rev.* **91**, 1071-1079.

Rakowsky, M. H., More, K. M., Kulikov, A. V., Eaton, G. R. and Eaton, S. S., 1995, "Time-domain electron paramagnetic resonance as a probe of electron-electron spin-spin interaction in spin-labeled low-spin iron porphyrins" *J. Amer. Chem. Soc.* **117**, 2049-2058.

Rhee, K.-H., Morris, E. P., Barber, J. and Kühlbrandt, W., 1998, "Three-dimensional structure of the plant photosystem II at 8 Å resolution" *Nature* **396**, 283-286.

Sahlin, M., Petersson, L., Graslund, A., Ehrenberg, A., Sjoberg, B. M. and Thelander, L., 1987, "Magnetic interaction between the tyrosyl free-radical and the anti-ferromagnetically coupled iron center in ribonucleotide reductase" *Biochemistry* **26**, 5541-5548.

Schiller, H., Dittmer, J., Iuzzolino, L., Dorner, W., Meyer-Klaucke, W., Sole, V. A., Nolting, H. F. and Dau, H., 1998, "Structure and orientation of the oxygen-evolving manganese complex of green algae and higher plants investigated by x-ray absorption linear dichroism spectroscopy on oriented photosystem II membrane particles" *Biochemistry* **37**, 7340-7350.

Shigemori, K., Hara, H., Kawamori, A. and Akabori, K., 1998, "Determination of distances from tyrosine D to Q_A and chlorophyll$_Z$ in photosystem II studied by '2+1' pulsed EPR" *Biochim. Biophys. Acta* **1363**, 187-198.

Smigel, M. D., Dalton, L. A. and Dalton, L. R., 1974, "Very slowly tumbling spin labels - Saturation recovery" *Chem. Phys.* **6**, 183-192.

Smith, I. D. and Pilbrow, J. R., 1974, "The determination of structural properties of dimeric transition metal ion complexes from EPR spectra" *Coord. Chem. Rev.* **13**, 173-278.

Stevenson, R. C., Sands, R. H., Singer, T. P. and Beinert, H., 1986, "Studies on the spin-spin interaction between flavin and iron-sulfur cluster in an iron-sulfur flavoprotein" *Biochim. Biophys. Acta* **869**, 81-88.

Stewart, D. H. and Brudvig, G. W., 1998, "Cytochrome b_{559} of photosystem II" *Biochim. Biophys. Acta* **1367**, 63-87.

Stewart, D. H., Cua, A., Chisholm, D. A., Diner, B. A., Bocian, D. F. and Brudvig, G. W., 1998, "Identification of histidine 118 in the D1 polypeptide of photosystem II as the axial ligand to chlorophyll Z" *Biochemistry* **37**, 10040-10046.

Styring, S. A. and Rutherford, A. W., 1988, "The microwave power saturation of SII slow varies with the redox state of the oxygen-evolving complex in photosystem II" *Biochemistry* **27**, 4915-4923.

Svensson, B., Etchebest, C., Tuffery, P., van Kan, P., Smith, J. and Styring, S., 1996, "A model for the photosystem II reaction center core including the structure of the primary donor P_{680}^{+}" *Biochemistry* **35**, 14486-14502.

Szalai, V. A. and Brudvig, G. W., 1996, "Formation and decay of the S3 EPR signal species in acetate-inhibited photosystem II" *Biochemistry* **35**, 1946-1953.

Szalai, V. A., Kühne, H., Lakshmi, K. V. and Brudvig, G. W., 1998a, "Characterization of the interaction between manganese and tyrosine Z in acetate-inhibited photosystem II" *Biochemistry* **37**, 13594-13603.

Szalai, V. A., Stone, D. A. and Brudvig, G. W., 1998b, in *Photosynthesis: Mechanisms and Effects* (Garab, G., Ed.) Kluwer Academic Publishers, Dordrecht, The Netherlands, pp 1403-1406.

Tan, X. L., Gultneh, Y., Sarneski, J. E. and Scholes, C. P., 1991, "EPR-ENDOR of the electronic-structure from two nitrogenously ligated Bis(Mu-Oxo)-MnIII-MnIV model complexes spectroscopically relevant to the multi-manganese center of photosystem II" *J. Amer. Chem. Soc.* **113**, 7853-7858.

Tang, X.-S., Randall, D. W., Force, D. A., Diner, B. A. and Britt, R. D., 1996, "Manganese-tyrosine interaction in the photosystem II oxygen-evolving complex" *J. Amer. Chem. Soc.* **118**, 7638-7639.

Un, S., Brunel, L.-C., Brill, T. M., Zimmermann, J.-L. and Rutherford, A. W., 1994, "Angular orientation of the stable tyrosyl radical within photosystem II by high-field 245-GHz electron-paramagnetic-resonance" *Proc. Natl. Acad. Sci. USA* **91**, 5262-5266.

Yachandra, V. K., Sauer, K. and Klein, M. P., 1996, "Manganese cluster in photosynthesis: Where plants oxidize water to dioxygen" *Chem. Rev.* **96**, 2927-2950.

Yoshii, T., Kawamori, A., Tonaka, M. and Akabori, K., 1999, "Relative positions of electron transfer components in photosystem II studied by '2+1' pulsed electron paramagnetic resonance: Y-D and Q(A)" *Biochim. Biophys. Acta* **1413**, 43-49.

Zheng, M. and Dismukes, G. C., 1996, "Orbital configuration of the valence electrons, ligand field symmetry, and manganese oxidation states of the photosynthetic water oxidizing complex: Analysis of the S_2 state multiline EPR signals" *Inorg. Chem.* **35**, 3307-3319.

Chapter 13

Photo-Induced Radical Pairs Investigated Using Out-Of-Phase Electron Spin Echo[1]

Sergei A. Dzuba[*] and Arnold J. Hoff[#]
[*]*Institute of Chemical Kinetics and Combustion, Russian Academy of Sciences, Novosibirsk, Russia*
[#]*Department of Biophysics, Leiden University, Leiden, The Netherlands*

Abstract: Electron spin echo experiment on photo-induced radical pairs in photosynthetic reaction centers produces the echo signal which is out-of-phase due to spin coherence. This signal is strongly modulated because of the electron spin-spin dipolar and exchange couplings in the pair. Fourier transformation of the time-domain modulation produces a "Pake doublet" from the two coupled spins, which allows the accurate determination of the distance between the two radicals. The precision with which the distance can be measured is 0.3 - 0.4 Å, for radicals separated in the pair by ~ 30 Å. This high precision provides a very sensitive tool for structural investigations in photosynthetic reaction centers. The method is applicable to spin-correlated radical pairs separated by 20 – 40 Å and possessing an EPR linewidth accessible for the full excitation by the microwave pulses.

1. INTRODUCTION

Radicals produced by illumination or ionizing radiation are generally formed in pairs, with the combined electron spin quantum number $S = 0$

[1] This chapter is based on an earlier paper (Hoff et al., 1998) a few sections of which are reproduced with permission from Elsevier Science

Biological Magnetic Resonance, Volume 19: Distance Measurements in Biological Systems by EPR, edited by Berliner et al. Kluwer Academic/Plenum Publishers, New York, 2000

(singlet state) or $S = 1$ (triplet state), depending on the excited state from which radical formation proceeds. The radicals are produced *coherently*, their phase relation evolving under the influence of local magnetic fields, due for example to hyperfine interactions, the difference in g-values, Δg, and magnetic interactions between the radicals, such as the dipolar and Heisenberg exchange interactions, D and J respectively.

The light-induced pairs in photosynthetic reaction centers (RC) were intensively studied using time-resolved electron paramagnetic resonance (EPR) spectroscopy (see, e.g. Hore, 1990, Angerhofer and Bittl, 1996). The obtained transient EPR spectra promoted the development of the spin-correlated radical pair (SCRP) model (Hore et al., 1987, Hore, 1990).

The first electron spin echo (ESE) experiment on SCRP in photosynthetic RCs was reported by Thurnauer and Norris, 1980. They found an abnormal phase behavior of the echo signal: in contradiction to the case of radicals in thermal equilibrium, where only normal *in-phase* echo exists, it was found to be *out-of-phase*. At the time there was no theory to explain this phenomenon. The theory of ESE of SCRP was developed by Salikhov et al., 1992 and, some time later, by Tang et al., 1994. The theory takes into account that EPR spectra in photosynthetic RCs are only moderately broadened (15 - 20 G for the total linewidth at X-band) implying full excitation of the EPR line by microwave pulses can be achieved in a normal ESE experimental setup and predicts that the ESE signal indeed must be out-of-phase.

Another important theoretical prediction was that signal must be strongly modulated, with modulation frequency determined by the D and J values. This modulation (ESE Envelope Modulation, or ESEEM) was indeed observed experimentally (Dzuba et al., 1995). It was shown that Fourier transformation of the time-domain modulation provides a direct way to determine the electron-electron dipolar and Heisenberg exchange interactions. The resulting spectrum is a "Pake doublet" from the two dipolar and exchange-coupled spins (with an antisymmetric shape with respect to the central frequency, due to spin polarization). A simple measurement of the frequency difference between the peaks and edges of the Pake spectrum yields the dipolar interaction, which in turn allows the accurate determination of the distance between the two radicals. This result initiated quite a few experimental studies of ESEEM of SCRP, aiming at structural investigation of photosynthetic RCs based on accurate measurement of the D value, thus allowing direct distance determination.

Other interesting phenomena that follow directly from the quantum-mechanical description of the spin motion in the radical pair are, for example, zero- and double-quantum coherence, which in principle provide further means of determining the magnetic interactions of the pair.

2. ELECTRON SPIN ECHO ENVELOPE MODULATION OF A PHOTOINDUCED RADICAL PAIR

2.1 Energy Levels of the Interacting Radical Pair

The simplified radical pair spin-Hamiltonian in the high-field approximation is usually written as (see, e. g., Hore, 1990)

$$\tan 2\phi = \frac{\Delta\omega}{2J + \frac{1}{3}D(3\cos^2\xi - 1)} \quad (1)$$

Here, μ_B is the Bohr magneton, B_0 is the magnetic field (defining the z-axis of the laboratory frame), g_1 and g_2 are the g-values of radical 1 and radical 2 respectively, $\boldsymbol{S}_1$ and $\boldsymbol{S}_2$ are their electron spin operators, A_{1j} and A_{2j} are their hyperfine coupling constants with the jth nucleus, $\boldsymbol{I}_j$ are nuclear spin operators, D represents the dipolar coupling, ξ is the angle between the line connecting the two species and the direction of the external magnetic field, $\boldsymbol{S} = \boldsymbol{S}_1 + \boldsymbol{S}_2$, and J is the value of the spin-exchange interaction.
The eigenvectors of the four radical pair states are given by (Fig. 1a)

$$\begin{aligned} |1\rangle &= |T_+\rangle \\ |2\rangle &= \cos\phi|S\rangle + \sin\phi|T_0\rangle \\ |3\rangle &= -\sin\phi|S\rangle + \cos\phi|T_0\rangle, \\ |4\rangle &= |T_-\rangle \end{aligned} \qquad \tan 2\phi = \frac{\Delta\omega}{2J + \frac{1}{3}D(3\cos^2\xi - 1)}, \quad (2)$$

where $\Delta\omega = \omega_1 - \omega_2$ is the difference in resonance frequencies for the two spins in the absence of dipolar and exchange interactions,

$$\Delta\omega = (g_1 - g_2)\mu_B B_0 + \sum_j (A_{1j} - A_{2j}) m_j \quad (3)$$

and m_j is the magnetic quantum number of nucleus j. The eigenvalues of the Hamiltonian (1) are

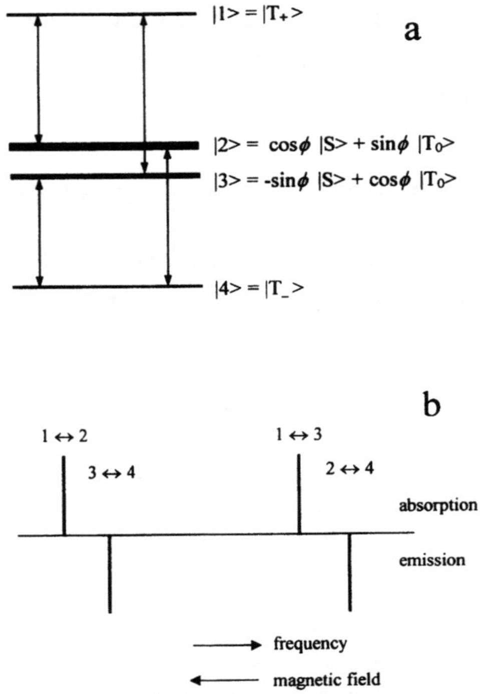

Figure 1. Energy levels of two coupled radicals (a) and the EPR stick spectra for a spin-polarized pair (b). Initially after the laser flash, only the two middle levels are populated.

$$\begin{aligned}
E_1 &= \tfrac{1}{2}(\omega_1+\omega_2)+\tfrac{1}{6}D(3\cos^2\xi-1)\\
E_2 &= J-\tfrac{1}{6}D(3\cos^2\xi-1)+\tfrac{1}{2}\sqrt{[2J+\tfrac{1}{3}D(3\cos^2\xi-1)]^2+(\Delta\omega)^2}\\
E_3 &= J-\tfrac{1}{6}D(3\cos^2\xi-1)-\tfrac{1}{2}\sqrt{[2J+\tfrac{1}{3}D(3\cos^2\xi-1)]^2+(\Delta\omega)^2}\\
E_4 &= -\tfrac{1}{2}(\omega_1+\omega_2)+\tfrac{1}{6}D(3\cos^2\xi-1).
\end{aligned} \tag{4}$$

Introducing the notation

$$\begin{aligned}
a &= \tfrac{1}{2}(\omega_1+\omega_2),\\
b &= J-\tfrac{1}{3}D(3\cos^2\xi-1),\\
c &= \tfrac{1}{2}\sqrt{[2J+\tfrac{1}{3}D(3\cos^2\xi-1)]^2+(\Delta\omega)^2},
\end{aligned} \tag{5}$$

the six possible energy differences are

$$E_{12} = a - b - c, \quad E_{13} = a - b + c,$$
$$E_{24} = a + b + c, \quad E_{34} = a + b - c, \qquad (6)$$
$$E_{14} = 2a, \qquad E_{23} = 2c.$$

The singlet character of eigenstates $|2\rangle$ and $|3\rangle$ (see (2)), and thus their initial populations, are given by $\cos^2\phi$ and $\sin^2\phi$, respectively. There is also a coherent superposition of $|2\rangle$ and $|3\rangle$ which oscillates at a frequency given by the energy difference:

$$E_{23} = \sqrt{[2J + \tfrac{1}{3}D(3\cos^2\xi - 1)]^2 + (\Delta\omega)^2}. \qquad (7)$$

This "zero-quantum coherence", is not directly detectable, but can be made observable by using a microwave pulse to convert it into single-quantum coherence (see below).

Not long after the radical-generating laser pulse, generally within less than 100 ns, the zero-quantum coherence will dephase as a result of inhomogeneous interactions in the ensemble of radical pairs, and by spin relaxation. The intensities of the EPR transitions are then given by the populations of states $|2\rangle$ and $|3\rangle$ (determined by their singlet characters) scaled by the probability of a transition to levels $|1\rangle$ and $|4\rangle$ (given by their triplet characters). All four transitions thus have intensities proportional to $\sin^2\phi\cos^2\phi$. In other words, the spectrum comprises two antiphase doublets with equal but opposite EPR amplitudes within each doublet (Fig. 1b).

2.2 Electron Spin Echo Formation

The formal solution of the time-dependent Schrödinger equation

$$\frac{d\rho}{dt} = -i[\mathcal{H}, \rho] = -i(\mathcal{H}\rho - \rho\mathcal{H}) \qquad (8)$$

is given by

$$\rho(t) = \exp(-i\mathcal{H}t)\rho(0)\exp(i\mathcal{H}t) \qquad (9)$$

where ρ is the density operator of the system. Hence, the time evolution of the radical pair subject to the pulse sequence (flash- t_0 - pulse1 - τ - pulse2 -

τ - echo) is given by a product of exponential operators depending on the appropriate Hamiltonians:

$$\rho(t_0,\tau) = \exp(-i\mathcal{H}\tau)\exp(-i\theta_2 S_x)\exp(-i\mathcal{H}\tau)\exp(-i\theta_1 S_x)\exp(-i\mathcal{H}t_0)\rho(0) \times \exp(i\mathcal{H}t_0)\exp(i\theta_1 S_x)\exp(i\mathcal{H}\tau)\exp(i\theta_2 S_x)\exp(i\mathcal{H}\tau) \tag{10}$$

where θ_1 and θ_2 are the flip angles of the first and second microwave pulses, respectively. We assume throughout that the microwave field is directed along the x-axis in the rotating frame and that the pulse duration is very short, so that one may ignore the evolution of the density matrix during the pulses, due to the interactions described by Eq. 1. For the periods of time after the flash and before the first microwave pulse, t_0, and between and after the pulses, τ, the Hamiltonian is given by Eq. 1. It is advantageous to transform the Hamiltonian to the eigenbasis $|1\rangle$, $|2\rangle$, $|3\rangle$, $|4\rangle$, so that the exponential operators $\exp(\pm i\mathcal{H}t)$ of Eqs. 9 and 10 are given by diagonal matrices determined by the eigenvalues in Eq. 4. The transformation matrix is given by

$$U = \begin{pmatrix} 1 & 0 & 0 & 0 \\ 0 & \cos\phi & \sin\phi & 0 \\ 0 & -\sin\phi & \cos\phi & 0 \\ 0 & 0 & 0 & 1 \end{pmatrix} \tag{11}$$

The spin operators S_x and S_y for an $S = 0, 1$ system in the $|S\rangle$, $|T_+\rangle$, $|T_0\rangle$, $|T_-\rangle$ basis are

$$S_x = \frac{1}{2}\begin{pmatrix} 0 & 0 & 1 & 0 \\ 0 & 0 & 0 & 0 \\ 1 & 0 & 0 & 1 \\ 0 & 0 & 1 & 0 \end{pmatrix}, \quad S_y = \frac{i}{2}\begin{pmatrix} 0 & 0 & -1 & 0 \\ 0 & 0 & 0 & 0 \\ 1 & 0 & 0 & -1 \\ 0 & 0 & 1 & 0 \end{pmatrix}. \tag{12}$$

To evaluate the exponential operators for the microwave pulses, we make use of the general relation

$$\exp(-i\theta S_x) = \mathbf{1} - (1-\cos\theta)S_x^2 - i\sin\theta S_x\,, \tag{13}$$

which is easily derived by Taylor expansion of the exponent, making use of the self-similarity relations $(S_x)^3=S_x$ and $(S_x)^4=(S_x)^2$; similar relations hold for S_y. Using (12), (13), one obtains

$$\exp(i\theta S_x) = \begin{pmatrix} \frac{1}{2}(1+\cos\theta) & 0 & \frac{i}{\sqrt{2}}\sin\theta & -\frac{1}{2}(1-\cos\theta) \\ 0 & 1 & 0 & 0 \\ \frac{i}{\sqrt{2}}\sin\theta & 0 & \cos\theta & \frac{i}{\sqrt{2}}\sin\theta \\ -\frac{1}{2}(1-\cos\theta) & 0 & \frac{i}{\sqrt{2}}\sin\theta & \frac{1}{2}(1+\cos\theta) \end{pmatrix} \tag{14}$$

Finally, the echo signal along the x-axis (out-of-phase) and along the y-axis (in-phase) are given by

$$\begin{aligned} M_x(t_0,\tau) &= \mathrm{Tr}\{S_x\rho(t_0,\tau)\}, \\ M_y(t_0,\tau) &= \mathrm{Tr}\{S_y\rho(t_0,\tau)\}. \end{aligned} \tag{15}$$

To simplify calculations, let us consider first the case of the $\pi/2$ - τ - π pulse sequence (see also the next section). Then the straightforward matrix multiplication in (10) results in the following simple relation for the transverse magnetization in the rotating frame (Tang et al., 1994):

$$\begin{aligned} M_x(t_0,\tau) &= \sin^2(2\phi)\cos(2\phi)[\cos(2c\tau)-1]\{\sin(2ct_0)\cos(2b\tau) \\ &\qquad - \cos(2\phi)[1-\cos(2ct_0)]\sin(2b\tau)\} \\ M_y(t_0,\tau) &= 0. \end{aligned} \tag{16}$$

This result predicts that for the spin-correlated radical pair only the out-of-phase echo appears. This is the reverse of the case for radicals with spin polarization in thermal equilibrium where only the normal in-phase echo exists.

For the radical pairs in photosynthetic reaction centers the c value is of the order of 10^8 rad/s. So if t_0 and $\tau >> 10$ ns (for τ this is always valid due to instrumental limitation), then Eq. 16 reduces to

$$M_x(t_0,\tau) = -\sin^2(2\phi)\cos^2(2\phi)\sin(2b\tau). \tag{17}$$

This equation means that the out-of-phase echo is modulated with the frequency $b = J - \frac{1}{3}D(3\cos^2\xi - 1)$. Note that the modulation frequency does not depend on the resonance frequencies ω_1 and ω_2 of the radicals. This means that all radical pairs having the same dipolar and spin-exchange couplings, independent of inhomogeneous broadening of the EPR lineshapes, will contribute to the ESEEM with the same modulation frequency.

2.3 Echo Amplitudes and Pulse Angle Dependences

The out-of-phase echo also has an abnormal pulse angle dependence. This was explored in detail by Tang et al., (1997), Hara et al., (1998) and Timmel et al., (1998). Tang et al., (1997) obtained analytical formulae in terms of the pulse angle, microwave phase, and the time interval between the laser pulse and the microwave pulses. They considered both singlet and triplet precursors. The phase dependence and the formation of the three pulse stimulated echo was also examined.

Timmel et al., (1998) explored the case of long delays after flash, $t_0 >> 10$ ns, which is the normal situation in experiment. In this case the eigenbasis (2) is more convenient for calculation. Then all coherences have decayed and the initial density matrix has only diagonal elements. Immediately after the flash

$$\rho_{11} = 0,\ \rho_{22} = \cos^2\phi,\ \rho_{33} = \sin^2\phi,\ \rho_{44} = 0. \tag{18}$$

It was then assumed that the populations are arbitrary (deviating from (18) by e.g. spin relaxation), and equal to p_1, p_2, p_3 and p_4, respectively, and that the turning angle θ_1 may be arbitrary (leaving for a while $\theta_2 = 180°$, see below). Furthermore we take both microwave pulses oriented along the x-axis in the rotating frame. Then, according to Timmel et al., 1998

$$\begin{aligned}
M_x &= P\ \sin 2\theta_1 \left\{S_0 \sin^2 2\phi + \cos 2\phi \left[\cos^2\phi\ S_{-1} - \sin^2\phi\ S_{+1}\right]\right\} \\
&\quad + \sin\theta_1 \left(p_2 - p_3\right)\sin^2 2\phi \cos 2\phi \left[S_0 - \tfrac{1}{2}S_{+1} - \tfrac{1}{2}S_{-1}\right] \\
M_y &= \left(p_4 - p_1\right)\sin\theta_1 \left\{C_0\ \sin^2 2\phi + \cos 2\phi \left[\cos^2\phi\ C_{-1} - \sin^2\phi\ C_{+1}\right]\right\} \\
P &= -\tfrac{1}{2}p_1 + p_2 \sin^2\phi + p_3 \cos^2\phi - \tfrac{1}{2}p_4 \\
S_q &= \sin\left[2(b + qc)\tau\right] \\
C_q &= \cos\left[2(b + qc)\tau\right],
\end{aligned} \tag{19}$$

where b and c are given by Eq. 5 and $q = 0, \pm 1$. In the absence of spin-lattice relaxation, the populations p_1 and p_4 are initially zero and remain zero until the first microwave pulse. Therefore, the in-phase y-echo for a spin-correlated radical pair is zero. In contrast, even when p_2 and p_3 are (approximately) equal, e.g. for a weakly coupled pair when the initial zero-quantum coherence has decayed, the out-of-phase x-echo has a definite amplitude given by the part of M_x that depends on $\sin 2\theta_1$. Note that this behavior contrasts with that of an equilibrated radical pair for which the out-of-phase echo is zero, while the in-phase echo is proportional to the Boltzmann population difference between the $T_\pm$ levels.

For intermediate-to-weak coupling, i.e. $(\Delta\omega)^2 \geq [2J + \frac{1}{3}D(3\cos^2\xi - 1)]^2$, the expressions in Eq. 19 can be simplified somewhat by observing that the higher frequency oscillations $S_{\pm 1}$ will generally dephase quickly compared with those of S_0, so that the corresponding terms in Eq. 19 can safely be dropped when observing echo signals beyond $\cong 100\,\text{ns}$, the experimental dead time of common spin-echo spectrometers. Then M_x is proportional to $S_0 \equiv \sin 2b\tau$ (cf. (17)). So one may conclude that the ESEEM does not depend on the pulse angles. Bittl and Zech (1997) have provided the experimental evidence supporting this important conclusion.

For the weak-coupling limit $\phi = 45°$, the echo amplitudes are simply

$$
\begin{aligned}
M_x &= \tfrac{1}{2}(-p_1 + p_2 + p_3 - p_4)\sin 2\theta_1\, S_0, \\
M_y &= (p_4 - p_1)\sin\theta_1\, C_0.
\end{aligned}
\tag{20}
$$

It is seen from Eq. 19 and even more clearly from Eq. 20 that the dependence of the echo signals of a spin-correlated radical pair on the microwave flip angle is quite unusual. While the in-phase y-echo has the normal $\sin\theta_1$ dependence, reflecting the Zeeman order of the spin system, the out-of-phase echo contains an unusual $\sin 2\theta_1$ dependence, dominant for weakly coupled pairs, which reflects the longitudinal two-spin order of the pair. It follows that the out-of-phase signal is maximal for $\theta_1 = 45°$, in contrast to normal spin-echo spectroscopy, which yields maximal amplitude for $\theta_1 = 90°$. This unusual dependence was indeed observed in experiment (Tang et al., 1997, Hara et al., 1998).

Finally, if θ_2 is allowed to deviate from 180°, a scaling factor of $\sin^4 \frac{1}{2}\theta_2$ must be included in (19) and (20) (Timmel et al., 1998). Note that $\theta_2 = 180°$ gives the optimum refocusing and the maximum echo amplitude.

2.4 Zero-Quantum and Double Quantum Coherences

The photogenerated radical pairs are formed in non-stationary states, i.e. in coherent superpositions of their eigenstates, which are given by the off-diagonal elements of the density matrix. The various interactions of the two electron spins cause the coherences to oscillate, allowing the system to evolve in time, e.g. from a singlet to a triplet (and back again). Through the effects of inhomogeneous interactions and random events such as nuclear spin flips, these oscillations will ultimately dephase and relax, rendering the ensemble average of the off-diagonal elements zero.

For the present system, in high magnetic field and in the absence of microwaves, the only off-diagonal elements that can be non-zero are $\rho_{23}(t) = \rho^*_{32}(t)$. Applying Eq. 9 at a time t_0 after a laser flash, it is easily seen that their time dependence is given by

$$\rho_{23}(t) = -\tfrac{1}{2}\sin 2\phi\ \exp(-2iE_{23}t_0) \tag{21}$$

with E_{23} given by Eq. 7. The oscillation with frequency E_{23} represents the zero-quantum coherence of the correlated spins. The first microwave pulse creates single-quantum coherences, i.e. transverse magnetization, given by the density matrix elements $\rho_{12}(t)$, $\rho_{13}(t)$, $\rho_{24}(t)$, $\rho_{34}(t)$ and their complex conjugates, which follow from

$$\rho(t) = \exp(-i\theta_1 S_x)\exp(-i\mathcal{H}t_0)\rho(0)\exp(i\mathcal{H}t_0)\exp(i\theta_1 S_x). \tag{22}$$

It turns out that these matrix elements depend not only on the populations of the four eigenstates given by the diagonal elements of $\rho(t_0)$, but also on $\rho_{23}(t_0)$ and $\rho_{32}(t_0)$. In other words, for times sufficiently short that exp($-iE_{23}t_0$) is not averaged to zero, the echo signal obtained after the second microwave pulse, will depend strongly on the time t_0 between the laser flash and the first microwave pulse. This is illustrated in Fig. 2. Note that the zero-quantum coherence is captured by the first microwave pulse, which may be applied shortly after the laser flash, and is then stored as single-quantum coherence. Therefore, pulsed microwave detection of zero-quantum coherence has a high time resolution (~10 ns, determined by the typical microwave pulselength), exceeding that of direct-detection continuous wave (cw) EPR, with the advantage that it is free from Torrey oscillations. Note that the traces in Fig. 2 represent the first experimental observation of the very rapidly decaying zero-quantum coherence in normal (protonated) photosynthetic reaction centers (Dzuba et al., 1996). Previously, zero-quantum coherence was detected by time-resolved EPR in deuterated reaction centers (Kothe et al., 1991).

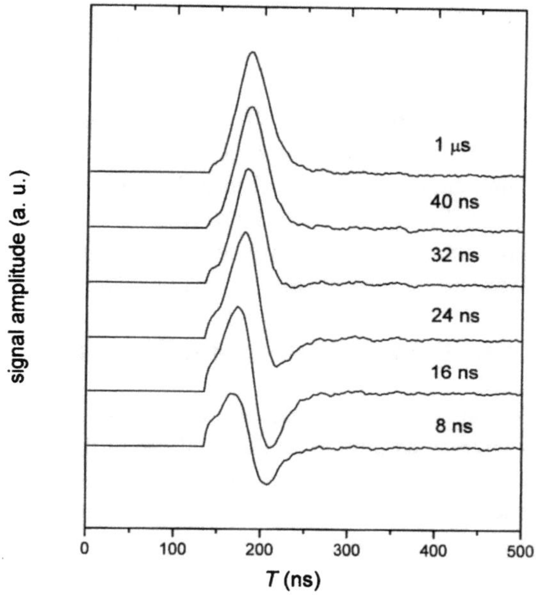

Figure 2. The out-of-phase echo signal as a function of the delay time t_0 between the laser flash and the first microwave pulse. The delay time τ was fixed at 184 ns (Dzuba et al. 1996), with permission from Elsevier Science.

Also, the first microwave pulse excites the double-quantum coherence, as noticed by Tang and Norris, 1995. I.e. the $\rho_{14}(t_0)$ matrix element is non-zero after the first pulse. Application of Eq. 22 leads to the following density matrix element immediately after the first microwave pulse:

$$\rho_{14} = \tfrac{1}{4}\sin^2 2\phi\left(1 - \cos E_{23}t_0\right). \tag{23}$$

When $t_0 = \tau$, this coherence is refocused as a second, double-quantum-coherence echo (Dzuba et al., 1996). It is positioned at the time delay 2τ after the second pulse (normal echo is at the time delay τ). It is again an out-of-phase echo. Qualitatively its appearance may be described as follows. After the first pulse the double-quantum coherence evolves with frequency $E_{14} = 2a$. The second microwave pulse, providing it is not exactly 180°, inverts the phase and transfers the double-quantum coherence into observable single-quantum coherences, which evolve at frequencies E_{12}, E_{13}, E_{24} and E_{34}. Collecting the phases, including that accumulated by the zero-quantum coherence before the first pulse, we obtain for the 1↔2 transition at time T after the second pulse

$$\varphi = E_{23}t_0 - E_{14}\tau + E_{12}T = 2ct_0 - 2a\tau + (a-b-c)T. \tag{24}$$

There is thus an echo when $T = 2\tau = 2t_0$, modulated again with frequency $2b$ when T, τ, and t_0 are varied simultaneously.

The double-quantum echo was observed experimentally for bacterial reaction centers (Dzuba et al., 1996). Hara et al., (1998) studied the double-quantum echo in photosystem I. Microwave phase dependence of the double quantum echo was treated theoretically in detail by Tang et. al. (1997).

2.5 Fourier Transformation and Distance Determination

Eqs. 17, 19 and 20 predict that the amplitude of the echo signal is proportional to sin2b so its amplitude is modulated at the frequency $2b$. When the radical pairs are randomly oriented, as in our example of a solid solution of photosynthetic reaction centers, and for full excitation, all frequencies corresponding to

$$\omega_{\text{mod}}(\xi) = 2b = 2J - \tfrac{2}{3}D(3\cos^2\xi - 1) \tag{25}$$

will be present. Averaging over a spherical distribution of orientations and sine Fourier transformation yields a "Pake spectrum" in frequency space. Because the echo amplitude also depends on $\Delta\omega$ (see e.g. Eqs. 17 and 2), to calculate the Pake spectrum we need in principle full information on the two g-tensors of the radicals and their relative orientation, and on the direction of the principal dipolar axis with respect to (one of) the g-tensors. Such information is indeed available for the photosynthetic reaction centers used as model system here. The peaks and edges of the Pake spectrum, however, are not influenced by the $\Delta\omega$-dependence of the echo amplitude. One may derive from Eq. 25 that they are given by (Dzuba et al., 1995)

$$\begin{aligned} f_{\perp} &= \pm\left(\tfrac{2}{3}D + 2J\right) \\ f_{||} &= \pm\left(-\tfrac{4}{3}D + 2J\right). \end{aligned} \tag{26}$$

Using Eq. 26, the values of D and J are immediately determined from the experimental spectrum. From D, in the point dipolar approximation, the distance can be calculated using

$$D/MHz = -\frac{78\times10^3}{\left(r/\overset{o}{A}\right)^3}. \tag{27}$$

The method of distance determination outlined here has now been used for determining the distance between electron donor and the electron acceptor in several photosynthetic reaction centers.

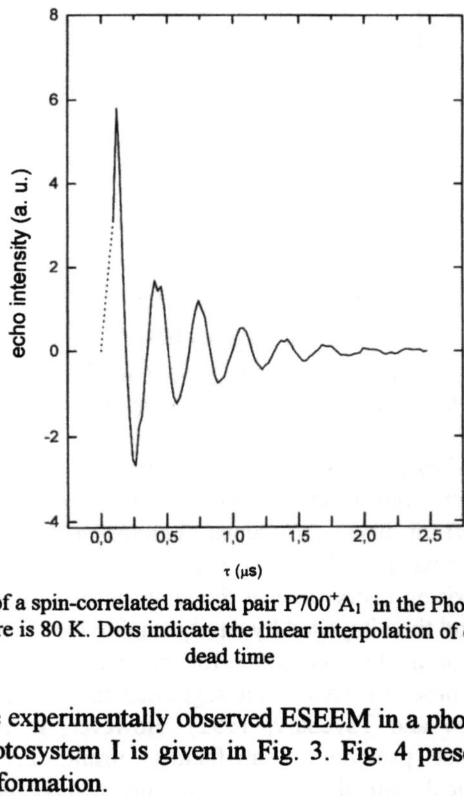

Figure 3. ESEEM of a spin-correlated radical pair $P700^+A_1^-$ in the Photosystem I reaction center. Temperature is 80 K. Dots indicate the linear interpolation of data missed in the dead time

Example of the experimentally observed ESEEM in a photoinduced radical pair $D^+Q_A^-$ in photosystem I is given in Fig. 3. Fig. 4 presents the result of sine Fourier transformation.

Some additional line broadening seen in Fig. 4 may be induced by transverse relaxation and by a distribution of distances in the pair (the latter is discussed also in section 3.4). The best way to take into account this additional broadening is to determine the distance via two steps. First, one measures D and J values from the peaks and edges of the Pake spectrum.

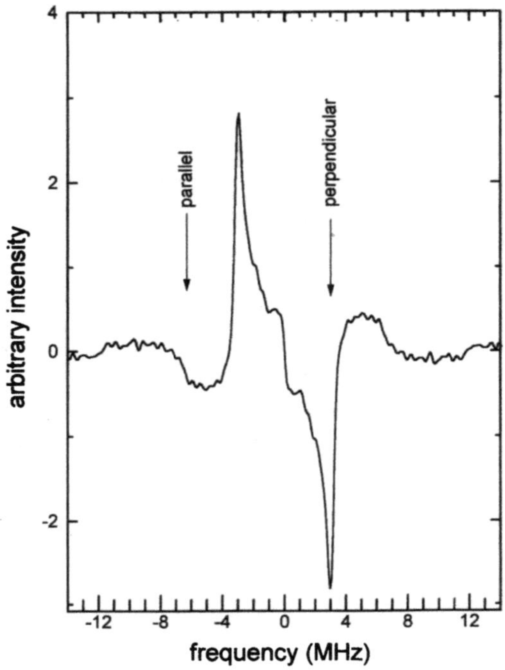

Figure 4. Sine Fourier transformation of the ESEEM given in Fig. 3. There are shown spectral positions for the canonical orientations of the pair respectively to the magnetic field

Secondly, one performs a precise computer simulation, convoluting the Pake spectrum with an appropriate Lorentzian broadening, and fitting the experimental spectrum, using the values obtained in the first step as a first approximation (Dzuba et al., 1995, Bittl and Zech, 1997).

One problem arising every time in the analysis of ESEEM is missing the data within the dead time imposed by cavity ringing (see Fig. 3). For ESEEM induced by electron-nuclear coupling for radicals in thermal equilibrium, many different approaches have been suggested to restore the missing data (see, e.g., Dikanov and Tsvetkov, 1992). However, in many cases these approaches do not provide a satisfactory result. For ESEEM of a photoinduced radical pair the situation in this respect is very favorable, because the initial time point is known precisely from the theory, i.e. the signal is equal to zero for $\tau=0$ (see Eqs. 16, 17, 19, 20). Normally, a simple linear or parabolic interpolation of data in the dead time provides a satisfactory result (Dzuba et al., 1997a). Nevertheless, in actual experiment one should try to diminish the dead time as much as possible, to improve the quality of the data. Bittl and Zech (1997) employed the "maximum entropy method" to reconstruct the ESEEM within the spectrometer dead time.

The precision with which the distance can be measured depends on the resolution of the frequency spectrum. (This resolution may become worse if the distances are distributed.) For well-resolved spectra, like those shown in Fig. 4, the typical uncertainty of the distance determination is quoted as 0.3 - 0.4 Å (Bittl and Zech, 1997, Dzuba et al., 1997b). Recently, Fursman and Hore (1999) explored in detail this precision, using the method of Cramér-Rao lower bounds. They proposed an optimum strategy for sampling the time-domain spin echo signal and noticed that the precision may be even better. Note that radical pairs are separated normally by ~ 30 Å, so the attained precision is extremely high, much higher than for all other known methods, including X-ray.

The electron spin density for radicals in the photosynthetic reaction centers is spread over many atoms. Let us evaluate the accuracy of the point dipole approximation assumed in (27). The variation ΔD is to first approximation proportional to $-3\Delta r/r^4$ where Δr is the variation of the distance when different atoms are considered. Δr averaged over all atoms is equal to zero if r refers to the center of the spin density distribution. To second order approximation the averaged value $<\Delta D> \propto 6<\Delta r^2>/r^5$. This value introduces a systematic positive error to the point dipole approximation, resulting in turn in a systematic negative error in the distance determination, $\delta r = -\frac{1}{3}<\Delta D>r^4 = -2<\Delta r^2>/r$. For radicals distributed within $<\Delta r^2>^{1/2} \approx 1.5$ Å, and separated by 30 Å, this systematic error is 0.15 Å. This rather small value may be further diminished, if the spin density distribution is precisely known and exact calculation of the dipolar coupling is performed.

2.6 Sequential Electron Transfer

Radical pairs in photosynthetic reaction centers are formed as a result of sequential electron transfer (see Section 3). Normally, the primary radical pair has a very short lifetime and therefore the secondary radical pair is investigated by ESEEM. In some cases, however, the primary pair lifetime may be long enough to have an influence on echo formation of the secondary radical pair.

Transfer of spin correlation between photoinduced radical pairs was considered in detail by Hore (1996). Jeschke and Bittl (1998) analyzed ESEEM in radical pairs undergoing secondary electron transfer. It was shown that an in-phase component of the echo should be observed also in this case. Analytical expressions have been derived for the ESEEM of a secondary pair, which revealed contributions of both the primary and secondary radical pair to the out-of-phase component of the signal, while only the secondary radical pair contributes to the in-phase component. Numerical calculations confirm that the frequency spectra depend strongly on the delay between generation of the primary pair and the echo formation. The primary and secondary radical give rise to spectral features at their own dipolar coupling frequencies. This in principle would allow study of primary and secondary radical pairs separately.

2.7 Contributions of Nuclear ESEEM

In addition to the echo amplitude modulations due to the spin-spin coupling, one would expect to observe also modulations due to nuclear ESEEM. It turns out, however, that such modulations contribute only slightly to the experimentally observed ESEEM of spin-correlated radical pairs (Dzuba et al., 1995, Bittl and Zech, 1997). Yet, for assessing possible introduction of error in the determination of D and J it is desirable to calculate quantitatively the contribution of nuclear ESEEM for the system of interest. This may be done by extending the Hamiltonian of Eq. 1 to include the anisotropic hyperfine and Zeeman interactions of a single spin-$\frac{1}{2}$ nucleus:

$$\begin{aligned} \mathcal{H} = \omega_1 S_{1z} + \omega_2 S_{2z} + \tfrac{1}{2} D(3\cos^2\xi - 1)(S_z^2 - \tfrac{1}{3}S^2) + J(\tfrac{1}{2} - 2\mathbf{S}_1.\mathbf{S}_2) \\ + a_s S_{1z} I_z + b_{ps} S_{1z} I_x - \omega_N I_z \end{aligned} \qquad (28)$$

with a_s and b_{ps} the secular and pseudo-secular coupling constants of the nucleus (axial symmetry of this coupling is assumed), with Larmor frequency ω_N, coupled to one of the two correlated electron spins. One may repeat the calculation, this time numerically, using a basis consisting of eight product states (Dzuba, 1997)

$$\begin{aligned} |1>=|T_+\alpha_N>, \quad |2>=|T_+\beta_N>, \quad |3>=|S\alpha_N>, \quad |4>=|T_0\alpha_N>, \\ |5>=|S\beta_N>, \quad |6>=|T_0\beta_N>, \quad |7>=|T_-\alpha_N>, \quad |8>=|T_-\beta_N>. \end{aligned} \qquad (29)$$

Alternatively, an analytical treatment is possible if the electrons are assumed to be weakly coupled (Timmel et al., 1998). In this case it is easily shown that the out-of-phase ESEEM signal is given simply by Eq. 20 supplemented by a term containing the normal nuclear modulation function:

$$M_x = \tfrac{1}{2}(-p_1 + p_2 + p_3 - p_4)\sin 2\theta_1 \sin 2b\tau \;\left[1 + f(t)\right] \qquad (30)$$

with

$$f(t) = 1 - \tfrac{1}{2}k\left[1 - \cos\omega_\alpha\tau - \cos\omega_\beta\tau + \tfrac{1}{2}\cos(\omega_\alpha - \omega_\beta)\tau + \tfrac{1}{2}\cos(\omega_\alpha + \omega_\beta)\tau\right]$$
$$k = (b_{ps}\omega_N / \omega_\alpha\omega_\beta)^2. \quad (31)$$

k is the modulation depth, and ω_α and ω_β are the NMR frequencies of a nucleus coupled to $m_S = +1/2$ and $-1/2$, electron spins, respectively:

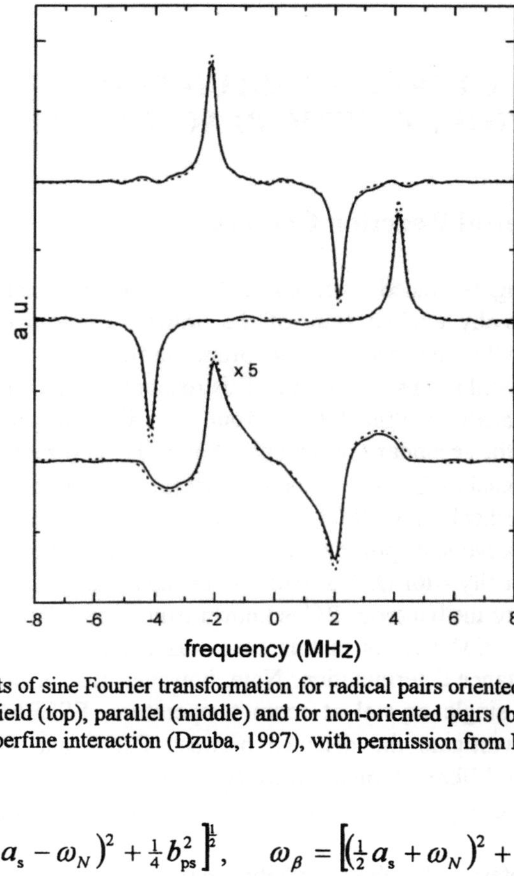

Figure 5. Results of sine Fourier transformation for radical pairs oriented perpendicular to the magnetic field (top), parallel (middle) and for non-oriented pairs (bottom). Dots: no anisotropic hyperfine interaction (Dzuba, 1997), with permission from Elsevier Science.

$$\omega_\alpha = \left[\left(\tfrac{1}{2}a_s - \omega_N\right)^2 + \tfrac{1}{4}b_{ps}^2\right]^{\frac{1}{2}}, \quad \omega_\beta = \left[\left(\tfrac{1}{2}a_s + \omega_N\right)^2 + \tfrac{1}{4}b_{ps}^2\right]^{\frac{1}{2}}. \quad (32)$$

Fig. 5 shows the result of the numerical calculation for oriented radical pairs and for a random distribution. It is seen that a few small extra features are present when the anisotropic hyperfine interaction is included, especially for the oriented sample, but that the peak positions of the Pake spectrum are not affected. The edges seem to move a little inward, which could have an effect on the determination of J. Overall, however, the effects are slight. We conclude that the determination of D is basically unaffected by the presence of nuclear ESEEM, while the value of J possibly could be somewhat overestimated.

Fursman and Hore (1999) considered the case when electron spins are coupled with two nuclei. The computer fitting of the data in the time domain showed that although the anisotropic hyperfine interactions change slightly the shape of the ESEEM spectra, they do not markedly shift frequencies of the parallel and perpendicular features that principally determine D.

3. STRUCTURAL INVESTIGATIONS IN PHOTOSYNTHETIC REACTION CENTERS[2]

3.1 Bacterial Reaction Centers

The primary light-induced reactions in RCs of photosynthetic bacteria are: (i) The electronically excited state of the primary electron donor P865, a bacteriochlorophyll dimer, produces the primary radical pair $P865^{+}I^{-}$, where I denotes a bacteriochlorophyll molecule. This reaction occurs within ~3 ps. (ii) Dark electron transfer within 200 ps from I^{-} to Q_A, the primary acceptor quinone, creates the secondary radical pair $P865^{+}Q_A^{-}$. (iii) Finally, within 200 μs the electron reaches Q_B, the secondary acceptor quinone. The lifetime of the secondary radical pair $P865^{+}Q_A^{-}$ is well suited for ESE experiment. Normally, to observe spin polarization, the native non-heme Fe^{2+} ion (which interacts magnetically with Q_A^{-}) is removed, or replaced by diamagnetic Zn^{2+}.

The structure of the bacterial RC is known from X-ray data. Therefore, the main goal of ESEEM studies in this case was mainly to evaluate the new approach for distance determination. Note, however, that X-ray studies were performed on a single crystal at room temperature. ESEEM provides the possibility to investigate the distance between $P865^{+}$ and Q_A^{-} at random orientation and at different temperatures. By comparison with the crystal data, the influence of charge separation (i.e. $P865^{+}$ and Q_A^{-} formation) may be investigated.

Reaction centers of photosynthetic purple bacterium *Rhodobacter sphaeroides* R26 were first investigated by Dzuba et al., 1995. A D value was found that was somewhat larger than that which could be expected from the X-ray studies. This data gives the center-to center distance between two species as $r = 28.1$ Å (Komiya et al., 1988), resulting in $D = -123$ μT (see Eq. 27), or $r = 29.2$ Å (Chang et al., 1991), resulting in $D = -106$ μT. Later, in

[2] See also chapter by Lakshmi and Brudwig of this volume

work of the same authors (Dzuba et al., 1997a), the origin of this difference was shown to arise from aging of the sample for the ESE experiments during prolonged storage at liquid nitrogen temperature. Note that only the ESEEM method of distance determination is sufficiently precise for indicating the effect of such aging. For freshly prepared samples it was found that $D = -115 \pm 5$ μT ($r = 28.9$ Å), which is in a good agreement with X-ray data. Zech et al., (1996) obtained for the same radical pair $D = 121$ μT ($r = 28.4$ Å).

The good agreement with the X-ray data implies that the point-dipole approximation used in Eq. 27 provides the same result as a purely geometric measurement between the two centers. This conclusion is not trivial because the spin density for both radicals, $P865^+$ and Q_A^-, is distributed in a rather large region (see also discussion in section 2.5). Also, the magnetic resonance measurement is carried out on reaction centers containing two charges, whereas the X-ray distance is for neutral reaction centers. Thus, the effect of electrostriction is very small.

One may conclude that ESEEM studies provide a reliable approach to measure precisely the distance between two electron transport cofactors. This opens the possibility to investigate RCs of plant photosynthesis where X-ray data do not exist or are incomplete.

3.2 Photosystem I Reaction Centers

The absorption of light in reaction centers of Photosystem I of green plants induces a charge separation between the donor chlorophyll *a* (P700) and the primary acceptor chlorophyll *a* (A_0). Then the electron is transferred to the secondary acceptor phylloquinone (A_1) on a picosecond time scale. For the $P700^+A_1^-$ radical pair a strong out-of-phase echo was observed (Moënne-Loccoz et al., 1994). So far, the structure of the Photosystem I reaction centers has been analyzed by X-ray with rather low resolution (~4 Å, Schubert et al., 1997). Therefore, an ESEEM study provides important information.

Zech et al., (1996, 1997b) found that for the $P700^+A_1^-$ radical pair a dipolar coupling $D = -170 \pm 4$ μT, indicating a distance of 25.4 ± 0.3 Å between the two cofactors. The same result, $D = -171 \pm 5$ μT, was obtained by Dzuba et al., (1997b) (see Fig. 3 for ESEEM data in the time domain and Fig. 4 for the Fourier transform in the frequency domain). Note the substantial decrease in comparison with the analogous distance in bacterial RCs (see above).

Dzuba et al., (1997b) and Zech et al., (1997b) also investigated RCs in which native phylloquinone was replaced by different artificial quinones. For all preparations, no change of the distance between the two cofactors was found within the experimental accuracy (±0.3 Å). It was concluded that the reconstituted quinones are bound in the native A_1 binding pocket.

Measurements on oriented samples such as single crystals also allow a determination of the orientation of the dipolar axis with respect to the crystal axes system. Photosystem I single crystals were investigated by Bittl et al., (1997). These crystals belong to the hexagonal space group $P6_3$ with two trimers of Photosystem I in the unit cell. The crystal was rotated in the plane perpendicular to the crystallographic *c* axis. The spectra obtained revealed a strong orientation dependence. This dependence was in agreement with the expected dependence for the spectral position of the Pake doublet, which provided an additional and important prove for the interpretation of the observed ESEEM as arising from the dipolar coupling between two unpaired electron spins in the radical pair. The orientation of the principal axis of the dipolar interaction, i.e., the axis connecting P700 and A_1, was determined. The angle between the dipolar coupling axis and the crystallographic *c* axis was found to be $27° \pm 5°$. Furthermore, the projection of the dipolar axis onto the crystallographic *a*, *b* plane was found to be parallel to the *a* axis. The results obtained allowed the determination of two possible locations of A_1 within the electron transfer chain of Photosystem I. These two positions are related to each other by the pseudo C_2 symmetry of the chlorophyll cofactors. One may conclude that the study of oriented samples is very informative from the point of view of obtaining the precise structural information. Schubert et al., (1997) used the results by Bittl et al., (1997) on single crystal measurements to assign the A_1 position in the X-ray electron density map.

3.3 Photosystem II Reaction Centers

Light excitation of Photosystem II reaction centers induces a charge separation between the donor chlorophyll *a* (P680) and the intermediate acceptor pheophytin *a*, which occurs in a few picoseconds. The electron is transferred, within a few hundred picosecond, to the primary electron acceptor plastoquinone Q_A, forming the secondary radical pair $P680^+Q_A^-$. Note that the X-ray crystal structure of Photosystem II reaction centers has not yet been obtained because of difficulties of crystallization. Therefore, ESEEM provides unique information, which so far cannot be obtained by other experimental methods.

The $P680^+Q_A^-$ pair was investigated using ESEEM by Zech et al., (1997c, 1999) and by Hara et al., (1997). They found a distance between the two cofactors $r = 27.4 \pm 0.3$ Å (Zech et al., 1997c), or $r = 27.2 \pm 1$ Å (Hara et al., 1997). Note that the decrease of accuracy, as compared with Photosystem I, is induced by poorer spectral resolution, arising, most probably, from distance dispersion (see also discussion in the next section). Also, the shortening of the lifetime of $P680^+Q_A^-$ pair may also decrease the spectral resolution (Zech et al., 1999).

$P680^+$ may be reduced by a tyrosine residue, Y_Z. As a result, the subsequent radical pair state, $Y_Z^{ox}Q_A^-$ state could be investigated (Zech et al., 1999). The dipolar coupling determined for this pair corresponds to a distance of 34 ± 1 Å.

3.4 The Energy Landscape of a Reaction Center Protein

It is now generally believed that the structure, dynamics and biological function of proteins are intimately linked to their so-called energy landscape, the complex, self-similar hierarchy of energy levels associated with the manifold of structures engendered by the set of amino acids building the protein (see, e.g., Thorn Leeson et al., 1997). It has been demonstrated that

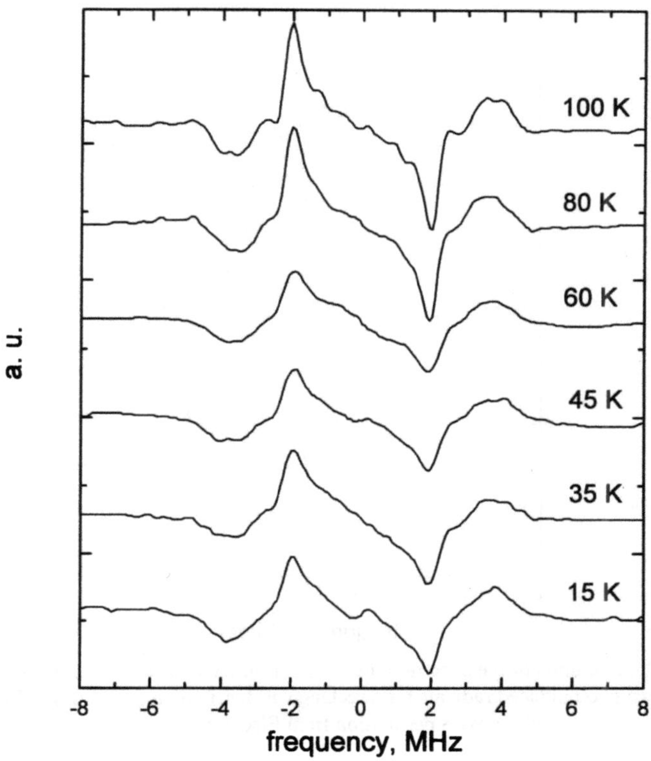

Figure 6. Pake spectra of the light-induced radical pair in bacterial photosynthetic reaction centers at various temperatures (Dzuba et al., 1997a), with permission from Elsevier Science.

similar features are found at physiological and cryogenic temperatures, down to a few millikelvin. The spectral diffusion and relaxation phenomena that reflect the properties of the energy landscape take place over an extraordinarily large time scale, spanning more than 18 orders of magnitude, from the sub-picosecond domain to that of a week and longer.

Although the concept of a hierarchical energy landscape seems now well established, much of the relevant data has been obtained for specific co-factor sites, such as the heme group in heme proteins, and Zn-protoporphyrin in myoglobin. Hence it could be argued that the measured energy landscape reflects local, not global properties of the protein. We have used the method of accurate distance determination described above to determine the distance and changes in distance for the primary donor and acceptor quinone cofactors in a photosynthetic reaction center protein, which are about 29 Å apart.

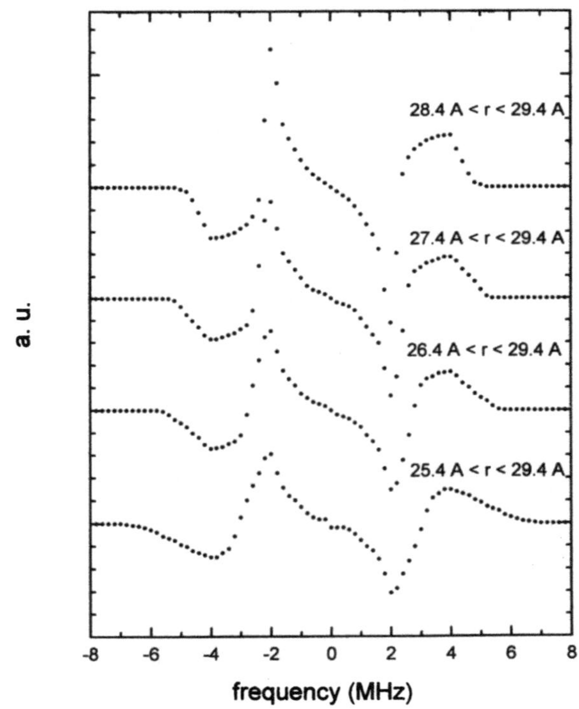

Figure 7. Simulation of the spectra of Fig. 6 for different distribution of the distance between two photoinduced radicals for a rectangular distribution function (Dzuba et al., 1997a), with permission from Elsevier Science.

Fig. 6 presents the resulting Pake spectra at various temperatures. The line broadening of the spectra appears to be discontinuous with temperature. For example, there is a clear transition between 35 and 45 K, and one more at 60 to 80 K. Above 100 K, the spectra were insensitive to temperature (data not given). It follows that the energy barriers in the energy landscape are indeed a *global* property of the protein. When cooling below such an energy barrier, the distance distribution induced by further cooling cannot anneal in the relatively short time allowed at each temperature (5 to 10 min.), and remains broad in a stepwise fashion, corresponding to the tiers of barrier heights.

The temperature-induced changes in the Pake spectrum could be quite well reproduced using a rectangular distance distribution (Fig. 7) for which the higher extreme is independent of temperature, whereas the low extreme moves closer to the high extreme with higher temperatures (compare with a tethered ball). At higher temperatures the distribution is quite narrow and the distance has a maximum value. Cooling below 100 K causes further contraction, but now, because exchange between substates in the energy landscape is inhibited, a broad distribution in distances results.

3.5 Kleinfeld effect

Rhodobacter sphaeroides R26 reaction centers frozen in the dark and under illumination exhibit quite different recombination kinetics of the $P865^+Q_A^-$ and $P865^+Q_B^-$ radical pairs (Kleinfeld et al., 1984). It was suggested that during illumination structural changes occur. Recently, X-ray studies were performed at cryogenic temperature for crystalline RCs frozen in the dark and under illumination (Stowell et al., 1997). For the charge-separated $P865^+Q_AQ_B^-$ state Q_B^- was found to be located approx. 5 Å away from the Q_B position in the charge-neutral DQ_AQ_B state, and to undergo a 180° propeller twist when moving from the former to the latter location. No significant change was detected for the Q_A position. This is perhaps not surprising, because light-induced structural changes for the Q_A position are expected for the $P865^+Q_A^-Q_B$ state only. Even then, the 2.2 Å resolution attained by Stowell et al., (1997) is probably not enough to detect these changes, which may be expected to be around 1 Å (Kleinfeld et al., 1988). ESEEM of the spin-correlated radical pairs allow us to attain much better accuracy. It was employed to assess a possible light-induced change in the distance between the P865 and Q_A cofactors (Zech et al., 1997a; Borovykh et al., 1998). For samples frozen under illumination a slight decrease in D value (see Fig. 8) was found implying an increase in the distance of 0.4 ± 0.2 Å (Borovykh et al., 1998). This result was obtained only for freshly prepared samples. For aged samples, the spectral resolution was not high enough to detect such a slight change. Probably, this was the reason why no distance change was detected by Zech et al., (1997a). The increased distance resolution (0.2 Å), as compared with that cited above (0.3 – 0.4 Å), refers only to the relative distance change and follows from the fact that very similar lineshapes are compared in Fig. 8.

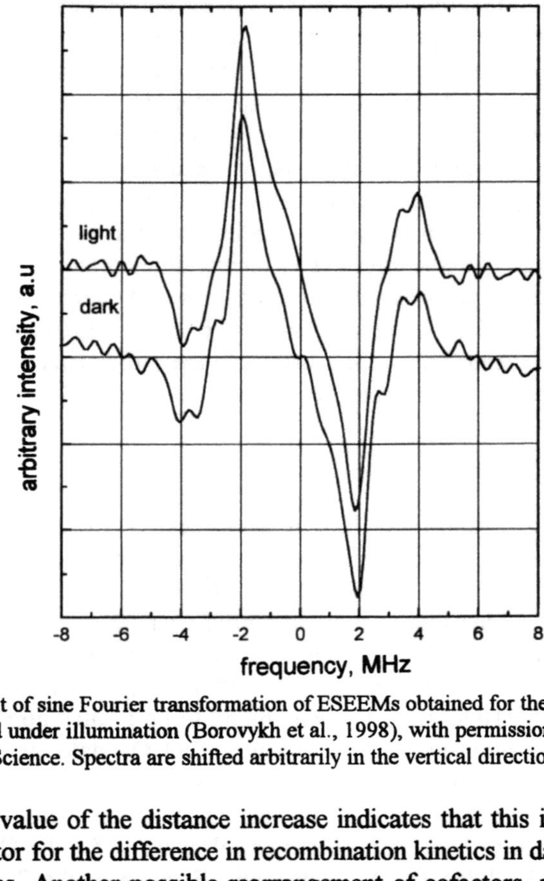

Figure 8. Result of sine Fourier transformation of ESEEMs obtained for the samples frozen in the dark and under illumination (Borovykh et al., 1998), with permission from Elsevier Science. Spectra are shifted arbitrarily in the vertical direction

The small value of the distance increase indicates that this increase is not the major factor for the difference in recombination kinetics in dark- and light-frozen samples. Another possible rearrangement of cofactors, rotation of the quinone, was ruled out earlier (Van den Brink et al., 1994). These results suggest that the major cause for the change in the recombination kinetics is not a rearrangement of the cofactors but rather a change in overall reorganization energy, for example because of frozen-in rearrangement of polar groups involved in charge separation.

4. CONCLUSION

ESEEM induced by the electron spin-spin dipolar interaction between radicals in a spin-correlated radical pair provides unique information that cannot be obtained by other experimental methods. The precision with which the distance between radicals can be measured is typically 0.3 - 0.4 Å, for radicals separated in the pair by ~ 30 Å. This high precision (see also Fursman and Hore, 1999) provides a very sensitive tool to perform structural investigations in photosynthetic reaction centers.

Another important advantage of the method is its physical clearness which in turn enhances the reliability of the information obtained. The spectrum in the frequency domain is a well-known "Pake doublet" from the two dipolar and exchange-coupled spins (with an antisymmetric shape with respect to the central frequency, due to spin polarization). A simple measurement of the frequency difference between the peaks and edges of the Pake spectrum yields a good first approximation for the dipolar interaction. This approximation may be improved by the subsequent precise computer simulation, based on convolution of the Pake "stick" spectrum with additional broadening induced by relaxation, etc.

The method is limited to photoinduced spin-correlated radical pairs separated by a distance of 20 - 40 Å. The lower limit comes from the condition that both radicals in the pair must be excited by the microwave pulse simultaneously, so D must be less than the pulse amplitude H_1. Typically, at the X-band H_1 is around 5 – 10 G (15 – 30 MHz). Then it follows from (27) that $r > 20$ Å. The upper limit is determined by the additional line broadening of the Pake doublet which determines the spectral resolution. For the broadening ~1 MHz (see Figs. 4, 6, 8) one may obtain from (27) that the condition $r < 40$ Å must hold.

ACKNOWLEDGEMENT

We are indebted to Drs. C.R. Timmel and C.E. Fursman for many stimulating discussions and to Prof. Yu.D. Tsvetkov and Dr. P.J. Hore for their interest and support over the years. The work carried out in Leiden was supported by TMR grant ERBFMRXCT980214, INTAS grant 93-2849ext and NWO grant 047-006.003.

REFERENCES

Angerhofer, A. and Bittl, R. (1996) Radicals and radical pairs in photosynthesis. *Photochem. Photobiol.* ***63***, 11-38.

Bittl, R. and Zech, S.G. (1997) Pulsed EPR study of spin coupled radical pairs in photosynthetic reaction centers: measurement of the distance between $P700^+$ and A_1^- in Photosystem I and between P865+ and Q_A^- in bacterial reaction center. *J. Phys. Chem. B* ***101***, 1429-1436.

Bittl, R., Zech, S.G., Fromme, P., Witt, H.T. and Lubitz, W. (1997) Pulsed EPR structure analysis of Photosystem I single crystals: localization of the phylloquinone acceptor. *Biochem.* ***36***, 12001-12004.

Borovykh, I.V., Dzuba, S.A., Proskuryakov, I.I., Gast P. and Hoff, A.J. (1998) Light-induced structural changes in photosynthetic reaction centers studied by ESEEM of spin-correlated $D^+Q_A^-$ radical pairs. *Biochim. Biophys. Acta* ***1363***, 182-186.

Chang, C.-H., El-Kabbani, O., Tiede, D., Norris J. and Schiffer M. (1991) Structure of the membrane-bound protein in photosynthetic reaction center from Rhodobacter sphaeroides. *Biochemistry,* ***30*** 5352.

Dikanov, S.A. and Tsvetkov, Yu.D. (1992) Electron spin echo modulation (ESEEM) spectroscopy CRC Press, Boca Raton.

Dzuba, S.A. (1997) Spin-correlated radical pairs in photosynthetic reaction centers: role of anisotropic hyperfine interaction as revealed by computational modeling. *Chem. Phys. Letters* ***278***, 333-340.

Dzuba, S.A., Gast P. and Hoff, A.J. (1995) ESEEM study of spin-spin interactions in spin-correlated $P^+Q_A^-$ pairs in the photosynthetic purple bacterium *Rhodobacter sphaerodes* R26. *Chem. Phys. Letters* ***236***, 595-602.

Dzuba, S.A., Bosch M.K. and Hoff, A.J. (1996) Electron spin echo detection of quantum beats and double-quantum coherence in spin-correlated radical pairs of protonated reaction centers *Chem. Phys. Letters* ***248***, 427-433.

Dzuba, S.A., Gast P. and Hoff, A.J. (1997a) Probing the energy landscape of bacterial photosynthetic reaction centers at cryogenic temperatures by ESEEM of spin-polarized $D^+Q_A^-$ radical pairs. *Chem. Phys. Letters* ***268***, 273-279.

Dzuba, S.A., Hara, H., Kawamori, A., Iwaki, M., Itoh S. and Tsvetkov, Yu.D. (1997b) Electron spin echo of spin-polarised radical pairs in intact and quinone-reconstituted plant Photosystem I reaction centers. *Chem. Phys. Letters* ***264***, 238-244.

Fursman, C.E. and Hore, P.J. (1999) Distance determination in spin-correlated radical pairs in photosynthetic reaction centers by electron spin echo envelope modulation. *Chem. Phys. Letters* ***303***, 593-600.

Hara, H., Dzuba, S.A., Kawamori, A., Akabori, K., Tomo, T., Satoh, K., Iwaki, M. and Itoh, S. (1997) The distance between P680 and Q_A in Photosystem II determined by ESEEM spectroscopy. *Biochim. Biophys. Acta* ***1322***, 77-85.

Hara, H., Tang, J., Kawamori, A., Itoh, S. and Iwaki, M. (1998) Anomalous pulse-angle and phase dependence of Hahn's electron spin echo and multiple-quantum echo of the spin-correlated radical pair $P700^+A_1^-$ in Photosystem I. *Appl. Magn. Reson.* ***14***, 367-379.

Hoff, A.J., Gast, P., Dzuba, S.A., Timmel, C.R., Fursman, C.E. and Hore, P.J. (1998) The nuts and bolts of distance determination and zero and double quantum coherence in photoinduced radical pairs. *Spectrochim. Acta A* ***54***, 2283-2293.

Hore, P.J. (1990) Analysis of polarized electron paramagnetic resonance spectra. in: *Advanced EPR, applications in biology and biochemistry,* (Hoff A.J., Ed.), pp. 405-440, Elsevier, Amsterdam.

Hore, P.J. (1996) Transfer of spin correlation between radical pairs in the initial steps of photosynthetic energy conversion. *Mol. Phys.* ***89***, 1195-1202.

Hore, P.J., Hunter, D.A., McKie, C.D. and Hoff, A.J., (1987) Electron paramagnetic resonance of spin-correlated radical pairs in photosynthetic reactions. *Chem. Phys. Letters* ***137***, 495-500.

Iwaki, M., Itoh, S., Hara, H. and Kawamori, A. (1998) Spin-polarized radical pair in Photosystem I reaction center that contains different quinones and fluorenones as the secondary electron acceptor. *J. Phys. Chem. B* ***102***, 10440-10445.

Jeschke, G. and Bittl, R. (1998) Electron spin echo envelope modulation by electronic spin-spin interactions in radical pairs undergoing electron transfer. *Chem. Phys. Letters* ***294***, 323-331.

Kleinfeld, D., Okamura, M.Y. and Feher, G. (1984) Electron transfer kinetics in photosynthetic reaction centers in the charge-separated state: evidence for light-induced structural changes. *Biochem.* ***23***, 5780-5786.

Komiya, H., Yeates, T.O., Rees, D.C., Allen, J.P. and Feher, J. (1988) Structure of the reaction center from Rhodobacter sphaeroides R-26. *Proc. Natl. Acad. Sci. US* ***85***, 9012-9016.

Kothe, G., Weber, S., Bittl, R., Ohmes, E., Thurnauer, M.C. and Norris, J.R., (1999) Transient EPR of light-induced radical pairs in plant photosystem I: observation of quantum beats. *Chem. Phys. Letters* ***186***, 474-480.

Moënne-Loccoz, P., Heathcote, P., Maclachlan, D.J., Berry, M.C., Davis, I.H. and Evans, M.C.W. (1994) Path of electron transfer in photosystem I: direct evidence of forward electron transfer from A_1 to $Fe\text{-}S_X$. *Biochem.* ***33***, 10037-10042.

Salikhov, K.M., Kandrashkin, Yu.E. and Salikhov, A.K. (1992) Peculiarities of free induction and primary spin echo signals for spin-correlated radicals pairs, *Appl. Magn. Reson.* ***3***, 199-217.

Schubert, W.-D., Klukas, O., Krauss, N., Saenger, W., Fromme, P., Witt, H.T. (1997) Photosystem I of Synechococcus elongatus at 4 Å resolution: comprehensive structure analysis. *J. Mol. Biol.* ***272***, 741-769.

Stowell, M.H.B., McPhillips, T.M., Rees, D.C., Soltis, S.M., Abresch, E., and Feher, G. (1997) Ligt-induced structural changes in photosynthetic reaction center: implication for mechanism of electron transfer. *Science* ***276***, 812-816.

Tang, J., Thurnauer, M.C. and Norris, J.R. (1994) Electron spin echo envelope modulation due to exchange and dipolar interactions in a spin-correlated radical pair. *Chem. Phys. Letters* ***219***, 283-290.

Tang, J. and Norris, J.R., (1995) Multiple-quantum EPR coherence in a spin-correlated radical pair system. *Chem. Phys. Letters* ***233***, 192-200.

Tang, J., Thurnauer, M.C., Kubo, A., Hara, H. and Kawamori, A. (1997) Anomalous pulse-angle and phase dependence of Hahn's electron spin echo and multiple-quantum echoes in a photoinduced spin-correlated radical pair. *J. Chem. Phys.* ***106***, 7471-7478.

Thorn Leeson, D., Wiersma, D.A., Fritsch, K. and Friedrich, J. (1997) The energy landscape of myoglobin: an optical study. *J. Phys. Chem. B* ***101***, 6331-6340.

Thurnauer, M.C. and Norris, J.R. (1980) An electron spin echo phase shift observed in photosynthetic algae. Possible evidence for dynamic radical pair interactions. *Chem. Phys. Letters* ***76***, 557-561.

Timmel, C.R., Fursman, C.E., Hoff A.J. and Hore, P.J. (1998) Spin-correlated radical pairs: microwave pulse effects on lifetimes, electron spin echo envelope modulations and optimum conditions for detection by electron spin echo spectroscopy. *Chem. Phys.* ***226***, 271-283.

Van den Brink, J.S., Hulsebosch, R.J., Gast, P., Hore, P.J. and Hoff, A.J. (1994) Q_A binding in reaction centers of the photosynthetic purple bacterium *Rhodobacter sphaeroides* R26 investigated with electron spin polarisation spectroscopy. *Biochemistry* ***33***, 13668-13677.

Zech, S.G., Lubitz W. and Bittl, R. (1996) Pulsed EPR experiments on radical pairs in photosynthesis: comparison of the donor acceptor distances in photosystem I and bacterial reaction centers. *Ber. Bunsenges. Phys. Chem.* ***100***, 2041-2044.

Zech, S.G., Bittl, R. Gardiner, A.T. and Lubitz W., (1997a) Transient and pulsed EPR spectroscopy on the radical pair state $P865^{+}Q_{A}^{-}$ to study light-induced changes in bacterial reaction centers. *Appl. Magn. Reson.* ***13***, 517-529.

Zech, S.G., van der Est, A.J. and Bittl, R., (1997b) Measurement of cofactor distances between $P700^{+}$ and A_{1}^{-} in native and quinone substituted photosystem I using pulsed electron paramagnetic resonance spectroscopy. *Biochem.* ***36***, 9774-9779.

Zech, S.G., Kurreck, J., Eckert, H.J., Renger, G., Lubitz W. and Bittl, R. (1997c) Pulsed EPR measurement of the distance between $P680^{+}$ and Q_{A}^{-} in Photosystem II. *FEBS Letters* ***414***, 454-456.

Zech, S.G., Kurreck, J., Renger, G., Lubitz W. and Bittl, R. (1999) Determination of the distance between Y_{Z}^{ox} and Q_{A}^{-} in photosystem II by pulsed EPR spectroscopy on light-induced radical pairs. *FEBS Letters* ***442***, 79-82.

Contents of Prior Related Volumes in this Series

Volume 8
Spin Labeling: Theory and Applications

VOLUME 13
EMR of Paramagnetic Molecules

VOLUME 14
Spin Labeling: The Next Millenium

Volume 18
In Vivo EPR (ESR): Theory and Applications

INDEX

MIX
Papier aus verantwortungsvollen Quellen
Paper from responsible sources
FSC® C105338

If you have any concerns about our products,
you can contact us on
ProductSafety@springernature.com

In case Publisher is established outside the EU,
the EU authorized representative is:
Springer Nature Customer Service Center GmbH
Europaplatz 3, 69115 Heidelberg, Germany

Printed by Libri Plureos GmbH
in Hamburg, Germany